Counterexamples in Markov Decision Processes

Series on Optimization and its Applications

Print ISSN: 2399-1593
Online ISSN: 2399-1607

(*formerly known as Imperial College Press Optimization Series* —
Print ISSN: 2041-1677)

Series Editor: Jean Bernard Lasserre *(LAAS-CNRS and Institute of Mathematics, University of Toulouse, France)*

Published

Vol. 6: *Counterexamples in Markov Decision Processes*
by Alexey Piunovskiy

Vol. 5: *Sparse Polynomial Optimization: Theory and Practice*
by Victor Magron and Jie Wang

Vol. 4: *The Moment-SOS Hierarchy: Lectures in Probability, Statistics, Computational Geometry, Control and Nonlinear PDEs*
by Didier Henrion, Milan Korda and Jean Bernard Lasserre

Vol. 3: *Genericity in Polynomial Optimization*
by Huy-Vui Hà and Tiến-Sơn Phạm

Vol. 2: *Examples in Markov Decision Processes*
by A. B. Piunovskiy

Vol. 1: *Moments, Positive Polynomials and Their Applications*
by Jean Bernard Lasserre

Series on Optimization and its Applications – Vol. 6

Counterexamples in Markov Decision Processes

Alexey Piunovskiy
University of Liverpool, UK

World Scientific

NEW JERSEY • LONDON • SINGAPORE • GENEVA • BEIJING • SHANGHAI • TAIPEI • CHENNAI

Published by

World Scientific Publishing Europe Ltd.
57 Shelton Street, Covent Garden, London WC2H 9HE
Head office: 5 Toh Tuck Link, Singapore 596224
USA office: 27 Warren Street, Suite 401-402, Hackensack, NJ 07601

Library of Congress Control Number: 2024053285

British Library Cataloguing-in-Publication Data
A catalogue record for this book is available from the British Library.

Series on Optimization and its Applications — Vol. 6
COUNTEREXAMPLES IN MARKOV DECISION PROCESSES

ISBN 978-1-80061-675-2 (hardcover)
ISBN 978-1-80061-676-9 (ebook for institutions)
ISBN 978-1-80061-677-6 (ebook for individuals)

For any available supplementary material, please visit
https://www.worldscientific.com/worldscibooks/10.1142/Q0494#t=suppl

Desk Editors: Soundararajan Raghuraman/Lai Fun Kwong

Typeset by Stallion Press
Email: enquiries@stallionpress.com

A small omission
of a required condition
may lead
to a large error of commission.

Guy Soulsby

Preface

Markov Decision Processes (MDP) is a branch of applied mathematics based on probability theory, optimal control, and mathematical analysis. Several books with counterexamples/paradoxes in probability [Stoyanov (1997); Szekely (1986)] and in analysis [Gelbaum and Olmsted (1964)] are in existence; it is therefore not surprising that MDP is also replete with unexpected counter-intuitive examples. The main goal of the current book is to collect such examples. Most of them are based on earlier publications; the remainder are new. This book should be considered as a complement to scientific monographs on MDP [Altman (1999); Bertsekas and Shreve (1978); Guo and Hernández-Lerma (2009); Hernández-Lerma and Lasserre (1996a, 1999); Kallenberg (2010); Piunovskiy (1997); Piunovskiy and Zhang (2020); Prieto-Rumeau and Hernández-Lerma (2012); Puterman (1994)]. It can also serve as a reference book to which one can turn for answers to curiosities that arise while studying or teaching MDP. All the examples are self-contained and can be read independently of each other. Concerning uncontrolled Markov chains, we mention the illuminating collection of examples in [Suhov and Kelbert (2008)].

A survey of meaningful applications is beyond the scope of the current book. The examples presented either lead to counter-intuitive solutions, or illustrate the importance of conditions in the known theorems. Not all examples are equally simple or complicated. Several examples are aimed at undergraduate students, whilst others will be of interest to professional researchers.

The book has five chapters in line with the main different types of MDP: the finite-horizon case, infinite horizon with total or discounted cost, and average cost over an infinite time interval; the fifth chapter is on continuous-time MDP. Some proofs of auxiliary assertions are postponed to the

Appendix. At the beginning of each chapter, the necessary piece of theory is briefly presented. Note that the known statements, with the references where they were proved, sometimes are presented not in the most general setting, e.g., under stronger conditions. The necessary knowledge on Borel spaces, measure theory, linear programming etc can be found in [Bertsekas and Shreve (1978)], in the appendix of [Piunovskiy and Zhang (2020)] and in many other monographs.

Compared with the first edition [Piunovskiy (2013)], multiple typos and inaccuracies are corrected and more details are provided making the text more reader friendly. Several new examples and the new chapter on continuous-time MDP are added making this book more comprehensive: now it contains more than 170 examples.

The following notations and conventions will often be used without explanation.

- $:=$ means 'equals by definition';
- $\mathbf{C}^\infty$ is the space of infinitely differentiable functions;
- $\mathbf{C}(\mathbf{X})$ is the space of continuous bounded functions on a (topological) space $\mathbf{X}$;
- for functions and mappings like $f(x)$ and f_x, formula $f = g$ means the point-wise equality;
- a Borel space is called topological in case a topology is fixed therein;
- $\mathbf{B}(\mathbf{X})$ is the space of bounded measurable functions on a (Borel) space $\mathbf{X}$;
- in discrete (finite or denumerable) spaces, the discrete topology is supposed to be fixed, unless specified otherwise;
- with some abuse of notation, we often say that μ is a measure on (a Borel space) $\mathbf{X}$ instead of saying that μ is a measure on the Borel σ-algebra $\mathcal{B}(\mathbf{X})$;
- $\mathbf{P}(\mathbf{X})$ is the space of probability measures on the (metrizable) space $\mathbf{X}$, equipped with the weak topology;
- If Γ is a subset of space $\mathbf{X}$ then Γ^c is the complement;
- $\mathbb{R}^N$ is the N-dimensional Euclidean space; $\mathbb{R} = \mathbb{R}^1$ is the straight line; in these spaces the Euclidean topology is supposed to be fixed, unless specified otherwise;
- $\mathbb{R}^* = [-\infty, +\infty]$ is the extended straight line;
- $\mathbb{I}\{statement\} = \begin{cases} 1, & \text{if the statement is correct;} \\ 0, & \text{if the statement is false} \end{cases}$ is the indicator function; $\mathbb{I}_A(x) := \mathbb{I}\{x \in A\}$;

- $\delta_a(dy)$ is the Dirac measure concentrated at point a: $\delta_a(\Gamma) = \mathbb{I}_\Gamma(a)$;
- for a measure μ on a discrete space $\mathbf{X}$, we write $\mu(x)$, not $\mu(\{x\})$ for singletons; similarly, for a measure μ on $\mathbf{X} \times \mathbf{A}$ we write $\mu(x \times \Gamma)$, not $\mu(\{x\} \times \Gamma)$ for $x \in \mathbf{X}$, $\Gamma \subset \mathbf{A}$; if $\mathbf{A}$ is also discrete, then a measure μ on singletons in $\mathbf{Z} = \mathbf{X} \times \mathbf{A}$ is denoted as $\mu(z) = \mu(x, a)$, not $\mu((x, a))$;
- If $r \in \mathbb{R}^*$ then $r^+ := \max\{0, r\}$, $r^- := \min\{0, r\}$;
- $\sum_{i=n}^{m} f_i := 0$ and $\prod_{i=n}^{m} f_i := 1$ if $m < n$;
- $\lfloor r \rfloor$ is the integer part, the maximal integer i such that $i \leq r$;
- the integral $\int_{(t_1,t_2]} F(t)dt$ is understood as $\int_{(t_1,\infty)} F(t)dt$ in case $t_2 = \infty$;
- $\pm\infty + c := \pm\infty$ for $c \in \mathbb{R}$, and $\infty - \infty := \infty$, unless specified otherwise.

Throughout the current book $\mathbf{X}$ is the state space, $\mathbf{A}$ is the action space, $p_t(dy|x, a)$ is the transition probability, $q(dy|x, a)$ is the transition rate, $c_t(x, a)$ are the cost functions or cost rates. DTMDP and CTMDP are the abbreviations for discrete-time and continuous-time Markov decision processes. On the figures illustrating DTMDPs, the arrows indicate the possible transitions, usually equipped with the corresponding probabilities (sometimes omitted if equal to 1). The similar concerns the figures on CTMDPs.

Normally, we denote random variables with capital letters (X), small letters (x) being used just for variables, arguments of functions, etc. Bold case ($\mathbf{X}$) is for spaces. All functions, mappings, and stochastic kernels are assumed to be Borel-measurable unless their properties are explicitly specified.

We say that a function on $\mathbb{R}^1$ with the values in a topological Borel space $\mathbf{A}$ is *piece-wise continuous* if there exists a sequence $y_i \in \mathbb{R}^*$ such that $\lim_{i\to\infty} y_i = \infty$; $\lim_{i\to-\infty} y_i = -\infty$, this function is continuous on each open interval (y_i, y_{i+1}), and there exists a right (left) limit as $y \to y_i + 0$ ($y \to y_{i+1} - 0$), $i = 0, \pm 1, \pm 2 \ldots$. A similar definition is accepted for real-valued piece-wise Lipschitz, and continuously differentiable functions.

If $\mathbf{X}$ is a measurable space and ν is a measure on it, then both formulae

$$\int_{\mathbf{X}} f(x) d\nu(x) \qquad \text{and} \qquad \int_{\mathbf{X}} f(x) \nu(dx)$$

denote the same integral of a real-valued function $f(\cdot)$ with respect to ν;

$$\int_{\mathbf{X}} f(x) d\nu(x) := \int_{\mathbf{X}} f^+(x) d\nu(x) + \int_{\mathbf{X}} f^-(x) d\nu(x) \qquad \text{with} + \infty - \infty := +\infty.$$

W.r.t. is the abbreviation for 'with respect to', a.s. means 'almost surely', and CDF means 'cumulative distribution function'.

We consider only minimization problems. When formulating theorems and examples published in books (articles) devoted to maximization, we always adjust the statements for our case without any special remarks.

It should be emphasized that the terminology in MDP is not entirely fixed. For example, very often strategies are called policies. There exist several slightly different definitions of a semi-continuous model, and so on.

A.B. Piunovskiy

Contents

Chapter 1

Finite-Horizon Models

In this chapter, we consider the model with the finite horizon T. In many examples, there is only one decision epoch: $T = 1$. After the very brief presentation of the necessary knowledge on MDPs in Sections 1.1–1.3, a number of examples are presented in Section 1.4.

1.1 Preliminaries

A decision maker is faced with the problem of influencing the behaviour of a stochastic system as it evolves over time. Decisions are made at discrete points in time referred to as *decision epochs* and denoted as $t = 1, 2, \ldots, T < \infty$. At each time epoch t, the system occupies a *state* $x \in \mathbf{X}$. The *state space* $\mathbf{X}$ can be either discrete (finite or denumerably infinite) or continuous (non-empty uncountable Borel subset of a complete, separable metric space, e.g. $\mathbb{R}^1$). $\mathcal{B}(\mathbf{X})$ is the Borel σ-algebra in $\mathbf{X}$. If the state at the time epoch t is considered as a random variable, it is denoted by a capital letter X_t; small letters x_t are just for possible values of X_t. Therefore, the behaviour of the system is described by a *stochastic (controlled) process*

$$X_0, X_1, X_2, \ldots, X_T.$$

In case of uncontrolled systems, the theory of Markov processes is well developed: the initial probability distribution $\mathsf{P}_0(dx)$ for X_0 is assumed to be given, and the dynamics is defined by *transition probabilities* $p_t(dy|x)$ which are measurable stochastic kernels on $\mathcal{B}(\mathbf{X})$ given $\mathbf{X}$. When $\mathbf{X}$ is finite and the process is time-homogeneous, those probabilities form a transition matrix with elements $p(j|i) = \mathsf{P}(X_{t+1} = j | X_t = i)$.

In the case of controlled systems, we assume that the *action space* $\mathbf{A}$ is given, which again can be an arbitrary Borel space (including the case of finite or denumerable $\mathbf{A}$). $\mathcal{B}(\mathbf{A})$ is the Borel σ-algebra in $\mathbf{A}$. As soon as the state X_{t-1} becomes known (equals x_{t-1}), the decision maker must choose an action/control $A_t \in \mathbf{A}$; in general this depends on all the realized values of $X_0, X_1, \ldots, X_{t-1}$ along with the past actions $A_1, A_2, \ldots, A_{t-1}$. Moreover, that decision can be randomized. The rigorous definition of a control strategy is given in the next section.

As a result of choosing action a at decision epoch t in state x, the decision maker loses $c_t(x,a)$ units, and the system state at the next decision epoch is determined by the probability distribution $p_t(dy|x,a)$. The function $c_t(x,a)$ is called the *one-step cost*. The *final/terminal* cost equals $C(x)$ when the final state $X_T = x$ is realized.

We assume that the *initial distribution* $\mathsf{P}_0(dx)$ of X_0 is given. Suppose a control strategy π is fixed (that is, the rule of choosing actions a_t; see the next section). Then the random sequence

$$X_0, A_1, X_1, A_2, X_2, \ldots, A_T, X_T$$

is well defined: there exists a single corresponding probability measure $\mathsf{P}^\pi_{\mathsf{P}_0}$ on the space of trajectories

$$(x_0, a_1, x_1, a_2, x_2, \ldots, a_T, x_T) \in \mathbf{X} \times (\mathbf{A} \times \mathbf{X})^T.$$

For example, if $\mathbf{X}$ is finite and the control strategy is defined by the map $a_t = \varphi_t(x_{t-1})$, then

$$\mathsf{P}^\varphi_{\mathsf{P}_0}(X_0 = i, A_1 = a_1, X_1 = j, A_2 = a_2, X_2 = k, \ldots, X_{T-1} = l, A_T = a_T, X_T = m)$$
$$= \mathsf{P}_0(\{i\})\mathbb{I}\{a_1 = \varphi_1(i)\}p_1(\{j\}|i, a_1)\mathbb{I}\{a_2 = \varphi_2(j)\}\ldots p_T(\{m\}|l, a_T).$$

Here and below, $\mathbb{I}\{\cdot\}$ is the indicator function. If $\mathbf{X}$ is discrete then transition probabilities $p_t(\cdot|x,a)$ are defined by the values on singletons $p_t(\{y\}|x,a)$. The same is for the initial distribution. In what follows we omit the parentheses for brevity and write $p_t(y|x,a)$.

Therefore, for a fixed control strategy π, the *total expected cost* equals $v^\pi_{\mathsf{P}_0} := \mathsf{E}^\pi_{\mathsf{P}_0}[W]$, where

$$W := \sum_{t=1}^{T} c_t(X_{t-1}, A_t) + C(X_T)$$

is the *total realized cost*. Here and below, $\mathsf{E}^\pi_{\mathsf{P}_0}$ is the mathematical expectation with respect to the probability measure $\mathsf{P}^\pi_{\mathsf{P}_0}$.

The aim is to find an *optimal* control strategy π^* solving the problem

$$v^\pi_{\mathsf{P}_0} = \mathsf{E}^\pi_{\mathsf{P}_0}\left[\sum_{t=1}^{T} c_t(X_{t-1}, A_t) + C(X_T)\right] \longrightarrow \inf_\pi. \tag{1.1}$$

$v^\pi_{\mathsf{P}_0}$ is called the *performance functional*, and $v^*_{\mathsf{P}_0} := \inf_\pi v^\pi_{\mathsf{P}_0}$.

Using the dynamic programming approach, under some technical conditions, one can prove the following statement. Suppose function $v_t(\cdot)$ on $\mathbf{X}$ satisfies the following equation

$$\begin{cases} v_T(x) = C(x); \\ v_{t-1}(x) = \inf_{a\in\mathbf{A}}\left\{c_t(x,a) + \int_X v_t(y)p_t(dy|x,a)\right\} \\ \qquad\quad = c_t(x,\varphi^*_t(x)) + \int_X v_t(y)p_t(dy|x,\varphi^*_t(x)); \qquad t = T, T-1, \dots, 1. \end{cases} \tag{1.2}$$

Then, the control strategy defined by the map $a_t = \varphi^*_t(x_{t-1})$ solves problem (1.1), i.e. it is *optimal*; $\inf_\pi v^\pi_{\mathsf{P}_0} = \int_{\mathbf{X}} v_0(x)\mathsf{P}_0(dx)$. Therefore, control strategies of the presented type are usually sufficient for solving standard problems. They are called *Markov selectors*. More about dynamic programming in Section 1.3.

1.2 Model Description

We now provide more rigorous definitions.

The *Markov Decision Process (MDP)* with a finite horizon is defined by the collection

$$\{\mathbf{X}, \mathbf{A}, T, p, c, C\},$$

where $\mathbf{X}$ and $\mathbf{A}$ are the state and action spaces (Borel); T is the *time horizon*; $p_t(dy|x,a)$, $t = 1, 2, \dots, T$, are measurable stochastic kernels on $\mathcal{B}(\mathbf{X})$ given $\mathbf{X}\times\mathbf{A}$; $c_t(x,a)$ are measurable functions on $\mathbf{X}\times\mathbf{A}$ with values on the straight-line $\mathbb{R} := (-\infty, +\infty)$; $C(x)$ is a measurable map $C(\cdot) : \mathbf{X} \to (\infty, +\infty)$. Sometimes, we allow the costs to be not finite. The model is called *homogeneous* if the transition probability p and the cost function $c(\cdot,\cdot)$ do not depend on time t. The model is called *finite* if the spaces $\mathbf{X}$ and $\mathbf{A}$ are finite. The model is called *discrete* if the spaces $\mathbf{X}$ and $\mathbf{A}$ are discrete, that is, finite or denumerable. If the state (or action) space is discrete, then, if needed, we always by default consider the discrete topology therein. Sometimes, the topologies in $\mathbf{X}$ and $\mathbf{A}$ are not important; otherwise, they are explicitly described.

The *space of trajectories* (or *histories*) up to decision epoch t is

$$\mathbf{H}_{t-1} := \mathbf{X}\times(\mathbf{A}\times\mathbf{X})^{t-1}, \quad t = 1, 2, \dots, T; \quad \mathbf{H} := \mathbf{X}\times(\mathbf{A}\times\mathbf{X})^T.$$

The space $\mathbf{H}$ plays the role of the *sample space*, sometimes called *canonical*; its elements are denoted as ω or h. All the random variables, like X_t and A_t

are just measurable mappings (projections) defined on $\mathbf{H}$, e.g., $X_t(\omega) = x_t$ and $A_t(\omega) = a_t$. A *control strategy* $\pi = \{\pi_t\}_{t=1}^T$ is a sequence of measurable stochastic kernels

$$\pi_t(da|x_0, a_1, x_1, \dots, a_{t-1}, x_{t-1}) = \pi_t(da|h_{t-1})$$

on $\mathcal{B}(\mathbf{A})$ given $\mathbf{H}_{t-1}$. In the above formula, we omit the brackets for $h_{t-1} = (x_0, a_1, x_1, \dots, a_{t-1}, x_{t-1})$. If a strategy π^{m} is defined by (measurable) stochastic kernels $\pi_t^{\mathrm{m}}(da|x_{t-1})$ then it will be called a *Markov* strategy. It is called *semi-Markov* if it has the form $\pi_t(da|x_0, x_{t-1})$. A Markov (semi-Markov) strategy is called *stationary* if none of the kernels $\pi^{\mathrm{s}}(da|x_{t-1})$ $(\pi^{\mathrm{s}}(da|x_0, x_{t-1}))$ depend on the time t. If, for any $t = 1, 2, \dots, T$, there exists a measurable mapping $\varphi_t(h_{t-1}) : \mathbf{H}_{t-1} \to \mathbf{A}$ such that $\pi_t(\Gamma|h_{t-1}) = \mathbb{I}\{\Gamma \ni \varphi_t(h_{t-1})\}$ for any $\Gamma \in \mathcal{B}(\mathbf{A})$, then the strategy is denoted by the symbol φ and is called a *selector* or *non-randomized* strategy. Selectors of the form $\varphi_t(x_{t-1})$ and $\varphi(x_{t-1})$ are called *Markov* and *stationary* respectively. Stationary semi-Markov strategies and semi-Markov (stationary) selectors are defined in the same way. In what follows, Π^{All} is the collection of all strategies, Π^{M} is the set of all Markov strategies, Π^{MN} is the set of all Markov selectors. In this connection, letter N corresponds to non-randomized strategies. Further, Π^{S} and Π^{SN} are the sets of all stationary strategies and of all stationary selectors.

We assume that the *initial* probability distribution $\mathsf{P}_0(dx)$ is fixed. If a control strategy π is fixed too, then there exists a unique probability measure $\mathsf{P}^\pi_{\mathsf{P}_0}$ on $\mathbf{H}$ such that, for all $\Gamma^X \in \mathcal{B}(\mathbf{H}_0) = \mathcal{B}(\mathbf{X})$,

$$\mathsf{P}^\pi_{\mathsf{P}_0}(\Gamma^X) = \mathsf{P}_0(\Gamma^X);$$

and, for each $t = 1, 2, \dots, T$ for all $\Gamma_i^X \in \mathcal{B}(\mathbf{X})$ $(i = 0, 1, \dots, t)$ and $\Gamma_i^A \in \mathcal{B}(\mathbf{A})$ $(i = 1, 2, \dots, t)$,

$$\begin{aligned} &\mathsf{P}^\pi_{\mathsf{P}_0}(X_0 \in \Gamma_0^X,\ A_1 \in \Gamma_1^A,\ \dots,\ X_{t-1} \in \Gamma_{t-1}^X,\ A_t \in \Gamma_t^A) \\ &\quad = \int_{\Gamma_0^X \times \Gamma_1^A \times \cdots \times \Gamma_{t-1}^X} \pi_t(\Gamma_t^A|x_0, a_1, \dots, x_{t-1}) \\ &\qquad \times \mathsf{P}^\pi_{\mathsf{P}_0}(X_0 \in dx_0,\ A_1 \in da_1,\ \dots,\ X_{t-1} \in dx_{t-1}); \end{aligned}$$

and

$$\begin{aligned} &\mathsf{P}^\pi_{\mathsf{P}_0}(X_0 \in \Gamma_0^X,\ A_1 \in \Gamma_1^A, \dots,\ X_t \in \Gamma_t^X) \\ &\quad = \int_{\Gamma_0^X \times \Gamma_1^A \times \cdots \times \Gamma_{t-1}^X \times \Gamma_t^A} p(\Gamma_t^X|x_{t-1}, a_t) \\ &\qquad \times \mathsf{P}^\pi_{\mathsf{P}_0}(X_0 \in dx_0,\ A_1 \in da_1, \dots,\ X_{t-1} \in dx_{t-1},\ A_t \in da_t). \end{aligned}$$

For details, see [Hernández-Lerma and Lasserre (1996a), (2.2.7)].

Here and below, with some less rigorous notation, we also denote $\mathsf{P}^\pi_{\mathsf{P}_0}(\cdot)$ the images of $\mathsf{P}^\pi_{\mathsf{P}_0}$ relative to projections of the types

$$\mathbf{H} \to \mathbf{H}_t \times \mathbf{A} =: \mathbf{G}_t, \quad \text{and } \mathbf{H} \to \mathbf{H}_t, \quad t = 0, 1, 2, \ldots, T-1. \tag{1.3}$$

$h_t = (x_0, a_1, x_1, \ldots, a_t, x_t)$ is the generic element of $\mathbf{H}_t$. When it is considered as the random element on $\mathbf{H}$, we use capital letter H_t, as usual.

Measures $\mathsf{P}^\pi_{\mathsf{P}_0}(\cdot)$ on $\mathbf{H}$ are called *strategic* measures; they form the space $\mathcal{P}_{\mathsf{P}_0} := \{\mathsf{P}^\pi_{\mathsf{P}_0}(\cdot) : \pi \in \Pi^{\mathrm{All}}\}$. The corresponding mathematical expectation is denoted as $\mathsf{E}^\pi_{\mathsf{P}_0}$. If the initial distribution $\mathsf{P}_0(dx) = \delta_{x_0}(dx)$ is degenerate, we use notations $\mathsf{P}^\pi_{x_0}$, $\mathcal{P}_{x_0}$ and $\mathsf{E}^\pi_{x_0}$. If the strategies are restricted (e.g., to Markov selectors Π^{MN}), then the notations like $\mathcal{P}^{\mathrm{MN}}_{\mathsf{P}_0}$ are in use.

Let $\mathcal{F}_T := \mathcal{B}(\mathbf{H})$ and let $\mathcal{F}_t$ be the σ-algebra in $\mathbf{H}$ defined as the pre-image of $\mathcal{B}(\mathbf{G}_t) = \mathcal{B}(\mathbf{H}_t \times \mathbf{A})$ with respect to (1.3), $t = 0, 1, 2, \ldots, T-1$. Now the trivial projections

$$\mathbf{H} \ni (x_0, a_1, x_1, \ldots, a_T, x_T) \to x_t \text{ and } \mathbf{H} \ni (x_0, a_1, x_1, \ldots, a_T, x_T) \to a_t$$

define the adapted stochastic process $\{X_t\}_{t=0}^T$, called *controlled process*, and the predictable (i.e., $\mathcal{F}_{t-1}$-measurable) stochastic process $\{A_t\}_{t=1}^T$, called *control process*, on the stochastic basis $(\mathbf{H}, \mathcal{B}(\mathbf{H}), \{\mathcal{F}_t\}_{t=0}^T, \mathsf{P})$, where the measure $\mathsf{P} \in \mathcal{P}_{\mathsf{P}_0}$ is under control. For fixed P, all the σ-algebras $\{\mathcal{F}_t\}_{t=0}^T$ are assumed to be completed. We use notation $(x_0, a_1, x_1, \ldots, a_T, x_T)$ for the elements of $\mathbf{H}$, in order to make the controlled and control processes to be adapted and predictable correspondingly. Notation $(x_0, a_0, x_1, \ldots, a_{T-1}, x_T)$ is also widely used by many authors.

For each $h \in \mathbf{H}$ the (realized) *total cost* equals

$$\begin{aligned} w(h) &= \sum_{t=1}^T c_t(x_{t-1}, a_t) + C(x_T) \\ &:= \sum_{t=1}^T c_t^+(x_{t-1}, a_t) + C^+(x_T) + \sum_{t=1}^T c_t^-(x_{t-1}, a_t) + C^-(x_T), \end{aligned}$$

where we put "$+\infty$" + "$-\infty$" := "$+\infty$". The map $W: \ h \to w(h)$ on the sample space $\mathbf{H}$ defines the random total cost, and the performance of a control strategy π is given by $v^\pi_{\mathsf{P}_0} = \mathsf{E}^\pi_{\mathsf{P}_0}[W]$. Here and below,

$$\begin{aligned} \mathsf{E}^\pi_{\mathsf{P}_0}[W] &:= \mathsf{E}^\pi_{\mathsf{P}_0}[W^+] + \mathsf{E}^\pi_{\mathsf{P}_0}[W^-]; \quad \text{"}+\infty\text{"} + \text{"}-\infty\text{"} := \text{"}+\infty\text{"}; \\ W^+ &:= \max\{0, W\}, \qquad W^- := \min\{0, W\}. \end{aligned} \tag{1.4}$$

The aim is to solve problem (1.1), i.e. to construct an *optimal* control strategy π^*. We say that a strategy $\hat\pi$ *dominates* strategy π is $v^{\hat\pi}_x \le v^\pi_x$

for all $x \in \mathbf{X}$, with the strong inequality for some $x \in \mathbf{X}$. Conventions on infinities and expectation are discussed in Subsection 1.4.7.

Sometimes it is assumed that action a can take values only in measurable subsets $\mathbf{A}(x)$ depending on the previous state $x \in \mathbf{X}$. In such cases, one can modify the cost function $c_t(\cdot,\cdot)$ by putting $c_t(x,a) = \infty$ if $a \notin \mathbf{A}(x)$. Another possibility is to put $p_t(dy|x,a) \equiv p_t(dy|x,\hat{a})$ and $c_t(x,a) \equiv c_t(x,\hat{a})$ for all $a \notin \mathbf{A}(x)$, where $\hat{a} \in \mathbf{A}(x)$ is a fixed point.

As mentioned in the future chapters, all similar definitions and constructions hold also for infinite-horizon models with $T = \infty$.

Remark 1.1. Any finite-horizon model can be considered as an infinite-horizon MDP with expected total cost. It is sufficient to introduce an artificial costless *cemetery* (that is, absorbing) state and to make the process absorbed there at the additional decision epoch $T+1$ with the cost $c_{T+1}(x,a) \equiv C(x)$ and without any future cost. The model becomes homogeneous if we incorporate the number of the decision epoch as the second component of the state: (x,t).

1.3 Dynamic Programming Approach

Bellman formulated his famous *principle of optimality (the Bellman principle)* as follows: "An optimal policy has the property that whatever the initial state and initial decision are, the remaining decisions must constitute an optimal policy with regard to the state resulting from the first decision." [Bellman (1957), Section 3.3].

The Bellman principle leads to the following equation

$$\begin{cases} v_T(x) = C(x); \\ v_{t-1}(x) = \inf_{a \in \mathbf{A}} \left\{ c_t(x,a) + \int_{\mathbf{X}} v_t(y) p_t(dy|x,a) \right\}, \end{cases} \tag{1.5}$$
$$t = T, T-1, \ldots, 1, \quad x \in \mathbf{X}.$$

Here

$$\int_{\mathbf{X}} v_t(y) p_t(dy|x,a) := \int_{\mathbf{X}} [v_t(y)]^+ p_t(dy|x,a) + \int_{\mathbf{X}} [v_t(y)]^- p_t(dy|x,a)$$

and "$+\infty$" + "$-\infty$" := "$+\infty$".

Below, we assume that this *optimality* (or *'Bellman'*) equation has a measurable solution $v_t(x)$.

Suppose the cost functions $c_t(\cdot,\cdot)$ and $C(\cdot)$ are bounded. Then $v_t(x)$ coincides with the *Bellman function*

$$v_t^*(x) := \inf_\pi \mathsf{E}_{\mathsf{P}_0}^\pi \left[\sum_{\tau=t+1}^{T} c_\tau(X_{\tau-1}, A_\tau) + C(X_T) | X_t = x \right]. \tag{1.6}$$

Moreover, a control strategy π^* is optimal in problem (1.1) if and only if, for all $t = 1, 2, \ldots, T$, the following equality holds $\mathsf{P}_{\mathsf{P}_0}^{\pi^*}$ – a.s.

$$v_{t-1}(X_{t-1}) = \int_{\mathbf{A}} \left\{ c_t(X_{t-1}, a) + \int_{\mathbf{X}} v_t(y) p_t(dy|X_{t-1}, a) \right\} \pi_t^*(da|H_{t-1}) \tag{1.7}$$

(here $H_{t-1} = (X_0, A_1, X_1, \ldots, A_{t-1}, X_{t-1})$ is a random history). The minimal expected cost equals

$$v_{\mathsf{P}_0}^* := v_{\mathsf{P}_0}^{\pi^*} = \inf_\pi v_{\mathsf{P}_0}^\pi = \int_{\mathbf{X}} v_0(x) \mathsf{P}_0(dx). \tag{1.8}$$

Suppose a history $(x_0, a_1, \ldots, x_\tau) = h_\tau \in \mathbf{H}_\tau$, $0 \le \tau \le T$ is fixed. Then we can consider the control process A_t and the controlled process X_t as developing on the time interval $\{\tau+1, \tau+2, \ldots, T\}$ which is empty if $\tau = T$. If a control strategy π (in the initial model) is fixed, then one can build the strategic measure on $\mathbf{H}$, denoted as $\mathsf{P}_{h_\tau}^\pi$, in the same way as was described above, satisfying the 'initial condition' $\mathsf{P}_{h_\tau}^\pi(h_\tau \times (\mathbf{A} \times \mathbf{X})^{T-\tau}) = 1$. The most important case is $\tau = 0$; then we have just $\mathsf{P}_{x_0}^\pi$. Recall that $\mathsf{P}_{x_0}^\pi$ is another notation for $\mathsf{P}_{\mathsf{P}_0}^\pi$ in the case when $\mathsf{P}_0(\cdot)$ is concentrated at point x_0. In reality, $\mathsf{P}_{h_\tau}^\pi(\cdot)$ coincides with the conditional probability $\mathsf{P}_{\mathsf{P}_0}^\pi(\cdot|H_\tau = h_\tau)$ for $\mathsf{P}_{\mathsf{P}_0}^\pi$-almost all h_τ if the measure $\mathsf{P}_{\mathsf{P}_0}^\pi$ on $\mathbf{H}_\tau$ has full support: **Supp** $\mathsf{P}_{\mathsf{P}_0}^\pi = \mathbf{H}_\tau$. Recall that the support **Supp** μ of a finite measure μ on $(\mathbf{Y}, \mathcal{B}(\mathbf{Y}))$, where $\mathbf{Y}$ is a topological space, is the unique closed set $F \subset \mathbf{Y}$ such that $\mu(\mathbf{Y} \setminus F) = 0$ and $\mu(F \cap G) > 0$ for every open set G such that $F \cap G \neq \emptyset$. We introduce $v_{h_\tau}^\pi := \mathsf{E}_{h_\tau}^\pi \left[\sum_{t=\tau+1}^{T} c_t(X_{t-1}, A_t) + C(X_T) \right]$ and call a control strategy π^* *uniformly optimal* if

$$v_{h_\tau}^{\pi^*} = \inf_\pi v_{h_\tau}^\pi =: v_{h_\tau}^* \text{ for all } h_\tau \in \bigcup_{t=0}^{T} \mathbf{H}_t.$$

In this connection, the Bellman function

$$v_{x_0}^* := \inf_\pi v_{x_0}^\pi = v_0^*(x_0) = v_0(x_0) \tag{1.9}$$

(cf (1.6)) represents the minimal possible cost on the total interval $\{0, 1, \ldots, T\}$, if started from $X_0 = x_0$. Recall that $v_t(\cdot)$ is the (measurable) solution to the Bellman equation (1.5), $v_t^*(\cdot)$ is given by (1.6), and we assumed that the cost functions are bounded. If the latter condition is not satisfied, then equalities (1.9) may be not valid: see, e.g., example from Subsection 1.4.2. One can say that the dynamic programming approach is successful if

$$v_t^*(x) = v_t(x), \qquad t = 0, 1, \ldots, T, \quad x \in \mathbf{X}.$$

Sometimes it is necessary to underline the time horizon T. In such cases, we use notations $V_x^{\pi,T} := v_x^\pi$ and $V_x^T := \inf_\pi v_x^\pi = v_x^*$.

Suppose function $v_{h_\tau}^* \neq \pm\infty$ is finite and $\varepsilon > 0$ is fixed. We call a control strategy π *uniformly ε-optimal* if $v_{h_\tau}^\pi \leq v_{h_\tau}^* + \varepsilon$ for all $h_\tau \in \bigcup_{t=0}^T \mathbf{H}_t$. Similarly, in the case when $\inf_\pi v_{\mathsf{P}_0}^\pi \neq \pm\infty$, we call the strategy π *ε-optimal* if $v_{\mathsf{P}_0}^\pi \leq \inf_\pi v_{\mathsf{P}_0}^\pi + \varepsilon$ (see (1.1)). Uniformly (ε)-optimal strategies are sometimes called *persistently (ε)-optimal* [Kertz and Nachman (1979)].

The dynamic programming approach leads to the following result. Suppose the cost functions $c_t(\cdot,\cdot)$ and $C(\cdot)$ are bounded and optimality equation (1.5) has a measurable solution $v_t(x)$. Then a control strategy π^* is uniformly optimal if and only if equality

$$v_{t-1}(x_{t-1}) = \int_{\mathbf{A}} \left\{ c_t(x_{t-1}, a) + \int_{\mathbf{X}} v_t(y) p_t(dy|x_{t-1}, a) \right\} \pi_t^*(da|h_{t-1}) \tag{1.10}$$

holds for all $t = 1, 2, \ldots, T$ and $h_{t-1} = (x_0, a_1, \ldots, x_{t-1}) \in \mathbf{H}_{t-1}$. In this case,

$$v_{h_t}^{\pi^*} = v_t(x_t), \quad \forall\ h_t = (x_0, a_1, \ldots, x_t), \quad t = 0, 1, \ldots, T. \tag{1.11}$$

Very often, the infimum in (1.5) is provided by a mapping $a = \varphi_t^*(x)$, so that Markov selectors form a sufficient class for solving problem (1.1). Another general observation is that, for an arbitrarily fixed initial distribution P_0, a uniformly optimal strategy is also optimal under the imposed above conditions, but not *vice versa*: see Subsection 1.4.3.

Note that in general, even if equality (1.7) (or (1.10)) holds, it can happen that the strategy π^* is not (uniformly) optimal: see Subsection 1.4.3.

Remark 1.2. We mainly study minimization problems. If one considers $v_{\mathsf{P}_0}^\pi \to \sup_\pi$ instead of (1.1), then all the statements remain valid if min and inf are replaced with max and sup. Simultaneously, the convention about the infinities should be modified: "$+\infty$" + "$-\infty$" := "$-\infty$".

1.4 Examples

The first five subsections demonstrate some pathological situations when only non-Markov strategies can be optimal, when the Bellman principle is violated etc. Subsection 1.4.6 is about the turnpike property of MDP. In Subsections 1.4.7 and 1.4.8, the possible conventions on infinities and the definition of a semi-continuous model are discussed. Possible "bad" properties of the Bellman function are presented in Subsections 1.4.9 and 1.4.10. In Subsections 1.4.11 and 1.4.12 some surprising properties of the so-called non-atomic MDPs are demonstrated, and the importance of all conditions in the known theorems is confirmed by counterexamples. Other unexpected results are provided in Subsections 1.4.13–1.4.17, with the last three of them being just real-life situations. Note also that the last example in Subsection 2.3.16 is in fact a finite-horizon model.

1.4.1 *Equality (1.7) is not necessary for optimality*

The first simple example based on [Bäuerle and Rieder (2011), Ex. 2.3.10] confirms that it is not necessary for A_t to provide the infimum in (1.5) for all t and all $x \in \mathbf{X}$.

Example. Let $T = 2$, $\mathbf{X} = \{0, 1\}$, $\mathbf{A} = \{0, 1\}$, $p_t(a|x, a) \equiv 1$, $c_t(x, a) \equiv -a$, $C(x) = 0$ (see Fig. 1.1). The initial distribution P_0 is arbitrarily fixed.

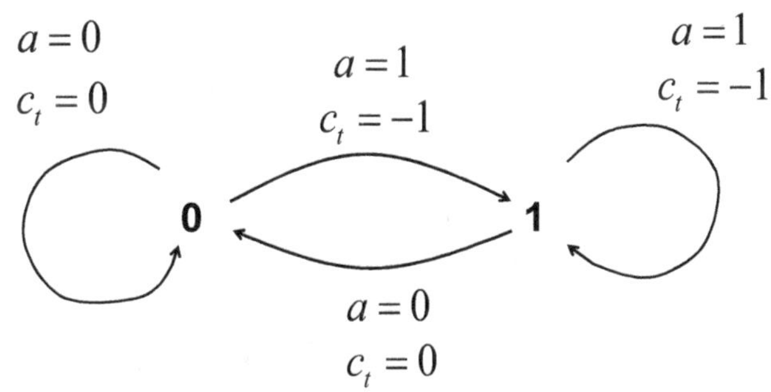

Fig. 1.1 Selector $\varphi_t(x) = \mathbb{I}\{t = 1\} + \mathbb{I}\{t > 1\}x$ is optimal, but not uniformly optimal.

Equation (1.5) has a solution $v_2(x) = 0$; $v_1(x) = -1$; $v_0(x) = -2$ and all actions providing the minima equal 1. Thus, selector $\varphi^1_t(x) \equiv 1$ is optimal for any initial distribution. But selector

$$\varphi_t(x) = \begin{cases} 1, & \text{if } t = 1; \\ 1, & \text{if } t = 2 \quad \text{and } x = 1; \\ 0, & \text{if } t = 2 \quad \text{and } x = 0 \end{cases}$$

is also optimal because state 0 will not be visited at time 1 and we do not meet it at decision epoch 2.

Incidentally, selector φ^1 is *uniformly* optimal whereas the above selector φ is not. □

The second example here shows that, in case the cost functions $c_t(\cdot,\cdot)$ and $C(\cdot)$ are not bounded, even if all states appear at all time steps with positive probability, then it is still not necessary for equality (1.7) to hold in order for a strategy π^* to be optimal.

Example. Let $T=2$, $\mathbf{X}=\{1,2,\ldots\}$, $\mathbf{A}=\{0,1\}$, $p_t(y|x,a)\equiv\mathbb{I}\{y=x\}$, $c_1(x,a)=2^x$, $c_2(x,a)=a$, $C(x)\equiv 0$, $\mathsf{P}_0(x)=(1/2)^x$: see Fig. 1.2.

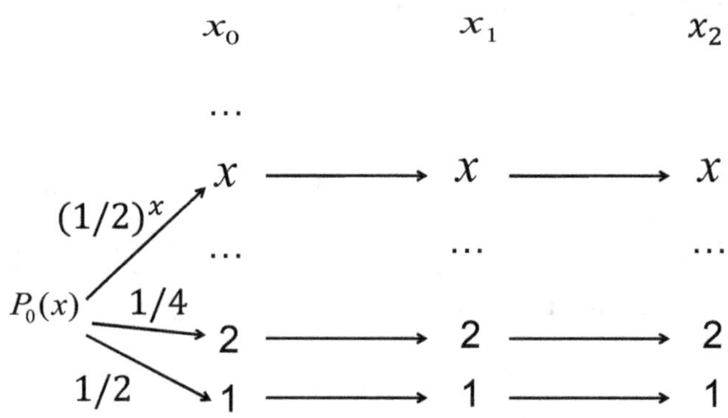

Fig. 1.2 Equality (1.7) is not necessary for optimality.

Equation (1.5) has a solution $v_2(x)\equiv 0$, $v_1(x)\equiv 0$, $v_0(x)=2^x$. For equality (1.7) to be valid, one needs to put $\pi_2^*(0|h_1)\equiv 1$. But, in this example, all strategies are equally optimal, even if $A_2=1$ (leading to $c_2(x,1)=1>c_2(x,0)=0$) because, for all strategies π,

$$\begin{aligned} v^\pi_{\mathsf{P}_0} &= \mathsf{E}^\pi_{\mathsf{P}_0}[c_1(X_0,A_1)] + \mathsf{E}^\pi_{\mathsf{P}_0}[c_2(X_1,A_2)+C(X_2)] \\ &= \sum_{x=1}^{\infty} 2^x(1/2)^x + \mathsf{E}^\pi_{\mathsf{P}_0}[c_2(X_1,A_2)+C(X_2)] = +\infty. \end{aligned}$$

□

1.4.2 *Markov or non-Markov strategy? Randomized or not? When is the Bellman principle violated? Unsuccessful dynamic programming approach*

It is well known that, for every control strategy π, there exists an *equivalent* Markov strategy π^{m} such that, for all $t=1,2,\ldots,T$

$$\mathsf{P}^{\pi^{\mathrm{m}}}_{\mathsf{P}_0}(X_{t-1}\in\Gamma^X, A_t\in\Gamma^A) = \mathsf{P}^{\pi}_{\mathsf{P}_0}(X_{t-1}\in\Gamma^X, A_t\in\Gamma^A)$$

and

$$\mathsf{P}_{\mathsf{P}_0}^{\pi^{\mathrm{m}}}(X_T \in \Gamma^X) = \mathsf{P}_{\mathsf{P}_0}^{\pi}(X_T \in \Gamma^X)$$

for any $\Gamma^X \in \mathcal{B}(\mathbf{X})$ and $\Gamma^A \in \mathcal{B}(\mathbf{A})$. (See Lemma 2.1 or [Derman and Strauch (1966)].) Therefore,

$$\begin{aligned} v_{\mathsf{P}_0}^{\pi} &= \sum_{t=1}^{T} \mathsf{E}_{\mathsf{P}_0}^{\pi}[c_t(X_{t-1}, A_t)] + \mathsf{E}_{\mathsf{P}_0}^{\pi}[C(X_T)] \qquad (1.12) \\ &= \sum_{t=1}^{T} \mathsf{E}_{\mathsf{P}_0}^{\pi^{\mathrm{m}}}[c_t(X_{t-1}, A_t)] + \mathsf{E}_{\mathsf{P}_0}^{\pi^{\mathrm{m}}}[C(X_T)] = v_{\mathsf{P}_0}^{\pi^{\mathrm{m}}} \end{aligned}$$

in the event that the sums of the type "$+\infty$" + "$-\infty$" do not appear. That is why the optimization in the class of all strategies can be replaced by the optimization in the class of Markov strategies, assuming the initial distribution is fixed.

The following example, published in [Piunovskiy (2009a)], shows that the requirement concerning the infinities is essential.

Example. Let $\mathbf{X} = \{0, \pm 1, \pm 2, \ldots\}$, $\mathbf{A} = \{0, -1, -2, \ldots\}$, $T = 3$, $\mathsf{P}_0(0) = 1$,

$$p_1(y|x,a) = \begin{cases} \frac{3}{|y|^2\pi^2}, & \text{if } y \neq 0; \\ 0, & \text{if } y = 0, \end{cases} \qquad p_2(0|x,a) = p_3(0|x,a) \equiv 1,$$

$c_1(x,a) \equiv 0$, $c_2(x,a) = x$, $c_3(x,a) = a$, $C(x) = 0$ (see Fig. 1.3).

Since actions A_1 and A_2 play no role, we shall consider only A_3.

The dynamic programming approach results in the following

$$v_3(x) = 0, \quad v_2(x) = -\infty, \quad v_1(x) = -\infty, \quad v_0(x) = -\infty.$$

Consider the Markov control strategy π^* with

$$\pi_3^*(0|x_2) = 0,\ \pi_3^*(a|x_2) = \frac{6}{|a|^2\pi^2} \text{ for } a < 0. \qquad (1.13)$$

Here equalities (1.10) hold because

$$\sum_{i=1}^{\infty} \frac{(-i) \times 6}{i^2\pi^2} = -\infty = v_2(x), \qquad x + v_2(0) = -\infty = v_1(x),$$

$$0 + \sum_{|y|=1}^{\infty} \frac{3}{|y|^2\pi^2} \cdot \text{“}-\infty\text{”} = -\infty = v_0(x).$$

On the other hand, for any Markov strategy π^{m}, $v_{\mathsf{P}_0}^{\pi^{\mathrm{m}}} = +\infty$. In fact, let $\hat{a} = \max\{j : \pi_3^{\mathrm{m}}(j|0) > 0\}$; $0 \geq \hat{a} > -\infty$, and consider the random

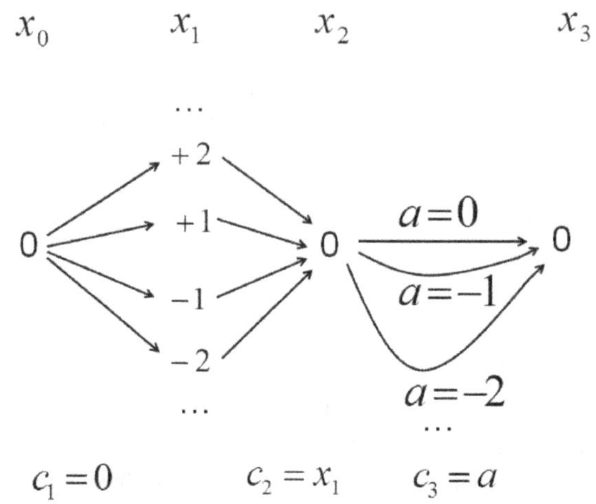

Fig. 1.3 Only a non-Markov randomized strategy can satisfy equalities (1.7) and (1.10) and be optimal and uniformly optimal.

variable $W^+ = (X_1 + A_3)^+$. It takes values $1, 2, 3, \ldots$ with probabilities not smaller than

$$p_1(1-\hat{a}|0,a)\pi_3^{\mathrm{m}}(\hat{a}|0) = \frac{3\pi_3^{\mathrm{m}}(\hat{a}|0)}{|1-\hat{a}|^2\pi^2},$$

$$p_1(2-\hat{a}|0,a)\pi_3^{\mathrm{m}}(\hat{a}|0) = \frac{3\pi_3^{\mathrm{m}}(\hat{a}|0)}{|2-\hat{a}|^2\pi^2},$$

$$p_1(3-\hat{a}|0,a)\pi_3^{\mathrm{m}}(\hat{a}|0) = \frac{3\pi_3^{\mathrm{m}}(\hat{a}|0)}{|3-\hat{a}|^2\pi^2},$$

$$\ldots$$

These expressions come from trajectories $(x_0 = 0, x_1 = i - \hat{a}, a_1, x_2 = 0, a_2 = \hat{a}, x_3 = 0)$. That means

$$\mathsf{E}_{\mathsf{P}_0}^{\pi^{\mathrm{m}}}[W^+] \geq \pi_3^{\mathrm{m}}(\hat{a}|0) \sum_{i=1}^{\infty} \frac{3i}{|i-\hat{a}|^2\pi^2} = +\infty$$

and $v_{\mathsf{P}_0}^{\pi^{\mathrm{m}}} = \mathsf{E}_{\mathsf{P}_0}^{\pi^{\mathrm{m}}}[W] = +\infty$. In particular, $v_{\mathsf{P}_0}^{\pi^*} = +\infty$.

At the same time, there exist optimal non-Markov strategies providing $v_{\mathsf{P}_0}^{\pi} = -\infty$. For example, put

$$a_3 = \varphi_3^*(x_1) = \begin{cases} -x_1, & \text{if } x_1 > 0; \\ 0, & \text{if } x_1 \leq 0. \end{cases} \tag{1.14}$$

Then $W = X_1 + A_3 = X_1^- \le 0$ and $\mathsf{E}^{\varphi}_{\mathsf{P}_0}[W] = \sum_{i=1}^{\infty}(-i)\frac{3}{i^2\pi^2} = -\infty$. Note that $x_0 = 0$; so $v^{\varphi^*}_{x_0} = \inf_\pi v^\pi_{x_0} = \inf_\pi v^\pi_{\mathsf{P}_0} = -\infty$, meaning that no-one Markov control strategy (including π^*) can be optimal or uniformly optimal. For the Markov strategy, equivalent to φ^*, formula (1.12) is not valid because one cannot represent the expectation of the sum as the sum of expectations.

The presented optimal control strategy φ^* does not satisfy either equalities (1.7), or (1.10). Indeed, for example, for the history $\hat{h}_2 = (0, a_1, 1, a_2, 0)$ with positive $\mathsf{P}^{\varphi^*}_{\mathsf{P}_0}$ probability, on the right-hand side of (1.7) and (1.10) we have

$$c_3(x_2 = 0, a_3 = \varphi_3(1)) + 0 = \varphi^*_3(1) = -1,$$

whereas on the left-hand side we have $v_2(0) = -\infty$. Since for this history $v^{\varphi^*}_{\hat{h}_2} = c_3(0, a_3) = -1$ and $\inf_\pi v^\pi_{\hat{h}_2} = -\infty$, the optimal control strategy φ^* is not uniformly optimal. This reasoning is correct for an arbitrary selector, meaning that non-randomized strategies cannot satisfy equalities (1.7) and (1.10) and cannot be uniformly optimal.

Therefore, only a non-Markov randomized strategy can satisfy the equalities (1.7) and (1.10) and be simultaneously optimal and uniformly optimal. As an example, take

$$\pi_3(j|x_1) = \begin{cases} \dfrac{6}{(x_1 + j - 1)^2\pi^2}, & \text{if } j \le -x_1 \text{ and } x_1 > 0; \\ \dfrac{6}{j^2\pi^2}, & \text{if } j < 0 \text{ and } x_1 < 0; \\ 0, & \text{otherwise.} \end{cases}$$

□

In the model investigated, for every optimal control strategy π^* we have $v^{\pi^*}_{x_0} = v_0(x_0)$. It can happen that this statement is false. Consider the following modification of the MDP being studied (see Fig. 1.4).

Example.

$$\mathbf{A} = \{0\}, \quad p_3(y|x, a) = \begin{cases} \dfrac{6}{|y|^2\pi^2}, & \text{if } y < 0; \\ 0, & \text{otherwise,} \end{cases} \qquad c_3(x, a) = 0, \quad C(x) = x.$$

Actually the process is not controlled and can be interpreted as a version of the previous MDP under a fixed Markov control strategy with distribution $\pi_3(\cdot|x) = p_3(\cdot|x, a)$. We know that the total expected cost here equals $+\infty$. Thus, in this modified model for the optimal control strategy (which is unique: $\varphi_t(x) \equiv 0$) we have $v^{\varphi}_{x_0} = v^*_{x_0} = v^*_0(x_0) = +\infty$. At the same time, the optimality equation (1.5) still gives $v_3(x) = x$, $v_2(x) = -\infty$,

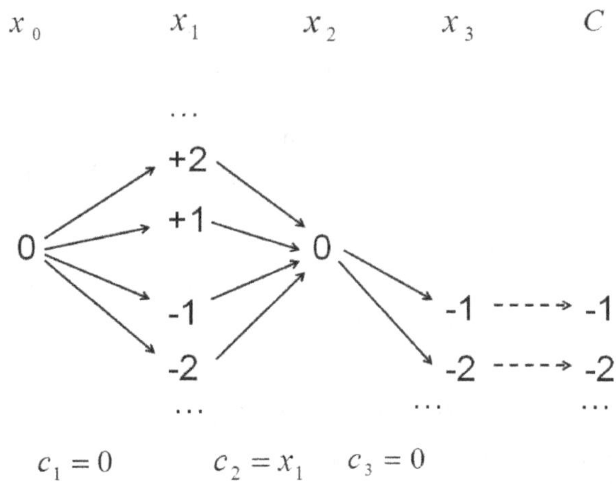

Fig. 1.4 $v_0(x_0) = -\infty < +\infty = v^{\varphi}_{x_0} = \inf_\pi v^\pi_{x_0}$.

$v_1(x) = -\infty$, and $v_0(x_0) = -\infty$: equalities (1.9) are not valid and the dynamic programming approach is not successful. □

Another similar example, illustrating that $v_0(x_0) = -\infty$ and $v_0^*(x_0) = v^*_{x_0} = \inf_\pi v^\pi_{x_0} = +\infty$ at some x_0, is presented in [Bertsekas and Shreve (1978), Section 3.2, Ex. 3]. See also [Hinderer (1977), Ex.3], where the maximization problem was studied and $v_t(x_0) = -\infty$, $v_t^*(x_0) = 0$ for $x_0 \geq 0$.

In the examples presented above, one should take into account the previous state(s) of the controlled process in order to control in the optimal way. This means, the Bellman principle about the sufficiency of the *current* state is violated.

Finally, let us present a very simple one-step model ($T = 1$) with a negative cost function, where only a randomized strategy is optimal.

Example. Let $\mathbf{X} = \{1\}$ be a singleton; $\mathbf{A} = \{1, 2, \ldots\}$; $c_1(x, a) = -2^{-a}$; $C(x) = 0$. For any selector $a = \varphi_1(x)$, $v^{\varphi}_{\mathsf{P}_0} = -2^a > -\infty$, but $\inf_\pi\{v^\pi_{\mathsf{P}_0}\} = -\infty$, and this infimum is attained, e.g., by the randomized strategy $\pi_1^*(a|x) = \left(\frac{1}{2}\right)^a$:

$$v^{\pi^*}_{\mathsf{P}_0} = \sum_{a=1}^{\infty} -2^a \left(\frac{1}{2}\right)^a = -\infty.$$

Compare with example from Subsection 2.3.19 with the infinite horizon. □

One can slightly modify the initial example (see Fig. 1.3) by ignoring the initial step and putting

$$\mathsf{P}_0(x) = \begin{cases} \dfrac{3}{x^2\pi^2}, & \text{if } x \neq 0; \\ 0, & \text{otherwise.} \end{cases}$$

The number of time moments decreases by 1 and $T = 2$. We still have that, for any Markov strategy π^{m}, $v^{\pi^{\mathrm{m}}}_{\mathsf{P}_0} = +\infty$, so that none of them is optimal. At the same time, the non-optimal strategy π^* as in (1.13) is now uniformly optimal.

1.4.3 *Optimal and uniformly optimal strategies*

As explained in Section 1.3, if the cost functions $c_t(\cdot,\cdot)$ and $C(\cdot)$ are bounded and optimality equation (1.5) has a measurable solution, then the dynamic programming approach is very fruitful. In this case, any uniformly optimal strategy is also optimal. On the other hand, suppose $\mathsf{P}_0(dx) = \delta_{x_0}(dx)$ and a state $\hat{x} \in \mathbf{X}$ cannot be reached under any strategy. Then, obviously, the actions in the isolated state $\hat{x}$ can be arbitrary, and an optimal strategy is not necessarily uniformly optimal.

In case the cost functions $c_t(\cdot,\cdot)$ and $C(\cdot)$ are not bounded (below or above), the situation becomes more complicated. Even if all states communicate, an optimal strategy can be not uniformly optimal. The trivial example is as follows:

Example. $T = 2$, $\mathbf{X} = \{1,2\}$, $\mathbf{A} = \{1,2\}$, $c_1(x,a) = -\infty$, $c_2(x,a) = a$, $C(x) = 0$; the initial distribution and the transition probabilities can be arbitrary. Now all strategies π are equally optimal, leading to $v^{\pi}_{\mathsf{P}_0} = -\infty$, but the stationary selector $\varphi(x) \equiv 2$ is not uniformly optimal. The 'bad' action 2 at the second decision epoch leads to $v^{\varphi}_{h_1} = 2 > 1 = \inf_\pi v^{\pi}_{h_1}$ (for all $h_1 = (x_0, a_1, x_1) \in \mathbf{H}_1$). □

Below, we present an example when a uniformly optimal strategy is not optimal and function $v_t(\cdot)$, the solution to (1.5), is finite. (Cf. the end of Subsection 1.4.2, where the non-optimal, but uniformly optimal strategy was also built.)

Firstly, we provide several statements whose proofs can be found in [Piunovskiy (2009a)].

Lemma 1.1. *Suppose the cost functions $c_t(\cdot,\cdot)$ and $C(\cdot)$ are finite-valued and the optimality equation (1.5) has a measurable solution. Then, for any control strategy π, for each $h_t = (x_0, a_1, \ldots, x_t) \in \mathbf{H}_t$, $t = 0, 1, \ldots, T$,* *inequality* $v^\pi_{h_t} \geq v_t(x_t)$ *is valid. (Note that function* $v_t(x)$ *can take values* $\pm\infty$.*)*

In the case when a strategy π^* *satisfies equality (1.10) and* $v^{\pi^*}_{h_t} < +\infty$ *for all* $h_t \in \mathbf{H}_t$, $t = 0, 1, \ldots, T$, *equality*

$$v^{\pi^*}_{h_t} \equiv v_t(x_t) = \inf_\pi v^\pi_{h_t}$$

is valid, so that π^* *is uniformly optimal.*

Corollary 1.1. *Under the conditions of Lemma 1.1, for each strategy* π,

$$v^\pi_{\mathsf{P}_0} \geq \int_{\mathbf{X}} v_0(x_0)\mathsf{P}_0(dx_0),$$

so that π^* *is optimal if* $v^{\pi^*}_{\mathsf{P}_0} = \int_{\mathbf{X}} v_0(x_0)\mathsf{P}_0(dx_0)$.

Corollary 1.2. *Under the conditions of Lemma 1.1, if a strategy* π^* *satisfies equality (1.10),* $v^{\pi^*}_{\mathsf{P}_0} < +\infty$, *and* $v^{\pi^*}_{h_t} < +\infty$ *for all* $h_t \in \mathbf{H}_t$, $t = 0, 1, \ldots, T$, *then the control strategy* π^* *is optimal and uniformly optimal.*

Example. Let $\mathbf{X} = \{\pm 1, \pm 2, \ldots\}$, $\mathbf{A} = \{0, 1\}$, $T = 1$,

$$\mathsf{P}_0(x) = \begin{cases} \dfrac{6}{|x|^2\pi^2}, & \text{if } x > 0; \\ 0, & \text{otherwise,} \end{cases}$$

$$p_1(y|x,1) = \mathbb{I}\{y = -x\}, \qquad p_1(y|x,0) = \begin{cases} 1/4, & \text{if } y = 2x; \\ 3/4, & \text{if } y = -2x; \\ 0, & \text{otherwise.} \end{cases}$$

$c_1(x,a) = x$, $C(x) = x$ (see Fig. 1.5).

The dynamic programming approach results in the following: $v_1(x) = x$, $v_0(x) = 0$, and the both actions provide the minimum in equation (1.5).

Consider action 1: $\varphi^1_1(x) = 1$. This control strategy φ^1 is uniformly optimal, since

$$v^{\varphi^1}_{x_0} = x_0 - x_0 = 0 = v_0(x_0) = \inf_\pi v^\pi_{x_0} = v^*_{x_0}.$$

It is also optimal by Corollary 1.1, because only trajectories $(x_0, a_1 = 1, x_1 = -x_0)$ are realized, for which $W = X_0 - X_0 = 0$, so that $v^{\varphi^1}_{\mathsf{P}_0} = 0 = \int_{\mathbf{X}} v_0(x)\mathsf{P}_0(dx)$.

Consider now action 0: $\varphi^0_1(x) = 0$. This control strategy φ^0 is also uniformly optimal, since

$$v^{\varphi^0}_{x_0} = x_0 + \frac{1}{4}(2x_0) + \frac{3}{4}(-2x_0) = 0 = v_0(x_0) = \inf_\pi v^\pi_{x_0} = v^*_{x_0}.$$

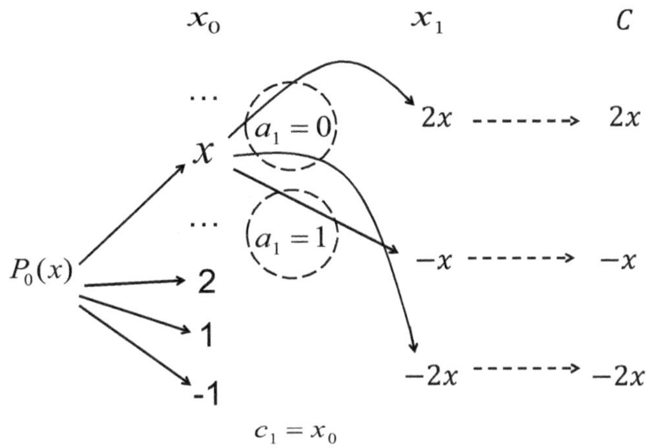

Fig. 1.5 Uniformly optimal stratwegy φ^0 is not optimal.

It also satisfies equality (1.10). However, it is not optimal, because

$$\mathsf{E}_{\mathsf{P}_0}^{\varphi^0}[W^+] = \sum_{x=1}^{\infty} 3x \cdot 1/4 \cdot \frac{6}{x^2\pi^2} = +\infty, \quad \mathsf{E}_{\mathsf{P}_0}^{\varphi^0}[W^-] = \sum_{x=1}^{\infty} (-x) \cdot 3/4 \cdot \frac{6}{x^2\pi^2} = -\infty,$$

so that $v_{\mathsf{P}_0}^{\varphi^0} = +\infty > v_{\mathsf{P}_0}^{\varphi^1} = 0$.

This example, first published in [Piunovskiy (2009a)], shows that the condition $v_{\mathsf{P}_0}^{\pi^*} < +\infty$ in Corollary 1.2 is important. □

1.4.4 *Martingales and the Bellman principle*

Recall that an adapted real-valued stochastic process $Y_0, Y_1, \dots$ on a stochastic basis $(\Omega, \mathcal{B}(\Omega), \{\mathcal{F}_t\}_{t=0}^T, \mathsf{P})$ is called a *martingale* if

$$\forall t = 0, 1, \dots, T-1 \qquad Y_t = \mathsf{E}[Y_{t+1}|\mathcal{F}_t] \quad \mathsf{P}\text{-a.s.}$$

If a control strategy π is fixed, we have the $\mathcal{F}_t$-adapted stochastic process X_t (the basis was constructed in Section 1.2) which makes possible constructing the following $\mathcal{F}_t$-adapted process, sometimes called the *estimating* process,

$$Y_t^\pi = \sum_{i=1}^{t} c_i(X_{i-1}, A_i) + v_t(X_t).$$

Here $v_t(\cdot)$ is a (measurable) solution to the optimality equation (1.5). If the cost functions $c_t(\cdot,\cdot)$ and $C(\cdot)$ are bounded from below and the function $v_t(\cdot)$ is finite-valued, then Y_t^π is a martingale if and only if the strategy π is optimal [Boel (1977)]; [Piunovskiy (1997), Remark 8]. See also Theorem 9.5.5 of [Hernández-Lerma and Lasserre (1999)].

The first example from Subsection 1.4.2 shows that the latter statement can be false. Take the optimal strategy φ^* defined by (1.14). Since

$$v_0(x) = v_1(x) = v_2(x) = -\infty,$$

we have

$$Y_2^{\varphi^*} = X_1 + v_2(X_2) = -\infty.$$

At the same time, $v_3(x) \equiv 0$ and

$$Y_3^{\varphi^*} = X_1 + A_3 = X_1^-,$$

so that $\mathsf{E}[Y_3^{\varphi^*}|\mathcal{F}_2] = X_1^- \neq Y_2^{\varphi^*}$.

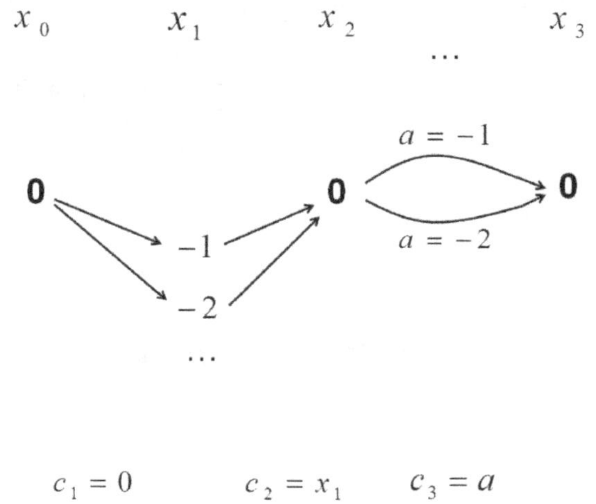

Fig. 1.6 Estimating process is not a martingale.

When we consider the strictly negative modification of this example, presented in Fig. 1.6, with $\mathbf{A} = \{-1,-2\}$, $p_1(y|x,a) = \frac{6}{|y|^2\pi^2}$, $p_2(0|x,a) = p_3(0|x,a) \equiv 1$, we still see that the optimal selector $\varphi_3^*(x_1) \equiv -1$ (like in the original example, the actions A_1 and A_2 play no role), providing $v_{\mathsf{P}_0}^{\varphi^*} = -\infty$, leads to the process $Y_t^{\varphi^*}$ which is not a martingale:

$$v_3(x) = 0, \quad v_2(x) = -2, \quad v_1(x) = x - 2, \quad v_0(x) = -\infty;$$

$$\mathsf{E}[Y_3^{\varphi^*}|\mathcal{F}_2] = X_1 - 1 \neq Y_2^{\varphi^*} = X_1 - 2.$$

The point is that the Bellman principle is violated: action $a_3 = -1$ is definitely not optimal at state $x_2 = 0$; nevertheless, it is optimal for the whole process on the time horizon $t = 0, 1, 2, 3$. The substantial negative cost $c_2(x_1, a_2) = x_1$ on the second step improves the performance up to $-\infty$.

□

1.4.5 *Constrained optimization*

Suppose we have two different cost functions ${}^1c_t(x,a)$ and ${}^2c_t(x,a)$, along with ${}^1C(x)$ and ${}^2C(x)$. Then every control strategy π results in two performance functionals ${}^1v^\pi_{\mathsf{P}_0}$ and ${}^2v^\pi_{\mathsf{P}_0}$ defined similarly to (1.1). Usually, objectives ${}^1v^\pi_{\mathsf{P}_0}$ and ${}^2v^\pi_{\mathsf{P}_0}$ are inconsistent, so that there does not exist a control strategy providing $\min_\pi {}^1v^\pi_{\mathsf{P}_0}$ and $\min_\pi {}^2v^\pi_{\mathsf{P}_0}$ simultaneously. One can construct the Pareto set corresponding to non-dominated control strategies. Another approach sets out the passage to a *constrained* problem:

$$ {}^1v^\pi_{\mathsf{P}_0} \to \inf_\pi; \qquad {}^2v^\pi_{\mathsf{P}_0} \le d, \tag{1.15}$$

where d is a pre-assigned constant. Strategies satisfying the inequality in (1.15) are called *admissible*. In such cases, the method of Lagrange multipliers has proved to be effective, although constructing an optimal strategy becomes more complicated [Piunovskiy (1997)]. The dynamic programming approach can also be useful here [Piunovskiy and Mao (2000)], but only after some preliminary work.

We present an example similar to [Haviv (1996)] showing that, in the constrained optimization, the Bellman principle fails to hold and the optimal control strategy can look strange.

Example. Let $\mathbf{X} = \{1,2,3,4\}$; $\mathbf{A} = \{1,2\}$; $T = 1$, $\mathsf{P}_0(1) = \mathsf{P}_0(2) = 1/2$;

$$p_1(y|1,a) = \mathbb{I}\{y=1\}; \qquad p_1(y|2,a) = \mathbb{I}\{y = 2+a\}.$$

Other transition probabilities play no role. The action in state 1 is immaterial.

Put

$$ {}^1c_1(x,a) = {}^2c_1(x,a) \equiv 0;$$

$$ {}^1C(x) = \begin{cases} 0, & \text{if } x = 1 \text{ or } 2; \\ 20, & \text{if } x = 3; \\ 10, & \text{if } x = 4; \end{cases} \qquad {}^2C(x) = \begin{cases} 0.2, & \text{if } x = 1 \text{ or } 2; \\ 0.05, & \text{if } x = 3; \\ 0.1, & \text{if } x = 4; \end{cases}$$

and consider problem (1.15) with $d = 0.125$ (see Fig. 1.7).

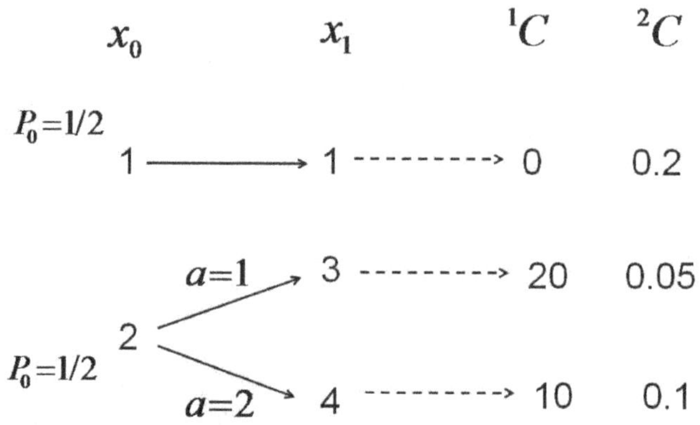

Fig. 1.7 Constrained optimization.

Intuition tells us that as soon as $X_0 = 2$ is realized, it is worth applying action $a = 2$ because that leads to the admissible value $^2C(X_1) = {}^2C(4) = 0.1 \leq 0.125$ and simultaneously to the minimal cost $^1C(X_1) = {}^1C(4) = 10$, when compared with $^1C(3) = 20$.

On the other hand, after taking into account the initial distribution, for such a control strategy we have:

$$^1v^{\pi}_{P_0} = 1/2 \cdot 10 = 5; \qquad ^2v^{\pi}_{P_0} = 1/2 \cdot 0.2 + 1/2 \cdot 0.1 = 0.15 > 0.125 = d,$$

meaning that the control strategy mentioned above is not admissible. One can see that the only admissible control strategy is $\varphi^*_1(2) = 1$ when $^1v^{\varphi^*}_{P_0} = 1/2 \cdot 20 = 10$, $^2v^{\varphi^*}_{P_0} = 1/2 \cdot 0.2 + 1/2 \cdot 0.05 = 0.125$. Therefore, in state 2 the decision maker should take into account not only the future dynamics, but also other trajectories ($X_0 = X_1 = 1$) that have already no chance of being realized; this means that the Bellman principle does not hold.

Suppose now that $^2C(1) = 0.18$. Then, action 2 in state 2 can be used but only with small probability. One should maximize that probability in order to minimize $^1v^{\pi}_{P_0}$. The solution to problem (1.15) is then given by

$$\pi^*_1(1|2) = 0.6, \qquad \pi^*_1(2|2) = 1 - \pi^*_1(1|2) = 0.4.$$

Remark 1.3. In the latter case, $^1v^{\pi^*}_{P_0} = 8$ and $^2v^{\pi^*}_{P_0} = 0.125$. Note that selector $\varphi_1(2) = 1$ provides $^1v^{\varphi_1}_{P_0} = 10$, $^2v^{\varphi_1}_{P_0} = 0.115$, and selector

$\varphi_2(2) = 2$ provides ${}^1v_{\mathsf{P}_0}^{\varphi_2} = 5$, ${}^2v_{\mathsf{P}_0}^{\varphi_2} = 0.14$. We see that no individual selector results in the same performance vector $({}^1v,\ {}^2v)$ as π^*. In a general MDP with finite horizon and finite spaces $\mathbf{X}$ and $\mathbf{A}$, *performance space* $\mathcal{V}_{\mathsf{P}_0} := \{({}^1v_{\mathsf{P}_0}^\pi,\ {}^2v_{\mathsf{P}_0}^\pi),\ \pi \in \Pi^{\text{All}}\}$ coincides with the (closed) convex hull of performance space $\mathcal{V}_{\mathsf{P}_0}^{\text{MN}} := \{({}^1v_{\mathsf{P}_0}^\varphi,\ {}^2v_{\mathsf{P}_0}^\varphi),\ \varphi \in \Pi^{\text{MN}}\}$. This statement can be generalized in many directions. Here, as usual, Π^{All} is the set of all strategies; Π^{MN} is the set of all Markov selectors. The details can be found in [Feinberg (1996)].

In the case when ${}^2C(1) \le 0.15$, constraint ${}^2v_{\mathsf{P}_0}^\pi \le d = 0.125$ becomes inessential, and the main objective ${}^1v_{\mathsf{P}_0}^\pi$ is minimized by the admissible control strategy $\varphi_1^*(2) = 2$. □

Note that, very often, the solution to a constrained MDP is given by a randomized Markov control strategy; however, there is still no reason to consider past-dependent strategies. Example 1 in [Frid (1972)], in the framework of a constrained discounted MDP, also shows that randomization is necessary. In general, the same is true also for the expected total costs [Altman (1999); Piunovskiy (1997)].

One can impose a constraint in a different way:

$$ {}^1v_{\mathsf{P}_0}^\pi \to \inf_\pi; \qquad {}^2W = \sum_{t=1}^{T} {}^2c_t(X_{t-1}, A_t) + {}^2C(X_T) \le d \ \ \mathsf{P}_{\mathsf{P}_0}^\pi - \text{a.s.} $$

After introducing the artificial random variables Z_t:

$$ Z_0 = 0, \qquad Z_t = Z_{t-1} + {}^2c_t(X_{t-1}, A_t), \quad t = 1, 2, \ldots, T, $$

one should modify the final cost:

$$ {}^1\tilde{C}(x, z) = \begin{cases} {}^1C(x), & \text{if } z + {}^2C(x) \le d; \\ +\infty, & \text{otherwise.} \end{cases} $$

In this new 'tilde' model, the Bellman principle holds. In the considered example (with the original data), $\tilde{\varphi}_1^*(x, z = 0) = 2$ is the optimal strategy leading to $\tilde{v}_0(x = 1, z = 0) = +\infty$, $\tilde{v}_0(x = 2, z = 0) = 10$. The infinite value shows that there are no admissible strategies if $x_0 = 1$, and $\tilde{v}_{\mathsf{P}_0}^* := \inf_\pi \tilde{v}_{\mathsf{P}_0}^\pi = +\infty$ in case $\mathsf{P}_0(1, 0) > 0$. If $\mathsf{P}_0(1, 0) = 0$, then $\tilde{v}_{\mathsf{P}_0}^* = 10$.

1.4.6 *Turnpike in finite MDPs*

In this subsection, we consider only homogeneous MDPs with finite state and action spaces $\mathbf{X}$ and $\mathbf{A}$, with transition probability $p(y|x, a)$, and with real-valued cost functions $C(\cdot)$ and $c(\cdot,\cdot)$, that is, homogeneous finite models. Recall that in such finite-horizon MDPs, there exists an optimal (not necessarily stationary) selector.

The turnpike property of an MDP can be formulated as follows.

If the time horizon is large enough, then the optimal control on the first several steps is also optimal for the infinite-horizon version of the model.

More about turnpikes for MDPs in [Lewis and Paul (2019); Piunovskiy (2023)].

First of all, we rewrite equation (1.5):

$$W^0(x) = C(x);$$
$$W^{k+1}(x) = \min_{a\in\mathbf{A}}\left\{c(x,a) + \sum_{y\in\mathbf{X}} W^k(y)p(y|x,a)\right\}, \quad k = 0,1,\ldots. \tag{1.16}$$

Clearly, $v_t(x) = W^{T-t}(x)$ for all $t = 0,1,\ldots,T$.

Equation (1.16) coincides with the value iteration for infinite-horizon MDPs with the expeted total cost and with the average cost: see (2.6) and (4.20). In the turnpike property, we have in mind the average cost MDP.

It is known that in finite models, there exists an AC-optimal stationary selector, e.g., the element φ^* of a canonical triplet: see Section 4.1.

Theorem 1.1. [Piunovskiy (2023), Thm. 1]. *Suppose the iterations (1.16) satisfy the requirement:* $\lim_{k\to\infty} sp(W^{k+1}(\cdot) - W^k(\cdot)) = 0$, *where*

$$sp(W(\cdot)) := \max_{x\in\mathbf{X}} W(x) - \min_{x\in\mathbf{X}} W(x)$$

is the so called "span-seminorm". Then there exists K such that, in the model with an arbitrary finite time horizon $T > K$, on the first steps $t = 1,2,\ldots,T-K$ the optimal selectors φ_t necessarily belong to the set of AC-optimal stationary selectors.

Here, with some abuse of notation, we call the particular mapping $\varphi_t(\cdot) : \mathbf{X} \to \mathbf{A}$ the selector on step t.

Sufficient conditions for the requirement $\lim_{k\to\infty} sp(W^{k+1}(\cdot)-W^k(\cdot)) = 0$ are provided in Section 5 of [Piunovskiy (2023)]. In particular, the following lemma holds true.

Lemma 1.2. *Suppose one of the following conditions is satisfied.*

- *For every stationary selector* $\varphi : \mathbf{X} \to \mathbf{A}$ *the transition matrix* $P(\varphi) := \{p(y|x,\varphi(x))\}_{(x,y)\in\mathbf{X}^2}$ *is irreducible and aperiodic, and* $p(x|x,\varphi(x)) > 0$ *for all* $x \in \mathbf{X}$.
- *For every AC-optimal stationary selector* φ^* *the matrix* $P(\varphi^*)$ *is aperiodic, and the optimal value of problem (4.1) (i.e., the element* $\rho(\cdot)$ *of a canonical triplet) does not depend on the initial state* x.

Then $\lim_{k\to\infty} sp(W^{k+1}(\cdot) - W^k(\cdot)) = 0$.

For the proof, see Corollary 1 and Lemma 6 of [Piunovskiy (2023)].

The following example, similar to Example 2 of [Piunovskiy (2023)], shows that, if $sp(W^{k+1}(\cdot) - W^k(\cdot))$ does not approach zero as $k \to \infty$ and φ^* is the unique AC-optimal stationary selector, then it can happen that, for each large enough T, for some initial states x_0, it is necessary to apply actions different from $\varphi^*(x_0)$ in order to gain the minimal expected cost over the finite interval $\{0, 1, 2, \ldots, T\}$.

Example. Let $\mathbf{X} = \{0, 1, \Delta\}$; $\mathbf{A} = \{\alpha, \beta, \gamma\}$;

$$p(1|0,\alpha) = p(0|1,\alpha) = 1; \quad p(1|0,\beta) = p(0|1,\beta) = 1 - \varepsilon;$$

$$p(0|0,\beta) = p(1|1,\beta) = \varepsilon;$$

$$p(\Delta|0,\gamma) = p(\Delta|1,\gamma) = p(\Delta|\Delta,\alpha) = p(\Delta|\Delta,\beta) = p(\Delta|\Delta,\gamma) = 1;$$

other transition probabilities equal zero. Δ is the absorbing state. Finally, put $C(x) \equiv 0$, $c(0,\alpha) = c(1,\alpha) = 0$, $c(0,\beta) = c(1,\beta) = h$, $c(0,\gamma) = -1$, $c(1,\gamma) = 0$, $c(\Delta, a) \equiv 2$. We assume that

$$\varepsilon \in (0,1); \; 0 < h < \varepsilon$$

See Fig. 1.8 for illustration.

As for the average cost version of this MDP, the only AC-optimal stationary selector is $\varphi^*(x) \equiv \alpha$, leading to $v_x^{\varphi^*} = \begin{cases} 0, & \text{if } x = 0, 1; \\ 2, & \text{if } x = \Delta : \end{cases}$ see Section 4.1. Of course, the actions in the cemetery Δ are immaterial. The canonical equations (4.2) give the canonical triplet $\langle \rho, h, \varphi^* \rangle$ with $\rho(0) = \rho(1) = 0$, $\rho(\Delta) = 2$, $h(\Delta) = 0$, $h(0) = h(1) \le -1$ is arbitrary, $\varphi^*(x) \equiv \alpha$. (The AC-optimal stationary selector φ^* is unique up to the values at Δ.) Conditions in Lemma 1.2 are not satisfied, as the stationary selector φ^* results in the periodic Markov chain if started from $x_0 = 0$ or $x_0 = 1$.

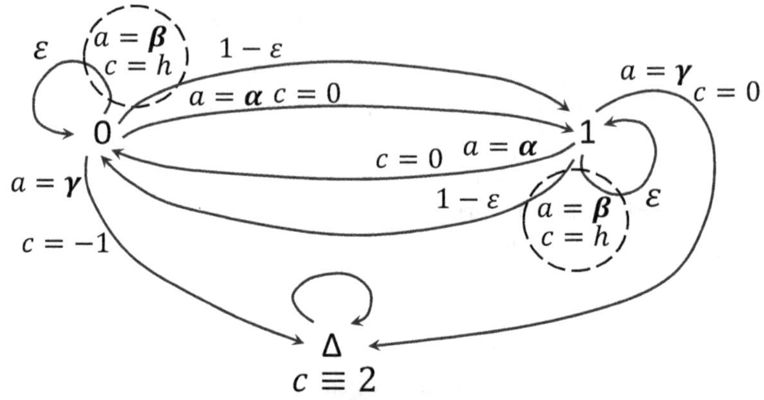

Fig. 1.8 Turnpike property fails.

Let us introduce the following two stationary selectors:

$$\varphi^o(0) = \alpha, \quad \varphi^o(1) = \beta, \quad \varphi^o(\Delta) = \alpha;$$

$$\varphi^e(0) = \beta, \quad \varphi^e(1) = \alpha, \quad \varphi^e(\Delta) = \alpha.$$

Neither of them is AC-optimal. But the both 'alternating' selectors

$$\varphi^1_t(x) = \begin{cases} \varphi^o(x), & \text{if } t \text{ is even;} \\ \varphi^e(x), & \text{if } t \text{ is odd;} \end{cases}$$

$$\varphi^2_t(x) = \begin{cases} \varphi^o(x), & \text{if } t \text{ is odd;} \\ \varphi^e(x), & \text{if } t \text{ is even} \end{cases}$$

are AC-optimal. To justify this, note that, starting from any initial state $x_0 \in \{0,1\}$, the (not AC-optimal) action β leading to the cost $h > 0$ is either never applied in the cases $\{x_0 = 0$ and $\varphi^2\}$ and $\{x_0 = 1$ and $\varphi^1\}$, or is applied L times, where the random variable L is geometric: $\mathsf{P}(L = l) = \varepsilon(1-\varepsilon)^{l-1}$, $l = 1, 2, \ldots$. After that, only the optimal action α is in use. Therefore,

$$0 \le \mathsf{E}^{\varphi^{1,2}}_{x_0}\left[\sum_{t=1}^{T} c(X_{t-1}, A_t)\right] \le \frac{h}{\varepsilon} \quad \text{and } v^{\varphi^{1,2}}_{x_0} = 0 = v^*_{x_0}. \tag{1.17}$$

The case of $x_0 = \Delta$ is trivial leading to $v^{\varphi^{1,2}}_{\Delta} = 2$. The situation is similar to the Parrondo's paradox, Subsection 4.3.33.

Lemma 1.3. *Let $w_1 = 0$ and $w_k = (1-\varepsilon)w_{k-1} - \varepsilon + h$ for $k \ge 2$. Then the following assertions hold.*

(a) $0 > w_{k-1} > w_k > -1 + \frac{h}{\varepsilon}$ *for all* $k \geq 2$.

(b) If $k \geq 1$ *is even, then, for* $W^k(\cdot)$ *as in (1.16),*

$$W^{k+1}(0) = -1, \quad W^{k+1}(1) = w_{k+1}, \quad W^{k+1}(\Delta) = 2(k+1),$$

and the minimum in (1.16) at $x \in \mathbf{X}$ *is provided by* $\varphi^o(x)$ *(only by* α *and* β *at* $x = 0$ *and* 1 *correspondingly).*

If $k \geq 1$ *is odd, then*

$$W^{k+1}(0) = w_{k+1}, \quad W^{k+1}(1) = -1, \quad W^{k+1}(\Delta) = 2(k+1),$$

and the minimum in (1.16) at $x \in \mathbf{X}$ *is provided by* $\varphi^e(x)$ *(only by* β *and* α *at* $x = 0$ *and* 1 *correspondingly).*

The index $o(e)$ *corresponds to the odd (even) value of* $k + 1$.

The proof is given in the appendix A.

Now suppose the time horizon $T \geq 2$ is even and return back to the Bellman function $v_t(x)$ and optimality equation (1.5). Since $v_1(x) = W^{T-1}(x)$, the actions providing the minimum in (1.5) at $t = 1$ are exactly those which provide the minimum in (1.16) at the odd value of $k = T - 1$, i.e., β and α for $x = 0$ and 1 correspondingly. We see that the first optimal selector is $\varphi_1 = \varphi^e \neq \varphi^*$, and, in case $x_0 = 0$, one has to apply action β which is different from $\varphi^*(0) = \alpha$.

If the time horizon $T \geq 2$ is odd, then the situation is symmetric: the first optimal selector is $\varphi_1 = \varphi^o \neq \varphi^*$, and, in case $x_0 = 1$, one has to apply action $\beta \neq \varphi^*(1) = \alpha$.

In any case, it is desirable to reach the state $X_{T-1} = 0$ in order to apply action $a = \gamma$ and obtain the negative cost $c(0, \gamma) = -1$. If, for example, the time horizon $T \geq 2$ is even, the initial state is $x_0 = 0$, and one applies only the actions $\varphi^*(x) \equiv \alpha$, then $X_{T-1} = 1$ and the total expected cost is 0. But applying $a = \beta$ on the odd steps t for $X_{t-1} = 0$ and on the even steps for $X_{t-1} = 1$ leads to the high enough probability to finish with $X_{T-1} = 0$ and obtain $c(0, \gamma) = -1$ on the last step T. The price for this is below $h/\varepsilon < 1$: see (1.17). Additional discussion of this example is in Subsection 4.3.16. □

1.4.7 *Conventions on expectation and infinities*

Some authors [Feinberg (1982), Section 4.1], [Feinberg (1987), Section 4], [Feinberg and Piunovskiy (2002)], [Puterman (1994), Section 5.2], while still accepting the rule "$+\infty$" + "$-\infty$" = "$+\infty$", suggest the following formula for calculating the performance functional:

$$v_{\mathsf{P}_0}^{\pi} = \mathsf{E}_{\mathsf{P}_0}^{\pi}\left[\sum_{t=1}^{T} c_t^+(X_{t-1}, A_t) + C^+(X_T)\right] + \mathsf{E}_{\mathsf{P}_0}^{\pi}\left[\sum_{t=1}^{T} c_t^-(X_{t-1}, A_t) + C^-(X_T)\right]. \tag{1.18}$$

In this situation, the value of $v_{\mathsf{P}_0}^{\pi}$ can only increase, compared with the convention (1.4), repeated below in (1.19). This means that most of the statements in Subsections 1.4.2 and 1.4.4 still hold. On the other hand, in the basic model presented in Fig. 1.3, any control strategy gives $v_{\mathsf{P}_0}^{\pi} = +\infty$ simply because

$$\mathsf{E}_{\mathsf{P}_0}^{\pi}\left[c_2^+(X_1, A_2)\right] = \mathsf{E}_{\mathsf{P}_0}^{\pi}\left[X_1^+\right] = +\infty.$$

(The same happens in the second example from Subsection 1.4.3.) Thus, any control strategy can be called optimal! But it seems intuitively clear that selector φ^* given in (1.14) is better than many other strategies because it compensates positive values of X_1. Similarly, it is natural to call the selector φ^1 optimal in the second example from Subsection 1.4.3.

If we accept convention (1.18), then it is easy to elaborate an example where the optimality equation (1.5) has a finite solution, but where, nevertheless, only a control strategy for which criterion (1.10) is violated, is optimal. The examples below, very similar to [Hinderer (1977), Ex. 1], appeared in [Piunovskiy (2009a)].

Example. Put $\mathbf{X} = \{0, 1, 2, \ldots\}$, $\mathbf{A} = \{0, 1\}$, $T = 2$, $\mathsf{P}_0(0) = 1$,

$$p_1(y|x, 0) = \mathbb{I}\{y = 0\}, \quad p_1(y|x, 1) = \begin{cases} \dfrac{6}{y^2\pi^2}, & \text{if } y > 0; \\ 0, & \text{if } y = 0, \end{cases}$$

$$p_2(y|x, a) = \mathbb{I}\{y = x\},$$

$$c_1(x, a) = 1 - a, \quad c_2(x, a) = x, \quad C(x) = -x.$$

Since action A_2 plays no role, we shall consider only A_1 (see Fig. 1.9).

The dynamic programming approach results in the following:

$$v_2(x) = C(x) = -x, \quad v_1(x) = x - x = 0, \quad v_0(x) = \min\{1 + 0,\ 0 + 0\} = 0,$$

and action $a_1 = 1$ provides this minimum. At the same time, for control strategy $\varphi_1(x_0) = 1$ we have

$$\mathsf{E}_{\mathsf{P}_0}^{\varphi}\left[c_2^+(X_1, A_2)\right] = \mathsf{E}_{\mathsf{P}_0}^{\varphi}\left[X_1\right] = +\infty,$$

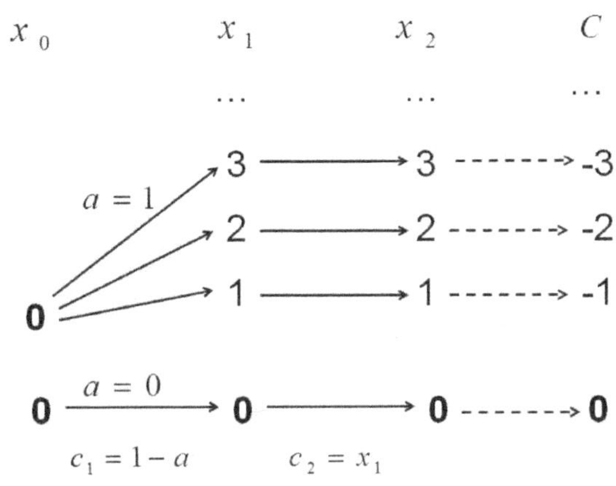

Fig. 1.9 $v_t(x)$ is finite, but $\inf_\pi v^\pi_{x_0} = v^{\varphi^*}_{x_0} = 1 > v_0(x_0) = 0$ provided formula (1.18) is accepted.

so that (1.18) gives $v^{\varphi}_{\mathsf{P}_0} = +\infty$. Hence, control strategy $\varphi^*_1(x_0) = 0$, resulting in $v^{\varphi^*}_{\mathsf{P}_0} = 1$, must be called optimal. In the opposite case, the control strategy $\varphi_1(x_0) = 1$ is optimal if we accept formula

$$v^\pi_{\mathsf{P}_0} = \mathsf{E}^\pi_{\mathsf{P}_0}\left[W^+\right] + \mathsf{E}^\pi_{\mathsf{P}_0}\left[W^-\right], \tag{1.19}$$

where $W = \sum_{t=1}^{T} c_t(X_{t-1}, A_t) + C(X_T)$ is the total realized cost. The big cost X_1 on the second step is totally compensated by the final (negative) cost $C(X_2) = -X_2 = -X_1$. □

The following example from [Feinberg (1982), Ex. 4.1] shows that a control strategy, naturally uniformly optimal under convention (1.19), is not optimal if we accept formula (1.18).

Example. Let $\mathbf{X} = \{0, 1, 2, \ldots\}$, $\mathbf{A} = \{1, 2, \ldots\}$, $T = 2$, $\mathsf{P}_0(0) = 1$, $p_1(y|x, a) = \mathbb{I}\{y = a\}$, $p_2(y|x, a) = \mathbb{I}\{y = 0\}$, $c_1(x, a) = -a$, $c_2(x, a) = x/2$, $C(x) = 0$ (see Fig. 1.10).

The dynamic programming approach (1.5) results in the following:

$$v_2(x) = C(x) = 0, \quad v_1(x) = x/2, \quad v_0(x) = \inf_{a \in \mathbf{A}} \{-a + a/2\} = -\infty.$$

No-one selector is optimal, and the randomized (Markov) strategy π^*, where

$$\pi^*_1(a|x_0) = \frac{6}{a^2\pi^2}, \quad \pi^*_2(a|x_2) \text{ is arbitrary,}$$

is uniformly optimal (and optimal) if we accept formula (1.19):

$$W = -\frac{a_1}{2}; \qquad v^{\pi^*}_{\mathsf{P}_0} = \mathsf{E}^{\pi^*}_{\mathsf{P}_0}[-a/2] = -\infty.$$

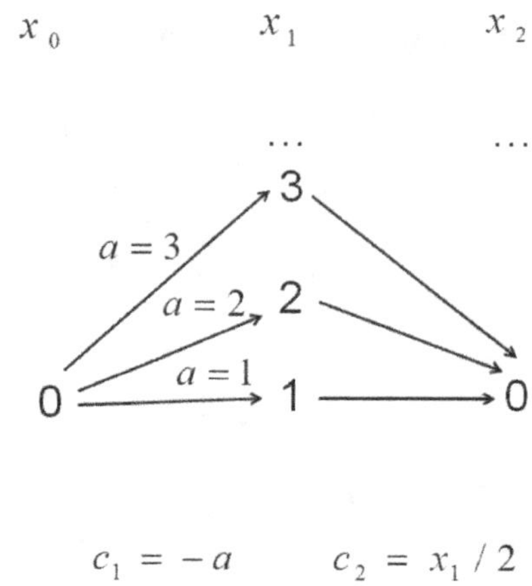

Fig. 1.10 No optimal strategies under convention (1.18).

On the other hand, if we accept convention (1.18), then

$$v_{x_0}^{\pi^*} = v_{\mathsf{P}_0}^{\pi^*} = \mathsf{E}_{\mathsf{P}_0}^{\pi^*}[c_1(x_0, a_1)] + \mathsf{E}_{\mathsf{P}_0}^{\pi^*}[c_2(x_1, a_2)] = -\infty + \infty = +\infty,$$

so that π^* (as well as any other control strategy) is not (uniformly) optimal. If $\mathsf{E}_{\mathsf{P}_0}^{\pi}[c_1(x_0, a_1)] = -\infty$, then $\mathsf{E}_{\mathsf{P}_0}^{\pi}[c_2(x_1, a_2)] = +\infty$, and hence, $v_{\mathsf{P}_0}^{\pi} = +\infty$. At the same time, for selectors $\varphi(x) = i \in \{1, 2, \dots\}$ we have $v_{\mathsf{P}_0}^{\varphi} = -i/2$, so $\inf_\pi v_{\mathsf{P}_0}^{\pi} = -\infty$. □

In the current book, we accept formula (1.19), as it was in the early articles (e.g., [Hinderer (1977)]), as well as, e.g., in [Hernández-Lerma and Lasserre (1999), (9.2.9),(9.3.1)].

We now discuss the possible conventions for infinity. Basically, if in (1.19) the expression "$+\infty$" + "$-\infty$" appears, then the random variable W is said to be not integrable. We have seen in Subsections 1.4.2, 1.4.3, and 1.4.4 that the convention

$$\text{“}+\infty\text{”} + \text{“}-\infty\text{”} = +\infty \tag{1.20}$$

leads to the violation of the Bellman principle and to other problems. One can show that all those principal difficulties also appear if we put "$+\infty$" + "$-\infty$" $= -\infty$. But convention (1.20) is still better because of the following.

Assume for a moment that "$+\infty$" + "$-\infty$" $= -\infty$. Then in the first example from Subsection 1.4.2 (Fig. 1.3), any strategy π provides $v_{\mathsf{P}_0}^{\pi} = -\infty$, so that all of them are equally optimal. But selector φ^* given by (1.14) seems better, and we want this to be mathematically confirmed. In a nutshell, if we meet "$+\infty$" + "$-\infty$" in (1.19), it is better to say that all such strategies are equally bad, rather than to accept that they are equally good.

Lemma 1.1 and Corollary 1.1 provided the lower boundary for the performance functional. That will not be the case if "$+\infty$" + "$-\infty$" $= -\infty$, as the following example shows.

Example. Let $\mathbf{X} = \{0, \pm 1, \pm 2, \ldots\}$, $\mathbf{A} = \{0, 1\}$, $T = 1$,

$$\mathsf{P}_0(x) = \begin{cases} \frac{6}{x^2\pi^2}, & \text{if } x > 0; \\ 0, & \text{otherwise,} \end{cases}$$

$$p_1(y|x,0) = \begin{cases} 1/4, & \text{if } y = 2x; \\ 3/4, & \text{if } y = -2x; \\ 0, & \text{otherwise,} \end{cases} \qquad p_1(y|x,1) = \mathbb{I}\{y = -x\},$$

$$c_1(x,a) = x + a, \qquad C(x) = x$$

(see Fig. 1.11).

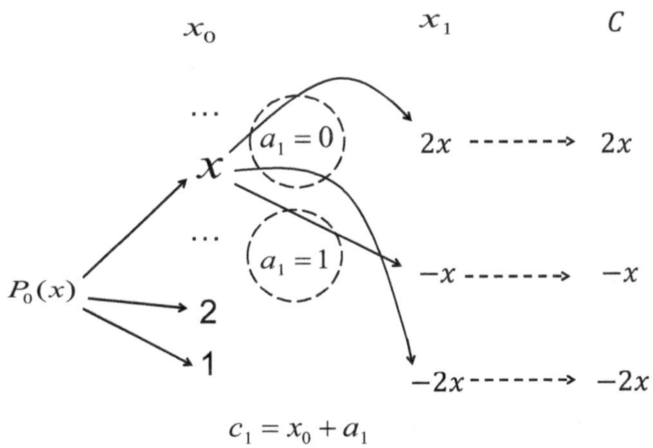

Fig. 1.11 No boundaries for $v^{\pi}_{\mathsf{P}_0}$.

The dynamic programming approach (1.5) results in the following:

$$v_1(x) = x, \qquad v_0(x) = 0,$$

and $a_1 = 0$ provides the minimum in the formula

$$v_0(x_0) = \min_a \{x_0 + a - x_0\} = 0.$$

At the same time, for control strategy $\varphi_1^0(x_0) \equiv 0$, we have $W = X_0 + X_1$ and

$$\mathsf{E}_{\mathsf{P}_0}^{\varphi^0}[W^+] = \sum_{i=1}^{\infty} 3i \cdot 1/4 \cdot \frac{6}{i^2\pi^2} = +\infty, \ \mathsf{E}_{\mathsf{P}_0}^{\varphi^0}[W^-] = \sum_{i=1}^{\infty} (-i) \cdot 3/4 \cdot \frac{6}{i^2\pi^2} = -\infty,$$

so that $v_{\mathsf{P}_0}^{\varphi^0} = -\infty < v_0(x_0) = 0$. Incidentally, for the selector $\varphi_1^1(x_1) = 1$ we have

$$W = X_0 + 1 + X_1 = X_0 + 1 - X_0 = 1,$$

and $v_{\mathsf{P}_0}^{\varphi^1} = 1 > v_0(x_0) = 0$. Thus, the solution to the optimality equation (1.5) provides no boundaries to the performance functional. □

Another simple example (in fact, of an uncontrolled MDP with $\mathbf{A} = \{\alpha\}$ being a singleton), when the dynamic programming approach results in the finite function $v_t(x)$, but, for the single control strategy π,

$$\mathsf{E}_{\mathsf{P}_0}^{\pi}[W^+] = +\infty \quad \text{and} \quad \mathsf{E}_{\mathsf{P}_0}^{\pi}[W^-] = -\infty,$$

was given in [Hinderer (1977), Ex. 2].

Everywhere else in this book we accept formula (1.20).

Remark 1.4. All the pathological situations in Subsections 1.4.2–1.4.7 appear only because we encounter expressions "$+\infty$" + "$-\infty$" when calculating expectations. That is why people impose the following conditions: for every strategy π and each degenerate initial distribution $\mathsf{P}_0(\cdot) = \delta_{x_0}(\cdot)$, $x_0 \in \mathbf{X}$,

$$\mathsf{E}_{\mathsf{P}_0}^{\pi}[c_t^+(x_{t-1}, a_t)] < +\infty, \ t = 1, 2, \ldots, T, \ \text{and} \ \mathsf{E}_{\mathsf{P}_0}^{\pi}[C^+(x_T)] < +\infty$$

or

$$\mathsf{E}_{\mathsf{P}_0}^{\pi}[c_t^-(x_{t-1}, a_t)] > -\infty, \ t = 1, 2, \ldots, T, \ \text{and} \ \mathsf{E}_{\mathsf{P}_0}^{\pi}[C^-(x_T)] > -\infty.$$

(See [Bertsekas and Shreve (1978), Section 8.1] and [Hernández-Lerma and Lasserre (1999), Section 9.3.2].)

To guarantee this, one can restrict oneself to 'negative' or 'positive' models with

$$c_t(x, a) \le K, \quad C(x) \le K, \ \text{or with} \ c_t(x, a) \ge -K, \quad C(x) \ge -K,$$

where $K > 0$. Another possibility is to consider 'contracting' models [Altman (1999), Section 7.7]; [Hernández-Lerma and Lasserre (1999), Section 8.3.2], where, for some positive function $\nu(\cdot)$ on $\mathbf{X}$ and constant $K > 0$,

$$\int_{\mathbf{X}} \nu(y) p_t(dy|x, a) \le K\nu(x) \ \text{and} \ \frac{|c_t(x, a)|}{\nu(x)} \le K, \quad \frac{|C(x)|}{\nu(x)} \le K$$

for all t, x, a.

1.4.8 *Definition of a semi-continuous model*

MDP is called *semi-continuous* if $\mathbf{X}$ and $\mathbf{A}$ are topological Borel spaces and the following condition is satisfied.

Condition 1.1.

(a) The action space $\mathbf{A}$ is compact,
(b) The transition probability $p_t(dy|x,a)$ is a (weakly) continuous stochastic kernel on $\mathbf{X}$ given $\mathbf{X} \times \mathbf{A}$, and
(c) The cost functions $c_t(\cdot,\cdot)$ and $C(\cdot)$ are lower semi-continuous and bounded below.

Some authors call a model semi-continuous if slightly different conditions are satisfied. More about that in Subsection 2.3.16. To avoid misunderstanding, we always give a reference to a specific condition.

In models, where Condition 1.1 is satisfied, the function $v_t(\cdot)$ is also lower semi-continuous and bounded below. Moreover, there exists a (measurable) selector $\varphi_t^*(x)$ providing the required minimum:

$$v_{t-1}(x) = c_t(x, \varphi_t^*(x)) + \int_{\mathbf{X}} v_t(y) p_t(dy|x, \varphi_t^*(x))$$

[Hernández-Lerma and Lasserre (1996a), Section 3.3.5], [Bertsekas and Shreve (1978), Statement 8.6], [Piunovskiy (1997), Section 1.1.4.1].

If the action space is not compact, or the transition probability is not continuous, or the cost functions are not lower semi-continuous, then trivial examples show that the desired selector φ^* may not exist. The following example based on [Feinberg (2002b), Ex. 6.2] confirms that the boundedness below of the cost functions is also important.

Example. Let $T = 1$, $\mathbf{X} = \{\Delta, 0, 1, 2, \ldots\}$ with the discrete topology. Suppose $\mathbf{A} = \{\infty, 1, 2, \ldots\}$, action ∞ being the one-point compactification of the sequence $1, 2, \ldots$, that is, $\lim_{a\to\infty} a = \infty$. We put $\mathsf{P}_0(0) = 1$;

$$\text{for } a \neq \infty, \qquad p_1(y|x,a) = \begin{cases} 1/a, & \text{if } y = a; \\ (a-1)/a, & \text{if } y = \Delta; \\ 0, & \text{otherwise;} \end{cases} \qquad p_1(\Delta|x,\infty) = 1,$$

$$c_1(x,a) \equiv 0, \qquad C(x) = \begin{cases} 1 - x, & \text{if } x \neq \Delta; \\ 0, & \text{if } x = \Delta \end{cases}$$

(see Fig. 1.12).

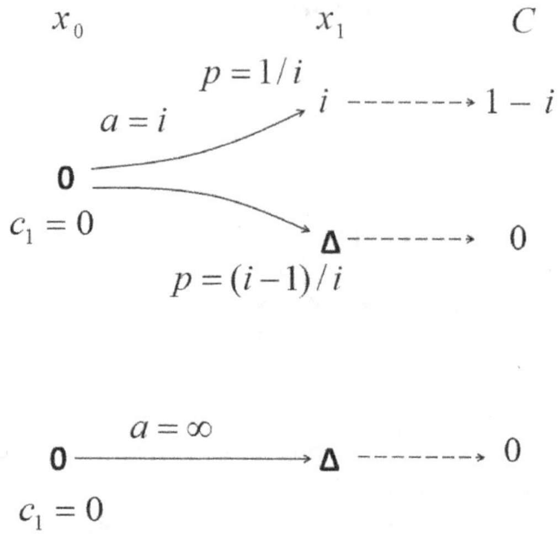

Fig. 1.12 Not a semi-continuous model.

This MDP satisfies Condition 1.1 apart from the requirement that the cost functions be bounded below. One can easily calculate

$$v_1(x) = C(x) = \begin{cases} 1 - x, & \text{if } x \neq \Delta; \\ 0, & \text{if } x = \Delta \end{cases}$$

$$v_0(x) = \min\left\{ \inf_{a \in \{1,2,\ldots\}} \left\{ \frac{1}{a}[1-a] \right\};\ 0 \right\} = \lim_{a\to\infty} \left(\frac{1}{a} - 1 \right) = -1$$

(in the second formula, zero corresponds to action ∞). But for any $a \in \mathbf{A}$,

$$c_1(0,a) + \sum_{y \in \mathbf{X}} v_1(y) p_1(y|0,a) > -1 = v_0(0),$$

and no-one strategy is optimal.

One can reformulate this model in the following way. $T = 1$, $\mathbf{X} = \{0, \Delta\}$, $\mathbf{A} = \{\infty, 1, 2, \ldots\}$, $\mathsf{P}_0(0) = 1$, $p_1(\Delta|x,a) \equiv 1$, $c_1(\Delta, a) = 0$, $c_1(0,a) = \frac{1}{a} - 1$ if $a \neq \infty$, $c_1(0,\infty) = 0$, $C(x) \equiv 0$. Now the cost function is bounded below, but it ceases to be lower semi-continuous: $\lim_{a\to\infty} c_1(0,a) = -1 < c_1(0,\infty) = 0$. □

In Subsection 2.3.24, it is demonstrated that one can always make the cost functions uniformly bounded (see Fig. 2.33); however, the time horizon will be infinite. According to Theorem 2.12, one can guarantee the existence of a uniformly optimal stationary selector only under additional conditions (i.e. if the state space is finite etc). See also Theorem 2.4: a uniformly optimal stationary selector exists in weakly semi-continuous *positive* homogeneous models with the total expected cost, where Condition 2.3 is satisfied. The presented above example shows that here the requirement $c_t(\cdot,\cdot) \geq 0$ is important. Note also that any finite-horizon MDP can be represented in the form of the infinite-horizon homogeneous model: see Remark 1.1.

The next example will show that the Bellman function $v_t^*(\cdot)$ may not necessarily be lower semi-continuous if the action space $\mathbf{A}$ is not compact, even when the infimum in the optimality equation (1.5) is attained at every $x \in \mathbf{X}$. Below, we assume that $\mathbf{X}$ and $\mathbf{A}$ are topological Borel spaces.

If the space $\mathbf{A}$ is not compact, it is convenient to impose the following condition: the cost function $c_t(\cdot,\cdot)$ is bounded below and *inf-compact* for any $x \in \mathbf{X}$, i.e. the set $\{a \in \mathbf{A} : \ c_t(x,a) \leq r\}$ is compact for each $r \in \mathbb{R}^1$.

Assume that the final cost $C(\cdot)$ is bounded below and lower semi-continuous. Now the infimum in equation

$$v_{t-1}(x) = \inf_{a \in \mathbf{A}} \left\{ c_t(x,a) + \int_{\mathbf{X}} v_t(y) p_t(dy|x,a) \right\}$$

(see (1.5)) is provided by a measurable selector $\varphi_t^*(x)$, if function $v_t(\cdot)$ is bounded below and lower semi-continuous. The function in the parentheses is itself bounded below, lower semi-continuous and inf-compact for any $x \in \mathbf{X}$ [Hernández-Lerma and Lasserre (1996a), Section 3.3.5]. To apply the mathematical induction method, what remains is to find conditions which guarantee that the function

$$v(x) = \inf_{a \in \mathbf{A}} \{f(x,a)\}$$

is lower semi-continuous for a bounded below, lower semi-continuous and inf-compact (for any $x \in \mathbf{X}$) function $f(\cdot)$. This problem was studied in [Luque-Vasquez and Hernandez-Lerma (1995)]: it is sufficient to require that the multifunction

$$x \to \mathbf{A}^*(x) = \{a \in \mathbf{A} : \ v(x) = f(x,a)\} \tag{1.21}$$

is *lower semi-continuous*; that is, the set $\{x \in \mathbf{X} : \ \mathbf{A}^*(x) \cap \Gamma \neq \emptyset\}$ is open for every open set $\Gamma \subseteq \mathbf{A}$. The following example from [Luque-Vasquez and Hernandez-Lerma (1995)] shows that this requirement is essential.

Example. Let $\mathbf{X} = \mathbb{R}^1$ and $\mathbf{A} = [0,\infty)$ be equipped with the standard (Euclidean) topology. Suppose

$$f(x,a) = \begin{cases} 1+a, & \text{if } x \leq 0 \text{ or } x > 0, \ 0 \leq a \leq \frac{1}{2x}; \\ (2+\frac{1}{x}) - (2x+1)a, & \text{if } x > 0, \ \frac{1}{2x} \leq a \leq \frac{1}{x}; \\ a - \frac{1}{x}, & \text{if } x > 0, \ a \geq \frac{1}{x} \end{cases}$$

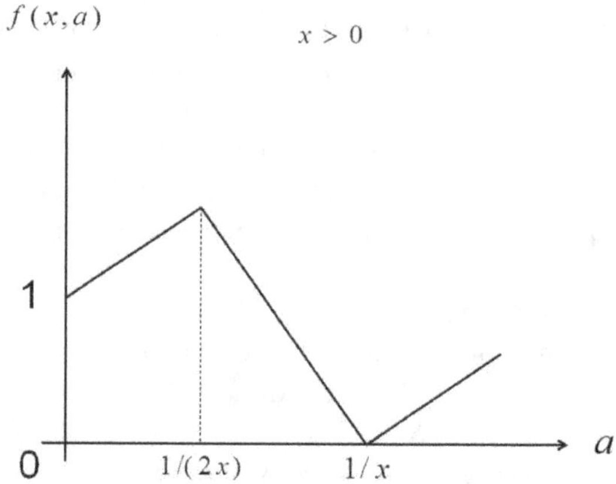

Fig. 1.13 Function $f(x,a)$.

(see Fig. 1.13).

Function $f(\cdot)$ is non-negative, continuous and inf-compact for any $x \in \mathbf{X}$. But multifunction (1.21) (it is actually a function, that is, the values are singletons) has the form

$$\mathbf{A}^*(x) = \begin{cases} \{0\}, & \text{if } x \leq 0; \\ \{\frac{1}{x}\}, & \text{if } x > 0, \end{cases}$$

and is not lower semi-continuous because the set $\Gamma := [0, r)$ is open in $\mathbf{A}$ for $r > 0$ and the set

$$\{x \in \mathbf{X} : \ \mathbf{A}^*(x) \in \Gamma\} = (-\infty, 0] \cup (1/r, \infty)$$

is not open in $\mathbf{X}$. The function $v(x) = \inf_{a \in \mathbf{A}}\{f(x,a)\} = \begin{cases} 1, & \text{if } x \leq 0; \\ 0, & \text{if } x > 0 \end{cases}$ is not lower semi-continuous; it is upper semi-continuous (see Theorem 1.2).

□

As for the other semi-continuous MDPs, let us also mention example from Subsection 2.3.16. The definition of a semi-continuous model given in that subsection differs from the one presented at the beginning of the current subsection. (See Condition 1.1.) However, in a similar way to Subsection 2.3.17, one can usually introduce different topologies in the space $\mathbf{A}$, so that one or two of the requirements (a,b,c) will be satisfied. Discussions of several slightly different semi-continuity conditions which guarantee the existence of (uniformly) optimal selectors can be found in [Hernández-Lerma and Lasserre (1996a), Section 3.3] and [Schäl (1975a)]. Recall that any finite-horizon model can be considered as an infinite-horizon MDP with expected total cost (see Remark 1.1).

1.4.9 *Nowhere differentiable function $v_t(x)$; discontinuous function $v_t(x)$*

It is known that a functional series can converge (absolutely) to a continuous, but nowhere differentiable function. As an example, take

$$f(x) := \sum_{i=0}^{\infty} \left(\frac{1}{2}\right)^i [\cos(7^i \cdot \pi x) + 1], \quad x \in \mathbb{R}$$

[Gelbaum and Olmsted (1964), Section 3.8]. Note that, for all $x \in \mathbb{R}$, $f_n(x) \geq f_{n-1}(x)$, where

$$f_n(x) := \sum_{i=0}^{n} \left(\frac{1}{2}\right)^i [\cos(7^i \cdot \pi x) + 1] \in \mathbf{C}^\infty;$$

$\mathbf{C}^\infty$ is the space of infinitely many times differentiable functions. It is also known that function

$$g(a) = \begin{cases} \exp\left[-\frac{1}{a^2} e^{-\frac{1}{(1-a)^2}}\right], & \text{if } 0 < a < 1; \\ 0, & \text{if } a = 0; \\ 1, & \text{if } a = 1 \end{cases} \tag{1.22}$$

is strictly increasing on $[0, 1]$ and belongs to $\mathbf{C}^\infty$, and $\left.\frac{dg}{da}\right|_{a=0} = \left.\frac{dg}{da}\right|_{a=1} = 0$.

Example. Let us put

$$c_1(x, a) = -f_{n-1}(x) - \left(\frac{1}{2}\right)^n [\cos(7^n \cdot \pi x) + 1]g(a - n + 1) \text{ if } a \in [n - 1, n], \tag{1.23}$$

where $n \in \{1, 2, \ldots\}$, $x \in \mathbf{X} := [0, 2]$, $a \in \mathbf{A} := [0, \infty)$. The spaces $\mathbf{X}$ and $\mathbf{A}$ are equipped with the standard (Euclidean) topology.

In the MDP $\{\mathbf{X}, \mathbf{A}, T = 1, p, c(\cdot,\cdot), C(\cdot) \equiv 0\}$ with arbitrarily fixed transition probability $p_1(dy|x, a)$ and initial distribution $\mathsf{P}_0(\cdot)$, we have

$$v_1(x) = C(x) = 0 \text{ and } v_0(x) = \inf_{a \in \mathbf{A}} c_1(x, a) = \lim_{a \to \infty} c_1(x, a) = f(x),$$

so that $c_1(\cdot,\cdot), C(\cdot) \in \mathbf{C}^\infty$, but function $v_0(\cdot)$ is continuous and nowhere differentiable (see Figs. 1.14–1.17). □

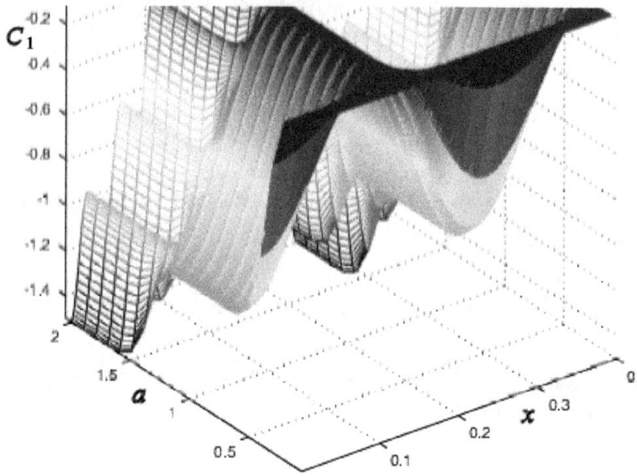

Fig. 1.14 Function (1.23) on the small area $x \in [0,\ 0.4]$, $a \in [0, 2]$.

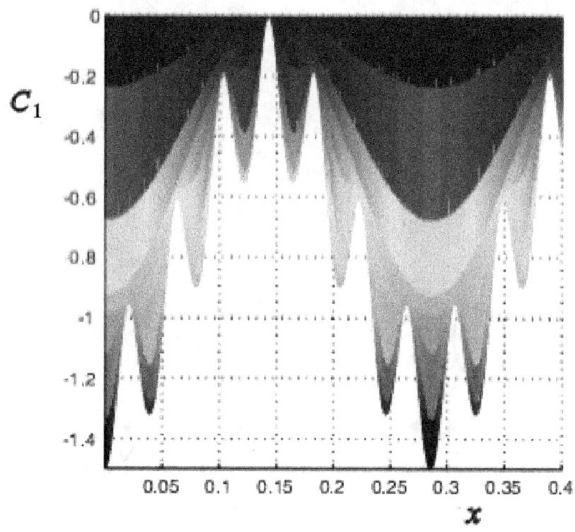

Fig. 1.15 Projection on the plane $x \times c_1$ of the function from Fig. 1.14.

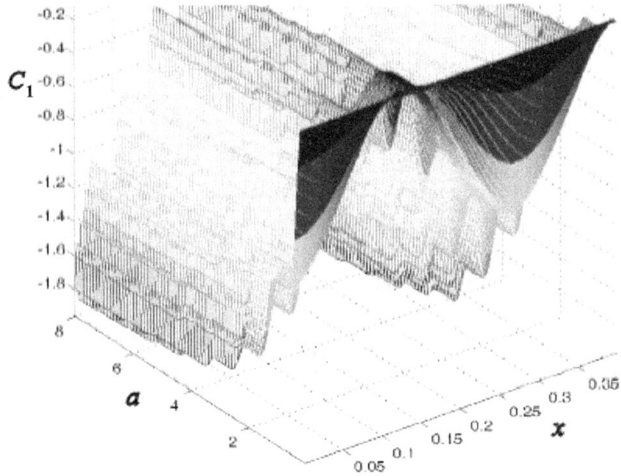

Fig. 1.16 Function (1.23) on the greater area $x \in [0,\ 0.4]$, $a \in [0, 8]$.

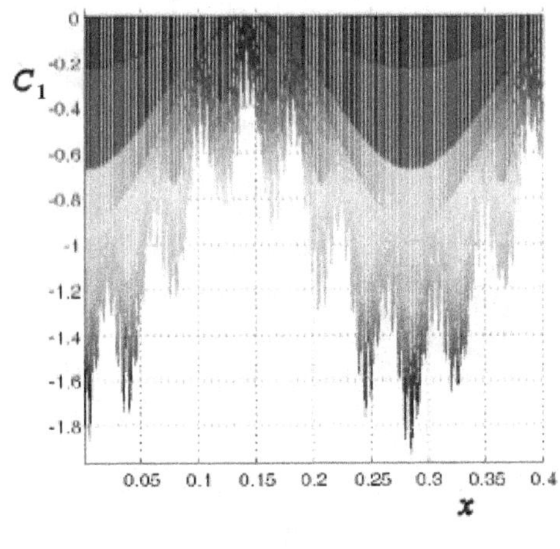

Fig. 1.17 Projection on the plane $x \times c_1$ of the function from Fig. 1.16.

One can easily construct a similar example where $v_0(\cdot)$ is discontinuous, although $c_1(\cdot,\cdot) \in \mathbf{C}^\infty$:

Example. $\mathbf{X} := [0,\infty)$, $\mathbf{A} := [0,\infty)$, $T = 1$, $p_1(dy|x,a)$ is arbitrarily fixed, $C(\cdot) \equiv 0$ and

$$c_1(x,a) = h_n(x)+[h_{n+1}(x)-h_n(x)]g(a-n+1), \text{ if } a \in [n-1,n],\ \ n \in \{1,2,\ldots\}, \tag{1.24}$$

where

$$h_n(x) = \begin{cases} 0, & \text{if } x \le 1-\frac{1}{n}; \\ g(1-n+nx), & \text{if } 1-\frac{1}{n} \le x \le 1; \\ 1, & \text{if } x \ge 1, \end{cases}$$

and the function $g(\cdot)$ is given by (1.22).

Now

$$v_1(x) = C(x) = 0 \text{ and } v_0(x) = \inf_{a\in\mathbf{A}} c_1(x,a) = \lim_{a\to\infty} c_1(x,a) = \begin{cases} 1, & \text{if } x \ge 1; \\ 0, & \text{if } x < 1 \end{cases} \tag{1.25}$$

(see Figs. 1.18 and 1.19). □

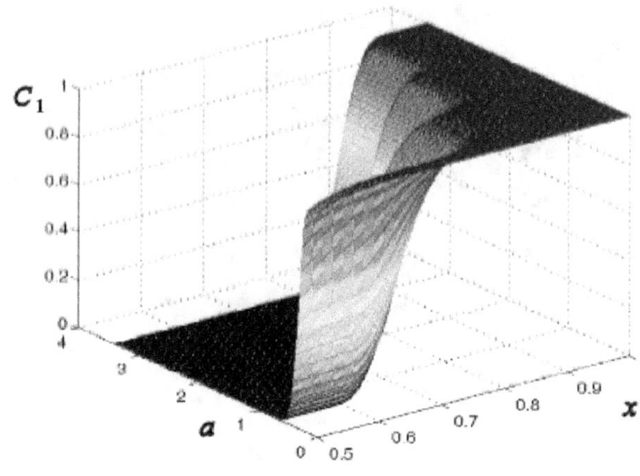

Fig. 1.18 Function (1.24) $c_1(x,a)$ on subset $0.5 \le x \le 1$, $a \in [0,3]$.

Remark 1.5. In general, Theorem 1.2 proved in [Bertsekas and Shreve (1978), Propositions 7.32 and 7.33] can be useful. Note that, if $\mathbf{A}$ is compact, then function $\inf_{a\in\mathbf{A}} c_1(x,a)$ is continuous provided $c_1(\cdot,\cdot)$ is continuous.

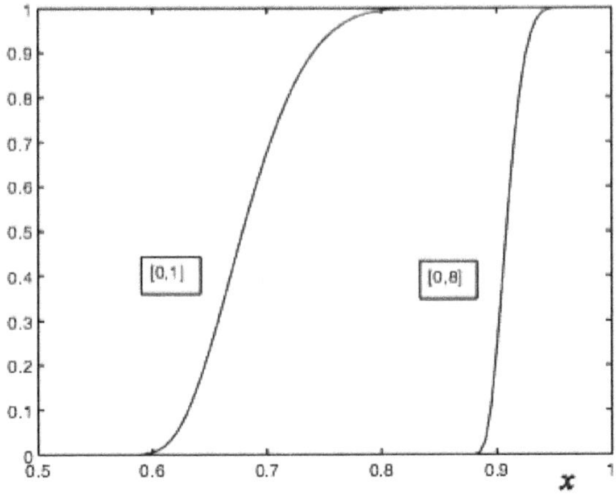

Fig. 1.19 Graphs of functions $\inf_{a\in[0,1]} c_1(x,a)$ and $\inf_{a\in[0,8]} c_1(x,a)$ for $c_1(x,a)$ shown in Fig. 1.18.

1.4.10 *Non-measurable Bellman function; no-one strategy is uniformly ε-optimal*

The dynamic programming approach is based on the assumption that optimality equation (1.5) has a measurable solution. The following example shows that the Bellman function $v_t^*(x)$ may be not Borel-measurable even in the simplest case when $T = 1$, $C(x) \equiv 0$ and the cost function $c_1(x,a)$ is measurable.

Example. Let $\mathbf{X} := [0,1]$, $\mathbf{A} := [0,1]$, $c_1(x,a) := -\mathbb{I}\{(x,a) \in B\}$, where B is a Borel subset of $\mathbf{X} \times \mathbf{A}$ with projection $B^1 := \{x \in \mathbf{X} : \exists a \in \mathbf{A} : (x,a) \in B\}$ that is not Borel (see Fig. 1.20). One can find the construction of such a set in [Dynkin and Yushkevich (1979), Appendix 2, Section 5]. The transition probability $p(dy|x,a)$ is arbitrarily fixed.

Now the Bellman function

$$v_0(x) = v_0^*(x) = \inf_{a\in\mathbf{A}} \{c_1(x,a)\} = -\mathbb{I}\{x \in B^1\}$$

is not Borel-measurable. A similar example can be also found in [Mine and Osaki (1970), Section 6.3].

Note that the Bellman function $v_t^*(x)$ is lower semi-analytical if the cost functions $c_t(\cdot,\cdot)$ and $C(\cdot)$ are all Borel-measurable and bounded below (or above) [Bertsekas and Shreve (1978), Corollary 8.2.1].

Suppose $\mathsf{P}_0(dx) := \delta_{x_0}(dx)$. If $x_0 \in B^1$, then any strategy $\pi^* =$

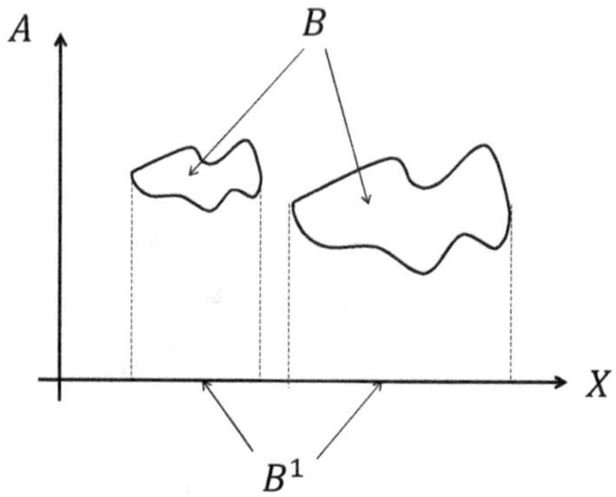

Fig. 1.20 Non-measurable Bellman function.

$\pi_1^*(da|x_0)$ with $\pi_1^*(\hat{\Gamma}|x_0) = 1$, where $\hat{\Gamma} := \{a \in \mathbf{A}: (x_0, a) \in B\}$, is optimal. If $x_0 \notin B^1$, then all strategies are equally optimal leading to $v^\pi_{\mathsf{P}_0} = 0$.

Now let us fix an arbitrary strategy $\pi = \pi_1(da|x_0)$. The set

$$\hat{B} := \{x_0 \in \mathbf{X}: v^\pi_{x_0} < 0\} = \left\{x_0 \in \mathbf{X}: \int_{\mathbf{A}} \mathbb{I}\{(x_0, a) \in B\}\pi_1(da|x_0) > 0\right\}$$

is a measurable subset of B^1 because the function

$$\lambda(x_0) := \int_{\mathbf{A}} \mathbb{I}\{(x_0, a) \in B\}\pi_1(da|x_0): \mathbf{X} \to [0, 1]$$

is measurable by Proposition 7.29 of [Bertsekas and Shreve (1978)]. Since B^1 is not measurable, $\hat{B}$ is a proper subset of B^1. Thus, there exists $\hat{x}_0 \in B^1 \setminus \hat{B}$ for which $\int_{\mathbf{A}} \mathbb{I}\{(\hat{x}_0, a) \in B\}\pi_1(da|\hat{x}_0) = 0$, leading to $v^\pi_{\hat{x}_0} = 0 > -1 = v_0^*(\hat{x}_0) = v^*_{\hat{x}_0}$. We conclude that the strategy π is not uniformly ε-optimal if $\varepsilon < 1$. □

A similar example was presented in [Blackwell (1965)] and [Strauch (1966), Ex. 4.1].

A very similar example was also presented in [Bertsekas and Shreve (1978), Section 8.2, Example 1]:

Example. Let $\mathbf{X} = [0, 1]$ and $\mathbf{A} = \mathcal{N}$, the Bair null space; $c_1(x, a) = -\mathbb{I}\{(x, a) \in B\}$, where B is a closed (hence Borel-measurable) subset of $\mathbf{X} \times \mathbf{A}$ with projection $B^1 = \{x \in \mathbf{X}: \exists a \in \mathbf{A}: (x, a) \in B\}$ that is not Borel. The construction of such closed B was described in Prop. 7.39 of [Bertsekas and Shreve (1978)]. The further reasoning is identical to that given above. □

In the both examples we have the 'two-dimensional' Borel set B whose projection B^1 is not Borel. Note that in the first case, $\mathbf{A}$ is compact and B is only measurable. On the other hand, in the second example B is closed, but $\mathbf{A}$ is not compact. We emphasize that one cannot have simultaneously closed B and compact $\mathbf{A}$, because in this case projection B^1 would have been closed: it is sufficient to apply the following theorem to the lower semi-continuous function $-\mathbb{I}\{(x,a) \in B\}$.

Theorem 1.2. [Bertsekas and Shreve (1978), Props. 7.32 and 7.33]. *Let $\mathbf{X}$ and $\mathbf{Y}$ be metrizable spaces, and let $f(\cdot): \mathbf{X} \times \mathbf{Y} \longrightarrow \mathbb{R}^*$ be given. Define*

$$g(x) := \inf_{y \in \mathbf{Y}} f(x,y).$$

(a) If $f(\cdot)$ is lower semi-continuous and $\mathbf{Y}$ is compact, then $g(\cdot)$ is lower semi-continuous and for every $x \in \mathbf{X}$ the infimum is attained by some $y \in \mathbf{Y}$. Furthermore, there exists a (Borel)-measurable function $\varphi: \mathbf{X} \longrightarrow \mathbf{Y}$ such that $f(x,\varphi(x)) = g(x)$ for all $x \in \mathbf{X}$.

(b) If $f(\cdot)$ is upper semi-continuous, then $g(\cdot)$ is also upper semi-continuous.

Another example, in which there are no uniformly ε-optimal selectors, but the Bellman function is measurable, can easily be built using the construction of the example from Subsection 3.3.8.

1.4.11 *Equivalent selectors in non-atomic MDPs*

Consider a one-step MDP with state and action spaces $\mathbf{X}$ and $\mathbf{A}$, initial distribution $\mathsf{P}_0(dx)$, and zero final cost, so that the transition probabilities play no role. We assume that $\mathbf{X}$ and $\mathbf{A}$ are topological Borel spaces. Suppose we have a finite collection of cost functions $\{\,^k c(\cdot,\cdot)\}_{k=1}^K$. In Remark 1.3 we saw that performance spaces

$$\mathcal{V}_{\mathsf{P}_0} := \{\{\,^k v^\pi_{\mathsf{P}_0}\}_{k=1}^K,\ \pi \in \Pi^{\mathrm{All}}\} \text{ and } \mathcal{V}^{\mathrm{MN}}_{\mathsf{P}_0} := \{\{\,^k v^\varphi_{\mathsf{P}_0}\}_{k=1}^K,\ \varphi \in \Pi^{\mathrm{MN}}\}$$

can be different: if $\mathbf{X}$ and $\mathbf{A}$ are finite, then $\mathcal{V}^{\mathrm{MN}}_{\mathsf{P}_0} \subset \mathbb{R}^K$ is finite because Π^{MN} is finite, but $\mathcal{V}_{\mathsf{P}_0} \subset \mathbb{R}^K$ is convex compact [Piunovskiy (1997), Section 3.2.2], equal to the convex hull of $\mathcal{V}^{\mathrm{MN}}_{\mathsf{P}_0}$ [Feinberg and Rothblum (2012), Thm.4.1]. On the other hand, according to [Feinberg and Piunovskiy (2002), Thm. 2.1], $\mathcal{V}_{\mathsf{P}_0} = \mathcal{V}^{\mathrm{MN}}_{\mathsf{P}_0}$ if the initial distribution $\mathsf{P}_0(dx)$ is non-atomic. See also [Feinberg and Piunovskiy (2010), Thm. 3.1] or [Feinberg and Piunovskiy (2019), Thm.2.3]. In other words, for any control strategy

π, there exists a Markov selector φ such that their performance vectors coincide:

$$\{\ ^k v^{\pi}_{\mathsf{P}_0}\}_{k=1}^K = \{\ ^k v^{\varphi}_{\mathsf{P}_0}\}_{k=1}^K.$$

We shall call such strategies π and φ *equivalent* w.r.t. $\{\ ^k c(\cdot,\cdot)\}_{k=1}^K$.

Recall that

$$\mathcal{P}_{\mathsf{P}_0} = \{\mathsf{P}^{\pi}_{\mathsf{P}_0}(\cdot),\ \pi \in \Pi^{\text{All}}\} \text{ and } \mathcal{P}^{\text{MN}}_{\mathsf{P}_0} = \{\mathsf{P}^{\varphi}_{\mathsf{P}_0}(\cdot),\ \varphi \in \Pi^{\text{MN}}\}$$

are the strategic measures spaces. In the general case, if $\mathbf{X}$ is a topological Borel space, $\mathbf{A}$ is compact and all transition probabilities $p_t(dy|x,a)$ are continuous stochastic kernels, then $\mathcal{P}_{\mathsf{P}_0}$ is convex compact in the *weak* topology [Schäl (1975a), Thm. 5.6]. That is the weakest topology with respect to which, for each $t = 0, 1, \ldots$, the mapping $\mathsf{P} \to \int_{\mathbf{H}_t} f(h_t) d\mathsf{P}(h_t)$ is continuous for each bounded continuous function $f(\cdot)$ on $\mathbf{H}_t$.

If, additionally, the initial distribution $\mathsf{P}_0(dx)$ is non-atomic, then $\mathcal{V}_{\mathsf{P}_0} = \mathcal{V}^{\text{MN}}_{\mathsf{P}_0}$, but the space $\mathcal{P}^{\text{MN}}_{\mathsf{P}_0}$ is usually not closed (in the weak topology), and hence $\mathcal{P}^{\text{MN}}_{\mathsf{P}_0} \neq \mathcal{P}_{\mathsf{P}_0}$, as the following example shows [Piunovskiy (1997), p.170].

Example. Let $\mathbf{X} = [0,1]$, $\mathbf{A} = \{1,2\}$, $\mathsf{P}_0(dx) = dx$, the Lebesgue measure, and for $n = 1, 2, \ldots$ put $\varphi^n(x) = \mathbb{I}\{x \in \Gamma_n\} + 2 \cdot \mathbb{I}\{x \in \Gamma^c_n\}$, where Γ_n is the set consisting of 2^{n-1} segments of the same length $\delta_n = \left(\frac{1}{2}\right)^n$:

$$\Gamma_n = [\delta_n, 2\delta_n] \cup [3\delta_n, 4\delta_n] \cup \ldots \cup [(2^n - 1)\delta_n, 2^n\delta_n = 1]$$

(see Fig. 1.21). Take an arbitrary bounded continuous function $c(x,a)$. Then

$$\sum_{a=1}^{2} \int_0^1 c(x,a)\mathbb{I}\{\varphi^n(x) = a\}dx = \int_{\Gamma_n} c(x,1)dx + \int_{\Gamma^c_n} c(x,2)dx$$

$$= \frac{1}{2}\int_0^1 c(x,1)dF^n_1(x) + \frac{1}{2}\int_0^1 c(x,2)dF^n_2(x),$$

where $F^n_1(\cdot)$ and $F^n_2(\cdot)$ are the cumulative distribution functions of uniform random variables on Γ_n and Γ^c_n correspondingly. Obviously, for all $x \in \mathbf{X}$, $F^n_1(x)$, $F^n_2(x) \to F(x) = x$ as $n \to \infty$, and the convergence is monotonic. Hence

$$\int_0^1 \hat{c}(x)dF^n_1(x), \quad \int_0^1 \hat{c}(x)dF^n_2(x) \to \int_0^1 \hat{c}(x)dx$$

for any continuous function $\hat{c}(\cdot)$. In particular,

$$\sum_{a=1}^{2} \int_0^1 c(x,a)\mathbb{I}\{\varphi^n(x) = a\}dx \to \frac{1}{2}\int_0^1 c(x,1)dx + \frac{1}{2}\int_0^1 c(x,2)dx$$

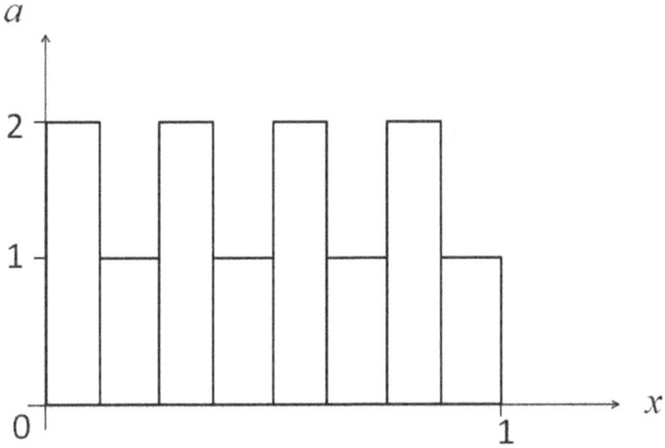

Fig. 1.21 Selector φ^n when $n = 3$.

$$= \sum_{a=1}^{2} \int_0^1 c(x,a)\pi^*(a|x)dx,$$

where $\pi^*(1|x) = \pi^*(2|x) = \frac{1}{2}$. This calculation confirms that, if we put $T = 1$ and, e.g., $p_1(dy|x,a) = \delta_0(dy)$, then the space $\mathcal{P}_{\mathsf{P}_0}$ is weakly closed. But the strategic measure $\mathsf{P}_{\mathsf{P}_0}^{\pi^*}$ cannot be generated by a selector. To prove this, suppose such a selector $\varphi(x)$ exists. Then, for function $g_1(x) := \mathbb{I}\{\varphi(x) = 1\}$, the following integrals must coincide:

$$\mathsf{E}_{\mathsf{P}_0}^{\pi^*}[g_1(X_0)\mathbb{I}\{A_1 = 1\}] = \int_0^1 g_1(x)\pi^*(1|x)dx = \frac{1}{2}\int_0^1 \mathbb{I}\{\varphi(x) = 1\}dx,$$

$$\mathsf{E}_{\mathsf{P}_0}^{\varphi}[g_1(X_0)\mathbb{I}\{A_1 = 1\}] = \int_0^1 g_1(x)\mathbb{I}\{\varphi(x) = 1\}dx = \int_0^1 \mathbb{I}\{\varphi(x) = 1\}dx,$$

meaning that $\int_0^1 I\{\varphi(x) = 1\}dx = 0$. Similarly, for $g_2(x) := \mathbb{I}\{\varphi(x) = 2\}$, we obtain $\int_0^1 \mathbb{I}\{\varphi(x) = 2\}dx = 0$. This contradiction implies that such a selector φ does not exist. The weak limit of $\{\mathsf{P}_{\mathsf{P}_0}^{\varphi^n}\}_{n=1}^{\infty}$ is not selector-generated. □

The following example shows that, even in the non-atomic case, the performance spaces $\mathcal{V}_{\mathsf{P}_0}$ and $\mathcal{V}_{\mathsf{P}_0}^{\mathrm{MN}}$ may be different if the collection of cost functions $\{\,{}^k c(\cdot,\cdot)\}_{k=1}^{\infty}$ is not finite

Example. Let $\mathbf{X}$ be an arbitrary topological Borel space and $\mathbf{A} = \{1, 2\}$. It is known [Parthasarathy (2005), Thm. 6.6] that there exists a sequence $\{\,^k f(x)\}_{k=1}^{\infty}$ of bounded uniformly continuous functions on $\mathbf{X}$ such that, if

$$\forall k = 1, 2, \ldots \qquad \int_{\mathbf{X}} {}^k f(x)\mu_1(dx) = \int_{\mathbf{X}} {}^k f(x)\mu_2(dx),$$

then the measures μ_1 and μ_2 coincide. Now put ${}^k c(x, a) := I\{a = 1\}\,{}^k f(x)$, take $\pi(1|x) = \pi(2|x) \equiv 1/2$ and suppose that there exists a selector φ equivalent to π w.r.t. $\{\,^k c(\cdot)\}_{k=1}^{\infty}$, that is,

$$\int_{\mathbf{X}} {}^k f(x)\pi(1|x)\mathsf{P}_0(dx) = \int_{\mathbf{X}} {}^k f(x)\mathbb{I}\{\varphi(x) = 1\}\mathsf{P}_0(dx), \quad k = 1, 2, \ldots.$$

We see that measures $\mu_1(dx) := \pi(1|x)\mathsf{P}_0(dx)$ and $\mu_2(dx) := \mathbb{I}\{\varphi(x) = 1\}\mathsf{P}_0(dx)$ on $\mathbf{X}$ must coincide. But for function $g_1(x) := \mathbb{I}\{\varphi(x) = 1\}$ we have

$$\int_{\mathbf{X}} g_1(x)\mu_1(dx) = \frac{1}{2}\int_{\mathbf{X}} \mathbb{I}\{\varphi(x) = 1\}\mathsf{P}_0(dx)$$

and

$$\int_{\mathbf{X}} g_1(x)\mu_2(dx) = \int_{\mathbf{X}} \mathbb{I}\{\varphi(x) = 1\}\mathsf{P}_0(dx),$$

so that $\int_{\mathbf{X}} \mathbb{I}\{\varphi(x) = 1\}\mathsf{P}_0(dx) = 0$. Similarly, when considering function $g_2(x) := \mathbb{I}\{\varphi(x) = 2\}$ we obtain $\int_{\mathbf{X}} \mathbb{I}\{\varphi(x) = 2\}\mathsf{P}_0(dx) = 0$. This contradiction shows that such a selector φ does not exist. □

1.4.12 *Strongly equivalent selectors in non-atomic MDPs*

Again, we consider a one-step MDP with state and action spaces $\mathbf{X}$ and $\mathbf{A}$, initial distribution $\mathsf{P}_0(dx)$, and zero final cost.

If we have an arbitrary collection $\{\,^\alpha c(\cdot,\cdot)\}_{\alpha\in\mathcal{A}}$ of cost functions, but of special form ${}^\alpha c(x, a) = {}^\alpha\rho(a)\cdot f(x)$, where all functions ${}^\alpha\rho(\cdot)$ are bounded (arbitrary in case $f(\cdot) \ge 0$ or $f(\cdot) \le 0$), then, as in Subsection 1.4.11, in the non-atomic case, for any control strategy π there exists a selector φ such that ${}^\alpha v^\pi_{\mathsf{P}_0} = {}^\alpha v^\varphi_{\mathsf{P}_0}$ for all $\alpha \in \mathcal{A}$ (see Lemma A.1). The latter statement can be reformulated as follows: if the function $f(\cdot)$ is fixed, then, for any control strategy π, there exists a selector φ such that the measures on $\mathbf{A}$

$$\nu^\pi(\Gamma) := \int_{\mathbf{X}} \pi(\Gamma|x)f(x)\mathsf{P}_0(dx) \text{ and } \nu^\varphi(\Gamma) := \int_{\mathbf{X}} I\{\Gamma \ni \varphi(x)\}f(x)\mathsf{P}_0(dx)$$

coincide (here $\Gamma \in \mathcal{B}(\mathbf{A})$). We call such strategies π and φ *strongly equivalent w.r.t.* $f(\cdot)$. This notion is important in the theory of mass transportation and for the so-called Monge–Kantorovich problems [Ball (2004)]; [Magaril-Il'yaev and Tikhomirov (2003), Section 12.2]. The generalized definition reads:

Definition 1.1. Suppose a collection of functions $\{\,^k f(x,a)\}_{k=1}^K$ is given. Two strategies π^1 and π^2 are called *strongly equivalent* w.r.t. $\{\,^k f(\cdot,\cdot)\}_{k=1}^K$ if, for an arbitrary bounded real measurable function $\rho(a)$ on $\mathbf{A}$, for all $k = 1, 2, \ldots, K$,

$$\begin{aligned} {}^k\nu^{\pi^1} &:= \int_{\mathbf{X}}\int_{\mathbf{A}} \rho(a)\ {}^k f(x,a)\pi^1(da|x)\mathsf{P}_0(dx) \\ = {}^k\nu^{\pi^2} &:= \int_{\mathbf{X}}\int_{\mathbf{A}} \rho(a)\ {}^k f(x,a)\pi^2(da|x)\mathsf{P}_0(dx). \end{aligned}$$

If π^1 and π^2 are equivalent w.r.t. $\{\,^k c(\cdot,\cdot)\}_{k=1}^K$ (as in Subsection 1.4.11), then they may be not strongly equivalent w.r.t. $\{\,^k c(\cdot,\cdot)\}_{k=1}^K$. On the other hand, if π^1 and π^2 are strongly equivalent w.r.t. $\{\,^k f(\cdot,\cdot)\}_{k=1}^K$, then they are equivalent w.r.t. any collection of measurable cost functions of the form $\{\,^\alpha c(x,a)\}_{\alpha\in\mathcal{A}} = \{\,^\gamma\rho(a)\cdot\,^k f(x,a)\}_{\gamma\in\Gamma;\ k=1,2,\ldots,K}$, where $\alpha = (\gamma, k)$, Γ is an arbitrary set, $\mathcal{A} := \Gamma \times \{1, 2, \ldots, K\}$ and, for each $\gamma \in \Gamma$ the function $^\gamma\rho(\cdot)$ is bounded (arbitrary in case every function $^k f(\cdot,\cdot)$ is either non-negative or non-positive).

Theorem 1.3. *Let initial distribution $\mathsf{P}_0(dx)$ be non-atomic and suppose one of the following conditions is satisfied:*

(a) *Action space $\mathbf{A}$ is finite and collection $\{\,^k f(x,a)\}_{k=1}^K$ is finite.*
(b) *Action space $\mathbf{A}$ is arbitrary, $K = 1$ and a single function $f(x) = \,^1 f(x)$ is given (independent of a).*

Then, for any control strategy π, there exists a selector φ, strongly equivalent to π w.r.t. $\{\,^k f(\cdot\}_{k=1}^K$.

Part (a) follows from [Feinberg and Piunovskiy (2002)]; see also [Feinberg and Piunovskiy (2010), Thm. 1]. For part (b), see Lemma A.1.

If the collection $\{\,^k f(\cdot,\cdot)\}_{k=1}^\infty$ is not finite, then assertion (a) is not valid (see Subsection 1.4.11). Now we will show that all the conditions in Theorem 1.3(b) are important.

Independence of function $f(\cdot)$ of a. This example is due to [Feinberg and Piunovskiy (2010), Ex. 2].

Example. **X** and **A** Let $\mathbf{X} = [0,1]$, $\mathbf{A} = [-1,+1]$, $\mathsf{P}_0(dx) = dx$, the Lebesgue measure, and $f(x,a) := 2x - |a|$. Consider a strategy $\pi(\Gamma|x) := \frac{1}{2}[\delta_x(\Gamma) + \delta_{-x}(\Gamma)]$ to be a mixture of two Dirac measures (see Fig. 1.22), and suppose there exists a selector φ strongly equivalent to π w.r.t. $f(\cdot,\cdot)$. Then, for any measurable bounded non-negative or non-positive function

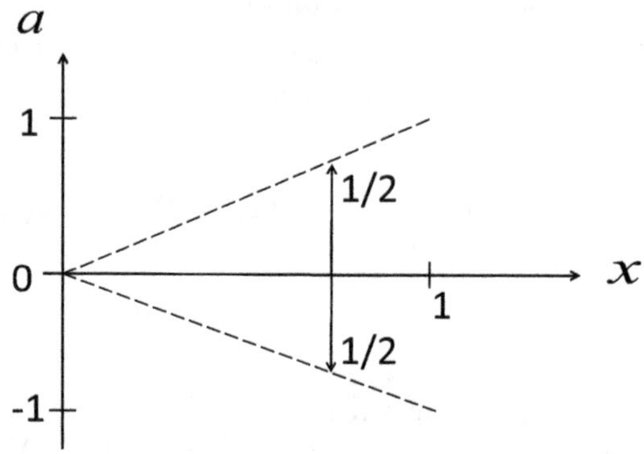

Fig. 1.22 Description of the strategy π.

$\rho(a)$, we must have

$$\begin{aligned}\int_0^1 \int_{-1}^1 \rho(a) f(x,a)\pi(da|x)dx &= \frac{1}{2}\int_0^1 [\rho(x)\cdot x + \rho(-x)\cdot x]dx \\ &= \int_0^1 \rho(\varphi(x))\mathbb{I}\{\varphi(x) > 0\}[2x - \varphi(x)]dx \\ &+ \int_0^1 \rho(\varphi(x))\mathbb{I}\{\varphi(x) \le 0\}[2x + \varphi(x)]dx.\end{aligned} \tag{1.26}$$

Consider $\rho(a) := a \cdot \mathbb{I}\{a > 0\}$. Then

$$\frac{1}{2}\int_0^1 x^2 dx = \int_0^1 \varphi(x)\mathbb{I}\{\varphi(x) > 0\}[2x - \varphi(x)]dx.$$

Hence

$$\int_0^1 \mathbb{I}\{\varphi(x) > 0\}[x - \varphi(x)]^2 dx = \int_0^1 \mathbb{I}\{\varphi(x) > 0\}x^2\, dx - \frac{1}{2}\int_0^1 x^2\, dx. \tag{1.27}$$

Consider $\rho(a) := a \cdot \mathbb{I}\{a \le 0\}$. Then

$$-\frac{1}{2}\int_0^1 x^2 dx = \int_0^1 \varphi(x)\mathbb{I}\{\varphi(x) \le 0\}[2x + \varphi(x)]dx.$$

Hence

$$\int_0^1 \mathbb{I}\{\varphi(x) \le 0\}[x + \varphi(x)]^2 dx = \int_0^1 \mathbb{I}\{\varphi(x) \le 0\}x^2\, dx - \frac{1}{2}\int_0^1 x^2\, dx. \tag{1.28}$$

If we add together the right-hand parts of (1.27) and (1.28), we obtain zero. Therefore,

$$\int_0^1 \mathbb{I}\{\varphi(x) > 0\}[x - \varphi(x)]^2 dx = \int_0^1 \mathbb{I}\{\varphi(x) \le 0\}[x + \varphi(x)]^2 dx = 0$$

and

$$\left.\begin{array}{ll} \varphi(x) = x, & \text{if } \varphi(x) > 0; \\ \varphi(x) = -x, & \text{if } \varphi(x) \le 0 \end{array}\right\} \ P_0\text{-a.s.} \tag{1.29}$$

Consider

$$\rho(a) := \mathbb{I}\{\varphi(a) = a\} \cdot \mathbb{I}\{a > 0\}/a > 0.$$

In equality (1.26) for $\min\{\rho(a), L\}$, where $L > 0$, the second term on the right-hand side is zero. Thus, after passing to the limit as $L \to \infty$, we obtain

$$\frac{1}{2}\int_0^1 \mathbb{I}\{\varphi(x) = x\}dx = \int_0^1 \mathbb{I}\{\varphi(x) > 0\}\frac{\mathbb{I}\{\varphi(\varphi(x)) = \varphi(x)\}}{\varphi(x)}[2x - \varphi(x)]dx$$

$$= \int_0^1 \mathbb{I}\{\varphi(x) > 0\}\mathbb{I}\{\varphi(\varphi(x)) = \varphi(x)\}\left[\frac{2x}{\varphi(x)} - 1\right] dx.$$

If $\varphi(x) = x$, then $\varphi(\varphi(x)) = \varphi(x)$. Hence,

$$\frac{1}{2}\int_0^1 \mathbb{I}\{\varphi(x) = x\}dx \ge \int_0^1 \mathbb{I}\{\varphi(x) > 0\}\mathbb{I}\{\varphi(x) = x\}\left[\frac{2x}{\varphi(x)} - 1\right] dx$$

$$= \int_0^1 \mathbb{I}\{\varphi(x) = x\}dx,$$

meaning that

$$\int_0^1 \mathbb{I}\{\varphi(x) = x\}dx = 0. \tag{1.30}$$

Consider

$$\rho(a) := \mathbb{I}\{\varphi(-a) = a\} \cdot \mathbb{I}\{a < 0\}/(-a).$$

The arguments similar to those presented above show that equality (1.26) implies

$$\frac{1}{2}\int_0^1 \mathbb{I}\{\varphi(x) = -x\}dx = \int_0^1 \mathbb{I}\{\varphi(x) < 0\}\frac{\mathbb{I}\{\varphi(-\varphi(x)) = \varphi(x)\}}{-\varphi(x)}[2x+\varphi(x)]dx$$

$$= \int_0^1 \mathbb{I}\{\varphi(x) < 0\}\mathbb{I}\{\varphi(-\varphi(x)) = \varphi(x)\}\left[\frac{2x}{-\varphi(x)} - 1\right] dx.$$

If $\varphi(x) = -x$, then $\varphi(-\varphi(x)) = \varphi(x)$. Hence

$$\frac{1}{2}\int_0^1 \mathbb{I}\{\varphi(x) = -x\}dx \geq \int_0^1 \mathbb{I}\{\varphi(x) < 0\}\mathbb{I}\{\varphi(x) = -x\}\left[\frac{2x}{-\varphi(x)} - 1\right]dx$$
$$= \int_0^1 \mathbb{I}\{\varphi(x) = -x\}dx,$$

meaning that

$$\int_0^1 \mathbb{I}\{\varphi(x) = -x\}dx = 0. \tag{1.31}$$

The contradiction obtained in (1.29), (1.30), (1.31) shows that the selector φ does not exist. □

One cannot have more than one function $f(x)$. This example is due to [Loeb and Sun (2006), Ex. 2.7].

Example. Let $\mathbf{X} = [0, 1]$, $\mathbf{A} = [-1, +1]$, $\mathsf{P}_0(dx) = dx$, the Lebesgue measure, and ${}^1f(x) \equiv 1$, ${}^2f(x) = 2x$. Consider the strategy $\pi(\Gamma|x) := \frac{1}{2}[\delta_x(\Gamma) + \delta_{-x}(\Gamma)]$ to be a mixture of two Dirac measures (see Fig. 1.22), and suppose there exists a selector φ strongly equivalent to π w.r.t. $\{{}^1f(\cdot), {}^2f(\cdot)\}$. Then, for any measurable bounded function $\rho(a)$, we must have

$$\int_0^1\int_{-1}^1 \rho(a)\pi(da|x)\,{}^1f(x)dx = \frac{1}{2}\int_0^1[\rho(x) + \rho(-x)]dx = \int_0^1 \rho(\varphi(x))dx; \tag{1.32}$$

$$\int_0^1\int_{-1}^1 \rho(a)\pi(da|x)\,{}^2f(x)dx = \frac{1}{2}\int_0^1[\rho(x)+\rho(-x)]2x\,dx = \int_0^1 \rho(\varphi(x))2x\,dx. \tag{1.33}$$

Consider $\rho(a) := a^2$. Then (1.32) implies

$$\int_0^1 [\varphi(x)]^2 dx = \int_0^1 x^2\,dx = 1/3.$$

Consider $\rho(a) := |a|$. Then (1.33) implies

$$\int_0^1 |\varphi(x)|2x\,dx = \int_0^1 x\cdot 2x\,dx = 2/3.$$

Therefore,

$$\int_0^1 [x - |\varphi(x)|]^2 dx = \int_0^1 x^2 dx - 2/3 + 1/3 = 0,$$

meaning that

$$\left.\begin{array}{ll}\varphi(x) = x, & \text{if } \varphi(x) > 0;\\ \varphi(x) = -x, & \text{if } \varphi(x) \leq 0\end{array}\right\} \mathsf{P}_0\text{-a.s.} \tag{1.34}$$

Consider $\rho(a) := \mathbb{I}\{\varphi(a) = a\}\mathbb{I}\{a > 0\}$. Equality (1.32) implies

$$\frac{1}{2}\int_0^1 \mathbb{I}\{\varphi(x) = x\}dx = \int_0^1 \mathbb{I}\{\varphi(\varphi(x)) = \varphi(x)\}\mathbb{I}\{\varphi(x) > 0\}dx.$$

If $\varphi(x) = x$, then $\varphi(\varphi(x)) = \varphi(x)$. Hence

$$\frac{1}{2}\int_0^1 \mathbb{I}\{\varphi(x) = x\}dx \geq \int_0^1 \mathbb{I}\{\varphi(x) = x\}dx,$$

meaning that

$$\int_0^1 \mathbb{I}\{\varphi(x) = x\}dx = 0. \tag{1.35}$$

Consider $\rho(a) := \mathbb{I}\{\varphi(-a) = a\}\mathbb{I}\{a \leq 0\}$. Equality (1.32) implies

$$\frac{1}{2}\int_0^1 \mathbb{I}\{\varphi(x) = -x\}dx = \int_0^1 \mathbb{I}\{\varphi(-\varphi(x)) = \varphi(x)\}\mathbb{I}\{\varphi(x) \leq 0\}dx.$$

If $\varphi(x) = -x$, then $\varphi(-\varphi(x)) = \varphi(x)$. Hence

$$\frac{1}{2}\int_0^1 \mathbb{I}\{\varphi(x) = -x\}dx \geq \int_0^1 \mathbb{I}\{\varphi(x) = -x\}dx,$$

meaning that

$$\int_0^1 \mathbb{I}\{\varphi(x) = -x\}dx = 0. \tag{1.36}$$

The contradiction obtained in (1.34), (1.35), (1.36) shows that selector φ does not exist. □

1.4.13 *More frequently used action is not better*

Example. Let $\mathbf{X} = \{-2, -1, +1, +2\}$, $\mathbf{A} = \{0, 1\}$, $T = 2$, the transition probability

$$p_1(y|x,a) \equiv p_1(y) = \begin{cases} p, & \text{if } y = +1; \\ q = 1 - p, & \text{if } y = -2; \\ 0, & \text{otherwise} \end{cases}$$

does not depend on x and a;

$$p_2(y|x,a) = \begin{cases} 1, & \text{if } y = x + a; \\ 0, & \text{otherwise.} \end{cases}$$

Finally, put $c_1(x,a) \equiv 0$, $c_2(x,a) \equiv 0$, $C(x) = \begin{cases} b, & \text{if } x = +1; \\ d, & \text{if } x = -1; \\ 0, & \text{otherwise,} \end{cases}$

where b and d are positive constants. (See Fig. 1.23.) The initial distribution P_0 is fixed arbitrarily.

Clearly, $\pi_1(a|x_0)$ can be arbitrary, and the control

$$a_2 = \varphi_2^*(x_1) = \begin{cases} 1, & \text{if } x = +1; \\ 0, & \text{if } x = -2 \end{cases}$$

is optimal. ($\pi_2(a|x_1)$ can be arbitrary at $x_1 = +2$ or -1 because, for each strategy π, $\mathsf{P}^\pi_{\mathsf{P}_0}(X_1 = +2) = \mathsf{P}^\pi_{\mathsf{P}_0}(X_1 = -1) = 0$.) Also, $v^*_{\mathsf{P}_0} := \min_\pi v^\pi_{\mathsf{P}_0} = 0$. When $p > q$, action $a_2 = 1$ is applied with higher probability, i.e. more frequently.

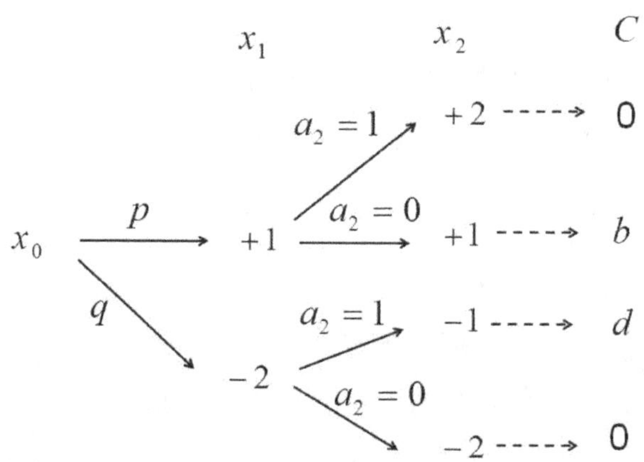

Fig. 1.23 Example 1.4.13.

Now suppose the decision maker cannot observe state X_1, but still has to choose action a_2. It turns out that $a_2 = 0$ is optimal if $b < qd/p$. Actually, we deal here with another MDP (with partial information) with the Bellman equation

$$v_0(x_0) = \min\{pb, qd\} = pb,$$

where the first (second) expression in the parentheses corresponds to $a_2 = 0$ (1). □

1.4.14 *Non-transitivity of the correlation*

Example. Let $\mathbf{X} = \{-1, 0, 1\}$, $\mathbf{A} = \{0, 1\}$, $T = 1$, $p_1(y|x, a) \equiv p_1(y|a)$ does not depend on x;

$$p_1(y|0) = \begin{cases} 1/3 - \varepsilon_0 - \varepsilon_-, & \text{if } y = 1; \\ 1/3 + \varepsilon_0, & \text{if } y = 0; \\ 1/3 + \varepsilon_-, & \text{if } y = -1, \end{cases}$$

where ε_0 and ε_- are small positive numbers; $p_1(y|1) \equiv 1/3$. Finally, put

$$c_1(x,a) \equiv 0,\ C(x) = \begin{cases} 1, & \text{if } x = 1; \\ 1+\delta, & \text{if } x = 0; \\ -1, & \text{if } x = -1, \end{cases} \quad \text{where } \delta > 0 \text{ is a small constant}$$

(see Fig. 1.24).

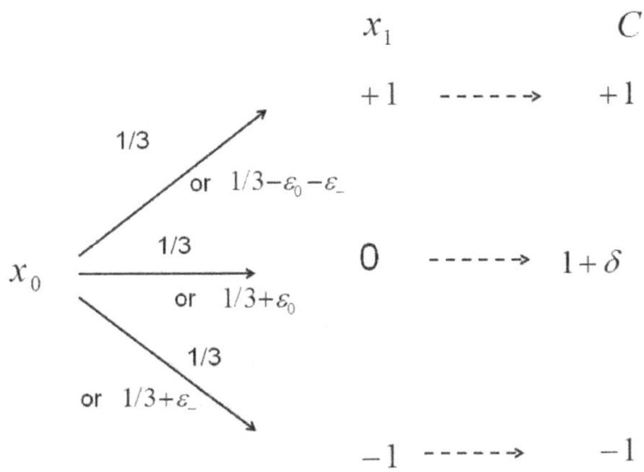

Fig. 1.24 Example 1.4.14.

The random variables X_1 and $W = C(X_1)$ do not depend on the initial distribution $\mathsf{P}_0(\cdot)$ which is arbitrarily fixed. Obviously, $Cov(X_1, W)$ is a continuous function of ε_0, ε_- and δ. If $\varepsilon_0 = \varepsilon_- = \delta = 0$, then (for any control strategy π) $\mathsf{E}^\pi_{\mathsf{P}_0}[X_1] = 0$, $\mathsf{E}^\pi_{\mathsf{P}_0}[W] = 1/3$ and

$$Cov(X_1, W) = \frac{1}{3}[1\cdot(1-1/3)] + \frac{1}{3}[0\cdot(1-1/3)] + \frac{1}{3}[(-1)\cdot(-1-1/3)] = 2/3$$

meaning that X_1 and W are positively correlated for small positive $\varepsilon_0, \varepsilon_-$ and δ. Under any randomized strategy $\pi_1(a|x_0)$, random variables A_1 and X_1 are also positively correlated for small positive $\varepsilon_0, \varepsilon_-$ and δ. To be more precise, for $p := \pi_1(1|x_0) \in (0,1)$ the straightforward calculation results in the following formula:

$$\mathrm{Cov}(A_1, X_1) = (p - p^2)(\varepsilon_0 + 2\varepsilon_-).$$

One might think that it is reasonable to minimize A_1 in order to obtain $\inf_\pi v^\pi_{\mathsf{P}_0} = \inf_\pi \mathsf{E}^\pi_{\mathsf{P}_0}[W]$, but it turns out that A_1 and W are *negatively* correlated if $\delta > 2\varepsilon_-/\varepsilon_0$: for $p := \pi_1(1|x_0) \in (0,1)$, we have

$$\mathrm{Cov}(A_1, W) = (p - p^2)(2\varepsilon_- - \delta\varepsilon_0) < 0.$$

In this case,

$$\begin{aligned} v_0(x_0) &= \min\{\ (1/3-\varepsilon_0-\varepsilon_-)+(1/3+\varepsilon_0)(1+\delta)-(1/3+\varepsilon_-);\\ &\qquad \tfrac{1}{3}+\tfrac{1}{3}(1+\delta)-\tfrac{1}{3}\}\\ &= \min\{\ \tfrac{1}{3}(1+\delta)+\delta\varepsilon_0-2\varepsilon_-;\ \ \tfrac{1}{3}(1+\delta)\}\\ &= \qquad 1/3+\delta/3, \end{aligned}$$

and the minimum is provided by $a_1^* = \varphi_1^*(x_0) \equiv 1$.

The property of being positively correlated is not necessarily transitive. This question was studied in [Langford *et al.* (2001)]. □

1.4.15 *Stock exchange; different utility functions*

Example. Suppose the investor would like to buy shares and can choose from two different types. Assume for simplicity that in a one-year period, the ith share price ($i = 1, 2$) can either double or drop to zero. In other words, if the initial capital is x_0 and the investments in the shares are $s^1 = a^1x_0$ and $s^2 = a^2x_0$ then, after one year, the capital will equal $x_0(1-a^1-a^2)+Y^1s^1+Y^2s^2$. Here $a := (a^1, a^2) \in \mathbf{A} = \{(a^1, a^2) :\ a^i \ge 0,\ a^1+a^2 \le 1\}$ is the decision how to split the initial capital x_0 and Y^1, Y^2 are random variables taking only values 2 and 0. The part $(1 - a^1 - a^2)$ of the initial capital is not invested and is not changed if we ignore the inflation. The state space is $\mathbf{X} := \{x \ge 0\} \subset \mathbb{R}$. Suppose the probabilities

$$p_{++} := \mathsf{P}\{Y^1 = Y^2 = 2\},\ \ p_{--} := \mathsf{P}\{Y^1 = Y^2 = 0\},$$

$$p_{+-} := \mathsf{P}\{Y^1 = 2, Y^2 = 0\},\ \ p_{-+} := \mathsf{P}\{Y^1 = -0, Y^2 = 2\},$$

$$p_{++} + p_{--} + p_{+-} + p_{-+} = 1$$

are known. Since the values of the capital at each time moment are proportional to x_0, one can fix $x_0 = 1$ without loss of generality.

We represent this model as MDP with $T = 1$, $\mathsf{P}_0(dx) = \delta_1(dx)$, $c_1(x,a) \equiv 0$ and the final reward ${}^1C(x) = x$. Unless most of examples, the target here is to *maximize* the expected final reward, equal to the capital at the end of the year: see Remark 1.2. One can say that here ${}^1C(x) = U(x) := x$ is the *utility function*. Another utility function ${}^2C(x)$ is considered later. The transition probabilities equal

$$\begin{aligned} p_1(\Gamma|x, a = (a^1, a^2)) = \ & p_{++}\delta_{x[2a^1+2a^2+(1-a^1-a^2)]}(\Gamma)\\ & +p_{--}\delta_{x[(1-a^1-a^2)]}(\Gamma)\\ & +p_{+-}\delta_{x[2a^1+(1-a^1-a^2)]}(\Gamma)\\ & +p_{-+}\delta_{x[2a^2+(1-a^1-a^2)]}(\Gamma). \end{aligned}$$

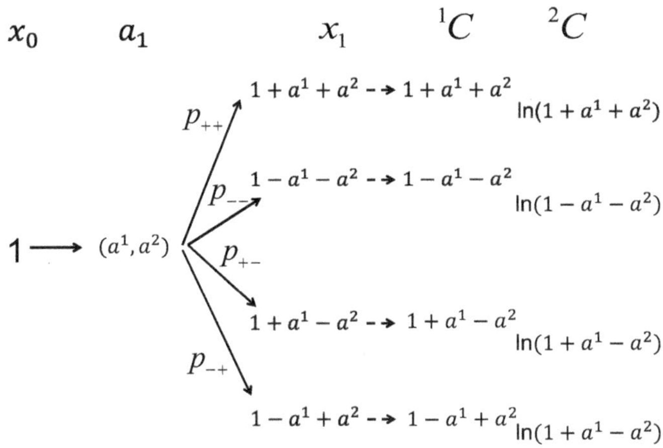

Fig. 1.25 Two different utility functions.

See Fig. 1.25.

Solution to the problem

$$\begin{aligned}
{}^1v^\pi_{P_0} &= \mathsf{E}^\pi_{P_0}[{}^1C(X_1)] = \int_{\mathbf{A}} \left[p_{++}(1 + a^1 + a^2) + p_{--}(1 - a_1 - a_2)\right. \\
&\quad \left. + p_{+-}(1 + a^1 - a^2) + p_{-+}(1 - a^1 + a^2)\right] \pi(d(a^1, a^2)|x_0 = 1) \\
&= \int_{\mathbf{A}} \left[1 + (p_{++} - p_{--})(a^1 + a^2) + (p_{+-} - p_{-+})(a^1 - a^2)\right] \\
&\quad \times \pi(d(a^1, a^2)|x_0 = 1) \to \max_\pi,
\end{aligned}$$

which is equivalent to the standard linear program

$$\begin{aligned}
&p_{++}(1 + a^1 + a^2) + p_{--}(1 - a_1 - a_2) \\
&+ p_{+-}(1 + a^1 - a^2) + p_{-+}(1 - a^1 + a^2) \to \max_{(a^1, a^2) \in \mathbf{A}},
\end{aligned}$$

is trivial. In case $p_{+-} \geq p_{-+}$ (i.e., if the price of shares 1 doubles with higher probability than the price of shares 2) it is given by the selector

$$\varphi^*_1(1) = \begin{cases} (1, 0), & \text{if } p_{++} - p_{--} + p_{+-} - p_{-+} > 0; \\ (0, 0), & \text{otherwise.} \end{cases}$$

In words, it is better to invest all the capital in shares 1 if they are at all profitable.

If $p_{+-} \le p_{-+}$, then the situation is symmetric:

$$\varphi_1^*(1) = \begin{cases} (0,1), & \text{if } p_{++} - p_{--} + p_{-+} - p_{+-} > 0; \\ (0,0), & \text{otherwise.} \end{cases}$$

Below we assume that $p_{+-} \ge p_{-+}$.

If $p_{++} - p_{--} + p_{+-} - p_{-+} > 0$ and $p_{--} + p_{-+} > 0$, then, under the optimal control strategy φ^*, there is a chance to finish with zero capital, i.e. with $X_1 = 0$, which is by far not good. In order to severely penalize the small values of X_1, people often introduce other utility functions $U(\cdot)$, e.g., ${}^2C(x) := U(x) = \ln(x)$. As usual, this utility function is concave, meaning that the *increase* of the wealth is more valuable if the wealth is small. Since $\mathsf{E}^\pi_{\mathsf{P}_0}[\ln(X_1)] \ne \ln(\mathsf{E}^\pi_{\mathsf{P}_0}[X_1])$, the solution will be different. Now we deal with the problem

$$\begin{aligned} {}^2v^\pi_{\mathsf{P}_0} &= \mathsf{E}^\pi_{\mathsf{P}_0}[{}^1C(X_1)] = \int_{\mathbf{A}} [p_{++}\ln(1+a^1+a^2) + p_{--}\ln(1-a_1-a_2) \\ &\quad + p_{+-}\ln(1+a^1-a^2) + p_{-+}\ln(1-a^1+a^2)]\, \pi(d(a^1,a^2)|x_0 = 1) \\ &\to \max_\pi, \end{aligned}$$

i.e., we need to maximize the following expression

$$p_{++}\ln(1+a^1+a^2)+p_{--}\ln(1-a^1-a^2)+p_{+-}\ln(1+a^1-a^2)+p_{-+}\ln(1-a^1+a^2)$$

with respect to $(a^1, a^2) \in \mathbf{A}$. Suppose

$$p_{++} > p_{--}, \qquad \frac{p_{++}}{p_{--}} > \frac{p_{+-}}{p_{-+}} = \max\left\{\frac{p_{-+}}{p_{+-}}; \frac{p_{+-}}{p_{-+}}\right\},$$

and all these probabilities are non-zero. Then the optimal solution is given by the selector

$$\begin{aligned} \varphi_1^*(1) = \Bigg(a^{1*} &= \frac{1}{2}\left[\frac{p_{++} - p_{--}}{p_{++} + p_{--}} + \frac{p_{+-} - p_{-+}}{p_{+-} + p_{-+}}\right], \\ a^{2*} &= \frac{1}{2}\left[\frac{p_{++} - p_{--}}{p_{++} + p_{--}} - \frac{p_{+-} - p_{-+}}{p_{+-} + p_{-+}}\right]\Bigg). \end{aligned}$$

Even if the shares exhibit identical properties ($p_{+-} = p_{-+}$), it is better to invest equal parts of the capital in both of them. A similar example was considered in [Szekely (1986), Section 3.6]. □

1.4.16 ***Secretary problem***

The classical secretary problem was studied in depth in, for example, [Puterman (1994), Section 4.6.4]. See also [Ross (1983), Chapter 1, Section

5] and [Suhov and Kelbert (2008), Section 1.11]. We shall consider only a very simple version here.

An employer seeks to hire an individual to fill a vacancy for a secretarial position. There are two candidates for this job, from two job centres. It is known that the candidate from the first centre is better with probability 0.6. The employer can, of course, interview the first candidate; however, immediately after this interview, he must decide whether or not to make the offer. If the employer does not offer the job, that candidate seeks employment elsewhere and is no longer eligible to receive an offer, so that the employer has to accept the second candidate, from the second job centre. As there is no reason for such an interview, the employer should simply make the offer to the first candidate. The aim is to maximize the probability of accepting the best candidate.

Now, suppose there is a third candidate from a third job centre. For simplicity, assume that the candidates can be ranked in only three ways:

- 1 is better than 2 and 2 is better than 3, the probability of this event being 0.3;
- 3 is better than 1 and 1 is better than 2, the probability is 0.3;
- 2 is better than 1 and 1 is better than 3, making the probability 0.4.

The first candidate is better than the second with probability of 0.6, but to maximize the probability of accepting the best candidate without interviews, the employer has to offer the job to the second candidate. There could be the following conversation between the employers:
– We have candidates from job centres 1 and 2. Who do you prefer?
– Of course, we'll hire the first one.
– Wait. Here is another application from job centre 3.
– Hm. In that case I prefer the candidate from the second job centre.

The employer can interview the candidates sequentially: the first one, the second and the third. At each step he can make an offer; if the first two candidates are rejected, the employer then has to hire the third one. Now the situation is similar to the classical case, and the problem can be formulated as an MDP [Puterman (1994), Section 4.6.4].

Example. The state space is $\mathbf{X} = \{b, \bar{b}, B, \bar{B}\}$. The state $b(\bar{b})$ means, the current candidate is the best (not the best) among those interviewed up to the current moment. The absorbing state $B(\bar{B})$ means, the offer is made, and the accepted candidate is the best (not the best) possible. The action space is $\mathbf{A} = \{O, R\}$, where $O(R)$ means 'make the offer to the current candidate (reject)'. $T = 3$, equal the number of decisions (make the offer or reject) to be made.

$\mathsf{P}_0(b) = 1$ because the first candidate is the best so far. If (s)he is accepted, then the new (absorbing) state is $B(\bar{B})$ with probability $p_1(B|b, O) = 0.3$ $(p_1(\bar{B}|b, O) = 0.7$: the first candidate is *a priori* the best with probability 0.3. If (s)he is rejected, then the new state is b with probability $p_1(b|b, R) = 0.4$, equal to the chance that 2 is better than 1; $p_1(\bar{b}|b, R) = 0.6$. If $x_1 = b$, then we have the third version of ranking candidates; so $p_2(B|b, O) = 1$. If $x_1 = \bar{b}$, then, obviously, $p_2(\bar{B}|\bar{b}, O) = 1$, and $p_2(b|\bar{b}|R) = p_2(\bar{b}|\bar{b}, R) = 0.5$ because here the first two versions of ranking candidates are equiprobable, hence the third candidate is the best one with probability 0.5. All the other non-zero transition probabilities equal 1 (see Fig.1.26). $C(B) = 0$, $C(\bar{B}) = 1$, and all the other costs are either zero, or immaterial. The target is to hire the best possible candidate.

The dynamic programming approach results in the following function $v_t(\cdot)$:

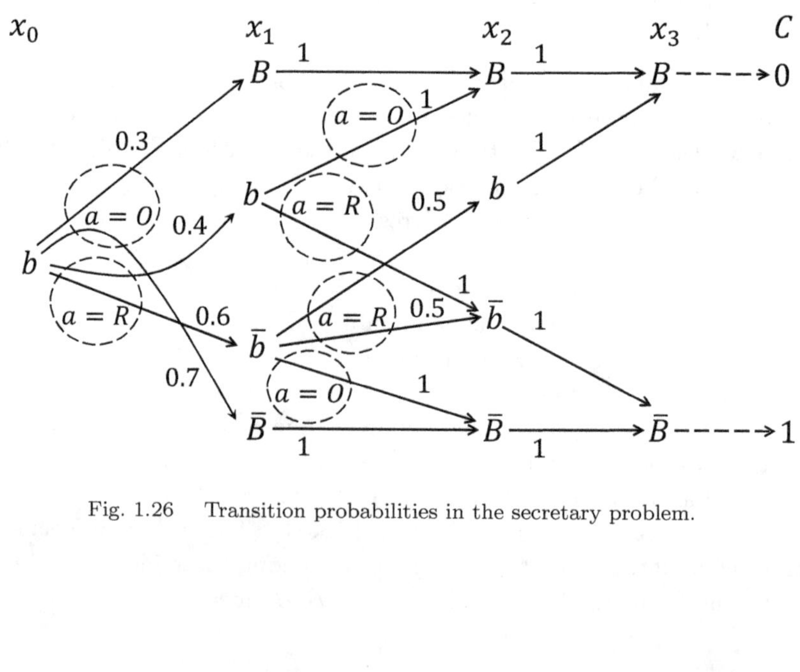

Fig. 1.26 Transition probabilities in the secretary problem.

$$v_3(B) = C(B) = 0, \quad v_3(\bar{B}) = C(\bar{B}) = 1;$$
$$v_2(B) = 0, \quad v_2(b) = 0, \quad v_2(\bar{b}) = 1, \quad v_2(\bar{B}) = 1;$$
$$v_1(B) = 0, \quad v_1(b) = 0, \quad v_1(\bar{b}) = 0.5, \quad v_1(\bar{B}) = 1;$$
$$v_0(b) = 0.3.$$

Other arguments of the function $v_t(\cdot)$ are immaterial, because those states cannot be realized. The minimal value $v^*_{\mathsf{P}_0} := \inf_\pi \mathsf{E}^\pi_{\mathsf{P}_0}[C(X_3)] = 0.3$ means that the highest probability of hiring the best ever candidate equals 0.7. The optimal strategy prescribes to reject the first candidate and accept the second one only if (s)he is better than the first.

Consider another sequence of interviews: the second job centre, the first and the third. Then the optimal control strategy prescribes the acceptance of candidate 2 (which can be done even without interviews). The probability of success equals 0.4. One can also investigate other sequences of interviews, and the optimal control strategy and probability of success can change again. □

1.4.17 *Voting*

Example. Suppose three magistrates investigate an accused person who is actually guilty. When making their decisions, the magistrates can make a mistake. To be specific, let p_i, $i = 1, 2, 3$ be the probability that magistrate i makes the (correct) decision that the accused is guilty; $q_i := 1 - p_i$. The final decision is in accordance with the majority among the three opinions. Suppose $p_1 > p_3 > p_2$. Is it not better for the less reliable magistrate 2 to share the opinion of the most reliable magistrate 1, instead of voting independently? Such a problem was discussed in [Szekely (1986), p.171].

To describe the situation mathematically, we make several assumptions. First of all, the magistrates make their decisions in sequence, one after another. We accept that magistrates 1 and 3 vote according to their personal estimates of the accused's guilt; magistrate 2 either follows the same rule (see Fig. 1.27) or shares the opinion of magistrate 1 (see Fig. 1.28), and (s)he makes his/her general decision at the very beginning. Put $T = 3$; $\mathbf{X} = \{(y, z), s_0\}$, where component $y \in \{-3, -2, \ldots, +3\}$ represents the current algebraic sum of decisions in favour of finding the accused guilty; $z \in \{\text{Own}, \text{Sh}\}$ denotes the general decision of magistrate 2 made initially; s_0 is the fictitious initial state. $\mathbf{A} = \{\text{Own}, \text{Sh}\}$, and action Own(Sh) means that magistrate 2 will make his/her own decision (will share the opinion of magistrate 1).

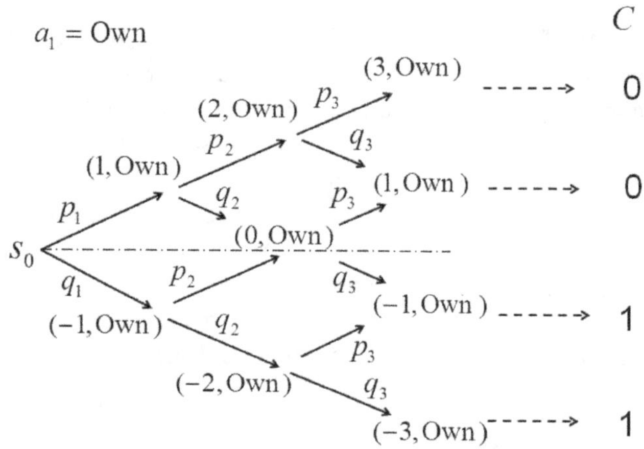

Fig. 1.27 Independent voting.

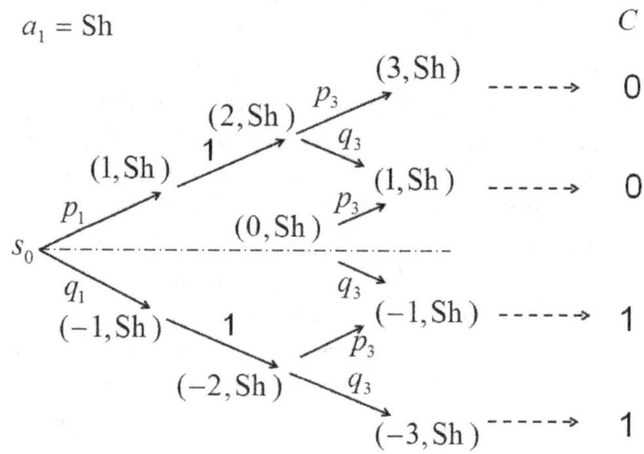

Fig. 1.28 Sharing opinion.

Now $x_0 = s_0$, i.e. $\mathsf{P}_0(x) = \mathbb{I}\{x = s_0\}$;

$$p_1((\hat{y}, \hat{z})|s_0, a) = \mathbb{I}\{\hat{z} = a\} \times \begin{cases} p_1, & \text{if } \hat{y} = 1; \\ q_1, & \text{if } \hat{y} = -1; \\ 0, & \text{otherwise,} \end{cases}$$

$$p_2((\hat{y}, \hat{z})|(y, \text{Own}), a) = \mathbb{I}\{\hat{z} = \text{Own}\} \times \begin{cases} p_2, & \text{if } \hat{y} = y + 1; \\ q_2, & \text{if } \hat{y} = y - 1; \\ 0, & \text{otherwise,} \end{cases}$$

$$p_2((\hat{y}, \hat{z})|(y, \text{Sh}), a) = \mathbb{I}\{\hat{z} = \text{Sh}\} \times \begin{cases} 1, & \text{if } \hat{y} = y + 1, \ y > 0; \\ 1, & \text{if } \hat{y} = y - 1, \ y < 0; \\ 0, & \text{otherwise,} \end{cases}$$

$$p_3((\hat{y}, \hat{z})|(y, z), a) = \mathbb{I}\{\hat{z} = z\} \times \begin{cases} p_3, & \text{if } \hat{y} = y + 1; \\ q_3, & \text{if } \hat{y} = y - 1; \\ 0, & \text{otherwise.} \end{cases}$$

We have presented only the significant values of the transition probability. Other transition probabilities are either zero or immaterial. If a state x cannot be reached by step t, then there is no reason to pay attention to $p_t(\hat{x}|x, a)$.

$$c_t(x, a) \equiv 0, \text{ for } v = (y, z), \ C((y, z)) = \begin{cases} 0, & \text{if } y \geq 0; \\ 1, & \text{if } y < 0, \end{cases} \quad C(s_0) \text{ is immaterial.}$$

The dynamic programming approach results in the following:

$$v_3(x) = C(x); \quad v_2((y, z)) = \begin{cases} 0, & \text{if } y = +2; \\ q_3, & \text{if } y = 0; \\ 1, & \text{if } y = -2 \end{cases}$$

(other values of y are of no interest).

$$v_1((1, \text{Own})) = q_2 q_3; \quad v_1((-1, \text{Own})) = p_2 q_3 + q_2;$$

$$v_1((1, \text{Sh})) = 0; \quad v_1((-1, \text{Sh})) = 1;$$

$$v_0(s_0) = \min\{p_1 q_2 q_3 + q_1(p_2 q_3 + q_2); \quad q_1\}.$$

The first (second) expression in parentheses corresponds to $a = \text{Own}$ (Sh). Other values of $x \in \mathbf{X}$ are of no interest.

Let $p_1 = 0.7$, $p_2 = 0.6$, $p_3 = 0.65$. Then

$$v_0(s_0) = p_1 q_2 q_3 + q_1(p_2 q_3 + q_2) = 0.281 < 0.35 = q_1.$$

Even the less reliable magistrate plays his role. If (s)he shares the opinion of the most reliable magistrate 1, then the total probability of making a mistake increases. Of course the situation changes if p_2 is too small: the critical value is $p_2 = \frac{p_1 q_3}{p_1 q_3 + q_1 p_3}$ (equal to 0.56 in the presented numerical example). □

Chapter 2

Homogeneous Infinite-Horizon Models: Expected Total Cost

In this chapter, we consider the model with the infinite horizon $T = \infty$. After the brief presentation of the necessary knowledge on such MDPs in Sections 2.1–2.2, a number of examples are presented in Section 2.3.

2.1 Homogeneous Undiscounted Model

The definitions of (Markov, stationary) strategies and selectors are the same as in Chapter 1. The sample space is now

$$\Omega := (\mathbf{X} \times \mathbf{A})^\infty = \{(x_0, a_1, x_1, \ldots)\}.$$

For a given initial distribution P_0 and control strategy π, the *strategic* measure $\mathsf{P}^\pi_{\mathsf{P}_0}$ on $\mathcal{B}(\Omega)$ is built in a similar way to that in Chapter 1; the rigorous construction is based on the Ionescu Tulcea Theorem [Bertsekas and Shreve (1978), Prop. 7.28], [Hernández-Lerma and Lasserre (1996a), Prop. C10], [Piunovskiy (1997), Thm. A1.11]. The space of all strategic measures (for a fixed initial distribution P_0) is denoted as $\mathcal{P}_{\mathsf{P}_0} := \{\mathsf{P}^\pi_{\mathsf{P}_0},\ \ \pi \in \Pi^{\text{All}}\}$; $\mathsf{E}^\pi_{\mathsf{P}_0}$ is the mathematical expectation w.r.t. $\mathsf{P}^\pi_{\mathsf{P}_0}$. If the initial distribution $\mathsf{P}_0(dx) = \delta_{x_0}(dx)$ is concentrated at the initial state $x_0 \in \mathbf{X}$, then we use notations $\mathsf{P}^\pi_{x_0}$, $\mathsf{E}^\pi_{x_0}$ and $\mathcal{P}_{x_0}$. The goal is to find an optimal control strategy π^* solving the problem

$$v^\pi_{\mathsf{P}_0} := \mathsf{E}^\pi_{\mathsf{P}_0}\left[\sum_{t=1}^{\infty} c(X_{t-1}, A_t)\right] \to \inf_\pi \tag{2.1}$$

for the fixed initial distribution P_0. As usual, $v^\pi_{\mathsf{P}_0}$ is called the *performance functional*. For a fixed $\varepsilon \geq 0$, a strategy π is called ε-optimal if $v^\pi_{\mathsf{P}_0} \leq \begin{cases} \inf_\pi v^\pi_{\mathsf{P}_0} + \varepsilon, & \text{if } \inf_\pi v^\pi_{\mathsf{P}_0} > -\infty; \\ -\frac{1}{\varepsilon}, & \text{if } \inf_\pi v^\pi_{\mathsf{P}_0} = -\infty. \end{cases}$ A 0-optimal strategy is called optimal; here $-\frac{1}{0}$ should be replaced by $-\infty$.

According to the following Lemma, usually it is sufficient to consider only Markov strategies.

Lemma 2.1. [Piunovskiy (1997), Lemma 2], [Dynkin and Yushkevich (1979), Chapter 3, Section 8]. *For every control strategy π, there exists a Markov strategy $\pi^{\rm m}$ such that for all $t = 1, 2, \dots$*

$$\mathsf{P}^{\pi}_{\mathsf{P}_0}(X_{t-1} \in \Gamma^{\mathbf{X}}, A_t \in \Gamma^{\mathbf{A}}) = \mathsf{P}^{\pi^{\rm m}}_{\mathsf{P}_0}(X_{t-1} \in \Gamma^{\mathbf{X}}, A_t \in \Gamma^{\mathbf{A}}) \qquad (2.2)$$

for any $\Gamma^{\mathbf{X}} \in \mathcal{B}(\mathbf{X})$ and $\Gamma^{\mathbf{A}} \in \mathcal{B}(\mathbf{A})$.

Assume that the following condition is satisfied. (It holds in all the examples of this chapter apart from Subsections 2.3.5 and 2.3.8.) The corresponding models are called *summable*.

Condition 2.1. For every control strategy π, either $\mathsf{E}^{\pi}_{\mathsf{P}_0}[\sum_{t=1}^{\infty} c^+(X_{t-1}, A_t)]$ or $\mathsf{E}^{\pi}_{\mathsf{P}_0}[\sum_{t=1}^{\infty} c^-(X_{t-1}, A_t)]$ is finite, so expressions '$\infty - \infty$' do not appear.

Similarly to Chapter 1, we accept formula

$$v^{\pi}_{\mathsf{P}_0} = \mathsf{E}^{\pi}_{\mathsf{P}_0}\left[W^+\right] + \mathsf{E}^{\pi}_{\mathsf{P}_0}\left[W^-\right], \qquad (2.3)$$

where

$$W(\omega) = \sum_{t=1}^{\infty} c(x_{t-1}, a_t) := \sum_{t=1}^{\infty} c^+(x_{t-1}, a_t) + \sum_{t=1}^{\infty} c^-(x_{t-1}, a_t)$$

with the usual convention $\infty - \infty := +\infty$. Under Condition 2.1, the result coincides with

$$v^{\pi}_{\mathsf{P}_0} = \mathsf{E}^{\pi}_{\mathsf{P}_0}\left[\sum_{t=1}^{\infty} c^+(x_{t-1}, a_t)\right] + \mathsf{E}^{\pi}_{\mathsf{P}_0}\left[\sum_{t=1}^{\infty} c^-(x_{t-1}, a_t)\right]. \qquad (2.4)$$

But, if Condition 2.1 is violated, the results (2.3) and (2.4) can be different as Example 2.3.5 shows.

Note that the cost function $c(\cdot, \cdot)$ and the transition probability $p(dy|x, a)$ do not depend on time. Such models are called *homogeneous*. As before, we write $\mathsf{P}^{\pi}_{x_0}$ if the initial distribution is concentrated at a single point $x_0 \in \mathbf{X}$. In this connection,

$$v^{\pi}_{x_0} := \mathsf{E}^{\pi}_{x_0}\left[\sum_{t=1}^{\infty} c(X_{t-1}, A_t)\right]$$

and

$$v^*_x := \inf_{\pi} v^{\pi}_x, \quad x \in \mathbf{X}$$

is the Bellman function (cf (1.9)). For a fixed $\varepsilon \geq 0$, we call a strategy π *uniformly ε-optimal* if, for all $x \in \mathbf{X}$, $v_x^\pi \leq \begin{cases} v_x^* + \varepsilon, & \text{if } v_x^* > -\infty; \\ -\frac{1}{\varepsilon}, & \text{if } v_x^* = -\infty. \end{cases}$ A uniformly 0-optimal strategy is called *uniformly optimal*. Here $-\frac{1}{0}$ should be replaced by $-\infty$. Note that this definition of the uniform optimality is slightly different from the one given in Chapter 1. In many books, uniformly (ε)-optimal strategies are called simply (ε)-optimal.

Remark 2.1. According to Lemma 2.1, Markov strategies are sufficient for solving optimization problems if the initial distribution is fixed. Since the uniform optimality concerns all possible initial states, it can happen that only a semi-Markov strategy is uniformly (ε-) optimal: see Subsections 2.3.13, 2.3.14 and Theorem 2.1.

The Bellman function v_x^* satisfies the *optimality* equation

$$v(x) = \inf_{a \in \mathbf{A}} \left\{ c(x,a) + \int_{\mathbf{X}} v(y) p(dy|x,a) \right\}, \quad x \in \mathbf{X}, \tag{2.5}$$

except for pathological situations similar to those described in Chapter 1, where expressions " $+\infty$" + " $-\infty$" appear: see Proposition 9.8 of [Bertsekas and Shreve (1978)]. In the finite-horizon case, the minimal expected cost (when starting from the state $x \in \mathbf{X}$) coincides with the solution of (1.5), except for those pathological cases. If $T = \infty$, the situation is different, because it is obvious that, if a function $v(\cdot)$ is a solution of (2.5), then $v(\cdot) + r$ is also a solution for any $r \in \mathbb{R}$. Moreover, as the example in Subsection 2.3.2 below shows, there can exist many other non-trivial solutions of equation (2.5). Thus, generally speaking, a solution of the optimality equation (2.5) provides no boundaries for the Bellman function.

A stationary selector φ is called *conserving* (or *thrifty*) if

$$v_x^* = c(x, \varphi(x)) + \int_{\mathbf{X}} v_y^* p(dy|x, \varphi(x)), \quad x \in \mathbf{X};$$

it is called *equalizing* if

$$\forall x \in \mathbf{X} \qquad \liminf_{T \to \infty} \mathsf{E}_x^\varphi \left[v_{X_T}^* \right] \geq 0$$

[Puterman (1994), Section 7.1.3]. In positive models, any strategy is equalizing.

More generally, a stationary strategy π^{s} is called conserving if

$$v_x^* = \int_{\mathbf{A}} \left[c(x,a) + \int_{\mathbf{X}} v_y^* p(dy|x,a) \right] \pi^{\mathrm{s}}(da|x), \quad x \in \mathbf{X}.$$

It is known that a conserving and equalizing stationary selector φ is uniformly optimal, i.e. $v_x^\varphi = v_x^*$, under very general conditions [Puterman (1994), Thm. 7.1.7]. Note also that, e.g., in *positive or negative models* (i.e., when $c(\cdot,\cdot) \geq 0$ or $c(\cdot,\cdot) \leq 0$) every uniformly optimal strategy is also optimal under any initial distribution.

Let the initial distribution P_0 be fixed. An MDP is called *absorbing* if there is an isolated state, say 0 or Δ, for which the controlled process is absorbed at time $T_0 := \min\{t > 0 : X_t = 0\}$, and for all π $\mathsf{E}^\pi_{\mathsf{P}_0}[T_0] < \infty$ [Altman (1999), Section 7.1]. In fact, there exists a constant M (of course dependent on P_0) such that $\mathsf{E}^\pi_{\mathsf{P}_0}[T_0] \leq M < \infty$ for all strategies π [Feinberg and Rothblum (2012), p.132]. All the future costs equal zero: $c(0,a) \equiv 0$. Absorbing models are considered in Subsections 2.3.2, 2.3.5, 2.3.9, 2.3.14, 2.3.17, 2.3.20, 2.3.24, 2.3.25, 2.3.31 and 2.3.34. Note that if an MDP is absorbing for one initial distribution, it may be non-absorbing for another initial distribution, even if all the states (apart from 0) communicate: see Remark 2.5 and also Subsection 2.3.25.

The examples in Subsections 2.3.3, 2.3.4, 2.3.11, 2.3.14 and 2.3.19 are from the area of *optimal stopping* in which, on each step, there exists the possibility of putting the controlled process in a special absorbing state (say 0, or Δ), sometimes called *cemetery*, with no future cost. Note that optimal stopping problems are not always about absorbing MDP: the absorption may be infinitely delayed, as in the examples in Subsections 2.3.4, 2.3.11 and 2.3.19.

Finally, note that, as explained in Section 3.1, discounted MDP is a special case of the (absorbing) model with the expected total cost.

2.2 Several Important Statements

Recall that the model is called discrete, if the spaces $\mathbf{X}$ and $\mathbf{A}$ are finite or denumerable. In this case, we usually assume that the topology in $\mathbf{X}$ and $\mathbf{A}$ is discrete. If the spaces $\mathbf{X}$ and/or $\mathbf{A}$ are general Borel, we sometimes call the model *Borel*. Below, we also assume that the model is summable.

Theorem 2.1. [Bertsekas and Shreve (1978), Props 9.19 and 9.20]. *In a positive (negative) model, for each $\varepsilon > 0$, there exists a uniformly ε-optimal Markov (semi-Markov) selector.*

The rather powerful method for constructing the Bellman function v_x^* is based on the so called *value iteration*. This algorithm works as follows:

$$v^0(x) \equiv 0;$$
$$v^{n+1}(x) = \inf_{a\in\mathbf{A}} \left\{ c(x,a) + \int_{\mathbf{X}} v^n(y)p(dy|x,a) \right\}, \quad n = 0,1,2,\ldots, \tag{2.6}$$
$$x \in \mathbf{X}.$$

(We leave aside the question of the measurability of functions $v^{n+1}(\cdot)$.)

Theorem 2.2. [Bertsekas and Shreve (1978), Prop. 9.14] and [Puterman (1994), Thm. 7.2.12]. *If the model is positive ($c(\cdot,\cdot) \geq 0$) or negative ($c(\cdot,\cdot) \leq 0$), then there exists the limit*

$$\lim_{n\to\infty} v^n(x) =: v^\infty(x), \quad x \in \mathbf{X} \tag{2.7}$$

and the convergence is monotonic (respectively, increasing or decreasing).

In negative models, $v^\infty(x) = v_x^*$ for all $x \in \mathbf{X}$, and $v^\infty(x)$ satisfies the optimality equation (2.5). Thus, the Bellman function v_x^* is the maximal non-positive solution to equation (2.5).

The similar statement, that is, the limit (2.7) exists and $v^\infty(x) = v_x^*$, holds for discounted models, if e.g., $\sup_{x\in\mathbf{X}} \sup_{a\in\mathbf{A}} |c(x,a)| < \infty$: see (3.3) [Bertsekas and Shreve (1978), Prop. 9.14], [Puterman (1994), Section 6.3].

The limit (2.7) also exists in positive models, but here

$$v^\infty(x) \leq v_x^*, \quad x \in \mathbf{X}. \tag{2.8}$$

Theorem 2.3. [Bertsekas and Shreve (1978), Prop. 9.16]. *Suppose the model is positive. Then $v^\infty(x) \equiv v_x^*$ if and only if*

$$v^\infty(x) = \inf_{a\in\mathbf{A}} \left\{ c(x,a) + \int_{\mathbf{X}} v^\infty(y)p(dy|x,a) \right\}, \quad x \in \mathbf{X}.$$

Moreover, $v^\infty(x) \equiv v_x^*$ if the action space $\mathbf{A}$ is finite [Bertsekas and Shreve (1978), Corollary 9.17.1].

According to [Feinberg (1996), Thm.4.2], [Bertsekas and Shreve (1978), Corollary 9.4.1], the Bellman function v_x^* is lower semianalytic. Since each lower semianalytic function on $\mathbf{X}$ with values in $[0,\infty]$, satisfying the optimality equation (2.5), is component-wise not smaller than v_x^* in positive models (see Proposition 9.10 of [Bertsekas and Shreve (1978)]), we conclude that v_x^* is the *minimal* lower semianalytic $[0,\infty]$-valued solution to (2.5) in case $c(\cdot,\cdot) \geq 0$. Similarly, in case $c(\cdot,\cdot) \leq 0$, the Bellman function v_x^* is the *maximal* lower semianalytic $[-\infty,0]$-valued solution to (2.5). In this connection, see also Theorems 7.2.3 and 7.3.2 of [Puterman (1994)] about discrete models.

Suppose the following conditions are satisfied:

Condition 2.2. [Puterman (1994), Assumptions 7.2.1, 7.2.2].

(a) $\forall x \in \mathbf{X}$, $\forall \pi \in \Pi^{\mathrm{All}}$

$$\mathsf{E}_x^\pi\left[\sum_{t=1}^{\infty} c^-(X_{t-1}, A_t)\right] > -\infty;$$

(b) $\forall x \in \mathbf{X}$, $\exists a \in \mathbf{A} \quad c(x,a) \le 0$.

Then, in such a discrete model, $v_x^* \le 0$ is the maximal non-positive finite-valued solution to equation (2.5) [Puterman (1994), Theorem 7.2.3(a)].

Remark 2.2. For a fixed stationary strategy $\pi^{\mathrm{s}}(da|x)$, one also can compute $v_x^{\pi^{\mathrm{s}}}$ using the value iteration

$$v^0(x) \equiv 0;$$
$$v^{n+1}(x) = \int_{\mathbf{A}} \left\{ c(x,a) + \int_{\mathbf{X}} v^n(y) p(dy|x,a) \right\} \pi^{\mathrm{s}}(da|x), \qquad n = 0, 1, 2, \ldots,$$
$$x \in \mathbf{X}.$$

In negative models, $\lim_{n\to\infty} v^n(x) =: v^\infty(x) = v_x^{\pi^{\mathrm{s}}}$, and in positive models $v^\infty(x) = v_x^{\pi^{\mathrm{s}}}$ if and only if

$$v^\infty(x) = \int_{\mathbf{A}} \left\{ c(x,a) + \int_{\mathbf{X}} v^\infty(y) p(dy|x,a) \right\} \pi^{\mathrm{s}}(da|x), \quad x \in \mathbf{X}.$$

For the proof, one can regard this case as the uncontrolled MDP with a dummy action Δ and with transition probability and cost function being $\int_{\mathbf{A}} p(dy|x,a)\pi^{\mathrm{s}}(da|x)$ and $\int_{\mathbf{A}} c(x,a)\pi^{\mathrm{s}}(da|x)$. After that, Theorems 2.2 and 2.3 are in use.

Let us introduce the following condition.

Condition 2.3. The spaces $\mathbf{X}$ and $\mathbf{A}$ are topological Borel spaces. The action space $\mathbf{A}$ is compact, the transition probability $p(dy|x,a)$ is *(weakly) continuous*, that is, $\int_{\mathbf{X}} u(y)p(dy|x,a)$ is continuous in (x,a) for each bounded continuous function $u(\cdot)$ on $\mathbf{X}$, and the cost function $c(\cdot,\cdot)$ is lower semi-continuous and bounded from below.

We will call the models, satisfying Condition 2.3 *weakly semi-continuous* and reserve the term 'semi-continuous' for the models satisfying Condition

2.4. Anyway, to avoid misunderstanding, we always give a reference to a specific condition when talking about semi-continuous models.

Theorem 2.4. [Bertsekas and Shreve (1978), Corollary 9.17.2]. *If a positive model is weakly semi-continuous (i.e., Condition 2.3 is satisfied), then $v^\infty(x) \equiv v_x^*$ and there exists a conserving stationary selector which is uniformly optimal.*

Some authors call a model *semi-continuous* if the following condition (slightly different from Condition 2.3) is satisfied.

Condition 2.4.

(a) The action space $\mathbf{A}$ is a compact topological Borel space;
(b) for each $x \in \mathbf{X}$ the cost function $c(x, a)$ is lower semi-continuous in a;
(c) for each $x \in \mathbf{X}$ function $\int_{\mathbf{X}} u(y)p(dy|x, a)$ is continuous in a for every (measurable) bounded function $u(\cdot)$.

If the state space $\mathbf{X}$ is finite or denumerable with the discrete topology, then Condition 2.4 is equivalent to Condition 2.3.

Theorem 2.5. [Schäl (1975a), Thm. 2.2 and Thm. 6.6]. *If the model is semi-continuous (i.e., Condition 2.4 is satisfied), the cost function $c(\cdot, \cdot)$ is bounded from below, $\mathsf{E}^\pi_{\mathsf{P}_0}\left[\sum_{t=1}^\infty c^-(X_{t-1}, A_t)\right] > -\infty$ for all $\pi \in \Pi^{\text{All}}$, and*

$$\inf_{T \ge n} \inf_{\pi \in \Pi^{\text{All}}} \sum_{t=n+1}^{T} \mathsf{E}^\pi_{\mathsf{P}_0}\left[c(X_{t-1}, A_t)\right] \to 0 \qquad \text{as } n \to \infty, \tag{2.9}$$

then there exists an optimal control strategy.

The solvability of problem (2.1) follows, e.g., if the space $\mathcal{P}_{\mathsf{P}_0}$ is compact in some metrizable topology and the mapping

$$\mathsf{P}^\pi_{\mathsf{P}_0} \to v^\pi_{\mathsf{P}_0} \tag{2.10}$$

is lower semi-continuous. In the space $\mathbb{R}^*$ of the values of $v^\pi_{\mathsf{P}_0}$ the Euclidean topology is fixed as usual.

Lemma 2.2. *If conditions 2.4(a,c) are satisfied (and $\mathbf{X}$ is also a topological Borel space), then the space $\mathcal{P}_{\mathsf{P}_0}$ of all strategic measures is compact in the weak topology.*

The proof follows from [Schäl (1975a), Thm. 6.6] and [Nowak (1988)]. One can also refer to Lemma 2.5.

The lower semi-continuity of the mapping (2.10) w.r.t. the weak topology in $\mathcal{P}_{\mathsf{P}_0}$ is a more delicate issue. For example, the following lemma was proved in [Jaśkiewicz and Nowak (2019), p.3133].

Lemma 2.3. *Suppose Condition 2.4 is satisfied and* $\mathbf{X}$ *is also a topological Borel space.*

Let the transition probability $p(dy|x,a)$ *be absolutely continuous w.r.t. some reference measure* ν *on* $\mathcal{B}(\mathbf{X})$ *with the Radon-Nikodym derivative* $\delta(x,a,y) := \frac{dp(\cdot|x,a)}{d\nu(\cdot)}$ *such that*

$$\lim_{k\to\infty} \int_{\mathbf{X}} |\delta(x,a_k,y) - \delta(x,a,y)|\nu(dy) = 0$$

for each $x \in \mathbf{X}$, *if* $\lim_{k\to\infty} a_k = a$.

Assume that there exists a measurable function $w(\cdot) : \mathbf{X} \to [0,\infty)$ *such that*

(a) $|c^-(x,a)| \le w(x)$ *for all* $(x,a) \in \mathbf{X} \times \mathbf{A}$;
(b) $\mathsf{E}^\pi_{\mathsf{P}_0}[w(X_n)] < \infty$ *for all* π *and* $n = 0,1,\dots$;
(c) $\lim_{n\to\infty} \sup_\pi \mathsf{E}^\pi_{\mathsf{P}_0}\left[\sum_{t=n}^\infty w(X_t)\right] = 0$;
(d) *for* $n = 0,1,\dots$ $\lim_{l\to\infty} \sup_\pi \mathsf{E}^\pi_{\mathsf{P}_0}[w(X_n)\mathbb{I}\{w(X_n) \ge l\}] = 0$.

Then the mapping (2.10) is lower semi-continuous w.r.t. the weak topology in $\mathcal{P}_{\mathsf{P}_0}$.

Theorem 2.6. [Strauch (1966), Thm. 8.3]. *If* $c(\cdot,\cdot) \ge 0$ *and there exists an optimal strategy, then there exists an optimal stationary selector.*

Definition 2.1. For a fixed initial distribution P_0 and control strategy π, the *occupation measure* $\eta^\pi_{\mathsf{P}_0}$ is the ($[0,\infty]$-valued) measure on $\mathcal{B}(\mathbf{X}\times\mathbf{A})$ given by the formula

$$\eta^\pi_{\mathsf{P}_0}(\Gamma^{\mathbf{X}}\times\Gamma^{\mathbf{A}}) := \sum_{t=0}^\infty \mathsf{P}^\pi_{\mathsf{P}_0}\{X_t \in \Gamma^{\mathbf{X}}, A_{t+1} \in \Gamma^{\mathbf{A}}\},\ \ \Gamma^{\mathbf{X}} \in \mathcal{B}(\mathbf{X}),\ \Gamma^{\mathbf{A}} \in \mathcal{B}(\mathbf{A}).$$

Proposition 2.1. *The set* $\mathcal{P}_{\mathsf{P}_0}$ *of all strategic measures (for a fixed initial distribution) is convex. The set* $\mathcal{D}_{\mathsf{P}_0} := \{\eta^\pi_{\mathsf{P}_0},\ \pi \in \Pi^{\mathrm{All}}\}$ *of all occupation measures for a fixed initial distribution* P_0 *is also a convex set in the cone of* $[0,\infty]$*-valued measures on* $\mathcal{B}(\mathbf{X}\times\mathbf{A})$.

The proof of the first statement can be found in Ch. 5, §5 of [Dynkin and Yushkevich (1979)], see also [Piunovskiy (1997), Thm. 8]. The second statement directly follows from the linearity of the mapping $\mathsf{P}^\pi_{\mathsf{P}_0} \to \eta^\pi_{\mathsf{P}_0}$ under a fixed P_0. For more details, see [Piunovskiy and Zhang (2020), Prop. 3].

If the initial distribution $\mathsf{P}_0(dx) = \delta_{x_0}(dx)$ is concentrated at the initial state $x_0 \in \mathbf{X}$, we use the notations $\eta^\pi_{x_0}$ and $\mathcal{D}_{x_0}$.

For any π, the occupation measure $\eta^\pi_{\mathsf{P}_0}$ satisfies equation

$$\eta(\Gamma \times \mathbf{A}) = \mathsf{P}_0(\Gamma) + \int_{\mathbf{X}\times\mathbf{A}} p(\Gamma|y,a)d\eta(y,a), \quad \Gamma \in \mathcal{B}(\mathbf{X}) \tag{2.11}$$

[Hernández-Lerma and Lasserre (1999), Lemma 9.4.3].

Usually (e.g. in positive and negative models),

$$v^\pi_{\mathsf{P}_0} = \int_{\mathbf{X}\times\mathbf{A}} c(x,a)d\eta^\pi_{\mathsf{P}_0}(x,a),$$

and investigation of MDP in terms of occupation measures (the so-called *convex analytic approach*) leads to the program

$$\int_{\mathbf{X}\times\mathbf{A}} c(y,a)d\eta(y,a) \ \to \ \inf_\eta \tag{2.12}$$

$$\text{subject to (2.11) and } \eta \geq 0,$$

traditionally called *Primal (Linear) Program.*

In the absorbing model with the costless cemetery 0, for each strategy π, $\eta^\pi_{\mathsf{P}_0}((\mathbf{X}\setminus\{0\})\times\mathbf{A}) = \mathsf{E}^\pi_{\mathsf{P}_0}[T_0] < \infty$. Thus, the occupation measure $\eta^\pi_{\mathsf{P}_0}$ satisfies the following equations

$$\eta((\mathbf{X}\setminus\{0\})\times\mathbf{A}) < \infty, \qquad \eta(\{0\}\times\mathbf{A}) = \infty,$$

and

$$\eta(\Gamma \times \mathbf{A}) = \mathsf{P}_0(\Gamma) + \int_{(\mathbf{X}\setminus\{0\})\times\mathbf{A}} p(\Gamma|y,a)d\eta(y,a), \quad \Gamma \in \mathcal{B}(\mathbf{X}\setminus\{0\}). \tag{2.13}$$

In this case, integration in (2.12) is over $(\mathbf{X}\setminus\{0\})\times\mathbf{A}$ and (2.11) is replaced by (2.13). With some abuse of notation, if the model is absorbing, we use the notation $\mathcal{D}_{\mathsf{P}_0}$ for the restrictions of occupation measures on $(\mathbf{X}\setminus\{0\})\times\mathbf{A}$.

Lemma 2.4. [Piunovskiy and Zhang (2024a), Lemma 2]. *Suppose the model is absorbing. If $\eta = \eta^\pi_{\mathsf{P}_0}$ is a (finite) occupation measure on $\mathcal{B}((\mathbf{X}\setminus\{0\})\times\mathbf{A})$, then $\eta = \eta^{\pi^s}_{\mathsf{P}_0}$, where π^s is the stationary strategy coming from the decomposition*

$$\eta(\Gamma_X \times \Gamma_A) = \int_{\Gamma_X} \pi^s(\Gamma_A|x)\eta(dx\times\mathbf{A}), \quad \Gamma_X \in \mathcal{B}((\mathbf{X}\setminus\{0\}), \ \Gamma_A \in \mathcal{B}(\mathbf{A}). \tag{2.14}$$

Note that the decomposition (2.14) can be also done for σ-finite measures η on $\mathcal{B}(\mathbf{X}\times\mathbf{A})$.

If $\pi^{\rm s}(da|y)$ is a stationary strategy, then $\eta^{\pi^{\rm s}}_{\mathsf{P}_0}(dx\times da)=\pi^{\rm s}(da|x)\eta^{\pi^{\rm s}}_{\mathsf{P}_0}(dx\times\mathbf{A})$, and the marginal measure $\hat\eta^{\pi^{\rm s}}_{\mathsf{P}_0}(dx):=\eta^{\pi^{\rm s}}_{\mathsf{P}_0}(dx\times\mathbf{A})$ on $\mathbf{X}$ is the minimal non-negative solution to equation

$$\hat\eta(\Gamma)=\mathsf{P}_0(\Gamma)+\int_{\mathbf{X}}\int_{\mathbf{A}}p(\Gamma|y,a)\pi^{\rm s}(da|y)\hat\eta(dy),\quad \Gamma\in\mathcal{B}(\mathbf{X}). \tag{2.15}$$

This solution to (2.15) can be built by successive approximations

$$\begin{aligned}\hat\eta_0(\Gamma)&=0,\quad \Gamma\in\mathcal{B}(\mathbf{X});\\ \hat\eta_{i+1}(\Gamma)&=\mathsf{P}_0(\Gamma)+\int_{\mathbf{X}}\int_{\mathbf{A}}p(\Gamma|y,a)\pi^{\rm s}(da|y)\hat\eta_i(dy),\\ &\Gamma\in\mathcal{B}(\mathbf{X}),\ i=0,1,\ldots:\end{aligned} \tag{2.16}$$

by induction, $\hat\eta_i(\Gamma)=\mathsf{E}^{\pi^{\rm s}}_{\mathsf{P}_0}\left[\sum_{t=0}^{i-1}\mathbb{I}\{X_t\in\Gamma\}\right]\to\hat\eta^{\pi^{\rm s}}_{\mathsf{P}_0}(\Gamma)$ as $i\to\infty$ for all $\Gamma\in\mathcal{B}(\mathbf{X})$. See also [Piunovskiy and Zhang (2020), pp.263-264].

For given P_0, p and $\pi^{\rm s}$, equation (2.15) w.r.t. $\hat\eta$ can have many solutions, but only the *minimal* non-negative solution gives the occupation measure: see Lemma 7.1 in the book [Altman (1999)], devoted to models with the countable space $\mathbf{X}$. Non-minimal solutions are usually phantom and do not correspond to any control strategy.

For countable state space $\mathbf{X}$, one can introduce the notion of *transient* models.

Definition 2.2. [Altman (1999), Section 7.1]. A control strategy π is called *transient* if

$$\hat\eta^{\pi}_{\mathsf{P}_0}(x):=\eta^{\pi}_{\mathsf{P}_0}(x\times\mathbf{A})=\sum_{t=0}^{\infty}\mathsf{P}^{\pi}_{\mathsf{P}_0}(X_t=x)<\infty\quad\text{for any } x\in\mathbf{X}. \tag{2.17}$$

In the case when state 0 is absorbing, we consider only $x\in\mathbf{X}\setminus\{0\}$ in (2.17). Sometimes the absorbing state is denoted as Δ.

An MDP is called *transient* if all its strategies are transient. Any absorbing MDP is also transient, but not *vice versa.*

In transient models, occupation measures $\eta^{\pi}_{\mathsf{P}_0}$ are finite on singletons but can only be σ-finite if there is no absorbing state. Occupation measures satisfy equation (2.11), but that equation can have phantom σ-finite solutions (see Subsection 2.3.20).

One can find the necessary knowledge about Borel spaces, semi-continuous functions, stochastic kernels etc in [Bertsekas and Shreve (1978)]. In particular, the following theorem is useful.

Theorem 2.7. [Bertsekas and Shreve (1978), Prop. 7.31]. *Let* $\mathbf{X}$ *and* $\mathbf{Y}$ *be separable metrizable spaces, let* $q(dy|x)$ *be a continuous stochastic kernel on* $\mathcal{B}(\mathbf{Y})$ *given* $\mathbf{X}$, *and let* $f(\cdot) : \mathbf{X} \times \mathbf{Y} \longrightarrow \mathbb{R}^*$ *be a measurable function. Define*

$$g(x) := \int_{\mathbf{Y}} f(x, y) q(dy|x).$$

(a) *If* $f(\cdot)$ *is lower semi-continuous and bounded below, then* $g(\cdot)$ *is lower semi-continuous and bounded below.*
(b) *If* $f(\cdot)$ *is upper semi-continuous and bounded above, then* $g(\cdot)$ *is upper semi-continuous and bounded above.*

Corollary 2.1. *Let* Y *be a topological Borel space and* Ξ *be a Borel set of uniformly bounded measures* η *on* Y, *equipped with the weak topology. Then the mapping*

$$\eta \to \int_Y c(y)\eta(dy)$$

on Ξ *is lower semi-continuous and bounded below if the function* $c(\cdot,\cdot) : Y \to \mathbb{R}^*$ *is lower semi-continuous and bounded below.*

The proof is given in Appendix A.

2.3 Examples

Most of the examples demonstrate that conditions in theorems on MDP are important. In particular, there are strong theorems for (weakly) semi-continuous models stating the existence of optimal solutions to problem (2.1): see, e.g., Theorems 2.4, 2.5 and 2.12. But if some additional requirements are not fulfilled, there may be no optimal strategies, as is shown in Subsections 2.3.16 and 2.3.24.

Examples in Subsection 2.3.1 are on mixed strategies. In Subsections 2.3.2–2.3.5, we discuss the solutions to the optimality equation and the connected sufficient/necessary conditions for a stationary selector to be uniformly optimal. Some counter-intuitive properties of combinations of strategies are presented in Subsection 2.3.6. Examples in Subsections 2.3.7–2.3.11 and 2.3.18 are about the strategy and value iterations. In particular example in Subsection 2.3.11 demonstrates that different initial functions $v^0(\cdot)$ can result in different limiting functions $v^\infty(\cdot)$ in the value iteration algorithm (2.6). In Subsections 2.3.12–2.3.15, we discuss the existence of uniformly ε-optimal and 'nearly uniformly optimal' stationary selectors. Semi-continuous models are studied in Subsections 2.3.16 and 2.3.17; see also 2.3.24. Example 2.3.18 presents a non-measurable Bellman function, and examples in Subsections 2.3.4, 2.3.19 show the necessity of randomized strategies. Properties of occupation measures are presented in Subsections 2.3.20–2.3.24. We compare absorbing and uniformly absorbing models in Subsection 2.3.25. Different topologies in the spaces of strategic and occupation measures are compared in Subsections 2.3.26 and 2.3.27. Constrained optimization is considered in Subsection 2.3.28. Some results when investigating real-life situations, like queues, gambling and optimal search, are given in Subsections 2.3.2, 2.3.16 and 2.3.29–2.3.33. Example in Subsection 2.3.34 is about the turnpike property of a random walk on the integer lattice.

Recall that any finite-horizon model is a special case of a homogeneous infinite-horizon model: see Remark 1.1. Thus, all the examples from Section 1.4 are relevant for the current chapter, too.

2.3.1 *Mixed strategies*

The space of all probability measures on the space of trajectories $\mathbf{H} = \mathbf{X} \times (\mathbf{A} \times \mathbf{X})^\infty$ is a Borel space (see Proposition 7.13 and Corollary 7.25.1 of [Bertsekas and Shreve (1978)]). In this connection, under a fixed initial distribution P_0, the spaces of strategic measures $\mathcal{P}^{\Pi}_{\mathsf{P}_0} = \{\mathsf{P}^\pi_{\mathsf{P}_0},\ \pi \in \Pi\}$ are known to be measurable and hence Borel for $\Pi = \Pi^{\text{All}} = \{\text{all strategies}\}$, $\Pi = \Pi^{\text{M}} = \{\text{Markov strategies}\}$, $\Pi = \Pi^{\text{S}} = \{\text{stationary strategies}\}$, and for Π^{AllN}, Π^{MN}, Π^{SN}, where letter N corresponds to non-randomized strategies (see [Feinberg (1996), Sections 2 and 3]). Below, we use the notation $\mathcal{P}_{\mathsf{P}_0}$ for $\Pi = \Pi^{\text{All}}$, $\mathcal{P}^{\text{N}}_{\mathsf{P}_0}$ for $\Pi = \Pi^{\text{AllN}}$ and so on.

Now, we say that a strategy π is *mixed* if, for some probability measure $\nu(d\mathsf{P})$ on $\mathcal{P}^{\text{N}}_{\mathsf{P}_0}$,

$$\mathsf{P}^\pi_{\mathsf{P}_0} = \int_{\mathcal{P}^{\text{N}}_{\mathsf{P}_0}} \mathsf{P}\,\nu(d\mathsf{P}). \tag{2.18}$$

Similarly, a Markov (stationary) strategy π is called *mixed* if

$$\mathsf{P}^\pi_{\mathsf{P}_0} = \int_{\mathcal{P}^{\text{MN}}_{\mathsf{P}_0}} \mathsf{P}\,\nu(d\mathsf{P}) \qquad \left(\mathsf{P}^\pi_{\mathsf{P}_0} = \int_{\mathcal{P}^{\text{SN}}_{\mathsf{P}_0}} \mathsf{P}\,\nu(d\mathsf{P})\right).$$

Incidentally, the space $\mathcal{P}_{\mathsf{P}_0}$ is convex and, for any probability measure ν on $\mathcal{P}_{\mathsf{P}_0}$,

$$\mathsf{P}^\nu := \int_{\mathcal{P}_{\mathsf{P}_0}} \mathsf{P}\nu(d\mathsf{P}) \in \mathcal{P}_{\mathsf{P}_0} : \tag{2.19}$$

see [Dynkin and Yushkevich (1979), Ch.5, Sec.5].

According to [Feinberg (1996), Thm.5.2], any general strategy $\pi \in \Pi^{\text{All}}$ is mixed, and any Markov strategy $\pi \in \Pi^{\text{M}}$ is mixed as well. (Examples 5.3 and 5.4 in [Feinberg (1996)] confirm that the measure ν here is not necessarily unique.) At the same time, the space $\mathcal{P}^{\text{M}}_{\mathsf{P}_0} := \{\mathsf{P}^\pi_{\mathsf{P}_0},\ \pi \in \Pi^{\text{M}}\}$ is usually not convex as the following simple example confirms.

Example. Let $\mathbf{X} = \{1, 2, \Delta\}$, where Δ is a cemetery; $\mathbf{A} = \{1, 2\}$; $\mathsf{P}_0(1) = 1$; $p(2|1, a) = p(\Delta|2, a) = p(\Delta|\Delta, a) \equiv 1$ with all the other transition probabilities being zero. The cost function $c(\cdot, \cdot)$ is immaterial. See Fig. 2.1.

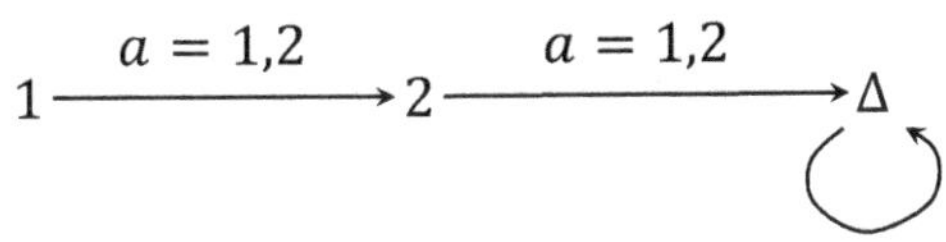

Fig. 2.1 The space of strategic measures coming from Markov strategies is not convex.

Let $\pi^{\text{m},1}_t(1|x) \equiv 1$ and $\pi^{\text{m},2}_t(2|x) \equiv 1$, $t = 1, 2, \ldots$. For these Markov strategies $\pi^{\text{m},1}$ and $\pi^{\text{m},2}$ we have

$$\begin{array}{ll} \mathsf{P}^{\pi^{\text{m},1}}_{\mathsf{P}_0}(A_1 = 1) = 1; & \mathsf{P}^{\pi^{\text{m},2}}_{\mathsf{P}_0}(A_1 = 2) = 1; \\ \mathsf{P}^{\pi^{\text{m},1}}_{\mathsf{P}_0}(X_1 = 2, A_2 = 1) = 1; & \mathsf{P}^{\pi^{\text{m},2}}_{\mathsf{P}_0}(X_1 = 2, A_2 = 2) = 1; \\ \mathsf{P}^{\pi^{\text{m},1}}_{\mathsf{P}_0}(A_1 = 1, A_2 = 1) = 1; & \mathsf{P}^{\pi^{\text{m},2}}_{\mathsf{P}_0}(A_1 = 2, A_2 = 2) = 1. \end{array}$$

For the convex combination $\mathsf{P} := \frac{1}{2}\mathsf{P}^{\pi^{\text{m},1}}_{\mathsf{P}_0} + \frac{1}{2}\mathsf{P}^{\pi^{\text{m},2}}_{\mathsf{P}_0}$ we have equalities

$$\begin{array}{ll} \mathsf{P}(A_1 = 1) = \frac{1}{2}; & \mathsf{P}(A_1 = 2) = \frac{1}{2}; \\ \mathsf{P}(X_1 = 2, A_2 = 1) = \frac{1}{2}; & \mathsf{P}(X_1 = 2, A_2 = 2) = \frac{1}{2}; \\ \mathsf{P}(A_1 = 1, A_2 = 1) = \frac{1}{2}. & \end{array}$$

This measure P, while being strategic by Proposition 2.1, is not generated by any Markov strategy under the initial distribution P_0. Indeed, that Markov strategy π^{m} (if exists) must be such that $\pi^{\text{m}}_1(1|1) = \pi^{\text{m}}_1(2|1) = \frac{1}{2}$ and $\pi^{\text{m}}_2(1|2) = \pi^{\text{m}}_2(2|2) = \frac{1}{2}$. But now $\mathsf{P}^{\pi^{\text{m}}}_{\mathsf{P}_0}(A_1 = 1, A_2 = 1) = \frac{1}{4}$. The obtained contradiction with $\mathsf{P}(A_1 = 1, A_2 = 1) = \frac{1}{2}$ confirms that the Markov strategy we are looking for does not exist. □

The following example from [Feinberg (1987), Remark 3.1] shows that not each stationary strategy is mixed.

Example. Let $\mathbf{X} = \{0\}$, $\mathbf{A} = \{0, 1\}$, $p(0|0, a) \equiv 1$, and consider the stationary randomized strategy $\pi^{\mathrm{s}}(a|0) = 0.5$, $a \in \mathbf{A}$. In this model, we have only two non-randomized stationary strategies $\varphi^0(0) = 0$ and $\varphi^1(0) = 1$. If

$$\mathsf{P}^{\pi^{\mathrm{s}}}_{\mathsf{P}_0} = \int_{\mathcal{P}^{\mathrm{SN}}_{\mathsf{P}_0}} \mathsf{P}\nu(d\mathsf{P}) = \alpha \mathsf{P}^{\varphi^0}_{\mathsf{P}_0} + (1-\alpha)\mathsf{P}^{\varphi^1}_{\mathsf{P}_0} \quad \text{for } \alpha \in [0, 1],$$

then measure $\mathsf{P}^{\pi^{\mathrm{s}}}_{\mathsf{P}_0}$ would have been concentrated on two trajectories $(x_0 = 0, a_1 = 0, x_1 = 0, a_2 = 0, \ldots)$ and $(x_0 = 0, a_1 = 1, x_1 = 0, a_2 = 1, \ldots)$ only, which is not the case:

$$\mathsf{P}^{\pi^{\mathrm{s}}}_{\mathsf{P}_0}(A_t = 0, A_{t+1} = 1) = \frac{1}{4} \qquad \forall\, t = 1, 2, \ldots.$$

□

The following example from [Piunovskiy (1997), Remark 34] shows that if φ^1 and φ^2 are two stationary selectors, then it can happen that equality

$$\mathsf{P}^{\pi}_{\mathsf{P}_0}(X_{t-1} \in \Gamma^X, A_t \in \Gamma^A) = \frac{1}{2}\mathsf{P}^{\varphi^1}_{\mathsf{P}_0}(X_{t-1} \in \Gamma^X, A_t \in \Gamma^A)$$
$$+\frac{1}{2}\mathsf{P}^{\varphi^2}_{\mathsf{P}_0}(X_{t-1} \in \Gamma^X, A_t \in \Gamma^A), \quad \Gamma^X \in \mathcal{B}(\mathbf{X}),\ \Gamma^A \in \mathcal{B}(\mathbf{A}),\ t = 1, 2, \ldots \tag{2.20}$$

holds for no-one stationary strategy π^{s}.

Example. Let $\mathbf{X} = \{1, \Delta\}$, $\mathbf{A} = \{1, 2\}$, $p(1|1, 1) = 1$, $p(\Delta|1, 2) = 1$, $p(\Delta|\Delta, a) \equiv 1$, with the other transition probabilities being zero (see Fig. 2.2).

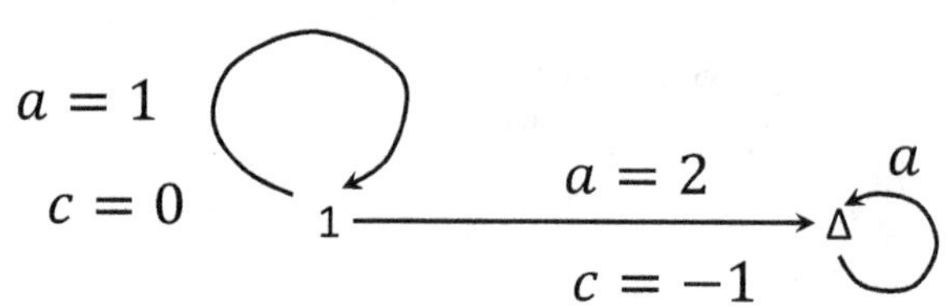

Fig. 2.2 Examples 2.3.1 and 2.3.3.

Suppose $\mathsf{P}_0(1) = 1$, $\varphi^1(x) \equiv 1$, $\varphi^2(x) \equiv 2$. If we take $t = 1$, $\Gamma^X = \{1\}$ and $\Gamma^A = \{1\}$, then (2.20) implies that $\pi_1(1|1) = \frac{1}{2}$ and

$$\mathsf{P}_1^\pi(X_1 = 1) = \frac{1}{2} = \frac{1}{2}\mathsf{P}_1^{\varphi^1}(X_1 = 1). \tag{2.21}$$

Now

$$\frac{1}{2}\mathsf{P}_1^{\varphi^1}(X_1 = 1,\ A_2 = 1) + \frac{1}{2}\mathsf{P}_1^{\varphi^2}(X_1 = 1,\ A_2 = 1)$$

$$= \frac{1}{2}\mathsf{P}_1^{\varphi^1}(X_1 = 1,\ A_2 = 1) = \frac{1}{2},$$

and it follows from (2.21) that we must put $\pi_2(1|1) = 1$ in order to have (2.20) for $t = 2$, $\Gamma^X = \{1\}$, $\Gamma^A = \{1\}$. Since $\pi_2(1|1) \neq \pi_1(1|1)$, equality (2.20) cannot hold for a stationary strategy $\pi^{\rm s}$. At the same time, that equality does hold for some Markov strategy: see Lemma 2.1. □

Suppose either $\mathsf{E}_{\mathsf{P}_0}^\pi\left[\sum_{t=1}^\infty c^+(X_{t-1}, A_t)\right]$ or $\mathsf{E}_{\mathsf{P}_0}^\pi\left[\sum_{t=1}^\infty c^-(X_{t-1}, A_t)\right]$ is uniformly bounded for all strategies π. Then the mapping $V : \mathcal{P}_{\mathsf{P}_0} \to \mathbb{R} \cup \{\infty\}$ (or $\mathbb{R} \cup \{-\infty\}$) defined by $V(\mathsf{P}_{\mathsf{P}_0}^\pi) := v_{\mathsf{P}_0}^\pi$, i.e., the performance functional, satisfies equality

$$V(\mathsf{P}^\nu) = \int_{\mathcal{P}_{\mathsf{P}_0}} V(\mathsf{P})\nu(d\mathsf{P})$$

for any probability measure ν on $\mathcal{P}_{\mathsf{P}_0}$. Here P^ν is defined in (2.19). More generally, one can consider $\mathbb{R}^*$-valued *concave* functionals satisfying inequality

$$V(\mathsf{P}^\nu) \geq \int_{\mathcal{P}_{\mathsf{P}_0}} V(\mathsf{P})\nu(d\mathsf{P}) \tag{2.22}$$

for all probability measures ν on $\mathcal{P}_{\mathsf{P}_0}$ such that the integral on the right is well defined.

Note that the given definition of a concave functional differs from the standard definition: a mapping $V : \mathcal{P}_{\mathsf{P}_0} \to \mathbb{R}$ is usually called concave if, for any $\mathsf{P}^1, \mathsf{P}^2 \in \mathcal{P}_{\mathsf{P}_0}$, $\forall \alpha \in [0, 1]$,

$$V(\alpha\mathsf{P}^1 + (1-\alpha)\mathsf{P}^2) \geq \alpha V(\mathsf{P}^1) + (1-\alpha)V(\mathsf{P}^2).$$

Theorem 2.8. [Feinberg (1982), Thm. 3.2]. *If the performance functional is concave in the sense (2.22) then, for any $\mathsf{P} \in \mathcal{P}_{\mathsf{P}_0}$, $\forall \varepsilon > 0$, there exists $\mathsf{P}^{\rm N} \in \mathcal{P}_{\mathsf{P}_0}^{\rm N}$ such that*

$$V(\mathsf{P}^{\rm N}) \leq \begin{cases} V(\mathsf{P}), & \text{if } V(\mathsf{P}) > -\infty; \\ -\frac{1}{\varepsilon}, & \text{if } V(\mathsf{P}) = -\infty. \end{cases}$$

The following example [Feinberg (1982), Ex. 3.1] shows that, if the mapping V is concave (in the usual sense), then Theorem 2.8 can fail.

Example. Let $\mathbf{X} = \{0\}$ be a singleton (there is no controlled process), so in this example the index P_0 is omitted. Put $\mathbf{A} = [0, 1]$ with the standard Euclidean topology and let

$$V(P) = \begin{cases} -1, & \text{if the marginal distribution of } A_1, \text{ i.e. } \pi_1(da|0), \text{ is} \\ & \text{absolutely continuous w.r.t. the Lebesgue measure;} \\ 0, & \text{otherwise.} \end{cases}$$

The mapping V is concave (in the usual sense), but for each $\mathsf{P}^{\mathrm{N}} \in \mathcal{P}^{\mathrm{N}}$ we have $V(\mathsf{P}^{\mathrm{N}}) = 0$, whereas $V(\mathsf{P}^{\pi}) = -1$, if $\pi_1(\cdot|0)$ is absolutely continuous w.r.t. the Lebesgue measure.

Inequality (2.22) is violated. To show this, introduce

$$\hat{\mathcal{P}} := \{\mathsf{P}^{\pi^{\alpha}} : \pi^{\alpha}(da|0) = \delta_{\alpha}(da), \ \alpha \in [0, 1]\}$$

and the measurable 1-1-correspondence

$$\alpha \longleftrightarrow \mathsf{P}^{\pi^{\alpha}} \tag{2.23}$$

between $[0, 1]$ and $\hat{\mathcal{P}}$. Let ν be the image on $\hat{\mathcal{P}}$ of the Lebesgue measure on $[0, 1]$ with respect to this correspondence. Note that, for each $\Gamma^A \in \mathcal{B}(\mathbf{A})$, $\alpha \in [0, 1]$, we have $\mathsf{P}^{\pi^{\alpha}}(A_1 \in \Gamma^A) = \mathbb{I}\{\alpha \in \Gamma^A\}$. Now, using (2.23), we have

$$\mathsf{P}^{\nu}(A_1 \in \Gamma^A) = \int_{\hat{\mathcal{P}}} \mathsf{P}(A_1 \in \Gamma^A)\nu(d\mathsf{P}) = \int_{[0,1]} \mathsf{P}^{\pi^{\alpha}}(A_1 \in \Gamma^A)d\alpha = Leb(\Gamma^A)$$

(the Lebesgue measure on the right). Thus, $V(\mathsf{P}^{\nu}) = -1$ while $V(\mathsf{P}) = 0$ for all $\mathsf{P} \in \hat{\mathcal{P}}$.

For each $\mathsf{P}^{\mathrm{N}} \in \mathcal{P}^{\mathrm{N}}$, $V(P^{\mathrm{N}}) = 0$, but in case $\pi_1(\Gamma^A|0) := Leb(\Gamma^A)$ we have $V(\mathsf{P}^{\pi}) = -1$: the statement of Theorem 2.8 fails to hold. □

2.3.2 *Multiple solutions to the optimality equation*

Consider a discrete-time *queueing* model. During each time period $(t, t+1]$, $t = 0, 1, \ldots$, there may be an arrival of a customer with probability $\lambda > 0$ or a departure from the queue (in case it is non-empty) with probability $\mu > 0$; $\lambda + \mu \leq 1$. State x means there are x customers in the queue. There is no control here, and we wish to compute the expected time for the queue to empty. A similar example was presented in [Altman (1999), Ex. 9.1].

Example. Let $\mathbf{X} = \{0, 1, 2, \ldots\}$, $\mathbf{A} = \{0\}$ (a dummy action), $p(0|0, a) = 1$,

$$\forall x > 0 \quad p(y|x, a) = \begin{cases} \lambda, & \text{if } y = x + 1; \\ \mu, & \text{if } y = x - 1; \\ 1 - \lambda - \mu, & \text{if } y = x; \\ 0, & \text{otherwise;} \end{cases} \qquad c(x, a) = \mathbb{I}\{x > 0\}.$$

The one-step cost equals 1 at all the positive states. (See Fig. 2.3.)

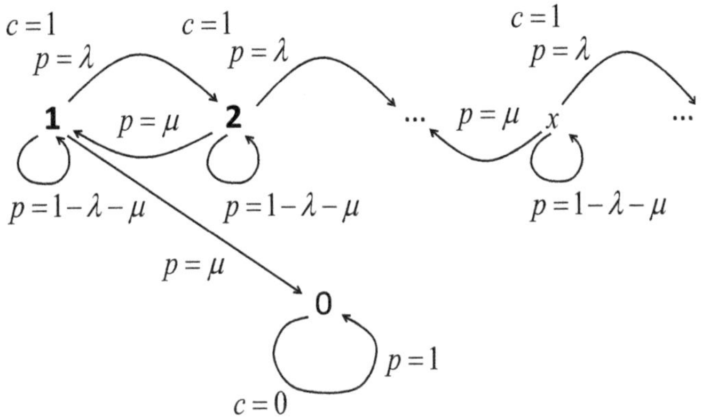

Fig. 2.3 Multiple solutions to the optimality equation.

Equation (2.5) has the form

$$v(x) = 1 + \mu v(x-1) + \lambda v(x+1) + (1-\lambda-\mu)v(x), \quad x > 0. \tag{2.24}$$

In the case when $\lambda \neq \mu$, the general solution to (2.24) is as follows:

$$v(x) = k_1 \left(\frac{\mu}{\lambda}\right)^x + k_2 + \frac{x}{\mu - \lambda}.$$

The Bellman function must satisfy the condition $v_x^* = 0$ at $x = 0$, so that one should put $k_2 = -k_1$. And even now, constant k_1 can be arbitrary. In fact, v_x^* is the *minimal* non-negative solution to (2.24), i.e. $k_1 = k_2 = 0$ in the case when $\mu > \lambda$. In this case, this MDP is absorbing (at 0) for each initial state, since v_x^* is the finite expected time to absorption when starting from $x \in \mathbf{X}$. It is absorbing (at 0) for any initial distribution P_0 with $\sum_{x=1}^{\infty} x\mathsf{P}_0(x) < \infty$. If $\sum_{x=1}^{\infty} x\mathsf{P}_0(x) = \infty$, then this MDP is not absorbing at the initial distribution P_0.

If $\mu < \lambda$, then equation (2.24) has no finite non-negative solutions. Here $v_x^* \equiv \infty$ for $x > 0$, and the model is not absorbing. We leave aside the case $\mu = \lambda$. □

2.3.3 *Finite model: multiple solutions to the optimality equation; conserving but not equalizing stationary selector*

Example. Let $\mathbf{X} = \{1, \Delta\}$, $\mathbf{A} = \{1, 2\}$, $p(1|1,1) = 1$, $p(\Delta|1,2) = 1$, $p(\Delta|\Delta, a) \equiv 1$, with the other transition probabilities being zero; $c(1,1) =$

0, $c(1,2) = -1$, $c(\Delta, a) \equiv 0$. See Fig. 2.2; similar examples were presented in [Dynkin and Yushkevich (1979), Chapter 4, Section 7], [Puterman (1994), Ex. 7.2.3 and 7.3.1] and [Kallenberg (2010), Ex. 4.1].

The optimality equation (2.5) is given by

$$\left.\begin{array}{l} v(1) = \min\{v(1); \quad -1 + v(\Delta)\}; \\ v(\Delta) = v(\Delta). \end{array}\right\} \tag{2.25}$$

Any pair of numbers satisfying $v(1) \le v(\Delta) - 1$ provides a solution. Conditions 2.2 are satisfied, so the Bellman function coincides with the maximal non-positive solution:

$$v_1^* = -1, \qquad v_\Delta^* = 0.$$

Any stationary selector is conserving, but selector $\varphi^1(x) \equiv 1$ is not equalizing: $\mathsf{E}_1^{\varphi^1}[v_{X_T}^*] \equiv v_1^* = -1$; $v_1^{\varphi^1} = 0$, $v_\Delta^{\varphi^1} = 0$. In the opposite case, the stationary selector $\varphi^2(x) \equiv 2$ is equalizing and hence uniformly optimal by [Puterman (1994), Thm. 7.1.7]: $v_1^{\varphi^2} = -1$, $v_\Delta^{\varphi^2} = 0$.

Remark 2.3. For a discounted model with discount factor $\beta \in (0,1)$ (Chapter 3), the optimality equation is given by (3.2). In that case, if the cost function $c(\cdot,\cdot)$ is bounded, it is known [Puterman (1994), Thm. 6.2.2] that $\tilde{v}(x) \ge v_x^*$ (or $\tilde{v}(x) \le v_x^*$) for all $x \in \mathbf{X}$, if function $\tilde{v}(\cdot)$ satisfies inequality

$$\tilde{v}(x) \ge (\text{ or } \le) \inf_{a \in \mathbf{A}} \left\{ c(x,a) + \beta \int_{\mathbf{X}} \tilde{v}(y) p(dy|x,a) \right\}, \qquad x \in \mathbf{X}.$$

In the current example (with the discount factor $\beta = 1$) this statement does not hold: one can take $\tilde{v}(1) = \tilde{v}(\Delta) = -3$ or $\tilde{v}(1) = 0$; $\tilde{v}(\Delta) = 2$.

□

2.3.4 *Single conserving stationary selector is not equalizing and not optimal*

Example. Let $\mathbf{X} = \{0,1,2,\ldots\}$, $\mathbf{A} = \{1,2\}$, $p(0|0,a) \equiv 1$, $c(0,a) \equiv 0$, $\forall x > 0$ $p(0|x,1) \equiv 1$, $p(x+1|x,2) \equiv 1$, with all other transition probabilities being zero; $c(x,1) = 1/x - 1$, $c(x,2) \equiv 0$. See Fig. 2.4; similar examples were presented in [Bäuerle and Rieder (2011), Ex. 7.4.4], [Bertsekas (2001), Ex. 3.4.4], [Dynkin and Yushkevich (1979), Chapter 6, Section 3], [Puterman (1994), Ex. 7.2.4], [Strauch (1966), Ex. 4.2] and in [van der Wal and Wessels (1984), Ex. 3.4]. See also Subsection 2.3.14.

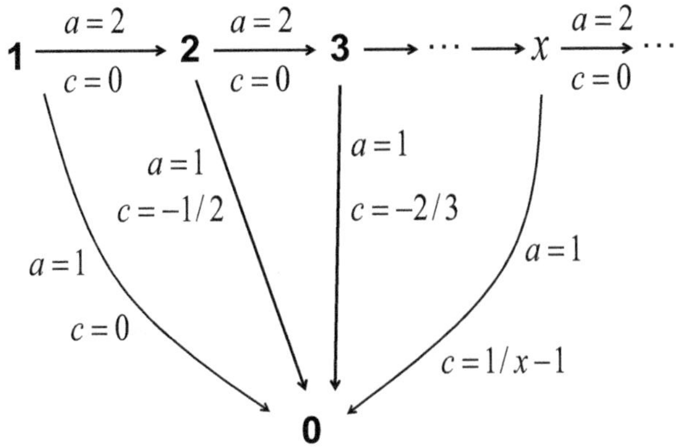

Fig. 2.4 No optimal strategies.

The optimality equation (2.5) is given by

$$v(0) = v(0);$$
$$v(x) = \min\{1/x - 1 + v(0), \quad v(x+1)\}, \qquad x > 0. \tag{2.26}$$

We are interested only in the solutions with $v(0) = 0$ because $v_0^* = 0$. Conditions 2.2 are satisfied, and v_x^* is the maximal non-positive solution to (2.26), i.e. $v_x^* \equiv -1$ for $x > 0$. By the way, for any $r \le -1$, $v(x) \equiv r$ for $x > 0$ is a solution to (2.26).

The stationary selector $\varphi^*(x) \equiv 2$ is the single conserving one, but

$$\lim_{t\to\infty} \mathsf{E}_x^{\varphi^*}\left[v_{X_t}^*\right] \equiv -1 \text{ for } x > 0,$$

so that it is not equalizing.

There exist no optimal strategies in this model for each initial disctribution $\mathsf{P}_0(\cdot)$ on $\{1, 2, \ldots\}$. The stationary selector φ^* indefinitely delays absorption in state 0, so that the decision maker receives no negative cost: $v_x^{\varphi^*} \equiv 0$.

Note that, for each $\varepsilon > 0$, the stationary selector

$$\varphi^\varepsilon(x) = \begin{cases} 2, & \text{if } x \le \varepsilon^{-1}; \\ 1, & \text{if } x > \varepsilon^{-1} \end{cases}$$

is uniformly ε-optimal: $\forall x > 0 \; v_x^{\varphi^\varepsilon} < \varepsilon - 1$.

Suppose now that there is an additional action 3 leading, from any state $x > 0$, directly to state 0 with cost -1: $p(0|x,3) \equiv 1$, $c(x,3) \equiv -1$. Now the stationary selector $\varphi(x) \equiv 3$ is conserving and equalizing, and hence is uniformly optimal by [Puterman (1994), Thm. 7.1.7]. It is also optimal for any initial distribution.

This example can also be adjusted for a discounted model; see Example 3.3.7. □

Consider the following example motivated by [Feinberg (2002b), Ex. 6.3].

Example. Let $\mathbf{X} = \{0, 1, 2, \ldots\}$; $\mathbf{A} = \{1, 2\}$; $p(0|0, a) \equiv 1$, $p(0|x, 1) \equiv 1$, $p(x+1|x, 2) = 1$ for $x \geq 1$, with other transition probabilities being zero; $c(0, a) \equiv 0$, for $x > 0$ $c(x, 1) = 2^{-x} - 1$, $c(x, 2) = -2^{-x}$ (see Fig. 2.5).

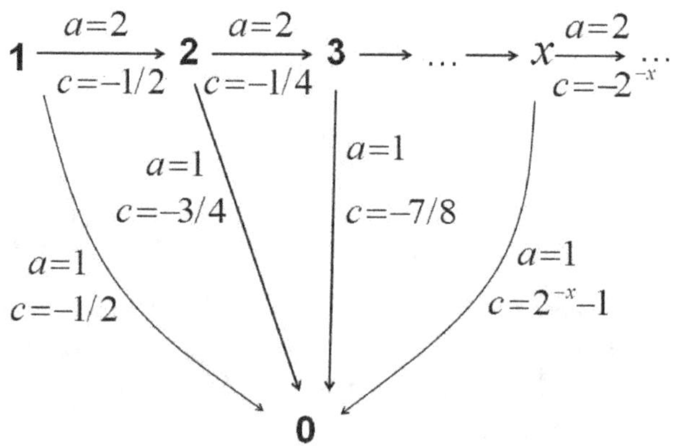

Fig. 2.5 No optimal strategies; stationary selectors are not dominating.

The Bellman function satisfies the optimality equation (2.5) and can be found by value iteration: see (2.6),(2.7):

$$v_x^* = \begin{cases} 0, & \text{if } x = 0; \\ -2^{-x+1} - 1, & \text{if } x > 0. \end{cases}$$

The stationary selector $\varphi^*(x) \equiv 2$ is the only conserving one, but it is not equalizing and not uniformly optimal because

$$v_x^{\varphi^*} = -2^{-x} - 2^{-(x+1)} - 2^{-(x+2)} - \cdots = -2^{-x+1}$$

for $x > 0$. There exist no optimal strategies in this model for each initial distribution $\mathsf{P}_0(\cdot)$ with $\mathsf{P}_0(0) < 1$ because, again, in order to be optimal, the strategy must indefinitely delay absorption in state 0.

Consider the following stationary randomized strategy: $\pi^{\mathrm{s}}(1|x) = \pi^{\mathrm{s}}(2|x) = 1/2$. One can compute $v_x^{\pi^{\mathrm{s}}} \equiv -1$ (for all $x > 0$), e.g. again using the value iteration: see Remark 2.2. Now, if φ is an arbitrary stationary selector, then either $\varphi = \varphi^*$ and $v_2^{\varphi} = -1/2 > v_2^{\pi^s} = -1$, or $\varphi(\hat{x}) = 1$ for some $\hat{x} > 0$, so that $v_{\hat{x}}^{\varphi} = 2^{-\hat{x}} - 1 > v_{\hat{x}}^{\pi^s} = -1$. We conclude that the strategy π^s cannot be dominated by any stationary selector. □

An example, similar to the above presented, is given in Subsection 2.3.14.

2.3.5 *Non-equalizing strategy is uniformly optimal*

In fact, the model below is uncontrolled (there exists only one control strategy $\pi_t(0|h_{t-1}) \equiv 1$). We intend to show that the Bellman function $v_x^* = v_x^{\pi}$ is finite and, for some values of x,

$$\liminf_{t\to\infty} \mathsf{E}_x^{\pi}\left[v_{X_t}^*\right] < 0. \tag{2.27}$$

Example. Let $\mathbf{X} = \{0, 1, 1', 1'', 2, 2', 2'', \ldots\}$, $\mathbf{A} = \{0\}$ (a dummy action),

$$\forall x > 0 \;\; p(y|x,a) = \begin{cases} p, & \text{if } y = x'; \\ 1-p, & \text{if } y = x''; \\ 0, & \text{otherwise,} \end{cases} \quad c(x,a) \equiv 0, \;\; p(x+1|x',a) = 1,$$

$c(x',a) = p^{-x}$, $p(0|x'',a) = 1$, $c(x'',a) = -\frac{p^{-x+1}}{1-p}$, $p(0|0,a) = 1$, $c(0,a) = 0$, where $p \in (0,1)$ is a fixed constant (see Fig. 2.6).

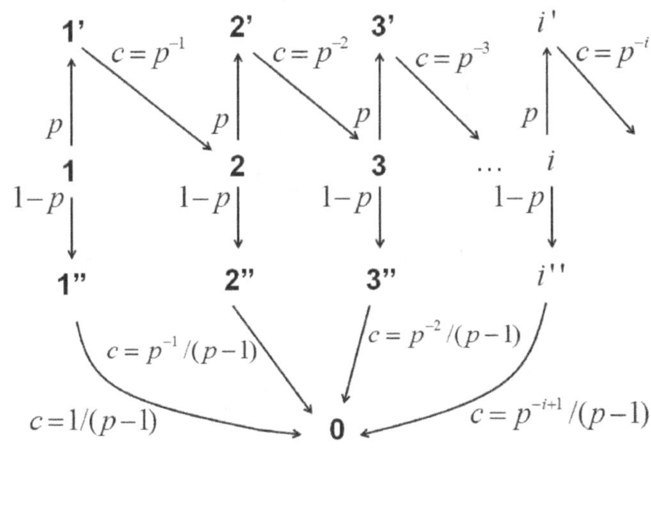

Fig. 2.6 Non-equalizing strategy is uniformly optimal.

This model is not summable because Condition 2.1 is violated: $\mathsf{E}_x^\pi\left[\sum_{t=1}^\infty c^+(X_{t-1},A_t)\right] = \infty$ and $\mathsf{E}_x^\pi\left[\sum_{t=1}^\infty c^-(X_{t-1},A_t)\right] = -\infty$ for all $x = 1,2,\ldots$: see the end of this subsection. But we will show that the Bellman function $v_x^* = v_x^\pi$ is well defined and finite for all $x \in \mathbf{X}$ if we accept formula $v_x^\pi = \mathsf{E}_x^\pi[W^+] + \mathsf{E}_x^\pi[W^-]$ for $w(h) = \sum_{t=1}^\infty c(x_{t-1},a_t)$. According to the description of the model, only the following trajectories are realized with the positive probability $p^n(1-p)$, when starting from initial state $X_0 = x \in \{1,2,\ldots\}$:

$$h = (x, x', (x+1), (x+1)', \ldots, (x+n), (x+n)'', 0, 0, \ldots) \qquad (n = 0,1,2,\ldots).$$

Hence, this MDP is absorbing, e.g., for each initial distribution P_0 with the finite support. The associated cost equals

$$w(h) = \sum_{j=0}^{n-1} p^{-(x+j)} - \frac{p^{-(x+n)+1}}{1-p} = -\frac{p^{-x+1}}{1-p}.$$

Therefore,

$$v_x^* = -\sum_{n=0}^\infty p^n(1-p)\frac{p^{-x+1}}{1-p} = -\frac{p^{-x+1}}{1-p}$$

and also

$$v_{x'}^* = v_{x''}^* = -\frac{p^{-x+1}}{1-p}.$$

Now, starting from $X_0 = x \in \{1,2,\ldots\}$,

$$X_{2t} = \begin{cases} x+t & \text{with probability } p^t; \\ 0 & \text{with probability } 1-p^t, \end{cases}$$

$$X_{2t+1} = \begin{cases} (x+t)'' & \text{with probability } p^t(1-p); \\ (x+t)' & \text{with probability } p^{t+1}; \\ 0 & \text{with probability } 1-p^t. \end{cases}$$

Therefore,

$$\mathsf{E}_x^\pi\left[v_{X_{2t}}^*\right] = \mathsf{E}_x^\pi\left[v_{X_{2t+1}}^*\right] = -\frac{p^{-x+1}}{1-p}.$$

Similar calculations are valid for $X_0 = x'$ and for $X_0 = x''$. Thus, inequality (2.27) holds.

The optimality equation is given by

$$v(0) = v(0);$$

for $x = 1, 2, \ldots$

$$v(x) = pv(x') + (1-p)v(x''),$$

$$v(x') = p^{-x} + v(x+1),$$

$$v(x'') = -\frac{p^{-x+1}}{1-p} + v(0).$$

We are interested only in the solutions with $v(0) = 0$ because $v_0^* = 0$. If we substitute the second and the third equations into the first one, we obtain

$$v(x) = pv(x+1), \quad x = 1, 2, \ldots.$$

The general solution is given by $v(x) = kp^{-x}$, and for v_x^* we have $k = -\frac{p}{1-p}$.

By the way, for $x = 1, 2, \ldots$ we have

$$\mathsf{E}_x^\pi\left[\sum_{t=1}^{\infty} c^+(X_{t-1}, A_t)\right] = \sum_{n=0}^{\infty} p^n(1-p)\sum_{j=0}^{n-1} p^{-(x+j)} = \infty$$

and similarly $\mathsf{E}_x^\pi\left[\sum_{t=1}^{\infty} c^-(X_{t-1}, A_t)\right] = -\infty$. □

2.3.6 *Combinations of strategies*

If the model is positive, i.e. $c(\cdot,\cdot) \geq 0$, and the cost function $c(\cdot,\cdot)$ is bounded, then the following theorem holds.

Theorem 2.9.

(a) *[Strauch (1966), Thm. 4.2] Let π and σ be two strategies and suppose $\exists n_0$: $\forall n > n_0$ $v_x^{\pi^n\sigma} \leq v_x^\sigma$ for all $x \in \mathbf{X}$. Here $\pi^n\sigma = \{\pi_1, \pi_2, \ldots, \pi_n, \sigma_{n+1} \ldots\}$ is the natural combination of the strategies π and σ. Then $v_x^\pi \leq v_x^\sigma$ for all $x \in \mathbf{X}$.*

(b) *[Strauch (1966), Cor. 9.2] Let φ^1 and φ^2 be two stationary selectors and put*

$$\hat{\varphi}(x) := \begin{cases} \varphi^1(x), & \text{if } v_x^{\varphi^1} \leq v_x^{\varphi^2}; \\ \varphi^2(x), & \text{otherwise.} \end{cases}$$

Then, for all $x \in \mathbf{X}$, $v_x^{\hat{\varphi}} \leq \min\{v_x^{\varphi^1},\ v_x^{\varphi^2}\}$.

Example from Subsection 2.3.4 (Fig. 2.4), shows that this theorem can fail for *negative* models, where $c(\cdot,\cdot) \leq 0$ (see [Strauch (1966), Examples 4.2 and 9.1]).

Statement (a). Let $\pi_t(2|h_{t-1}) \equiv 1$ and $\sigma_t(1|h_{t-1}) \equiv 1$ be stationary selectors. Then $v_0^\sigma = 0$; for $x > 0$ $v_x^\sigma = 1/x - 1$; $v_0^{\pi^n\sigma} = 0$ and $v_x^{\pi^n\sigma} = 1/(x+n) - 1$ for $x > 0$. Therefore, $v_x^{\pi^n\sigma} \le v_x^\sigma$ for all n, but $v_x^\pi = 0 > v_x^\sigma$ for $x > 1$.

Statement (b). For $x > 0$ let

$$\varphi^1(x) = \begin{cases} 1, & \text{if } x \text{ is odd;} \\ 2, & \text{if } x \text{ is even;} \end{cases} \qquad \varphi^2(x) = \begin{cases} 1, & \text{if } x \text{ is even;} \\ 2, & \text{if } x \text{ is odd.} \end{cases}$$

Then, for positive odd $x \in \mathbf{X}$,

$$v_x^{\varphi^1} = \frac{1}{x} - 1; \qquad v_x^{\varphi^2} = \frac{1}{x+1} - 1,$$

so that $\hat{\varphi}(x) = \varphi^2(x) = 2$.
For positive even $x \in \mathbf{X}$,

$$v_x^{\varphi^1} = \frac{1}{x+1} - 1; \qquad v_x^{\varphi^2} = \frac{1}{x} - 1,$$

so that $\hat{\varphi}(x) = \varphi^1(x) = 2$ (for $x = 0$, $v_0^\pi = 0$ for any strategy π).
Now, for all $x > 0$, we have $\hat{\varphi}(x) = 2$ and $v_x^{\hat{\varphi}} \equiv 0 > \min\{v_x^{\varphi^1}, v_x^{\varphi^2}\} = \frac{1}{x+1} - 1$. $\square$

2.3.7 *When strategy iteration is not successful*

The basic strategy iteration algorithm constitutes a paraphrase of [Puterman (1994), Section 7.2.5].

1. Set $n = 0$ and select a stationary selector φ^0 arbitrarily enough.

2. Obtain $w^n(x) := v_x^{\varphi^n}$.

3. Choose $\varphi^{n+1} : \mathbf{X} \to \mathbf{A}$ such that

$$c(x, \varphi^{n+1}(x)) + \int_{\mathbf{X}} w^n(y) p(dy|x, \varphi^{n+1}(x))$$

$$= \inf_{a \in \mathbf{A}} \left\{ c(x,a) + \int_{\mathbf{X}} w^n(y) p(dy|x,a) \right\},$$

   setting $\varphi^{n+1}(x) = \varphi^n(x)$ whenever possible.

4. If $\varphi^{n+1} = \varphi^n$, then stop and set $\varphi^* = \varphi^n$. Otherwise, increment n by 1 and return to step 2.

Theorem 2.10. [Puterman (1994), Thm. 7.2.16]. *In negative models with finite spaces* $\mathbf{X}$ *and* $\mathbf{A}$ *and with finite values of* v_x^π *at any* π *and* $x \in \mathbf{X}$*, the strategy iteration algorithm terminates and returns a uniformly optimal strategy.*

Example in Subsection 2.3.4 (Fig. 2.4) shows that this algorithm does not always terminate if the state space $\mathbf{X}$ is not finite, even if the action space is finite. Indeed, choose $\varphi^0(x) \equiv 2$; then $w^0(x) = v_x^{\varphi^0} \equiv 0$. Now $\varphi^1(x) = 1$ if $x > 1$ and $\varphi^1(0) = \varphi^1(1) = 2$. Therefore,

$$w^1(x) = v_x^{\varphi^1} = \begin{cases} \frac{1}{x} - 1, & \text{if } x > 1; \\ -\frac{1}{2}, & \text{if } x = 1; \\ 0, & \text{if } x = 0. \end{cases}$$

Now, for $x \geq 1$, we have

$$c(x,2) + w^1(x+1) = \frac{1}{x+1} - 1 < c(x,1) + w^1(0) = \frac{1}{x} - 1,$$

so that $\varphi^2(x) \equiv 2$ for all $x \in \mathbf{X}$, and the strategy iteration algorithm will cycle between these two stationary selectors φ^0 and φ^1. This is not surprising, because Example 2.3.4 illustrates that there are no optimal strategies at all. □

Now we modify the *example from Subsection 2.3.3* (see Fig. 2.7): we put $c(1,2) = +1$. This is a simplified version of Example 7.3.4 from [Puterman (1994)].

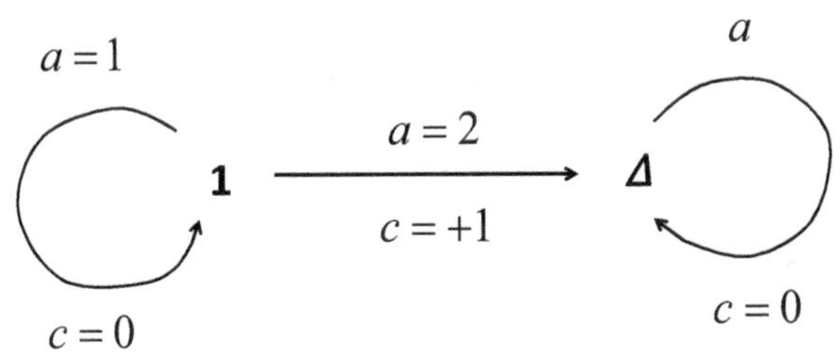

Fig. 2.7 Strategy iteration returns a sub-optimal strategy.

The optimality equation

$$v(1) = \min\{v(1),\ 1 + v(\Delta)\};$$
$$v(\Delta) = v(\Delta)$$

– compare with (2.25) – again has many solutions. We deal now with the positive model, in which the minimal non-negative solution $v(1) = 0$, $v(\Delta) = 0$ coincides with the Bellman function v_x^*, and the stationary selector $\varphi^*(x) \equiv 1$ is conserving, equalizing and uniformly optimal. If we apply the strategy iteration algorithm to selector $\varphi^0(x) \equiv 2$, we see that

$$w^0(x) = \begin{cases} +1, & \text{if } x = 1; \\ 0, & \text{if } x = \Delta. \end{cases}$$

Hence, φ^0 provides the minimum on step 3 in the following expressions corresponding to $x = 1$ and $x = \Delta$:

$$\begin{array}{ll} \min\{w^0(1), & +1 + w^0(\Delta)\}; \\ \min\{w^0(\Delta), & w^0(\Delta)\}. \end{array}$$

Thus, the strategy iteration algorithm terminates and returns the stationary selector $\varphi^0(x) \equiv 2$ which is not uniformly optimal: $v_1^{\varphi^0} = 1 > v_1^* = 0$. Condition $c(\cdot,\cdot) \le 0$ is important in Theorem 2.10. Note that, in the discounted case, the strategy iteration algorithm is powerful much more often [Puterman (1994), Section 6.4]. □

2.3.8 *When value iteration is not successful: unstable models*

Some authors call a Markov Decision Process *stable* if the limit

$$\lim_{n\to\infty} v^n(x) = v_x^* \tag{2.28}$$

exists and coincides with the Bellman function [Kallenberg (2010), p. 112]. Below, we present three MDPs which are not stable.

Example. Let $\mathbf{X} = \{0,1,2\}$, $\mathbf{A} = \{1,2\}$, $p(0|0,a) = p(0|1,a) \equiv 1$, $p(2|2,1) = 1$, $p(1|2,2) = 1$, $c(0,a) \equiv 0$, $c(1,a) \equiv 1$, $c(2,1) = 0$, $c(2,2) = -2$ (see Fig. 2.8; a similar example was presented in [Kallenberg (2010), Ex. 4.1, p.112]).

Obviously, $v_0^* = 0$, $v_1^* = 1$, $v_2^* = -1$, but value iteration leads to the following:

$$\begin{array}{lll} v^0(0) = 0 & v^0(1) = 0 & v^0(2) = 0 \\ v^1(0) = 0 & v^1(1) = 1 & v^1(2) = -2 \\ v^2(0) = 0 & v^2(1) = 1 & v^2(2) = -2 \\ \dots & \dots & \dots \end{array}$$

□

In the following examples, the limit (2.7) does not exist at all.

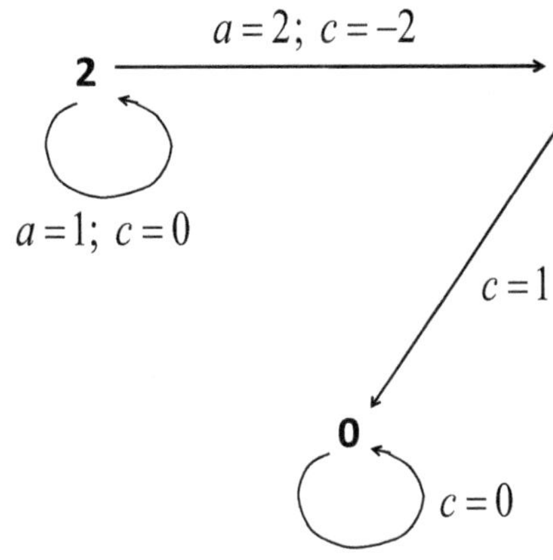

Fig. 2.8 Unstable MDP.

Example. Let $\mathbf{X} = \{\Delta, 0, 1, 2, \dots\}$, $\mathbf{A} = \{0, 1, 2, \dots\}$, $p(\Delta|\Delta, a) \equiv 1$, $p(2a+1|0,a) = 1$, $p(\Delta|1,0) = 1$, for $a > 0$ $p(\Delta|1,a) = p(0|1,a) \equiv 1/2$, for $x > 1$ $p(x-1|x,a) \equiv 1$. All the other transition probabilities are zero. Let $c(\Delta, a) \equiv 0$, $c(0,a) \equiv 12$, $c(1,0) = 1$, for $a > 1$ $c(1,a) \equiv -4$, for $x > 1$ $c(x,a) \equiv 0$ (see Fig. 2.9).

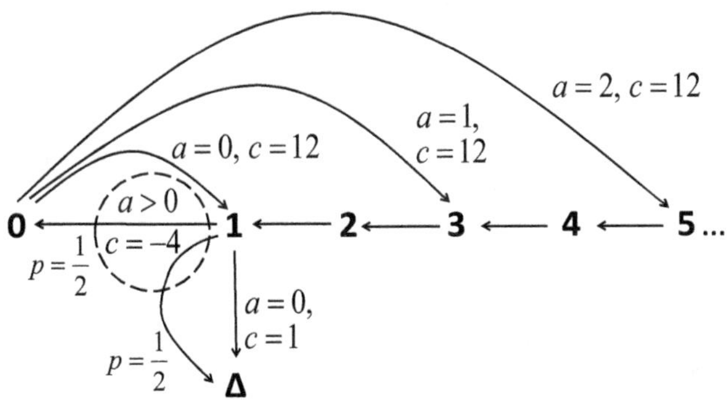

Fig. 2.9 Unstable MDP, no limits.

This MDP is absorbing at Δ starting from each initial state, the performance functional $v^\pi_{\mathsf{P}_0}$ and the Bellman function v^*_x are well defined and finite. Since any one cycle $1 \to 0 \to 2a+1 \to 2a \to \cdots \to 1$ leads to an increment of the performance, one should put $\varphi^*(1) = 0$, so that

$$v_0^* = 13, \qquad \text{for } x > 0 \ \ v_x^* = 1, \qquad \text{and } v_\Delta^* = 0.$$

On the other hand, the value iteration gives the following values:

x	0	1	2	3	4	5 ...
$v^0(x)$	0	0	0	0	0	0
$v^1(x)$	12	−4	0	0	0	0
$v^2(x)$	8	1	−4	0	0	0
$v^3(x)$	12	0	1	−4	0	0
$v^4(x)$	8	1	0	1	−4	0
$v^5(x)$	12	0	1	0	1	−4 ...
...						...

□

The third example is based on [Whittle (1983), Chapter 25, Section 5].

Example. Let $\mathbf{X} = \{0, 1, 2\}$, $\mathbf{A} = \{1, 2\}$, $p(0|0, a) \equiv 1$, $p(0|1, 1) = 1$, $p(2|1, 2) = 1$, $p(1|2, a) \equiv 1$, the other transition probabilities being zero; $c(0, a) \equiv 0$, $c(1, 1) = 1$, $c(1, 2) = 3$, $c(2, a) \equiv -3$. See Fig. 2.10.

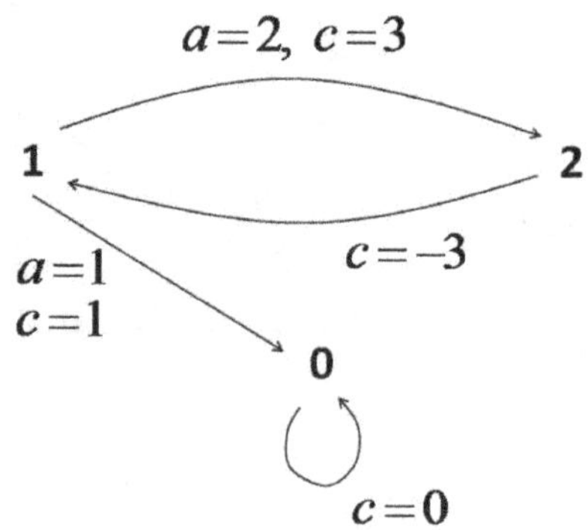

Fig. 2.10 Unstable MDP, no limits.

Here Condition 2.1 is violated if $\mathsf{P}_0(0) < 1$. For the stationary selector $\varphi(x) \equiv 2$ the loop $h = (1, 2, 1, 2, \ldots)$ having a positive probability leads to $\mathsf{E}^{\varphi}_{\mathsf{P}_0}[\sum_{t=1}^{\infty} c^+(X_{t-1}, A_t)] = \infty$ and $\mathsf{E}^{\varphi}_{\mathsf{P}_0}[\sum_{t=1}^{\infty} c^-(X_{t-1}, A_t)] = -\infty$. According to the basic conventions in Section 2.1, we have to put $w(h) = +\infty$, when calculating $v^{\varphi}_{\mathsf{P}_0}$ by using formula (2.3), leading to $v^{\varphi}_{\mathsf{P}_0} = +\infty$. Thus, selector φ is not (uniformly) optimal because, e.g., for selector $\varphi^*(x) \equiv 1$, $v^{\varphi^*}_{\mathsf{P}_0}$ and $v^{\varphi^*}_x$ are finite for all $x \in \mathbf{X}$ and all initial distributions $\mathsf{P}_0(\cdot)$.

In this situation, one can call a strategy π^* optimal if it minimizes $\limsup_{\beta\to 1-} v^{\pi,\beta}_{\mathsf{P}_0}$ (see Section 3.1; β is the discount factor). In [Whittle (1983)], such strategies, which additionally satisfy equation

$$\lim_{T\to\infty}\left|\mathsf{E}^{\pi^*}_{\mathsf{P}_0}\left[\sum_{t=1}^{T} c(X_{t-1},A_t)\right] - \mathsf{E}^{\pi^*}_{\mathsf{P}_0}\left[\sum_{t=1}^{\infty} c(X_{t-1},A_t)\right]\right| = 0, \tag{2.29}$$

are called *transient-optimal.* It is assumed that all the mathematical expectations in (2.29) are well defined.

For any $\beta \in [0,1)$, the stationary selector $\varphi^*(x) \equiv 1$ is uniformly optimal in problem (3.1) and satisfies (2.29), because the controlled process X_t is absorbed at 0 in no more than two steps. Thus, selector φ^* is transient-optimal. The discounted Bellman function equals

$$v^{*,\beta}_x = v^{\varphi^*,\beta}_x = \begin{cases} 0, & \text{if } x=0;\\ 1, & \text{if } x=1;\\ \beta-3, & \text{if } x=2,\end{cases}$$

so that in this example one should put

$$v^*_x = \lim_{\beta\to 1-} v^{*,\beta}_x = \begin{cases} 0, & \text{if } x=0;\\ 1, & \text{if } x=1;\\ -2, & \text{if } x=2.\end{cases}$$

But the value iteration leads to the following:

$$v^0(x) \equiv 0;$$

$$v^1(x) = \begin{cases} 0, & \text{if } x=0;\\ 1, & \text{if } x=1;\\ -3, & \text{if } x=2,\end{cases} \qquad v^2(x) = \begin{cases} 0, & \text{if } x=0;\\ 0, & \text{if } x=1;\\ -2, & \text{if } x=2,\end{cases}$$

$$v^3(x) = \begin{cases} 0, & \text{if } x=0;\\ 1, & \text{if } x=1;\\ -3, & \text{if } x=2,\end{cases}$$

and so on. At every step, the optimal action in state 1 switches from 1 to 2 and back. Each one cycle 1-2-1 results in the cost equal to 3-3=0, and the decision maker can apply action 2 in state 1 any finite number of times without influencing the total expected cost coming from the jump $1 \to 0$ under action 1. Recall that selector $\varphi(x) \equiv 2$ is not (uniformly) optimal.

□

2.3.9 *When value iteration is not successful: positive model I*

The following example shows that inequality (2.8) can be strict for some $x \in \mathbf{X}$.

Example. Let $\mathbf{X} = \{0, 1, 2, \ldots\}$, $\mathbf{A} = \{1, 2, \ldots\}$, $p(0|0, a) \equiv 1$, $p(0|2, a) \equiv 1$, $p(x-1|x, a) \equiv 1$ for $x > 2$, and

$$p(y|1, a) = \begin{cases} 1, & \text{if } y = 1 + a; \\ 0, & \text{otherwise.} \end{cases}$$

All the other transition probabilities are zero. Let $c(2, a) \equiv 1$, with all other costs being zero (see Fig. 2.11). Versions of this example were presented in [Bäuerle and Rieder (2011), Ex. 7.2.4], [Bertsekas and Shreve (1978), Chapter 9, Ex. 1], [Dynkin and Yushkevich (1979), Chapter 4, Section 6], [Puterman (1994), Ex. 7.3.3] and [Strauch (1966), Ex. 6.1]. This MDP is absorbing if, e.g., $\mathsf{P}_0(x) = 1$ for some $x = 2, 3, \ldots$, but it is not absorbing if $\mathsf{P}_0(1) = 1$, as the expected time to absorption in 0 is unbounded from above for different control strategies.

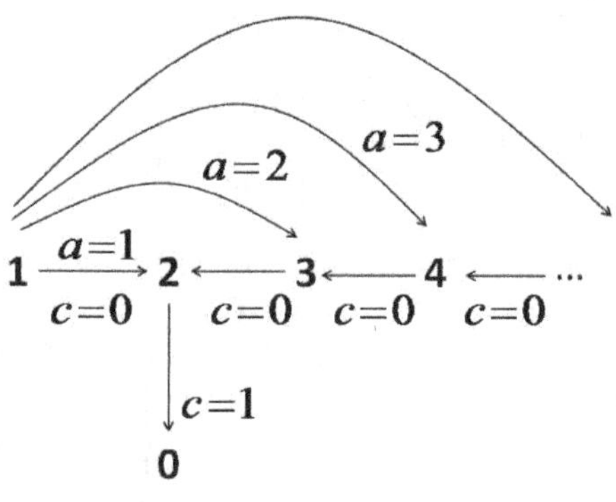

Fig. 2.11 Value iteration does not converge to the Bellman function.

The optimality equation (2.5) takes the form

$$\begin{aligned} v(0) &= v(0); \\ v(1) &= \inf_{a \in \mathbf{A}} \{v(1+a)\}; \\ v(2) &= 1 + v(0); \\ v(x) &= v(x-1), \text{ if } x > 2. \end{aligned}$$

The minimal non-negative solution $v(x)$ coincides with the Bellman function v^*_x and can be built using the following reasoning: $v(0) = 0$, hence $v(2) = 1$ and $v(x) = 1$ for all $x > 2$; therefore, $v(1) = 1$.

The value iteration results in the following sequence

$$\begin{aligned}
&v^0(x) \equiv 0;\\
&v^1(0) = 0;\quad v^1(2) = 1;\qquad\qquad\qquad v^1(x) \equiv 0 \text{ for } x > 2 \text{ and } v^1(1) = 0;\\
&v^2(0) = 0;\quad v^2(2) = 1;\quad v^2(3) = 1;\quad v^2(x) \equiv 0 \text{ for } x > 3 \text{ and } v^2(1) = 0;
\end{aligned}$$

and so on.

On step $n = 1, 2, \ldots$ the minimum with respect to $a \in \mathbf{A}$ in the formula for $v^{n+1}(1)$ is provided by $a > n + 1$.

Eventually $\lim_{n\to\infty} v^n(0) = 0$, $\lim_{n\to\infty} v^n(x) = 1$ for $x \geq 2$, but $v^n(1) = 0$ for each n, so that $v^\infty(1) = \lim_{n\to\infty} v^n(1) = 0 < 1 = v^*_1$. If $x = 1$, we have the strict inequality

$$v^\infty(1) = 0 < 1 = \inf_{a\in\mathbf{A}} \{v^\infty(1 + a)\}.$$

The above presented model is not weakly semi-continuous (Condition 2.3 fails to hold) since the action space $\mathbf{A}$ is not compact in the discrete topology. But it becomes so if we add a point ∞ to $\mathbf{A}$ as the one-point compactification of the set $\{1, 2, \ldots\}$ with the discrete topology: $\lim_{a\to\infty} a = \infty$. To make the transition probability and the cost function correspondingly weakly continuous and continuous, one has to add ∞ to $\mathbf{X}$ (making the state space also compact) and to put $p(\infty|\infty, a) \equiv 1$, $c(\infty, a) \equiv 0$. Now $v^*_\infty = 0$, $v^*_1 = 0$ (one can send the process to ∞ using action ∞), and $v^\infty(x) \equiv v^*_x$. □

Examples from Subsection 2.3.16 show that the statement of Theorem 2.4 can fail to hold if the model is not positive.

2.3.10 *When value iteration is not successful: positive model II*

Example 2.3.9 showed that the following statement can fail if the model is not negative:

Statement 1. $\exists\, v^\infty(x) = \lim_{n\to\infty} v^n(x) = v^*_x$, $x \in \mathbf{X}$.

It is also useful to look at the strategies that provide the infimum in the value iteration (2.6). Namely, for $x \in \mathbf{X}$, put

$$\Gamma_n(x) := \left\{ a \in \mathbf{A} :\ \ c(x, a) + \int_{\mathbf{X}} v^n(y) p(dy|x, a) = v^{n+1}(x) \right\};$$

$$\Gamma_\infty(x) := \{a \in \mathbf{A} : \quad a \text{ is an accumulation point}$$

$$\text{of some sequence } a_n \text{ with } a_n \in \Gamma_n(x)\}$$

(we assume that $\mathbf{A}$ is a topological space), and

$$\Gamma^*(x) := \left\{a \in \mathbf{A} : \quad c(x,a) + \int_{\mathbf{X}} v_y^* p(dy|x,a) = v_x^*\right\}$$

(the set of conserving actions at $x \in \mathbf{X}$).

One can expect that usually the following statement holds.

<u>Statement 2.</u> $\Gamma_\infty(x) \subseteq \Gamma^*(x)$ for all $x \in \mathbf{X}$.

Sufficient conditions for statements 1 and 2 were discussed in [Schäl (1975b)]. Below, we present a slight modification of Example 7.1 from [Schäl (1975b)] which shows that statements 1 and 2 can fail separately or simultaneously.

Example. Let $\mathbf{X} = [0,2] \times \{1,2,\ldots\}$, $\mathbf{A} = [0,2]$, and we consider the natural Euclidean topology in $[0,2]$ and the product topology in $\mathbf{X}$. For $x = (y,k)$ with $k \geq 2$, we put $p((y,k+1)|(y,k),a) \equiv 1$; $p((a,2)|(y,1),a) \equiv 1$ with other transition probabilities being zero (see Fig. 2.12).

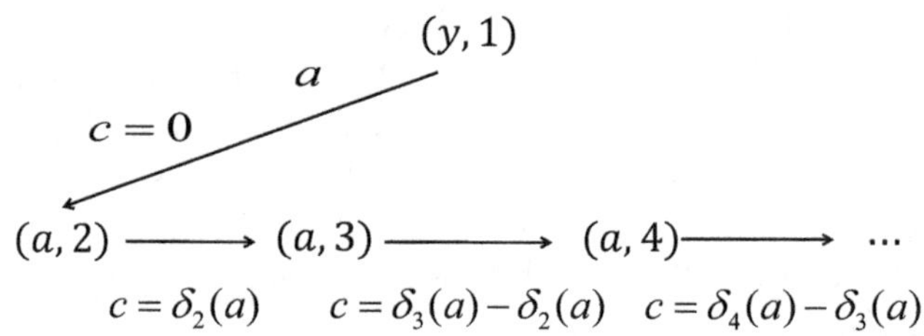

Fig. 2.12 Value iteration can be unsuccessful.

To describe the one-step cost $c(\cdot,\cdot)$, we need to introduce functions $\delta_n(y)$ on $[0,2]$, $n = 1,2,\ldots$ Suppose positive numbers $c_2 \leq c_3 \leq \cdots \leq d \leq b$ are fixed; $c_\infty := \lim_{i\to\infty} c_i$. Let $\delta_1(y) \equiv 0$. For $n \geq 2$, we put

$$\delta_n(y) = \begin{cases} b, & \text{if } y = 0; \\ c_n, & \text{if } 0 < y \leq 1/n; \\ b, & \text{if } 1/n < y < 1; \\ d, & \text{if } 1 \leq y \leq 2 \end{cases}$$

(see Fig. 2.13). Now $c((y,1)) \equiv 0$ and, for $x = (y,k)$ with $k \geq 2$, $c(x,a) \equiv \delta_k(y) - \delta_{k-1}(y)$.

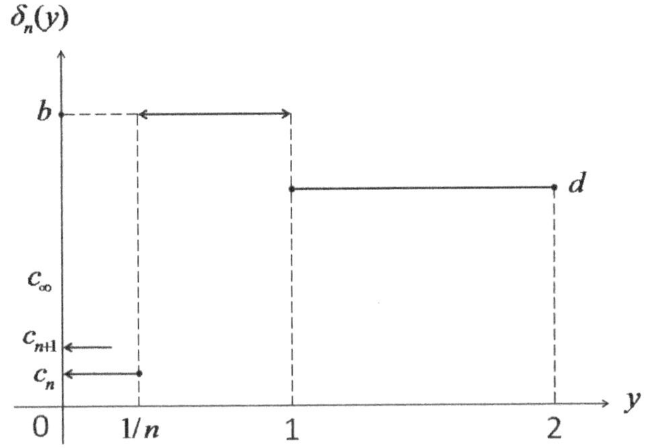

Fig. 2.13 Construction of the cost function.

Value iteration (2.6) results in the following table:

x	$(y,1)$	$(y,2)$	$(y,3)$	$(y,4)$	$\ldots$
$v^0(x)$	0	0	0	0	$\ldots$
$v^1(x)$	0	$\delta_2(y)$	$\delta_3(y)-\delta_2(y)$	$\delta_4(y)-\delta_3(y)$	$\ldots$
$v^2(x)$	$\inf_a \delta_2(a)=c_2$	$\delta_3(y)$	$\delta_4(y)-\delta_2(y)$	$\delta_5(y)-\delta_3(y)$	$\ldots$
$v^3(x)$	$\inf_a \delta_3(a)=c_3$	$\delta_4(y)$	$\delta_5(y)-\delta_2(y)$	$\delta_6(y)-\delta_3(y)$	$\ldots$
$\ldots$	$\ldots$	$\ldots$	$\ldots$	$\ldots$	$\ldots$

For $x=(y,k)$ with $k\ge 2$,

$$v^*_x = \lim_{n\to\infty} v^n(x) = \lim_{i\to\infty} \delta_i(y) - \delta_{k-1}(y) = \begin{cases} b, & \text{if } 0\le y<1; \\ d, & \text{if } 1\le y\le 2 \end{cases} - \delta_{k-1}(y),$$

and $\Gamma_\infty(x)=\Gamma^*(x)=\mathbf{A}$: all actions $a\in\mathbf{A}$ provide the infimum in the Bellman equation and in (2.6) on the steps $n=0,1,\ldots$.

Since for $x=(y,2)$,

$$\lim_{n\to\infty} v^n(x) = v^*_x = \lim_{i\to\infty}\delta_i(y) = \begin{cases} b, & \text{if } 0\le y<1; \\ d, & \text{if } 1\le y\le 2, \end{cases}$$

and $d\le b$, we conclude that $v^*_{(y,1)}=d$ and $\Gamma^*((y,1))=[1,2]$ when $d<b$. If $d=b$ then $\Gamma^*((y,1))=[0,2]=\mathbf{A}$.

At the same time, $\lim_{n\to\infty} v^n((y,1)) = \lim_{n\to\infty} c_n = c_\infty$ and $\Gamma_\infty((y,1))=\{0\}$ because $\Gamma_n((y,1))=(0,1/n]$.

Therefore,

- If $c_\infty < d < b$ then Statements 1 and 2 both fail.
- If $c_\infty = d = b$ then Statements 1 and 2 both hold.
- If $c_\infty = d < b$ then Statement 1 holds, but Statement 2 fails.
- If $c_\infty < d = b$ then Statement 1 fails, but Statement 2 holds.

□

2.3.11 *Value iteration and stability in optimal pure stopping models*

The optimal pure *stopping model* has the action space $\mathbf{A} = \{s, n\}$, where s (n) means the decision to stop (not to stop) the process. Let Δ be a specific absorbing state (cemetery) meaning that the process is stopped: for $x \neq \Delta$, $p(\Gamma|x,s) = \mathbb{I}\{\Gamma \ni \Delta\}$ $p(\mathbf{X} \setminus \{\Delta\}|x,n) = 1$; $p(\Gamma|\Delta,a) \equiv \mathbb{I}\{\Gamma \ni \Delta\}$ $c(\Delta,a) \equiv 0$.

Now, equation (2.5) takes the form

$$v(x) = \min\{c(x,n) + \int_{\mathbf{X}} v(y)p(dy|x,n);\ \ c(x,s)\}, \quad x \in \mathbf{X} \setminus \{\Delta\};$$

$$v(\Delta) = 0.$$

In this framework, the traditional value iteration (2.6) is often replaced with calculation of V_x^T, the minimal expected total cost incurred if started in state $x \in \mathbf{X} \setminus \{\Delta\}$ and maximum of T steps are allowed before stopping, that is, one investigates the finite horizon model.

Function V_x^T satisfies equations

$$V_x^0 = c(x,s);$$
$$V_x^{T+1} = \min\{c(x,n) + \int_{\mathbf{X}} V_y^T p(dy|x,n);\ \ c(x,s)\}, \quad T = 0,1,2,\ldots$$

Definition 2.3. [Ross (1983), p. 53]. The optimal stopping problem is called *stable* if

$$\lim_{T\to\infty} V_x^T = v_x^*.$$

In the following example, published in [Ross (1983), Chapter III, Ex. 2.1a], we present an unstable problem for which the traditional value iteration algorithm (2.6) provides $v^n(x) \to v^*_x$ as $n \to \infty$, and thus is stable in the sense of (2.28).

Example. Let $\mathbf{X} = \{\Delta, 0, \pm 1, \pm 2, \dots\}$; $\mathbf{A} = \{s, n\}$; for $x \neq \Delta$, $p(x + 1|x, n) = p(x - 1|x, n) = 1/2$ with other transition probabilities being zero, $c(x, n) \equiv 0$, $c(x, s) = x$ (see Fig. 2.14).

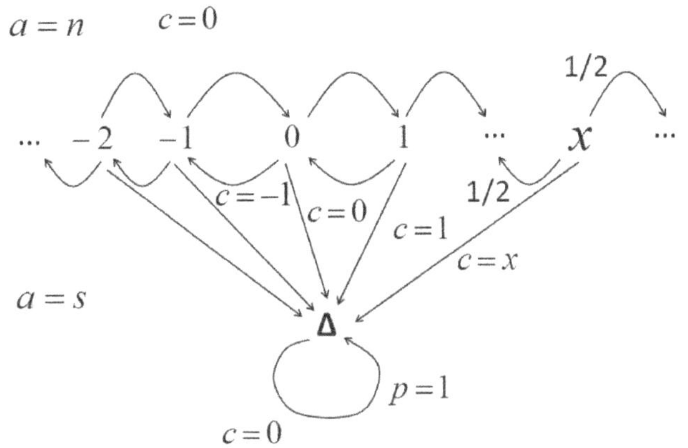

Fig. 2.14 Non-stable optimal stopping problem.

One can check by induction that $V_x^T = x$:

$$V_x^0 = x; \quad V_x^{T+1} = \min\left\{x, \ \frac{1}{2}V_{x-1}^T + \frac{1}{2}V_{x+1}^T\right\} = x, \quad x \in \mathbf{X} \setminus \{\Delta\}.$$

On the other hand, value iteration (2.6) results in the following:

$$v^0(x) = 0,$$

$$v^1(x) = \min\{x; \ 0\} = \begin{cases} x, & \text{if } x \leq 0; \\ 0, & \text{if } x > 0, \end{cases}$$

$$v^2(x) = \begin{cases} x, & \text{if } x < 0; \\ -1/2, & \text{if } x = 0; \\ 0, & \text{if } x > 0, \end{cases}$$

$$v^3(x) = \begin{cases} x, & \text{if } x < -1; \\ -5/4, & \text{if } x = -1; \\ -1/2, & \text{if } x = 0; \\ -1/4, & \text{if } x = 1; \\ 0, & \text{if } x > 1, \end{cases}$$

and so on, leading to $v^\infty(x) = \lim_{n\to\infty} v^n(x) = -\infty$ for all $x \in \mathbf{X} \setminus \{\Delta\}$.

We see that different initial functions result in different limiting functions in the value iteration algorithm (2.6).

To show that $v_x^* = -\infty$ for $x \in \mathbf{X}\setminus\{\Delta\}$, consider the stationary selector

$$\varphi^N(x) = \begin{cases} n, & \text{if } x > -N; \\ s, & \text{if } x \le -N, \end{cases}$$

where $N > 0$. For each $x \in \mathbf{X} \setminus \{\Delta\}$, $v_x^{\varphi^N} \le -N$ because the random walk under consideration is (null-)recurrent, so that state $-N$ will be reached from any initial state $x > -N$. Therefore, $\inf_{N>0} v_x^{\varphi^N} = -\infty$ and $v_x^* = -\infty$. At the same time, $\lim_{T\to\infty} V^T(x) = x > -\infty$.

Note that there is obviously no reason to stop the process at positive states: if the chain is never stopped then the total cost equals zero. Hence we can replace $c(x, s)$ with zeroes at $x > 0$ and obtain a negative model. The standard value iteration (2.6) for this modified model provides the same functions $v^n(x)$ as above, and $v^\infty(x) = v_x^*$, as is usual for negative models. □

2.3.12 *Stationary ε-optimal strategy does not exist (positive model)*

This example was published in [Dynkin and Yushkevich (1979), Chapter 6, Section 6].

Example. Let $\mathbf{X} = \{0, 1, 2\}$, $\mathbf{A} = \{1, 2, \ldots\}$, $p(0|0, a) = p(0|1, a) \equiv 1$, $p(1|2, a) = \left(\frac{1}{2}\right)^a$, $p(2|2, a) = 1 - \left(\frac{1}{2}\right)^a$, with all other transition probabilities being zero. We put $c(0, a) \equiv 0$, $c(1, a) \equiv 1$, $c(2, a) \equiv 0$, $\mathsf{P}_0(2) = 1$. See Fig. 2.15; a similar example was presented in [Puterman (1994), Ex. 7.3.2].

In order to construct the Bellman function v_x^*, we put $v(0) = 0$, and from the optimality equation, we obtain $v(1) = 1$,

$$v(2) = \inf_{a\in\mathbf{A}} \left\{ \left(\frac{1}{2}\right)^a + \left(1 - \left(\frac{1}{2}\right)^a\right) v(2) \right\}.$$

Any value $v(2) \in [0, 1]$ satisfies the last equation, and the minimal solution, which is of our interest, is $v(2) = 0$. The function $v(x) = v_x^*$ coincides

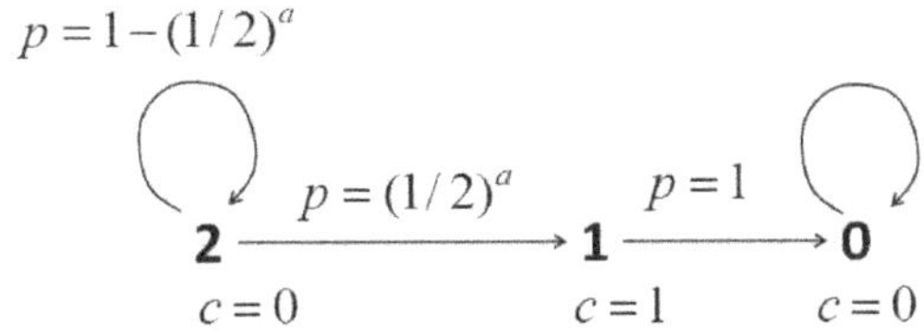

Fig. 2.15 Stationary ε-optimal strategy does not exist.

with the Bellman function because of the following. The value iteration (2.6) results exactly in this function $v(\cdot) = v^\infty(\cdot)$, and $v^\infty(\cdot)$ satisfies the optimality equation. (See Theorem 2.3.)

For a fixed integer $m \geq 0$, the non-stationary selector $\hat{\varphi}_t^m(2) = m + t$ provides the total cost equal to zero (when starting from $x_0 = 2$) with probability $\prod_{t=1}^{\infty}\left(1 - \left(\frac{1}{2}\right)^{m+t}\right)$, and equal to one with the complementary probability, so that

$$v_2^{\hat{\varphi}^m} = 1 - \prod_{t=1}^{\infty}\left(1 - \left(\frac{1}{2}\right)^{m+t}\right).$$

The last expression approaches 0 as $m \to \infty$, because $\prod_{t=1}^{\infty}\left(1 - \left(\frac{1}{2}\right)^{m+t}\right)$ is the tail of the converging infinite product $\prod_{t=1}^{\infty}\left(1 - \left(\frac{1}{2}\right)^{t}\right) > 0$. This convergence holds, since the series $\sum_{t=1}^{\infty} \ln\left[1 - \left(\frac{1}{2}\right)^t\right]$ converges to a finite (negative) number because of the comparison test:

$$\lim_{t\to\infty} \frac{\ln\left[1 - \left(\frac{1}{2}\right)^t\right]}{\left(\frac{1}{2}\right)^t} = 1 \quad \text{and} \quad \sum_{t=1}^{\infty}\left(\frac{1}{2}\right)^t = 1.$$

Hence, $\lim_{m\to\infty} \prod_{t=1}^{\infty}\left(1 - \left(\frac{1}{2}\right)^{m+t}\right) = 1$ and $\inf_{m\geq 0} v_2^{\hat{\varphi}^m} = 0$.

At the same time, for any stationary selector φ (and also for any stationary randomized strategy) we have $v_2^{\varphi} = 1 > 0 = \inf_{m\in\{0,1,2,\dots\}} v_2^{\hat{\varphi}^m}$.

Note that, when starting from $x_0 = 2$ under any control strategy π, the probability of absorption at 0 equals v_2^π because it coincides with the probability of having the total cost equal to one. As was shown above, $v_2^{\hat{\varphi}^m}$

is close to zero for big values of m, so that the expected time to absorption is infinity. The considered MDP is not absorbing at 0 if $\mathsf{P}_0(2) > 0$. A little more about this MDP in Subsection 2.3.17. A very similar example is presented in Subsection 5.3.5 when investigating the induced DTMDP (see Fig. 5.9). □

We present another simple example for which a stationary ε-optimal strategy does not exist.

Example. Let $\mathbf{X} = \{1\}$, $\mathbf{A} = \{1, 2, \ldots\}$, $p(1|1, a) \equiv 1$, $c(1, a) = \left(\frac{1}{2}\right)^a$. The model is positive, and the minimal non-negative solution to equation (2.5), which has the form

$$v(1) = \inf_{a \in \mathbf{A}} \left\{ \left(\frac{1}{2}\right)^a + v(1) \right\},$$

equals $v(1) = v^\infty(1) = v_1^* = 0$. No-one stationary selector is conserving. Moreover, for any stationary strategy π^{s}, there is the same positive cost on each step, equal to $\sum_{a \in \mathbf{A}} \pi^{\mathrm{s}}(a|1) \left(\frac{1}{2}\right)^a$, meaning that $v_1^{\pi^{\mathrm{s}}} = \infty$. At the same time, for each $\varepsilon > 0$, there is a non-stationary ε-optimal selector. For instance, one can put $\varphi_1(1)$ equal to any $a_1 \in \mathbf{A}$ such that $c(1, a_1) < \frac{\varepsilon}{2}$, $\varphi_2(1)$ equal to any $a_2 \in \mathbf{A}$ such that $c(1, a_2) < \frac{\varepsilon}{4}$, and so on.

The same reasoning remains valid for any other positive cost function, such that $\inf_{a \in \mathbf{A}} c(1, a) = 0$. □

In the presented examples one can add ∞ to the action space $\mathbf{A}$ as the one point compactification (considering the discrete topology at the beginning in $\mathbf{X}$ and $\mathbf{A}$). After adjusting the cost function and the transition probability, one can obtain the weakly semi-continuous model (see Condition 2.3) with the uniformly optimal selector $\varphi(x) \equiv \infty$.

2.3.13 *Stationary uniformly ε-optimal selector does not exist (negative model)*

Example. Let $\mathbf{X} = \{\ldots, -2, -1, 0, 1, 2, \ldots\}$, $\mathbf{A} = \{1, 2, \ldots\}$. For all $i > 0$, $p(-i+1| -i, a) = p(1| -i, a) \equiv \frac{1}{2}$, $p(i|i, a) \equiv 1$, $p(a|0, a) = 1$. All other transition probabilities are zero. We put $c(0, a) = -a$, with all other values of the cost function being zero. See Fig. 2.16; a similar example was presented in [Ornstein (1969), p. 564], see also [Bertsekas and Shreve (1978), Chapter 8, Ex. 2]. Clearly, only actions in state 0 are important.

Obviously, $v_x^* = \begin{cases} 0, & \text{if } x > 0; \\ -\infty & \text{if } x \le 0. \end{cases}$: for each negative initial state, there is a positive probability to reach 0, where $v_0^* = -\infty$. However, for any stationary selector φ, if $\hat{a} = \varphi(0)$ then $v_x^\varphi = -\left(\frac{1}{2}\right)^{|x|} \hat{a}$ for $x \le 0$, so that, for any

$\varepsilon > 0$, selector φ is not uniformly ε-optimal. If we put $\varphi(x_0, 0) \geq 2^{|x_0|}/\varepsilon$, then this semi-Markov selector will be uniformly ε-optimal (see Theorem 2.1). Note also that, for the stationary randomized strategy $\pi(a|x) = \frac{6}{a^2\pi^2}$, we have $v_0^\pi = -\infty$. Hence, $v_x^\pi = -\infty$ for all $x \leq 0$, and this strategy is uniformly optimal. □

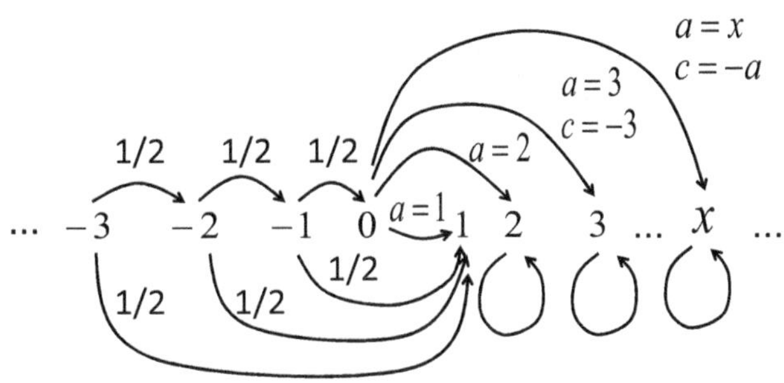

Fig. 2.16 Stationary uniformly ε-optimal selector does not exist.

In the example given in Subsection 2.3.14 the Bellman function is finite, but again unbounded.

We now present a very special example from [Ornstein (1969)], where $v_x^* \equiv -1$, but, for any stationary selector φ, there is a state $\hat{x} \in \mathbf{X}$ such that $v_{\hat{x}}^\varphi > -(1/2)$. In fact, according to the proof, 1/2 can be replaced here by any preliminary fixed positive number.

Example. Let $\mathbf{A} = \{1, 2, \ldots\}$; the state space $\mathbf{X}$ and the transition probability will be defined inductively. See Fig. 2.17 for illustration. Note that the space $\mathbf{X}$ will be not Borel, but it will be clear that the optimization problem (2.1) is well defined.

Let $C_1 := \{y_0\} \subset \mathbf{X}$ and let 0 and g be two points in $\mathbf{X}$; $p(0|0,a) \equiv p(0|g,a) \equiv 1$, $c(g,a) \equiv -1$, with all the other values of the cost function being zero. $p(g|y_0,a) = 1 - (1/2)^a$, $p(0|y_0,a) = (1/2)^a$. Obviously, $v_{y_0}^* = -1$: consider $\varphi(y_0) \to \infty$. State g is the 'goal', and state 0 is the 'cemetery'.

Suppose we have built all the subsets $C_\beta \subset \mathbf{X}$ for $\beta < \alpha$, where $\alpha, \beta \in \Omega$ which is the collection of all the ordinals up to (and excluding) the first uncountable one. (Or, more simply, Ω is the first uncountable ordinal.) Suppose also that $v_x^* = -1$ for all $x \in \bigcup_{\beta<\alpha} C_\beta$. We shall build C_α such that $C_\alpha \cap C_\beta = \emptyset$ for all $\beta < \alpha$ and $v_x^* = -1$ for all $x \in C_\alpha$. Eventually we

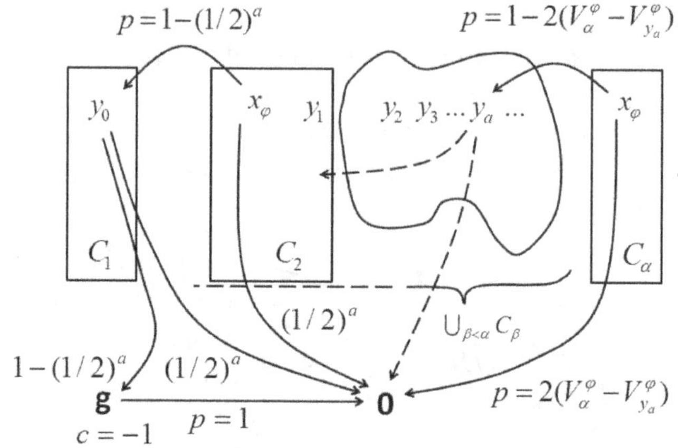

Fig. 2.17 Stationary uniformly ε-optimal selector does not exist.

shall put

$$\mathbf{X} := \bigcup_{\alpha<\Omega} C_\alpha \cup \{0, g\}.$$

Note that, for each $x \in C_\alpha$, there will be a sequence $y_a \in \bigcup_{\beta<\alpha} C_\beta$, $a = 1, 2, \ldots$ such that only transition probabilities $p(y_a|x, a)$ and $p(0|x, a)$ are positive.

For each stationary selector φ on $\bigcup_{\beta<\alpha} C_\beta$, i.e. for each mapping $\varphi: \bigcup_{\beta<\alpha} C_\beta \to \mathbf{A}$, such that

$$V_\alpha^\varphi := \sup_{y\in\bigcup_{\beta<\alpha} C_\beta} v_y^\varphi \le -\frac{1}{2},$$

we will introduce one point x_φ in C_α. The transition probabilities from x_φ are defined as follows:

- if $v_{\hat{y}}^\varphi = V_\alpha^\varphi$ for some $\hat{y} \in \bigcup_{\beta<\alpha} C_\beta$, then $p(\hat{y}|x_\varphi, a) = 1 - \left(\frac{1}{2}\right)^a$ and $p(0|x_\varphi, a) = \left(\frac{1}{2}\right)^a$;
- otherwise, pick a sequence $y_a \in \bigcup_{\beta<\alpha} C_\beta$, $a = 1, 2, \ldots$, such that $V_\alpha^\varphi - v_{y_a}^\varphi < 1/2$ and $\lim_{a\to\infty} v_{y_a}^\varphi = V_\alpha^\varphi$, and put $p(y_a|x_\varphi, a) = 1 - 2(V_\alpha^\varphi - v_{y_a}^\varphi)$ and $p(0|x_\varphi, a) = 2(V_\alpha^\varphi - v_{y_a}^\varphi)$.

Starting from $x_0 \in \mathbf{X}$, under any control strategy π, the probability of each trajectory $x_0, a_1, x_1, \ldots$ equals the product $P_0(x_0)\pi_1(a_1|x_0)p(x_1|x_0, a_1)\cdots$, simply because both π_t and p are atomic. Every one (random) trajectory

$$X_0 = x_0 \neq 0, A_1, X_1, \ldots$$

is absorbed at 0 at some finite $T > 0$ (directly, or through the state g) $\mathsf{P}^{\pi}_{x_0}$-a.s., with the tail being $X_T = 0, A_{T+1}, 0, \ldots$. This is true because, for each $t = 0, 1, 2, \ldots$, if $X_t, X_{t+1} \neq 0, g$, then $X_t \in C_{\alpha_t}$ and $X_{t+1} \in C_{\alpha_{t+1}}$ with $\alpha_{t+1} < \alpha_t$, i.e., the ordinals corresponding to $X_t \neq 0, g$ *decrease* with time t. After the minimal ordinal α^* among $\{\alpha_0, \alpha_1, \alpha_2, \ldots\}$ is reached at some $\tilde{T} < \infty$, $X_{\tilde{T}+1} \in \{0, g\}$. (Recall that the ordinals are well-ordered, so the minimal ordinal α^* exists.) This reasoning holds $\mathsf{P}^{\pi}_{x_0}$-a.s.

Hence, the optimization problem (2.1) is well defined.

Let us show that, for an arbitrarily fixed $\varepsilon > 0$, for each $x \in \bigcup_{\alpha<\Omega} C_\alpha \cup \{g\}$, there is a stationary selector φ^* such that $v^{\varphi^*}_x < -1 + \varepsilon$. For y_0 and g this statement is obvious. Suppose it is valid for all $x \in \bigcup_{\beta<\alpha} C_\beta$ and consider $\hat{x} = x_\varphi \in C_\alpha$. ($\varphi$ is a selector on $\bigcup_{\beta<\alpha} C_\beta$ with $V^{\varphi}_{\alpha} \leq -\frac{1}{2}$.) Let us choose an action $a^* \in \mathbf{A}$ such that some point $y \in \bigcup_{\beta<\alpha} C_\beta$ is reached in one step with a probability bigger than $1-\varepsilon/2$. This can be done because either $y = \hat{y} \in \bigcup_{\beta<\alpha} C_\beta$ with $v^{\varphi}_{\hat{y}} = V^{\varphi}_{\alpha}$ and $p(\hat{y}|x_\varphi, a) = 1 - \left(\frac{1}{2}\right)^a$, or, otherwise, there is a sequence $y_a \in \bigcup_{\beta<\alpha} C_\beta$, $a = 1, 2, \ldots$ with $\lim_{a\to\infty} v^{\varphi}_{y_a} = V^{\varphi}_{\alpha}$, and $p(y_a|x_\varphi, a) = 1 - 2(V^{\varphi}_{\alpha} - v^{\varphi}_{y_a})$. According to the induction supposition, for $y \in \bigcup_{\beta<\alpha} C_\beta$, there exists a stationary selector φ^* such that $v^{\varphi^*}_y < -1+\varepsilon/2$. The latter inequality will not change if we put $\varphi^*(\hat{x}) := a^*$ because $\hat{x} \in C_\alpha$ and all the states in all (random) trajectories $X_0 = y, A_1, X_1, \ldots$ belong to $\bigcup_{\beta<\alpha} C_\beta \cup \{0, g\}$ $\mathsf{P}^{\varphi^*}_y$-a.s. Now $v^{\varphi^*}_{\hat{x}} < (1 - \varepsilon/2)(-1 + \varepsilon/2) < -1 + \varepsilon$. (Note that $v^{\varphi^*}_z \leq 0$ for all $z \in \mathbf{X}$.) The desired statement follows by the transfinite induction.

Since $\varepsilon > 0$ was arbitrary, we conclude that, for each $x \in \bigcup_{\alpha<\Omega} C_\alpha \cup \{g\}$, $v^*_x := \inf_\pi v^\pi_x = -1$ (because $v^*_x \geq -1$).

Now, taking any stationary selector φ, we shall prove by contradiction that $V^{\varphi}_{\alpha} > -(1/2)$ for some $\alpha < \Omega$.

Suppose $\forall \alpha < \Omega$ $V^{\varphi}_{\alpha} \leq -(1/2)$. Then the function

$$h(\alpha) := \sup_{y \in \bigcup_{\beta\leq\alpha} C_\beta} v^{\varphi}_y$$

exhibits the following properties: h is (obviously) non-positive and non-decreasing; for each $\alpha \in \Omega$,

$$\sup_{\gamma<\alpha} h(\gamma) = \sup_{y \in \bigcup_{\beta<\alpha} C_\beta} v^{\varphi}_y = V^{\varphi}_{\alpha} \leq -1/2 < 0,$$

and $h(\alpha) > \sup_{\gamma<\alpha} h(\gamma)$. Below, we prove the latter inequality.

For a fixed α, consider the restriction of φ to $\bigcup_{\beta<\alpha} C_\beta$ and take the point $x_\varphi \in C_\alpha$ as constructed above. We have

- either $v^{\varphi}_{x_\varphi} = \left[1-(1/2)^{\varphi(x_\varphi)}\right] v^{\varphi}_{\hat{y}} > V^{\varphi}_{\alpha}$ when $v^{\varphi}_{\hat{y}} = V^{\varphi}_{\alpha}$ (recall that $V^{\varphi}_{\alpha} < 0$),
- or $v^{\varphi}_{x_\varphi} = p(y_a|x_\varphi, a)v^{\varphi}_{y_a} = [1-2(V^{\varphi}_{\alpha} - v^{\varphi}_{y_a})]v^{\varphi}_{y_a}$ otherwise, where $a = \varphi(x_\varphi)$, $y_a \in \bigcup_{\beta<\alpha} C_\beta$.

In the latter case, $v^{\varphi}_{y_a} < V^{\varphi}_{\alpha} \le -(1/2)$, so that $2v^{\varphi}_{y_a} < -1$ and $v^{\varphi}_{x_\varphi} = v^{\varphi}_{y_a} - 2v^{\varphi}_{y_a}(V^{\varphi}_{\alpha} - v^{\varphi}_{y_a}) > v^{\varphi}_{y_a} + V^{\varphi}_{\alpha} - v^{\varphi}_{y_a} = V^{\varphi}_{\alpha}$. Hence, in any case,

$$h(\alpha) \ge v^{\varphi}_{x_\varphi} > V^{\varphi}_{\alpha} = \sup_{\gamma<\alpha} h(\gamma).$$

But, according to Theorem A.1, $h(\alpha) = 0$ for some $\alpha < \Omega$, which contradicts the assumption $V^{\varphi}_{\alpha+1} \le -(1/2)$. Therefore, $\exists \alpha < \Omega$: $V^{\varphi}_{\alpha} > -(1/2)$, and there is $\hat{x} \in \bigcup_{\beta<\alpha} C_\beta \subset \mathbf{X}$ such that $v^{\varphi}_{\hat{x}} > -(1/2)$. We emphasize that Theorem A.1 is not applicable if Ω is replaced with a countable ordinal. □

The considered model can be called *gambling*, because, for each state (e.g. amount of money), when using a gambling action $a \in \mathbf{A}$, the gambler either reaches his or her goal (state g), loses everything (state 0), or moves to one of a countable number of new states. The objective is to maximize the probability of reaching the goal. We emphasize that the state space $\mathbf{X}$ is uncountable in the current example. For gambling models with countable $\mathbf{X}$, for each $\varepsilon > 0$ there is a stationary selector φ such that $v^{\varphi}_{x} \le (1-\varepsilon)v^{*}_{x}$ for all $x \in \mathbf{X}$ [Ornstein (1969), Thm. B], see also [Puterman (1994), Thm. 7.2.7]. Note that we reformulate all problems as minimization problems.

Other gambling examples are given in Subsections 2.3.29 and 2.3.30.

2.3.14 *Finite-action negative model where a stationary uniformly ε-optimal selector does not exist*

Example. Let $\mathbf{X} = \{0, 1, 2, \ldots\}$, $\mathbf{A} = \{1, 2\}$, $p(0|x, 1) = 1$, $p(0|0, a) \equiv 1$, i.e. state 0 is absorbing; $p(0|x, 2) = p(x+1|x, 2) = 1/2$ for $x > 0$. All the other transition probabilities are zero. We put $c(0, a) = 0$, $c(x, 1) = 1 - 2^x$; $c(x, 2) = 0$ for $x > 0$. See Fig. 2.18; a similar example was presented in [Puterman (1994), Ex. 7.2.5] and in [Altman (1999), Section 9.8]. See also Subsections 2.3.15 and 2.3.24.

Remark 2.4. This example can be reformulated as a gamble [Dynkin and Yushkevich (1979), Chapter 6, Section 6]. State x means 2^x units in hand, and if the decision is to continue ($a = 2$) then the capital is either doubled or lost with equal probability. If the decision is to stop ($a = 1$) then the player has to pay one unit. The goal is to maximize the final wealth.

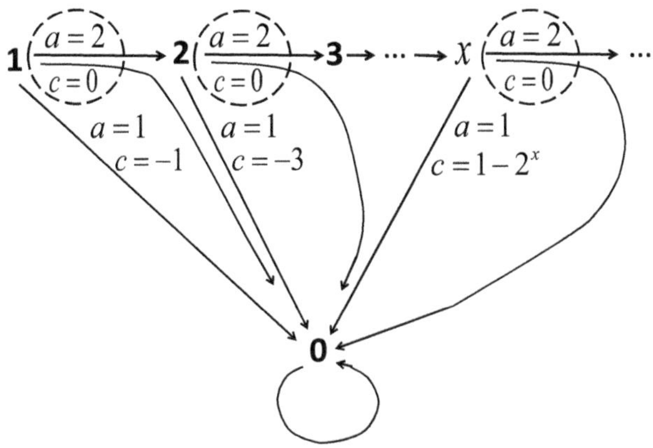

Fig. 2.18 Stationary uniformly ε-optimal selector does not exist.

The optimality equation (2.5) is given by

$$v(0) = v(0); \quad \text{for } x > 0 \quad v(x) = \min\{1 - 2^x + v(0);\ 0.5v(x+1) + 0.5v(0)\},$$

and the Bellman function v_x^* will be built using Theorem 2.2.

Equations (2.6) take the form

$$v^0(x) \equiv 0; \quad v^n(0) \equiv 0; \quad v^{n+1}(x) = \min\left\{1 - 2^x,\ \frac{1}{2}v^n(x+1)\right\},\ x = 1, 2, \ldots$$

The solution is given by formulae

$$v^n(0) \equiv 0; \quad v^n(x) = \frac{1}{2^{n-1}} - 2^x, \quad x = 1, 2, \ldots:$$

the proof is by induction starting from $v^1(x) = 1 - 2^x$ for $x = 1, 2, \ldots$.

Therefore, $v_0^* = 0$,

$$v_x^* = v^\infty(x) := \lim_{n\to\infty} v^n(x) = -2^x, \quad x = 1, 2, \ldots,$$

and the Bellman function v_x^* satisfies the optimality equation (2.5):

$$v_x^* = \min\left\{1 - 2^x;\ \ \frac{1}{2}v_{x+1}^*\right\}, \qquad x = 1, 2, \ldots.$$

Note that, for any constant $K \geq 1$, $Kv_x^* = -K2^x$ is also a solution to the optimality equation for $x = 1, 2, \ldots$.

Below, we assume that the initial state x_0 is not zero. The stationary selector $\varphi^2(x) \equiv 2$ is conserving but not equalizing: $\forall x > 0,\ t$, $\mathsf{E}_x^{\varphi^2}[v_{X_t}^*] = -2^x$. It is far from being optimal because $v_x^{\varphi^2} \equiv 0 > v_x^* = -2^x$. In a similar way to Example 2.3.4, this selector indefinitely delays the ultimate absorption in state 0. In this model, there are no optimal strategies, because, for an optimal strategy, we must have the equation $v^\pi_x = -2^x$ which only holds if $\pi_1(2|x) = 1$ (control $A_1 = 1$ is excluded because it leads to the cost $1 - 2^x > v^*_x = -2^x$), and $v^\pi_x = 0.5 v^\pi_{h_1=(x,2,x+1)}$. Thus, we must have $v^\pi_{(x,2,x+1)} = 2^{x+1}$. The same reasoning applies to state $x+1$ after history $h_1 = (x, 2, x+1)$ is realized, and so on. Therefore, the only candidate for the optimal strategy is φ^2, but we already know that it is not optimal.

If φ is an arbitrary stationary selector, different from φ^2, then $\varphi(\hat{x}) = 1$ for some $\hat{x} > 0$, and $v^\varphi_{\hat{x}} = 1 - 2^{\hat{x}} > v^*_{\hat{x}} = -2^{\hat{x}}$. Hence, for $\varepsilon < 1$, a stationary uniformly ε-optimal selector does not exist. On the other hand, for each $\varepsilon > 0$, there exists a non-stationary Markov uniformly ε-optimal selector $\varphi_t(x)$. Indeed, put $\varphi_t(y) = 2$ for all $1 \le t \le n$ and $\varphi_t(y) = 1$ for $t > n$, where $n \in \{0, 1, 2, \ldots\}$ is such that $n \ge -\frac{\ln \varepsilon}{\ln 2}$. Then, starting from $x > 0$, the controlled process will reach the state $x + n$ in n steps with probability $\left(\frac{1}{2}\right)^n$ with the further absorption in 0, leading to

$$v^\varphi_x = \left(\frac{1}{2}\right)^n (1 - 2^{x+n}) \le -2^x + \varepsilon.$$

The same selector can be represented as a stationary semi-Markov selector $\varphi(x_0, y) = \begin{cases} 2, & \text{if } y - x_0 < n; \\ 1, & \text{otherwise.} \end{cases}$ (See also Theorem 2.1.)

At the same time, for an arbitrary $\varepsilon > 0$, after we put $\tilde{\varepsilon} := \min\left\{\varepsilon, \frac{1}{2}\right\} \le \varepsilon$, the stationary randomized strategy $\tilde{\pi}^s(1|y) = \delta := \frac{\tilde{\varepsilon}}{2 - \tilde{\varepsilon}}$; $\tilde{\pi}^s(2|y) = 1 - \delta$ (with $\delta \in (0,1)$) is uniformly ε-optimal. To show this, note that the trajectory of the form $(x, 2, x+1, 2, \ldots, x+n, 1, 0, a_{n+1}, 0, \ldots)$, starting from $x > 0$, is realized with probability $\left[\frac{1-\delta}{2}\right]^n \delta$ and leads to the cost $(1 - 2^{x+n})$. All the other trajectories, when only action 2 is in use, result in zero cost. Therefore,

$$v^{\tilde{\pi}^s}_x = \sum_{n=0}^{\infty} \left(\frac{1-\delta}{2}\right)^n \delta(1 - 2^{x+n}) = -2^x + \frac{2\delta}{1+\delta} = -2^x + \tilde{\varepsilon} \le v^*_x + \varepsilon.$$

The MDP thus considered is semi-continuous (see Condition 2.4); more about such models in Subsection 2.3.16.

We also remark that this model is *absorbing* for each initial distribution. To justify this, note that, for each control strategy π and each initial state $x_0 \in \{1, 2, \ldots\}$, $\mathsf{P}^\pi_{x_0}(T_0 = k) \le \left(\frac{1}{2}\right)^{k-1}$, $k = 1, 2, \ldots$: if $T_0 := \min\{t > 0 : X_t = 0\} = k$, then the state $x_0 + k - 1$ was reached, and the probability of the latter event is below $\left(\frac{1}{2}\right)^{k-1}$. Thus,

$$\mathsf{E}^\pi_{x_0}[T_0] \le \sum_{k=1}^{\infty} k \left(\frac{1}{2}\right)^{k-1} = 4.$$

One can find more properties of the considered MDP in Subsection 2.3.24. □

2.3.15 *Nearly uniformly optimal selectors in negative models*

If the state space $\mathbf{X}$ is countable, then the following statement holds.

Theorem 2.11. [Ornstein (1969), Thm. C]. *Suppose the model is negative and $v_x^* > -\infty$ for all $x \in \mathbf{X}$. Then, for each $\varepsilon > 0$, there is a stationary selector φ such that*

$$v_x^\varphi \leq v_x^* + \varepsilon |v_x^*|, \quad x \in \mathbf{X}. \tag{2.30}$$

One can call such a stationary selector 'nearly uniformly optimal'.

The following example, based on [van der Wal and Wessels (1984), Ex. 7.1], shows that this theorem cannot be essentially improved. In other words, if $\mathbf{X} = \{0, 1, 2, \ldots\}$, and if we replace (2.30) with

$$v_x^\varphi \leq v_x^* + \varepsilon |v_x^*| \delta(x), \quad x \in \mathbf{X}, \tag{2.31}$$

where $0 < \delta(x) \leq 1$ is a fixed (perhaps, ε-dependent) function with $\lim_{x\to\infty} \delta(x) = 0$, then Theorem 2.11 fails to hold.

Suppose the desired sequence $\delta(x)$ exists for a fixed $\varepsilon > 0$, and put $\gamma_x := \sqrt{\delta(x)}$. Below, we describe and investigate the model having in hand a sequence γ_x $(x = 0, 1, 2, \ldots)$ such that $\gamma_x \in [0, 1]$ and $\liminf_{x\to\infty} \gamma_x = 0$. We intend to show that inequality (2.31) is not valid for this model.

Example. Let $\mathbf{X} = \{0, 1, 2, \ldots\}$, $\mathbf{A} = \{1, 2\}$, $p(0|0, a) \equiv 1$, $p(0|x, 1) \equiv 1$, $p(x + 1|x, 2) = \frac{1+\gamma_x}{2(1+\gamma_{x+1})}$, $p(0|x, 2) = \frac{1+2\gamma_{x+1}+\gamma_x}{2(1+\gamma_{x+1})}$ for $x > 0$. All the other transition probabilities are zero. Note that $\frac{1+\gamma_x}{2(1+\gamma_{x+1})} \leq 1$ because $\gamma_x \leq 1 \leq 1 + 2\gamma_{x+1}$. We put $c(0, a) = 0$, $c(x, 2) \equiv 0$ and $c(x, 1) = -2^x$ for $x > 0$. See Fig. 2.19, compare with example from Subsection 2.3.14.

The optimality equation (2.5) looks as follows:

$$v(0) = v(0); \quad \text{clearly, we must put } v(0) = 0;$$

$$v(x) = \min\left\{-2^x, \frac{1+\gamma_x}{2(1+\gamma_{x+1})} v(x+1)\right\} \quad \text{for } x > 0,$$

and the Bellman function v_x^* will be built using Theorem 2.2.

Equations (2.6) take the form

$$v^0(x) \equiv 0; \quad v^n(0) \equiv 0;$$

$$v^{n+1}(x) = \min\left\{-2^x, \frac{1+\gamma_x}{2(1+\gamma_{x+1})} v^n(x+1)\right\}, \quad x = 1, 2, \ldots$$

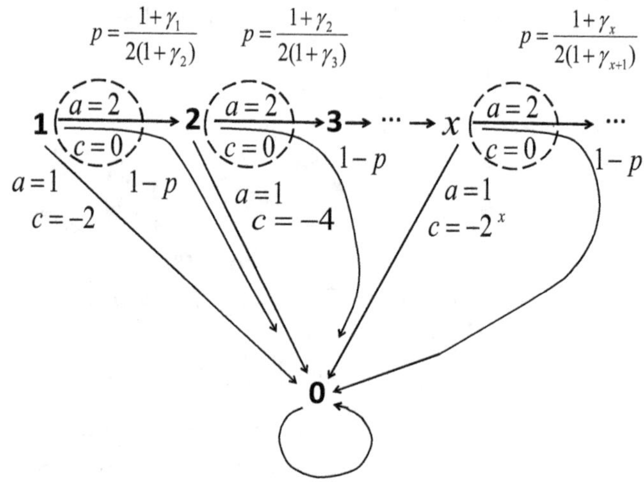

Fig. 2.19 Example 2.3.15.

The solution is given by formulae

$$v^n(0) \equiv 0; \quad v^n(x) = -(1+\gamma_x)2^x \frac{1}{1+\min_{x \le y < x+n} \gamma_y}, \quad x = 1, 2, \ldots :$$

the proof is by induction starting from $v^1(x) = -2^x$ for $x = 1, 2, \ldots$. If the given formula is correct for some $n \ge 1$, then, for $x = 1, 2, \ldots$,

$$\begin{aligned} v^{n+1}(x) = \min \Bigg\{ -2^x, \quad & \frac{1+\gamma_x}{2(1+\gamma_{x+1})} \left[-(1+\gamma_{x+1})2^{x+1} \right] \\ & \times \frac{1}{1+\min_{x+1 \le y < x+1+n} \gamma_y} \Bigg\} \\ = -(1+\gamma_x)2^x \max & \left\{ \frac{1}{1+\gamma_x}, \quad \frac{1}{1+\min_{x+1 \le y < x+1+n} \gamma_y} \right\} \\ = -(1+\gamma_x)2^x & \max_{x \le y < x+n+1} \frac{1}{1+\gamma_y} \\ = -(1+\gamma_x)2^x & \frac{1}{1+\min_{x \le y < x+n+1} \gamma_y}. \end{aligned}$$

The presented formula is valid for $n+1$ and hence for all $n = 0, 1, 2, \ldots$. Therefore,

$$v_x^* = v^\infty(x) = -2^x(1+\gamma_x) \frac{1}{1+\inf_{y \ge x} \gamma_y} = -2^x(1+\gamma_x), \quad x = 1, 2, \ldots,$$

because $\liminf_{x\to\infty}\gamma_x = 0 \Longrightarrow \inf_{y\ge x}\gamma_y = 0$.

If a stationary selector φ in the studied model satisfies (2.31), then $\varphi(x) = 1$ for infinitely many $x \in \mathbf{X}$. Otherwise, $v_x^\varphi = 0$ for all sufficiently large x, but, since $\lim_{x\to\infty}\delta(x) = 0$, $\delta(x) \le \frac{1}{2\varepsilon}$ for all sufficiently large x. Thus, for such values of x,

$$v_x^* + \varepsilon|v_x^*|\delta(x) = -2^x(1+\gamma_x)[1-\varepsilon\delta(x)] \le -2^{x-1}(1+\gamma_x),$$

and inequality (2.31) is violated.

For the values $x_1, x_2, \dots$ with $\lim_{i\to\infty} x_i = \infty$ and $\varphi(x_i) = 1$, we have $v_{x_i}^\varphi = -2^{x_i}$, and

$$\frac{v_{x_i}^\varphi - v_{x_i}^*}{|v_{x_i}^*|\delta(x_i)} = \frac{\gamma_{x_i}}{\delta(x_i)(1+\gamma_{x_i})} = \frac{1}{\sqrt{\delta(x_i)}+\delta(x_i)} \ge \frac{1}{2\sqrt{\delta(x_i)}}$$

because $\delta(\cdot) \le 1$. The right-hand side approaches $+\infty$ as $x_i \to \infty$, and cannot remain smaller than ε for all x_i, meaning that inequality (2.31) is violated and sequence $\delta(\cdot)$ does not exist. □

2.3.16 *Semi-continuous models and the blackmailer's dilemma*

As was mentioned in Section 2.2, very powerful results are known for weakly semi-continuous models, when Condition 2.3 is satisfied. (See Theorem 2.4.)

Theorem 2.12. **[Cavazos-Cadena** *et al.* **(2000), Thm. 3.1].** *If the model is semi-continuous (that is, Condition 2.4 is satisfied), the state space* $\mathbf{X}$ *is finite, Condition 2.2(a) is satisfied, and, for each stationary selector, the controlled process* X_t *has a single positive recurrent class (unichain model), then there exists a uniformly optimal stationary selector.*

Example in Subsection 2.3.14 shows that, for denumerable $\mathbf{X}$, the statement of Theorem 2.12 can fail. That model is semi-continuous (i.e., Condition 2.4 is satisfied), Condition 2.2(a) is satisfied, and $\{0\}$ is the single positive recurrent class. A more complicated example illustrating the same ideas can be found in [Cavazos-Cadena *et al.* (2000), Section 4]; see also Subsection 1.4.8.

We now show that the requirement $c(\cdot,\cdot) \ge 0$ in Theorem 2.4 and the unichain condition in Theorem 2.12 are important. Similar examples were published in [Cavazos-Cadena *et al.* (2000), Ex. 3.1] and in [van der Wal and Wessels (1984), Ex. 3.5].

Example. Let $\mathbf{X} = \{0,1\}$; $\mathbf{A} = [0,1]$ equipped with the Euclidean topology; $p(0|0,a) \equiv 1$, $p(1|0,a) \equiv 0$, $p(y|1,a) = \begin{cases} a, & \text{if } y = 0; \\ 1-a, & \text{if } y = 1, \end{cases}$ $c(0,a) \equiv 0$,

$c(1,a) = -a(1-a)$. Here, for $\varphi(x) \equiv 0$, both states are positive recurrent. See Fig. 2.20.

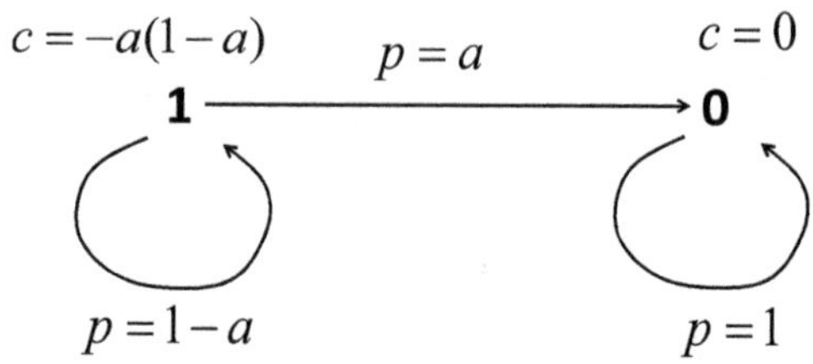

Fig. 2.20 No optimal selectors in a semi-continuous model.

The optimality equation (2.5) looks as follows:

$$v(0) = v(0);$$
$$v(1) = \inf_{a \in \mathbf{A}} \{-a(1-a) + av(0) + (1-a)v(1)\},$$

and the Bellman function v_x^* will be built using Theorem 2.2.

For the functions $v^n(\cdot)$ (see (2.6)) one can prove by induction that

$$v^n(0) = 0, \quad -1 < v^n(1) \le 0, \text{ and } v^{n+1}(1) = -\frac{(v^n(1)-1)^2}{4}, \; n = 0, 1, \ldots.$$

Like in every negative model, $v^n(1)$ decreases with $n \uparrow \infty$ and the limit $v^\infty(1) \in [-1, 0]$ satisfies equation

$$v^\infty(1) = -\frac{(v^\infty(1)-1)^2}{4},$$

that is, $v^\infty(1) = v_1^* = -1$ and, obviously, $v^\infty(0) = v_0^* = 0$.

The stationary selector $\varphi^*(x) \equiv 0$ is the single conserving one, but

$$\lim_{t \to \infty} \mathsf{E}_1^{\varphi^*}[v_{X_t}^*] = v_1^* = -1,$$

so that it is not equalizing and not optimal if $x_0 = 1$. Indeed, $v_1^{\varphi^*} = 0$, and, for each stationary selector $\varphi(x) \equiv \hat{a} > 0$,

$$v_1^\varphi = -\hat{a}(1-\hat{a})\mathsf{E}_1^\varphi[T_0],$$

where T_0, the time to absorption in state 0, has a geometric distribution: $\mathsf{P}_1^\varphi(T_0 = k) = (1-\hat{a})^{k-1}\hat{a}$, $k = 1, 2, \ldots$. Thus, $v_1^\varphi = -(1-\hat{a})$ and $\inf_{\hat{a} \in \mathbf{A}} v_1^\varphi = -1 = v_1^*$.

If $\mathsf{P}_0(1) > 0$, then there are no optimal strategies in this model, unless it is semi-continuous, i.e., Condition 2.4 is satisfied. To show this, fix an arbitrary Markov strategy $\hat{\pi}$ different from $\varphi^*(x) \equiv 0$. (No need to look at non-Markov strategies due to Lemma 2.1.) We know that

$$\inf_\pi v^\pi_{\mathsf{P}_0} = \inf_\pi \mathsf{E}^\pi_{\mathsf{P}_0}\left[\sum_{t=1}^{\infty} c(X_{t-1}, A_t)\right] = \int_{\mathbf{X}} v^*_x \mathsf{P}_0(dx) = -\mathsf{P}_0(1),$$

and we will show that $v^{\hat{\pi}}_{\mathsf{P}_0} > -\mathsf{P}_0(1)$. Let $T \geq 1$ be the first time moment when $\hat{\pi}_T(\{0\}|x_{T-1}) < 1$. Then

$$v^{\hat{\pi}}_{\mathsf{P}_0} = \mathsf{P}_0(1)\int_{[0,1]}[-a(1-a)]\hat{\pi}_T(da|x_{T-1})$$
$$+\mathsf{E}^{\hat{\pi}}_{\mathsf{P}_0}\left[\sum_{t=T+1}^{\infty} c(X_{t-1}, A_t)|X_T = 1\right]\mathsf{P}^{\hat{\pi}}_{\mathsf{P}_0}(X_T = 1).$$

Since

$$\mathsf{E}^{\hat{\pi}}_{\mathsf{P}_0}\left[\sum_{t=T+1}^{\infty} c(X_{t-1}, A_t)|X_T = 1\right] \geq v^*_1 = -1$$

and

$$\mathsf{P}^{\hat{\pi}}_{\mathsf{P}_0}(X_T = 1) = \mathsf{P}_0(1)\int_{[0,1]} p(1|1,a)\hat{\pi}_T(da|x_{T-1}),$$

we conclude that

$$v^{\hat{\pi}}_{\mathsf{P}_0} \geq \mathsf{P}_0(1)\left\{\int_{[0,1]}[-a(1-a)]\hat{\pi}_T(da|x_{T-1}) - \int_{[0,1]}(1-a)\hat{\pi}_T(da|x_{T-1})\right\}$$
$$= \mathsf{P}_0(1)\int_{[0,1]}(a^2-1)\hat{\pi}_T(da|x_{T-1}) = -\mathsf{P}_0(1) + \int_{(0,1]} a^2\hat{\pi}_T(da|x_{T-1})$$
$$> -\mathsf{P}_0(1).$$

□

The next example, based on [Bertsekas (1987), Section 6.4, Ex. 2], where Condition 2.4 is satisfied, also illustrates that the unichain assumption and Condition 2.2(a) in Theorem 2.12 are important. Moreover, it shows that Theorem 2.6 does not hold in negative models.

Example. Let $\mathbf{X} = \{1, \Delta\}$; $\mathbf{A} = [0,1]$ with the Euclidean topology; $p(\Delta|\Delta, a) \equiv 1$, $p(\Delta|1,a) = 1 - p(1|1,a) = a^2$, other transition probabilities being zero; $c(\Delta, a) \equiv 0$, $c(1,a) = -a$ (see Fig. 2.21).

"We may regard a as a demand made by a blackmailer, and state 1 as the situation where the victim complies. State Δ is the situation where the victim refuses to yield to the blackmailer's demand. The problem then can be seen as one whereby the blackmailer tries to maximize his total

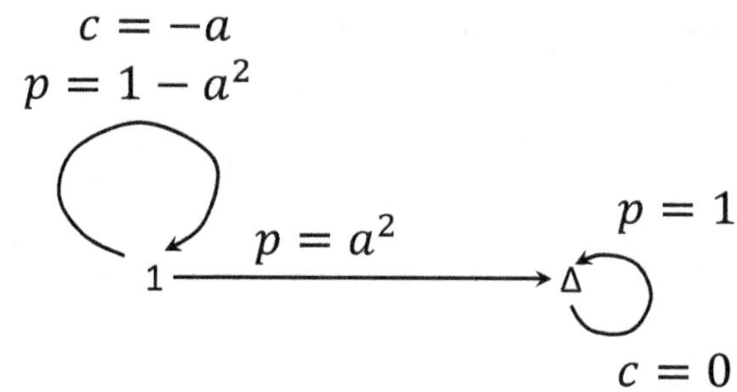

Fig. 2.21 Blackmailer's dilemma.

gain by balancing his desire for increased demands with keeping his victim compliant." [Bertsekas (1987), p. 254].

The model is negative, clearly, $v^*_\Delta = 0$, and, according to Theorem 2.2, $v^*_1 = v^\infty(1) = \lim_{n\to\infty} v^n(1)$, where

$$v^0(1) = 0;\ v^{n+1}(1) = \inf_{a\in[0,1]}\{-a + (1-a^2)v^n(1)\},\ n = 0,1,\ldots.$$

(See (2.6).) Obviously, $v^1(1) = -1$. Since $v^n(1)$ decreases with $n \uparrow \infty$, the above infimum is attained at $a = -\frac{1}{2v^n(1)}$ ($n = 1,2,\ldots$) and equals $v^n(1) + \frac{1}{4v^n(1)}$. After passing to the limit as $n \to \infty$ in equality

$$v^{n+1}(1) = v^n(1) + \frac{1}{4v^n(1)},$$

we see that $v^\infty(1) = v^*_1 = -\infty$. From another viewpoint, for the stationary selector $\varphi(x) \equiv \hat{a} > 0$, $v^\varphi_1 = -\hat{a}\mathsf{E}^\varphi_1[T_0]$, where T_0, the time to absorption in state Δ, has a geometric distribution: $\mathsf{P}^\varphi_1(T_0 = k) = (1-\hat{a}^2)^{k-1}\hat{a}^2$, $k = 1,2,\ldots$. Thus, $v^\varphi_1 = -\frac{1}{\hat{a}}$, and $\inf_{\hat{a}\in(0,1]}\left(-\frac{1}{\hat{a}}\right) = -\infty$.

Everyone stationary selector $\varphi(x) < 1$ is conserving, but $\mathsf{E}^\varphi_1[v^*_{X_T}] = -\infty$ for all $T = 0,1,2,\ldots$ because $\mathsf{P}^\varphi_1(X_T = 1) > 0$. If $\mathsf{P}_0(1) > 0$, then no-one stationary selector is optimal: for $\varphi(x) \equiv \hat{a} > 0$, $v^\varphi_{\mathsf{P}_0} = -\frac{1}{\hat{a}}\mathsf{P}_0(1) > \inf_\pi v^\pi_{\mathsf{P}_0} = -\infty$ (see the above calculations), and note that, for $\varphi(x) \equiv 0$, we have $v^\varphi_{\mathsf{P}_0} = 0$. More generally, the similar reasoning shows that no-one stationary randomized strategy is optimal if $\mathsf{P}_0(1) > 0$. We also emphasize that actions in state Δ are immaterial.

Suppose $x_0 = 1$, and consider the non-stationary selector $\varphi^*_t(x) = \sqrt{1 - e^{-1/t^2}}$. Clearly,

$$\begin{aligned} v^{\varphi^*}_1 &= c(1,\varphi^*_1(1)) + \mathsf{P}^{\varphi^*}_1(X_1 = 1)\mathsf{E}^{\varphi^*}_1\left[\sum_{t=2}^\infty c(X_{t-1},A_t)|X_1 = 1\right] \\ &= -\varphi^*_1(1) + [1-(\varphi^*_1(1))^2]\{-\varphi^*_2(1) + [1-(\varphi^*_2(1))^2]\{\cdots\}\}. \end{aligned}$$

First of all, notice that $Q := \prod_{t=1}^{\infty}[1 - \varphi_t^*(1))^2] > 0$, because

$$\sum_{t=1}^{\infty} \ln[1 - (\varphi_t^*(1))^2] = -\sum_{t=1}^{\infty} \frac{1}{t^2} < \infty.$$

Now, $v_1^{\varphi^*} \le -\varphi_1^*(1)Q - \varphi_2^*(1)Q - \ldots$ because $\prod_{t=1}^{K}[1 - \varphi_t^*(1)]^2$ decreases to Q with $K \uparrow \infty$. Therefore, $v_1^{\varphi^*} \le -Q \cdot \sum_{t=1}^{\infty} \varphi_t^*(1)$. But

$$\sum_{t=1}^{\infty} \varphi_t^*(1) = \sum_{t=1}^{\infty} \sqrt{1 - e^{-1/t^2}} = +\infty$$

because $\sum_{t=1}^{\infty} \frac{1}{t} = +\infty$ and

$$\lim_{t\to\infty} \frac{\sqrt{1 - e^{-1/t^2}}}{\frac{1}{t}} = \sqrt{\lim_{\varepsilon\to 0} \frac{1 - e^{-\varepsilon}}{\varepsilon}} = 1.$$

Therefore, $v_1^{\varphi^*} = -\infty$ and selector φ^* is uniformly optimal and optimal for each initial distribution P_0. Any actions taken in state Δ play no role. Another remark about the blackmailer's dilemma appears at the end of Subsection 4.3.3. □

In the presented examples, the *polytope* condition is satisfied: for each $x \in \mathbf{X}$ the set $\Pi(x) := \{p(0|x,a), p(1|x,a), \ldots, p(m|x,a) | a \in \mathbf{A}\}$ has a finite number of extreme points. (Here we assume that the state space $\mathbf{X} = \{0, 1, \ldots, m\}$ is finite.) It is known that in such MDPs with the *average* cost, an optimal stationary selector exists, if the model is semi-continuous [Cavazos-Cadena *et al.* (2000)]. The situation is different for MDPs with the expected total cost.

The both models investigated so far in this subsection are also weakly semi-continuous (i.e., Condition 2.3 is satisfied; the topology in $\mathbf{X}$ is discrete). The statement of Theorem 2.4 does not hold because those models are not positive.

As was explained in Section 2.2, if the statements of Lemmas 2.2 and 2.3 are valid, then there exists an optimal strategy. Moreover, that strategy can be found within the class of stationary strategies: see [Jaśkiewicz and Nowak (2019), p. 3133]. The following example shows that requirement (d) in Lemma 2.3 is important. It again confirms that Theorem 2.4 only holds for positive models.

Example. Let $\mathbf{X} = \{\Delta, 0, 1, 2, \ldots\}$, $\mathbf{A} = \left(\{\frac{1}{n}\}_{n=1}^{\infty}\right) \cup \{0\}$. The space $\mathbf{X}$ ($\mathbf{A}$) is equipped with the discrete (Euclidean) topology, so that $\mathbf{A}$ is compact.

$$p(y|0,a) = \begin{cases} 1 - \frac{1}{n+1}, & \text{if } y = 1 \text{ and } a = \frac{1}{n}; \\ \frac{1}{n+1}, & \text{if } y = n+1 \text{ and } a = \frac{1}{n}; \\ 1, & \text{if } y = 1 \text{ and } a = 0; \end{cases}$$
$$p(\Delta|x,a) \equiv 1 \text{ for } x = \Delta, 1, 2, \ldots;$$

other transition probabilities are zero. $c(x,a) \equiv -x$ for $x = 0,1,2,\ldots$; $c(\Delta,a) \equiv 0$. See Fig. 2.22 for illustration. The maximization version of this example was published in [Jaśkiewicz and Nowak (2019), Ex. 7.1].

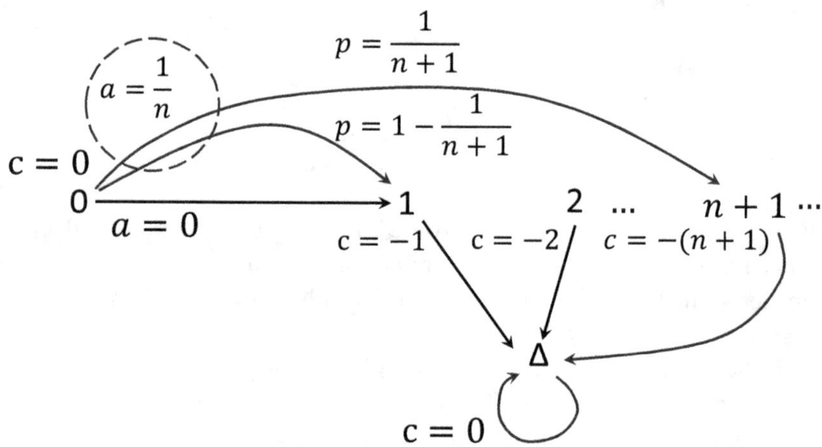

Fig. 2.22 Semi-continuous negative model.

The optimality equation (2.5) looks as follows:

$$v(\Delta) = v(\Delta);$$
$$v(0) = \min\left\{ \inf_{n=1,2,\ldots} \left\{ \left(1 - \frac{1}{n+1}\right) v(1) + \frac{1}{n+1} v(n+1) \right\},\quad v(1) \right\};$$
$$v(n) = -n + v(\Delta), \quad n = 1, 2, \ldots.$$

The maximal non-positive solution is

$$v(\Delta) = 0; \quad v(n) = -n \text{ for } n = 1, 2, \ldots; \quad v(0) = -2$$

and coincides with the Bellman function v_x^* by Theorem 2.2, see also the paragraph below Theorem 2.3.

If $\mathsf{P}_0(0) > 0$, then there are no optimal strategies: $v_0^\pi > -2 = v_0^*$ for all strategies π. In what follows, we fix the initial state $x_0 = 0$.

This MDP is (weakly) semi-continuous: Conditions 2.3 and 2.4 are satisfied and the space $\mathcal{P}_0$ is compact in the weak topology by Lemma 2.2.

Let us check the conditions in Lemma 2.3.

Condition 2.1 is satisfied because the cost function $c(\cdot,\cdot)$ is non-positive and continuous.

One can take ν as an arbitrary probability distribution on $\mathcal{B}(\mathbf{X})$ with $\nu(x) > 0$ for all $y \in \mathbf{X}$: $\delta(x,a,y) = p(y|x,a)/\nu(y)$;

$$\sum_{y\in\mathbf{X}} |\delta(x,a_k,y) - \delta(x,a,y)|\nu(y) = \sum_{y\in\mathbf{X}} |p(y|x,a_k) - p(y|x,a)| \to 0$$

for each $x \in \mathbf{X}$ if $a_k \to a$ as $k \to \infty$.

The minimal possible function $w(\cdot) : \mathbf{X} \to [0,\infty)$, which satisfies requirement (a), is $w(x) = x$ for $x = 0,1,\ldots$ and $w(\Delta) = 0$. Since

$$\sum_{y\in\mathbf{X}} w(y)p(y|0,a) \le \sup_{n=1,2,\ldots} \left\{\left(1 - \frac{1}{n+1}\right) + 1\right\} = 2,$$

requirement (b) is fulfilled for all π for $n = 1$ and obviously for all other $n = 0,2,3,\ldots$: $X_n \equiv \Delta$ for $n \ge 2$. Requirement (c) also is obviously fulfilled because $w(\Delta) = 0$. As for Item (d), let us fix $n = 1$. Then, for $l \ge 2$, we have

$$\sum_{y\in\mathbf{X}} w(y)\mathbb{I}\{w(y) \ge l\}p\left(y|0, \frac{1}{l-1}\right) = w(l)\cdot\frac{1}{l} = 1$$

leading to

$$\sup_{\pi} \mathsf{E}_0^{\pi}[w(X_1)\mathbb{I}\{w(X_1) \ge l\}] \ge 1,$$

and requirement (d) is not fulfilled. Obviously, the increase of function $w(\cdot)$ cannot make Item (d) valid. Lemma 2.3 is not applicable in this example.

The mapping (2.10) is not lower semi-continuous. To show this, consider the sequence of stationary selectors $\varphi^n(x) \equiv \frac{1}{n}$, $n = 1,2,\ldots$. We have $\mathsf{P}_0^{\varphi^n} \to \mathsf{P}_0^{\varphi^0}$ as $n \to \infty$, where $\varphi^0(x) \equiv 0$, but $v_0^{\varphi^n} = -2 + \frac{1}{n+1}$ and $v_0^{\varphi^0} = -1 > \lim_{n\to\infty} v_0^{\varphi^n} = -2$.

The considered model is uniformly absorbing at Δ. (See Definition 2.8.) Since the model is semi-continuous and the space $\mathcal{P}_0$ is weakly compact by Lemma 2.2, the space of occupation measures $\mathcal{D}_0 = \{\eta_0^{\pi}$ on $(\mathbf{X}\setminus\{\Delta\})\times\mathbf{A}$, $\pi \in \Pi^{\mathrm{All}}\}$ is weakly compact according to Theorem 2.14(a). (Recall Lemma 2.5(a) and the well known fact that the continuous image of a compact is compact.) At the same time, the mapping $\eta \to \sum_{(x,a)\in(\mathbf{X}\setminus\{\Delta\})\times\mathbf{A}} c(x,a)\eta(x,a)$ from $\mathcal{D}_0$ to $\mathbb{R}$ is not lower semi-continuous:

for the sequence of stationary selectors $\varphi^n(x) \equiv \frac{1}{n}$ as above, we have $\eta_0^{\varphi^n} \to \eta_0^{\varphi^0}$ (where $\varphi^0(x) \equiv 0$), but

$$\lim_{n\to\infty} \sum_{(x,a)\in(\mathbf{X}\setminus\{\Delta\})\times\mathbf{A}} c(x,a)\eta_0^{\varphi^n}(x,a)$$
$$= \lim_{n\to\infty} v_0^{\varphi^n} = \lim_{n\to\infty} \left\{-2 + \frac{1}{n+1}\right\} = -2$$

and $\sum_{(x,a)\in(\mathbf{X}\setminus\{\Delta\})\times\mathbf{A}} c(x,a)\eta_0^{\varphi^0}(x,a) = -1 > -2$. Theorem 2.5 and Corollary 2.1 are not applicable since the cost function $c(\cdot,\cdot)$ is not bounded below. Note that all the other conditions in Theorem 2.5 are satisfied.

Finally, this example again confirms that Theorem 2.4 fails to hold for negative models. □

2.3.17 *Not a semi-continuous model*

If the model is not semi-continuous then the existence of optimal strategies becomes more problematic. Moreover, the following example [van der Wal and Wessels (1984), Ex. 4.1] shows that no-one stationary selector is uniformly ε-optimal for $\varepsilon < 1$.

Example. Let $\mathbf{X} = \{0,1,2\}$, $\mathbf{A} = (0,1]$ with the Euclidean toology. Note that $\mathbf{A}$ is not compact. Put $p(0|0,a) = p(0|1,a) \equiv 1$, $p(1|2,a) = a$, $p(2|2,a) = 1-a$; all the other transition probabilities are zero; $c(1,a) \equiv 1$, $c(0,a) = c(2,a) \equiv 0$. See Fig. 2.23. In fact, this is a slight modification of the example in Subsection 2.3.12.

Reasoning similarly to Subsection 2.3.12 results in the Bellman function $v_0^* = v_2^* = 0$, $v_1^* = 1$. For any one stationary selector, $\varphi(2) > 0$, so that the controlled process will be absorbed at 0. Thus, $v_2^\varphi = 1 > v_2^* = 0$, and φ is not uniformly ε-optimal for $\varepsilon < 1$. Here, no-one stationary selector is conserving, i.e., providing the infimum in the optimality equation

$$v_2^* = \inf_{a\in\mathbf{A}} \{av_1^* + (1-a)v_2^*\}.$$

Recall, as was explained in Subsection 2.3.12, this MDP is not absorbing (at 0) for $\mathsf{P}_0(2) > 0$. □

Now we present a unichain absorbing model with finite space $\mathbf{X}$ which is not semi-continuous. This example is trivial but can help understanding several topological issues.

Example. Let $\mathbf{X} = \{0,1,2\}$; $\mathbf{A} = \{a^1_\infty, a^2_\infty, 1, 2, \ldots\}$; $p(0|0,a) \equiv 1$,

$$p(2|1,a) = \begin{cases} 1/2, & \text{if } a = a^1_\infty; \\ 1/7, & \text{if } a = a^2_\infty; \\ 1/2 - (1/3)^a, & \text{if } a = 1,2,\ldots, \end{cases}$$

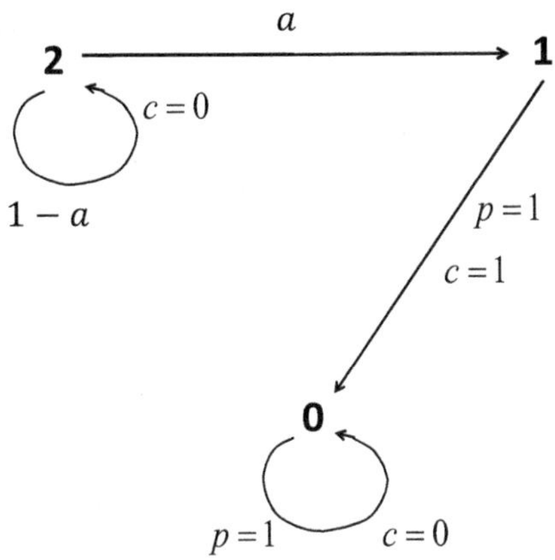

Fig. 2.23 No ε-optimal selectors for $\varepsilon < 1$.

$p(0|1,a) = 1 - p(2|1,a)$, $p(1|2,a) \equiv 1$; all the other transition probabilities are zero; $c(0,a) = c(2,a) \equiv 0$,

$$c(1,a) = \begin{cases} D, & \text{if } a = a^1_\infty; \\ -8/7, & \text{if } a = a^2_\infty; \\ -1 + 1/a, & \text{if } a = 1, 2, \dots, \end{cases}$$

where $D \in (-1, 0]$. 0 is the costless cemetery. (See Fig. 2.24).

This MDP is absorbing for any initial distribution because, under each strategy, the transition $1 \to 0$ occurs with the probability bigger (or equal) than 1/2.

This model is negative. Clearly, $v^*_0 = 0$, and, according to Theorem 2.2, $v^*_{1,2} = v^\infty(1,2) = \lim_{n\to\infty} v^n(1,2)$, where

$$\begin{aligned} v^0(1,2) &= 0; \\ v^{n+1}(2) &= v^n(1); \\ v^{n+1}(1) &= \inf_{a \in \mathbf{A}} \{-1 + 1/a + [1/2 - (1/3)^a] v^n(2);\ D + 1/2 \cdot v^n(2); \\ &\quad -8/7 + 1/7 \cdot v^n(2)\}, \quad n = 0, 1, \dots . \end{aligned}$$

(See (2.6).) The first expression in the parentheses decreases with a, so its infimum is $-1 + 1/2 \cdot v^n(2) < D + 1/2 \cdot v^n(2)$. Thus, we investigate the formula

$$v^{n+1}(1) = \min\{-1 + 1/2 \cdot v^{n-1}(1); \quad -8/7 + 1/7 \cdot v^{n-1}(1)\}$$

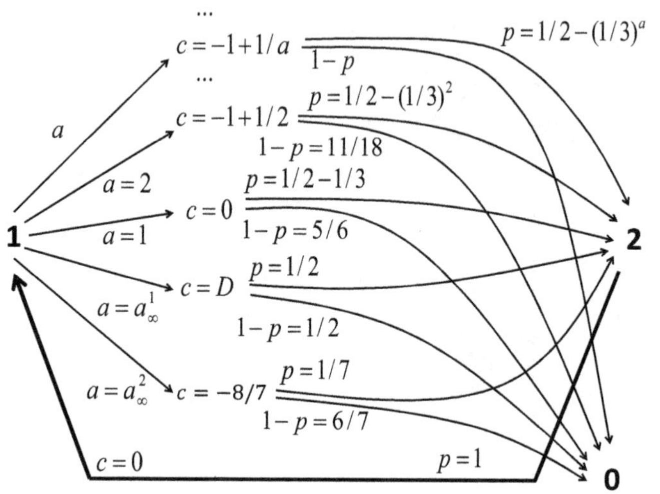

Fig. 2.24 Not a semi-continuous absorbing model.

with the initial values $v^0(1) = 0$, $v^1(1) = -8/7$.

If, for some $n \geq 1$, $v^{n,n-1}(1) \geq -2$, then

$$-1 + 1/2 \cdot v^{n-1}(1) \geq -2 \text{ and } -8/7 + 1/7 \cdot v^{n-1}(1) \geq -8/7 - 2/7 > -2.$$

Therefore, $v^\infty(1) \geq -2$, and $v^\infty(1) = v_1^*$ comes from the equation

$$v_1^* = \min\{-1 + 1/2 \cdot v_1^*; \quad -8/7 + 1/7 \cdot v_1^*\}, \quad v_1^* \geq -2.$$

Assume for a moment that $v_1^* = -8/7 + 1/7 \cdot v_1^*$. Then $v_1^* = -4/3 \geq -2$, but

$$-1 + 1/2 \cdot v_1^* = -5/3 < -8/7 + 1/7 \cdot v_1^* = v_1^* = -4/3.$$

Therefore, we have to accept that $v_1^* = -1 + 1/2 \cdot v_1^*$ leading to $v_1^* = -2$. Note that now $-8/7 + 1/7 \cdot v_1^* = -10/7 > v_1^* = -2$.

As the result, $v_0^* = 0$, $v_1^* = -2$ and $v_0^* = 0$.

No-one stationary selector φ is conserving, satisfying the optimality equation

$$v_1^* = c(1, \varphi(1)) + \sum_{y \in \mathbf{X}} v_y^* p(y|x, \varphi(1)).$$

If $\mathsf{P}_0(0) < 1$, then there are no optimal strategies in this model: in state 1, for big enough $a = 1, 2, \ldots$, action $a + 1$ is better than $b = 1, 2, \ldots, a$, a_∞^1 and a_∞^2.

The only reasonable topology in the state space $\mathbf{X}$ is the discrete one. One can introduce the topology in the action space $\mathbf{A}$ in different ways:

(a) Suppose the topology is *discrete*: all singletons are open sets, hence all subsets are open (and simultaneously closed). Then any function is certainly continuous, Conditions 2.4(b,c) are satisfied, but $\mathbf{A}$ is not compact. Thus, the model is not semi-continuous and not weakly semi-continuous, as Condition 2.3 is violated either.

(b) To make $\mathbf{A}$ compact, we can say that all singletons $\{a\}$ in $\mathbf{A}$, except for a^1_∞, are open, along with their complements $\mathbf{A}\setminus\{a\}$, and consider the coarsest topology containing all those open sets. In other words, we simply accept that sequence $1, 2, \dots$ in $\mathbf{A}$ converges to a^1_∞, or one can interpret a^1_∞ as 0, $a = i$ as $1/i$, $a = a^2_\infty$ as 2, and consider the trace on $\mathbf{A}$ of the standard topology in $\mathbb{R}^1$. Now Condition 2.4(c) is satisfied, but the cost function $c(1, a)$ is not lower semi-continuous in a because $D > -1$. Condition 2.3 fails to hold either. Note that if $D \le -1$, then this construction leads to a semi-continuous model, and $\varphi(x) \equiv a^1_\infty$ is the uniformly optimal stationary selector. In this case, Condition 2.3 is satisfied as well.

(c) We can make $\mathbf{A}$ compact in a different way by announcing that all singletons $\{a\}$ in $\mathbf{A}$, except for a^2_∞, are open, along with their complements $\mathbf{A} \setminus \{a\}$. Now Condition 2.4(b) is satisfied, but part (c) is violated: in the case when $u(2) = 1$ and $u(0) = 0$ we have $\sum_{y\in\mathbf{X}} u(y)p(y|1, a) = (1/2) - (1/3)^a$ does not converge to $\sum_{y\in\mathbf{X}} u(y)p(y|1, a^2_\infty) = 1/7$ as $a = 1, 2, \dots$ increases (equivalently, as $a \to a^2_\infty$). Hence, this topology in $\mathbf{A}$ again does not result in a semi-continuous model; Condition 2.3 fails to hold either. Like previously, if we had 1/2 instead of 1/7, then the both conditions 2.4 and 2.3 would have been satisfied, and $\varphi(x) \equiv a^2_\infty$ would have been the uniformly optimal stationary selector.

□

2.3.18 *Positive model with the non-measurable Bellman function, where $v^\infty(\cdot)$ is measurable*

As was mentioned above, many examples from Chapter 1 can be modified as infinite-horizon models: see Remark 1.1. In particular, Example 1.4.10 demonstrates that the Bellman function can be non-measurable. Below, we present the example published in [Bertsekas and Shreve (1978), Section 9.5, Ex. 2].

Example. Let Ξ be the (countable) set of finite sequences ζ of positive integers and G the set of all functions $g : \Xi \to \{0, 1\}$. Then G can be regarded as the countable Cartesian product of copies of $\{0,1\}$ indexed by $\zeta \in \Xi$. Let $\{0,1\}$ have the discrete topology and G the product topology, so G is a (topological) Borel space [Bertsekas and Shreve (1978), Prop.7.4]. The set

$$\mathcal{G} := \{g \in G : \exists \{i_1, i_2, \ldots\} \text{ with } i_k \in \{1,2,\ldots\},\ k = 1,2,\ldots, \quad (2.32)$$
$$\text{such that } g(\{i_1, i_2, \ldots, i_K\}) = 0 \text{ for all } K = 1,2,\ldots\} \subset G$$

is known to be analytic, but not (Borel)-measurable [Bertsekas and Shreve (1978), p.233].

Let the state space be $\mathbf{X} := G \times (\Xi \cup \{0\})$ and put $\mathbf{A} := \{1,2,\ldots\}$. The topology in $(\Xi \cup \{0\})$ and in $\mathbf{A}$ is discrete, and $\mathbf{X}$ is equipped with the product topology, hence it is a Borel space. The transitions are deterministic: if $x_{t-1} = (g_{t-1}, \zeta_{t-1}) \in \mathbf{X}$ and $a_t \in \mathbf{A}$ is selected, then $x_t = f(x_{t-1}, a_t) := (g_{t-1}, \zeta_t)$ with

$$\zeta_t := \begin{cases} \{a_t\}, & \text{if } \zeta_{t-1} = 0; \\ \zeta_{t-1}, a_t, & \text{if } \zeta_{t-1} \in \Xi. \end{cases}$$

Here and below, with some abuse of notation, $\zeta, a := \{i_1, i_2, \ldots, i_K, a\}$ for $\zeta = \{i_1, i_2, \ldots, i_K\}$.

The cost function is given by

$$c(x = (g,\zeta), a) := \begin{cases} g(\{a\}), & \text{if } \zeta = 0; \\ g(\zeta, a), & \text{if } \zeta \in \Xi. \end{cases}$$

See Fig. 2.25.

Obviously, there exists $x \in \mathbf{X}$ such that the Bellman function $v_x^* = 0$: e.g., if $x = (g^0, \hat{\zeta})$ with $g^0(\zeta) \equiv 0$ for all $\zeta \in \Xi$, and arbitrarily fixed $\hat{\zeta} \in \Xi \cup \{0\}$.

Suppose the Bellman function v_x^* is (Borel)-measurable. Then the set $\{(g,0) : g \in G,\ v^*_{(g,0)} = 0\} = \{x \in \mathbf{X} :\ v_x^* = 0\} \cap \{x \in \mathbf{X} :\ x = (g,0),\ g \in G\}$ must be measurable. But the set $\{g \in G :\ v^*_{(g,0)} = 0\}$ coincides with $\mathcal{G}$ (see (2.32)) because there must be such a_1 that $c((g,0), a_1) = g(\{a_1\}) = 0$ and, on each step $t > 1$, there must be a possibility to choose a_t with $g(\zeta_{t-1}, a_t) = 0$. Since $\mathcal{G}$ is not measurable, the Bellman function v_x^* is not measurable.

At the same time, $v^0(\cdot) \equiv 0$ is certainly a measurable function on $\mathbf{X}$ and, if, for some $n \geq 0$, $v^n(\cdot)$ is a measurable function on $\mathbf{X}$, then

$$v^{n+1}(x) = \inf_{a \in \mathbf{A}} \{c(x,a) + v^n(f(x,a))\}, \quad x \in \mathbf{X}$$

is a measurable function on $\mathbf{X}$ because $c(\cdot,\cdot)$ and $f(\cdot)$ are measurable mappings, and $\mathbf{A}$ is countable. Thus, $v^\infty(\cdot) = \lim_{n\to\infty} v^n(\cdot) : \mathbf{X} \to [0,\infty]$ is measurable. In this positive model $v^\infty(x) \leq v_x^*$, $x \in \mathbf{X}$, but inequality is strict for some x: see (2.8). □

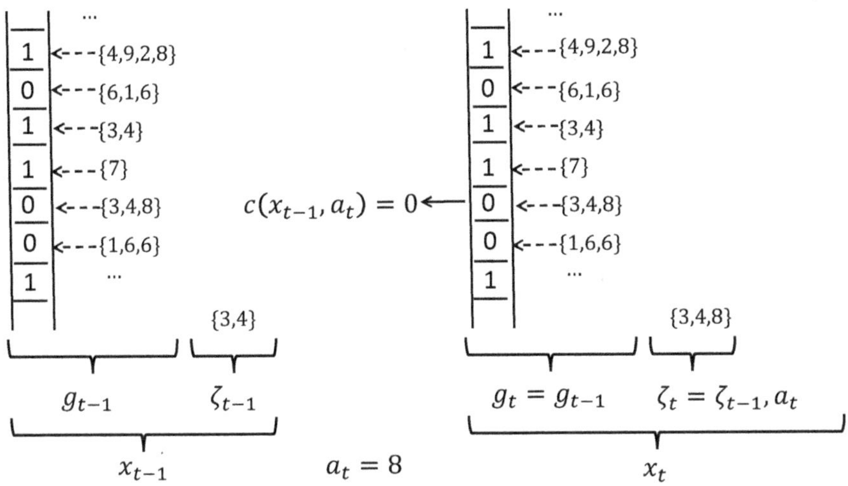

Fig. 2.25 Example of transition and cost.

2.3.19 *Randomized strategy is better than any selector (finite action space)*

The following example, published in [Bertsekas and Shreve (1978), Section 9.6, Ex. 3] shows that the statement of Theorem 2.6 does not hold for negative models.

Example. Let $\mathbf{X} = \{0, 1, 2, \ldots\}$; $\mathbf{A} = \{1, 2\}$; $p(0|0, a) \equiv 1$, $p(0|x, 1) \equiv 1$ for all $x \geq 0$, $p(x+1|x, 2) \equiv 1$ for all $x > 0$; all the other transition probabilities are zero; $c(0, a) \equiv 0$, $c(x, 1) = -2^x$ and $c(x, 2) \equiv 0$ for $x > 0$ (see Fig. 2.26).

Obviously, $v_0^* = 0$ and, for $x > 0$, $v_x^* = -\infty$ because $v_x^* \leq -2^x$; hence (if one applies action 2 in state x) $v_x^* \leq -2^{x+1}$; hence (if one applies action 2 in state $x+1$) $v_x^* \leq -2^{x+2}$, and so on. One can also use Theorem 2.2 to build $v^\infty(x) = v_x^*$ using value iteration (2.6).

The only conserving stationary selector is $\varphi^*(x) \equiv 2$, but it is not equalizing and not optimal, since $v_x^{\varphi^*} \equiv 0$.

If $\varphi_t(h_{t-1})$ is an arbitrary selector and $x_0 > 0$ is the initial state then

- either $\varphi_t(x_0, a_1, \ldots, x_{t-1}) \equiv 2$ for all $t > 0$, so that $v_{x_0}^\varphi = 0$,
- or $\varphi_t(x_0, a_1, \ldots, x_{t-1}) = 1$ for some (first) value of $t \geq 1$, in which case
  $v_{x_0}^\varphi = -2^{-(x_0+t-1)} > -\infty.$

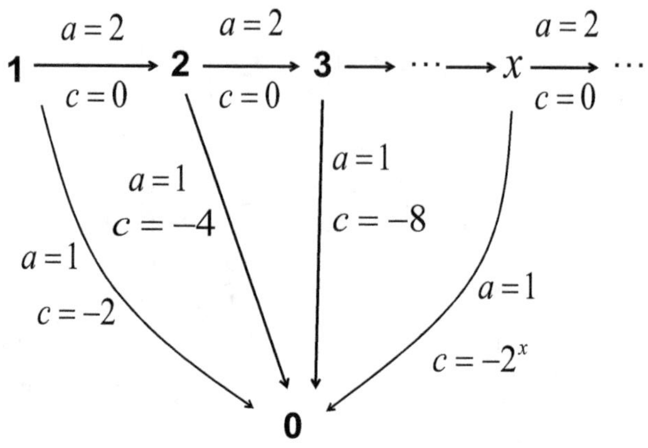

Fig. 2.26 Only a randomized strategy is (uniformly) optimal.

Thus, no-one selector is uniformly optimal.

In the case when $\mathsf{P}_0(x) = \left(\frac{1}{2}\right)^x$ for $x > 0$, the stationary selector $\varphi(x) \equiv 1$ is optimal (but certainly not uniformly optimal):

$$v^{\varphi}_{\mathsf{P}_0} = \sum_{x=1}^{\infty} \mathsf{P}_0(x)(-2^x) = -\infty.$$

Similarly, the randomized stationary strategy $\pi^*(1|x) = \pi^*(2|x) := \frac{1}{2}$ provides, for each $x > 0$,

$$\begin{aligned} v^{\pi^*}_x &= -2^{x-1} + \frac{1}{2} v^{\pi^*}_{x+1} \\ &= -2^{x-1} - 2^{x-1} + \left(\frac{1}{4}\right) v^{\pi^*}_{x+2} = \cdots = -k2^{x-1} + \left(\frac{1}{2}\right)^k v^{\pi^*}_{x+k}, \\ & k = 1, 2, \ldots \end{aligned}$$

Passing to the limit as $k \to \infty$, we see that $v^{\pi^*}_x = -\infty$ (note that $v^{\pi^*}_y \le 0$). Therefore, the π^* strategy is uniformly optimal. It is also optimal for any initial distribution P_0:

$$v^{\pi^*}_{\mathsf{P}_0} = \int_{\mathbf{X}} v^{\pi^*}_x \mathsf{P}_0(dx) = \begin{cases} -\infty, & \text{if } P_0(0) < 1; \\ 0, & \text{if } \mathsf{P}_0(0) = 1. \end{cases}$$

□

At the end of Subsection 2.3.4, one can find another example of a stationary (randomized) strategy π^{s} such that, for any stationary selector φ, there is an initial state $\hat{x}$, for which $v^{\pi^{\mathrm{s}}}_{\hat{x}} < v^{\varphi}_{\hat{x}}$.

2.3.20 *Occupation measures: phantom solutions*

The following example shows that equation (2.15) can have many non-minimal (phantom) solutions.

Example. Let $\mathbf{X} = \{0, 1, 2, \dots\}$, $\mathbf{A} = \{0\}$ (a dummy action). In reality, the model under consideration is uncontrolled, as there exists only one control strategy π^{s}. We put

$$p(0|0,a) \equiv 1, \qquad \forall x > 0 \;\; p(y|x,a) = \begin{cases} p_+, & \text{if } y = x+1; \\ p_-, & \text{if } y = x-1; \\ 0, & \text{otherwise,} \end{cases}$$

where $p_+ + p_- = 1$, $p_+ < p_-$, $p_+, p_- \in [0,1]$ are arbitrary numbers. The cost function does not play any role. See Fig. 2.27.

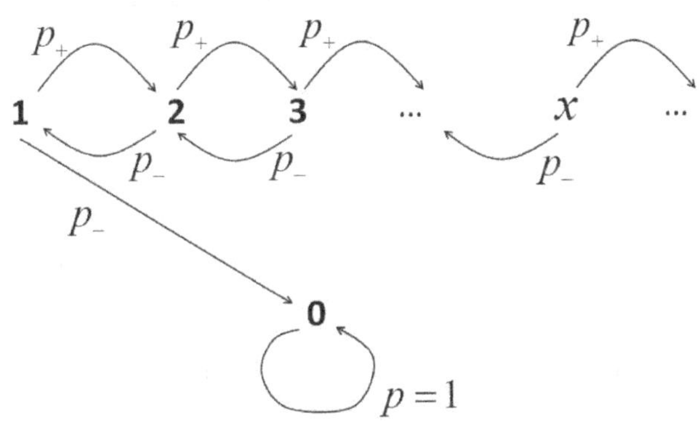

Fig. 2.27 Phantom solutions to equation (2.15).

This model is known to be absorbing at 0 when starting from each initial state: see Remark 2.5. To give more details, assume that $\mathsf{P}_0(1) = 1$ and $p_+ > 0$. (The case $p_+ = 0$ is trivial.) Approximations (2.16) on $\mathbf{X} \setminus \{0\} = \{1, 2, \dots\}$ look as follows:

$$\begin{aligned} \hat{\eta}_0(x) &= 0, \quad x = 1, 2, \dots; \\ \hat{\eta}_{i+1}(x) &= p_+\hat{\eta}_i((x-1)) + p_-\hat{\eta}_i(x+1), \qquad x > 1; \\ \hat{\eta}_{i+1}(1) &= 1 + p_-\hat{\eta}_i(2). \end{aligned}$$

For all $x = 1, 2, \ldots$, $\hat{\eta}_0(x) \le \frac{1}{p_+}\left(\frac{p_+}{p_-}\right)^x$ and, if $\hat{\eta}_i(x) \le \frac{1}{p_+}\left(\frac{p_+}{p_-}\right)^x$, then

$$\hat{\eta}_{i+1}(x) \le p_+\left[\frac{1}{p_+}\left(\frac{p_+}{p_-}\right)^{x-1}\right] + p_-\left[\frac{1}{p_+}\left(\frac{p_+}{p_-}\right)^{x+1}\right] = \frac{1}{p_+}\left(\frac{p_+}{p_-}\right)^x, \quad x > 1;$$

$$\hat{\eta}_{i+1}(1) \le 1 + p_-\left[\frac{1}{p_+}\left(\frac{p_+}{p_-}\right)^2\right] = \frac{1}{p_+}\left(\frac{p_+}{p_-}\right).$$

Therefore, $\hat{\eta}_1^{\pi^s}(x) \le \frac{1}{p_+}\left(\frac{p_+}{p_-}\right)^x$ for all $x = 1, 2, \ldots$, and the expected time to absorption

$$\mathsf{E}_1^{\pi^s}[T_0] = \eta_1^{\pi^s}((\mathbf{X} \setminus \{0\}) \times \mathbf{A}) = \hat{\eta}_1^{\pi^s}(\mathbf{X} \setminus \{0\}) \le \frac{1}{p_+}\sum_{x=1}^{\infty}\left(\frac{p_+}{p_-}\right)^x < \infty.$$

Here, as usual, $T_0 := \min\{t > 0 : X_t = 0\}$. Hence, the model is absorbing for $\mathsf{P}_0(1) = 1$. State 1 communicates with any other state $x > 1$, $\mathsf{E}_x^{\pi^s}[T_1] < \infty$ for $T_1 := \min\{t \ge 0 : X_t = 1\}$. Thus, the model is absorbing for $\mathsf{P}_0(x) = 1$ with $x > 1$, as well.

Equation (2.15) on $\mathbf{X} \setminus \{0\}$ (equivalently, equation (2.13)) is expressed as follows:

$$\hat{\eta}(x) = \mathsf{P}_0(x) + p_+\hat{\eta}(x-1) + p_-\hat{\eta}(x+1), \qquad x > 1;$$
$$\hat{\eta}(1) = \mathsf{P}_0(1) + p_-\hat{\eta}(2).$$

For $\mathsf{P}_0(1) = 1$, any function η_x of the form

$$\eta_x = d\left[1 - \left(\frac{p_+}{p_-}\right)^x\right] + \frac{1}{p_+}\left(\frac{p_+}{p_-}\right)^x, \quad x \ge 1$$

provides a solution, and the minimal non-negative solution corresponds to $d = 0$. (Cf the above presented approximations). Negative values of d result in $\eta_x < 0$ for large values of x.

Remark 2.5. One can recognize that the investigated model is a special case of the example from Subsection 2.3.2, where one should put $\lambda + \mu = 1$ and denote $p_+ = \lambda$ and $p_- = \mu$. Now $v_x^* = \frac{x}{p_- - p_+}$. If $x = 1$, then v_1^* is the total expected time to absorption at 0 when starting from 1. It equals the total value of the occupation measure on $\mathcal{B}(\mathbf{X} \setminus \{0\})$:

$$\sum_{x=1}^{\infty}\eta_x = \frac{1}{p_+}\frac{\frac{p_+}{p_-}}{1 - \frac{p_+}{p_-}} = \frac{1}{p_- - p_+}.$$

As was mentioned in Subsection 2.3.2, this model is absorbing at 0 for each initial state provided $p_+ < p_-$, but it is not absorbing for such initial distributions that $\sum_{x=1}^{\infty} x\mathsf{P}_0(x) = \infty$.

Now put $p_+ = 0$; $p_- = 1$. Then equation (2.15) on $\mathbf{X} \setminus \{0\}$ takes the form

$$\hat{\eta}(x) = \mathsf{P}_0(x) + \hat{\eta}(x+1), \qquad x \geq 1.$$

The minimal non-negative solution is given by

$$\hat{\eta}(x) = \sum_{i=x}^{\infty} \mathsf{P}_0(i), \qquad x \geq 1,$$

which is the unique finite solution: $\hat{\eta}(\mathbf{X} \setminus \{0\}) < \infty$. At the same time, one can obviously add any constant to this solution, and equation (2.15) remains satisfied. □

Definition 2.4. In an absorbing model with cemetery 0, a measure μ on $\mathcal{B}((\mathbf{X} \setminus \{0\}) \times \mathbf{A})$ is called *invariant* if

$$\mu(\Gamma \times \mathbf{A}) = \int_{(\mathbf{X}\setminus\{0\})\times\mathbf{A}} p(\Gamma|y,a) d\mu(y,a), \quad \Gamma \in \mathcal{B}(\mathbf{A}).$$

Proposition 2.2. [Dufour and Prieto-Rumeau (2024a), Prop. 3.8]. *Suppose the model, under the initial distribution* P_0*, is absorbing with cemetery 0. Then the following assertions hold.*

(a) If μ is a finite invariant measure, then, for every strategy π, the measure $\eta^{\pi}_{\mathsf{P}_0}$ is such that $\eta^{\pi}_{\mathsf{P}_0} \perp \mu$ on $\mathcal{B}((\mathbf{X} \setminus \{0\}) \times \mathbf{A})$, i.e., for all $\Gamma_X \in \mathcal{B}(\mathbf{X} \setminus \{0\})$, $\Gamma_A \in \mathcal{B}(\mathbf{A})$,

$$\eta^{\pi}_{\mathsf{P}_0}(\Gamma_X \times \Gamma_A) \times \mu(\Gamma_X \times \Gamma_A) = 0.$$

(b) For every finite on $\mathcal{B}((\mathbf{X} \setminus \{0\}) \times \mathbf{A})$ measure η satisfying equation (2.13), there exist a unique occupation measure $\eta^{\pi^{\mathrm{s}}}_{\mathsf{P}_0}$, generated by some stationary strategy π^{s}, and a unique finite invariant measure μ, such that $\eta = \eta^{\pi^{\mathrm{s}}}_{\mathsf{P}_0} + \mu$.

The examples in the current subsection show that here the requirement for the invariant measure to be finite is important.

Let us present a simple example illustrating Proposition 2.2.

Example. Put $\mathbf{X} = \{0, 1, 2\}$, $\mathbf{A}$ is an arbitrary Borel space, $\mathsf{P}_0(1) = 1$,

$$p(0|0,a) = p(0|1,a) = p(2|2,a) \equiv 1,$$

other transition probabilities being zero. See Fig. 2.28.

Equation (2.13) takes the form

$$\eta(1 \times \mathbf{A}) = 1 + 0;$$
$$\eta(2 \times \mathbf{A}) = 0 + \eta(2 \times \mathbf{A}).$$

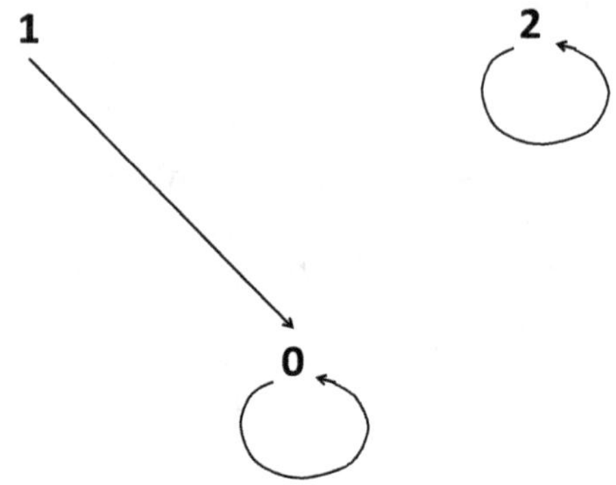

Fig. 2.28 Illustration to Proposition 2.2.

For each control strategy π,

$$\eta^\pi_{\mathsf{P}_0}(x\times\mathbf{A}) = \begin{cases} 1, & \text{if } x=1; \\ 0, & \text{if } x=2; \\ \infty, & \text{if } x=0, \end{cases}$$

and a measure μ is finite and invariant if and only if $\mu(2\times\mathbf{A})=k$ for some $k\ge 0$ and $\mu(1\times\mathbf{A})=0$. The both parts of Proposition 2.2 are obviously valid. □

2.3.21 *Occupation measures in transient models*

The following example [Feinberg and Sonin (1996), Ex. 4.3] shows that if π^{s} is a stationary strategy defined by (2.14) for a σ-finite occupation measure $\eta^\pi_{\mathsf{P}_0}$ in a transient MDP, then it can happen that $\eta^{\pi^{\mathrm{s}}}_{\mathsf{P}_0}\neq\eta^\pi_{\mathsf{P}_0}$. One can only claim that $\hat\eta^{\pi^{\mathrm{s}}}_{\mathsf{P}_0}(x)\le\hat\eta^\pi_{\mathsf{P}_0}(x)$, for all $x\in\mathbf{X}$, where, as usual, $\hat\eta^\pi_{\mathsf{P}_0}(x):=\eta^\pi_{\mathsf{P}_0}(x\times\mathbf{A})$ is the marginal (see [Altman (1999), Thm. 8.1]). Thus, conditions in Lemma 2.4 (absorbing model and finiteness of $\eta^\pi_{\mathsf{P}_0}$) are important.

Example. Let $\mathbf{X}=\{0,1,2,\ldots\}$, $\mathbf{A}=\{f,b\}$. State 0 is absorbing: $p(0|0,a)\equiv 1$, $c(0,a)\equiv 0$. The model will be transient, not absorbing, i.e. for each control strategy, formula (2.17) holds, but condition $\mathsf{E}^\pi_{\mathsf{P}_0}[T_0]<\infty$ is not guaranteed for $T_0:=\min\{t>0:\ X_t=0\}$. We put

$$p(x-1|x,b)=\gamma_x \quad \text{and } p(0|x,b)=1-\gamma_x \quad \text{for } x\ge 2;$$

$$p(2|1,b)=\gamma_2; \qquad p(0|1,b)=1-\gamma_2;$$

$$p(x+1|x,f) = \gamma_{x+1} \qquad \text{and } p(0|x,f) = 1-\gamma_{x+1} \quad \text{for } x \geq 1.$$

Here $0 < \gamma_x \leq \gamma_{x+1} < 1$, $x = 1,2,3,\ldots$ are numbers such that

$$\gamma_1 \prod_{j=2}^{\infty} \gamma_j^{2^j+1} > 0.9. \tag{2.33}$$

$\gamma_1 \in (0,1)$ does not appear in the description of this MDP, but is needed in the future. Other transition probabilities are zero (see Fig. 2.29). The cost function does not play any role. We assume that $\mathsf{P}_0(x) = \mathbb{I}\{x=1\}$.

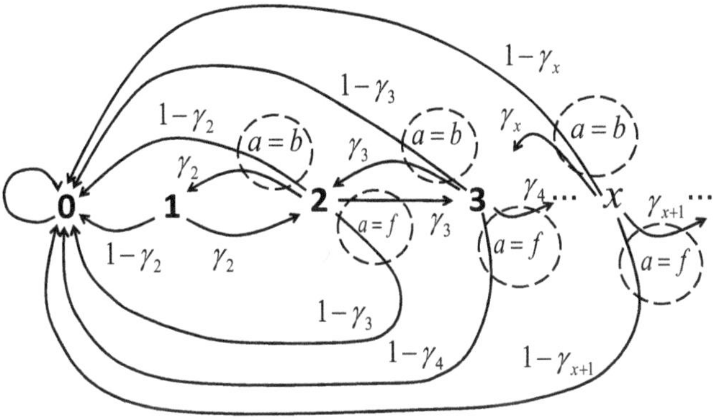

Fig. 2.29 $\eta_{\mathsf{P}_0}^{\pi^{\mathrm{s}}} \neq \eta_{\mathsf{P}_0}^{\pi}$ for π^{s} given by equation (2.14).

First of all, this MDP is transient, since for each strategy π and for each state $x \in \mathbf{X} \setminus \{0\}$, the probability of returning to state x is bounded above by γ_{x+1}, so that

$$\hat{\eta}_1^{\pi}(x) = \eta_1^{\pi}(x \times \mathbf{A}) \leq 1 + \gamma_{x+1} + \gamma_{x+1}^2 + \cdots = \frac{1}{1-\gamma_{x+1}} < \infty.$$

We consider the following control strategy π:

$$\pi_t(b|x_0, a_1, \ldots, x_{t-1}) = \begin{cases} 1, & \text{if } m = \sum_{n=0}^{t-1} I\{x_n = x_{t-1}\} \leq 2^{x_{t-1}-1}; \\ 0, & \text{otherwise}; \end{cases}$$

$$\pi_t(f|x_0, a_1, \ldots, x_{t-1}) = 1 - \pi_t(b|x_0, a_1, \ldots, x_{t-1}).$$

In words, if the process visits state $j \geq 1$ for the mth time, then the strategy π selects action b if $m \leq 2^{j-1}$, and action f otherwise.

For this strategy, starting from $x_0 = 1$, the controlled process will never be absorbed in 0 if the following events realize.

(a) $X_1 = 2$ with probability γ_2.
(b) The cycles $2 \to 1 \to 2 \to 1 \to 2$ (under the action b in state 2 applied $2^{2-1} = 2$ times) are realized; the probability is $(\gamma_2^2)^2$. Combining with (a), we obtain the probability $\gamma_2 \cdot \gamma_2^4 = \gamma_2^{s^2+1}$.
(c) After that, the action f is applied in state 2, and the next state is 3 with probability γ_3.
(d) The cycles $3 \to 2 \to \ldots \to 3$ (under the action b in state 3 applied $2^{3-1} = 4$ times) are realized; the probability is $(\gamma_3^2)^4$. Combining with (c), we obtain the probability $\gamma_3 \cdot \gamma_3^8 = \gamma_3^{2^3+1}$.
(e) And so on.

Therefore, according to (2.33),

$$\mathsf{P}_1^\pi(T_0 = \infty) = \prod_{j=2}^{\infty} \gamma_j^{2^j+1} > 0.9,$$

where $T_0 := \min\{t > 0 : \ X_t = 0\}$. We have proved that the model is not absorbing.

Note that

- state 1 can appear not more than three times: $x_0 = 1$, $X_2 = 1$, $X_4 = 1$; after that, the action f is applied in state 2.
- state $x \geq 2$ can appear not more than $1 + 2^{x-1} + 2^x$ times: first time as the transition $x - 1 \to x$, plus 2^{x-1} times after action b in state x (and action f in state $x - 1$) is in use, plus 2^x times after action b in state $x + 1$ (and action f in state x) is in use.

Therefore, $\hat{\eta}_1^\pi(x) \leq 3 \cdot 2^{x-1} + 1$ for $x \geq 2$. If $T_0 = \infty$, then state $x \geq 2$ appears exactly $3 \cdot 2^{x-1} + 1$ times. Hence, for $x \geq 2$,

$$\begin{aligned} 3 \cdot 2^{x-1} + 1 \geq \hat{\eta}_1^\pi(x) &\geq \mathsf{P}_1^\pi(T_0 = \infty) \mathsf{E}_1^\pi \left[\sum_{t=0}^{\infty} \mathbb{I}\{X_t = x\} | T_0 = \infty \right] \quad (2.34) \\ &= \mathsf{P}_1^\pi(T_0 = \infty) \cdot [3 \cdot 2^{x-1} + 1] > 0.9[3 \cdot 2^{x-1} + 1]. \end{aligned}$$

The marginal occupation measure $\hat{\eta}_1^\pi$ is σ-finite on $\{1, 2, \ldots\}$.

Note also that, under the condition $T_0 = \infty$, action f is applied in state $x \geq 2$ exactly $1 + 2^x$ times: first time after the last transition $x - 1 \to x$ plus $2^{(x+1)-1}$ times after the transitions $x + 1 \to x$ under action b. (After that, action f is applied in state $x + 1$, and state x will never be visited.) By the way, if $T_0 = \infty$, then in state $x \geq 2$ action b is applied 2^{x-1} times, and the state x appears $2^{x-1}+2^x+1=3\cdot 2^{x-1}+1$ times in total. Now

$$\begin{aligned}\eta_1^\pi(x\times f) &\geq \mathsf{P}_1^\pi(T_0=\infty)\mathsf{E}_1^\pi\left[\sum_{t=0}^{\infty}\mathbb{I}\{X_t=x,A_{t+1}=f\}|T_0=\infty\right]\\ &= \mathsf{P}_1^\pi(T_0=\infty)[2^x+1],\quad x\geq 2,\end{aligned}$$

and, using (2.34), we conclude that the stationary control strategy $\pi^{\rm s}$ from (2.14) satisfies inequality

$$\pi^{\rm s}(f|x)=\frac{\eta_1^\pi(x\times f)}{\hat\eta_1^\pi(x)}\geq\frac{2^x+1}{3\cdot 2^{x-1}+1}\mathsf{P}_1^\pi(T_0=\infty),\quad x\geq 2.$$

Similar calculations show that

$$\pi^{\rm s}(b|x)=\frac{\eta_1^\pi(x\times b)}{\hat\eta_1^\pi(x)}\geq\frac{2^{x-1}}{3\cdot 2^{x-1}+1}\mathsf{P}_1^\pi(T_0=\infty),\quad x\geq 2.$$

Below, we use the notation $\lambda_x:=\gamma_{x+1}\pi^{\rm s}(f|x)$, $\mu_x:=\gamma_x\pi^{\rm s}(b|x)$ for $x\geq 2$, and $\lambda_1:=\gamma_2$ for brevity. Note that $0<\lambda_x<1$ for all $x\geq 1$ and

$$0<\mu_x=\gamma_x(1-\pi^{\rm s}(f|x))\leq\gamma_{x+1}(1-\pi^{\rm s}(f|x)<1-\lambda_x,\quad x\geq 2.$$

Moreover, for $x\geq 2$,

$$\lambda_x\geq\gamma_1\pi^{\rm s}(f|x)>\frac{2^x+1}{3\cdot 2^{x-1}+1}\cdot 0.9>\lim_{x\to\infty}\frac{2^x+1}{3\cdot 2^{x-1}+1}\cdot 0.9=\frac{2}{3}\cdot 0.9=\frac{3}{5}$$

and

$$\mu_x<1-\lambda_x<\frac{2}{5},$$

so that

$$\sum_{j=2}^{\infty}\frac{\mu_2\mu_3\dots\mu_j}{\lambda_2\lambda_3\dots\lambda_j}<\sum_{j=1}^{\infty}\left(\frac{2}{3}\right)^j<\infty.$$

Now, according to [Altman (1999), Lemma 7.1], the occupation measure $\hat\eta_1^{\pi^{\rm s}}(x)$ is the minimal non-negative solution to equation (2.15), which takes the form

$$\begin{aligned}&\hat\eta_1^{\pi^{\rm s}}(1)=1+\mu_2\hat\eta_1^{\pi^{\rm s}}(2);\\ &\hat\eta_1^{\pi^{\rm s}}(x)=\lambda_{x-1}\hat\eta_1^{\pi^{\rm s}}(x-1)+\mu_{x+1}\hat\eta_1^{\pi^{\rm s}}(x+1),\qquad x\geq 2.\end{aligned}$$

According to Lemma A.2,

$$\hat\eta_1^{\pi^{\rm s}}(1)\leq 1+\sum_{j=2}^{\infty}\left(\frac{\mu_2\cdots\mu_j}{\lambda_2\cdots\lambda_j}\right);$$

$$\hat\eta_1^{\pi^{\rm s}}(x)\leq\frac{1}{\lambda_x}\left[1+\sum_{j=1}^{\infty}\frac{\mu_{x+1}\dots\mu_{x+j}}{\lambda_{x+1}\dots\lambda_{x+j}}\right]<\frac{1}{3/5}\left[1+\sum_{j=1}^{\infty}\left(\frac{2}{3}\right)^j\right]=5,\quad x\geq 2.$$

But, according to (2.34), the function $\hat\eta_1^\pi(\cdot)$ is unbounded from above and $\hat\eta_1^\pi(x)>0.9\cdot 7=6.3>\hat\eta_1^{\pi^{\rm s}}(x)$ for $x\geq 2$. □

2.3.22 *Occupation measures and proper strategies*

In this subsection, we consider MDPs with a discrete state space and with the costless cemetery Δ.

Definition 2.5. [Bertsekas (2018), (7)]. A Markov strategy π^{m} is called *proper* at $x_0 \in \mathbf{X}$, if

$$v_{x_0}^{\pi^{\mathrm{m}}} = \mathsf{E}_{x_0}^{\pi^{\mathrm{m}}}\left[\sum_{t=1}^{\infty} c(X_{t-1}, A_t)\right] < \infty \quad \text{and } \eta_{x_0}^{\pi^{\mathrm{m}}}((\mathbf{X} \setminus \{\Delta\}) \times \mathbf{A}) < \infty.$$

The set of all proper at $x_0 \in \mathbf{X}$ strategies is denoted as $\Pi_{x_0}^p$.

If the cost function $c(\cdot, \cdot) \geq \delta > 0$ is separated from zero, then, for any fixed $x \in \mathbf{X}$, one can take into account only the proper at x strategies when solving problem (2.1). In particular, $\hat{v}_x := \inf_{\pi \in \Pi_x^p} v_x^\pi = v_x^*$, the function $\hat{v}_x$ satisfies optimality equation (2.5), and, if a stationary selector φ^* is such that

$$\hat{v}_x = c(x, \varphi^*(x)) + \sum_{y \in \mathbf{X}} \hat{v}_y p(y|x, \varphi^*(x)), \quad \forall x \in \mathbf{X}, \tag{2.35}$$

then φ^* is uniformly optimal [Puterman (1994), Thm.7.1.7]. Moreover, if a stationary selector φ^* satisfies equality (2.35) and $\varphi^* \in \Pi^p := \bigcap_{x \in \mathbf{X}} \Pi_x^p$, then

$$v_x^{\varphi^*} = \hat{v}_x = \inf_{\pi \in \Pi^p} v_x^\pi \qquad \forall x \in \mathbf{X}.$$

Properties of proper strategies and function $\inf_{\pi \in \Pi^p} v_x^\pi$ in a more general situation were investigated in [Bertsekas (2018)].

Below, we also need the 'perturbed' Bellman function

$$V^\delta(x) := \inf_{\pi \in \Pi^{\mathrm{All}}} \{v_x^\pi + \delta \eta_x^\pi((\mathbf{X} \setminus \{\Delta\}) \times \mathbf{A})\}, \quad \delta > 0, \ x \in \mathbf{X}.$$

Proposition 2.3. [Bertsekas (2020), Props. 5 and 6]. *Suppose the model is positive and either there is $\delta > 0$ such that*

$$c(x, a) + \delta + \sum_{y \in \mathbf{X}} V^\delta(y) p(y|x, a) < \infty \ \ \forall x \in \mathbf{X}, \ a \in \mathbf{A}, \tag{2.36}$$

or the number of elements of the set $Y_{x,a} := \{y : \ p(y|x, a) > 0\}$ is uniformly bounded. Then the following assertions hold.

- *Function $\hat{v}_x$ satisfies optimality equation (2.5).*
- *If a stationary selector belongs to Π^p and satisfies equality (2.35), then*

$$v_x^{\varphi^*} = \hat{v}_x = \inf_{\pi \in \Pi^p} v_x^\pi \qquad \forall x \in \mathbf{X}.$$

In what follows, we present an example, which demonstrates that requirement (2.36) (or boundedness of $Y_{x,a}$), which was missing in [Bertsekas (2018)], is important. A very similar example appeared in [Bertsekas (2020)].

Example. Let $\mathbf{X} = \{\Delta, 0, 1, \ldots\}$ and $\mathbf{A} = \{1, 2\}$; $p(\Delta|\Delta, a) = p(\Delta|1, a) \equiv 1$; $p(x-1|x, a) \equiv 1$ for $x > 1$; $p(y|0, 1) = p_y$, where the probability distribution p_y on $\{1, 2, \ldots\}$ is such that $\sum_{y=1}^{\infty} y p_y = \infty$; $p(0|0, 2) = p(\Delta|0, 2) = 1/2$. All the other transition probabilities are zero. Finally, $c(x, a) = \mathbb{I}\{x = 0, a = 2\}$. See Fig. 2.30 for illustration.

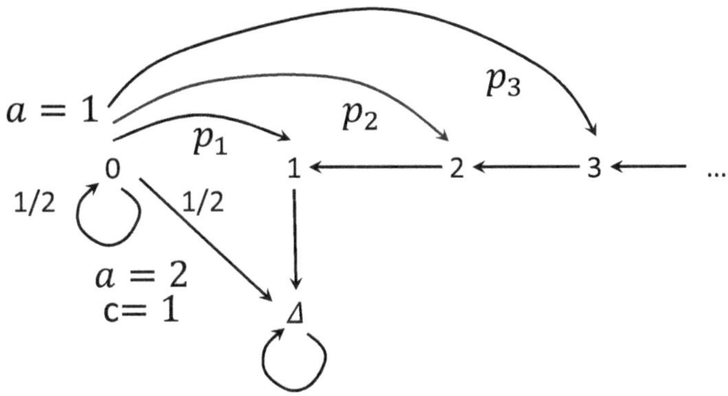

Fig. 2.30 Proper strategies.

All (Markov) strategies are proper at all $x_0 \neq 0$, but only the strategy π^{m} satisfying $\pi^{\mathrm{m}}_t(1|0) \equiv 1$ is proper at $x_0 = 0$. Therefore,

$$\hat{v}_x = 0 \quad \text{for } x \neq 0 \qquad \text{and } \hat{v}_0 = 1,$$

and function $\hat{v}_x$ does not satisfy equation (2.5) at $x = 0$:

$$\hat{v}_0 = 1 \neq \min_{a \in \mathbf{A}} \left\{ c(0, a) + \sum_{y \in \mathbf{X}} \hat{v}_y p(y|0, a) \right\}$$
$$= c(0, 1) + \sum_{y \in \mathbf{X}} \hat{v}_y p(y|0, 1) = 0.$$

Conditions in Proposition 2.3 are violated: $V^{\delta}(x) = \delta x$ for $x > 0$, so that, if $x = 0$ and $a = 1$, then

$$c(0,1)+\delta+\sum_{y\in\mathbf{X}}V^\delta(y)p(y|0,1)=\delta+\delta\sum_{y=1}^{\infty}yp_y=\infty.$$

□

Another simple example shows that, even if the statement of Proposition 2.3 is valid and the model is positive, then it can happen that $\hat{v}_x \neq v^*_x$ for some $x \neq \Delta$.

Example. Let $\mathbf{X}=\{\Delta,1\}$, $\mathbf{A}=\{1,2\}$, $p(\Delta|\Delta,a)\equiv 1$, $p(1|1,1)=p(\Delta|1,2)=1$, all the other transition probabilities being zero, and $c(\Delta,a)\equiv 0$, $c(1,1)=0$, $c(1,2)=1$. See Fig. 2.31 for illustration.

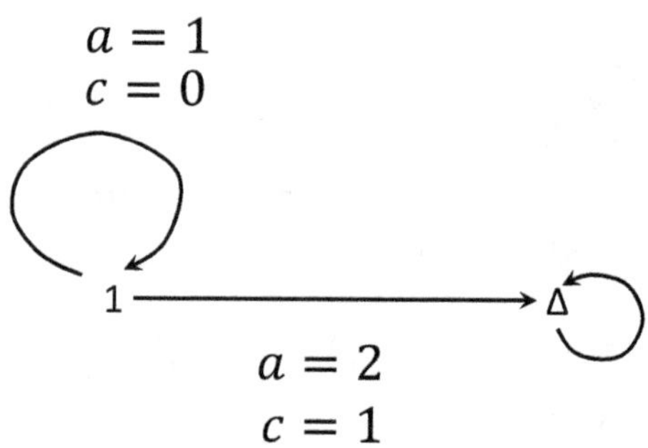

Fig. 2.31 Proper strategies are not sufficient.

Here $Y_{x,a}\subset\mathbf{X}$ is a finite set. Moreover, requirement (2.36) is also satisfied. It hods for $x=\Delta$ because $c(\Delta,a)=0$ and $V^\delta(\Delta)=0$. If $x=1$, then $\eta^\pi_1(\{1\}\times\mathbf{A})<\infty$ if and only if the expected time to absorption at Δ, starting from $x_0=1$, is finite. (See the paragraph below (2.12).) In other words,

$$\eta^\pi_1((\mathbf{X}\setminus\{\Delta\})\times\mathbf{A})=\eta^\pi_1(\{1\}\times\mathbf{A})<\infty\Longleftrightarrow v^\pi_1=1, \tag{2.37}$$

and $V^\delta(1)=1+\delta$ for $\delta>0$: the minimal value of $\eta^\pi_1(\{1\}\times\mathbf{A})$ equals 1 and is provided by the stationary selector $\varphi(x)\equiv 2$. Therefore,

$$c(1,a)+\delta+\sum_{y\in\mathbf{X}}V^\delta(y)p(y|x,a)<\infty \quad \forall a\in\mathbf{A}.$$

For each proper at $x_0 = 1$ strategy π^{m} we have $v_1^{\pi^{\mathrm{m}}} = 1$: see (2.37). Therefore, $\hat{v}_1 = 1$ and, obviously, $\hat{v}_\Delta = 0$: all Markov strategies are proper at $x_0 = \Delta$. Optimality equation (2.5) holds for the function $\hat{v}_x$:

$$1 = \hat{v}_1 = \min\{0 + \hat{v}_1; \quad 1 + \hat{v}_\Delta\}; \quad 0 = \hat{v}_\Delta = 0 + \hat{v}_\Delta.$$

The both stationary selectors $\varphi^1(x) \equiv 1$ and $\varphi^2(x) \equiv 2$ satisfy equality (2.35), but only selector φ^2 belongs to Π^p, and $v_x^{\varphi^2} = \hat{v}_x = \inf_{\pi \in \Pi^p} v_x^\pi$ for all $x \in \mathbf{X}$.

At the same time, $v_1^* = v_1^{\varphi^1} = 0 < \hat{v}_1$. We see that, when solving problem (2.1), one can miss the optimal solution if restricted to proper strategies, even if the model is positive. □

2.3.23 *Occupation measures and duality*

If an MDP with a discrete state space is transient (possibly with an absorbing state Δ) and *positive* then the value of the primal linear program (2.12), which takes the form

$$\begin{aligned} &\int_{(\mathbf{X}\setminus\{\Delta\})\times\mathbf{A}} c(y,a)d\eta(y,a) \to \inf_\eta \\ &\text{subject to} \qquad\qquad\qquad\qquad\qquad\qquad\qquad\qquad (2.38) \\ &\eta(\Gamma\times\mathbf{A}) = \mathsf{P}_0(\Gamma) + \int_{(\mathbf{X}\setminus\{\Delta\})\times\mathbf{A}} p(\Gamma|y,a)d\eta(y,a), \quad \Gamma \in \mathcal{B}(\mathbf{X}\setminus\{\Delta\}), \end{aligned}$$

coincides with $\inf_\pi v_{\mathsf{P}_0}^\pi = \inf_\pi \int_{\mathbf{X}\setminus\{\Delta\}} v_x^\pi \mathsf{P}_0(dx)$ [Altman (1999), Thm. 8.5]. Moreover, this statement holds also for a general MDP if the state space $\mathbf{X}$ is arbitrary Borel, $c(\cdot,\cdot) \in \mathbb{R}$, the transition probability $p(dy|x,a)$ is absolutely continuous w.r.t. some probability measure $q(dy)$ on $\mathbf{X}$, the action space $\mathbf{A}$ is finite, and the optimal value of program (2.12) is finite [Dufour and Piunovskiy (2013), Thm. 4.2].

The following example shows that the imposed conditions are important.

Example. Let $\mathbf{X} = \{\ldots,-2,-1,0,1,2,\ldots\}$, $\mathbf{A} = \{a\}$: the model is actually uncontrolled. We put $p(x+1|x,a) \equiv 1$ for all $x \in \mathbf{X}$, $c(x,a) = -\mathbb{I}\{x=0\}$, $\mathsf{P}_0(0) = 1$. This model is transient (but not absorbing), see Fig. 2.32.

Since the model is uncontrolled, we omit the argument a in the primal linear program (2.38):

$$-\eta(0) \to \inf_\eta \tag{2.39}$$

$$\text{subject to } \eta(x) = \mathbb{I}\{x=0\} + \eta(x-1) \text{ for } x \in \mathbf{X},\ \eta(\cdot) \geq 0.$$

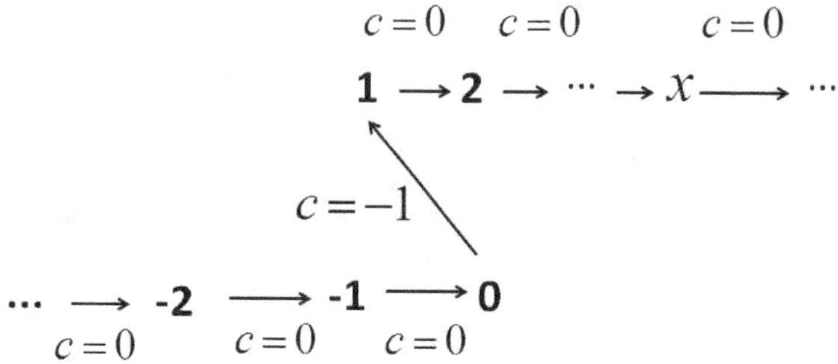

Fig. 2.32 Phantom solutions to the linear programs in duality.

The optimal value equals $-\infty$: for each $k \geq 0$, $\eta(x) := \begin{cases} k, & \text{if } x < 0; \\ k+1, & \text{if } x \geq 0 \end{cases}$ satisfies the constraint in (2.39) with $-\eta(0) = -(k+1) \to \infty$ as $k \to \infty$. But, obviously, $v_x^* = -\mathbb{I}\{x \leq 0\}$ and $\inf_\pi v_{\mathsf{P}_0}^\pi = \sum_{x \in \mathbf{X}} v_x^* \mathsf{P}_0(x) = -1$.

In a general MDP with a countable state space, the *Dual Linear Program* looks as follows [Altman (1999), p. 123]:

$$\sum_{x \in \mathbf{X}} \mathsf{P}_0(x) \tilde{v}(x) \to \sup_{\tilde{v}} \tag{2.40}$$
$$\text{subject to } \tilde{v}(x) \leq c(x,a) + \sum_{y \in \mathbf{X}} p(y|x,a)\tilde{v}(y), \quad x \in \mathbf{X},\ a \in \mathbf{A}.$$

In an arbitrary negative Borel model, if $\tilde{v}(x) \leq 0$ and $\tilde{v}(x) \leq c(x,a) + \int_{\mathbf{X}} \tilde{v}(y) p(dy|x,a)$, then $\tilde{v}(x) \leq v_x^*$ [Bertsekas and Shreve (1978), Prop. 9.10]. Since in negative models the Bellman function v_x^* is the *maximal* non-positive semianalytic solution to equation (2.5), it is not surprising that the optimal value of the program (2.40), with the additional requirement $\tilde{v}(x) \leq 0$, is provided by v_x^*.

For the example presented above (see Fig. 2.32) the *Dual Linear Program* looks as follows:

$$\tilde{v}(0) \to \sup_{\tilde{v}} \tag{2.41}$$
$$\text{subject to } \tilde{v}(x) \leq -\mathbb{I}\{x = 0\} + \tilde{v}(x+1) \text{ for all } x \in \mathbf{X}$$

and has the optimal value $+\infty$, if we do not require that $\tilde{v}(\cdot) \leq 0$.

One can rewrite programs (2.39) and (2.41) in the following form:

$$\left.\begin{array}{l}\text{Primal Linear Program: } \sup_{\tilde{v}} L(\eta,\tilde{v}) \to \inf_{\eta\ge 0},\\ \text{where } L(\eta,\tilde{v}) := -\eta(0) + \sum_{x\in\mathbf{X}} [\mathbb{I}\{x=0\} + \eta(x-1) - \eta(x)]\tilde{v}(x);\\ \text{Dual Linear Program: } \inf_{\eta\ge 0} \tilde{L}(\eta,\tilde{v}) \to \sup_{\tilde{v}},\\ \text{where } \tilde{L}(\eta,\tilde{v}) := \tilde{v}(0) + \sum_{x\in\mathbf{X}} \eta(x)[-\mathbb{I}\{x=0\} + \tilde{v}(x+1) - \tilde{v}(x)]\end{array}\right\} \quad (2.42)$$

correspondingly. In spite of their titles, linear programs (2.42) do not yet make a dual pair because $L(\eta,\tilde{v}) \neq \tilde{L}(\eta,\tilde{v})$. To make the Lagrangeans $L = \tilde{L}$ coincident, we must impose conditions making the series $\sum_{i=-\infty}^{\infty} \eta(i)|v(i)|$ and $\sum_{i=-\infty}^{\infty} \eta(i)|v(i+1)|$ convergent. For example, restrict ourselves with absolutely summable functions $\tilde{v}(\cdot) \in l^1$ and σ-finite measures η, uniformly bounded on singletons. Then the primal optimal value, like previously, is $-\infty$, and the dual linear program (2.41) has no feasible solutions leading to the optimal value $-\infty$. If we consider uniformly bounded functions $\tilde{v}(\cdot)$ and finite measures η, then the primal linear program (2.39) has no feasible solutions leading to the optimal value $+\infty$, and the dual optimal value, like previously, is $+\infty$. Finally, let us consider such functions $\tilde{v}(\cdot)$, that $\sum_{x>0} |\tilde{v}(x)| < \infty$ and $\sup_{x\le 0} |\tilde{v}(x)| < \infty$, and such measures η, that $\sup_{x>0} \eta(x) < \infty$ and $\sum_{x\le 0} \eta(x) < \infty$. Then the both programs (2.39) and (2.41) are feasible and have the coincident optimal value -1. The solution to the primal program is provided by $\eta^*(x) := \begin{cases} 1, & \text{if } x > 0,\\ 0, & \text{if } x \le 0,\end{cases}$ and the solution to the dual program is provided by $v^*_x := -\mathbb{I}\{x \le 0\}$. □

In specific examples one should accurately specify the proper classes of measures and functions to be considered in the primal and dual programs.

2.3.24 *Occupation measures in absorbing models*

Definition 2.6. Assuming that the spaces $\mathbf{X}$ and $\mathbf{A}$ are topological Borel spaces, the ws^∞-topology in the space of all strategic measures $\mathcal{P}_{\mathsf{P}_0} = \{\mathsf{P}^\pi_{\mathsf{P}_0},\ \pi \in \Pi^{\text{All}}\}$ is the coarsest topology rendering the mapping $\mathsf{P} \to \int_{\mathbf{H}_T} f(h_T) d\mathsf{P}(h_T)$ continuous for each $0 \le T < \infty$, for each measurable bounded function $f(h_T) = f(x_0, a_1, x_1, \ldots, a_T, x_T)$, continuous w.r.t. $(a_1, a_2, \ldots, a_T)$ under arbitrarily fixed $x_0, x_1, \ldots, x_T$ (i.e., for each Carathéodory function $f(\cdot)$: see Definition 2.9.)

Clearly, the weak topology in general is weaker than the ws^∞-topology.

Lemma 2.5. *Suppose the spaces* $\mathbf{X}$ *and* $\mathbf{A}$ *are topological Borel spaces and conditions 2.4(a,c) are satisfied. Then the following assertions hold.*

(a) *[Nowak (1988), Thm. 1], [Kurushima et al. (2018), Thm. 3.1]. The* ws^∞*-topology is equivalent to the weak topology in the space of all strategic measures* $\mathcal{P}_{\mathsf{P}_0}$.

(b) *[Schäl (1975a), Thm. 6.6]. The space of all strategic measures* $\mathcal{P}_{\mathsf{P}_0}$ *is compact in the* ws^∞*-topology.*

Example. Consider again the MDP described in Subsection 2.3.14 and fix the discrete topology in the spaces $\mathbf{X}$ and $\mathbf{A}$. Suppose the initial distribution P_0 is arbitrarily fixed. Since the model is semi-continuous (see Condition 2.4), the space of all strategic measures $\mathcal{P}_{\mathsf{P}_0}$ is compact in the weak topology by Lemma 2.5. Now the mapping $\mathsf{P} \to \int_{\mathbf{H}} s(h_T)\mathsf{P}(dh_T)$ is continuous, hence lower semi-continuous on the compact $\mathcal{P}_{\mathsf{P}_0}$ for any bounded function $s(\cdot)$ on $\mathbf{H}_T$, meaning that, in the finite-horizon version of this MDP, there is an optimal strategy. But, as we already know, there are no optimal strategies in the entire infinite-horizon model. Thus, the statement of Theorem 2.5 does not hold.

We emphasize that the cost function $c(\cdot,\cdot)$ is not bounded from below and requirement (2.9) is violated for each initial state $x > 0$:

$$\inf_{T\geq n}\inf_{\pi\in\Pi^{\mathrm{All}}}\sum_{t=n+1}^{T}\mathsf{E}_x^\pi[c(X_{t-1},A_t)] \leq \inf_{\pi\in\Pi^{\mathrm{All}}}\mathsf{E}_x^\pi[c(X_n,A_{n+1})]$$

$$\leq \mathsf{E}_x^{\varphi^n}[c(X_n,A_{n+1})] = \left(\frac{1}{2}\right)^n(1-2^{x+n}) \leq 1-2^x$$

does not go to zero as $n\to\infty$. Here φ^n is the following Markov selector:

$$\varphi_t^n(y) = \begin{cases} 2, & \text{if } t\leq n; \\ 1, & \text{if } t > n. \end{cases}$$

On the other hand, one can use Theorem 8.2 from [Altman (1999)].

Definition 2.7. [Altman (1999), Def. 7.4]. In a model with a topological Borel action space $\mathbf{A}$ and a countable (or finite) state space $\mathbf{X}$ with the costless cemetery 0, a function $\mu:\ \mathbf{X}\to[1,\infty)$ is said to be a *uniform Lyapunov function* if (i)–(iii) below hold.

(i) $1+\sum_{y\in\mathbf{X}\setminus\{0\}} p(y|x,a)\mu(y) \leq \mu(x), \quad x\in\mathbf{X}, a\in\mathbf{A}.$

(ii) For each $x \in \mathbf{X}$, the mapping $a \to \sum_{y\in\mathbf{X}\setminus\{0\}} p(y|x,a)\mu(y)$ is continuous.

(iii) For each $x \in \mathbf{X}$ and each stationary selector φ, $\lim_{t\to\infty} \mathsf{E}_x^\varphi[\mu(X_t) \times \mathbb{I}\{T_0 > t\}] = 0$. As usual, here $T_0 := \min\{t > 0 : X_t = 0\}$.

Theorem 2.13. [Altman (1999), Thm. 8.2(ii)]. *Suppose the action space* $\mathbf{A}$ *is compact, the state space* $\mathbf{X}$ *is denumerable (or finite), equipped with the discrete topology, with the costless cemetery 0. If there exists a uniform Lyapunov function* $\mu(\cdot)$*, then the space of all occupation measures* $\mathcal{D}_{\mathsf{P}_0} = \{\eta^\pi_{\mathsf{P}_0},\ \pi \in \Pi^{\text{All}}\}$ *on* $(\mathbf{X}\setminus\{0\})\times\mathbf{A}$*, under an arbitrarily fixed initial distribution* P_0*, is convex compact in the weak topology.*

In the MDP under investigation,

$$\mu(y) := \begin{cases} 1, & \text{if } y = 0; \\ 2, & \text{if } y \neq 0 \end{cases}$$

is the uniform Lyapunov function: assertions (i)–(ii) of Definition 2.7 are obvious, and assertion (iii) follows from equality

$$\lim_{t\to\infty} \mathsf{E}_x^\varphi[\mathbb{I}\{T_0 > t\}] = \mathsf{P}_x^\varphi(T_0 = \infty) = 0.$$

(See the end of Subsection 2.3.14).

Consider the case of $\mathsf{P}_0(1) = 1$. The space of all occupation measures $\mathcal{D}_1 = \{\eta_1^\pi,\ \pi \in \Pi^{\text{All}}\}$ on $(\mathbf{X}\setminus\{0\})\times\mathbf{A}$ is convex compact (in the weak topology), but the mapping

$$\eta_1^\pi \to \sum_{(x,a)\in(\mathbf{X}\setminus\{0\})\times\mathbf{A}} c(x,a)\eta_1^\pi(x,a) \tag{2.43}$$

is *not* lower semi-continuous. To show this, consider the sequence of stationary selectors

$$\varphi^n(x) = \begin{cases} 1, & \text{if } x \geq n, \\ 2, & \text{if } x < n; \end{cases} \qquad n = 1, 2, \ldots.$$

It is easy to see that

$$\eta_1^{\varphi^n}(x,a) = \begin{cases} \left(\frac{1}{2}\right)^{x-1}, & \text{if } 0 < x < n, \quad a = 2; \\ \left(\frac{1}{2}\right)^{n-1}, & \text{if } x = n, \quad a = 1; \\ 0, & \text{otherwise.} \end{cases}$$

Therefore,

$$\sum_{(x,a)\in(\mathbf{X}\setminus\{0\})\times\mathbf{A}} c(x,a)\eta_1^{\varphi^n}(x,a) = (1-2^n)\left(\frac{1}{2}\right)^{n-1}$$

$$= \left(\frac{1}{2}\right)^{n-1} - 2 \to -2 \quad \text{as } n \to \infty.$$

For the stationary selector $\varphi^\infty(x) \equiv 2$, we have $\eta_1^{\varphi^\infty}(x,a) = \left(\frac{1}{2}\right)^{x-1}\mathbb{I}\{a = 2\}$, $x > 0$. Thus, for any *bounded* continuous function $s(x,a)$ on $(\mathbf{X}\setminus\{0\})\times \mathbf{A}$ (for an arbitrary bounded function in the discrete case)

$$\left|\sum_{(x,a)\in(\mathbf{X}\setminus\{0\})\times\mathbf{A}} s(x,a)\eta_1^{\varphi^\infty}(x,a) - \sum_{(x,a)\in(\mathbf{X}\setminus\{0\})\times\mathbf{A}} s(x,a)\eta_1^{\varphi^n}(x,a)\right|$$

$$= \left|\sum_{a\in\mathbf{A}} s(n,a)\eta_1^{\varphi^\infty}(n,a) - \sum_{a\in\mathbf{A}} s(n,a)\eta_1^{\varphi^n}(n,a)\right.$$

$$\left. + \sum_{x>n}\sum_{a\in\mathbf{A}} s(x,a)\eta_1^{\varphi^\infty}(x,a) - \sum_{x>n}\sum_{a\in\mathbf{A}} s(x,a)\eta_1^{\varphi^n}(x,a)\right|$$

$$\le K\left[\left(\frac{1}{2}\right)^{n-1} + \left(\frac{1}{2}\right)^{n-1}\right] + K\sum_{x>n}\left(\frac{1}{2}\right)^{x-1} \to 0 \text{ as } n\to\infty$$

if $|s(x,a)| \le K$. Therefore, $\lim_{n\to\infty}\eta_1^{\varphi^n} = \eta_1^{\varphi^\infty}$ in the weak topology, but

$$\sum_{(x,a)\in(\mathbf{X}\setminus\{0\})\times\mathbf{A}} c(x,a)\eta_1^{\varphi^\infty}(x,a) = 0$$

$$> -2 = \lim_{n\to\infty}\sum_{(x,a)\in(\mathbf{X}\setminus\{0\})\times\mathbf{A}} c(x,a)\eta_1^{\varphi^n}(x,a),$$

and mapping (2.43) is not lower semi-continuous. Note that mapping (2.43) would have been lower semi-continuous if function $c(x,a)$ were lower semi-continuous and *bounded below*: see Corollary 2.1. Recall that the MDP is absorbing, as was shown in Subsection 2.3.14, so the space of all occupation measures $\{\eta_1^\pi,\ \pi\in\Pi^{\mathrm{All}}\}$ on $(\mathbf{X}\setminus\{0\})\times\mathbf{A}$ is uniformly bounded. □

Now, we slightly modify the model from Subsection 2.3.14 in such a way that the one-step cost $\hat{c}(x,a)$ becomes bounded (and continuous). A similar trick was demonstrated in [Altman (1999), Section 7.3].

Example. We introduce artificial states $1', 2' \ldots$, so that the cost $c(i,1) = 1 - 2^i$ will be accumulated owing to the long stay in state i'. In other words, $\mathbf{X} = \{0,1,1',2,2',\ldots\}$, $\mathbf{A} = \{1,2\}$, $p(0|0,a) \equiv 1$, for $i > 0$ $p(i'|i,1) = 1$, $p(0|i,2) = p(i+1|i,2) = 1/2$, $p(i'|i',a) = \frac{2^i-2}{2^i-1}$, $p(0|i',a) = \frac{1}{2^i-1}$; all the other transition probabilities are zero. We put $\hat{c}(i,a) \equiv 0$ for $i \ge 0$, $\hat{c}(i',a) \equiv -1$ for all $i \ge 1$. See Fig. 2.33.

Only actions in states $1, 2, \ldots$ play a role; as soon as $A_{t+1} = 1$ and $X_t = i$, the process jumps to state $X_{t+1} = i'$ and remains there for T_i

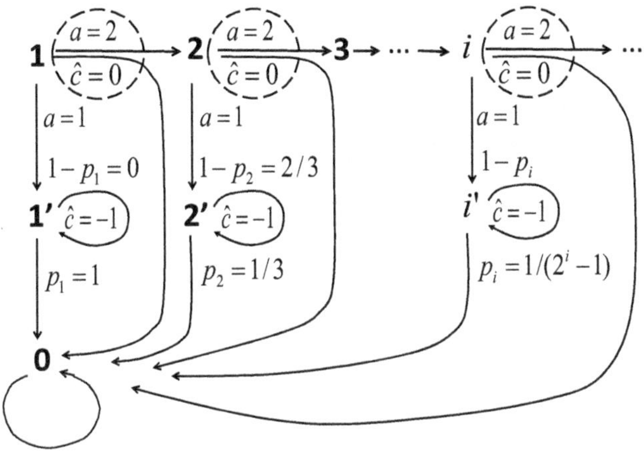

Fig. 2.33 How to make the cost function bounded.

time units, where T_i is geometrically distributed with parameter $p_i = \frac{1}{2^i-1}$. Since $\hat{c}(i',a) = -1$, the total expected cost, up to absorption at zero from state i', equals $\frac{-1}{p_i} = 1 - 2^i$, meaning that this modified model essentially coincides with the MDP from Subsection 2.3.14. Function $\hat{c}(\cdot,\cdot)$ is now uniformly bounded.

Let us show that, for each initial state $x \neq 0$, the modified MDP is absorbing at 0. To do so, suppose the cost function is $d(x,a) := -\mathbb{I}\{x \neq 0\}$ and construct the corresponding Bellman function V_x^* using Theorem 2.2. We use capital letters V_x^* here to distinguish from v_x^* used in Subsection 2.3.14. $V_0^* = 0$ and, for $x = i'$ $(i = 1, 2, \ldots)$, equation $V_{i'}^* = -1 + (1-p_i)V_{i'}^*$ implies that $V_{i'}^* = 1 - 2^i$. Next, for $x = i = 1, 2, \ldots$, equation $V_i^* = -1 + \min\{1 - 2^i, \frac{1}{2}V_{i+1}^*\}$ can be solved by successive approximations:

$$V^0(i) = 0; \; V^{n+1}(i) = -1 + \min\left\{1 - 2^i, \; \frac{1}{2}V^n(i+1)\right\}, \; i = 1, 2, \ldots,$$

$n = 0, 1, \ldots$: one can prove by induction w.r.t. n that

$$V^n(i) = -2 + \left(\frac{1}{2}\right)^{n-2} - 2^i, \qquad i = 1, 2, \ldots, \; n = 1, 2, \ldots.$$

Thus,

$$V_i^* = V^\infty(i) := \lim_{n\to\infty} V^n(i) = -2 - 2^i, \quad i = 1, 2, \ldots$$

and, for each control strategy π, for each $i = 1, 2, \dots$ and $x = i, i'$, we have inequality

$$V_x^\pi := \mathsf{E}_x^\pi \left[\sum_{t=1}^{\infty} d(X_{t-1}, A_t) \right] = -\mathsf{E}_x^\pi \left[\sum_{t=1}^{\infty} \mathbb{I}\{X_{t-1} \neq 0\} \right]$$
$$= -\eta_x^\pi((\mathbf{X} \setminus \{0\}) \times \mathbf{A}) \geq \begin{cases} -2 - 2^i, & \text{if } x = i; \\ 1 - 2^i, & \text{if } x = i'. \end{cases}$$

Therefore, after we fix $\mathsf{P}_0(x) = 1$ for $x = i$ or $x = i'$ $(i > 0)$, we see that

$$\eta_x^\pi((\mathbf{X} \setminus \{0\}) \times \mathbf{A}) = \mathsf{E}_x^\pi[T_0] \leq 2 + 2^i$$

for all strategies π, where, as usual, $T_0 := \min\{t > 0 : \ X_t = 0\}$. Hence, this MDP is absorbing (at 0).

Like previously, below $\mathsf{P}_0(1) = 1$. Recall that $\mathcal{D}_1$ is the space of the restrictions of occupation measures η_1^π $(\pi \in \Pi^{\mathrm{All}})$ on $(\mathbf{X} \setminus \{0\}) \times \mathbf{A}$. All the measures $\eta \in \mathcal{D}_1$ are uniformly bounded: $\eta_1^\pi((\mathbf{X} \setminus \{0\}) \times \mathbf{A}) \leq 4$. Thus, by Corollary 2.1, the mapping

$$\eta_1^\pi \to \sum_{(x,a) \in (\mathbf{X} \setminus \{0\}) \times \mathbf{A}} \hat{c}(x, a) \eta_1^\pi(x, a)$$

is bounded below and lower semi-continuous in the weak topology. Since, like in Subsection 2.3.14, there are no optimal strategies (i.e., no optimal occupation measures), we must conclude that the space of occupation measures $\mathcal{D}_1$ is not compact in the weak topology.

Let us prove explicitly the latter statement keeping in mind that the topologies in the spaces $\mathbf{X}$ and $\mathbf{A}$ are discrete. Consider the sequence $\eta_1^{\varphi^n}$, $n = 1, 2, \dots$ of measures from $\mathcal{D}_1$, generated by the stationary selectors

$$\varphi^n(x) := \begin{cases} 1, & \text{if } x = i \text{ or } x = i' \text{ with } i \geq n; \\ 2, & \text{otherwise.} \end{cases}$$

Recall that $\mathsf{P}_0(1) = 1$. Then $\eta_1^{\varphi^n}(\{i'\} \times \mathbf{A}) = 0$ as soon as $n > i$ $(i, n = 1, 2, \dots)$. Therefore, if $\eta_1^{\varphi^{n_j}} \to \eta$ for some subsequence n_j, $j = 1, 2, \dots$, then $\eta(\{i'\} \times \mathbf{A}) = 0$ for all $i = 1, 2, \dots$. The only occupation measure, satisfying this requirement, comes from the stationary selector $\varphi^\infty(i) \equiv 2$, $i = 1, 2, \dots$. (The values $\varphi^\infty(i')$ and $\varphi^\infty(0)$ are immaterial.) But, for such a selector, for

$$d(x, a) = \begin{cases} 1, & \text{if } x \in \{1', 2', \dots\}; \\ 0, & \text{otherwise,} \end{cases}$$

we have $\sum_{x\in(\mathbf{X}\setminus\{0\})\times\mathbf{A}} d(x,a)\eta_1^{\varphi^\infty}(x,a) = 0$. At the same time, for $\eta_1^{\varphi^{n_j}}$ we have

$$\sum_{x\in(\mathbf{X}\setminus\{0\})\times\mathbf{A}} d(x,a)\eta_1^{\varphi^{n_j}}(x,a) = \eta_1^{\varphi^{n_j}}((n_j)'\times\mathbf{A})$$
$$= \left(\frac{1}{2}\right)^{n_j-1}\cdot(2^{n_j}-1) = 2 - \left(\frac{1}{2}\right)^{n_j-1} :$$

$\left(\frac{1}{2}\right)^{n_j-1}$ is the probability to reach the state n_j (starting from $x_0 = 1$), and $(2^{n_j}-1)$ is the expected time spent in the state n_j. Thus,

$$\lim_{j\to\infty} \sum_{x\in(\mathbf{X}\setminus\{0\})\times\mathbf{A}} d(x,a)\eta_1^{\varphi^{n_j}}(x,a) = 2$$
$$\neq 0 = \sum_{x\in(\mathbf{X}\setminus\{0\})\times\mathbf{A}} d(x,a)\eta_1^{\varphi^\infty}(x,a).$$

No-one subsequence $\eta_1^{\varphi^{n_j}}$ is weakly convergent in $\mathcal{D}_1$. The proof is completed.

One can show that, in the modified model with an arbitrarily fixed initial state $x_0 = i > 0$, requirement (2.9) is violated, while all the other conditions of Theorem 2.5 are satisfied. Thus, in that theorem equation (2.9) is important.

The modified model is again semi-continuous (see Condition 2.4), and the space of all strategic measures $\mathcal{P}_1 = \{\mathsf{P}_1^\pi,\ \pi\in\Pi^{\mathrm{All}}\}$ is again compact in the weak and ws^∞-topology: see Lemma 2.5. (Recall that $\mathsf{P}_0(1) = 1$.) But the mapping $\mathsf{P}_1^\pi \to \eta_1^\pi$ from $\mathcal{P}_1$ to $\mathcal{D}_1$ is not continuous. In the space $\mathcal{D}_1$ we fix the weak topology, as usual. For the sequence of stationary selectors φ^n as above, we have $\mathsf{P}_1^{\varphi^n} \to \mathsf{P}_1^{\varphi^\infty}$ (as $n\to\infty$) for $\varphi^\infty(x)\equiv 2$ but, as was shown, the sequence $\eta_1^{\varphi^n}$ does not converge weakly to $\eta_1^{\varphi^\infty}$.

In the modified model, the uniform Lyapunov function $\mu(\cdot)$ does not exit (see Definition 2.7); therefore, Theorem 2.13 is not applicable. Indeed, the function μ must satisfy the following conditions.

- For each i' ($i = 1,2,\dots$), $1 + (1-p_i)\mu(i') \le \mu(i')$, so that $\mu(i') \ge \frac{1}{p_i} = 2^i - 1$.
- For each $i = 1,2,\dots$ (under action 1), $1 + \mu(i') \le \mu(i)$, so that $\mu(i) \ge 2^i$.

Now, for each $x = i = 1,2,\dots$, for the stationary selector $\varphi^\infty(x)\equiv 2$, we have

$$\mathsf{E}_i^\varphi[\mu(X_t)\mathbb{I}\{T_0 > t\}] \ge \left(\frac{1}{2}\right)^t\cdot 2^{i+t} = 2^i.$$

Requirement (iii) fails to hold. □

A similar example was considered in [Piunovskiy and Zhang (2024b), Ex. 2].

2.3.25 *Occupation measures in uniformly absorbing models*

Definition 2.8. [Feinberg and Piunovskiy (2019), Def. 3.6]. Let the initial distribution P_0 be fixed. An absorbing (at 0) MDP is called *uniformly absorbing* if

$$\lim_{n\to\infty} \sup_{\pi\in\Pi^{\mathrm{All}}} \mathsf{E}^\pi_{\mathsf{P}_0}\left[\sum_{t=n}^{\infty} \mathbb{I}\{t < T_0\}\right] \tag{2.44}$$

$$= \lim_{n\to\infty} \sup_{\pi\in\Pi^{\mathrm{All}}} \mathsf{E}^\pi_{\mathsf{P}_0}\left[\sum_{t=n}^{\infty} \mathbb{I}\{X_t \in (\mathbf{X}\setminus\{0\})\}\right] = 0.$$

As usual, here $T_0 := \min\{t > 0 : \ X_t = 0\}$.

If equation (2.44) is valid, then the model is certainly absorbing because, for each control strategy π,

$$\mathsf{E}^\pi_{\mathsf{P}_0}[T_0] = \mathsf{E}^\pi_{\mathsf{P}_0}\left[\sum_{t=0}^{\infty} \mathbb{I}\{X_t \in (\mathbf{X}\setminus\{0\})\}\right] \le n - 1 + \varepsilon.$$

Here, for a fixed $\varepsilon > 0$, $n \ge 1$ is such that $\sup_{\pi\in\Pi^{\mathrm{All}}} \mathsf{E}^\pi_{\mathsf{P}_0}\left[\sum_{t=n}^{\infty} \mathbb{I}\{X_t \in (\mathbf{X}\setminus\{0\})\}\right] \le \varepsilon$.

Definition 2.9. If $\mathbf{Y}$ and $\mathbf{A}$ are (topological) Borel spaces, then a function $f(\cdot,\cdot): \mathbf{Y}\times\mathbf{A} \to \mathbb{R}$ is called a *Carathéodory* function if $f(\cdot, a)$ is measurable for each $a \in \mathbf{A}$ and $f(y,\cdot)$ is continuous for each $y \in \mathbf{Y}$.

The coarsest topology in the space of finite measures on $\mathcal{B}(\mathbf{Y}\times\mathbf{A})$, for which the mappings $\mu \to \int_{\mathbf{Y}\times\mathbf{A}} f(y,a)d\mu(y,a)$ are continuous for all bounded Carathéodory functions $f(\cdot,\cdot)$, is called the *ws-topology*.

Recall that the *weak* topology in the space of finite measures on $\mathcal{B}(\mathbf{Y}\times\mathbf{A})$ is the coarsest topology for which the mappings $\mu \to \int_{\mathbf{Y}\times\mathbf{A}} f(y,a)d\mu(y,a)$ are continuous for all bounded continuous functions $f(\cdot,\cdot)$.

The *strong* topology in the space of finite measures on $\mathcal{B}(\mathbf{Y})$ is the coarsest topology for which the mappings $\mu \to \int_{\mathbf{Y}} f(y)d\mu(y)$ are continuous for all bounded measurable functions $f(\cdot)$.

Theorem 2.14. *Suppose* $\mathbf{X}$ *and* $\mathbf{A}$ *are topological Borel spaces, the initial distribution* P_0 *is fixed and the model is absorbing (at 0).*

(a) *If the model is semi-continuous (i.e., Conditions 2.4(a,c) are satisfied), then the mapping* $\mathsf{P}^\pi_{\mathsf{P}_0} \to \eta^\pi_{\mathsf{P}_0}$ *from* $\mathcal{P}_{\mathsf{P}_0}$ *(with the weak or* ws^∞*-topology) to* $\mathcal{D}_{\mathsf{P}_0} := \{\eta^\pi_{\mathsf{P}_0}$ *on* $(\mathbf{X} \setminus \{0\}) \times \mathbf{A},\ \pi \in \Pi^{\mathrm{All}}\}$ *(with the weak topology) is continuous if and only if the model is uniformly absorbing.*

(b) *If* $\mathbf{A}$ *is compact and the transition probability* p *is a continuous stochastic kernel, then the mapping* $\mathsf{P}^\pi_{\mathsf{P}_0} \to \eta^\pi_{\mathsf{P}_0}$ *from* $\mathcal{P}_{\mathsf{P}_0}$ *(with the weak topology) to* $\mathcal{D}_{\mathsf{P}_0} := \{\eta^\pi_{\mathsf{P}_0}$ *on* $(\mathbf{X}\setminus\{0\})\times\mathbf{A},\ \pi \in \Pi^{\mathrm{All}}\}$ *(with the weak topology) is continuous if and only if the model is uniformly absorbing.*

For the proof see [Piunovskiy and Zhang (2024b), Thm. 2]. See also Lemma 2.5(a): in a semi-continuous model, the ws^∞-topology in $\mathcal{P}_{\mathsf{P}_0}$ is equivalent to the weak topology. According to Lemma 2.5(b), under the conditions of Theorem 2.14(a), in a uniformly absorbing model the space $\mathcal{D}_{\mathsf{P}_0}$ is weakly compact as the continuous image of the compact $\mathcal{P}_{\mathsf{P}_0}$. Note also that, under the conditions of part (b) of Theorem 2.14, the space $\mathcal{P}_{\mathsf{P}_0}$ is compact in the (metrizable) weak topology by Lemma 2.7 of [Dufour and Prieto-Rumeau (2024b)] meaning that in this case in a uniformly absorbing model the space $\mathcal{D}_{\mathsf{P}_0}$ is also compact in the weak topology.

Theorem 2.15. *Suppose the initial distribution* P_0 *is fixed and the model is semi-continuous (i.e., Conditions 2.4(a,c) are satisfied) and absorbing (at 0). Then the following assertions are valid.*

(a) *The space* $\mathcal{D}_{\mathsf{P}_0}$ *is compact in the ws-topology if and only if the model is uniformly absorbing.*

(b) *The space* $\mathcal{D}_{\mathsf{P}_0}$ *is compact in the weak topology if and only if the model is uniformly absorbing.*

Statement (a) was proved in [Dufour and Prieto-Rumeau (2024a), Thm. 4.10]. Statement (b) was later proved in [Dufour and Prieto-Rumeau (2024b), Thms. 3.2 and 3.5].

Example. As was shown, the modified model in Subsection 2.3.24 (see Fig. 2.33) with $\mathsf{P}_0(1) = 1$ is semi-continuous in the sense of Condition 2.4, and absorbing (at 0), but the mapping $\mathsf{P}^\pi_{\mathsf{P}_0} \to \eta^\pi_{\mathsf{P}_0}$ is not continuous w.r.t. the topologies as in Theorem 2.14(a), and the space $\mathcal{D}_{\mathsf{P}_0}$ is not (weakly) compact. Note also that the weak and ws-topologies in $\mathcal{D}_{\mathsf{P}_0}$ coincide in that example just because the model is discrete. Let us show explicitly that the model is not uniformly absorbing.

Consider again the stationary selectors

$$\varphi^n(x) := \begin{cases} 1, & \text{if } x = i \text{ or } x = i' \text{ with } i \geq n; \\ 2, & \text{otherwise} \end{cases}$$

for $n = 1, 2, \ldots$. The occupation measure of the selector φ^n (recall that $x_0 = 1$) on $(\mathbf{X} \setminus \{0\}) \times \mathbf{A}$ is given by

$$\eta_1^{\varphi^n}(x,a) = \begin{cases} \left(\frac{1}{2}\right)^{i-1}, & \text{if } x = i < n \text{ and } a = 2 \\ & \text{or } x = n \text{ and } a = 1; \\ \left(\frac{1}{2}\right)^{n-1} \cdot \frac{1}{p_n} = 2 - \left(\frac{1}{2}\right)^{n-1}, & \text{if } x = n' \text{ and } a = 1; \\ 0, & \text{otherwise:} \end{cases}$$

the states $x = i \leq n$ are reached with probability $\left(\frac{1}{2}\right)^{i-1}$, and the controlled process spends one time unit in those states; the states $x = i'$ with $i < n$ and the states $x = i, i'$ with $i > n$ are never reached. For the state $x = n'$, note that $\left(\frac{1}{2}\right)^{n-1}$ is the probability that $X_{n-1} = n$ (the complementary probability $1 - \left(\frac{1}{2}\right)^{n-1}$ is for $X_{n-1} = 0$). After that, $X_n = n'$ and $X_{n+t} = n'$ with probability $(1 - p_n)^t$ leading to

$$\eta_1^{\varphi^n}(\{n'\} \times \mathbf{A}) = \left(\frac{1}{2}\right)^{n-1} [1 + (1 - p_n) + (1 - p_n)^2 + \ldots] = \left(\frac{1}{2}\right)^{n-1} \cdot \frac{1}{p_n}.$$

For each $n \geq 1$,

$$\sup_{\pi \in \Pi^{\mathrm{All}}} \mathsf{E}_1^\pi \left[\sum_{t=n}^{\infty} \mathbb{I}\{X_t \in (\mathbf{X} \setminus \{0\})\} \right] \geq \mathsf{E}_1^{\varphi^n} \left[\sum_{t=n}^{\infty} \mathbb{I}\{X_t = n'\} \right]$$

$$= \eta_1^{\varphi^n}(\{n'\} \times \mathbf{A}) = 2 - \left(\frac{1}{2}\right)^{n-1} \to 2 \text{ as } n \to \infty.$$

Thus, the modified model in Subsection 2.3.24 with $\mathsf{P}_0(1) = 1$ is not uniformly absorbing. □

Remark 2.6. In absorbing models, the space $\mathcal{D}_{\mathsf{P}_0}$ coincides with the space of occupation measures coming from stationary strategies by Lemma 2.4. In this connection, Lemma 4.7 of [Feinberg and Rothblum (2012)], claiming that $\mathcal{D}_{\mathsf{P}_0}$ is weakly compact in a semi-continuous absorbing model, is inaccurate: see Theorem 2.15. The modified model from Subsection 2.3.24, discussed above, also confirms this.

In absorbing, but not uniformly absorbing models most of the statements in Subsection 2.3.24 are valid.

Proposition 2.4. [Altman (1999), p.107]. *Suppose the state space* $\mathbf{X}$ *is denumerable or finite, endowed with the discrete topology, and 0 is the* *isolated absorbing state. Assume also that the model is semi-continuous in the sense of Condition 2.3 or 2.4. (Here the requirements about the cost function $c(\cdot,\cdot)$ are immaterial, and the other requirements in the mentioned conditions are equivalent.) If there exists a uniform Lyapunov function $\mu(\cdot)$ as in Definition 2.7, then the MDP is uniformly absorbing (at 0) for each initial state x_0.*

It can happen that, in the framework of Proposition 2.4, for some initial distribution P_0, the MDP model is not absorbing (at 0) for P_0. Below we present the example published in [Piunovskiy and Zhang (2024b), Ex. 1] to illustrate this; see also Remark 2.5.

Example. Consider the MDP model with $\mathbf{X} = \{0, 1, \dots\}$, endowed with the discrete topology, and $\mathbf{A}$ being a singleton, so that we will omit the argument $a \in \mathbf{A}$ everywhere in this example. Put $p(0|x) = \frac{1}{2^x} = 1 - p(x|x)$ for all $x \in \{1, 2, \dots\}$ and $p(0|0) = 1$. The cost function is immaterial. The

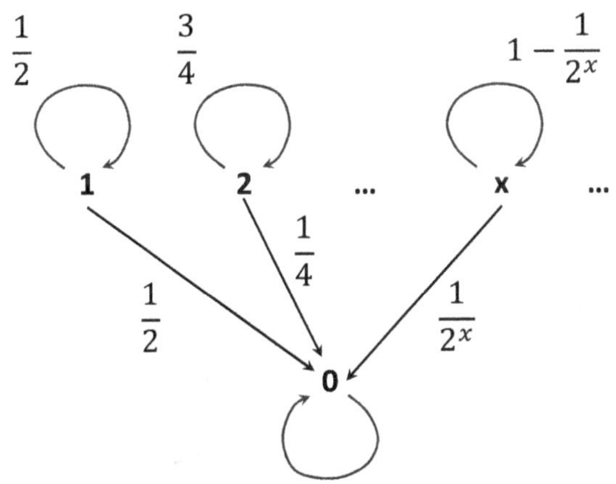

Fig. 2.34 Non-absorbing MDP with a uniform Lyapunov function.

graphical illustration is on Fig. 2.34. There is only one strategy say π in this model. Then for each $x_0 \in \{1, 2, \dots\}$,

$$\mathsf{E}_{x_0}^{\pi}[T_0] = 2^{x_0}.$$

Hence, this MDP model is absorbing, in fact uniformly absorbing (at 0) for every initial state x_0. It has a uniform Lyapunov function given by

$\mu(0) = 1$ and $\mu(x) = 2^x$ for $x \geq 1$. Indeed, conditions (i,ii) in Definition 2.7 can be easily checked to be satisfied by $\mu(\cdot)$, and for (iii), note that, for each $x \geq 1$,

$$\mathsf{E}_x^\pi[\mu(X_t)\mathbb{I}\{T_0 > t\}] = \mathsf{E}_x^\pi[2^x \mathbb{I}\{X_t = x\}] = 2^x \mathsf{P}_x^\pi(X_t = x)$$
$$= 2^x(1 - \frac{1}{2^x})^t \to 0$$

as $t \to \infty$, and the convergence also takes place for $x = 0$. Now, if we take the geometric initial distribution on $\{1, 2, \dots\}$: $\mathsf{P}_0(x) = \frac{2}{5}\left(\frac{3}{5}\right)^{x-1}$, then $\mathsf{E}_{\mathsf{P}_0}^\pi[T_0] = \sum_{x=1}^\infty \frac{2}{5}(\frac{3}{5})^{x-1}2^x = \infty$. □

We required the absorbing state to be an isolated point. If this is not the case, then the statements of Theorem 2.15 may be false as the following example, presented in [Dufour and Prieto-Rumeau (2024b), Ex. 3.12], confirms.

Example. Consider the MDP with $\mathbf{X} = \{-1, 0\} \cup \{1/n\}_{n=1}^\infty$ and $\mathbf{A} = \{0\} \cup \{1/n\}_{n=1}^\infty$. The spaces $\mathbf{X}$ and $\mathbf{A}$ are endowed with the topologies generated by the Euclidean metric on $\mathbb{R}$ (the traces of the Euclidean topology). Assume that $x_0 = -1$. Put $p(a|-1, a) \equiv 1$ and, for $x \neq -1$, $p(0|x, a) \equiv 1$ with other transition probabilities being zero. The cost function is immaterial. For illustration, see Fig. 2.35.

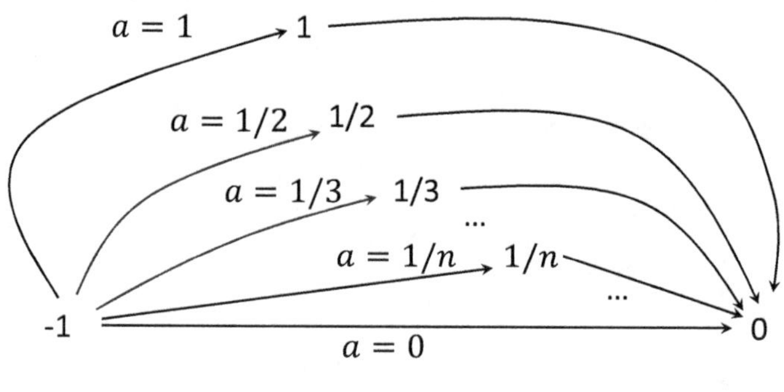

Fig. 2.35 Non-isolated absorbing state.

It is obvious that this MDP is uniformly absorbing (at 0): for each strategy π, for all $t \geq 2$ $\mathsf{E}_{-1}^\pi[\mathbb{I}\{X_t \in (\mathbf{X} \setminus \{0\})\}] = 0$. Conditions 2.4(a,c) are satisfied, but the absorbing state 0 is not an isolated point in $\mathbf{X}$. The space of occupation measures $\mathcal{D}_{-1}$ is not weakly compact because the sequence of occupation measures $\eta_{-1}^{\varphi^n}$ on $\mathcal{B}((\mathbf{X} \setminus \{0\}) \times \mathbf{A})$, corresponding to the stationary selectors $\varphi^n(x) \equiv 1/n$, has the form

$$\eta_{-1}^{\varphi^n}(-1, 1/n) = 1; \quad \eta_{-1}^{\varphi^n}(1/n, 1/n) = 1, \quad n = 1, 2, \ldots,$$

and this sequence of measures has no (weak) limiting points in the space of finite measures on $\mathcal{B}((\mathbf{X} \setminus \{0\}) \times \mathbf{A})$. Note that the trace of the Euclidean topology on $\mathbf{X} \setminus \{0\}$ is just the discrete topology. Since the weak topology in $\mathcal{D}_{-1}$ is weaker than the ws-topology, $\mathcal{D}_{-1}$ is also not compact in the ws-topology. Incidentally, the requirement of the absorbing state to be isolated was missing in [Feinberg and Rothblum (2012)]. □

2.3.26 *Topological properties of the space $\mathcal{P}_{\mathsf{P}_0}$ of strategic measures*

The first example in Subsection 1.4.11 shows that the space of strategic measures generated by selectors is usually not closed in the weak topology.

According to Lemma 2.5, under appropriate conditions, the weak and ws^∞-topologies in the space $\mathcal{P}_{\mathsf{P}_0}$ of strategic measures are equivalent, and $\mathcal{P}_{\mathsf{P}_0}$ is compact. The following example (reported by Dr. Y.Zhang in a private communication) shows that situation is different if one replaces Condition 2.4(c) with the requirement that the transition probability p is a continuous stochastic kernel. (The latter requirement is also popular in the literature on MDP: see Condition 2.3 and [Bertsekas and Shreve (1978), Def. 8.7], [Hernández-Lerma *et al.* (2023), Thm. 3.13(b)].)

Example. Let $\mathbf{X} = \mathbf{A} = [0, 1]$ endowed with the Euclidean topology, $p(dy|x, a) = \delta_a(dy)$; the cost function $c(\cdot, \cdot)$ is immaterial. See Fig. 2.36.

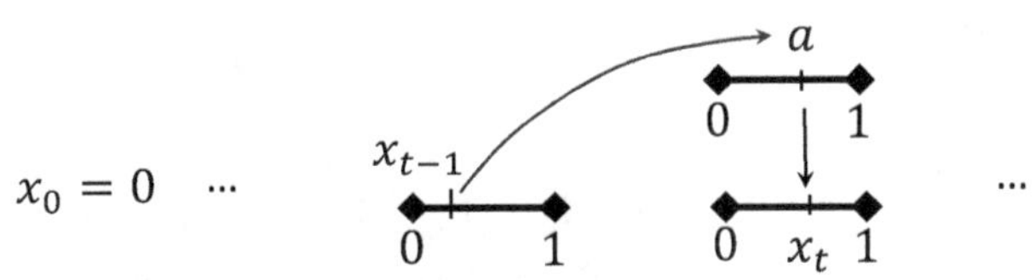

Fig. 2.36 Weak and ws^∞-topologies in $\mathcal{P}_{\mathsf{P}_0}$ are different.

The spaces $\mathbf{X}$ and $\mathbf{A}$ are compact and the transition probability p is a continuous stochastic kernel:

$$\int_{\mathbf{X}} g(y) p(dy|x, a) = g(a)$$

is the continuous function on $\mathbf{X} \times \mathbf{A}$ for each continuous function $g(\cdot)$ on $\mathbf{X}$. Therefore, the space $\mathcal{P}_{\mathsf{P}_0}$ of strategic measures is weakly compact for each fixed initial distribution P_0 by Theorem 5.6 of [Schäl (1975a)]. At the same time, Condition 2.4(c) is not satisfied: e.g., for $g(x) = \mathbb{I}\{x = 1/2\}$, function $\int_{\mathbf{X}} g(y)p(da|x,a) = \mathbb{I}\{a = 1/2\}$ is not continuous. In what follows, we demonstrate that some sequence $\{\mathsf{P}_n\}_{n=1}^{\infty}$ from $\mathcal{P}_{\mathsf{P}_0}$ is convergent to $\mathsf{P} \in \mathcal{P}_{\mathsf{P}_0}$ in the weak topology, but has no convergent in $\mathcal{P}_{\mathsf{P}_0}$ subsequences w.r.t. the ws^{∞}-topology.

Consider the sequence $\{\pi^{s,n}\}_{n=1}^{\infty}$ of stationary strategies $\pi^{s,n}(\Gamma|x) = \mu_n(\Gamma) := n\ Leb(\Gamma \cap [0, 1/n])$, $\Gamma \in \mathcal{B}(\mathbf{A})$, where *Leb* is the Lebesgue measure. The measures μ_n converge to $\mu_\infty := \delta_0$ in the weak topology on $\mathbf{P}(\mathbf{A})$ (probability measures on $\mathbf{A}$) as $n \to \infty$ because, for each continuous bounded function $g(\cdot)$ on $\mathbf{A}$, $\int_{\mathbf{A}} g(a)\mu_n(da) = n\int_0^{1/n} g(a)da \to g(0)$ as $\to \infty$. Consequently, for each $t \geq 1$, the product measure $\mu_n(da_1)\ldots\mu_n(da_t)$ converges to $\mu_\infty(da_1)\ldots\mu_\infty(da_t) = \delta_0(da_1)\ldots\delta_0(da_t)$ w.r.t. the weak topology on $\mathbf{P}(\mathbf{A}^t)$ as $n \to \infty$. Let $\pi^{s,\infty}$ be the stationary strategy given by $\pi^{s,\infty}(da|x) := \mu_\infty(da) = \delta_0(da)$. Note that, for each strategy $\pi^{s,n}$ $(n = 1, 2, \ldots, \infty)$, $X_i = A_i$ $\mathsf{P}_0^{\pi^{s,n}}$-a.s. for all $i = 1, 2, \ldots$ and the random variables A_i are independent and identically distributed. Therefore, for each bounded continuous function $f(\cdot)$ on $\mathbf{H}_t$,

$$\begin{aligned} &\int_{\mathbf{H}_t} f(h_t)\mathsf{P}_0^{\pi^{s,n}}(dh_t \times (\mathbf{A}\times\mathbf{X})^\infty) \\ =\ &\int_{\mathbf{A}^t} f(0, a_1, a_1, a_2, a_2, \ldots, a_t, a_t)\mu_n(da_1)\mu_n(da_2)\ldots\mu_n(da_t) \to \\ (\text{as } n\to\infty)\ &\int_{\mathbf{A}^t} f(0, a_1, a_1, a_2, a_2, \ldots, a_t, a_t)\mu_\infty(da_1)\mu_\infty(da_2)\ldots\mu_\infty(da_t) \\ =\ & f(0, 0, \ldots, 0) = \int_{\mathbf{H}_t} f(h_t)\mathsf{P}_0^{\pi^{s,\infty}}(dh_t \times (\mathbf{A}\times\mathbf{X})^\infty). \end{aligned}$$

To summarize, $\mathsf{P}_0^{\pi^{s,n}} \to \mathsf{P}_0^{\pi^{s,\infty}}$ w.r.t. the weak topology as $n \to \infty$.

Now we will show that the sequence $\{\mathsf{P}_0^{\pi^{s,n}}\}_{n=1}^{\infty}$ does not contain convergent subsequences in $\mathcal{P}_0$ w.r.t. the ws^{∞}-topology. Suppose for contradiction that a subsequence $\{\mathsf{P}_0^{\pi^{s,n_k}}\}_{k=1}^{\infty}$ is convergent w.r.t. the ws^{∞}-topology. Since the ws^{∞}-topology is finer than the weak topology on $\mathbf{P}(\Omega)$, the limit of $\{\mathsf{P}_0^{\pi^{s,n_k}}\}_{k=1}^{\infty}$ must coincide with $\mathsf{P}_0^{\pi^{s,\infty}}$. Consider the function

$$f(x, a, y) := \mathbb{I}\{y > 0\}, \quad \forall (x, a, y) \in \mathbf{X}\times\mathbf{A}\times\mathbf{X}.$$

It is a Carathéodory function, i.e., for each $a \in \mathbf{A}$, $f(x, a, y)$ is measurable in $(x, y) \in \mathbf{X}\times\mathbf{X}$ and, for each $(x, y) \in \mathbf{X}\times\mathbf{X}$, $f(x, a, y)$ is continuous in

$a \in \mathbf{A}$. Similarly to the above calculations,

$$\int_{\mathbf{H}_1} f(h_1)\mathsf{P}_0^{\pi^{s,n_k}}(dh_1 \times (\mathbf{A}\times\mathbf{X})^\infty) = \int_{\mathbf{A}} f(0,a_1,a_1)\mu_{n_k}(da_1)$$
$$= \int_0^{1/n_k} \mathbb{I}\{a>0\}n_k\, da \equiv 1.$$

On the other hand,

$$\int_{\mathbf{H}_1} f(h_1)\mathsf{P}_0^{\pi^{s,\infty}}(dh_1 \times (\mathbf{A}\times\mathbf{X})^\infty) = \int_{\mathbf{A}} f(0,a_1,a_1)\mu_{\infty}(da_1)$$
$$= \int_{\mathbf{A}} \mathbb{I}\{a>0\}\ \delta_0(da) = 0 \neq 1.$$

Thus, $\mathsf{P}_0^{\pi^{s,\infty}}$ is not the limit of $\{\mathsf{P}_0^{\pi^{s,n_k}}\}_{k=1}^\infty$, and we conclude that the sequence $\{\mathsf{P}_0^{\pi^{s,n}}\}_{n=1}^\infty$ does not contain convergent subsequences in $\mathcal{P}_0$ w.r.t. the ws^∞-topology. The proof is completed.

We showed that the weak and ws^∞-topologies in $\mathcal{P}_0$ are different. Moreover, the space $\mathcal{P}_0$ is weakly compact, but not compact w.r.t. the ws^∞-topology. For the last assertion, note that, in case of the ws^∞-topology, $\mathcal{P}_0$ is relatively compact in the space $\mathbf{P}(\Omega)$ of probability measures on $\mathcal{B}(\Omega)$ if and only if it is relatively sequentially compact therein [Schäl (1975a), Remark 4.4.]. □

2.3.27 *Topological properties of the space $\mathcal{D}_{\mathsf{P}_0}$ of occupation measures*

In this subsection we consider the absorbing (at Δ) model, assume that $\mathbf{X}$ and $\mathbf{A}$ are topological Borel spaces and compare the main topologies in the space $\mathcal{D}_{\mathsf{P}_0}$ of measures on $\mathcal{B}((\mathbf{X}\setminus\{\Delta\})\times\mathbf{A})$, namely, the weak and the ws-topology: see Definition 2.9. In general, the weak topology is weaker than the ws-topology, and they coincide in the case of a discrete model with discrete topologies in $\mathbf{X}$ and $\mathbf{A}$.

Proposition 2.5. [Zhang and Zheng (2024), Prop. 3.1]. *If the initial state $x_0 \in \mathbf{X}$ is fixed and the model is semi-continuous (in the sense of Conditions 2.4(a,c)) and absorbing (at Δ), then a sequence $\{\eta\}_{n=1}^\infty \subset \mathcal{D}_{x_0}$ converges to some $\eta \in \mathcal{D}_{x_0}$ in the weak topology if and only if $\lim_{n\to\infty}\eta_n = \eta$ in the ws-topology.*

As was explained in [Zhang and Zheng (2024)], in the framework of Proposition 2.5, if the model is uniformly absorbing, then the weak and ws-topologies coincide. The following example, presented in

[Zhang and Zheng (2024), Ex. 3.1], shows that the situation becomes different if Condition 2.4(c) is replaced by the requirement that the transition probability p is a continuous stochastic kernel.

Example. Let $\mathbf{X} = (\{0,1\} \times [0,1]) \cup \{\Delta\}$ and $\mathbf{A} = [0,1]$ with the Euclidean topology in $[0,1]$, the discrete topology in $\{0,1\}$, and the product topology in $\{0,1\} \times [0,1]$; Δ is the isolated cemetery. The cost function $c(\cdot,\cdot)$ is immaterial and, for all $x \in [0,1]$,

$$p((1,a)|(0,x),a) = p(\Delta|(1,x),a) = p(\Delta|\Delta,a) \equiv 1,$$

with all the other transition probabilities being zero. Here and below, the elements of $\mathbf{X}$ are denoted as $\bar{x}$, and the generic notation of the elements of $\mathbf{X} \setminus \{\Delta\}$ is $(k,x) \in \{0,1\} \times [0,1]$; $\bar{x}_0 = (0,0)$ is the initial state. (See Fig. 2.37.)

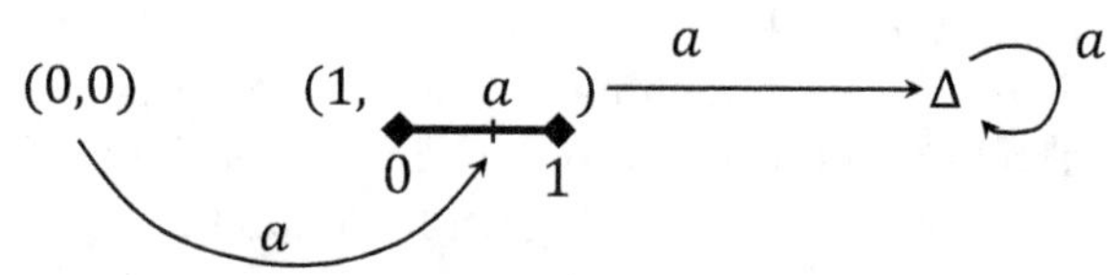

Fig. 2.37 Weak and *ws*-topologies are different.

This MDP is obviously uniformly absorbing. Condition 2.4(c) is not satisfied because, for the measurable function $u((k,x)) := \mathbb{I}\{x > 0\}$, $u(\Delta) := 0$, we have

$$\int_{\mathbf{X}} u((k,y))p(d(k,y)|(0,0),a) = u((1,a)) = \mathbb{I}\{a > 0\},$$

which is not a continuous function of $a \in \mathbf{A}$. On the other hand, for each bounded continuous function $u(\cdot)$ on $\mathbf{X}$ (and omitting the trivial case of $\bar{x} = \Delta$), we have

$$\int_{\mathbf{X}} u(\bar{y})p(d\bar{y}|\bar{x},a) = \int_{\mathbf{X}} u(\bar{y})p(d\bar{y}|(n,x),a) = \begin{cases} u(\Delta), & \text{if } n = 1; \\ u((1,a)), & \text{if } n = 0, \end{cases}$$

which is continuous in $(\bar{x}, a)$. Hence, the transition probability p is a continuous stochastic kernel.

Now consider the sequence of Markov strategies $\{\pi^{\mathrm{m},i}\}_{i=1}^{\infty}$, where $\pi_1^{\mathrm{m},i}(da|\bar{x}) \equiv i\mathbb{I}\{a \in [0,1/i]\}da$ (the Lebesgue measure); $\pi_t^{\mathrm{m},i}(da|\bar{x}) \equiv \mu(da)$ for all $t \geq 2$ for some fixed measure μ on $\mathcal{B}(\mathbf{A})$. Let $\pi^{\mathrm{m},\infty}$ be the Markov strategy given by $\pi_1^{\mathrm{m},\infty}(da|\bar{x}) \equiv \delta_0(da)$; $\pi_t^{\mathrm{m},\infty}(da|\bar{x}) \equiv \mu(da)$ for all $t \geq 2$.

Proposition 2.6.

(a) *The sequence of occupation measures* $\{\eta^{\pi^{\mathrm{m},i}}_{(0,0)}\}_{i=1}^{\infty}$ *converges to* $\eta^{\pi^{\mathrm{m},\infty}}_{(0,0)}$ *as* $i \to \infty$ *in the weak topology.*

(b) *The sequence* $\{\eta^{\pi^{\mathrm{m},i}}_{(0,0)}\}_{i=1}^{\infty}$ *does not contain any convergent subsequence in* $\mathbf{P}((\mathbf{X} \setminus \{\Delta\}) \times \mathbf{A})$ *with respect to the ws-topology.*

The proof can be found in [Zhang and Zheng (2024)] and is presented in Appendix A for consistency.

As was explained after Theorem 2.14, in this example the space $\mathcal{D}_{(0,0)}$ is weakly compact. At the same time, it is not compact in the ws-topology because, otherwise, according to Proposition 8.10.64 of [Bogachev (2007)] it would have been metrizable and thus sequentially ws-compact contradicting part (b) of Proposition 2.6. □

The next example shows that, in the framework of Proposition 2.5, the weak and strong topologies in the space of marginal occupation measures $\{\hat{\eta}(dx) := \eta(dx \times \mathbf{A}),\ \eta \in \mathcal{D}_{x_0}\}$ are not equivalent. We shall also show that the space $\mathcal{D}_{x_0}$ is not weakly compact.

Example. Let $\mathbf{X} = \{1, b_1, b_2, \ldots\} \cup \{\Delta\}$, where $b_n := \sum_{j=1}^{n} \frac{1}{2^j}$ for $n = 1, 2, \ldots$, and $\mathbf{A} = \{1, 2, 3\}$. The topology in $\{1, b_1, b_2, \ldots\}$ is the relative Euclidean, Δ is the isolated point, and the topology in $\mathbf{A}$ is discrete. The cost function $c(\cdot,\cdot)$ is immaterial, $x_0 = b_1 = \frac{1}{2}$ is the initial state, and

$$
\begin{aligned}
&p(\{\Delta\}|\Delta, a) = p(\{\Delta\}|1, a) \equiv 1; \quad p(\{\Delta\}|b_1, 1) = 1;\\
&p(\{b_n\}|b_n, 1) = 1 - \frac{1}{2^{n-2}}; \quad p(\{\Delta\}|b_n, 1) = \frac{1}{2^{n-2}} \quad \forall n \geq 2;\\
&p(\{\Delta\}|b_n, 2) = p(\{b_{n+1}\}|b_n, 2) \equiv \frac{1}{2}, \quad \forall n \geq 1;\\
&p(\{\Delta\}|b_n, 3) = p(\{1\}|b_n, 3) \equiv \frac{1}{4}; \quad p(\{b_{n+1}\}|b_n, 3) = \frac{1}{2}, \quad \forall n \geq 1
\end{aligned}
$$

with all the other transition probabilites being zero. See Fig. 2.38.

This MDP satisfies all the conditions of Proposition 2.5. Indeed, Conditions 2.4(a,c) are satisfied because the topology in $\mathbf{A}$ is discrete. To show that this model is absorbing, we need to prove that

$$
\begin{aligned}
\sup_{\pi} \mathsf{E}^{\pi}_{b_1}[T_\Delta] &= \sup_{\pi} \eta^{\pi}_{b_1}((\mathbf{X} \setminus \{\Delta\}) \times \mathbf{A}) = \sup_{\pi} \mathsf{E}^{\pi}_{b_1}\left[\sum_{t=1}^{\infty} \mathbb{I}\{X_t \neq \Delta\}\right]\\
&= -\inf_{\pi} \mathsf{E}^{\pi}_{b_1}\left[\sum_{t=1}^{\infty} -\mathbb{I}\{X_t \neq \Delta\}\right] < \infty,
\end{aligned}
$$

where $T_\Delta := \min\{t \geq 0 :\ X_t = \Delta\}$. Let us introduce the cost function $c(x, a) := -\mathbb{I}\{x \neq \Delta\}$. Then, using Theorem 2.2, we see that the Bellman

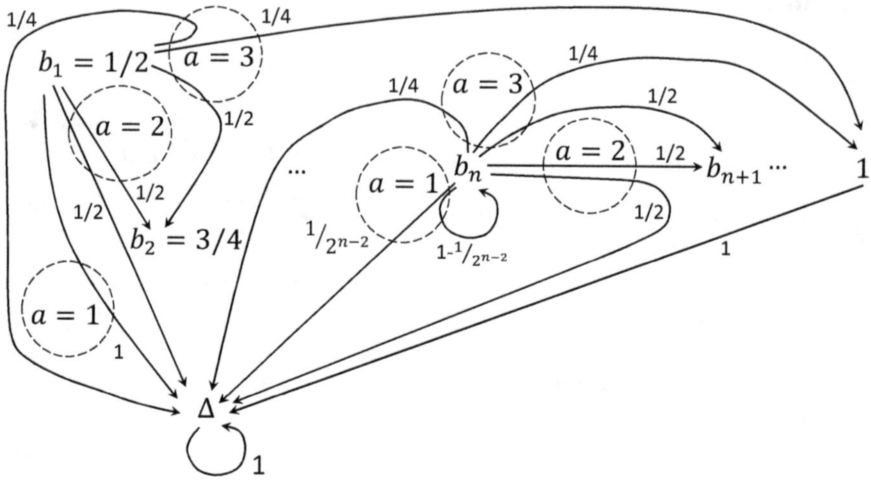

Fig. 2.38 Weak and strong topologies in the space of marginal occupation measures are different.

function $v_x^* := \inf_\pi \mathsf{E}_x^\pi \left[\sum_{t=1}^\infty -\mathbb{I}\{X_t \neq \Delta\}\right]$ is the maximal non-positive solution to the optimality equation (2.5) looking as follows:

$$v(\Delta) = 0; \qquad v(1) = -1;$$
$$v(b_n) = -1 + \min\left\{\mathbb{I}\{n > 2\}\left(1 - \frac{1}{2^{n-2}}\right) v(b_n),\ \frac{1}{2}v(b_{n+1}),\right.$$
$$\left.\frac{1}{2}v(b_{n+1}) - \frac{1}{4}\right\},\ n = 1, 2, \ldots.$$

It is easy to check by direct substitution that function

$$v(b_n) := -\left(\frac{1}{2} + 2 + 2^{n-2}\right), \quad n = 1, 2, \ldots$$

satisfies the presented equation meaning that $0 \geq v_{b_n}^* \geq v(b_n) > -\infty$. Thus, $\sup_\pi \mathsf{E}_{b_1}^\pi[T_\Delta] < \infty$.

One can show that this model is not uniformly absorbing [Zhang and Zheng (2024), Prop. 3.3(b)].

Now consider the following sequence of stationary selectors:

$$\varphi^i(x) := 2\mathbb{I}\{x \in \{b_1, \ldots, b_i\}\} + \mathbb{I}\{x \in \{\Delta, 1, b_{i+1}, b_{i+2}, \ldots\}\}, \quad i = 1, 2, \ldots.$$

The corresponding occupation measures (on $\mathcal{B}((\mathbf{X} \setminus \{\Delta\}) \times \mathbf{A})$) have the form

$$\eta^{\varphi^i}_{1/2}(x,a) = \begin{cases} \left(\frac{1}{2}\right)^{j-1}, & \text{if } x = b_j \in \{b_1, b_2, \ldots, b_i\},\ a = 2; \\ \left(\frac{1}{2}\right)^{i} \cdot 2^{i-1} & \\ = \frac{1}{2}, & \text{if } x = b_{i+1},\ a = 1; \\ 0, & \text{otherwise}, \end{cases} \qquad i = 1, 2, \ldots$$

In the presented expression, 2^{i-1} is the expected time spent by the controlled process X_t in state b_{i+1} (under action $a = 1$) after that state is reached.

Let us show that the sequence of finite measures $\{\eta^{\varphi^i}_{1/2}\}_{i=1}^{\infty}$ has no subsequence convergent in the weak topology (hence, in the ws-topology) to a measure from $\mathcal{D}_{1/2}$. Assume for contradiction that $\eta^{\varphi^{i_j}}_{1/2} \to \eta \in \mathcal{D}_{1/2}$ as $j \to \infty$. For continuous functions $u_n(x,a) := \mathbb{I}\{x = b_n,\ a = 1\}$ on $\{1, b_1, b_2, \ldots\} \times \mathbf{A}$

$$\lim_{j\to\infty} \sum_{(x,a)\in\{1,b_1,b_2,\ldots\}\times\mathbf{A}} u_n(x,a)\eta^{\varphi^{i_j}}_{1/2}(x,a) = 0 = \eta(b_n, 1), \quad n = 1, 2, \ldots$$

For $u(x,a) := \mathbb{I}\{a = 1\}$

$$\lim_{j\to\infty} \sum_{(x,a)\in\{1,b_1,b_2,\ldots\}\times\mathbf{A}} u(x,a)\eta^{\varphi^{i_j}}_{1/2}(x,a) = 1/2$$

$$= \sum_{(x,a)\in\{1,b_1,b_2,\ldots\}\times\mathbf{A}} u(x,a)\eta(x,a);$$

hence $\eta(1,1) = 1/2$. From equation (2.13) we have

$$1/2 \le \eta(1 \times \mathbf{A}) = \sum_{n=1}^{\infty} p(\{1\}|b_n, 3)\eta(b_n, 3),$$

so that $\eta(b_{\hat{n}}, 3) > 0$ for some $\hat{n} = 1, 2, \ldots$. But for the continuous function $u := \mathbb{I}\{x = b_{\hat{n}}, a = 3\}$

$$\lim_{j\to\infty} \sum_{(x,a)\in\{1,b_1,b_2,\ldots\}\times\mathbf{A}} u(x,a)\eta^{\varphi^{i_j}}_{1/2}(x,a) = 0$$

$$\neq \sum_{(x,a)\in\{1,b_1,b_2,\ldots\}\times\mathbf{A}} u(x,a)\eta(x,a) = \eta(b_{\hat{n}}, 3) > 0.$$

The obtained contradiction confirms that the weak limit $\lim_{j\to\infty} \eta^{\varphi^{i_j}}_{1/2}$ (if exists) cannot belong to $\mathcal{D}_{1/2}$. Thus the space $\mathcal{D}_{1/2}$ is not weakly compact (hence, not ws-compact), consistently with Theorem 2.15(b): recall that the model is not uniformly absorbing.

Now introduce the stationary selector $\varphi(x) \equiv 3$ and show that the sequence of marginal occupation measures $\hat{\eta}^{\varphi^i}_{1/2}$ converges weakly to $\hat{\eta}^{\varphi}_{1/2}$, but does not contain any strongly convergent subsequence in $\{\hat{\eta}(dx) := \eta(dx \times \mathbf{A}),\ \eta \in \mathcal{D}_{1/2}\}$. The occupation measure of φ looks as following:

$$\eta^{\varphi}_{1/2}(x,a) = \begin{cases} \frac{1}{2}, & \text{if } x = 1,\ a = 3; \\ \left(\frac{1}{2}\right)^{n-1}, & \text{if } x = b_n,\ a = 3; \\ 0, & \text{otherwise.} \end{cases}$$

The first expression here follows from the straightforward calculation

$$\begin{aligned} \eta^{\varphi}_{1/2}(1,3) &= \mathsf{P}^{\varphi}_{1/2}(X_1 = 1) + \mathsf{P}^{\varphi}_{1/2}(X_1 \notin \{\Delta, 1\},\ X_2 = 1) \\ &\quad + \mathsf{P}^{\varphi}_{1/2}(X_1, X_2 \notin \{\Delta, 1\},\ X_3 = 1) + \dots \\ &= \frac{1}{4} + \frac{1}{2} \cdot \frac{1}{4} + \frac{1}{2^2} \cdot \frac{1}{4} + \dots = \frac{1}{2}; \end{aligned}$$

the second expression can be justified similarly.

For an arbitrarily fixed bounded continuous function $f(\cdot)$ on $\{1, b_1, b_2, \dots\}$ we have

$$\begin{aligned} &\left| \sum_{x \in \{1, b_1, b_2, \dots\}} f(x)\hat{\eta}^{\varphi}_{1/2}(x) - \sum_{x \in \{1, b_1, b_2, \dots\}} f(x)\hat{\eta}^{\varphi^i}_{1/2}(x) \right| \\ &= \left| f(1)\hat{\eta}^{\varphi}_{1/2}(1) + \sum_{n=1}^{\infty} f(b_n)\hat{\eta}^{\varphi}_{1/2}(b_n) \right. \\ &\qquad \left. - \sum_{j=1}^{i} f(b_j)\hat{\eta}^{\varphi^i}_{1/2}(b_j) - f(b_{i+1})\hat{\eta}^{\varphi^i}_{1/2}(b_{i+1}) \right| \\ &= \left| f(1)\hat{\eta}^{\varphi}_{1/2}(1) - f(b_{i+1})\hat{\eta}^{\varphi^i}_{1/2}(b_{i+1}) + \sum_{n=i+1}^{\infty} f(b_n)\hat{\eta}^{\varphi}_{1/2}(b_n) \right|, \\ &\le \left| \frac{1}{2} f(1) - \frac{1}{2} f(b_{i+1}) \right| + \sum_{n=i+1}^{\infty} |f(b_n)| \hat{\eta}^{\varphi}_{1/2}(b_n). \end{aligned}$$

The second equality holds because $\hat{\eta}^{\varphi^i}_{1/2}(b_j) = \hat{\eta}^{\varphi}_{1/2}(b_j) = \left(\frac{1}{2}\right)^{j-1}$ for all $b_j \in \{b_1, b_2, \dots, b_i\}$. The very last term approaches zero when $i \to \infty$ as the tail of the convergent series. Since $b_{i+1} = \sum_{j=1}^{i+1} \frac{1}{2^j} \to 1$ as $i \to \infty$ and the function $f(\cdot)$ is continuous, we conclude that $\lim_{i\to\infty} \left|\frac{1}{2} f(1) - \frac{1}{2} f(b_{i+1})\right| = 0$ and hence $\lim_{i\to\infty} \sum_{x \in \{1, b_1, b_2, \dots\}} f(x)\hat{\eta}^{\varphi^i}_{1/2}(x) = \sum_{x \in \{1, b_1, b_2, \dots\}} f(x)\hat{\eta}^{\varphi}_{1/2}(x)$, so that $\hat{\eta}^{\varphi^i}_{1/2} \to \hat{\eta}^{\varphi}_{1/2}$ (as $i \to \infty$) in the weak topology.

Assume for contradiction that a subsequence $\{\hat{\eta}_{1/2}^{\varphi^{i_j}}\}_{j=1}^{\infty}$ has a strong limit $\hat{\eta}$ as $j \to \infty$ in $\{\hat{\eta}(dx) := \eta(dx \times \mathbf{A}),\ \eta \in \mathcal{D}_{1/2}\}$. Since $\hat{\eta}_{1/2}^{\varphi^i} \to \hat{\eta}_{1/2}^{\varphi}$ (as $i \to \infty$) in the weak topology and the strong topology is finer than the weak topology, $\hat{\eta} = \hat{\eta}_{1/2}^{\varphi}$. On the other hand, consider the bounded measurable function $g(x) := \mathbb{I}\{x = 1\}$ on $\{1, b_1, b_2, \ldots\}$.

$$\sum_{x \in \{1, b_1, b_2, \ldots\}} g(x)\hat{\eta}_{1/2}^{\varphi^{i_j}}(x) = 0 \text{ and } \sum_{x \in \{1, b_1, b_2, \ldots\}} g(x)\hat{\eta}_{1/2}^{\varphi}(x) = \frac{1}{2}$$

meaning that $\hat{\eta} \neq \hat{\eta}_{1/2}^{\varphi}$. The obtained contradiction confirms that the sequence $\{\hat{\eta}_{1/2}^{\varphi^i}\}_{i=1}^{\infty}$ does not contain any strongly convergent subsequence in $\{\hat{\eta}(dx) := \eta(dx \times \mathbf{A}),\ \eta \in \mathcal{D}_{1/2}\}$. $\square$

2.3.28 *Convex combinations of occupation measures*

In this subsection, we consider an MDP with the costless cemetery Δ: $p(\Delta|\Delta, a) \equiv 1$.

According to Proposition 2.1, the set $\mathcal{D}_{\mathsf{P}_0}$ of all occupation measures (under a fixed initial distribution P_0) is convex. It is easy to see that the set of all occupation measures with $\eta_{\mathsf{P}_0}^{\pi}((\mathbf{X} \setminus \{\Delta\}) \times \mathbf{A}) < \infty$, denoted below as $\mathcal{D}_{\mathsf{P}_0}^{f}$, is also convex: see Proposition 3 of [Piunovskiy and Zhang (2024a)] for more details. As usual, if $\mathsf{P}(dx) = \delta_{x_0}(dx)$, then the notation $\mathcal{D}_{x_0}^{f}$ is in use.

In the current subsection, we assume that the initial distribution P_0 is fixed and, for simplicity, degenerate: $\mathsf{P}_0(dx) = \delta_{x_0}(dx)$, $x_0 \neq \Delta$. The case of arbitrary P_0 is in fact not more general: one can add a fictitious isolated initial state $\circ$ to $\mathbf{X}$, put $p(dx|\circ, a) \equiv \mathsf{P}_0(dx)$, $p(\{\circ\}|x, a) \equiv 0$ for $x \in \mathbf{X}$, $c(\circ, a) \equiv 0$, and fix the initial state in the modified model as $x_0 := \circ$.

In view of the above, if $\eta_{x_0}^{\pi^1}, \eta_{x_0}^{\pi^2}, \ldots, \eta_{x_0}^{\pi^L}$ is a finite collection of occupation measures from $\mathcal{D}_{x_0}$ or from $\mathcal{D}_{x_0}^{f}$, then, for a set $\alpha_1, \alpha_2, \ldots, \alpha_L$ of non-negative weights with $\sum_{l=1}^{L} \alpha_l = 1$, the convex combination $\sum_{l=1}^{L} \alpha_l \eta_{x_0}^{\pi^l} = \eta_{x_0}^{\pi}$ is an occupation measure from $\mathcal{D}_{x_0}$ or from $\mathcal{D}_{x_0}^{f}$ correspondingly, for some strategy $\pi \in \Pi^{\text{All}}$. For brevity, we call this strategy π a *$\mathcal{D}$-mixture* of the strategies $\pi^1, \pi^2, \ldots, \pi^L$.

If $\mathsf{P}_{x_0}^{\pi} = \sum_{l=1}^{L} \alpha_l \mathsf{P}_{x_0}^{\pi^l}$, then π is a $\mathcal{D}$-mixture of $\pi_1, \pi_2, \ldots, \pi_L$. On the other hand, since the same occupation measure may be generated by different control strategies, if π is a $\mathcal{D}$-mixture of $\pi_1, \pi_2, \ldots, \pi_L$, that is, $\eta_{x_0}^{\pi} = \sum_{l=1}^{L} \alpha_l \eta_{x_0}^{\pi^l}$, then it can easily happen that $\mathsf{P}_{x_0}^{\pi} \neq \sum_{l=1}^{L} \alpha_l \mathsf{P}_{x_0}^{\pi^l}$. The following simple example illustrates this.

Example. Let $\mathbf{X} = \{\Delta, 1\}$, $\mathbf{A} = \{\alpha, \beta, \gamma\}$, $p(\Delta|\Delta, a) \equiv 1$, $p(1|1, \alpha) = p(1|1, \beta) = p(\Delta|1, \gamma) = 1$, all the other transition probabilities begin zero. The cost function $c(\cdot, \cdot)$ is of no importance. Assume that $x_0 = 1$. (See Fig. 2.39.)

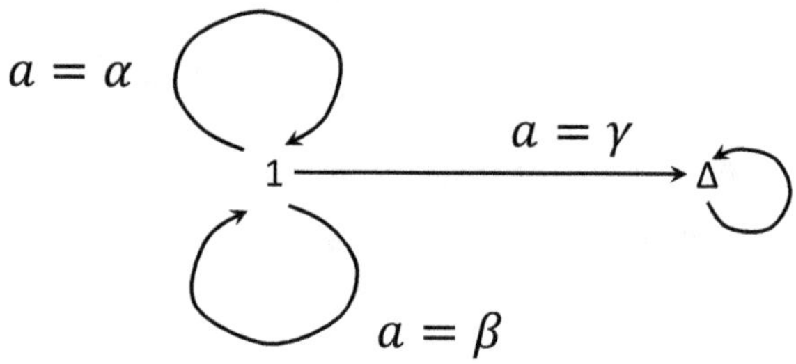

Fig. 2.39 Multiple strategies generating the same occupation measure.

Consider the following control strategies:

$$\pi^\alpha : \pi_1^\alpha(\alpha|h_0) = \pi_2^\alpha(\alpha|h_1) = 1,\ \pi_3^\alpha(\gamma|h_2) = \pi_4^\alpha(\gamma|h_3) = \ldots = 1;$$
$$\pi^\beta : \pi_1^\beta(\beta|h_0) = \pi_2^\beta(\beta|h_1) = 1,\ \pi_3^\beta(\gamma|h_2) = \pi_4^\alpha(\gamma|h_3) = \ldots = 1.$$

The strategic measure $\mathsf{P}_1^{\pi^\alpha}$ is concentrated on the sequences ω of the form $(1, \alpha, 1, \alpha, 1, \gamma, \Delta, \gamma, \Delta, \ldots)$, and the strategic measure $\mathsf{P}_1^{\pi^\beta}$ is concentrated on the sequences ω of the form $(1, \beta, 1, \beta, 1, \gamma, \Delta, \gamma, \Delta, \ldots)$. The occupation measures $\eta_1^{\pi^\alpha}$ and $\eta_1^{\pi^\beta}$ are as follows:

$$\eta_1^{\pi^\alpha}(1 \times \alpha) = 2,\ \ \eta_1^{\pi^\alpha}(1 \times \beta) = 0,\ \ \eta_1^{\pi^\alpha}(1 \times \gamma) = 1,$$
$$\eta_1^{\pi^\alpha}(\Delta \times \alpha) = 0,\ \eta_1^{\pi^\alpha}(\Delta \times \beta) = 0,\ \eta_1^{\pi^\alpha}(\Delta \times \gamma) = \infty;$$
$$\eta_1^{\pi^\beta}(1 \times \alpha) = 0,\ \ \eta_1^{\pi^\beta}(1 \times \beta) = 2,\ \ \eta_1^{\pi^\beta}(1 \times \gamma) = 1,$$
$$\eta_1^{\pi^\beta}(\Delta \times \alpha) = 0,\ \eta_1^{\pi^\beta}(\Delta \times \beta) = 0,\ \eta_1^{\pi^\beta}(\Delta \times \gamma) = \infty.$$

Consider the convex combination $\eta := \frac{1}{2}\eta_1^{\pi^\alpha} + \frac{1}{2}\eta_1^{\pi^\beta}$:

$$\eta(1 \times \alpha) = 1,\ \ \eta(1 \times \beta) = 1,\ \ \eta(1 \times \gamma) = 1,$$
$$\eta(\Delta \times \alpha) = 0,\ \eta(\Delta \times \beta) = 0,\ \eta(\Delta \times \gamma) = \infty.$$

Firstly, $\eta = \eta_1^\pi$, where the strategy π, given by

$$\pi_1(\alpha|1) = \frac{1}{2},\ \pi_1(\beta|1) = \frac{1}{2},\ \pi_2(\alpha|1, \alpha, 1) = 1,\ \pi_2(\beta|1, \beta, 1) = 1,$$
$$\pi_3(\gamma|h_2) = \pi_4(\gamma|h_3) = \ldots = 1$$

(other kernels $\pi_2(a|h_1)$ are immaterial), is a $\mathcal{D}$-mixture of π^α and π^β, in which case $\mathsf{P}_1^\pi = \frac{1}{2}\mathsf{P}_1^{\pi^\alpha} + \frac{1}{2}\mathsf{P}_1^{\pi^\beta}$. In terms of Subsection 2.3.1, the strategy π is mixed over π^α and π^β. But the strategy $\hat{\pi}$, defined as

$$\hat{\pi}_1(\alpha|h_0) = \hat{\pi}_1(\beta|h_0) = \frac{1}{2}, \quad \hat{\pi}_2(\alpha|h_1) = \hat{\pi}_2(\beta|h_1) = \frac{1}{2},$$
$$\hat{\pi}_3(\gamma|h_2) = \hat{\pi}_4(\gamma|h_3) = \ldots = 1$$

is also a $\mathcal{D}$-mixture of π^α and π^β because $\eta = \eta_1^{\hat{\pi}}$. At the same time, $\mathsf{P}_1^{\hat{\pi}} \neq \frac{1}{2}\mathsf{P}_1^{\pi^\alpha} + \frac{1}{2}\mathsf{P}_1^{\pi^\beta}$, since

$$\mathsf{P}_1^{\hat{\pi}}((1,\alpha,1,\beta,1,\gamma,\Delta,\ldots)) = \frac{1}{4}$$

and

$$\mathsf{P}_1^{\pi^\alpha}((1,\alpha,1,\beta,1,\gamma,\Delta,\ldots)) = \mathsf{P}_1^{\pi^\beta}((1,\alpha,1,\beta,1,\gamma,\Delta,\ldots)) = 0.$$

□

Convex combinations of occupation measures and $\mathcal{D}$-mixtures of strategies appear in constrained optimization considered below.

Suppose there are several cost functions $c_0(x,a), c_1(x,a), \ldots, c_J(x,a)$, $J \geq 1$, $c_j(\Delta,a) \equiv 0$, and consider the *constrained* MDP under a fixed initial state $x_0 \in \mathbf{X} \setminus \{\Delta\}$, where the problem is as follows:

$$\mathsf{E}_{x_0}^\pi\left[\sum_{t=1}^\infty c_0(X_{t-1},A_t)\right] \to \inf_\pi \tag{2.45}$$
$$\text{subject to} \quad \mathsf{E}_{x_0}^\pi\left[\sum_{t=1}^\infty c_j(X_{t-1},A_t)\right] \leq d_j, \ j=1,2,\ldots,J.$$

Here $d_1, d_2, \ldots, d_J$ are fixed numbers.

Definition 2.10.

(a) A strategy π is called *feasible* if all the constraints in (2.45) are satisfied; it is called *feasible with a finite value* if, additionally, $-\infty < \mathsf{E}_{x_0}^\pi\left[\sum_{t=1}^\infty c_0(X_{t-1},A_t)\right] < \infty$; it is called *optimal* if it solves problem (2.45).

(b) A constrained MDP is called *semi-continuous* if the spaces $\mathbf{X}$ and $\mathbf{A}$ are topological Borel spaces, the action space $\mathbf{A}$ is compact, the transition probability $p(dy|x,a)$ is (weakly) continuous, and the cost functions $c_j(x,a)$, $j=0,1,\ldots,J$ are lower semi-continuous.

Theorem 2.16. [Piunovskiy and Zhang (2024a), Thm. 2]. *Suppose the constrained MDP is weakly semi-continuous in the sense of Condition 2.3, $c_j(\cdot,\cdot) \geq 0$ for all $j = 0, 1, \ldots, J$, and there exists a feasible strategy π with a finite value. Furthermore, assume that, for each strategy π such that $\eta^\pi_{x_0} \notin \mathcal{D}^f_{x_0}$, there is some $\tilde{j} \in \{0, 1, \ldots, J\}$, possibly depending on π, satisfying $\int_{\mathbf{X}\times\mathbf{A}} c_{\tilde{j}}(x,a)\eta^\pi_{x_0}(dx \times da) = \infty$. Recall that $\mathcal{D}^f_{x_0} := \{\eta^\pi_{x_0} : \pi \in \Pi^{\text{All}}, \ \eta^\pi_{x_0}((\mathbf{X} \setminus \{\Delta\}) \times \mathbf{A}) < \infty\}$.*

Then there exists an optimal strategy in problem (2.45) in the form of a $\mathcal{D}$-mixture of $J+1$ stationary selectors.

Let us emphasize that discounted MDP is a special case of the (uniformly absorbing) model with the expected total cost, where $\mathcal{D}^f_{x_0} = \mathcal{D}_{x_0}$: see Section 3.1. Thus, the corresponding version of Theorem 2.16 holds for discounted MDPs, too.

The above theorem asserts the existence of an optimal strategy in the form of a $\mathcal{D}$-mixture of $J+1$ stationary selectors. It does not claim that *every* optimal strategy can be represented as a $\mathcal{D}$-mixture of finitely many stationary selectors. Below, we adapt Example 3.3.3 of [Kallenberg (1983)] to demonstrate this.

Example. Consider the MDP with $\mathbf{X} = \{\Delta, 0, 1, 2\}$, Δ being the costless cemetery, $\mathbf{A} = \{0, 1\}$, $p(1|1,0) = 1$, $p(2|1,1) = 1$, $p(2|2,0) = 1$, $p(2|2,1) = p(\Delta|2,1) = \frac{1}{2}$, $p(1|0,a) = p(2|0,a) \equiv \frac{1}{2}$, $p(\Delta|\Delta,a) \equiv 1$, all the other transition probabilities being zero. The state and action spaces are endowed with their discrete topologies. Let $x_0 = 0$. Let $c_0(x,a) \equiv 0$, $c_1(x,a) = 1$ for $x = 1, 2$, and $c_1(0,a) \equiv 0$. Let $J = 1$ and $d_1 = 3$. See Fig. 2.40 for illustration.

Any feasible strategy will be optimal, and any non-absorbing strategy π will be unfeasible with $\mathsf{E}^\pi_{x_0}\left[\sum_{n=0}^\infty c_1(X_n, A_{n+1})\right] = \infty$. Here, we call a strategy π non-absorbing if $\mathsf{E}^\pi_0[T_0] = \infty$, where, as usual, $T_0 := \min\{t > 0 : \ X_t = \Delta\}$. All the conditions in Theorem 2.16 are satisfied.

The class Φ of absorbing stationary selectors is specified by $\varphi(1) = \varphi(2) = 1$: the state 0 is essentially uncontrolled, and $\varphi(0)$ is immaterial as well as $\varphi(\Delta)$. In what follows, we put $\pi_t(da|h_{t-1}) \equiv \delta_0(da)$ for all $t = 1, 2, \ldots$ and for all histories h_{t-1} with $x_{t-1} \in \{0, \Delta\}$. Clearly, $\mathcal{D}^f_0$ is a proper subset of $\mathcal{D}_0$.

Consider the stationary strategy defined by $\pi^s(0|1) = \pi^s(1|1) = \frac{1}{2}$; $\pi^s(1|2) = 1$. Then, for each $\varphi \in \Phi$,

$$\mathsf{E}^\varphi_0\left[\sum_{n=0}^\infty c_1(X_n, A_{n+1})\right] =: W_1(0,\varphi) = \frac{1}{2}(1+2) + \frac{1}{2}2 = \frac{5}{2}$$

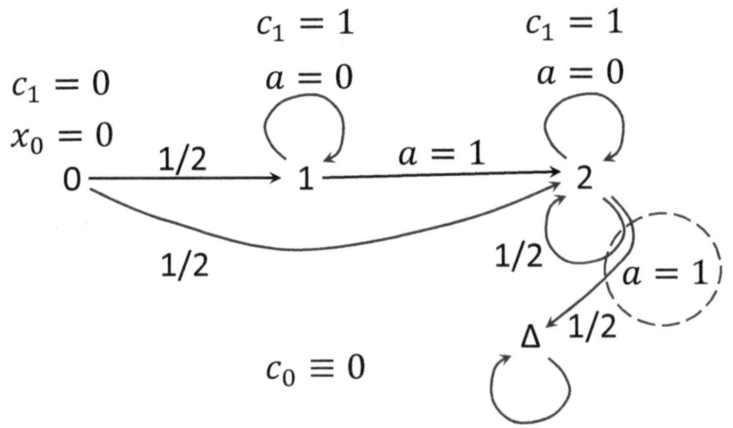

Fig. 2.40 Optimal strategy which is not a mixture of stationary selectors.

and

$$\mathsf{E}_0^{\pi^s}\left[\sum_{n=0}^{\infty} c_1(X_n, A_{n+1})\right] =: W_1(0, \pi^s) = \frac{1}{2}(2+2) + \frac{1}{2}2 = 3.$$

Therefore, π^s and all stationary selectors $\varphi \in \Phi$ are feasible and thus optimal.

On the other hand, $W_1(0, \pi^s) > W_1(0, \varphi)$ for all $\varphi \in \Phi$, so that the occupation measure $\eta_0^{\pi^s}$ cannot be represented as the convex combination of the occupation measures of stationary selectors from Φ. Of course, $\eta_0^{\pi^s}$ cannot be represented as a convex combination of occupation measures of non-absorbing stationary selectors together with the ones from Φ, either.

□

2.3.29 *Bold strategy in gambling is not optimal (house limit)*

Suppose a gambler wishes to obtain at least a certain fortune, say 100, in a primitive casino. He can stake any amount of fortune in his possession but no more than he possesses, and he gains his stake with probability w and loses his stake with the complementary probability $\bar{w} = 1 - w$. How much should the gambler stake every time so as to maximize his chance of eventually obtaining 100?

It is known that the gambler should play boldly in case the casino is unfair ($w < 1/2$): $a = \varphi(x) = \min\{x, \ (100 - x)\}$, where $x \in (0, 100)$ is the current value of the fortune [Dubins and Savage (1965)], [Bertsekas (1987), Section 6.6]. It is worth mentioning that the function $F(x)$, $x \in [0, 100]$, the probability of eventually obtaining 100 if the gambler plays boldly in an unfair casino, exhibits the following properties:

- $F(x)$ increases at all points $x \in [0, 100]$ from 0 to 1 and is continuous;
- $F'(x) = 0$ at all points except for the set of the null Lebesgue measure.

For the proof, see [Billingsley (1983)].

Suppose now that there is a house limit $h \in (0, 100)$. It is known [Wilkins (1972)] that, if $w < 1/2$ and $h = \frac{100}{n}$ for some $n = 1, 2, \ldots$, then the gambler should still play boldly, that is, $a = \varphi^{\text{b}}(x) = \min\{x,\ 100 - x,\ h\}$. But as is shown in [Heath *et al.* (1972)], if $h \in \left(\frac{100}{n+1}, \frac{100}{n}\right)$ for some integer $n \geq 3$, then the bold strategy is not optimal for all w sufficiently close to zero.

Example. One can model this game as an MDP in the following way. $\mathbf{X} = [0, 200)$, $\mathbf{A} = (0, 100)$, and state 0 (meaning the game is over) is absorbing with no future costs. For $0 < x < 100$,

$$p(dy|x, a) = w\delta_{x+a}(dy) + \bar{w}\delta_{\max\{0,(x-a)\}}(dy);$$

$$c(x, a) = \begin{cases} 0, & \text{if } a \leq \min\{x, h\}; \\ +\infty, & \text{if } a > \min\{x, h\}. \end{cases}$$

Infinity corresponds to the stakes which are forbidden. For $x \geq 100$ (i.e., when the goal is reached),

$$p(dy|x, a) = \delta_0(dy), \qquad c(x, a) = -1.$$

Minimizing the total expected cost means the maximization of the chance to eventually obtain 100 or more. See Fig. 2.41.

Below, we assume that $h = 22 \in \left(\frac{100}{5}, \frac{100}{4}\right)$ is an integer, and we consider the finite model $\mathbf{X} = \{0, 1, 2, \ldots, 198\}$, $\mathbf{A} = \{1, 2, \ldots, 99\}$. Following [Heath *et al.* (1972)], we shall prove that the bold strategy is not optimal for small values of w.

Consider the initial state $x_0 = 37$ and action $a = 20$. After the first decision, starting from the new states 57 or 17, the strategy is bold. We call this strategy 'timid' and denote it as $\hat{\varphi}$. We intend to prove that $v_{37}^{\hat{\varphi}} < v_{37}^{\varphi^{\text{b}}}$ for small enough w. (Remember that the performance functional to be minimized equals minus the probability of success.) The main point

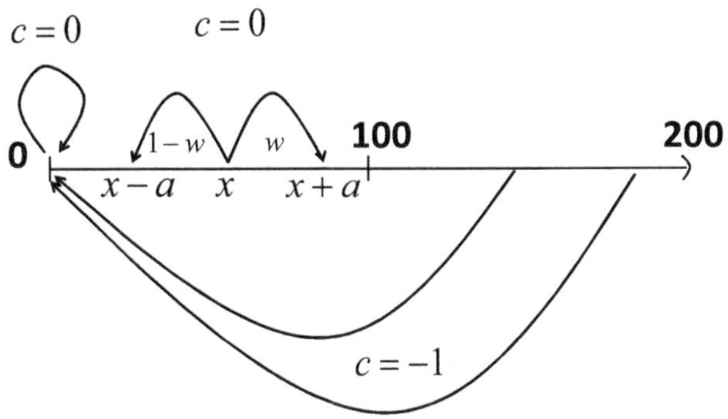

Fig. 2.41 Gambling with the house limit.

is that, starting from $x = 17$, it is possible to reach 100 in 4 steps, but it is impossible to do so starting from $x = 15$, i.e. after losing the bold stake $a = 22$.

When playing boldly, the gambler can win in no more than four successful plays in only the following three cases:

fortune	37	59	81	100		
result of the game		win	win	win		
fortune	37	59	37	59	81	100
result of the game		win	loss	win	win	win
fortune	37	59	81	59	81	100
result of the game		win	win	loss	win	win

Thus, the chance to obtain 100 or more, if the bold gambler wins less than five plays in total, is $w^3 + 2w^4\bar{w}$. When using the timid strategy $\hat{\varphi}$, the value of the fortune will change (decrease by 2), but there will still be no other ways to reach 100 in four or fewer successful plays, apart from the aforementioned additional path

$$37 \to 17 \to 34 \to 56 \to 78 \to 100.$$

Thus, the chance to obtain 100 or more, if the 'timid' gambler wins less than five plays in total, is $w^3 + 3w^4\bar{w}$.

One can estimate the number of plays $M(k)$ such that at the end the bold gambler reaches 0 or 100 experiencing at most k winning bets.

$$M(0) \le 9 \text{ (if starting from } x = 99);$$
$$M(1) \le 8 + 1 + M(0) = 18 :$$

If the first nine plays were unfavourable, then $M(1) \leq 9$. If one of the first nine plays was favourable, then, afterwards, the game is over after less than $M(0)$ plays. And so on

$$M(2) \leq 8 + 1 + M(1) \leq 27;$$
$$M(3) \leq 36; \qquad M(4) \leq 45.$$

Thus, after 46 plays, either the bold gambler wins less than five plays (and in this case the game is finished with the chance to obtain 100 or more being $w^3 + 2w^4\bar{w}$), or (s)he wins five or more plays and gets the further chance to obtain 100 or more. The probability of the latter event is

$$\sum_{i=5}^{46} {}^{46}C_i w^i \bar{w}^{46-i} \leq \sum_{i=0}^{46} {}^{46}C_i w^5 = 2^{46} w^5.$$

Summing up, we conclude that the chance to obtain 100 or more, when playing boldly, is below $w^3 + 2w^4\bar{w} + 2^{46}w^5$, that is,

$$v_{37}^{\varphi^b} \geq -(w^3 + 2w^4\bar{w} + 2^{46}w^5).$$

But, as was shown above, for the timid strategy we have

$$v_{37}^{\hat{\varphi}} \leq -(w^3 + 3w^4\bar{w}),$$

and $v_{37}^{\hat{\varphi}} < v_{37}^{\varphi^b}$ if w is sufficiently small.

The presented calculations are very rough: e.g., for $w = 0.01$ we have $w^3 + 3w^4\bar{w} - (w^3 + 2w^4\bar{w} + 2^{46}w^5) \approx 0.99 \cdot 10^{-8} - 70.37 \cdot 10^2 < 0$. Nevertheless, the numerical calculations for $w = 0.01$ give the following values (φ^* is the optimal stationary selector and $v_x^{\varphi^*} = v_x^*$ is the Bellman function).

x	36	37	38	39
$v_x^{\varphi^b}$	-102041×10^{-11}	-102061×10^{-11}	-102071×10^{-11}	-103070×10^{-11}
$v_x^{\varphi^*}$	-102060×10^{-11}	-103031×10^{-11}	-103050×10^{-11}	-103070×10^{-11}
$\varphi^b(x)$	22	22	22	22
$\varphi^*(x)$	20	19	18	22

The graph of $\varphi^*(x)$ is presented in Fig. 2.42. □

2.3.30 *Bold strategy in gambling is not optimal (inflation)*

The problem under study is due to [Chen *et al.* (2004)], where the situation is described as follows.

A gambler with a (non-random) initial fortune x_0 less than 1, wants to buy a house which sells today for 1. Owing to inflation, the price of the house tomorrow will be $\frac{1}{\beta} \geq 1$, and will continue to go up at this

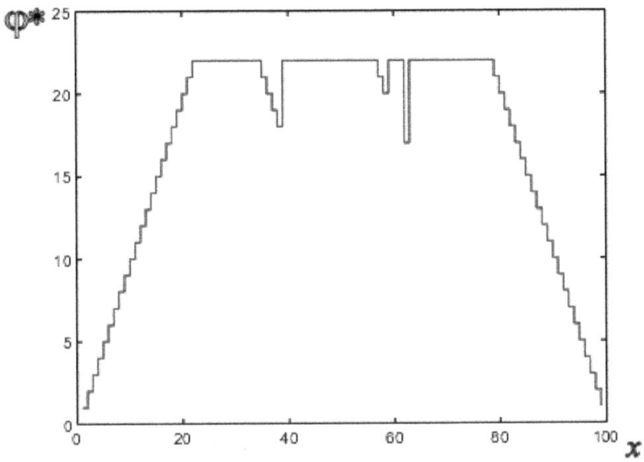

Fig. 2.42 Optimal strategy in gambling with a house limit.

rate each day, so as to become $\left(\frac{1}{\beta}\right)^n$ on the nth day. Once each day, the gambler can stake any amount a of the fortune in his possession, but no more than he possesses, in a primitive casino. If he makes a bet, he gains $r > 0$ times his stake with probability $w \in (0,1)$ and loses his stake with the complementary probability $\bar{w} = 1 - w$. To say differently, if the return on the winning state is ka, then $ra = ka - a$, so $k = r + 1 > 1$. How much should the gambler stake each day so as to maximize his chance of eventually catching up with inflation and being able to buy the house?

It is known that, if there is no inflation ($\beta = 1$), the gambler should play boldly in case the casino is unfair (that is if $w < \frac{1}{1+r}$): $a = \varphi(x) = \max\{x, (1-x)/r\}$, where x is the current value of fortune, since there is no other strategy that offers a higher probability of reaching the goal [Dubins and Savage (1965), Chapter 5]. The presence of inflation would intuitively motivate the gambler to recognize the time value of his fortune, and we would suspect that the gambler should again play boldly. However, in this example we show that bold play is not necessarily optimal. Below, we assume that $\beta < 1$.

To construct the mathematical model, it is convenient to accept that the house price remains at the same level 1, and express the current fortune in terms of the actual house price. In other words, if the fortune today is x, the stake is a, and the gambler wins, then his fortune increases to $(x+ra)\beta$. If he loses, the value becomes $(x-a)\beta$. We assume that $(r+1)\beta > 1$, because otherwise the gambler can never reach his goal.

Example. Therefore, $\mathbf{X} = [0,(r+1)\beta)$, $\mathbf{A} = [0,1)$, and state 0, meaning the game is over, is absorbing with no future costs. For $0 < x < 1$,

$$p(dy|x,a) = w\delta_{(x+ra)\beta}(dy) + \bar{w}\delta_{\max\{0,(x-a)\beta\}}(dy);$$

$$c(x,a) = \begin{cases} 0, & \text{if } a \le x; \\ +\infty, & \text{if } a > x. \end{cases}$$

For $x \ge 1$ (that is, when the goal is reached),

$$p(dy|x,a) = \delta_0(dy); \qquad c(x,a) = -1$$

(see Fig. 2.43). Recall that we deal with minimization problems; the value $c(x,a) = +\infty$ prevents the stakes bigger than the current fortune. Minimizing the total expected cost means the maximization of the chance to eventually buy the house.

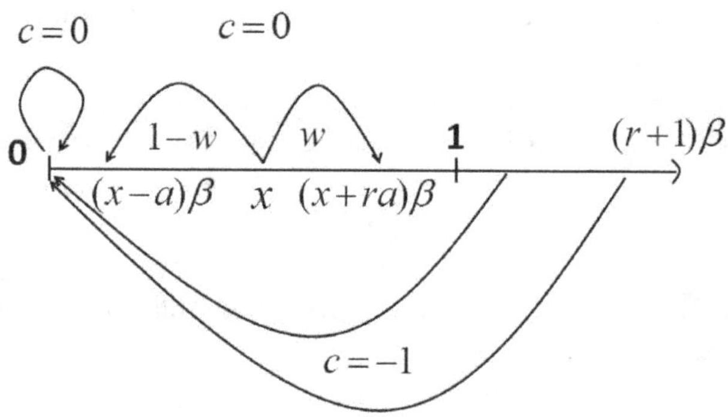

Fig. 2.43 Gambling in the presence of inflation.

Now we are in the framework of problem (2.1), and the Bellman equation (2.5) is expressed as follows:

$$v(x) = \inf_{a\in[0,x]} \{wv((x+ra)\beta) + \bar{w}v((x-a)\beta)\} \text{ if } x < 1;$$

$$v(x) = -1 \text{ if } x \ge 1; \qquad v(0) = 0.$$

Because of inflation, the bold strategy looks slightly different:

$$\varphi^b(x) = \min\left\{x, \frac{1-\beta x}{r\beta}\right\}.$$

Function $v_x^{\varphi^b}$ satisfies equation

$$v_x^{\varphi^b} = wv_{(x+r\varphi^b(x))\beta}^{\varphi^b} + \bar{w}v_{(x-\varphi^b(x))\beta}^{\varphi^b}, \text{ if } x < 1;$$

$$v_x^{\varphi^b} = -1 \text{ if } x \geq 1; \qquad v_0^{\varphi^b} = 0.$$

Therefore,

$$\begin{aligned} v_x^{\varphi^b} &= -1, \text{ if } x \geq 1, \\ v_x^{\varphi^b} &= -w + \bar{w}v_{\frac{\beta(r+1)x-1}{r}}^{\varphi^b}, \text{ if } \frac{1}{(r+1)\beta} \leq x < 1, \\ v_x^{\varphi^b} &= wv_{(r+1)x\beta}^{\varphi^b}, \text{ if } 0 < x < \frac{1}{(r+1)\beta}, \\ v_x^{\varphi^b} &= 0, \text{ if } x = 0. \end{aligned} \tag{2.46}$$

Suppose $r > 1$ and let

$$B = \begin{cases} \max\left\{\frac{1}{r+1}, \ r^{-(1+1/K)}\right\}, & \text{if } K \geq 1; \\ \frac{1}{r+1}, & \text{if } K = 0, \end{cases}$$

where $K := \left\lfloor \frac{\ln(w)}{\ln(\bar{w})} \right\rfloor$ is the integer part. Assume also that $\frac{1}{r+1} < \beta < \min\left\{1, \frac{1}{B(r+1)}\right\}$, so that

$$\gamma := \frac{1}{(r+1)\beta} \in (B, 1).$$

Finally, we fix a positive integer m such that $r\gamma^m < 1 - \gamma$. Note that $1 - \gamma\beta = 1 - \frac{1}{r+1} > 0$ and

$$\gamma \sum_{i=0}^{\infty} (r\gamma^m)^i = \frac{\gamma}{1 - r\gamma^m} < 1. \tag{2.47}$$

If $x = \gamma$ then the second equation in (2.46) implies that $v_\gamma^{\varphi^b} = -w$ and, as the result of the third equation, for all $i = 0, 1, 2, \ldots$,

$$v_{\gamma^i}^{\varphi^b} = -w^i.$$

If $x = \gamma + r\gamma^{i+1} < 1$, then the second equation in (2.46) implies that

$$v_{\gamma+r\gamma^{i+1}}^{\varphi^b} = -w + \bar{w}v_{\gamma^i}^{\varphi^b} = -w - \bar{w}w^i,$$

and, because of the third equation, for all $j = 0, 1, 2, \ldots$,

$$v^{\varphi^b}_{\gamma^{j+1}+r\gamma^{j+i+1}} = -w^j[w + \bar{w}w^i] = -w^{j+1} - \bar{w}w^{i+j}.$$

If we continue in the same way, we see that, in a rather general case,

$$v^{\varphi^b}_x = -\sum_{l=0}^{k} \bar{w}^l w^{n_l - l}, \text{ if } x = \sum_{l=0}^{k} r^l \gamma^{n_l} \tag{2.48}$$

(see [Chen *et al.* (2004), Thm. 2]). In particular, this expression for $v^{\varphi^b}_x$ under $0 < x < \gamma = \frac{1}{(r+1)\beta}$ follows directly from (2.46). So, for

$$0 < \hat{x} := \gamma^2[1 + r\gamma^m + r^2\gamma^{2m} + \cdots + r^{K+2}\gamma^{(K+2)m}] < \gamma \cdot \frac{\gamma}{1 - r\gamma^m} < \gamma$$

we have $0 < \hat{x} < (1 - \beta\hat{x})/(r\beta)$ and

$$v^{\varphi^b}_{\hat{x}} = -w^2[1 + \bar{w}w^{m-1} + \bar{w}^2 w^{2(m-1)} + \cdots + \bar{w}^{K+2} w^{(K+2)(m-1)}].$$

We intend to show that

$$\inf_{a \in [0, \hat{x}]} \left\{ w v^{\varphi^b}_{(\hat{x}+ra)\beta} + \bar{w} v^{\varphi^b}_{(\hat{x}-a)\beta} - v^{\varphi^b}_{\hat{x}} \right\} < 0. \tag{2.49}$$

To do so, take

$$\begin{aligned}
\hat{a} &:= \hat{x} - \gamma\{\gamma^m + r\gamma^{2m} + \cdots + r^K \gamma^{(K+1)m}\}/\beta \\
&= \left(\gamma - \frac{1}{\beta r}\right)\gamma[1 + r\gamma^m + (r\gamma^m)^2 + \ldots + (r\gamma^m)^{K+1}] + \frac{\gamma}{\beta r} + \gamma^2(r\gamma^m)^{K+2} \\
&> \frac{-\gamma^2}{r} \cdot \frac{1}{1 - r\gamma^m} + \frac{\gamma}{\beta r} + \gamma^2(r\gamma^m)^{K+2}.
\end{aligned}$$

The inequality holds because $\gamma - \frac{1}{\beta r} = -\frac{\gamma}{r}$. Thus,

$$\hat{x} > \hat{a} > \frac{\gamma}{\beta r}\left[1 - \frac{\gamma\beta}{1 - r\gamma^m}\right] > 0,$$

since $1 - r\gamma^m > \gamma > \beta\gamma$. Now

$$(\hat{x} - \hat{a})\beta = \gamma^{m+1} + r\gamma^{2m+1} + \cdots + r^K \gamma^{(K+1)m+1},$$

so that, since,

$$0 < (\hat{x} - \hat{a})\beta < \frac{\gamma}{r}\left[\frac{1}{1 - r\gamma^m} - 1\right] < \frac{1}{r} - \frac{\gamma\beta}{r} = \frac{1}{r+1} < \gamma,$$

we have, according to (2.48),

$$v^{\varphi^b}_{(\hat{x}-\hat{a})\beta} = -w^2[w^{m-1} + \bar{w}w^{2(m-1)} + \cdots + \bar{w}^K w^{(K+1)(m-1)}].$$

Now

$$\gamma < (\hat{x} + r\hat{a})\beta = \gamma(1 + (r\gamma^m)^{K+2}) < 1.$$

The last inequality holds according to (2.47).

Thus, according to (2.46),

$$v^{\varphi^b}_{(\hat{x}+r\hat{a})\beta} = -w + \bar{w}v^{\varphi^b}_{r^{K+1}\gamma^{(K+2)m}} \leq -w + \bar{w}v^{\varphi^b}_{\gamma^{(K+2)m-K}},$$

because the function $v^{\varphi^b}_x$ decreases with x [Chen *et al.* (2004), Thm. 1] and because $r^{K+1}\gamma^K \geq 1$ (recall that $\gamma > B \geq r^{-(1+1/K)}$ if $K \geq 1$).

Since, by (2.48), $v^{\varphi^b}_{\gamma^{(K+2)m-K}} = -w^{(K+2)m-K}$ and $0 < w < 1$, we have:

$$\begin{aligned}
&wv^{\varphi^b}_{(\hat{x}+r\hat{a})\beta} + \bar{w}v^{\varphi^b}_{(\hat{x}-\hat{a})\beta} - v^{\varphi^b}_{\hat{x}} \\
&\leq -w^2 + w\bar{w}[-w^{(K+2)m-K}] \\
&\quad -\bar{w}w^2[w^{m-1} + \bar{w}w^{2(m-1)} + \cdots + \bar{w}^K w^{(K+1)(m-1)}] \\
&\quad +w^2[1 + \bar{w}w^{m-1} + \bar{w}^2 w^{2(m-1)} + \cdots + \bar{w}^{K+2} w^{(K+2)(m-1)}] \\
&= \bar{w}w^{(K+2)(m-1)+2}[\bar{w}^{K+1} - w] < 0,
\end{aligned}$$

because $w > \bar{w}^{K+1}$ according to the definition of K: $K + 1 > \frac{\ln(w)}{\ln(\bar{w})}$. Inequality (2.49) is proved. Thus, having $\hat{x}$ in hand, making stake $\hat{a}$ and playing boldly afterwards is better than just playing boldly. For more about this model see [Chen *et al.* (2004)]. □

2.3.31 *Search strategy for a moving target*

Example. Suppose that a moving object is located in one of two possible locations 1 or 2, and moves according to the Markov chain with the transition matrix $\begin{bmatrix} 1-q_1 & q_1 \\ q_2 & 1-q_2 \end{bmatrix}$, where $q_1, q_2 \in (0,1)$. The initial probability of being in location 1, $x_0 \in [0,1]$, is given. The current position is unknown, but on each step the decision maker can calculate the *a posteriori* probability x of the object to be in location 1. Based on this information, the objective is to discover the object in the minimum expected time. Similar problems were studied in [Ross (1983), Chapter III, Section 5].

In fact, we have a model with incomplete state information, and the *a posteriori* probability x is the so-called *sufficient statistic* which plays the role of the state in the MDP under study: see [Bertsekas and Shreve (1978), Chapter 10]. Obviously, x can take values in the segment $[0,1]$, so that the state space is $\mathbf{X} = [0,1] \cup \{\Delta\}$, where Δ means that the object is discovered and the search process stopped. We put $\mathbf{A} = \{1,2\}$. The action

a means looking at the corresponding location. The transition probability (for the sufficient statistic x) and the cost function are given by

$$p(dy|x,a) = \begin{cases} (1-x)\delta_{q_2}(dy), & \text{if } x \in [0,1] \text{ and } a = 1; \\ x\delta_{1-q_1}(dy), & \text{if } x \in [0,1] \text{ and } a = 2; \end{cases}$$

$$p(\Delta|x,a) = \begin{cases} x, & \text{if } x \in [0,1] \text{ and } a = 1; \\ 1-x, & \text{if } x \in [0,1] \text{ and } a = 2; \end{cases}$$

$$p(\Delta|\Delta,a) \equiv 1; \qquad c(x,a) = \begin{cases} 1, & \text{if } x \in [0,1]; \\ 0, & \text{if } x = \Delta. \end{cases}$$

See Fig. 2.44.

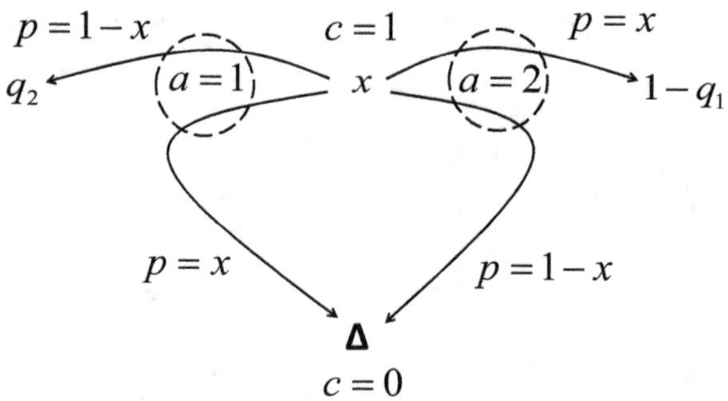

Fig. 2.44 Optimal search.

It seems plausible that the optimal strategy is simply to search the location that gives the highest probability of finding the object:

$$\varphi(x) = \begin{cases} 1, & \text{if } x > 1/2; \\ 2, & \text{if } x \le 1/2. \end{cases} \tag{2.50}$$

We shall show that this strategy is not optimal unless $q_1 = q_2$ or $q_1+q_2 = 1$.

The optimality equation (2.5) can be written as follows

$$v(x) = 1+\min\{(1-x)v(q_2),\ xv(1-q_1)\} \text{ for } x \in [0,1]; \quad v(\Delta) = 0. \tag{2.51}$$

The model under consideration is positive, and, starting from $t = 1$, $X_t \in \{q_2,\ 1-q_1,\ \Delta\}$. Thus, we first solve equation (2.51) for $x \in \{q_2, 1-q_1\}$. The following assertions can be easily verified; see Fig. 2.45.

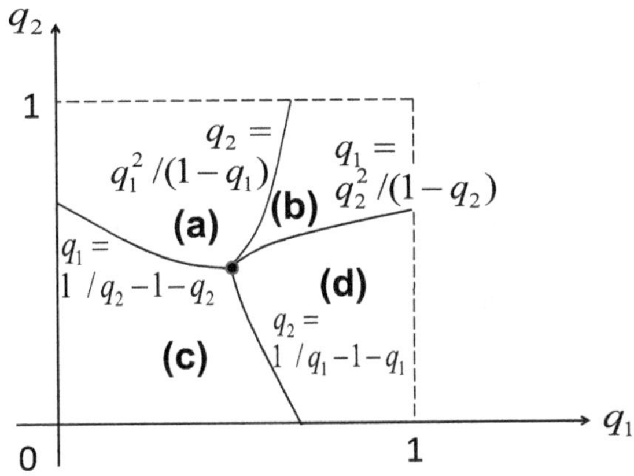

Fig. 2.45 Structure of the Bellman function.

(a) If

$$q_1 \geq \frac{1}{q_2} - 1 - q_2 \text{ and } q_2 \geq \frac{q_1^2}{1-q_1}$$

then

$$v(q_2) = \frac{1}{q_2}; \qquad v(1-q_1) = 1 + \frac{q_1}{q_2},$$

and the stationary selector $\varphi^*(q_2) = \varphi^*(1-q_1) = 1$ provides the minimum in (2.51).

(b) If

$$q_1 \leq \frac{q_2^2}{1-q_2} \text{ and } q_2 \leq \frac{q_1^2}{1-q_1}$$

then

$$v(q_2) = \frac{1}{q_2}; \qquad v(1-q_1) = \frac{1}{q_1},$$

and the stationary selector $\varphi^*(q_2) = 1$, $\varphi^*(1-q_1) = 2$ provides the minimum in (2.51).

(c) If

$$q_1 \leq \frac{1}{q_2} - 1 - q_2 \text{ and } q_2 \leq \frac{1}{q_1} - 1 - q_1$$

then

$$v(q_2) = \frac{1+q_2}{1-q_1q_2}; \qquad v(1-q_1) = \frac{1+q_1}{1-q_1q_2},$$

and the stationary selector $\varphi^*(q_2) = 2$, $\varphi^*(1-q_1) = 1$ provides the minimum in (2.51).

(d) If

$$q_1 \geq \frac{q_2^2}{1-q_2} \text{ and } q_2 \geq \frac{1}{q_1} - 1 - q_1$$

then

$$v(q_2) = 1 + \frac{q_2}{q_1}; \qquad v(1-q_1) = \frac{1}{q_1},$$

and the stationary selector $\varphi^*(q_2) = \varphi^*(1-q_1) = 2$ provides the minimum in (2.51).

For small (large) values of x, the minimum in (2.51) is provided by the second (first) term: $\varphi^*(x) = 2$ ($\varphi^*(x) = 1$) but the critical value of x is not necessarily equal to 1/2. The stationary selector (2.50) is optimal if and only if the situation is symmetric: $v(q_2) = v(1-q_1)$. In cases (a) and (d), we must have $q_1 + q_2 = 1$, and in cases (b) and (c) we see that $q_1 = q_2$.

A good explanation of why the strategy (2.50) is not optimal in the asymmetric case is given in [Ross (1983), Chapter III, Section 5.3]. Suppose $q_1 = 0.99$ and $q_2 = 0.5$. Then we have case (d):

$$\frac{q_2^2}{1-q_2} = 0.5 \leq 0.99 = q_1 \text{ and } \frac{1}{q_1} - 1 - q_1 \approx -0.9799 \leq 0.5 = q_2;$$

$$v(q_2) \approx 1.505 \text{ and } v(1-q_1) \approx 1.010.$$

The critical value of x in (2.51), when $(1-x)v(q_2) = xv(1-q_1)$, equals

$$x^* = \frac{v(q_2)}{v(1-q_1) + v(q_2)} \approx \frac{1.505}{2.515} \approx 0.598.$$

Thus, the optimal stationary selector has the form

$$\varphi(x) = \begin{cases} 1, & \text{if } x > 0.598; \\ 2, & \text{if } x \leq 0.598. \end{cases}$$

If, e.g., $x = 0.55$, then an immediate search of location 1 will discover the object with probability 0.55, whereas a search of location 2 discovers it with probability 0.45. However, an unsuccessful search of location 2 leads to a near-certain discovery at the next stage (because $q_1 = 0.99$ is the probability of the object to be in the second location), whereas an unsuccessful search of location 1 results in complete uncertainty as to where it will be at the time of the next search. It is optimal to search location 2, unless $x = 0.55 > 0.5$. □

2.3.32 *Job search*

We present a slightly simplified version of the model described in [McCall and McCall (2008), §8.5]:

> "An individual is seeking employment in a market in which jobs are homogeneous in all nonwage attributes. Since he/she does not know which wage is being offered by which prospective employer, but only the distribution function F of wages, he/she proceeds by randomly applying for positions, one at a time. For each application submitted there is a cost incurred. The problem, then, is that of determining when to terminate the search process and accept a job offer in order to maximize the expected wage net of search costs."

Example. Assume that the distribution F is concentrated on a finite interval $[0, K]$, has density $f(\cdot)$, and $\int_0^K yf(y)dy > k$, where $k > 0$ is the cost for each application. Here and below, $dy = Leb(dy)$ is the Lebesgue measure. We formulate this search model as the following optimal stopping problem: $\mathbf{X} = [0, K] \cup \{\Delta\}$, where the cemetery Δ means the search process is stopped; $\mathbf{A} = \{s, n\}$, where s (n) means the decision to stop, i.e., accept the current job offer (not to stop).

$$\begin{aligned} c(x,a) &= \begin{cases} k, & \text{if } a = n; \\ -x, & \text{if } a = s \end{cases} \qquad \text{for } x \neq \Delta; \\ p(dy|x,n) &= f(y)dy \qquad \text{for } x \neq \Delta; \\ p(\Delta|x,s) &= 1 \qquad \text{for } x \neq \Delta; \\ \text{and } c(\Delta, a) &\equiv 0, \quad p(\Delta|\Delta, a) \equiv 1. \end{aligned}$$

The endless job search is clearly not optimal leading to the infinite cost. Thus, action s will be used. Now we modify this MDP and make it positive by putting $c(x, s) = K - x$ for $x \neq \Delta$. The performance functional increases by the constant $K > 0$, and the optimal strategies in the initial and modified problem will be the same. The initial state is $x_0 = 0$. See Fig. 2.46 for illustration.

The optimality equation (2.5) takes the form

$$\begin{aligned} v(x) &= \min\left\{k + \int_0^K v(y)f(y)dy; \quad K - x\right\}, \qquad x \in [0, K]; \\ v(\Delta) &= v(\Delta) \end{aligned}$$

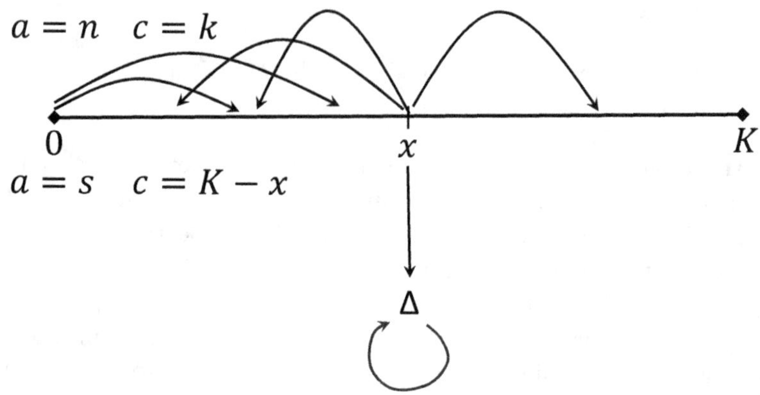

Fig. 2.46 Job search.

and can be solved by value iteration (2.6):

$$v^n(\Delta) \equiv 0;$$
$$v^0(x) = 0 \quad \text{for } x \in [0, K];$$
$$v^{n+1}(x) = \min\left\{k + \int_0^K v^n(y) f(y)dy; \quad K - x\right\} \quad \text{for } x \in [0, K].$$

On each step n, function $v^n(\cdot)$ has the form as on Fig. 2.47:

$$v^n(x) = \begin{cases} C^n, & \text{if } x < K - C^n; \\ K - x, & \text{if } x \geq K - C^n; \end{cases}$$
$$C^0 = 0;$$
$$C^{n+1} = k + C^n \int_0^{K-C^n} f(y)dy + \int_{K-C^n}^K (K - y) f(y)dy, \quad n = 0, 1, \ldots.$$

Since $C^0 = 0 < C^1 = k < K$ (recall that $k < \int_0^K y f(y)dy$) and

$$\frac{d}{dC}\left[C \int_0^{K-C} f(y)dy + \int_{K-C}^K (K - y) f(y)dy\right]$$
$$= \int_0^{K-C} f(y)dy - C f(K - C) + C f(K - C) \geq 0,$$

we conclude that $0 = C^0 < C^1 \leq C^2 \leq \ldots$. Let $C^\infty := \lim_{n\to\infty} C^n$. Since $z^n := K - C^n$ decrease with n starting from $z^0 = K$, $C^\infty \in (0, K)$.

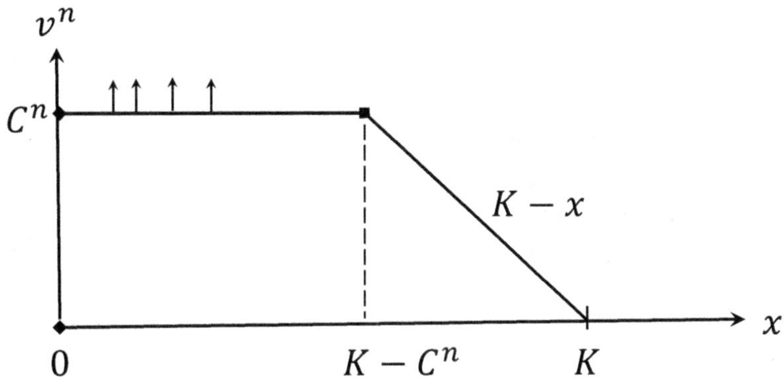

Fig. 2.47 Shape of functions $v^n(\cdot)$. The value of C^n increases with n.

Function

$$v^\infty(x) = \begin{cases} C^\infty, & \text{if } x < K - C^\infty; \\ K - x, & \text{if } x \geq K - C^\infty \end{cases}$$

on $[0, K]$ satisfies optimality equation (2.5):

$$v^\infty(x) = \min\{C^\infty, \quad K - x\}$$

and coincides with the Bellman function v^*_x by Theorem 2.3. (Of course, $v^\infty(\Delta) = v^*_\Delta = 0$.) The critical value C^∞ satisfies equation

$$C^\infty = k + C^\infty \int_0^{K-C^\infty} f(y)dy + \int_{K-C^\infty}^K (K - y)f(y)dy.$$

As claimed in [McCall and McCall (2008), p. 228], the equivalent equation

$$\int_{z^\infty}^K (y - z^\infty)f(y)dy = k$$

for $z^\infty := \lim_{n\to\infty} z^n = K - C^\infty$ has a unique solution.

The optimal strategy (stationary selector φ^*) looks as follows:

$$\varphi^*(x) = \begin{cases} n, & \text{if } x < K - C^\infty = z^\infty; \\ s, & \text{if } x \geq K - C^\infty = z^\infty \end{cases} \quad \text{for } x \in [0, K],$$

or, in words, accept the job offer X_n if and only if $X_n \geq z^\infty$. This selector is conserving and equalizing; its optimality follows, e.g., from Theorem 2.4.

□

2.3.33 *Three-way duel ('Truel')*

The sequential truel is a game that generalizes the simple duel. Three marksmen, A, B, and C, have accuracies $0 < \alpha < \beta < \gamma \leq 1$. Because of this disparity the marksmen agree that A shall shoot first, followed by B, followed by C, this sequential rotation continuing until only one man (the winner) remains standing. When all three men are standing, the active marksman must decide who to shoot at. Note, it is not forbidden to intentionally miss, i.e., shoot at the sky or not shoot at all. Every marksman wants to maximize his probability of winning the game. We assume that every player behaves in the same way under identical circumstances. Obviously, nobody will intentionally miss if only two men are standing (otherwise, the probability of winning is zero for the player who decides not to shoot). The following notation will be used, just for brevity: $p_B(ABC)$ is the probability that B wins the game if all three men are standing and A shoots first; $p_C(BC)$ is the probability that C wins the game if only B and C are standing and B shoots first, and so on. In what follows, $\bar{\alpha} = 1 - \alpha$, $\bar{\beta} = 1 - \beta$ and $\bar{\gamma} = 1 - \gamma$.

Suppose for the moment that it is not allowed to miss intentionally, and consider the behaviour of the marksmen B and C if all three men are standing. The probability $p_C(AC)$ satisfies equation

$$p_C(AC) = \bar{\alpha}[\gamma + \bar{\gamma} p_C(AC)]:$$

C wins with the probability γ if A does not hit him; the probability of reaching the same state AC equals $\bar{\alpha}\bar{\gamma}$. Thus

$$p_C(AC) = \frac{\bar{\alpha}\gamma}{\alpha + \bar{\alpha}\gamma}.$$

Similarly,

$$p_C(BC) = \frac{\bar{\beta}\gamma}{\beta + \bar{\beta}\gamma}.$$

Since $\alpha < \beta$, $p_C(AC) > p_C(BC)$ because the function $\frac{(1-x)\gamma}{x+(1-x)\gamma}$ decreases with $x \geq 0$. Now it is clear that in state CAB, the marksman C will shoot B as the situation is the same, i.e., ABC, if he misses, but if he hits the target, it is better to face A afterwards rather than B who is stronger.

By a similar argument,

$$p_B(AB) = \frac{\bar{\alpha}\beta}{\alpha + \bar{\alpha}\beta}; \quad p_B(CB) = \frac{\bar{\gamma}\beta}{\gamma + \bar{\gamma}\beta},$$

and, in state BCA, marksman B will shoot C.

Below, we assume that marksmen B and C behave as described above, and we allow A to miss intentionally. Now, in the case of marksman A (when he is first to shoot), the game can be modelled by the following MDP.

Example. $\mathbf{X} = \{A, AB, AC, ABC, \Delta\}$, where Δ means the game is over (for A). $\mathbf{A} = \{0, \hat{b}, \hat{c}\}$, where 0 means 'intentionally miss', $\hat{b}(\hat{c})$ means 'shoot B(C)'. In states AB and AC the both actions $\hat{b}$ and $\hat{c}$ mean 'shoot the standing partner' (do not miss intentionally).

$$p(ABC|ABC, a) = \begin{cases} \bar{\beta}\bar{\gamma}, & \text{if } a = 0; \\ \bar{\alpha}\bar{\beta}\bar{\gamma}, & \text{if } a \neq 0, \end{cases}$$

$$p(AB|ABC, a) = \begin{cases} \beta, & \text{if } a = 0; \\ \bar{\alpha}\beta, & \text{if } a = \hat{b}; \\ \alpha\bar{\beta} + \bar{\alpha}\beta, & \text{if } a = \hat{c}, \end{cases}$$

$$p(AC|ABC, a) = \begin{cases} \bar{\beta}\gamma, & \text{if } a = 0; \\ \alpha\bar{\gamma} + \bar{\alpha}\bar{\beta}\gamma, & \text{if } a = \hat{b}; \\ \bar{\alpha}\bar{\beta}\gamma, & \text{if } a = \hat{c}, \end{cases}$$

$$p(A|ABC, a) \equiv 0, \quad p(\Delta|ABC, a) = \begin{cases} 0, & \text{if } a = 0; \\ \alpha\gamma, & \text{if } a = \hat{b}; \\ \alpha\beta, & \text{if } a = \hat{c}, \end{cases}$$

$$p(AB|AB, a) = \begin{cases} \bar{\beta}, & \text{if } a = 0; \\ \bar{\alpha}\bar{\beta}, & \text{if } a \neq 0, \end{cases}$$

$$p(A|AB, a) = \begin{cases} 0, & \text{if } a = 0; \\ \alpha, & \text{if } a \neq 0, \end{cases} \quad p(\Delta|AB, a) = \begin{cases} \beta, & \text{if } a = 0; \\ \bar{\alpha}\beta, & \text{if } a \neq 0, \end{cases}$$

$$p(AC|AC, a) = \begin{cases} \bar{\gamma}, & \text{if } a = 0; \\ \bar{\alpha}\bar{\gamma}, & \text{if } a \neq 0, \end{cases}$$

$$p(A|AC, a) = \begin{cases} 0, & \text{if } a = 0; \\ \alpha, & \text{if } a \neq 0, \end{cases} \quad p(\Delta|AC, a) = \begin{cases} \gamma, & \text{if } a = 0; \\ \bar{\alpha}\gamma, & \text{if } a \neq 0, \end{cases}$$

$$p(\Delta|A, a) \equiv 1, \qquad p(\Delta|\Delta, a) \equiv 1,$$

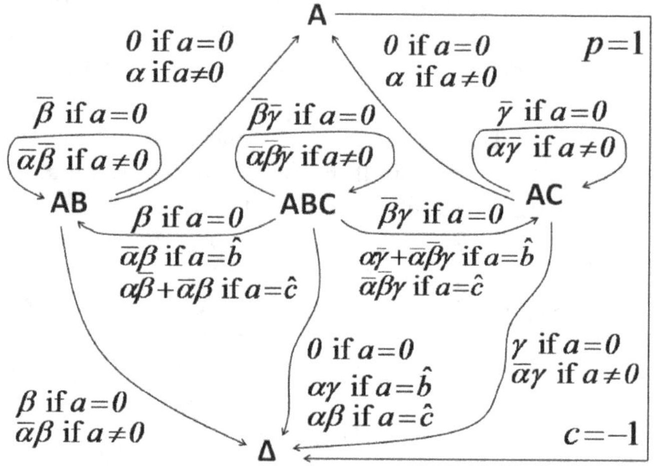

Fig. 2.48 Truel. The arrows are marked with the transition probabilities.

$$c(x,a) = \begin{cases} -1, & \text{if } x = A; \\ 0, & \text{otherwise,} \end{cases} \qquad \mathsf{P}_0(ABC) = 1.$$

All the other transition probabilities are zero. See Fig. 2.48.

After we define the MDP in this way, the modulus of the performance functional $|v^\pi_{\mathsf{P}_0}|$ coincides with the probability for A to win the game, and minimization $v^\pi_{\mathsf{P}_0} \to \inf_\pi$ means the maximization of that probability. The optimality equation (2.5) is given by

$$\begin{aligned}
v(\Delta) &= 0; \\
v(A) &= -1 + v(\Delta) = -1; \\
v(AB) &= \min\{\bar\beta\; v(AB) + \beta\; v(\Delta);\quad \bar\alpha\bar\beta\; v(AB) + \alpha\; v(A) + \bar\alpha\beta\; v(\Delta)\}; \\
v(AC) &= \min\{\bar\gamma\; v(AC) + \gamma\; v(\Delta);\quad \bar\alpha\bar\gamma\; v(AC) + \alpha\; v(A) + \bar\alpha\gamma\; v(\Delta)\}; \\
v(ABC) &= \min\{\bar\beta\bar\gamma\; v(ABC) + \beta\; v(AB) + \bar\beta\gamma\; v(AC); \\
&\qquad \bar\alpha\bar\beta\bar\gamma\; v(ABC) + \bar\alpha\beta\; v(AB) + (\alpha\bar\gamma + \bar\alpha\bar\beta\gamma) v(AC) + \alpha\gamma\; v(\Delta); \\
&\qquad \bar\alpha\bar\beta\bar\gamma\; v(ABC) + (\alpha\bar\beta + \bar\alpha\beta) v(AB) + \bar\alpha\bar\beta\gamma\; v(AC) + \alpha\beta\; v(\Delta)\}.
\end{aligned}$$

Therefore, $v(\Delta) = 0$, $v(A) = -1$, $v(AB) = -\frac{\alpha}{\beta+\alpha\bar\beta}$, $v(AC) = -\frac{\alpha}{\gamma+\alpha\bar\gamma}$, and actions $\varphi^*(AB) = \hat b$, $\varphi^*(AC) = \hat c$ are optimal as expected (for marksman A).

Firstly, note that in state ABC, action $\hat{b}$ is not optimal. To show this, compare the actions $\hat{b}$ and $\hat{c}$: the difference of the resulting costs equals

$$\delta = \alpha\bar{\gamma}\ v(AC) - \alpha\bar{\beta}\ v(AB) = \alpha^2 \left[\frac{\bar{\beta}}{\beta + \alpha\bar{\beta}} - \frac{\bar{\gamma}}{\gamma + \alpha\bar{\gamma}} \right].$$

The function $\frac{1-x}{x+\alpha(1-x)}$ decreases with $x \geq 0$, so that $\delta > 0$ because $\gamma > \beta$.

Now we rewrite the last optimality equation as follows:

$$\begin{aligned} v(ABC) &= \min\left\{ \bar{\beta}\bar{\gamma}\ v(ABC) - \frac{\beta\alpha}{\beta+\alpha\bar{\beta}} - \frac{\bar{\beta}\gamma\alpha}{\gamma+\alpha\bar{\gamma}}; \right. \\ &\qquad \left. \bar{\alpha}\bar{\beta}\bar{\gamma}\ v(ABC) - \frac{\alpha(\alpha\bar{\beta}+\bar{\alpha}\beta)}{\beta+\alpha\bar{\beta}} - \frac{\alpha\bar{\alpha}\bar{\beta}\gamma}{\gamma+\alpha\bar{\gamma}} \right\} \\ &=: T \circ v(ABC). \end{aligned}$$

Since $\bar{\beta}\bar{\gamma} \in (0,1)$ and $\bar{\alpha}\bar{\beta}\bar{\gamma} \in (0,1)$, the introduced operator $T: \mathbb{R} \to \mathbb{R}$ is a contraction in the standard metric $|x|$; the details are similar to the proof of Prop. 7.2.9 in [Hernández-Lerma and Lasserre (1999)]. Thus, the presented above optimality equation has the unique finite solution which can be obtained as follows.

- Compute
$$v^0(ABC) = -\alpha\frac{\beta\gamma + \alpha\beta\bar{\gamma} + \beta\bar{\beta}\gamma + \alpha(\bar{\beta})^2\gamma}{(\beta+\alpha\bar{\beta})(\gamma+\alpha\bar{\gamma})(\beta+\gamma\bar{\beta})}.$$
- Compute
$$v^{\hat{c}}(ABC) = -\alpha\frac{\alpha\bar{\beta}\gamma + \bar{\alpha}\beta\gamma + \alpha^2\bar{\beta}\bar{\gamma} + \alpha\bar{\alpha}\beta\bar{\gamma} + \bar{\alpha}\bar{\beta}\gamma\beta + \alpha\bar{\alpha}(\bar{\beta})^2\gamma}{(\beta+\alpha\bar{\beta})(\gamma+\alpha\bar{\gamma})(1-\bar{\alpha}\bar{\beta}\bar{\gamma})}.$$
- $v(ABC) = \min\{v^0(ABC),\ \ v^{\hat{c}}(ABC)\}$.
- The corresponding action 0 or $\hat{c}$ is optimal.

Suppose $v(ABC) = v^0(ABC)$, and the optimal action for the marksman A at the beginning is $\varphi^*(ABC) = 0$ (deliberately miss). Along with the probability $p_A(ABC)$ that marksman A wins the truel (which equals $-v(ABC)$), it is not hard to calculate the similar probabilities for B and C:

$$p_B(ABC) = \bar{\beta}\bar{\gamma}\ p_B(ABC) + \beta\ p_B(AB),$$

so that $p_B(ABC) = \dfrac{\bar{\alpha}\beta^2}{(\beta+\alpha\bar{\beta})(\beta+\gamma\bar{\beta})}$, and

$$p_C(ABC) = \bar{\beta}\bar{\gamma}\ p_C(ABC) + \bar{\beta}\gamma\ p_C(AC),$$

so that $p_C(ABC) = \dfrac{\bar{\alpha}\bar{\beta}\gamma^2}{(\gamma+\alpha\bar{\gamma})(\beta+\gamma\bar{\beta})}$.

Take $\alpha = 0.3$, $\beta = 0.5$ and $\gamma = 0.6$. Then $v^0(ABC) \approx -0.445$, $v^{\hat{c}}(ABC) \approx -0.370$, and $\varphi^*(ABC) = 0$. Marksman A intentionally misses his shot as soon as the state is ABC, that is, all the marksmen are standing. He wins the game with probability $p_A(ABC) = -v(ABC) \approx 0.445$; the other marksmen B and C win the game with probabilities $p_B(ABC) \approx 0.337$ and $p_C(ABC) \approx 0.219$ correspondingly. The order of shots is more important than their accuracies. The decision to miss intentionally allows A to wait until the end of the duel between B and C; after that he has the advantage of shooting first.

Now take $\alpha = 0.1$, $\beta = 0.15$ and $\gamma = 0.99$. Then $v^0(ABC) \approx -0.149$, $v^{\hat{c}}(ABC) \approx -0.171$, and $\varphi^*(ABC) = \hat{c}$. The dual with B is preferable (compared with C) for marksman A, even if B shoots first, because B is almost as weak as A, but marksman C is very strong. The first goal for the both marksmen A and B is to hit C. Probability to win the whole game for A is $p_A(ABC) = -v(ABC) \approx 0.171$.

The probabilities to win the game for the marksmen B and C, that is, $p_B(ABC)$ and $p_C(ABC)$ come from equations

$$p_B(ABC) = \bar{\alpha}\bar{\beta}\bar{\gamma}p_B(ABC) + \alpha p_B(BA) + \bar{\alpha}\beta p_B(AB);$$
$$p_C(ABC) = \bar{\alpha}\bar{\beta}\bar{\gamma}p_B(ABC) + \bar{\alpha}\bar{\beta}\gamma p_C(AC).$$

Expressions for $p_B(AB) = \frac{\bar{\alpha}\beta}{\alpha+\bar{\alpha}\beta}$ and $p_C(AC) = \frac{\bar{\alpha}\gamma}{\alpha+\bar{\alpha}\gamma}$ have been already obtained. The formula for $p_B(BA)$ comes from equation

$$p_B(BA) = \beta + \bar{\beta}\bar{\alpha}p_B(BA) \Longrightarrow p_B(BA) = \frac{\beta}{1-\bar{\alpha}\bar{\beta}}.$$

Direct substitution gives the answer:

$$p_B(ABC) \approx 0.142 \quad \text{and} \quad p_C(ABC) \approx 0.686.$$

□

The reader can find more about truels in [Kilgour (1975)].

2.3.34 *Random walk*

This absorbing MDP model was investigated in [Piunovskiy (2023)].

Example. Suppose

$$\mathbf{A} = \{a_1, a_2, \ldots, a_N\}$$

is the finite action space and, for each $a \in \mathbf{A}$, let $Z(a)$ be the random variable taking values $m = 1, 2, \ldots, M$ with probabilities $q_m(a)$. The states of the random walk are integer numbers. If action $a \in \mathbf{A}$ is chosen in the state $i \geq 0$, then the new state is just $j = i - Z(a)$. Another choice of action $a \in \mathbf{A}$ at the next decision epoch leads to the independent random variable $Z(a)$. The negative states of the random walk are costless and absorbing. When starting from the initial state $x_0 \geq 0$, the states smaller than $-M$ cannot be reached. Therefore, we accept that the state space of this MDP is

$$\mathbf{X} := \{-M, -M+1, -M+2, \ldots\}.$$

All the states $-M, -M+1, \ldots, -1$ are absorbing, with zero cost. See Fig. 2.49. Although the states $x > x_0$ cannot be reached, it is convenient to keep the infinite space $\mathbf{X}$ because we will consider arbitrary (large) initial states x_0.

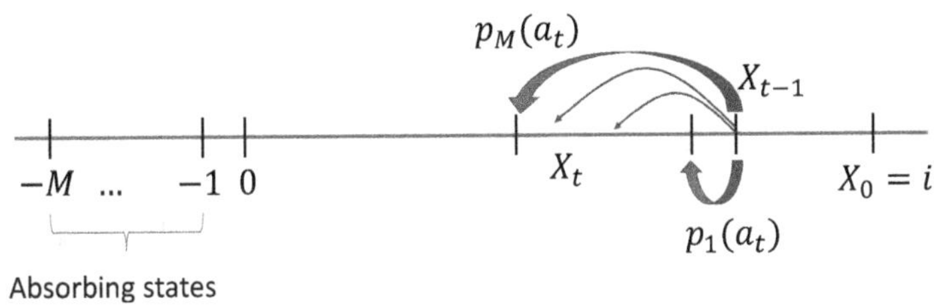

Fig. 2.49 Random walk.

The transition probability is given by

$$p(j|i,a) = \begin{cases} q_m(a), & \text{if } j = i - m, \quad m = 1, 2, \ldots, M; \\ 0, & \text{otherwise.} \end{cases}$$

The associated (expected) cost in states $i \geq 0$ equals $c(i,a) = C^a$ and is i-independent. For example, if $c_{Z(a)}(a)$ is the cost associated with the action $a \in \mathbf{A}$ and the value of $Z(a)$, then

$$C^a = \sum_{m=1}^{M} c_m(a) q_m(a).$$

Let us introduce the following notations: $L^a := \mathbb{E}[Z(a)] = \sum_{m=1}^{M} m q_m(a) \geq 1$; $c_* := \min_{a \in \mathbf{A}} \frac{C^a}{L^a}$, and

$$\mathbf{A}_* := \{a \in \mathbf{A} : \frac{C^a}{L^a} = c_*\}.$$

For a fixed $a \in \mathbf{A}$, for a large initial state $x_0 = i$, the total expected number of time steps up to the absorption equals $\approx \frac{i}{L^a}$. Thus, the total expected cost is $\approx i\frac{C^a}{L^a}$. To put it different, the expected cost, coming from passing through one point on the lattice, equals $\approx \frac{C^a}{L^a}$. Therefore, it is natural to expect that the following **conjecture** is valid

$$\begin{gathered}\text{There is such } 0 \le I < \infty \text{ that, for each } i \ge I, \\ \text{(assuming that } x_0 \ge i) \text{ the optimal actions belong to } \mathbf{A}_*.\end{gathered} \tag{2.52}$$

One can recognize the *turnpike* property of the MDP under study, formulated in Subsection 1.4.6, which we repeat here for convenience:

> *If the time horizon is large enough, then the optimal control on the first several steps is also optimal for the infinite-horizon version of the model.*

We will show that this conjecture may be not valid in some situations. But firstly, let us present a couple of conditions sufficient for the conjecture to hold.

Denote $\mathbf{S} := \{0, 1, \ldots, M-1\}$ and let $Q(a)$ be the following stochastic matrix describing a Markov chain on $\mathbf{S}$: $Q(a)$

$$:= \begin{pmatrix} Q_{0,0}(a) = q_M(a) & Q_{0,1}(a) = q_{M-1}(a) & \ldots\ Q_{0,M-1}(a) = q_1(a) \\ Q_{1,0}(a) = q_1(a) & Q_{1,1}(a) = q_M(a) & \ldots\ Q_{1,M-1}(a) = q_2(a) \\ \ldots & \ldots & \ldots\ldots \\ Q_{M-1,0}(a) = q_{M-1}(a) & Q_{M-1,1}(a) = q_{M-2}(a) & \ldots\ Q_{M-1,M-1}(a) = q_M(a) \end{pmatrix},$$

Proposition 2.7. [Piunovskiy (2023), Cor. 2,3]. *Suppose one of the following conditions is satisfied.*

(a) $q_M(a) > 0$ *for all* $a \in \mathbf{A}$ *and, for any two points* $i, j \in \mathbf{S}$, *there exists a path* $i_0 = i \to i_1 \to \ldots \to i_N = j$ *in* $\mathbf{S}$ *such that, for any* $a_0, a_1, \ldots, a_{N-1} \in \mathbf{A}$,

$$Q_{i_0,i_1}(a_0)Q_{i_1,i_2}(a_1)\ldots Q_{i_{N-1},i_N}(a_{N-1}) > 0.$$

(b) $\mathbf{A}_* = \{a_*\}$ *is a singleton,* $q_M(a_*) > 0$, *and the matrix* $Q(a_*)$ *is irreducible and aperiodic.*

Then the conjecture (2.52) is valid.

Suppose the random walk described at the beginning of the current subsection is slightly modified: there is additional isolated costless state $\Delta \notin \mathbf{X}$ (cemetery) and, for $i, j \in \mathbf{X}$,

$$p(j|i,a) = \begin{cases} \beta q_m(a), & \text{if } j = i-m,\ m = 1,2,\ldots,M; \\ 0 & \text{for other } i,j \in \mathbf{X}; \end{cases}$$
$$p(\Delta|i,a) \equiv (1-\beta), \quad i \in \mathbf{X};$$
$$p(\Delta|\Delta,a) \equiv 1; \qquad c(\Delta,a) \equiv 0,$$

where $\beta \in (0,1)$; the cost function $c(i,a) = C^a$ is the same as before for $i \in \mathbf{X}$.

One can recognize here the discounted version of the original random walk. Denote $c^* := \min_{a\in\mathbf{A}} C^a$ and $\mathbf{A}^* := \{a \in \mathbf{A} : C^a = c^*\}$. Below, we assume that the conjecture (2.52) in the original random walk is valid; the corresponding value of I is denoted below as I_1.

The optimality equation for the modified (discounted) version looks like follows

$$v^\beta(i) = \min_{a\in\mathbf{A}} \left\{ C^a + \beta \sum_{m=1}^{M} v^\beta(i-m) q_m(a) \right\} \qquad \text{for } i \geq 0; \tag{2.53}$$
$$v^\beta(i) = 0 \quad \text{for } i = -M, -M+1, \ldots, -1$$

and again can be solved successively for $i = 0, 1, \ldots$. Actions $a \in \mathbf{A}$ at state $i \in \mathbf{X}$ are optimal if and only if they provide the minimum in (2.53). Actions $a \in \mathbf{A}$ in the original random walk are optimal at state $i \in \mathbf{X}$ if and only if they provide the minimum in (2.53) at $\beta = 1$ (cf (2.54) below and the corresponding explanation). Since, for each $i \in \mathbf{X}$, $v^\beta(i)$ is a continuous function of β, for an *arbitrarily* fixed $I_2 > I_1$, there is $\beta_0 \in (0,1)$, close enough to 1, such that, for all $\beta \in [\beta_0, 1)$, for all $i \in [I_1, I_2]$, the minimum in (2.53) is provided only by $a \in \mathbf{A}_*$, i.e., the optimal actions in the discounted random walk belong to $\mathbf{A}_*$. We obtain the turnpike property for $[I_1, I_2]$. In what follows, $\beta \in [\beta_0, 1)$ is fixed.

The turnpike theory for discounted MDPs is well developed: see [Puterman (1994), §6.8]. For the modified (discounted with $\beta \in [\beta_0, 1)$) random walk the following statement is valid.

There is such $0 \leq I_3 < \infty$ that, for each $i \geq I_3$,
(assuming that $x_0 \geq i$) the optimal actions belong to $\mathbf{A}^*$.

The detailed proof can be found in [Piunovskiy (2021)]: see Thm. 2 therein, where the maximization version of the optimal control problem was studied. Since that proof is rather long, let us describe only the main idea. Since $\beta \in (0,1)$, the effect of the initial condition $v^\beta(i) = 0$ at $i = -M, -M+1, \ldots, -1$ on the values of $v^\beta(i)$ at $i \geq 0$ becomes less and less important as i increases, and one can show hat $\lim_{i\to\infty} |v^\beta(i) - \hat{v}^\beta(i)| = 0$, where the function $\hat{v}^\beta(i)$ satisfies the same equation (2.53), but with any other fixed initial condition. In particular, if we put $\hat{v}^\beta(i) = \frac{c^*}{1-\beta}$ at $i = -M, -M+1, \ldots, -1$, then $\hat{v}^\beta(i) \equiv \frac{c^*}{1-\beta}$ for all $i \in \mathbf{X}$. After that, one can accurately complete the proof keeping in mind that the minimum in (2.53) for $\hat{v}^\beta(i)$, $i \geq 0$, is provided only by $a \in \mathbf{A}^*$.

To summarize, if conjecture (2.52) in the original (undiscounted) random walk is valid, then, for β sufficiently close to 1, one can observe two turnpikes: on a big interval $[I_1, I_2]$, the optimal actions belong to $\mathbf{A}_*$, and for $i \in [I_3 > I_2, \infty)$ the optimal actions belong to $\mathbf{A}^*$: see Fig. 2.50. □

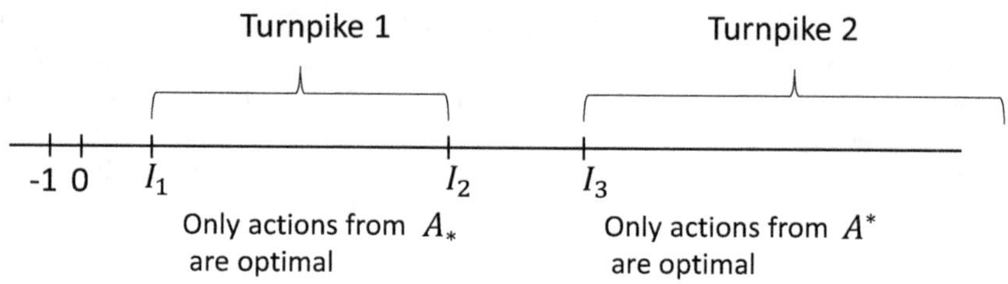

Fig. 2.50 Turnpikes for $\beta \approx 1$.

Now, consider the following specific example to show that the conjecture (2.52) may be false.
Example.

$$\mathbf{A} := \{a_1, a_2\},\ M := 3,\ q_2(a_1) = 1,\ q_2(a_2) = 1 - \varepsilon,\ q_3(a_2) = \varepsilon,$$

where $\varepsilon \in (0, 1)$. Other q-probabilities are zero. Let $C^{a_1} := -2$ and $C^{a_2} := h \in (-2 - \varepsilon, -2)$. The cost function $c(i, a) = C^a$ is negative, so in fact the problem is about maximizing rewards. Now

$$L^{a_1} = 2,\ L^{a_2} = 2 + \varepsilon,\ \frac{C^{a_1}}{L^{a_1}} = -1,\ \frac{C^{a_2}}{L^{a_2}} = \frac{h}{2 + \varepsilon} > -1,\ c_* = -1,\ \mathbf{A}_* = \{a_1\}.$$

The optimality equation (2.5) looks as follows:

$$v(i) = \min_{a \in \mathbf{A}} \left\{ C^a + \sum_{m=1}^{3} v(i - m) q_m(a) \right\}, \quad i \geq 0; \tag{2.54}$$
$$v(i) = 0, \quad i = -3, -2, -1.$$

It can be solved successively for $i = 0, 1, \ldots$, resulting in the Bellman function v_i^*, $i \in \mathbf{X}$. Clearly, the solution satisfies inequality

$$(i + 1)h = (i + 1) \min_{a \in \mathbf{A}} \{C^a\} \leq v_i^* \leq 0.$$

Recall that $C^a < 0$. The rigorous proof is by induction. Note also that, if the stationary selector $\varphi^*(i)$ provides the minimum in (2.54), then $v_i^{\varphi^*}$ can be again constructed successively for $i = 0, 1, \ldots$, and $v_i^{\varphi^*} = v_i^*$. The rigorous proof is again by induction w.r.t. i. Finally, if, at some state $0 \le i \le x_0$, there is a positive probability of using an action which does not provide the minimum in (2.54), then such a strategy cannot be optimal.

The first several values of v_i^* are directly calculated as follows, keeping in mind that $h < -2$.

- $v_0^* = v_1^* = \min\{-2, h\} = h$;
- $v_2^* = \min\{-2+v_0^* = -2+h;\ \ h+(1-\varepsilon)v_0^*+\varepsilon v_{-1}^* = h+(1-\varepsilon)h\} = -2+h$ because

$$\frac{-2}{1-\varepsilon} = -2[1+\varepsilon+\varepsilon^2+\ldots] < -2-\varepsilon < h \Longrightarrow -2 < (1-\varepsilon)h.$$

- $v_3^* = \min\{-2+v_1^* = -2+h;\ \ h+(1-\varepsilon)v_1^*+\varepsilon v_0^* = 2h\} = 2h.$

Further properties of the function v_i^* are given in the following lemma.

Lemma 2.6.

(a) For all even values of $i > 0$, $v_i^* = -i + h$, and the minimum in (2.54) is provided by a_1 only.

(b) For all odd values of $i \ge 3$,

$$v_i^* = (1-\varepsilon)v_{i-2}^* + (1+\varepsilon)h - \varepsilon(i-3),$$

and the minimum in (2.54) is provided by a_2 only.

The proof is given in Appendix A.

Conjecture (2.52) is false: the unique optimal action a_2 at all odd states $i \ge 3$ does not belong to $\mathbf{A}_*$. Proposition 2.7 is not applicable because $q_3(a_*) = q_3(a_1) = 0$. □

Chapter 3

Homogeneous Infinite-Horizon Models: Discounted Cost

In this chapter, we consider the special case of uniformly absorbing model, the discounted MDP. After the brief presentation of the necessary knowledge on such MDPs in Sections 3.1 and 3.2, a number of examples are given in Section 3.3.

3.1 Preliminaries

This chapter is about the following problem:

$$v^{\pi}_{\mathsf{P}_0} = \mathsf{E}^{\pi}_{\mathsf{P}_0}\left[\sum_{t=1}^{\infty} \beta^{t-1} c(X_{t-1}, A_t)\right] \to \inf_{\pi}, \tag{3.1}$$

where $\beta \in (0,1)$ is the *discount factor*. In all the examples in Section 3.3, $v^{\pi}_{\mathsf{P}_0}$ is well defined and expression $\infty - \infty$ does not appear. As usual, $v^{\pi}_{\mathsf{P}_0}$ is called the *performance functional*. Like previously, we use the notations $v^{\pi}_{x_0}$ and $v^{*}_{x_0} := \inf_{\pi} v^{\pi}_{x_0}$ if $\mathsf{P}_0(dx) = \delta_{x_0}(dx)$. The discounted model is a particular case of an MDP with the expected total cost. To see this, modify the transition probability $p(dy|x,a) \to \beta p(dy|x,a)$, introduce the costless cemetery $\Delta \notin \mathbf{X}$, adjust the state space $\mathbf{X} \to \mathbf{X} \cup \{\Delta\}$ and put $p(\{\Delta\}|\Delta, a) \equiv 1$, $p(\{\Delta\}|x,a) := (1-\beta)$, $c(\Delta, a) \equiv 0$. Investigation of problem (3.1) is now equivalent to the investigation of the modified (uniformly absorbing) model, with finite, totally bounded expected absorption time: $\mathsf{E}^{\pi}_{\mathsf{P}_0}[T_0] = \frac{1}{1-\beta}$, where $T_0 := \min\{t \geq 0 : X_t = \Delta\}$ as usual. Therefore, all the examples in the current chapter can also be considered as the examples on the homogeneous infinite-horizon models with the expected total cost. Nevertheless, discounted models traditionally constitute a special area in MDP. One can also say that the MDP is with the expected total cost, but the time horizon $T = T_0$ is the geometric random variable: $\mathsf{P}(T = l) = \beta^{l-1}(1-\beta)$ for $l = 1, 2, \ldots$, and the terminal cost $C(\cdot)$ is zero.

The optimality equation (2.5) takes the form

$$v(x) = \inf_{a \in \mathbf{A}} \left\{ c(x,a) + \beta \int_{\mathbf{X}} v(y) p(dy|x,a) \right\}. \tag{3.2}$$

All definitions, notations and other details are similar to those introduced in Chapter 2 and earlier, and all the main theorems from Chapter 2 apply. Incidentally, a stationary selector φ is called *equalizing* if

$$\forall x \in \mathbf{X} \qquad \liminf_{t\to\infty} \mathsf{E}_x^\varphi \left[\beta^t\, v^*_{X_t}\right] \geq 0.$$

A stationary strategy π^s is called *conserving* if

$$v_x^* = \int_{\mathbf{A}} \left[c(x,a) + \beta \int_{\mathbf{X}} v_y^* p(dy|x,a)\right] \pi^s(da|x), \quad x \in \mathbf{X}.$$

Sometimes we use notation $v_{\mathsf{P}_0}^{\pi,\beta}$, $v_x^{*,\beta}$ etc. if it is important to underline the dependence on the discount factor β.

If $\sup_{(x,a)\in\mathbf{X}\times\mathbf{A}} |c(x,a)| < \infty$ (and in many other cases when the investigation is performed, e.g. in weighted normed spaces; see [Hernández-Lerma and Lasserre (1999), Chapter 8]), the solution to equation (3.2) is unique in the space of bounded functions (respectively, in the space of functions with the finite weighted norm), coincides with the Bellman function $v_x^* := \inf_\pi v_x^\pi$, and can be built using the *value iteration algorithm* (cf (2.6)): $v(x) = v_x^* = v^\infty(x) := \lim_{n\to\infty} v^n(x)$, where

$$v^0(x) \equiv 0;$$
$$v^{n+1}(x) = \inf_{a\in\mathbf{A}} \left\{c(x,a) + \beta\int_{\mathbf{X}} v^n(y)p(dy|x,a)\right\}, \quad n = 0,1,2,\ldots \tag{3.3}$$

(we leave aside the question of the measurability of $v^{n+1}(\cdot)$). For justification see [Hernández-Lerma and Lasserre (1996a), p. 51], [Feinberg (2002b), Subsection 5.5.1] and [Bertsekas and Shreve (1978), Prop. 9.14]. In many other cases, e.g., if the model is positive or negative, there also exists the limit $v^\infty(x) := \lim_{n\to\infty} v^n(x)$, $x \in \mathbf{X}$. Note also that, if the cost function $c(\cdot,\cdot)$ is bounded below or above, one can pass to the positive or negative model correspondingly by adding/subtracting a constant K to the cost function: the performance functional will increase or decrease by the constant $\frac{K}{1-\beta}$.

Definition 3.1. If the limit $v^\infty(x)$ exists, the model is called *stable* in case $v^\infty(x) = v_x^*$, $x \in \mathbf{X}$.

3.2 Several Important Statements

Theorem 3.1.

(a) *[Bertsekas and Shreve (1978), Prop. 9.13].*
If $\sup_{(x,a)\in\mathbf{X}\times\mathbf{A}} |c(x,a)| < \infty$, then a stationary control strategy $\hat\pi^s$ is uniformly optimal if and only if

$$v_x^{\hat\pi^s} = \inf_{a\in\mathbf{A}} \left\{c(x,a) + \beta \int_{\mathbf{X}} v_y^{\hat\pi^s} p(dy|x,a)\right\}, \qquad x \in \mathbf{X}. \tag{3.4}$$

In negative models, this statement holds even for $\beta = 1$ and without any other restrictions on the cost function.

(b) *[Bertsekas and Shreve (1978), Prop. 9.12]. If $\sup_{(x,a)\in\mathbf{X}\times\mathbf{A}} |c(x,a)| < \infty$, then a stationary control strategy π^s is uniformly optimal if and only if it is conserving. In positive models, this statement holds even for $\beta = 1$ and without any other restrictions on the cost function.*

According to Lemma 2.1, one can restrict to Markov strategies.

Theorem 3.2.

(a) *[Bertsekas and Shreve (1978), Prop. 9.14].*
If the model is negative (i.e., $c(\cdot,\cdot) \le 0$) or $\sup_{(x,a)\in\mathbf{X}\times\mathbf{A}} |c(x,a)| < \infty$, then $v^\infty(x) = v_x^*$.

(b) *[Bertsekas and Shreve (1978), Cor. 9.17.2].*
Suppose the model is weakly semi-continuous (see Condition 2.3). If it is positive or $\sup_{(x,a)\in\mathbf{X}\times\mathbf{A}} |c(x,a)| < \infty$, then the Bellman function v_x^* is lower semi-continuous, $v_x^* = v^\infty(x)$ and there exists a uniformly optimal stationary selector.

Definition 3.2. For fixed initial distribution P_0 and control strategy π, the *occupation measure* $\eta^\pi_{\mathsf{P}_0}$ is the finite measure on $\mathbf{X}\times\mathbf{A}$ given by the formula

$$\eta^\pi_{\mathsf{P}_0}(\Gamma^{\mathbf{X}} \times \Gamma^{\mathbf{A}}) := (1-\beta)\sum_{t=0}^{\infty} \beta^{t-1} \mathsf{P}^\pi_{\mathsf{P}_0}\{X_t \in \Gamma^{\mathbf{X}}, A_{t+1} \in \Gamma^{\mathbf{A}}\},$$

$$\Gamma^{\mathbf{X}} \in \mathcal{B}(\mathbf{X}), \quad \Gamma^{\mathbf{A}} \in \mathcal{B}(\mathbf{A}).$$

$\mathcal{D}_{\mathsf{P}_0} := \{\eta^\pi_{\mathsf{P}_0},\ \pi \in \Pi^{\mathrm{All}}\}$ is the space of all occupation measures under the fixed initial distribution P_0. As usual, we use notations $\eta^\pi_{x_0}$ and $\mathcal{D}_{x_0}$ if $\mathsf{P}(dx) = \delta_{x_0}(dx)$ is a Dirac measure.

Usually (e.g. in positive and negative models and in the case when $\sup_{x\in\mathbf{X},\ a\in\mathbf{A}} |c(x,a)| < \infty$),

$$v^\pi_{\mathsf{P}_0} = \frac{1}{1-\beta}\int_{\mathbf{X}\times\mathbf{A}} c(x,a)\, d\eta^\pi_{\mathsf{P}_0}(x,a), \tag{3.5}$$

and investigation of the MDP in terms of occupation measures (the so-called *convex analytic approach*) is fruitful, especially in constrained problems.

Theorem 3.3. [Piunovskiy (1997), Lemmas 24, 25, p.142].
Suppose the initial distribution P_0 is fixed.

(a) *A finite measure η on $\mathbf{X}\times\mathbf{A}$ is an occupation measure if and only if it satisfies the equation*

$$\eta(\Gamma \times \mathbf{A}) = (1-\beta)\mathsf{P}_0(\Gamma) + \beta\int_{\mathbf{X}\times\mathbf{A}} p(\Gamma|y,a)d\eta(y,a)$$

for all $\Gamma \in \mathcal{B}(\mathbf{X})$.

(b) *The space of all occupation measures is convex and, for any strategy* π, *there exists a stationary strategy* π^s *such that* $\eta^{\pi}_{\mathsf{P}_0} = \eta^{\pi^s}_{\mathsf{P}_0}$

(c) *The strategy* π^s *in (b) is (any version of) the stochastic kernel in the decomposition*

$$\eta^{\pi}_{\mathsf{P}_0}(\Gamma^{\mathbf{X}} \times \Gamma^{\mathbf{A}}) = \int_{\Gamma^{\mathbf{X}}} \pi^s(\Gamma^{\mathbf{A}}|x)\eta^{\pi}_{\mathsf{P}_0}(dx \times \mathbf{A})$$

and, for a stationary strategy π^s,

$$\eta^{\pi^s}_{\mathsf{P}_0}(\Gamma^{\mathbf{X}} \times \Gamma^{\mathbf{A}}) = \int_{\Gamma^{\mathbf{X}}} \pi^s(\Gamma^{\mathbf{A}}|x)\eta^{\pi^s}_{\mathsf{P}_0}(dx \times \mathbf{A}).$$

Suppose an occupation measure $\eta^{\pi}_{\mathsf{P}_0}$ is not extreme: $\eta^{\pi}_{\mathsf{P}_0} = \alpha\eta^{\pi^1}_{\mathsf{P}_0} + (1-\alpha)\eta^{\pi^2}_{\mathsf{P}_0}$, where $\eta^{\pi^1}_{\mathsf{P}_0} \neq \eta^{\pi^2}_{\mathsf{P}_0}$, $\alpha \in (0,1)$. On the one hand, by Theorem 3.3(b), $\eta^{\pi}_{\mathsf{P}_0}$ is generated by a stationary strategy π^s: $\eta^{\pi}_{\mathsf{P}_0} = \eta^{\pi^s}_{\mathsf{P}_0}$. On the other hand, according to Lemma 2.1, there exists a Markov strategy π^m for which equality (2.2) holds, and hence $\eta^{\pi}_{\mathsf{P}_0} = \eta^{\pi^m}_{\mathsf{P}_0}$. In a typical situation, π^m is non-stationary: see Subsection 2.3.1, the example after formula (2.20). Therefore, very often the same occupation measure $\eta^{\pi}_{\mathsf{P}_0}$ can be generated by many different strategies. This was underlined after Lemma 24 in [Piunovskiy (1997)].

Definition 3.3. [Kallenberg (2010), Section 4.1], [Puterman (1994), Subsection 5.4.3]. A strategy π^* is called *Blackwell optimal* if, for some $\beta_0 \in (0,1)$, $v^{\pi^*,\beta}_x = v^{*,\beta}_x$ for all $x \in \mathbf{X}$ and all $\beta \in [\beta_0, 1) \neq \emptyset$. Such strategies were introduced in [Blackwell (1962), p. 721].

If the model is finite, then a Blackwell optimal strategy does exist and can be found in the form of a stationary selector [Blackwell (1962), Thm. 5], [Puterman (1994), Thm. 10.1.4].

Definition 3.4. [Blackwell (1962), p. 721]. A strategy π^* is called *nearly optimal* if

$$\lim_{\beta \to 1-} [v^{\pi^*,\beta}_x - v^{*,\beta}_x] = 0, \qquad x \in \mathbf{X}.$$

3.3 Examples

In this section, examples and counterexamples illustrate the theory of discounted MDPs which is well developed in [Altman (1999); Bertsekas and Shreve (1978); Dynkin and Yushkevich (1979); Hernández-Lerma and Lasserre (1996a, 1999); Kallenberg (2010); Piunovskiy (1997); Puterman (1994)].

Subsections 3.3.1 and 3.3.2 are about the standard optimal control of linear systems. They illustrate the stability issues and show that the optimality equation (3.2) can have (phantom) solutions different from the Bellman function.

The value iteration algorithm is discussed in Subsections 3.3.3–3.3.5: sometimes it does not converge and does not provide uniformly optimal selectors. Counterexamples 3.3.6 and 3.3.7 show that stationary selectors, looking as the natural candidates for the optimality, may be far not optimal. Similarly to Subsections 1.4.10 and 2.3.18, the Bellman function is non-measurable in Subsection 3.3.8. Several unexpected properties of myopic strategies and of the models with partial information are presented in Subsections 3.3.9 and 3.3.10. Example 3.3.11 illustrates the linear properties of occupation measures and stationary strategies. Constrained problems are discussed in Subsections 3.3.12–3.3.14. Some non-standard discounted problems (e.g., several different discount factors, the non-linear utility function etc) are investigated in Subsections 3.3.15–3.3.17. The Blackwell-type optimality concepts are illustrated in Subsections 3.3.18–3.3.23. Subsection 3.3.24 is about the approach to MDPs with the expected average cost via the vanishing discount factor.

3.3.1 *Stable and unstable controllers in linear systems*

MDPs with finite-dimensional Euclidean spaces $\mathbf{X}$ and $\mathbf{A}$ are often defined by *system equations* of the form

$$X_t = bX_{t-1} + gA_t + d\zeta_t, \tag{3.6}$$

where $\{\zeta_t\}_{t=1}^{\infty}$ is a sequence of *disturbances*, i.e. mutually independent random vectors with $E[\zeta_t] = 0$ and $E[\zeta_t \zeta_t'] = I$ (the identity matrix). In what follows, all vectors are columns; b, g, and d are the given matrices of appropriate dimensionalities, and the dash means transposition. The transition probability can be defined through the density function of ζ_t: if, e.g., all the random variables ζ_t are identically distributed with the density $h(\cdot)$, then $p(\Gamma|x,a) = \int_\Gamma h(y - bx - ga)dy$. In the framework of dynamical systems (3.6), selectors are called (feedback) *controllers/controls*.

Definition 3.5. A system (process) X_t is called *stable* for a selector φ_t if, in the absence of disturbances ζ_t, the state $x_t = bx_{t-1} + g\varphi_t(x_{t-1})$ tends to zero as $t \to \infty$ (for all or several initial states x_0); see [Bertsekas (2005), p. 153]. Likewise, the controller $\varphi_t(X_{t-1})$ is called *stable* if $\varphi_t(x_{t-1}) \to 0$ in the absence of disturbances. The same definition applies to particular components of X_t and $\varphi_t(X_{t-1})$ in the multiple dimensional case.

The following example, similar to [Bertsekas (2005), Ex. 5.3.1], shows that the stabilizing control can itself be unstable.

Example. Let $\mathbf{X} = \mathbb{R}^2$, $\mathbf{A} = \mathbb{R}^1$ with the Euclidean topologies, $b = \begin{bmatrix} 1 & -2 \\ 0 & 0 \end{bmatrix}$, $g = \begin{bmatrix} 1 \\ 1 \end{bmatrix}$, $d = \begin{bmatrix} 1 \\ 0 \end{bmatrix}$, $\zeta_t \in \mathbb{R}^1$ and assume that all the random variables ζ_t are identically distributed. The system equation for $X = \begin{pmatrix} X^1 \\ X^2 \end{pmatrix}$,

$$\begin{cases} X_t^1 = X_{t-1}^1 - 2X_{t-1}^2 + A_t + \zeta_t \\ X_t^2 = A_t \end{cases}$$

can be rewritten as

$$X_t^1 = X_{t-1}^1 - 2A_{t-1} + A_T + \zeta_t.$$

We put $c(x, a) = (x^1)^2$. The discount factor $\beta \in (0, 1)$ is arbitrarily fixed. See Fig. 3.1.

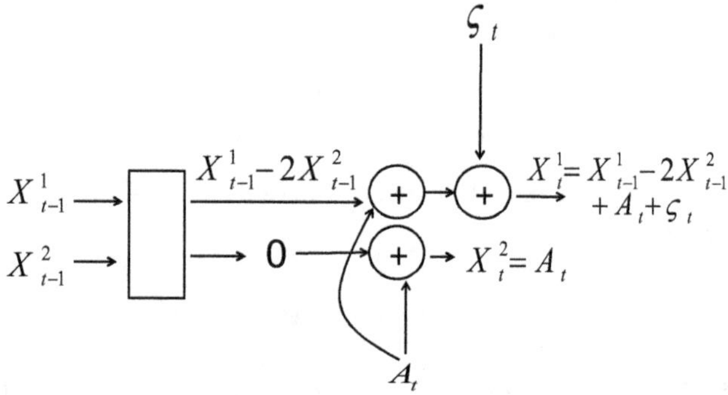

Fig. 3.1 Unstable controller.

The optimality equation takes the form

$$v(x^1, x^2) = \inf_{a \in \mathbf{A}} \left\{ (x^1)^2 + \beta \ \mathsf{E}[v(x^1 - 2x^2 + a + \zeta, a)] \right\},$$

where the expectation E corresponds to the random variable ζ. Using the standard convention $\mathsf{E}[\zeta] = 0$, $\mathsf{E}[\zeta^2] = 1$, one can calculate the minimal non-negative solution

$$v(x^1, x^2) = (x^1)^2 + \frac{\beta}{1 - \beta},$$

which coincides with the Bellman function v_x^* because the model is positive: see Theorem 2.3 and the paragraph below it. According to Theorem 3.1(b), the uniformly optimal stationary selector is given by the formula

$$\varphi^*(x) = 2x^2 - x^1,$$

so that $X_t^1 = \zeta_t$ for all $t = 1, 2, \ldots$. The first component X_t^1 is stable for φ^* and each initial state $x_0 = (x_0^1, x_0^2) \in \mathbf{X}$.

At the same time, for the initial state $x_0 = (x_0^1, x_0^2) \in \mathbf{X}$, the feedback control φ^* results in the sequence

$$A_1 = 2x_0^2 - x_0^1,$$
$$A_2 = 2A_1 - \zeta_1,$$
$$A_3 = 4A_1 - 2\zeta_1 - \zeta_2,$$

and, in general,

$$A_t = \varphi^*(X_{t-1}) = 2^{t-1}A_1 - \sum_{i=1}^{t-1} 2^{t-i-1}\zeta_i,$$

meaning that the optimal controller $\varphi^*(X_{t-1})$ (as well as the second component X_t^2 of the state) is unstable if $x_0^1 \neq 2x_0^2$. Moreover, it is even 'discounted'-unstable: in the absence of disturbances, $\lim_{t\to\infty} \beta^{t-1}A_t \neq 0$ for $\beta \geq 1/2$, if $A_1 = 2x_0^2 - x_0^1 \neq 0$.

Now, consider the selector

$$\varphi_t(x) = \begin{cases} x^1, & \text{if } t \text{ is even;} \\ 0, & \text{if } t \text{ is odd.} \end{cases}$$

Direct calculations lead to the following expressions:

$$X_t^1 = \begin{cases} 2(\zeta_{t-2} + \zeta_{t-1}) + \zeta_t, & \text{if } t \text{ is even;} \\ \zeta_{t-1} + \zeta_t, & \text{if } t \text{ is odd} \end{cases} \quad X_t^2 = \begin{cases} \zeta_{t-2} + \zeta_{t-1}, & \text{if } t \text{ is even;} \\ 0, & \text{if } t \text{ is odd} \end{cases}$$

for all $t \geq 3$, so that all the processes X_t^1, X_t^2 and the controller $\varphi_t(X_{t-1})$ are stable for each initial state $x_0 \in \mathbf{X}$.

One can find an acceptable solution to this control problem by taking into account all the variables; that is, we put

$$c(x,a) = k_1(x^1)^2 + k_2(x^2)^2 + k_3a^2 \qquad k_1, k_2, k_3 > 0.$$

Now we can use the well developed theory of linear–quadratic control, see for example [Piunovskiy (1997), Subsection 1.2.2.5].

For an arbitrarily fixed discount factor $\beta \in (0,1)$, the uniformly optimal stationary selector and the Bellman function are given by the formulae

$$\varphi^*(x) = -\frac{\beta g' f b}{k_3 + \beta g' f g}x; \qquad v_x^* = x' f x + q,$$

where $q = \frac{\beta Sp(d' f d)}{1-\beta} = \frac{\beta f_{11}}{1-\beta}$ and $f = \begin{bmatrix} f_{11} & f_{12} \\ f_{21} & f_{22} \end{bmatrix}$ (a symmetric matrix) is the unique positive-definite solution to the equation

$$f = \beta b' f b + \begin{bmatrix} k_1 & 0 \\ 0 & k_2 \end{bmatrix} - \frac{\beta^2 b' f g g' f b}{k_3 + \beta g' f g}.$$

For the proof, see Theorem 4 of [Piunovskiy (1997)].

Since

$$\begin{aligned} v^*_{x_0} &= \lim_{T\to\infty} \mathsf{E}^{\varphi^*}_{x_0} \left[\sum_{t=1}^{T} \beta^{t-1} c(X_{t-1}, A_t) + \beta^T v^*_{X_T} \right] \\ &= \lim_{T\to\infty} \mathsf{E}^{\varphi^*}_{x_0} \left[\sum_{t=1}^{\infty} \beta^{t-1} c(X_{t-1}, A_t) \right], \end{aligned}$$

$\lim_{T\to\infty} \mathsf{E}^{\varphi^*}_{x_0} \left[\beta^T (X'_T f X_T + q)\right] = 0$. After we diagonolize the matrix f: $f = Q'\Lambda Q = Q^{-1}\Lambda Q$ with $\Lambda = \begin{bmatrix} \lambda_1 & 0 \\ 0 & \lambda_2 \end{bmatrix}$ ($\lambda_i > 0$ being the eigenvalues of f), we see that

$$\begin{aligned} &\lim_{T\to\infty} \mathsf{E}^{\varphi^*}_{x_0} \left[\beta^T (QX_t)'\Lambda(QX_T)\right] = 0 \\ \Longrightarrow &\lim_{T\to\infty} \mathsf{E}^{\varphi^*}_{x_0} \left[\beta^T QX_T\right] = Q \lim_{T\to\infty} \mathsf{E}^{\varphi^*}_{x_0} \left[\beta^T X_T\right] = 0 \\ \Longrightarrow &\, Q^{-1}Q \lim_{T\to\infty} \mathsf{E}^{\varphi^*}_{x_0} \left[\beta^T X_T\right] = \lim_{T\to\infty} \mathsf{E}^{\varphi^*}_{x_0} \left[\beta^T X_T\right] = 0. \end{aligned}$$

If there are no disturbances, then all the formulae and statements survive (apart from q which now equals zero) for any $\beta \in (0,1)$. Hence, the system (process) is stable for φ^* and each initial state $x_0 \in \mathbf{X}$, as well as the controller $\varphi^*(X_{t-1})$. □

It can happen that system (3.6) can be made stable only for some initial states x_0 [Bertsekas (2001), Ex. 3.1.1]. The next example confirms this.

Example. Let $\mathbf{X} = \mathbb{R}^1$, $\mathbf{A} = [-3,3]$, $b = 3$, $g = 1$, $d = 1$ and suppose there are no disturbances ($\zeta_t \equiv 0$):

$$X_t = 3X_{t-1} + A_t.$$

If $|x_0| < 3/2$ then we can put $A_t = \varphi_t(X_{t-1}) = -3\,\mathrm{sign}(X_{t-1})$ for $t = 1,2,\ldots$, up to the moment when $|X_\tau| \le 1$. Afterwards, applying $A_t = \varphi_t(X_{t-1}) = -3X_{t-1}$ for $t = \tau+1, \tau+2, \ldots$, we finish with $X_t \equiv 0$, so the system and this controller $\varphi_t(X_{t-1})$ are stable under this selector. The system is unstable under any control strategy if $|x_0| \ge 3/2$.

Now let

$$\zeta_t = \begin{cases} +1 & \text{with probability } 1/2 \\ -1 & \text{with probability } 1/2 \end{cases}$$

and consider the performance functional

$$v_x^\pi = \mathsf{E}_x^\pi \left[\sum_{t=1}^{\infty} \beta^{t-1} |X_{t-1}| \right]$$

with $\beta = 1/2$.

Firstly, let us show that, for any control strategy π, $v_{x_0}^\pi = \infty$ if $|x_0| > 1$. In the case when $x_0 > 1$, there is a positive probability of having $\zeta_1 = \zeta_2 = \cdots = \zeta_\tau = 1$, up to the moment when $X_\tau > 4$: the sequence

$$X_1 \geq 3x_0 - 3 + 1; \quad X_2 \geq 3X_1 - 3 + 1, \quad \ldots$$

approaches $+\infty$. Thereafter,

$$X_{\tau+1} \geq 3X_\tau - 3 - 1 = 2X_\tau + (X_\tau - 4) > 2X_\tau > 4$$

and, for all $i = 1, 2, \ldots$,

$$X_{\tau+i+1} \geq 3X_{\tau+i} - 4 = 2X_{\tau+i} + (X_{\tau+i} - 4) > 2X_{\tau+i}$$

by induction, meaning that $v_{x_0}^\pi = \infty$. The reasoning for $x_0 < -1$ is similar. Hence $v_{x_0}^* = \infty$ for $|x_0| > 1$.

If $|x_0| \leq 1$ then $v_{x_0}^* = |x_0| + \frac{\beta}{1-\beta} = |x_0| + 1$, and the stationary selector $\varphi^*(x) = -3x$ is optimal. (Note that $\varphi^*(X_t) \in [-3, 3]$ $P_x^{\varphi^*}$-almost surely). The reader can independently investigate the optimality equation. □

3.3.2 *Phantom solutions of the optimality equation*

Example. Let $\mathbf{X} = \mathbb{R}^1$, $\mathbf{A} = \mathbb{R}^1$, $p(\Gamma|x, a) = \int_\Gamma h(y - bx)dy$, $\Gamma \in \mathcal{B}(\mathbf{X})$, where $h(\cdot)$ is a fixed density function with $\int_{-\infty}^{\infty} yh(y)dy = 0$ and $\int_{-\infty}^{\infty} y^2 h(y)dy = 1$.

Remark 3.1. One can represent the evolution of the process as the following *system equation:*

$$X_t = bX_{t-1} + \zeta_t,$$

where $\{\zeta_t\}_{t=1}^\infty$ is a sequence of mutually independent random variables with the probability density $h(\cdot)$ (see Fig. 3.2).

We put $c(x, a) = J_{AA}a^2 + J_{XX}x^2$, where $J_{AA}, J_{XX} > 0$. The described model is a special case of linear–quadratic systems [Piunovskiy (1997), Chapter 4].

The optimality equation (3.2) takes the form

$$v(x) = \inf_{a \in \mathbf{A}} \left\{ J_{AA}a^2 + J_{XX}x^2 + \beta \int_{-\infty}^{\infty} h(y - bx)v(y)dy \right\}$$

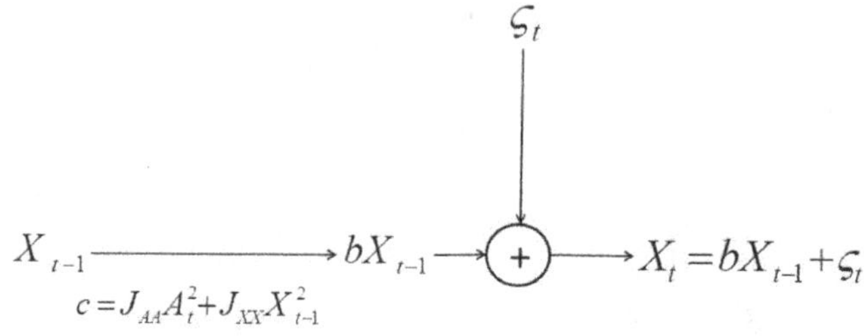

Fig. 3.2 Phantom solutions of the optimality equation.

and, if $\beta b^2 \neq 1$, has a finite solution which does not depend on $h(\cdot)$:

$$v(x) = \frac{J_{XX}}{1-\beta b^2} x^2 + \frac{\beta J_{XX}}{(1-\beta)(1-\beta b^2)}. \tag{3.7}$$

The stationary selector $\varphi^*(x) \equiv 0$ provides the infimum in the optimality equation.

Value iterations give the following:

$$v^n(x) = f_n x^2 + q_n,$$

where

$$f_n = \begin{cases} J_{XX}\frac{1-(\beta b^2)^n}{1-\beta b^2}, & \text{if } \beta b^2 \neq 1; \\ nJ_{XX}, & \text{if } \beta b^2 = 1 \end{cases}$$

$$q_n = \beta f_{n-1} + \beta q_{n-1}, \qquad q_0 = 0.$$

In the case when $\beta b^2 < 1$ we really have the point-wise convergence $\lim_{n\to\infty} v^n(x) = v^\infty(x) = v(x)$ and $v(x) = v_x^*$. But if $\beta b^2 \geq 1$, then $v^\infty(x) = \infty$, and one can prove that $v_x^* = \infty \neq v(x)$.

Note that the system is not stable if $|b| > 1$ (i.e. $\lim_{t\to\infty} |X_t| = \infty$ if $X_0 \neq 0$ and $\zeta_t \equiv 0$; see Definition 3.5), and nevertheless the MDP is well defined and the Bellman function is finite if the discount factor $\beta < \frac{1}{b^2}$ is small enough. This example first appeared in [Piunovskiy (1997), Subsection 1.2.2.3]. □

Another example is similar to the one presented in Subsection 2.3.2, see Fig. 2.3.

Example. Let $\mathbf{X} = \{0, 1, 2, \ldots\}$, $\mathbf{A} = \{0\}$ (a dummy action), $p(0|0,a) = 1$,

$$\forall x > 0 \quad p(y|x,a) = \begin{cases} \lambda, & \text{if } y = x+1; \\ \mu, & \text{if } y = x-1; \\ 1-\lambda-\mu, & \text{if } y = x; \\ 0, & \text{otherwise;} \end{cases} \qquad c(x,a) = I\{x > 0\}.$$

The one-step cost equals 1 for all positive states. For simplicity, we take $\lambda+\mu=1$, $\beta\in(0,1)$ is arbitrary. Now the optimality equation (3.2) takes the form

$$v(x)=1+\beta\mu v(x-1)+\beta\lambda v(x+1),\qquad x>0,$$

and has the following general solution, satisfying the obvious condition $v(0)=0$:

$$v(x)=\frac{1}{1-\beta}-\left[\frac{1}{1-\beta}+C\right]\gamma_1^x+C\gamma_2^x,\quad C\in\mathbb{R},$$

where

$$\gamma_{1,2}=\frac{1\pm\sqrt{1-4\beta^2\lambda\mu}}{2\beta\lambda},$$

and one can show that $0<\gamma_1<1$, $\gamma_2>1$.

The Bellman function v_x^* coincides with the minimal positive solution corresponding to $C=0$ and, in fact, is the only bounded solution. See Theorem 2.3 and the paragraph below it. □

Another nice example was presented in [Bertsekas (1987), Section 5.4, Ex. 2]:

Example. let $\mathbf{X}=[0,\infty)$, $\mathbf{A}=\{a\}$ (a dummy action), $p(x/\beta|x,a)=1$ with all the other transition probabilities zero. Put $c(x,a)\equiv 0$. Now the optimality equation (3.2) takes the form

$$v(x)=\beta v(x/\beta),$$

and is satisfied by any linear function $v(x)=kx$. But the Bellman function $v_x^*\equiv 0$ coincides with the unique bounded solution corresponding to $k=0$.

□

Other simple examples, in which the cost function $c(\cdot,\cdot)$ is bounded and the optimality equation has unbounded phantom solutions, can be found in [Hernández-Lerma and Lasserre (1996a), p. 51] and [Feinberg (2002b), Ex. 6.4]. See also Example 3.3.6.

3.3.3 *When value iteration is not successful: positive model*

It is known that, if $c(\cdot,\cdot)\le 0$, or $\sup_{(x,a)\in\mathbf{X}\times\mathbf{A}}|c(x,a)|<\infty$, or the action space $\mathbf{A}$ is finite and $c(\cdot,\cdot)\ge 0$, then $v^\infty(x)=v_x^*$ [Bertsekas and Shreve (1978), Prop. 9.14, Corollary 9.17.1], so that the MDP is stable: see Definition 3.1. A positive MDP is stable if and only if

$$v^\infty(x)=\inf_{a\in\mathbf{A}}\left\{c(x,a)+\beta\int_{\mathbf{X}}v^\infty(y)p(dy|x,a)\right\}$$

[Bertsekas and Shreve (1978), Prop. 9.16]. Below, we present a positive discounted MDP which is not stable; the idea is similar to Example 2.3.9.

Example. Let $\mathbf{X}=\{\Delta,0,(n,i)\ :\ n=1,2,\ldots;\ i=1,2,\ldots,n\}$; $\mathbf{A}=\{1,2,\ldots\}$; $p(\Delta|\Delta,a)\equiv 1$, $p(\Delta|(n,1),a)\equiv 1$, $p((n,i-1)|(n,i),a)\equiv 1$ for all

$i = 2, 3, \ldots, n$, and $p((a,a)|0,a) \equiv 1$, with all the other transition probabilities being zero. We put $\beta = 1/2$, $c((n,1),a) \equiv 2^n$; all the other costs equal zero. (See Fig. 3.3).

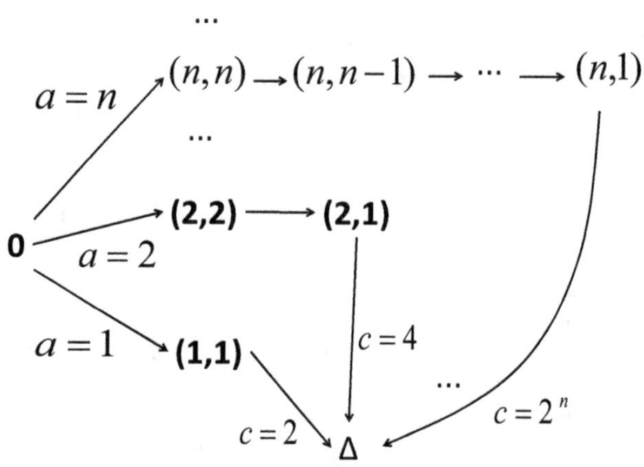

Fig. 3.3 Value iteration does not converge to the Bellman function.

The optimality equation (3.2) takes the form

$$v(\Delta) = \frac{1}{2}v(\Delta); \qquad v((n,1)) = 2^n + \frac{1}{2}v(\Delta);$$

$$v((n,i)) = \frac{1}{2}v((n,i-1)), \quad i = 2,3,\ldots,n; \qquad v(0) = \inf_{a\in\mathbf{A}}\left\{\frac{1}{2}v((a,a))\right\},$$

and has a single finite solution $v(\Delta) = 0$,

$$v((n,n)) = 2, \quad v((n,i)) = 2^{n-i+1}, \; i = 1,2,\ldots,n, \quad n = 1,2,\ldots;$$

$v(0) = 1$.

Value iteration results in the following sequence:

$$v^0(0) = v^0(\Delta) = 0, \qquad v^0((i,j)) \equiv 0;$$

$$v^1((1,1)) = 2, \;\; v^1((2,1)) = 4, \;\; v^1((3,1)) = 8, \ldots, \; v^1((n,1)) = 2^n, \ldots,$$

and these values remain unchanged in the further calculations. The remainder values of $v^1(\cdot)$ are zero. $v^1(0) = 0$ because $v^0((1,1)) = 0$.

$$v^2((2,2)) = 2, \;\; v^2((3,2)) = 4, \ldots, \; v^2((n,2)) = 2^{n-1}, \ldots,$$

and these values remain unchanged in the further calculations. $v^2(0) = 0$ because $v^1((2,2)) = 0$.

And so on.

We see that $v^\infty(\Delta) = v(\Delta) = 0$ and

$$v^\infty((n,i)) = v((n,i)) = 2^{n-i+1} \quad \text{for all } i = 1, 2, \ldots, n,$$

but

$$v^\infty(0) = 0 < 1 = \inf_{a \in \mathbf{A}} \left\{ \frac{1}{2} v^\infty((a,a)) \right\}.$$

Sufficient conditions for the equality $v^\infty(x) = v_x^*$, $x \in \mathbf{X}$ in positive models are also given in [Hernández-Lerma and Lasserre (1996a), Lemma 4.2.8]. In the current example, assuming that the spaces $\mathbf{X}$ and $\mathbf{A}$ are equipped with the discrete topology, all of those conditions are satisfied apart from the assumption that, for any $x \in \mathbf{X}$, the cost function $c(x, \cdot)$ is inf-compact: e.g., for $x = (2,2)$, $\{a \in \mathbf{A} : c((2,2), a) \equiv 0 \leq 1\} = \mathbf{A}$ is not compact. □

Another example in which value iteration does not converge to the Bellman function in a positive model can be found in [Bertsekas (2001), Exercise 3.1].

3.3.4 *Value iteration and convergence of stationary selectors*

Suppose the state and action spaces $\mathbf{X}$ and $\mathbf{A}$ are finite and the cost function $c(\cdot, \cdot)$ is finite-valued. Then the optimality equation (3.2) has a single bounded solution $v(x) = v_x^*$ coincident with the Bellman function, and any stationary selector from the set of *conserving (or thrifty)* selectors

$$\Phi^* := \left\{ \varphi : \mathbf{X} \to \mathbf{A} : \ c(x, \varphi(x)) + \beta \sum_{y \in \mathbf{X}} v(y) p(y|x, \varphi(x)) = v(x) \right\} \tag{3.8}$$

is uniformly optimal. For justification, see [Puterman (1994), Thms. 6.2.5(a) and 6.2.7].

Let us introduce the sets

$$\Phi_n^* := \left\{ \varphi : \mathbf{X} \to \mathbf{A} : \ c(x, \varphi(x)) + \beta \sum_{y \in \mathbf{X}} v^n(y) p(y|x, \varphi(x)) = v^{n+1}(x) \right\}, \tag{3.9}$$

$$n = 0, 1, 2, \ldots.$$

It is known that for all sufficiently large n, $\Phi_n^* \subseteq \Phi^*$ [Puterman (1994), Thm. 6.8.1]. The following example published in [Puterman (1994), Ex. 6.8.1] illustrates that the inclusion $\Phi_n^* \subset \Phi^*$ may be proper for all $n \geq 0$.

Example. Let $\mathbf{X} = \{1, 2, 3, 4, 5\}$, $\mathbf{A} = \{1, 2\}$, $\beta = 3/4$; $p(2|1,1) = p(4|1,2) = 1$, $p(3|2,a) = p(2|3,a) = p(5|4,a) = p(4|5,a) \equiv 1$, with other transition probabilities being zero; $c(1,1) = 10$, $c(1,2) = 8$, $c(2,a) \equiv 8$, $c(3,a) \equiv 142/9$, $c(4,a) = c(5,a) \equiv 12$ (see Fig. 3.4).

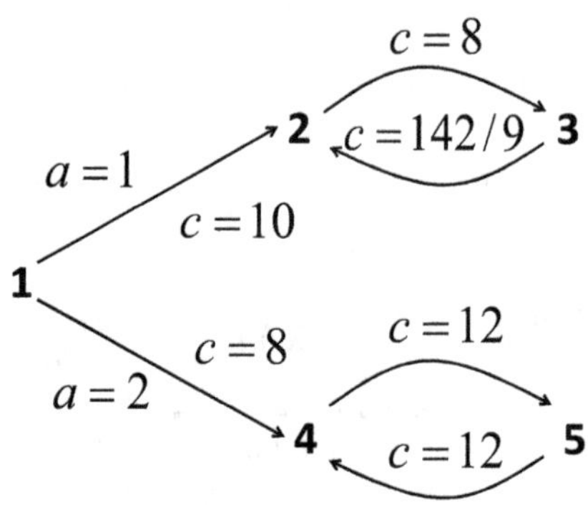

Fig. 3.4 Erratic value iterations.

The value iterations take the form:

$$\begin{aligned} v^0(x) &\equiv 0 \quad \text{and} \\ v^{n+1}(1) &= \min\{10 + 3/4 \cdot v^n(2); \quad 8 + 3/4 \cdot v^n(4)\}; \\ v^{n+1}(2) &= 8 + 3/4 \cdot v^n(3); \\ v^{n+1}(3) &= 142/9 + 3/4 \cdot v^n(2); \\ v^{n+1}(4) &= 12 + 3/4 \cdot v^n(5); \\ v^{n+1}(5) &= 12 + 3/4 \cdot v^n(4) \\ &\quad \text{for } n = 0, 1, 2, \ldots. \end{aligned}$$

Obviously,

$$v^n(4) = v^n(5) = 48\left[1 - \left(\frac{3}{4}\right)^n\right], \quad n = 0, 1, \ldots;$$

$$v^{n+1}(2) = \frac{119}{6} + \frac{9}{16}v^{n-1}(2), \quad n = 1, 2, \ldots.$$

Since $v^0(2) = 0$ and $v^1(2) = 8$, we conclude that, for all $n = 0, 1, \ldots,$

$$v^n(2) = \begin{cases} \frac{136}{3} - 34\left(\frac{3}{4}\right)^{n-1}, & \text{if } n \text{ is even;} \\ \frac{136}{3} - \frac{112}{3} \cdot \left(\frac{3}{4}\right)^{n-1}, & \text{if } n \text{ is odd.} \end{cases}$$

The rigorous proof can be done by induction. $v(x) = v_x^* = \lim_{n\to\infty} v^n(x)$ by Theorem 3.2.

Now, Φ^* coincides with the set of all stationary selectors: it is sufficient to look at state 1 only:

$$v(2) = \lim_{n\to\infty} v^n(2) = \frac{136}{3}; \qquad v(4) = \lim_{n\to\infty} v^n(4) = 48.$$

If $a = 1$, then

$$c(1,a) + \beta \sum_{y \in \mathbf{X}} v(y)p(y|1,a) = 10 + \frac{3}{4} \cdot v(2) = 44 = v(1);$$

if $a = 2$, then

$$c(1,a) + \beta \sum_{y \in \mathbf{X}} v(y)p(y|1,a) = 8 + \frac{3}{4} \cdot v(4) = 44 = v(1).$$

As for Φ_n^*, we see that, for fixed n, $\varphi \in \Phi_n^*$ if and only if

$$\varphi(1) = \begin{cases} 1, & \text{if } \Delta^n \le 0; \\ 2, & \text{if } \Delta^n \ge 0, \end{cases}$$

where

$$\Delta^n = 2 + \frac{3}{4}[v^n(2) - v^n(4)] = \begin{cases} 2 \cdot \left(\frac{3}{4}\right)^n, & \text{if } n \text{ is even;} \\ -\left(\frac{3}{4}\right)^{n-1}, & \text{if } n \text{ is odd.} \end{cases}$$

Therefore, the stationary selector $\varphi^1(x) \equiv 1$ belongs to Φ_n^* if and only if n is odd, and the selector $\varphi^2(x) \equiv 2$ belongs to Φ_n^* if and only if n is even. □

If either of the spaces $\mathbf{X}$ or $\mathbf{A}$ is not finite, it can easily happen that $\Phi_n^* \cap \Phi^* = \emptyset$ for all $n = 0, 1, 2, \ldots$. The next subsection illustrates this point.

3.3.5 *Value iteration in countable models*

Example. Let $\mathbf{X} = \{1, 2, 3\}$, $\mathbf{A} = \{0, 1, 2, \ldots\}$, $\beta = 1/2$,

$$p(2|1,a) = \begin{cases} \frac{1}{2} - \left(\frac{1}{2}\right)^a, & \text{if } a > 0; \\ \frac{1}{2}, & \text{if } a = 0, \end{cases} \qquad p(3|1,a) = 1 - p(2|1,a),$$

$p(2|2,a) = p(3|3,a) \equiv 1$, with other transition probabilities being zero; $c(1,a) = \left(\frac{1}{2} - \left(\frac{1}{2}\right)^a\right)^2$, $c(2,a) \equiv 1$, $c(3,a) \equiv 2$ (see Fig. 3.5). One can say that 0 is the one-point compactification of the set $\{1, 2, \ldots\}$. After this observation, the model is semi-continuous, i.e., Conditions 2.3 and 2.4 are satisfied.

For $x = 2$ and $x = 3$, value iteration gives

$$v^n(2) = 2 - \left(\frac{1}{2}\right)^{n-1}; \qquad v^n(3) = 4 - \left(\frac{1}{2}\right)^{n-2}.$$

We consider state $x = 1$ and firstly calculate the following infimum

$$\inf_{a>0} \left\{ \left(\frac{1}{2} - \left(\frac{1}{2}\right)^a\right)^2 + \frac{1}{2}\left[\left(\frac{1}{2} - \left(\frac{1}{2}\right)^a\right)\left(2 - \left(\frac{1}{2}\right)^{n-1}\right)\right.\right.$$

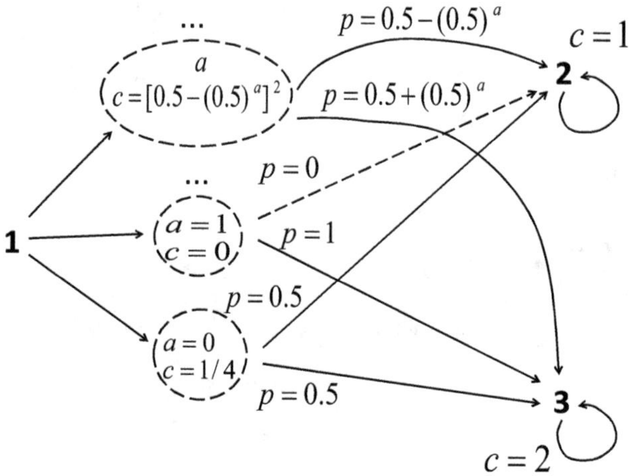

Fig. 3.5 Value iteration, denumerable action space.

$$+\left(\frac{1}{2}+\left(\frac{1}{2}\right)^a\right)\left(4-\left(\frac{1}{2}\right)^{n-2}\right)\Bigg]\Bigg\}.$$

Writing $\gamma := \left(\frac{1}{2}\right)^a$, we see that the infimum w.r.t. γ in the expression

$$\left(\frac{1}{2}-\gamma\right)^2+\frac{1}{2}\left[\left(\frac{1}{2}-\gamma\right)\left(2-\left(\frac{1}{2}\right)^{n-1}\right)+\left(\frac{1}{2}+\gamma\right)\left(4-\left(\frac{1}{2}\right)^{n-2}\right)\right]$$

is attained at $\gamma = \left(\frac{1}{2}\right)^{n+1}$ and equals $\frac{7}{4}-\frac{3}{2}\cdot\left(\frac{1}{2}\right)^n-\left(\frac{1}{2}\right)^{2n+2}$. Since

$$c(1,0)+\beta p(2|1,0)v^n(2)+\beta p(3|1,0)v^n(3)=\frac{7}{4}-\frac{3}{2}\cdot\left(\frac{1}{2}\right)^n,$$

we conclude that, for each $n = 0, 1, 2, \ldots$, the action $n+1$ provides the infimum in the formula

$$v^{n+1}(1) = \inf_{a\in\mathbf{A}}\{c(1,a)+\beta p(2|1,a)v^n(2)+\beta p(3|1,a)v^n(3)\}$$

$$=\frac{7}{4}-\frac{3}{2}\cdot\left(\frac{1}{2}\right)^n-\left(\frac{1}{2}\right)^{2n+2}.$$

Following (3.9),

$$\Phi_n^* = \{\varphi : \quad \varphi(1) = n+1\} \text{ for all } n = 0, 1, 2, \ldots.$$

Now

$$v^\infty(1) = \lim_{n\to\infty} v^n(1) = \frac{7}{4};\ v^\infty(2) = \lim_{n\to\infty} v^n(2) = 2;\ v^\infty(3) = \lim_{n\to\infty} v^n(3) = 4;$$

$v^\infty(x) = v(x) = v^*_x$ (see Section 3.1), and

$$\inf_{a\in\{1,2,\dots\}} \{c(1,a) + \beta p(2|1,a)v(2) + \beta p(3|1,a)v(3)\}$$

$$= \inf_{a\in\{1,2,\dots\}} \left\{ \left(\frac{1}{2} - \left(\frac{1}{2}\right)^a\right)^2 + p(2|1,a) + 2p(3|1,a)\right\} = \frac{7}{4}.$$

To see this, using the previous notation $\gamma := \left(\frac{1}{2}\right)^a$ for $a = 1, 2, \dots$, one can compute

$$\inf_{\gamma>0} \left\{ \left(\frac{1}{2} - \gamma\right)^2 + \frac{1}{2} - \gamma + 2\left(\frac{1}{2} + \gamma\right)\right\} = \frac{7}{4};$$

this infimum is attained at $\gamma = 0$, corresponding to no-one action $a > 0$. At the same time,

$$c(1,0) + p(2|1,0) + 2p(3|1,0) = \frac{7}{4}.$$

Therefore, following (3.8),

$$\Phi^* = \{\varphi^* :\quad \varphi^*(1) = 0\} \cap \Phi^*_n = \emptyset$$

for all $n = 0, 1, 2, \dots$. Note that the conserving stationary selectors φ^* are uniformly optimal by Theorem 3.1(b). □

The same situation can occur if the state space is infinite.

Example. Take $\mathbf{X} = \{0, 1, 2, \dots\}$, $\mathbf{A} = \{1, 2\}$, $\beta = 1/2$, $p(0|0,a) \equiv 1$; for $x > 0$ put $p(x|x,1) \equiv 1$ and $p(x-1|x,2) \equiv 1$, with other transition probabilities being zero; $c(0,a) \equiv -1$; for $x > 0$ put $c(x,1) \equiv 0$ and $c(x,2) = \left(\frac{1}{4}\right)^x$ (see Fig. 3.6).

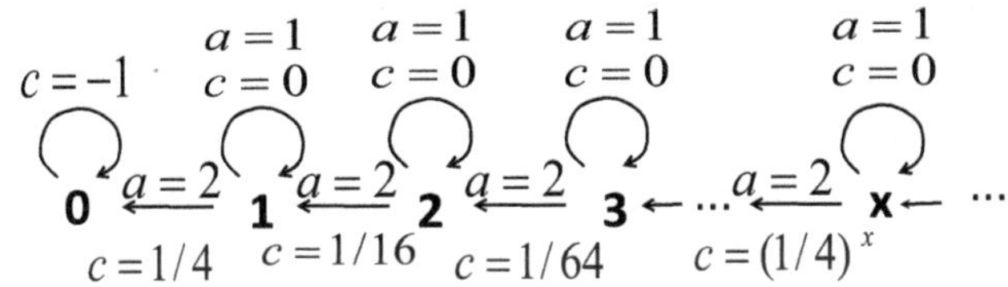

Fig. 3.6 Value iteration, denumerable state space.

Value iteration gives the following table:

x	0	1	2	3	4 ...
$v^0(x)$	0	0	0	0	0
$v^1(x)$	-1	0	0	0	0
$v^2(x)$	$-3/2$	$-1/4$	0	0	0
$v^3(x)$	$-7/4$	$-1/2$	$-1/16$	0	0
$v^4(x)$	$-15/8$	$-5/8$	$-3/16$	$-1/64$	0
...					...

In general,

$$v^n(0) = -2 + \left(\frac{1}{2}\right)^{n-1},$$

and for $x > 0$, $n > 0$ we have

$$v^n(x) = \min\left\{\frac{1}{2}v^{n-1}(x); \quad \left(\frac{1}{4}\right)^x + \frac{1}{2}v^{n-1}(x-1)\right\}$$

$$= \begin{cases} 0, & \text{if } n \leq x; \\ -\left(\frac{1}{2}\right)^x - \left(\frac{1}{4}\right)^x + \left(\frac{1}{2}\right)^{n-1}, & \text{if } n \geq x+1. \end{cases}$$

The minimum is provided by the following stationary selectors:

$$\Phi_n^* = \{\varphi : \quad \text{for } x > 0, \quad \varphi(x) = 1 \text{ if } x \geq n, \text{ and } \varphi(x) = 2 \text{ if } x < n\}.$$

At the same time, for all $x \in \mathbf{X}$,

$$\lim_{n\to\infty} v^n(x) = v^\infty(x) = v_x^* = v(x) = -\left(\frac{1}{2}\right)^x - \left(\frac{1}{4}\right)^x$$

(see Section 3.1) and

$$\Phi^* = \{\varphi^* : \text{ for } x > 0, \ \varphi^*(x) \equiv 2\}; \quad \Phi_n^* \cap \Phi^* = \emptyset \quad \text{for all } n = 0, 1, \ldots.$$

The conserving stationary selectors φ^* are uniformly optimal by Theorem 3.1(b). □

3.3.6 *Non-optimal stationary selector $\hat{\varphi}$ for which $v_x^{\hat{\varphi}}$ solves the optimality equation*

Theorem 3.1(a) allows to check the uniform optimality of a strategy without calculating the Bellman function. The following lemma provides a slight generalization. A very similar statement was proved in

[Hernández-Lerma and Lasserre (1996a), Thm. 4.2.3(c)], but under some additional conditions.

Lemma 3.1. *Suppose* $c(\cdot,\cdot) \ge K > -\infty$ *and* $\hat{\pi}$ *is a strategy such that for all* $x \in \mathbf{X}$ $v_x^{\hat{\pi}}$ *is finite and equation (3.4) is satisfied.*

Assume that, for each strategy π *and each* $x_0 \in \mathbf{X}$, $\mathsf{E}_{x_0}^{\pi}[v_{X_t}^{\hat{\pi}}]$ *is finite for all* $t = 0, 1, \ldots$ *and*

$$\lim_{T\to\infty} \beta^T \mathsf{E}_{x_0}^{\pi}[v_{X_T}^{\hat{\pi}}] = 0. \tag{3.10}$$

Then $\hat{\pi}$ *is uniformly optimal.*

The proof is given in Appendix A.

Example 3.1.4 of [Bertsekas (2001)] presents a non-optimal strategy $\hat{\pi}$ in a positive model satisfying equation (3.4), for which $v_x^{\hat{\pi}} = \infty$ and $\lim_{T\to\infty} \beta^T \mathsf{E}_x^{\pi}\left[v_{X_T}^{\hat{\pi}}\right] = \infty$ for some states x and some strategies π. Below, we provide another illustrative example where all functions are finite, which confirms that requirement (3.10) is important. It also confirms that one cannot omit the condition $\sup_{(x,a)\in\mathbf{X}\times\mathbf{A}} |c(x,a)| < \infty$ in Theorem 3.1(a).

Example. Let $\mathbf{X} = \{0, 1, 2 \ldots\}$, $\mathbf{A} = \{1, 2\}$, $p(0|0,a) \equiv 1$, $c(0,a) \equiv 0$. For $x > 0$ we put $p(x+1|x,1) = p(0|x,1) \equiv 1/2$, $p(x+1|x,2) \equiv 1$, other transition probabilities being zero; $c(x,1) = 2^x$, $c(x,2) \equiv 1$. The discount factor $\beta = 1/2$ (see Fig. 3.7).

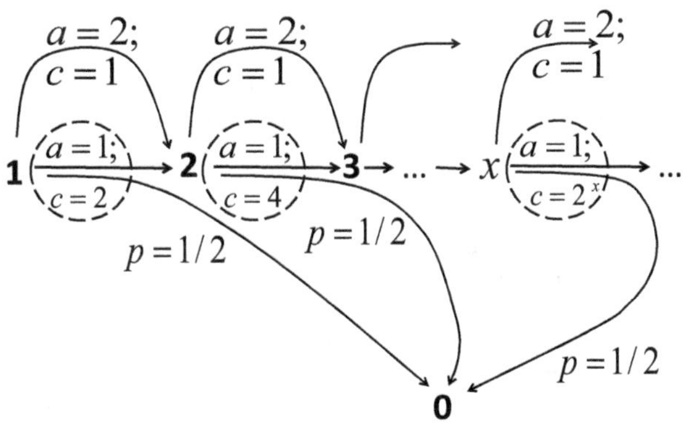

Fig. 3.7 Selector $\hat{\varphi}(x) \equiv 1$ is not optimal.

The optimality equation (3.2) is given by

$$v(0) = \frac{1}{2}v(0);$$

$$\text{for } x > 0 \;\; v(x) = \min\{2^x + \frac{1}{4}v(x+1) + \frac{1}{4}v(0);\;\; 1 + \frac{1}{2}v(x+1)\},$$

and has the minimal non-negative solution

$$v(x) = v_x^* \equiv 2 \quad \text{for } x > 0; \qquad\qquad v(0) = 0,$$

coincident with the Bellman function by [Puterman (1994), Thm. 7.3.2]. The stationary selector $\varphi^*(x) \equiv 2$ is conserving and equalizing and hence uniformly optimal; see Theorem 3.1(b).

Now look at the stationary selector $\hat{\varphi}(x) \equiv 1$. The performance functional $v_x^{\hat{\varphi}}$ is given by $v_x^{\hat{\varphi}} = 2^{x+1}$ (for $x > 0$) and satisfies equation (3.4):

$$1 + \frac{1}{2}v_{x+1}^{\hat{\varphi}} = 1 + 2^{x+1} > 2^{x+1} = v_x^{\hat{\varphi}} = 2^x + \frac{1}{4}v_{x+1}^{\hat{\varphi}}.$$

Equation (3.10) is violated for selector φ^*:

$$\beta^T \mathsf{E}_x^{\varphi^*}\left[v_{X_T}^{\hat{\varphi}}\right] = \beta^T\ 2^{x+T+1} = 2^{x+1},$$

and the selector $\hat{\varphi}$ is certainly non-optimal. All the conditions in Lemma 3.1 are satisfied for $\hat{\varphi}$ apart from equation (3.10). In particular, for each strategy π, $\mathsf{E}_x^\pi\left[v_{X_t}^{\hat{\varphi}}\right] \le \mathsf{E}_x^{\varphi^*}\left[v_{X_t}^{\hat{\varphi}}\right] = 2^{x+t+1}$. By the way,

$$\beta^T \mathsf{E}_x^{\hat{\varphi}}\left[v_{X_T}^{\hat{\varphi}}\right] = 2^{x+1-T} \to 0 \text{ as } T \to \infty.$$

We see that the both functions $v_x^* = v_x^{\varphi^*}$ and $v_x^{\hat{\varphi}}$ solve the optimality equation (3.2), but only v_x^* is the minimal non-negative solution. Note that equation (3.2) has many other solutions, e.g. $v(x) = 2 + k\ 2^x$ with $k \in [0, 1/2]$. □

3.3.7 *Single conserving strategy is not equalizing and not optimal*

This example is based on the same ideas as Example 2.3.4, and is very similar to the example described in [Hordijk and Tijms (1972)].

Example. Let $\mathbf{X} = \{0, 1, 2, \ldots\}$, $\mathbf{A} = \{1, 2\}$, $p(0|0, a) \equiv 1$, $c(0, a) \equiv 0$, $\forall x > 0$ $p(0|x, 1) \equiv 1$, $p(x+1|x, 2) \equiv 1$, with all other transition probabilities being zero; $c(x, 1) = 1 - 2^x$, $c(x, 2) \equiv 0$, $\beta = 1/2$ (see Fig. 3.8).

Like any other negative MDP, this model is stable in the sense of Definition 3.1: see Theorem 3.2(a). The optimality equation (3.2) is given by

$$v(0) = \frac{1}{2}v(0);$$

$$\text{for } x > 0 \qquad v(x) = \min\left\{1 - 2^x + \frac{1}{2}v(0);\ \frac{1}{2}v(x+1)\right\},$$

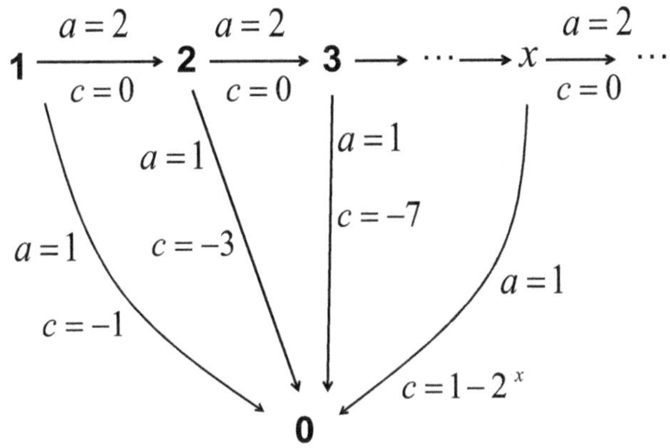

Fig. 3.8 No optimal strategies.

and has the maximal non-positive solution $v(0) = 0$; $v(x) = -2^x = v_x^*$ for $x > 0$, which coincides with the Bellman function. It can be built by the value iteration.

The stationary selector $\varphi^*(x) \equiv 2$ is the single conserving strategy at $x > 0$, but

$$\lim_{t\to\infty} E_x^{\varphi^*}\left[\beta^t\, v_{X_t}^*\right] = -2^x < 0,$$

so that it is not equalizing and not optimal. Note that equality (3.4) is violated for φ^* because $v_x^{\varphi^*} \equiv 0$.

There exist no optimal strategies in this model, but for an arbitrarily fixed $\varepsilon > 0$ the selector

$$\varphi_t^\varepsilon(x) = \begin{cases} 2, & \text{if } t \le 1 - \frac{\ln\varepsilon}{\ln 2}; \\ 1, & \text{otherwise} \end{cases}$$

is uniformly ε-optimal: for $\varepsilon \le 1$

$$v_x^{\varphi^\varepsilon} = \left(1 - 2^{x + \lfloor 1 - \frac{\ln\varepsilon}{\ln 2}\rfloor}\right)\left(\frac{1}{2}\right)^{\lfloor 1 - \frac{\ln\varepsilon}{\ln 2}\rfloor} \le 2^{-\frac{\ln\varepsilon}{\ln 2}} - 2^x = \varepsilon - 2^x$$

($\lfloor\cdot\rfloor$ is the integer part), and for $\varepsilon > 1$ $\varphi_t^\varepsilon(x) \equiv 1$ leading to $v_x^{\varphi^\varepsilon} = 1 - 2^x \le \varepsilon - 2^x$. □

Another trivial example of a negative MDP where a conserving strategy is not optimal can be found in [Bertsekas (2001), Ex. 3.1.3].

3.3.8 *Non-measurable Bellman function; no-one strategy (or selector) is uniformly ε-optimal*

Consider the model described in Subsection 1.4.10 (see Fig. 1.20), put $p(x|x,a) \equiv 1$ and fix the discount factor $\beta \in (0,1)$ arbitrarily.

For any $x \in \mathbf{X} \setminus B^1$, $v_x^* \equiv 0$ and, for any fixed $\hat{x} \in B^1$, $v_{\hat{x}}^* \equiv \frac{-1}{1-\beta}$: it is sufficient to take the stationary selector $\hat{\varphi}(x) \equiv \hat{a}$ with $\hat{a}$ such that $(\hat{x}, \hat{a}) \in B$. Thus, the Bellman function v_x^* is not measurable.

Now, consider an arbitrary control strategy π. Since $\pi_1(da|x)$ is a measurable stochastic kernel, $\int_{\mathbf{A}} c(x,a)\pi_1(da|x)$ is also measurable and

$$\left\{x \in \mathbf{X} : \quad \int_{\mathbf{A}} c(x,a)\pi_1(da|x) < 0\right\} \subseteq B^1$$

is a Borel subset of $\mathbf{X}$; hence there is $y \in B^1$ such that $\int_{\mathbf{A}} c(y,a)\pi_1(da|y) = 0$, meaning that, even if on the next steps the cost equals -1,

$$v_y^\pi \geq -\beta - \beta^2 - \cdots = -\frac{\beta}{1-\beta} = \frac{-1}{1-\beta} + 1 = v_y^* + 1,$$

i.e. the strategy π is not uniformly ε-optimal for any $\varepsilon < 1$.

The same example was presented in [Blackwell (1965), Ex. 2] and in [Puterman (1994), Ex. 6.2.3].

In the next example, first published in [Dynkin and Yushkevich (1979), Chapter 6, Section 7], the Bellman function $v_x^* \equiv -2$ is well defined, but no-one selector is uniformly ε-optimal.

Example. Let $\mathbf{X} = [0,1]$, $\mathbf{A} = [0,1]$ and let $Q \subset \mathbf{X} \times \mathbf{A}$ be a Borel subset such that for each $x \in \mathbf{X}$ there exists $a \in \mathbf{A}$ such that $(x,a) \in Q$ and Q does not contain graphs of measurable maps $\mathbf{X} \to \mathbf{A}$. (Such a subset was constructed in [Dynkin and Yushkevich (1979), App. 3, Section 3].) We put $c(x,a) = -\mathbb{I}\{(x,a) \in Q\}$, $p(\{0\}|x,a) \equiv 1$, and $\beta = 1/2$; other transition probabilities are zero. See Fig. 3.9.

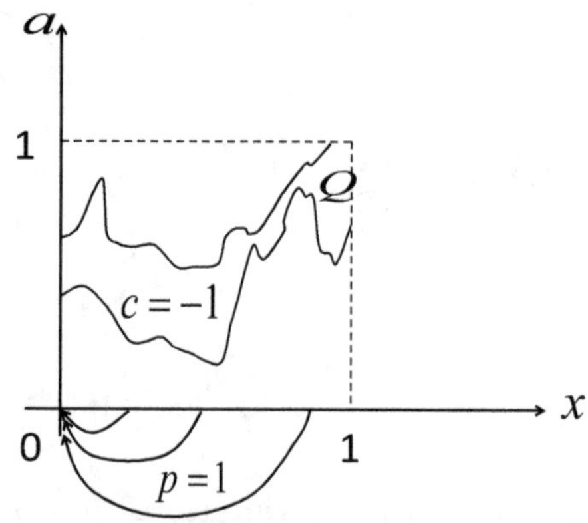

Fig. 3.9 No uniformly ε-optimal selectors.

For any fixed initial state $x_0 \in \mathbf{X}$ one can deterministically choose $\{A_t\}_{t=1}^\infty$ such that $(x_0, A_1) \in Q$ and $(0, A_t) \in Q$ for all $t > 1$; so $v_{x_0}^* \equiv -2$ and there exists an optimal control strategy. But for any selector φ,

$$v_x^\varphi = c(x, \varphi_1(x)) + \mathsf{E}_x^\varphi\left[\sum_{t=2}^{\infty} \beta^{t-1} c(X_t, A_t)\right] \geq c(x, \varphi_1(x)) - 1$$

(the second term, equal to the total discounted cost, starting from state $X_1 = 0$, cannot be smaller than -1). Since $\varphi_1(x)$ is a measurable map $\mathbf{X} \to \mathbf{A}$, there is a point $y \in \mathbf{X}$ such that $(y, \varphi_1(y)) \notin Q \Longrightarrow c(y, \varphi_1(y)) = 0$, and $v_y^\varphi \geq -1$. Thus, the selector φ is not uniformly ε-optimal if $\varepsilon < 1$. □

3.3.9 *Myopic strategies*

Definition 3.6. If the cost function $c(x, a)$ really depends on $a \in \mathbf{A}$, then a stationary strategy $\pi^s(da|x)$, uniformly optimal in the homogeneous one-step model ($T = 1$) with $C(x) \equiv 0$, is called *myopic*. In other words, $\int_{\mathbf{A}} c(x, a)\pi^s(da|x) \equiv \inf_{a \in \mathbf{A}}\{c(x, a)\}$.

When the cost function $c(x)$ does not depend on the action a, a stationary strategy π^s is called *myopic* if $\int_{\mathbf{A}} \int_{\mathbf{X}} p(dy|x, a)c(y)\pi^s(da|x) \equiv \inf_{a \in \mathbf{A}} \int_{\mathbf{X}} p(dy|x, a)c(y)$.

It is known that a myopic strategy is uniformly optimal in the discounted MDP with an arbitrary discount factor $\beta \in (0, 1)$, if it is uniformly optimal in the finite-horizon case with any time horizon $T < \infty$, without final cost ($C(x) \equiv 0$), and without discounting ($\beta = 1$). See [Piunovskiy (2006), Lemma 2.1]. The converse assertion is not always true, as the following example, published in the above article, shows.

Example. Let $\mathbf{X} = \{1, 2, 3\}$, $\mathbf{A} = \{1, 2\}$, $p(1|1, 1) = 1$, $p(2|1, 2) = 1$, $p(3|2, a) = p(3|3, a) \equiv 1$, with other transition probabilities being zero; $c(1, 1) = -2$, $c(1, 2) = -3$, $c(2, a) \equiv 0$, $c(3, a) \equiv -3$ (see Fig. 3.10). The discount factor $\beta \in (0, 1)$ is arbitrarily fixed.

One can check that the following bounded function satisfies the optimality equation (3.2):

$$v(3) = -\frac{3}{1-\beta}; \qquad v(2) = -\frac{3\beta}{1-\beta}; \qquad v(1) = -3 - \frac{3\beta^2}{1-\beta}.$$

The stationary selector $\varphi^*(x) \equiv 2$ is uniformly optimal independently of the value of β, and $v_x^* \equiv v_x^{\varphi^*} \equiv v(x)$. (See Section 3.1.) Obviously, the stationary selector φ^* is myopic.

At the same time, if $T = 2$ then, in this non-discounted finite-horizon model without final cost,

$$v_2(x) = C(x) \equiv 0, \quad v_1(x) = \begin{cases} -3, & \text{if } x = 1; \\ 0, & \text{if } x = 2; \\ -3, & \text{if } x = 3, \end{cases} \quad v_0(x) = \begin{cases} -5, & \text{if } x = 1; \\ -3, & \text{if } x = 2; \\ -6, & \text{if } x = 3, \end{cases}$$

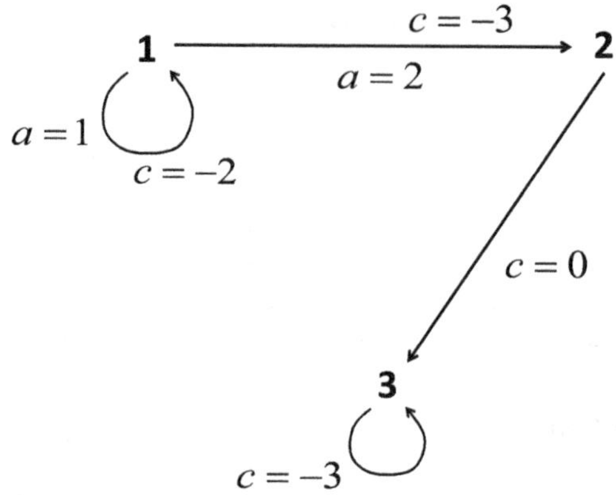

Fig. 3.10 Optimal myopic strategy.

the selector

$$\varphi_t(1) = \begin{cases} 1, & \text{if } t = 1; \\ 2, & \text{if } t = 2, \end{cases} \qquad \varphi_t(2) = \varphi_t(3) \equiv 2$$

is uniformly optimal, and the myopic selector φ^* is *not* optimal for the initial state $x_0 = 1$. □

3.3.10 *Model with partial information*

Suppose the transition probabilities depend on the unknown parameter $\theta \in \{\theta_1, \theta_2\}$, and the decision maker knows only the initial probability $q_0 = \mathsf{P}\{\theta = \theta_1\}$. It is known that the *a posteriori* probability $q_t = P\{\theta = \theta_1 | X_0, X_1, \dots, X_t\}$ is a sufficient statistic which plays the role of the component of the state in the MDP under study [Bertsekas and Shreve (1978), Chapter 10]. This means that one has to switch to the state space $\mathbf{X} \times [0,1] = \{(x,q)\}$, adjust the cost function and solve this new MDP model with complete information: the dynamics of the component q_t is calculated based on the observations of the component x_t. The initial value q_0 is assumed to be X_0-independent. More about MDPs with partial information in [Bertsekas and Shreve (1978), Chapter 10] and [Dynkin and Yushkevich (1979), Chapter 8].

The Bellman function is denoted as $v^*_{(x,q)}$ like usually. If the value of θ is known, i.e., $q_0 \in \{0,1\}$, then we deal with the model with complete information leading to the corresponding Bellman functions $v^*_{x|\theta_1}$ or $v^*_{x|\theta_2}$. It is more or less obvious that $qv^*_{x|\theta_1} + (1-q)v^*_{x|\theta_2} \le v^*_{(x,q)}$: if θ is known, the decision maker applies the corresponding best possible control strategy. In case θ is not known, there is always the following compromise: try to improve the knowledge of θ or try to minimize the performance functional based on the current knowledge of θ. It can happen that $q \in (0,1)$, but the optimal control strategy is such that q remains unchanged, i.e., no new information about θ arrives: attempts to obtain that information are too expensive. The following example published in [Bertsekas (2005), Ex. 6.1.6] confirms this point.

Example. Let $\mathbf{X} = \{1, 2\}$ and $\mathbf{A} = \{1, 2\}$. Assume that the one-step costs $k(x, y)$ depend on the current (x) and next (y) states only; $k(1, 1) = 0$, $k(1, 2) = 1$, $k(2, 1) = 1$. The transitions $2 \to 2$ never occur: $p(2|2, a) \equiv 0$. Independently of θ,

$$p(1|1, 1) = p(2|1, 1) = 0.5; \qquad p(1|2, a) \equiv 1.$$

The transition probability $p^\theta(1|1, 2)$ depends on the value of θ:

$$p^\theta(1|1, 2) = \begin{cases} 0.6, & \text{if } \theta = \theta_1; \\ 0.3, & \text{if } \theta = \theta_2 \end{cases} \qquad p^\theta(2|1, 2) = 1 - p^\theta(1|1, 2)$$

(see Fig. 3.11). The discount factor $\beta \in (0, 1)$ is arbitrarily fixed. Actions in state 2 play no role.

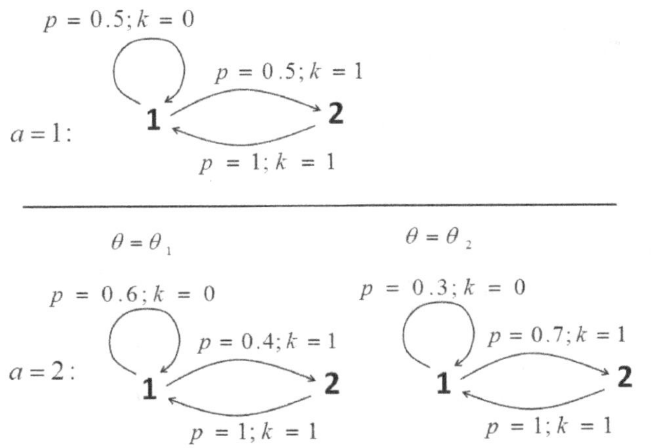

Fig. 3.11 MDP with partial information.

The *a posteriori* probability q_t can change only if $X_{t-1} = 1$ and $A_t = 2$:

$$\text{if } X_t = 1 \text{ then } \; q_t = \frac{0.6q_{t-1}}{0.6q_{t-1} + 0.3(1 - q_{t-1})} = \frac{0.6q_{t-1}}{0.3 + 0.3q_{t-1}};$$

$$\text{if } X_t = 2 \text{ then } \; q_t = \frac{0.4q_{t-1}}{0.4q_{t-1} + 0.7(1 - q_{t-1})} = \frac{0.4q_{t-1}}{0.7 - 0.3q_{t-1}}.$$

The transition probabilities and the cost function for the model with complete information and state space $\mathbf{X} \times [0, 1]$ are defined by the following formulae (see Fig. 3.12 for illustration):

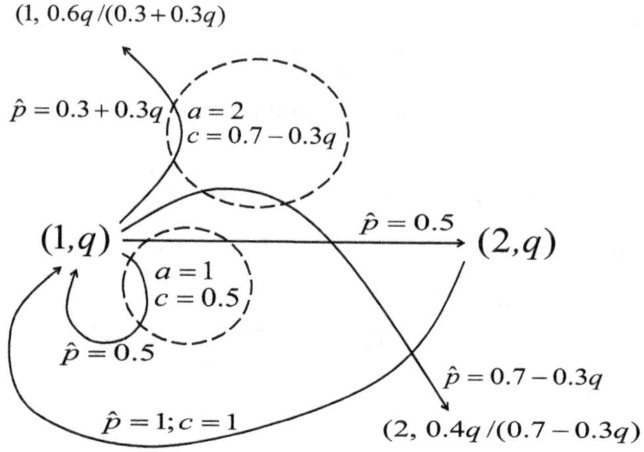

Fig. 3.12 Equivalent MDP with complete information.

$$\hat{p}((1,q)|(1,q),1) = \hat{p}((2,q)|(1,q),1) = 0.5; \quad \hat{p}((1,q)|(2,q),a) \equiv 1;$$

$$\hat{p}\left(\left(1, \frac{0.6q}{0.3+0.3q}\right)\bigg|(1,q),2\right) = 0.6q + 0.3(1-q) = 0.3+0.3q;$$

$$\hat{p}\left(\left(2, \frac{0.4q}{0.7-0.3q}\right)\bigg|(1,q),2\right) = 0.4q + 0.7(1-q) = 0.7-0.3q;$$

(other transition probabilities are zero);

$$c((1,q),1) = 0.5; \quad c((2,q),a) \equiv 1; \quad c((1,q),2) = 0.4q + 0.7(1-q) = 0.7-0.3q.$$

The optimality equation (3.2) takes the form

$$\begin{aligned} v((1,q)) = \min\Big\{&0.5 + \beta[0.5\ v((1,q)) + 0.5\ v((2,q))]; \\ &0.7 - 0.3q + \beta\left[(0.3+0.3q)v\left(\left(1, \frac{0.6q}{0.3+0.3q}\right)\right)\right. \\ &\left.+(0.7-0.3q)v\left(\left(2, \frac{0.4q}{0.7-0.3q}\right)\right)\right]\Big\}; \\ v((2,q)) = 1 + &\beta\ v((1,q)); \end{aligned}$$

$v((x,q)) = v^*_{(x,q)}$ is the single bounded solution, and there exists an optimal stationary selector $\varphi^*((x,q))$ because the action space $\mathbf{A}$ is finite [Bertsekas and Shreve (1978), Corollary 9.17.1].

We will show that there are positive values of q for which $\varphi^*((1,q)) = 1$. If $q \in (0,1)$ then $\frac{0.6q}{0.3+0.3q}, \frac{0.4q}{0.7q-0.3q} \in (0,1)$, and we can concentrate on the model with complete information and with state space $\mathbf{X} \times (0,1)$, for which φ^* is also an optimal strategy. For the stationary selector $\varphi((x,q)) \equiv 2$, we have

$$v^{\varphi}_{(1,q)} = \frac{0.4(1+\beta)q}{1-0.6\beta-0.4\beta^2} + \frac{0.7(1+\beta)(1-q)}{1-0.3\beta-0.7\beta^2};$$

$$v^{\varphi}_{(2,q)} = 1 + \beta\, v^{\varphi}_{(1,q)} :$$

the performance functional equals $\frac{0.4(1+\beta)}{1-0.6\beta-0.4\beta^2}$ $\left(\frac{0.7(1+\beta)}{1-0.3\beta-0.7\beta^2}\right)$ if $\theta = \theta_1$ (θ_2). Now

$$v^{\varphi}_{(1,q)} - \{0.5 + \beta[0.5v^{\varphi}_{(1,q)} + 0.5v^{\varphi}_{(2,q)}]\} =$$

$$(1+\beta)\,[1-0.5\beta(1+\beta)]\left[\frac{0.4q}{1-0.6\beta-0.4\beta^2} + \frac{0.7(1-q)}{1-0.3\beta-0.7\beta^2}\right] - 0.5(1+\beta).$$

This last function of q is continuous, and it approaches the positive value

$$\frac{0.7(1+\beta)(1-0.5\beta(1+\beta))}{1-0.3\beta-0.7\beta^2} - 0.5(1+\beta) = \frac{0.2(1-\beta^2)}{1-0.3\beta-0.7\beta^2} > 0$$

as $q \to 0$. Therefore, in the model under investigation, the selector $\varphi((x,q)) \equiv 2$ is not optimal by Theorem 3.1(a) if $q > 0$ is small enough. Since actions in state 2 play no role, we conclude that, for some positive values of q, it is reasonable to apply action $a = 1$ in state 1: $\varphi^*((1,q)) = 1$. But thereafter the value of q remains unchanged, and the decision maker applies action $a = 1$ every time. The attempt to apply action 2 in state 1 in order to better estimate θ leads to the expected one step cost $c((1,q),2) = 0.7 - 0.3q$ which is too large compared with $c((1,q),1) = 0.5$ if q is small. □

3.3.11 *Occupation measures and stationary strategies*

The space $\mathcal{D}_{\mathsf{P}_0}$ of occupation measures under a fixed initial distribution P_0 is convex, generated by stationary strategies by Theorem 3.3(b). According to (3.5), the performance functional $v^{\pi}_{\mathsf{P}_0}$ is linear on the space of occupation measures in general enough situations, e.g., when $\sup_{(x,a)\in\mathbf{X}\times\mathbf{A}} |c(x,a)| < \infty$. The space of stationary strategies is also convex, that is, for stationary strategies π^{s^1} and π^{s^2}, the stochastic kernel $\alpha\pi^{s^1}(da|x) + (1-\alpha)\pi^{s^2}(da|x)$ with $\alpha \in [0,1]$ defines the third stationary strategy. But the map $\pi^{s} \to \eta^{\pi^{s}}_{\mathsf{P}_0}$ and the function $v^{\pi^{s}}_{\mathsf{P}_0} : \pi^{s} \to \mathbb{R}^1$ may be non-affine. The following example confirms this.

Example. Let $\mathbf{X} = \{1,2\}$, $\mathbf{A} = \{0,1\}$, $p(1|1,0) = 1$, $p(2|1,1) = 1$, $p(1|2,a) \equiv 1$, with other transition probabilities being zero; $\mathsf{P}_0(1) = 1$, $c(1,a) \equiv 0$, $c(2,a) \equiv 1$ (see Fig. 3.13).

Since the values of $\pi^{s}(a|2)$ play no role, one can fix, for example, $\pi^{s}(0|2) = 1$ and describe any stationary strategy π^{s} with the number $\gamma = \pi^{s}(0|1)$. Now, for

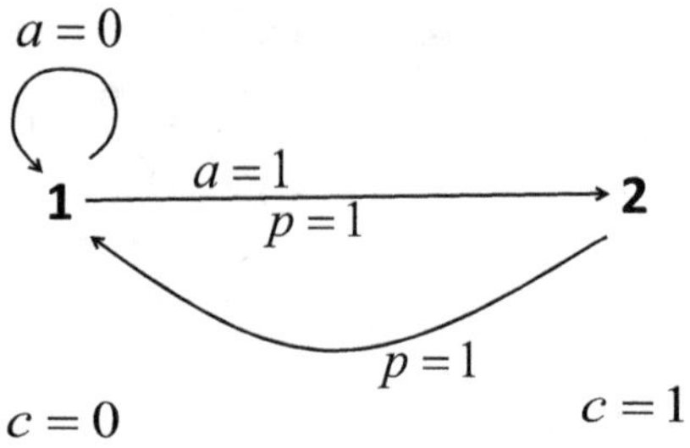

Fig. 3.13 Performance functional is non-convex w.r.t. to stationary strategies.

$\alpha \in [0,1]$, the strategy $\pi^{s^3}(a|x) = \alpha\pi^{s^1}(a|x) + (1-\alpha)\pi^{s^2}(a|x)$ corresponds to $\gamma^3 = \alpha\gamma^1 + (1-\alpha)\gamma^2$, where γ^1 and γ^2 describe strategies π^{s^1} and π^{s^2}. (Of course, $\pi^{s^3}(0|2) = 1$.) The linearity w.r.t. stationary strategies π^s corresponds to the linearity in γ, and the convex space of stationary strategies with $\pi^s(0|2) = 1$ is isomorphic to the segment $[0,1] \ni \gamma$.

For a fixed value of γ, the marginal $\hat{\eta}^\gamma(x) := \eta^\gamma_{\mathsf{P}_0}(x \times \mathbf{A})$ satisfies equations

$$\hat{\eta}^\gamma(1) = (1-\beta) + \beta[\gamma\hat{\eta}^\gamma(1) + \hat{\eta}^\gamma(2)];$$
$$\hat{\eta}^\gamma(2) = \beta(1-\gamma)\hat{\eta}^\gamma(1).$$

The index γ corresponds to the stationary strategy π^s defined by γ: $\eta^\gamma_{\mathsf{P}_0} := \eta^{\pi^s}_{\mathsf{P}_0}$. For justification, see Theorem 3.3:

$$\sum_{a=0}^{1} p(1|1,a)\eta^\gamma_{\mathsf{P}_0}(1,a) + \sum_{a=0}^{1} p(1|2,a)\eta^\gamma_{\mathsf{P}_0}(2,a) = \pi^s(0|1)\hat{\eta}^\gamma(1) + \hat{\eta}^\gamma(2).$$

Therefore,

$$\hat{\eta}^\gamma(1) = \frac{1-\beta}{1-\beta\gamma-\beta^2(1-\gamma)}, \qquad \hat{\eta}^\gamma(2) = \frac{(\beta-\beta\gamma)(1-\beta)}{1-\beta\gamma-\beta^2(1-\gamma)}, \tag{3.11}$$

and the mapping $\gamma \to \eta^\gamma_{\mathsf{P}_0}$ (or, equivalently, the map $\pi^s \to \eta^{\pi^s}_{\mathsf{P}_0}$) is not affine. Similarly, the function $v^{\pi^s}_{\mathsf{P}_0} = \hat{\eta}^\gamma(2) = \frac{(\beta-\beta\gamma)(1-\beta)}{1-\beta\gamma-\beta^2(1-\gamma)}$ is non-affine w.r.t γ (and w.r.t π^s).

One can encode occupation measures using the value of $\delta = \hat{\eta}^\gamma(1)$. The corresponding value of γ can be calculated using the first formula in (3.11):

$$\gamma = 1 + \frac{1}{\beta} - \frac{1}{\beta\delta}.$$

In order to have $\gamma \in [0,1]$, one can fix δ only from the interval $\left[\frac{1}{1+\beta}, 1\right]$. The corresponding occupation measure is given by

$$\eta^{\gamma}_{P_0}(1,0) = \gamma\delta = \delta - \frac{1-\delta}{\beta}; \quad \eta^{\gamma}_{P_0}(1,1) = (1-\gamma)\delta = \frac{1-\delta}{\beta}; \quad \hat{\eta}^{\gamma}(2) = 1-\delta.$$

The separate values $\eta^{\gamma}_{P_0}(2,0)$ and $\eta^{\gamma}_{P_0}(2,1)$ are of no importance but, if needed, they can be defined by

$$\eta^{\gamma}_{P_0}(2,0) = (1-\delta)\varepsilon, \quad \eta^{\gamma}_{P_0}(2,1) = (1-\delta)(1-\varepsilon),$$

where $\varepsilon \in [0,1]$ is an arbitrary number corresponding to $\pi^{s}(0|2)$. (Recall that we agreed to consider only the stationary strategies with $\pi^{s}(0|2) = 1$, i.e., $\varepsilon = 1$.) Now the performance functional $v^{\pi^{s}}_{P_0} = \hat{\eta}^{\gamma}(2) = 1-\delta$ is affine w.r.t. δ (and w.r.t. $\eta^{\gamma}_{P_0}$ which is affine w.r.t. δ).

In accordance with Theorem 3.3(b), stationary strategies are sufficient for solving this trivial optimal control problem $v^{\pi}_{P_0} \to \inf_{\pi}$: the solution is given by the maximal value of δ:

$$\delta = 1, \ \gamma = 1, \ \pi^{s^*}(0|1) = 1 \ (\pi^{s^*}(a|2) \text{ is immaterial}), \ v^*_1 = 0$$

and, obviously, $v^*_2 = 1$. □

3.3.12 *Constrained optimization and the Bellman principle*

Suppose we have two cost functions ${}^{1}c(\cdot,\cdot)$ and ${}^{2}c(\cdot,\cdot)$. Then every control strategy π results in two performance functionals ${}^{1}v^{\pi}_{P_0}$ and ${}^{2}v^{\pi}_{P_0}$ defined according to (3.1), the discount factor β being the same. The *constrained* problem looks like

$$ {}^{1}v^{\pi}_{P_0} \to \inf_{\pi} \qquad {}^{2}v^{\pi}_{P_0} \le d, \tag{3.12}$$

where d is a chosen number. Strategies satisfying the above inequality are called *admissible*. The following example, first published in [Altman and Shwartz (1991a), Ex. 5.3] shows that, in this framework, the optimal strategy depends on the initial distribution and, moreover, the Bellman principle fails to hold.

Example. Let $\mathbf{X} = \{1,2\}$; $\mathbf{A} = \{1,2\}$; $p(1|x,a) \equiv 0.1$; $p(2|x,a) \equiv 0.9$; $\beta = 0.1$;

$${}^{1}c(1,a) \equiv 0 \quad {}^{1}c(2,a) = \begin{cases} 1, & \text{if } a = 1; \\ 0, & \text{if } a = 2 \end{cases}$$

$${}^{2}c(1,a) \equiv 1 \quad {}^{2}c(2,a) = \begin{cases} 0, & \text{if } a = 1; \\ 0.1, & \text{if } a = 2 \end{cases}$$

(see Fig. 3.14).

Let $x_0 = 1$ and take $d = \frac{91}{90}$, the minimal value of ${}^{2}v^{\pi}_{1}$, i.e. the value of the Bellman function ${}^{2}v^{*}_{1}$ associated with the cost ${}^{2}c(\cdot,\cdot)$. The actions in state 1 play no role, and the optimal action in state 2 is 1 leading to ${}^{2}c(2,1) = 0$. In state 2

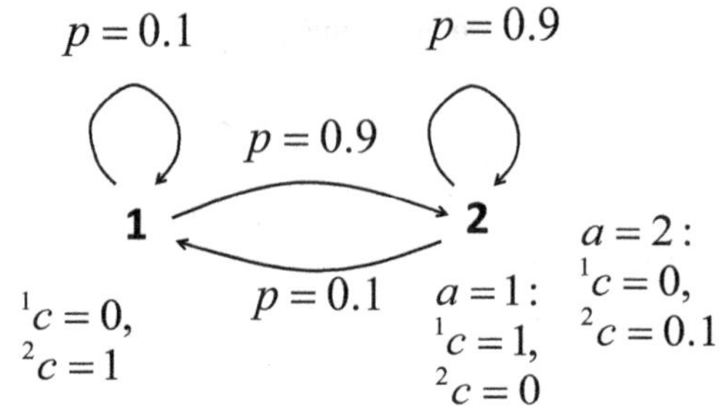

Fig. 3.14 Constrained problem.

one has to apply action 1 because otherwise the constraint in (3.12) is violated. Therefore, the stationary selector $\varphi^1(x) \equiv 1$ solves problem (3.12).

Suppose now that the initial state is $x_0 = 2$ and consider the stationary selector $\varphi^2(x) \equiv 2$. This strategy solves the unconstrained problem ${}^1v_2^\pi \to \inf_\pi$: ${}^1v_2^{\varphi^2} = {}^1v_2^* = 0$. In the constrained problem (3.12), ${}^2v_2^{\varphi^2} = \frac{109}{900} < d = \frac{910}{900}$. So, the stationary selector φ^2 is admissible and hence optimal. Therefore, the optimal strategy depends on the initial state. The Bellman principle fails to hold because the optimal actions in state 2 at all the decision epochs depend on the value of x_0.

Suppose $x_0 = 1$ and suppose we have the observed trajectory

$$X_0 = 1 \to X_1 = 2 \to X_2 = 2.$$

The optimal strategy φ^1 prescribes the actions $A_1 = A_2 = A_3 = \cdots = 1$. At the same time, at decision epoch 3, we know the current value of the accumulated realized second-type cost:

$$ {}^2c(X_0, 1) + \beta\ {}^2c(X_1, 1) = 1 + 0 = 1, \tag{3.13}$$

so that, if we apply selector φ^2 starting from the decision epoch 3, then the future expected second-type cost equals

$$\beta^2 \cdot {}^2v_2^{\varphi^2} = 0.01 \cdot \frac{109}{900} \approx 0.0012$$

which, together with one unit from (3.13), makes $1.0012 < d = \frac{91}{90}$. Thus, why do we not use selector φ^2 resulting in a smaller value for the main functional ${}^1v_1$ in this situation? The answer is: if we do that, then, after many repeated experiments, the average value of ${}^2v_1$ will be greater than d. □

3.3.13 *Constrained optimization, Lagrange multipliers and stationary selectors*

When solving constrained problems like (3.12), the following version of the Kuhn-Tucker Theorem is useful. For simplicity assume that either the model is positive/negative, or $\sup_{(x,a)\in \mathbf{X}\times\mathbf{A}} |c(x,a)| < \infty$.

Proposition 3.1. *Suppose the performance functionals* ${}^1v^{\pi}_{\mathsf{P}_0}$ *and* ${}^2v^{\pi}_{\mathsf{P}_0}$ *are finite for any strategy* π *and* ${}^2v^{\hat\pi}_{\mathsf{P}_0} < d$ *for some strategy* $\hat\pi$ *(the so called Slater condition). Let*

$$L(\pi,\lambda) = {}^1v^{\pi}_{\mathsf{P}_0} + \lambda({}^2v^{\pi}_{\mathsf{P}_0} - d), \qquad \lambda \ge 0$$

be the Lagrange function and assume that function $L(\cdot,1)$ *is bounded from below.*

Then the following assertions hold.

(a) *There exists* $\lambda^* \ge 0$ *such that*

$$g(\lambda^*) = \sup_{\lambda\ge 0} g(\lambda),$$

where $g(\lambda) := \inf_\pi L(\pi,\lambda)$ *is called dual functional.*

(b) *A strategy* π^* *is a solution to problem (3.12) if and only if*

$$L(\pi^*,\lambda^*) = \min_\pi L(\pi,\lambda^*), \quad {}^2v^{\pi^*}_{\mathsf{P}_0} \le d \quad \text{and} \quad \lambda^*\cdot({}^2v^{\pi^*}_{\mathsf{P}_0} - d) = 0.$$

The last equality here is called condition of complementary slackness.

More about the Lagrange approach can be found in [Piunovskiy (1997), Section 2.3]: see Theorem 13 and Lemma 9 therein. See also Subsection B.2.3 of [Piunovskiy and Zhang (2020)].

Example. Let us solve the constrained problem (3.12) for the MDP described in Subsection 3.3.12 (see Fig. 3.14) under $x_0 = 1$ for $d = 1.016$. One can build the dual functional $g(\lambda)$ after solving the unconstrained problem (3.1) for the cost function ${}^1c(\cdot,\cdot) + \lambda\,{}^2c(\cdot,\cdot)$. The optimality equation (3.2) has the form

$$\begin{aligned} v(1) &= \lambda + 0.01\,v(1) + 0.09\,v(2);\\ v(2) &= \min\{1 + 0.01\,v(1) + 0.09\,v(2);\\ &\qquad\qquad 0.1\lambda + 0.01\,v(1) + 0.09\,v(2)\}. \end{aligned}$$

If $\lambda < 10$, then

$$v(2) = 0.1\lambda + 0.01\,v(1) + 0.09\,v(2) = v(1) - 0.9\,\lambda$$

and equation

$$0.99\,v(1) = \lambda + 0.09[v(1) - 0.9\,\lambda]$$

gives $v(1) \approx 1.021\ \lambda$. Similarly, if $\lambda \geq 10$, then $v(1) \approx 0.1 + 1.011\ \lambda$. Therefore,

$$g(\lambda) \approx \begin{cases} 1.021\ \lambda - 1.016\ \lambda = 0.005\ \lambda, & \text{if } \lambda < 10; \\ 0.1 + 1.011\ \lambda - 1.016\ \lambda = 0.1 - 0.005\ \lambda, & \text{if } \lambda \geq 10, \end{cases}$$

and $\lambda^* = 10$. For this value of λ, $L(\pi, \lambda^*)$ does not depend on π and equals ≈ 0.05.

No-one stationary selector is optimal:

$${}^2v_1^{\varphi^2} = \frac{0.919}{0.9} \approx 1.021 > d = 1.016 \text{ and } {}^2v_1^{\varphi^1} = \frac{91}{90} \approx 1.011$$

which violates the complementary slackness condition. The stationary selectors $\varphi^1(x) \equiv 1$ and $\varphi^2(x) \equiv 2$ were introduced in Subsection 3.3.12. Recall that actions in state 1 are immaterial. The answer to problem (3.12) is given by the mixture of selectors φ^1 and φ^2, that is, such a strategy π that $\mathsf{P}_1^\pi = \frac{1}{2}\mathsf{P}_1^{\varphi^1} + \frac{1}{2}\mathsf{P}_1^{\varphi^2}$ (see Subsection 2.3.1): ${}^2v_1^\pi = 1.016 = d$, and π is optimal by Proposition 3.1(b). □

We see that stationary selectors are not sufficient for solving constrained problems. In general, it is usually sufficient to restrict to mixtures of $k+1$ stationary selectors, where k is the number of constraints-inequalities: see Theorem 2 of [Piunovskiy and Zhang (2024a)] and references provided in that article.

The dual functional $g(\lambda) := \inf_\pi L(\pi, \lambda)$ is obviously helpful for solving constrained problems, so its properties are of the great importance. It is known that $g(\cdot)$ is concave. If, for example, the cost functions ${}^1c(\cdot)$ and ${}^2c(\cdot)$ are bounded below then the dual functional $g(\cdot)$ is continuous for $\lambda > 0$, like any other concave function with the full domain ($\forall \lambda \geq 0\ g(\lambda) > -\infty$); see [Rockafellar (1970), Thm. 10.1]. Incidentally, the article [Frid (1972)] contains a minor mistake: on p. 189, the author claims that the dual functional $g(\cdot)$ is continuous on $[0, \infty)$. The following example shows that functional $g(\cdot)$ may be discontinuous at zero (see [Sennott (1991), Ex. 3.1]).

Example. Let $\mathbf{X} = \{0, 1, 2, \ldots\}$; $\mathbf{A} = \{0, 1\}$; $p(0|0,0) = 1$, $p(1|0,1) = 1$, $p(i+1|i,a) \equiv 1$ for all $i = 1, 2, \ldots$, with other transition probabilities being zero. We put

$${}^1c(x,a) = \begin{cases} 1, & \text{if } x = 0; \\ 0, & \text{if } x > 0; \end{cases} \qquad {}^2c(x,a) = 2^x;$$

$\beta = 1/2$; $\mathsf{P}_0(0) = 1$ and $d \geq 2$ (see Fig. 3.15). Only the actions in state $x = 0$ play any role.

Since ${}^2v_0^\pi = +\infty$ for any strategy except for those which apply action $a = 0$ in state $x = 0$, we conclude that

$$g(\lambda) = {}^1v_0^{\varphi^0} + \lambda({}^2v_0^{\varphi^0} - 1) = 2 + \lambda$$

if $\lambda > 0$. Here $\varphi^0(x) \equiv 0$. If $\lambda = 0$, then $\min_\pi {}^1v_0^\pi = 1$ is provided by the stationary selector $\varphi^1(x) \equiv 1$, and $g(0) = 1$. Hence, the dual functional is discontinuous at zero. The solution to the constrained problem (3.12) is provided

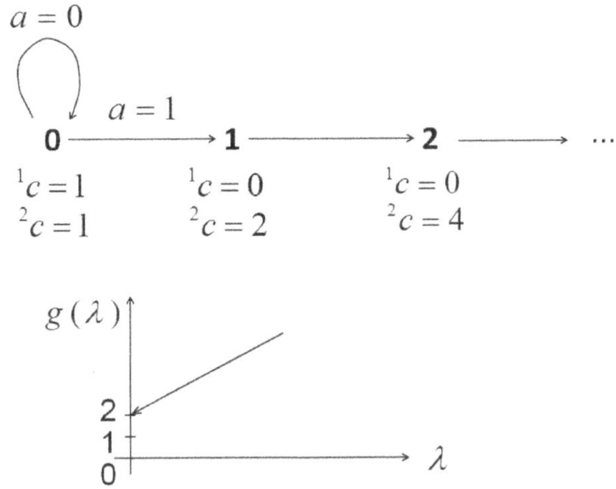

Fig. 3.15 Discontinuous dual functional.

by the selector φ^0. The statement of Proposition 3.1 is not valid because the performance functional ${}^2v_0^\pi$ is not finite, e.g. for the selector φ^1; λ^* maximizing the dual functional does not exist.

When the cost functions ${}^1c(\cdot,\cdot)$ and ${}^2c(\cdot,\cdot)$ are bounded below and ${}^2v_{\mathsf{P}_0}^{\hat\pi} < \infty$ for a strategy $\hat\pi$ providing $\min_\pi {}^1v_{\mathsf{P}_0}^\pi$, the dual functional is also continuous at $\lambda = 0$: the proof of Lemma 3.5 [Sennott (1991)] also holds in the general situation (not only for countable $\mathbf{X}$). In the presented example $\min_\pi {}^1v_{\mathsf{P}_0}^\pi$ is provided by the stationary selector $\varphi^1(x) \equiv 1$, and ${}^2v_0^{\varphi^1} = +\infty$. $\square$

The following example shows that the *Slater* condition ${}^2v_{\mathsf{P}_0}^{\hat\pi} < d$ is also important in Proposition 3.1.

Example. Let $\mathbf{X} = [1,\infty) \cup \{0\}$; $\mathbf{A} = [1,\infty) \cup \{0\}$; $p(\Gamma^{\mathbf{X}}|0,a) = I\{a \in \Gamma^{\mathbf{X}}\}$, $p(\Gamma^{\mathbf{X}}|x,a) \equiv \mathbb{I}\{x \in \Gamma^{\mathbf{X}}\}$ for $x \geq 1$,

$$ {}^1c(x,a) = \begin{cases} 2, & \text{if } x = 0,\ a = 0; \\ 0, & \text{if } x = 0,\ a \geq 1; \\ 1 - \frac{1}{x}, & \text{if } x \geq 1, \end{cases} \qquad {}^2c(x,a) = \begin{cases} 0, & \text{if } x = 0; \\ \frac{1}{x^2}, & \text{if } x \geq 1, \end{cases} $$

$\beta = 1/2$, $\mathsf{P}_0(\{0\}) = 1$ and $d = 0$ (see Fig. 3.16).

Only the actions in state $x = 0$ play any role, and only the strategies π which apply action $a = 0$ in state $x = 0$ are admissible leading to ${}^2v_0^\pi = 0 \leq d$. When calculating the dual functional $g(\cdot)$, without loss of generality, we need consider only stationary selectors φ or, to be more precise, only the values $\varphi(0) = a$. Now

$$ L(\varphi,\lambda) = {}^1v_0^\varphi + \lambda\, {}^2v_0^\varphi = \begin{cases} 4, & \text{if } a = 0; \\ 1 - \frac{1}{a} + \frac{\lambda}{a^2}, & \text{if } a \geq 1, \end{cases} $$

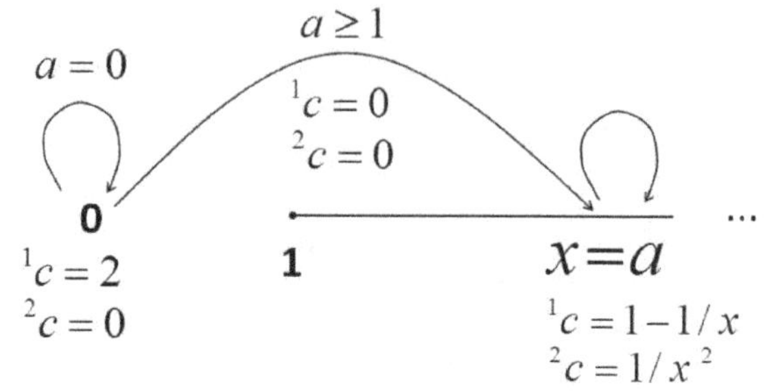

Fig. 3.16 No optimal values of the Lagrange multiplier λ.

and

$$g(\lambda) = \inf_{\varphi} L(\varphi, \lambda) = \begin{cases} \lambda, & \text{if } \lambda < \frac{1}{2}; \\ 1 - \frac{1}{4\lambda}, & \text{if } \lambda \geq \frac{1}{2} \end{cases}$$

(see Fig. 3.17). The optimal value of $a = \varphi(0)$ providing the infimum equals $a^* = \begin{cases} 1, & \text{if } \lambda < \frac{1}{2}; \\ 2\lambda, & \text{if } \lambda \geq \frac{1}{2}. \end{cases}$ Obviously, $\sup_{\lambda \geq 0} g(\lambda) = 1$, but there is no-one λ that provides this supremum. The statement of Proposition 3.1 fails to hold because the Slater condition is violated. □

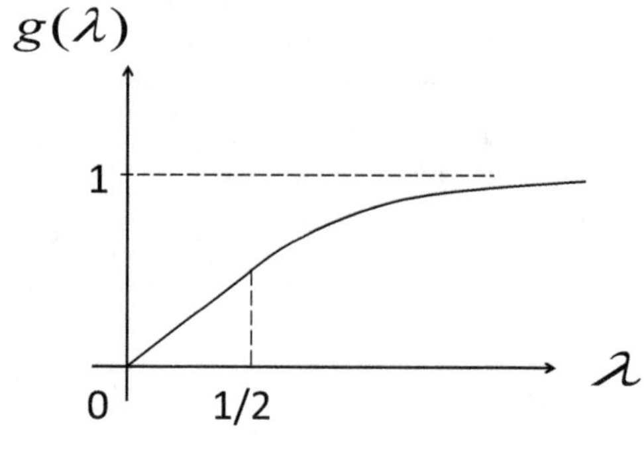

Fig. 3.17 Dual functional.

3.3.14 *Constrained optimization: multiple solutions*

Consider again the constrained problem (3.12). Subsection 3.3.12 demonstrated that, here, the Bellman principle can fail to hold. Below, we show that a solution to a simple constrained problem can be obtained by uncountably many different stationary strategies. Incidentally, in terms of occupation measures, the constrained problem

$$\int_{\mathbf{X}\times\mathbf{A}} {}^1c(x,a)d\eta^{\pi}_{\mathsf{P}_0}(x,a) \to \inf_{\pi}; \quad \int_{\mathbf{X}\times\mathbf{A}} {}^2c(x,a)d\eta^{\pi}_{\mathsf{P}_0}(x,a) \le (1-\beta)d$$

usually has a solution $\eta^{\pi}_{\mathsf{P}_0}$ which is a non-extreme occupation measure, and thus is generated by many different control strategies: see the end of Section 3.1.

Example. Let $\mathbf{X} = \{0,1,2,\ldots\}$, $\mathbf{A} = \{0,1\}$, $p(x+1|x,a) \equiv 1$, with all other transition probabilities being zero. We put ${}^1c(x,a) = a$, ${}^2c(x,a) = 1-a$, $\beta = 1/4$, $\mathsf{P}_0(0) = 1$ (see Fig. 3.18). Clearly, ${}^1v^{\pi}_0 = \frac{1}{1-\beta} - {}^2v^{\pi}_0$: on each step ${}^1c(X_{t-1},A_t) + {}^2c(X_{t-1},A_t) = 1$.

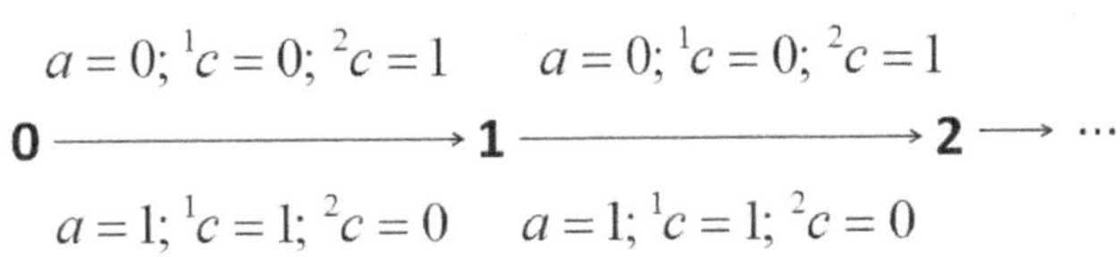

Fig. 3.18 Continuum constrained-optimal strategies.

Consider the constrained problem

$${}^1v^{\pi}_0 \to \inf_{\pi}; \qquad {}^2v^{\pi}_0 \le 1/2.$$

Any strategy, for which ${}^2v^{\pi}_0 = 1/2$, is optimal, and we intend to build uncountably many stationary optimal strategies.

Let $M \subseteq \{1,2,\ldots\}$ be a non-empty set, put $q_M := \sum_{i\in M}\beta^i \le 1/3$, and consider the following stationary strategy:

$$\pi^{s,M}(a|x) := \begin{cases} \frac{1}{2} - q_M, & \text{if } x = 0,\ a = 0; \\ \frac{1}{2} + q_M, & \text{if } x = 0,\ a = 1; \\ 1, & \text{if } x \in M \text{ and } a = 0 \\ & \text{or if } x \notin M,\ x \ne 0 \text{ and } a = 1; \\ 0, & \text{otherwise.} \end{cases}$$

Clearly,

$${}^2v^{\pi^{s,M}}_0 = \frac{1}{2} - q_M + \sum_{t=2}^{\infty}\beta^{t-1}\mathbb{I}\{X_{t-1} \in M\} = \frac{1}{2} - q_M + q_M = \frac{1}{2},$$

and all these strategies $\pi^{s,M}$ are optimal.

This example is similar to the one developed by J.Gonzalez-Hernandez (unpublished communication). □

3.3.15 *Two discount factors and* (N, ∞)*-stationary selectors*

Suppose there are two discount factors $1 > \beta_1 > \beta_2 > 0$ and two cost functions ${}^1c(\cdot,\cdot)$, ${}^2c(\cdot,\cdot)$, and consider the performance functional

$$v^\pi_{P_0} = \mathsf{E}^\pi_{P_0}\left[\sum_{t=1}^{\infty}\{\beta_1^{t-1}\cdot {}^1c(X_{t-1},A_t) + \beta_2^{t-1}\cdot {}^2c(X_{t-1},A_t)\}\right] \to \inf_\pi . \tag{3.14}$$

If the model is finite then the following reasoning helps to solve problem (3.14).

Since $\beta_1 > \beta_2$, the main impact at big values of t comes from the first term ${}^1c(\cdot,\cdot)$, meaning that, starting from some $t = N \geq 1$, the optimal strategy (stationary selector ψ) is the one that solves the problem

$$\mathsf{E}^\pi_{P_0}\left[\sum_{t=1}^{\infty}\beta_1^{t-1}\cdot {}^1c(X_{t-1},A_t)\right] \to \inf_\pi .$$

(If there are several different optimal selectors then one should choose the one that minimizes the cost of the second type.) As a result, we know the value

$$R(x) = \mathsf{E}^\psi_{P_0}\left[\sum_{t=N}^{\infty}\{\beta_1^{t-1}\cdot{}^1c(X_{t-1},A_t) + \beta_2^{t-1}\cdot{}^2c(X_{t-1},A_t)\}\,|X_{N-1}=x\right].$$

We still need to solve the (non-homogeneous) finite-horizon problem with the one-step cost $c_t(x,a) = \beta_1^{t-1}\cdot {}^1c(x,a) + \beta_2^{t-1}\cdot {}^2c(x,a)$ and final cost $R(x)$, which leads to a Markov selector $\varphi_t(x)$ in the decision epochs $t \in \{1,2,\ldots,N-1\}$. The resulting selector

$$\varphi^*_t(x) = \begin{cases} \varphi_t(x), & \text{if } t < N; \\ \psi(x), & \text{if } t \geq N \end{cases} \tag{3.15}$$

solves problem (3.14). More about the weighted discounted criteria can be found in [Feinberg and Shwartz (1994)].

Definition 3.7. Selectors looking like (3.15) with finite N are called (N,∞)-*stationary*.

The following examples, similar to those published in [Feinberg and Shwartz (1994)], show that this method does not work if the model is not finite.

Example. Let $\mathbf{X} = \{0\}$, $\mathbf{A} = [0,1]$, $\beta_1 = 1/2$, ${}^1c(x,a) = a^2$, $\beta_2 = 1/4$, ${}^2c(x,a) = a^2 - a$. The only optimal strategy is

$$\varphi^*_t(x) = \frac{1}{2+2^t}$$

which minimizes the one-step cost $(1/2)^{t-1}a^2 + (1/4)^{t-1}(a^2 - a)$, so that an (N,∞)-stationary optimal selector does not exist. □

Example. Let $\mathbf{X} = \{0,1,2,\ldots\} \cup \{\Delta\}$, $\mathbf{A} = \{1,2\}$, $\mathsf{P}_0(0) = 1$,

$$p(i|0,a) = (1/2)^{i+1} \text{ for all } i \geq 0, \quad a \in \mathbf{A},$$

$$p(\Delta|\Delta,a) = p(\Delta|i,a) \equiv 1 \text{ for all } i \geq 1,$$

with all other transition probabilities being zero. We put $\beta_1 = 1/2$, $\beta_2 = 1/4$, ${}^1c(0,a) = {}^2c(0,a) = {}^1c(\Delta,a) = {}^2c(\Delta,a) \equiv 0$. For $i \geq 1$,

$${}^1c(i,a) = \begin{cases} 0; & \text{if } a = 1; \\ 3 \cdot 2^{-i}, & \text{if } a = 2 \end{cases} \qquad {}^2c(i,a) = \begin{cases} 1, & \text{if } a = 1; \\ 0, & \text{if } a = 2 \end{cases}$$

(see Fig. 3.19).

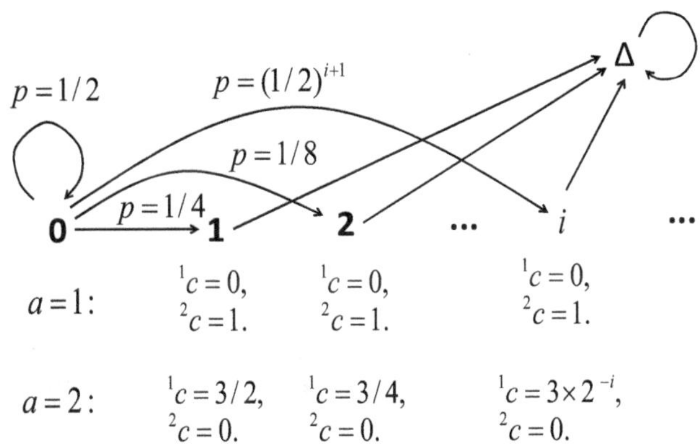

Fig. 3.19 No (N,∞)-stationary optimal selectors.

Note that the process can take any value $x \geq 1$ at any moment $t \geq 1$, and the one-step discounted cost, which is non-zero only once, equals

$$\beta_1^t \cdot {}^1c(x,a) + \beta_2^t \cdot {}^2c(x,a) = \begin{cases} (1/2)^{2t}, & \text{if } a = 1; \\ 3 \cdot (1/2)^{t+x}, & \text{if } a = 2. \end{cases}$$

Therefore, the optimal strategy (Markov selector) in states $x \geq 1$ is unique and is defined by

$$\varphi_t^*(x) = \begin{cases} 1, & \text{if } x - t - \log_2 3 < 0; \\ 2, & \text{if } x - t - \log_2 3 > 0. \end{cases}$$

The actions in states 0 and Δ are immaterial. There is no $N < \infty$ for which there exists an optimal (N,∞)-stationary selector. □

3.3.16 *Non-constant discounting*

A natural generalization of the MDP with discounted cost (3.1) is as follows:

$$v^{\pi}_{\mathsf{P}_0} = \mathsf{E}^{\pi}_{\mathsf{P}_0}\left[\sum_{t=1}^{\infty} f(t)c(X_{t-1}, A_t)\right] \to \inf_{\pi}, \tag{3.16}$$

where function $f(\cdot)$ decreases to zero rapidly enough.

Definition 3.8. A function $f: \{1, 2, \ldots\} \to \mathbb{R}$ is called *exponentially representable* if it is not identical zero and there exist sequences $\{d_k\}_{k=1}^{\infty}$ and $\{\gamma_k\}_{k=1}^{\infty}$ such that $\{\gamma_k\}_{k=1}^{\infty}$ is positive, strictly decreasing and $\gamma_1 < 1$; $f(t) = \sum_{k=1}^{\infty} d_k \gamma_k^t$, and this sum converges absolutely after some $N < \infty$.

The standard case, when $f(t) = \beta^{t-1}$, corresponds to $d_1 = 1/\beta$, $d_k = 0$ for $k \geq 1$, $\gamma_1 = \beta$, and $\{\gamma_k\}_{k=2}^{\infty}$ is an arbitrary positive decreasing sequence. If function $f(\cdot)$ is exponentially representable, then (3.16) is a specific infinite version of the weighted discounted cost considered in Subsection 3.3.15:

$$v^{\pi}_{\mathsf{P}_0} = \mathsf{E}^{\pi}_{\mathsf{P}_0}\left[\sum_{t=1}^{\infty}\sum_{k=1}^{\infty} \beta_k^{t-1}\, {}^k c(X_{t-1}, A_t)\right]$$

(here $\beta_k := \gamma_k$ and ${}^k c(x,a) := \gamma_k d_k c(x,a)$).

Theorem 3.4. [Carmon and Shwartz (2009), Thm. 1.5]. *If the model is finite and function $f(\cdot)$ is exponentially representable, then there exists an optimal (N, ∞)-stationary selector.*

If function $f(\cdot)$ is not exponentially representable, then this statement can be false even if $f(\cdot)$ monotonically decreases to zero, as the following example confirms.

Example. Suppose $f(t) = \beta^{t-1} \cdot h(t)$ with $\beta = 1/4$, $h(t) = \begin{cases} 1, & \text{if } t \text{ is odd;} \\ 1/2, & \text{if } t \text{ is even.} \end{cases}$
This function $f(\cdot)$ is not exponentially representable, because the necessary condition [Carmon and Shwartz (2009), Lemma 3.1.]

$$\exists \gamma \in (0,1): \quad \lim_{t\to\infty} \gamma^{-t} f(t) = c \neq 0 \text{ and } c < \infty$$

is violated.

Let $\mathbf{X} = \{1, 2, 3\}$, $\mathbf{A} = \{0, 1\}$, $p(1|1, a) = a$, $p(2|1, a) = 1 - a$, $p(3|2, a) = p(1|3, a) \equiv 1$, with other transition probabilities being zero. We put $c(1, a) \equiv 1$, $c(2, a) \equiv 5/4$, $c(3, a) \equiv 0$ and $\mathsf{P}_0(1) = 1$ (see Fig. 3.20). Note that the transitions are deterministic.

To solve problem (3.16), we deal with a standard discounted MDP where the one-step cost depends on the value of $h(t)$; it is reasonable to extend the states by incorporating the values of $h(t) \in \{1, 1/2\}$. Thus, we put

$$\tilde{\mathbf{X}} = \{(1,1),\ (1,1/2),\ (2,1),\ (2,1/2),\ (3,1),\ (3,1/2)\},$$

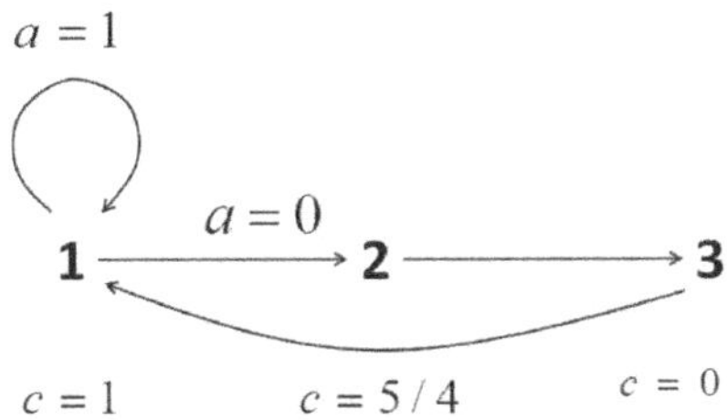

Fig. 3.20 Non-constant discounting.

the action space **A** remains the same,

$$
\begin{aligned}
&\tilde{p}((1,1/2)|(1,1),a) = \tilde{p}((1,1)|(1,1/2),a) = a,\\
&\tilde{p}((2,1/2)|(1,1),a) = \tilde{p}((2,1)|(1,1/2),a) = 1-a,\\
&\tilde{p}((3,1/2)|(2,1),a) = \tilde{p}((3,1)|(2,1/2),a)\\
&= \tilde{p}((1,1/2)|(3,1),a) = \tilde{p}((1,1)|(3,1/2),a) \equiv 1,
\end{aligned}
$$

and other transition probabilities in the auxiliary tilde-model are zero. Finally,

$$\tilde{c}((1,1),a) \equiv 1, \quad \tilde{c}((1,1/2),a) \equiv 1/2, \quad \tilde{c}((2,1),a) \equiv 5/4,$$

$$\tilde{c}((2,1/2),a) \equiv 5/8, \quad \tilde{c}((3,1),a) = \tilde{c}((3,1/2).a) \equiv 0$$

(see Fig. 3.21).

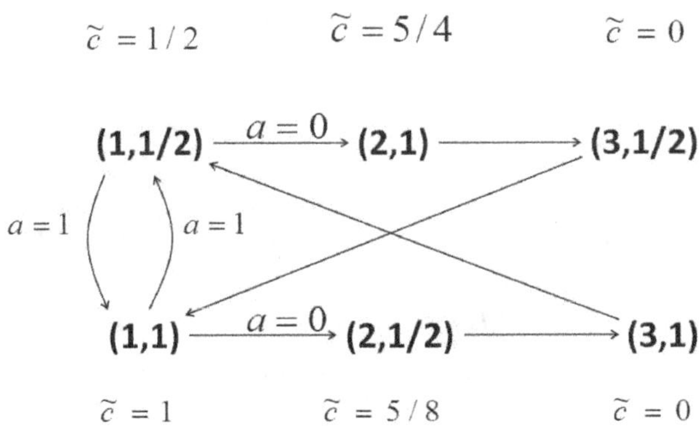

Fig. 3.21 Non-constant discounting, auxiliary model.

The optimality equation (3.2) is given by

$$\begin{aligned} v(1,1) &= \min\{1+\beta v(2,1/2);\quad 1+\beta v(1,1/2)\}; \\ v(1,1/2) &= \min\{1/2+\beta v(2,1);\quad 1/2+\beta v(1,1)\}; \\ v(2,1) &= 5/4+\beta v(3,1/2); \qquad v(2,1/2) = 5/8+\beta v(3,1); \\ v(3,1) &= \beta v(1,1/2); \qquad v(3,1/2) = \beta v(1,1) \end{aligned}$$

and has the solution

$$v(1,1) = \frac{298}{255}; \qquad v(1,1/2) = \frac{202}{255};$$

$$v(2,1) = \frac{2699}{2040}; \qquad v(2,1/2) = \frac{172}{255};$$

$$v(3,1) = \frac{101}{510}; \qquad v(3,1/2) = \frac{149}{510}.$$

With some abuse of notation, we write $v(1,1)$ instead of $v((1,1))$ etc. In state $(1,1)$, action $a=0$ is optimal; in state $(1,1/2)$, action $a=1$ is optimal. Therefore, if the initial state is $(1,1)$, corresponding to the initial state $x_0=1$ in the original model, then only the following sequences of actions are optimal in the both models:

t	1	2	3	4	5	6	...
State of the auxiliary tilde-model	$(1,1)$	$(2,\frac{1}{2})$	$(3,1)$	$(1,\frac{1}{2})$	$(1,1)$	$(2,\frac{1}{2})$	...
State of the original model	1	2	3	1	1	2	...
Action	0	0 or 1	0 or 1	1	0	0 or 1	...

In the original model, in state 1, the optimal actions always switch from 1 to 0 and back, and a (N,∞)-stationary optimal selector does not exist. □

3.3.17 *Risk-sensitive criteria*

Suppose $c(\cdot,\cdot)$ is not the cost but the reward function. Then the target is to maximize the expected discounted reward

$$\mathsf{E}^{\pi}_{\mathsf{P}_0}\left[\sum_{t=1}^{\infty}\beta^{t-1}c(X_{t-1},A_t)\right] = \mathsf{E}^{\pi}_{\mathsf{P}_0}[R] \to \sup_{\pi}. \tag{3.17}$$

We assume that all the expressions are well defined.

The *risk-sensitive* criterion looks as

$$U^{-1}\left(\mathsf{E}^{\pi}_{\mathsf{P}_0}\left[U\left(\sum_{t=1}^{\infty}\beta^{t-1}c(X_{t-1},A_t)\right)\right]\right)\to\sup_{\pi},\tag{3.18}$$

where $U:\mathbb{R}\to\mathbb{R}$ is a continuous strictly increasing function, which can be called 'utility' if it is concave. We emphasize that criterion (3.17) corresponds to linear functions $U(r)=kr$ with $k\neq 0$ and is called *risk-neutral.*

One of the most common utility functions is $U(r):=\frac{-1}{\gamma}e^{-\gamma r}$ with $\gamma>0$. For other functions $U(\cdot)$ and the rationale of introducing them see [Bäuerle and Rieder (2014)]. Note that, if $U(r):=\frac{-1}{\gamma}e^{-\gamma r}$ with $\gamma>0$, then, for the random variable R, for small γ,

$$U^{-1}\left(\mathsf{E}^{\pi}_{\mathsf{P}_0}[U(R)]\right)\approx\mathsf{E}^{\pi}_{\mathsf{P}_0}[R]-\frac{\gamma}{2}\mathsf{V}ar^{\pi}_{\mathsf{P}_0}[R]$$

(see [Bäuerle and Rieder (2014), p. 107]). This expression can be justified by the Taylor expansion of the exponent and logarithm. Therefore, when solving problem (3.18), one takes into account not only the expectation as in (3.17), but also the variance of the (random) discounted reward, which can be called 'risk' of the decision. Criteria like (3.18) are popular in financial mathematics [Bäuerle and Rieder (2011)]. In the current subsection, we put $\gamma=1$ and rewrite problem (3.18) as follows

$$-\ln\left[-\mathsf{E}^{\pi}_{\mathsf{P}_0}\left[-\exp\left[-\sum_{t=1}^{\infty}\beta^{t-1}c(X_{t-1},A_t)\right]\right]\right]\to\sup_{\pi},$$

or, equivalently,

$$\mathsf{E}^{\pi}_{\mathsf{P}_0}\left[\exp\left[-\sum_{t=1}^{\infty}\beta^{t-1}c(X_{t-1},A_t)\right]\right]\to\inf_{\pi}.\tag{3.19}$$

Differently from the risk-neutral criterion (3.17), stationary selectors are not sufficient for solving problem (3.19) even in finite models, as the following example confirms.

Example. Let $\mathbf{X}=\{1,2,3\}$, $\mathbf{A}=\{1,2\}$, $p(2|1,1)=p(3|1,1)=1/2$, $p(2|1,2)=0.9$, $p(3|1,2)=0.1$, $p(1|2,a)=p(1|3,a)\equiv 1$ with other transition probabilities being zero; $c(1,1)=0$, $c(1,2)=1$, $c(2,a)\equiv 0$, $c(3,a)\equiv 8$; $\beta=1/2$; $\mathsf{P}_0(1)=1$. See Fig. 3.22. The similar example was published in [Jaquette (1976)].

Clearly, for any control strategy π, $X_{2n}=1$ (P^{π}_1-a.s.) for all $n=0,1,2,\ldots$. Therefore, when restricting to Markov strategies π^{m} (see Lemma 2.1), on the left-hand side of (3.19) we have

$$\prod_{n=0}^{\infty}\mathsf{E}^{\pi^{\mathrm{m},n}}_1\left[\exp(-\beta^{2n}c(X_0,A_1)-\beta^{2n+1}c(X_1,A_2))\right],$$

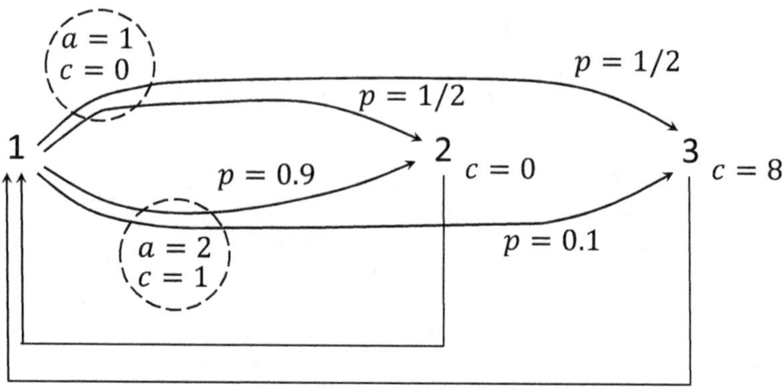

Fig. 3.22 No stationary optimal selectors.

where $\pi_t^{\mathrm{m},n}(da|x_{t-1}) = \pi_{t+2n}^{\mathrm{m}}(da|x_{t+2n-1})$ is the shifted strategy. One can minimize the particular terms

$$\mathsf{E}_1^{\pi^{\mathrm{m},n}}\left[\exp(-\beta^{2n}c(X_0,A_1) - \beta^{2n+1}c(X_1,A_2))\right]$$

separately. Since the actions in states 2 and 3 are immaterial, we conclude that the actions $A_{2n+1} =: a$ in state 1 on the steps $1, 3, 5, \ldots$ (corresponding to $n = 0, 1, 2, \ldots$) must minimize expression

$$V_n(a) := \mathsf{E}[\exp(-\beta^{2n}c(X_0 = 1, a) - \beta^{2n+1}\tilde{c}(X_1))], \quad a \in \mathbf{A}.$$

Here the expectation is with respect to the transition probability $p(y|1,a)$ and $\tilde{c}(y) := \begin{cases} 0, & \text{if } y = 2; \\ 8, & \text{if } y = 3. \end{cases}$ Thus, denoting $s := \beta^{2n}$, we have

$$\begin{aligned} \min_{a \in \mathbf{A}} V_n(a) &= \min\left\{e^{-sc(1,1)}\left[\frac{1}{2}e^{-s\beta\tilde{c}(2)} + \frac{1}{2}e^{-s\beta\tilde{c}(3)}\right];\right. \\ &\qquad \left. e^{-sc(1,2)}\left[0.9e^{-s\beta\tilde{c}(2)} + 0.1e^{-s\beta\tilde{c}(3)}\right]\right\} \\ &= \min\left\{\frac{1}{2} + \frac{1}{2}e^{-4s};\quad 0.9e^{-s} + 0.1e^{-5s}\right\}. \end{aligned}$$

- If $s = 1$ (i.e., $n = 0$), then

$$\frac{1}{2} + \frac{1}{2}e^{-4} \approx 0.509 > 0.9e^{-1} + 0.1e^{-5} \approx 0.332,$$

and the optimal action on step 1 is $a = A_1 = 2$.

- If $s = \frac{1}{4}$ (i.e., $n = 1$), then

$$\frac{1}{2} + \frac{1}{2}e^{-1} \approx 0.684 < 0.9e^{-\frac{1}{4}} + 0.1e^{-\frac{5}{4}} \approx 0.729,$$

and the optimal action on step 3 is $a = A_3 = 1$.

Let us show that on the further steps $5, 7, \ldots$, i.e., when $s = \frac{1}{16}, \frac{1}{64}, \ldots$ (corresponding to $n = 2, 3, \ldots$), the optimal actions are $A_5 = A_7 = \ldots = 1$. The difference

$$f(s) := \left[0.9e^{-s} + 0.1e^{-5s}\right] - \left[\frac{1}{2} + \frac{1}{2}e^{-4s}\right]$$

is such that, for $0 \le s \le \frac{1}{16}$,

$$\frac{df}{ds} = -0.9e^{-s} - 0.5e^{-5s} + 2e^{-4s} \ge -0.9 - 0.5 + 2e^{-\frac{1}{4}} \approx 0.158.$$

So, function $f(\cdot)$ increases on the interval $[0, \frac{1}{16}]$, and hence $f(s) > f(0) = 0$ for $s \in (0, \frac{1}{16}]$: the only optimal action is $a = 1$.

Since only actions in state 1 are important, one can say a little less rigorously, that there are only two stationary selectors $\varphi^1(x) \equiv 1$ and $\varphi^2(x) \equiv 2$. In the presented example, only the non-stationary Markov selector $\varphi_t = \begin{cases} \varphi^2, & \text{if } t = 1; \\ \varphi^1, & \text{otherwise} \end{cases}$ is optimal if $x_0 = 1$. As was proved in [Jaquette (1976), Thm. 1] , if the MDP is finite, then there is a uniformly optimal Markov selector of the form $\varphi_t \equiv \hat{\varphi}$ for all $t = T, T+1, \ldots$ at some $T < \infty$, where $\hat{\varphi}$ is a stationary selector. If the model is not finite, then it can happen that the only uniformly optimal Markov selector φ_t is such that, for each T, there is $t > T$ with $\varphi_t \neq \varphi_T$ [Brau-Rojas *et al.* (1998), Ex. 4]. □

3.3.18 *Nearly optimal strategy is not Blackwell optimal*

One of the approaches to problem (2.1) is the study of discounted problem (3.1) when β approaches 1. A nearly optimal strategy (see Definition 3.4) is also uniformly optimal in problem (2.1) (under Condition 2.1):

$$\forall x \in \mathbf{X}\ \forall \pi\ v_x^\pi = \lim_{\beta \to 1-} v_x^{\pi,\beta} \ge \lim_{\beta \to 1-} v_x^{*,\beta} = \lim_{\beta \to 1-} v_x^{\pi^*,\beta} = v_x^{\pi^*}.$$

Any Blackwell optimal strategy (see Definition 3.3) is obviously nearly optimal. The converse is not true, even in finite models, as the following example shows (see also [Blackwell (1962), Ex. 1]).

Example. Let $\mathbf{X} = \{0, 1\}$, $\mathbf{A} = \{1, 2\}$, $p(1|1,1) = p(0|1,1) = 1/2$, $p(0|1,2) = 1$, $p(0|0,a) \equiv 1$ other transition probabilities being zero, $c(1,a) = -a$, $c(0,a) \equiv 0$ (see Fig. 3.23).

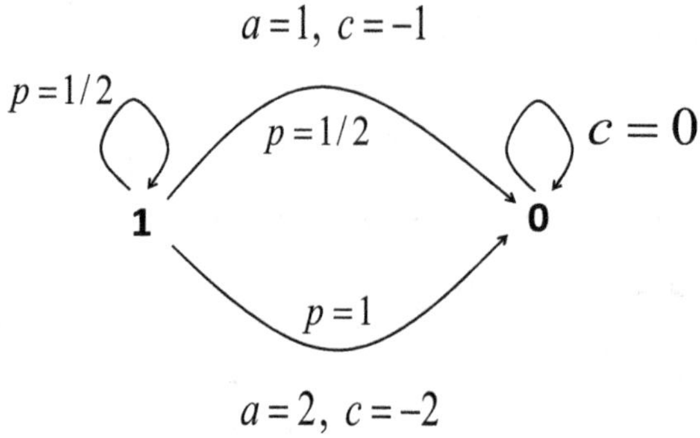

Fig. 3.23 Nearly optimal strategy is not Blackwell optimal.

The optimality equation (3.2) is given by

$$v^\beta(0) = \beta v^\beta(0),$$
$$v^\beta(1) = \min\{-1 + \frac{1}{2}\beta(v^\beta(0) + v^\beta(1)); \quad -2 + \beta v^\beta(0)\},$$

so that, for any $\beta \in (0,1)$, for stationary selector $\varphi^2(x) \equiv 2$ we have $v^\beta(0) = v_0^{*,\beta} = v_0^{\varphi^2,\beta} = 0$, $v^\beta(1) = v_1^{*,\beta} = v_1^{\varphi^2,\beta} = -2$, so that φ^2 is uniformly optimal and Blackwell optimal. At the same time, for selector $\varphi^1(x) \equiv 1$ we have

$$v_0^{\varphi^1,\beta} = 0, \quad v_1^{\varphi^1,\beta} = -\frac{2}{2-\beta},$$

so that φ^1 is nearly optimal, but certainly not Blackwell optimal:

$$\forall \beta < 1 \quad v_1^{\varphi^1,\beta} = \frac{-2}{2-\beta} > -2 = v_1^{*,\beta}.$$

The both stationary selectors φ^1 and φ^2 are uniformly optimal in problem (2.1) since they are conserving and equalizing: see Section 2.1. □

3.3.19 *Blackwell optimal strategies and opportunity loss*

If the time horizon T is finite, one can evaluate a control strategy π using the *opportunity loss* (or regret) $\mathsf{E}_x^\pi\left[\sum_{t=1}^T c(X_{t-1}, A_t)\right] - V_x^T$, where as usual $V_x^T := \inf_\pi \mathsf{E}_x^\pi\left[\sum_{t=1}^T c(X_{t-1}, A_t)\right]$. In the infinite-horizon models, the goal may be to find a strategy that solves the problem

$$\limsup_{T\to\infty}\left\{\mathsf{E}_x^\pi\left[\sum_{t=1}^T c(X_{t-1}, A_t)\right] - V_x^T\right\} \to \inf_\pi \text{ for all } x \in \mathbf{X}, \tag{3.20}$$

that is an *opportunity-loss optimal* strategy. The following example, based on [Flynn (1980), Ex. 3], shows that it can easily happen that a Blackwell optimal strategy does not solve problem (3.20) and, *vice versa*, a strategy, minimizing the opportunity loss, may not be Blackwell optimal.

Example. Let $\mathbf{X} = \{0,1,2,2',3,3'\}$, $\mathbf{A} = \{1,2,3\}$, $p(1|0,1) = p(2|0,2) = p(3|0,3) = 1$, $p(1|1,a) \equiv 1$, $p(2'|2,a) = p(2|2',a) \equiv 1$, $p(3'|3,a) = p(3|3',a) \equiv 1$, other transition probabilities being zero; $c(0,1) = 1/2$, $c(0,2) = -1$, $c(0,3) = 1$, $c(2,a) \equiv 2$, $c(2',a) \equiv -2$, $c(3,a) \equiv -2$, $c(3',a) \equiv 2$ (see Fig. 3.24). Note, in the infinite-horizon version of this MDP with the (undiscounted) total cost, Condition 2.1 is not satisfied for $x_0 = 2$ or 3.

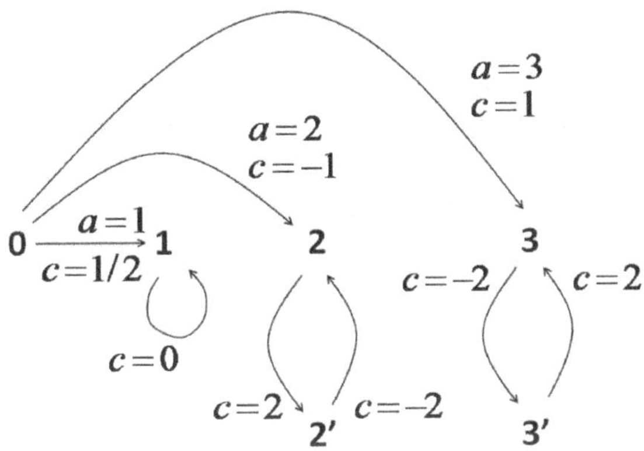

Fig. 3.24 Opportunity loss versus Blackwell optimality.

We only need study the initial state state 0, and only the stationary selectors $\varphi^1(x) \equiv 1$, $\varphi^2(x) \equiv 2$ and $\varphi^3(x) \equiv 3$ need be considered.

Clearly,

$$v_1^{\varphi^1,\beta} = v_1^{\varphi^2,\beta} = v_1^{\varphi^3,\beta} = v_1^{*,\beta} = 0,$$

and the identical values

$$v_2^{\varphi^1,\beta} = v_2^{\varphi^2,\beta} = v_2^{\varphi^3,\beta} = v_1^{*,\beta} = \frac{2}{1+\beta}$$

come from equation (3.2):

$$v(2) = 2 + \beta v(2'), \quad v(2') = -2 + \beta v(2).$$

Similarly,

$$v_3^{\varphi^1,\beta} = v_3^{\varphi^2,\beta} = v_3^{\varphi^3,\beta} = v_3^{*,\beta} = \frac{-2}{1+\beta}.$$

Thus,

$$v_0^{\varphi^1,\beta} = 1/2, \quad v_0^{\varphi^2,\beta} = -1 + \frac{2\beta}{1+\beta}, \quad v_0^{\varphi^3,\beta} = 1 - \frac{2\beta}{1+\beta},.$$

Since $v_0^{\varphi^2,\beta} < 0$ and $v_0^{\varphi^3,\beta} > 0$ for $\beta < 1$, only the stationary selector φ^2 is Blackwell optimal.

At the same time, direct calculations result in the following table:

T:	1	2	3	4	5	...
V_0^T:	-1	-1	-1	-1	-1	...
$\mathsf{E}_0^{\varphi^1}\left[\sum_{t=1}^T c(X_{t-1},A_t)\right]$:	$\frac{1}{2}$	$\frac{1}{2}$	$\frac{1}{2}$	$\frac{1}{2}$	$\frac{1}{2}$	...
$\mathsf{E}_0^{\varphi^2}\left[\sum_{t=1}^T c(X_{t-1},A_t)\right]$:	-1	1	-1	1	-1	...
$\mathsf{E}_0^{\varphi^3}\left[\sum_{t=1}^T c(X_{t-1},A_t)\right]$:	1	-1	1	-1	1	...

So, the left-hand side of (3.20) at $x = 0$ equals 3/2, 2 and 2 at φ^1, φ^2 and φ^3 correspondingly. The infimum is provided by φ^1, and any mixture of those stationary selectors results in the value bigger than 3/2. Thus, only selector φ^1 solves problem (3.20).

Note that all three selectors φ^1, φ^2 and φ^3 are equally AC-optimal (see Chapter 4). □

3.3.20 *Blackwell optimal and n-discount optimal strategies*

Definition 3.9. [Puterman (1994), Subsections 5.4.3 and 10.1.1]. A strategy π^* is *n-discount optimal* for some $n \geq -1$ if

$$\lim_{\beta\to 1-}\sup\ (1-\beta)^{-n}[v_x^{\pi^*,\beta} - v_x^{\pi,\beta}] \leq 0$$

for all $x \in \mathbf{X}$, π.

If a strategy is n-discount optimal for $n > -1$, then it is $(n-1)$-discount optimal. The proof follows directly from the definition.

Remark 3.2. Since

$$\limsup_{\beta\to 1-}[v_x^{\pi^*,\beta} - v_x^{\pi,\beta}] \leq \limsup_{\beta\to 1-}[v_x^{\pi^*,\beta} - v_x^{*,\beta}] + \limsup_{\beta\to 1-}[v_x^{*,\beta} - v_x^{\pi,\beta}],$$

we conclude that any nearly optimal strategy is 0-discount optimal. The converse is not true: see Subsection 4.3.25.

In finite models, if π^* is a Blackwell optimal strategy, then it is n-discount optimal for any $n \geq -1$. This statement was proved in [Puterman (1994), Thm. 10.1.5] for stationary Blackwell strategies, but the proof survives for arbitrary strategies, too. In finite models, if a stationary strategy is n-discount optimal for any $n \geq -1$, then it is Blackwell optimal [Puterman (1994), Thm. 10.1.6].

The following example, similar to [Puterman (1994), Ex. 10.1.1] and [Hordijk and Yushkevich (2002), Ex. 7.1], shows that a strategy can be n-discount optimal for all $n < m$, but not Blackwell optimal and not n-discount optimal for $n \geq m$.

Example. Let $\mathbf{X} = \{0, 1, 2, \ldots, m+1\}$, $\mathbf{A} = \{1, 2\}$, $p(1|0,1) = 1$, $p(i+1|i,a) \equiv 1$ for all $i = 1, 2, \ldots, m$, $p(m+1|m+1,a) \equiv 1$, $p(m+1|0,2) = 1$, with other transition probabilities being zero. We put $c(0,2) = 0$, $c(0,1) = 1$, $c(i,a) \equiv (-1)^i \frac{m!}{i!(m-i)!}$ for $i = 1, 2, \ldots, m$; $c(m+1,a) = 0$. Fig. 3.25 illustrates the case $m = 4$.

Here, there are only two essentially different stationary selectors $\varphi^1(x) \equiv 1$ and $\varphi^2(x) \equiv 2$ (actions in states $1, 2, \ldots, m+1$ play no role). For each $\beta \in (0,1)$, for all strategies π, we have

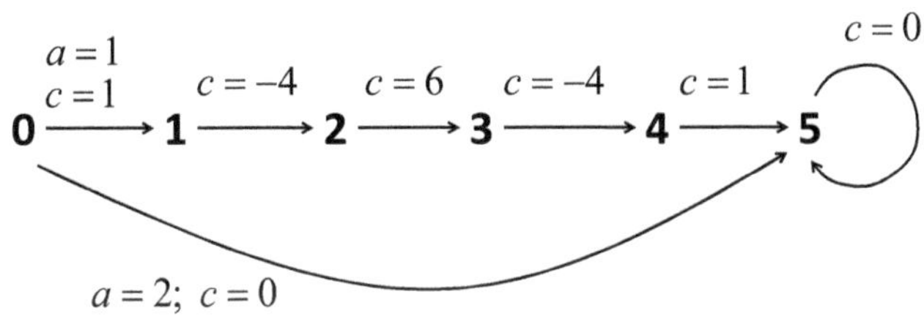

Fig. 3.25 3-discount optimal strategy is not Blackwell optimal ($m = 4$).

$$\begin{aligned}
v_{m+1}^{\pi,\beta} &\equiv v_{m+1}^{*,\beta} = 0, \quad v_m^{\pi,\beta} \equiv v_m^{*,\beta} = (-1)^m, \\
v_{m-1}^{\pi,\beta} &\equiv v_{m-1}^{*,\beta} = (-1)^{m-1} m + (-1)^m \beta, \\
v_{m-2}^{\pi,\beta} &\equiv v_{m-2}^{*,\beta} = (-1)^{m-2} \frac{m(m-1)}{2} + (-1)^{m-1} m\beta + (-1)^m \beta^2, \ldots, \\
v_1^{\pi,\beta} &\equiv v_1^{*,\beta} = -m + \frac{m(m-1)}{2}\beta + \cdots + (-1)^{m-2} \frac{m(m-1)}{2} \beta^{m-3} \\
&\quad + (-1)^{m-1} m\beta^{m-2} + (-1)^m \beta^{m-1}, \\
v_0^{\varphi^1,\beta} &= 1 - m\beta + \frac{m(m-1)}{2}\beta^2 + \ldots + (-1)^m \beta^m = (1-\beta)^m, \\
v_0^{\varphi^2,\beta} &= 0 < v_0^{\varphi^1,\beta},
\end{aligned}$$

and φ^2 is the only uniformly optimal strategy: $v_x^{*,\beta} = v_x^{\varphi^2,\beta}$. We see that φ^2 is Blackwell optimal, and φ^1 is not, although all the strategies are equally optimal in the undiscounted (total expected cost) version of this MDP, corresponding to $\beta = 1$. At the same time, φ^1 is n-discount optimal for all $n = -1, 0, 1, 2, \ldots, m-1$:

for each strategy π

$$\lim_{\beta\to 1-}(1-\beta)^{-n}[v_0^{\varphi^1,\beta}-v_0^{\pi,\beta}]\le \lim_{\beta\to 1-}(1-\beta)^{-n}[v_0^{\varphi^1,\beta}]=0,\ \text{if } n<m.$$

If $n\ge m$, then

$$\lim_{\beta\to 1-}(1-\beta)^{-n}[v_0^{\varphi^1,\beta}-v_0^{\varphi^2,\beta}]=\lim_{\beta\to 1-}(1-\beta)^{-n}[v_0^{\varphi^1,\beta}]=\begin{cases}1, & \text{if } n=m;\\ \infty, & \text{if } n>m.\end{cases}$$

Calculation for the initial state $x>0$ is trivial. □

The next example shows that, if the model is not finite, then a stationary strategy, which is not Blackwell optimal, can still be n-discount optimal for any $n\ge -1$.

Example. Let $\mathbf{X}=\{\Delta,0,1,2,\ldots\}$, $\mathbf{A}=\{1,2\}$, $p(1|0,1)=1$, $p(i+1|i,a)\equiv 1$ for all $i=1,2,\ldots$, $p(\Delta|0,2)=1$, $p(\Delta|\Delta,a)\equiv 1$, with other transition probabilities zero. We put $c(0,2)=0$, $c(\Delta,a)\equiv 0$, $c(0,1)=1/e:=C_0$ and, similarly, $c(i,a)\equiv C_i$, where C_i is the i-th coefficient in the Taylor expansion

$$g(\beta)=e^{-\frac{1}{(1-\beta)^2}}=\sum_{j=0}^{\infty}C_j\beta^j$$

(see Fig. 3.26). Since the function g is holomorphic everywhere except for the unique singular point $\beta=1$, this series converges absolutely for all $\beta\in(0,1)$ [Priestley (1990), Taylor's Theorem, p. 69].

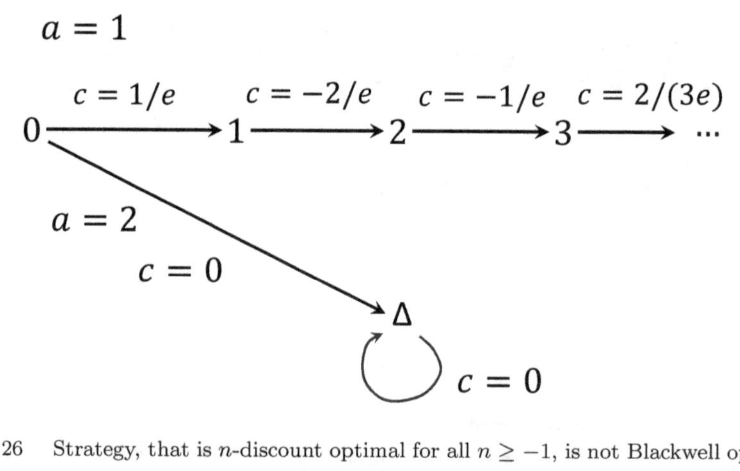

Fig. 3.26 Strategy, that is n-discount optimal for all $n\ge -1$, is not Blackwell optimal.

Here again we have only two essentially different stationary selectors $\varphi^1(x)\equiv$

1 and $\varphi^2(x) \equiv 2$ (actions in states $\Delta, 1, 2, \ldots$ play no role). For each $\beta \in (0,1)$, for all strategies π, we have

$$\begin{aligned}
v_\Delta^{\pi,\beta} &\equiv v_\Delta^{*,\beta} = 0, \\
v_i^{\pi,\beta} &\equiv v_i^{*,\beta} = \sum_{j=i}^{\infty} C_j \beta^{j-i} \quad \text{for } i = 1, 2, \ldots \\
v_0^{\varphi^1,\beta} &= \sum_{j=0}^{\infty} C_j \beta^j = g(\beta) = e^{-\frac{1}{(1-\beta)^2}}, \\
v_0^{\varphi^2,\beta} &= 0 < v_0^{\varphi^1,\beta},
\end{aligned}$$

and φ^2 is the only uniformly optimal strategy: $v_x^{*,\beta} = v_x^{\varphi^2,\beta}$. We see that φ^2 is Blackwell optimal, and φ^1 is not. At the same time, φ^1 is n-discount optimal for all $n = -1, 0, 1, 2, \ldots$: for each strategy π

$$\lim_{\beta \to 1-} (1-\beta)^{-n} [v_0^{\varphi^1,\beta} - v_0^{\pi,\beta}] \le \lim_{\beta \to 1-} (1-\beta)^{-n} [v_0^{\varphi^1,\beta}] = 0.$$

The case of initial state $x \neq 0$ is trivial. □

3.3.21 *No Maitra optimal strategies*

A.Maitra (1965) suggested the following definitions, similar to, but weaker than, the Blackwell optimality. By the way, recall that Blackwell optimal strategies exist in finite MDPs [Blackwell (1962), Thm. 5], [Puterman (1994), Thm. 10.1.4], [Kallenberg (2010), Thm. 4.1].

Definition 3.10. [Maitra (1965)]. A strategy is called *good* (we try to avoid the over-used term 'optimal') if, for each $x \in \mathbf{X}$, there is $\beta_0(x) < 1$ such that $v_x^{\pi^*,\beta} = v_x^{*,\beta}$ for all $\beta \in [\beta_0(x), 1)$.

In the same article the following even weaker definition appears.

Definition 3.11. A strategy π^* is (*Maitra*) optimal if, for any strategy π, for each $x \in \mathbf{X}$ there exists $\beta_0(x,\pi) \in (0,1)$ such that $v_x^{\pi^*,\beta} \le v_x^{\pi,\beta}$ for all $\beta \in [\beta_0(x,\pi), 1)$.

Let us emphasize that some authors call Maitra optimal strategies as Blackwell optimal, and the strategies in Definition 3.3 as *strongly* Blackwell optimal: [Hordijk and Spieksma (1989); Hordijk and Yushkevich (2002)].

Clearly, every Blackwell optimal strategy is automatically a 'good' one, and any 'good' strategy is automatically Maitra optimal.

If the MDP is finite, then there is a finite set of stationary selectors. Suppose π^* is a Maitra optimal strategy. One can put $\hat{\beta}_0 := \max_{x \in \mathbf{X}, \varphi \in \Pi^{\text{SN}}} \beta_0(x, \varphi) < 1$, where Π^{SN} is the set of all stationary selectors. As the result, for all $x \in \mathbf{X}$, $\varphi \in \Pi^{\text{SN}}$, $\beta \in (\hat{\beta}_0, 1)$,

$$v_x^{\pi^*,\beta} \le v_x^{\varphi,\beta} \Longrightarrow v_x^{\pi^*,\beta} \le \min_{\varphi \in \Pi^{\text{SN}}} v_x^{\varphi,\beta} = \min_{\pi \in \Pi^{\text{ALL}}} v_x^{\pi,\beta} = v_x^{*,\beta}.$$

(Recall that, in finite MDPs, there exists a uniformly optimal stationary selector: see Theorem 3.2(b) or [Kallenberg (2010), Cor. 3.3].) Thus, the Maitra optimal strategy π^* is also Blackwell optimal.

If the state space is not finite, then it can happen that there is no Blackwell optimal strategy, but the 'good' strategy (in the sense of Definition 3.10) does exist: see Subsection 3.3.22. In Subsection 3.3.23 we provide an example showing that, if the action space is not finite, then it can happen that there are no Blackwell optimal and 'good' strategies, but the Maitra optimal strategy does exist. Another example confirming that, in general, the Blackwell optimality is stronger than the Maitra optimality can be found in [Hordijk and Yushkevich (2002)]: see Counter-example 1 therein.

If the state space $\mathbf{X}$ is countable, then the sufficient condition for the existence of a Maitra optimal strategy is given in the following theorem.

Theorem 3.5. [Hordijk and Yushkevich (2002), Thm. 7.7]. *Suppose the MDP is semi-continuous (i.e., Condition 2.4 is satisfied), the cost function $c(\cdot,\cdot)$ is bounded and continuous in $a \in \mathbf{A}$ for each $x \in \mathbf{X}$. Assume that, for each stationary selector φ, the t-th convolution $P^t(\varphi) = \left[P^t_{y|x}(\varphi)\right]_{(x,y)\in\mathbf{X}^2}$ of the transition matrix $p(y|x,\varphi(x))$ converges, as $t \to \infty$, to a limiting stochastic matrix $Q(\varphi) = \left[Q_{y|x}(\varphi)\right]_{(x,y)\in\mathbf{X}^2}$ geometrically fast (the uniform ergodicity condition):*

$$\sup_{f(\cdot):\sup_{x\in\mathbf{X}}|f(x)|\le 1} \left\{ \sup_{x\in\mathbf{X}} \left| \sum_{y\in\mathbf{X}} f(y)\left(P^t_{y|x}(\varphi) - Q_{y|x}(\varphi)\right) \right| \right\} \le C\gamma^t$$

for some $C < \infty$ and $\gamma \in (0,1)$.

Then there exists a Maitra optimal strategy.

The following example, similar, but simpler than [Maitra (1965), p. 246], shows that, if $\mathbf{X}$ is not finite, a Maitra optimal strategy may not exist.

Example. Let $\mathbf{X} = \{\Delta, 0, 1, 2, \dots\}$, $\mathbf{A} = \{0, 1\}$, $p(\Delta|\Delta, a) \equiv 1$, for $i \in \{0, 1, 2, \dots\}$ $p(\Delta|i, 1) \equiv 1$, $p(i+1|i, 0) \equiv 1$, with all other transition probabilities being zero; $c(\Delta, a) \equiv 0$, for $i \in \{0, 1, 2, \dots\}$, $c(i, 0) \equiv 0$, $c(i, 1) = -1 + 2^{-i}$. (See Fig. 3.27).

Suppose there exists a Maitra optimal strategy π^* which, without loss of generality, can be chosen as Markov: see Lemma 2.1. Consider $x_0 = 0$.

Obviously, $\pi^* \ne \varphi^0$, where $\varphi^0(x) \equiv 0$, because any other strategy π with $\pi_1(0|0) > 0$ provides $v_0^{\pi,\beta} < 0$ and $v_0^{\varphi^0,\beta} = 0$ at each discount factor $\beta \in (0,1)$. Therefore, there exists the minimal state $0 \le n-1 < \infty$ (equivalently, the minimal decision epoch $1 \le n < \infty$) such that $\pi^*_n(1|n-1) = p > 0$. Let $\pi^*_{n+1}(1|n) =: q \in [0,1]$ and consider the following Markov strategy $\hat\pi$:

$$\begin{aligned} \hat\pi_t(0|x) &\equiv 1 \quad \text{for all } t = 1, 2, \dots, n; \\ \hat\pi_{n+1}(0|x) &\equiv (1-p)(1-q); \\ \hat\pi_t(0|x) &\equiv \pi^*_t(0|x), \quad t = n+2, n+3, \dots. \end{aligned}$$

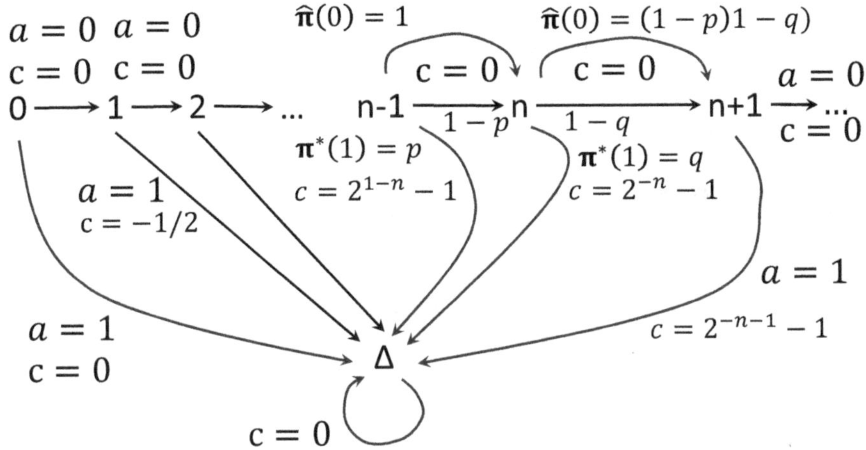

Fig. 3.27 No Maitra optimal strategies.

We have equality $\mathsf{P}_0^{\pi^*}(H_t) = \mathsf{P}_0^{\hat{\pi}}(H_t)$ for all $t = 0, 1, \dots, n-1$, where, as usual, $H_t = (X_0, A_1, X_1, \dots, A_t, X_t)$. Moreover

$$\mathsf{P}_0^{\pi^*}(X_{n+1}) = \mathsf{P}_0^{\hat{\pi}}(X_{n+1}) = \begin{cases} (1-p)(1-q), & \text{if } X_{n+1} = n+1; \\ 1-(1-p)(1-q), & \text{if } X_{n+1} = \Delta; \\ 0, & \text{otherwise,} \end{cases}$$

and hence $\mathsf{P}_0^{\pi^*}(X_t, A_{t+1}, X_{t+1}, \dots) = \mathsf{P}_0^{\hat{\pi}}(X_t, A_{t+1}, X_{t+1}, \dots)$ for all $t \geq n+1$. Therefore, in order to compare $v_0^{\pi^*,\beta}$ and $v_0^{\hat{\pi},\beta}$, we consider only the following difference:

$$\begin{aligned} \Pi^\beta &:= \mathsf{E}_0^{\pi^*,\beta}[\beta^{n-1}c(X_{n-1}, A_n) + \beta^n c(X_n.A_{n+1})] \\ &\quad - \mathsf{E}_0^{\hat{\pi},\beta}[\beta^{n-1}c(X_{n-1}, A_n) + \beta^n c(X_n.A_{n+1})] \\ &= \beta^{n-1}[p(2^{1-n}-1) + \beta(1-p)q(2^{-n}-1)] \\ &\quad - \beta^{n-1}[0 + \beta(1-(1-p)(1-q))(2^{-n}-1)]. \end{aligned}$$

If $\beta = 1$, then

$$\begin{aligned} \Pi^1 &= \left(\frac{1}{2}\right)^{n-1}\left\{p + \frac{(1-p)q}{2} - \frac{1}{2}[1-(1-p)(1-q)]\right\} \\ &\quad - p - (1-p)q + [1-(1-p)(1-q)] = \left(\frac{1}{2}\right)^n p > 0. \end{aligned}$$

Thus $\Pi^\beta > 0$ for all β close to 1, and the strategy π^* is not Maitra optimal.

Theorem 3.5 is not applicable because the uniform ergodicity condition is not satisfied for the stationary selector $\varphi^0(x) \equiv 0$: $P^t(\varphi^0)$ converges to the zero matrix as $t \to \infty$.

In this example, the uniformly optimal selector exists for each $\beta \in [0,1)$ by Theorem 3.2, but this MDP with total cost (when $\beta = 1$) has no uniformly optimal strategy because the infimum $\inf_\pi v_x^{\pi,1} = -1$ for $x = 0,1,2,\ldots$ is not achieved. □

3.3.22 *"Good" strategy which is not Blackwell optimal*

Example. Let $\mathbf{X} = \{1,2,\ldots\} \cup \{1',2',\ldots\}$; $\mathbf{A} = \{1,2\}$; $p(i|i,1) = 1$, $p(i'|i,2) = 1$, $p(i'|i',a) \equiv 1$ with all other transition probabilities being zero; $c(i,1) = 0$, $c(i,2) = -1$, $c(i',a) \equiv \frac{1}{i}$. See Fig. 3.28. Clearly, only actions in states $i \in \{1,2,\ldots\}$ are important.

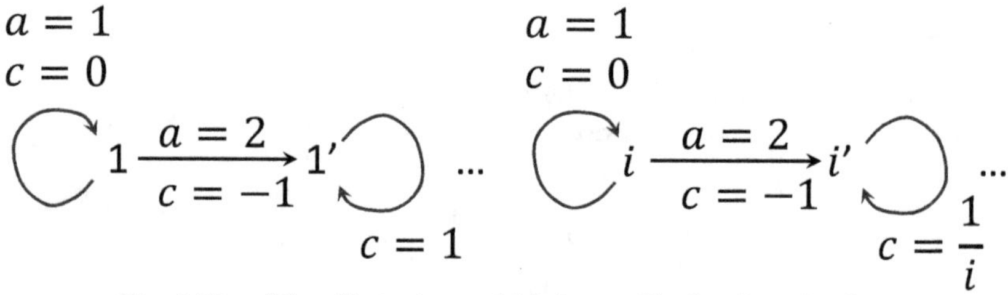

Fig. 3.28 "Good" strategy which is not Blackwell optimal.

For a fixed $\beta \in (0,1)$, the optimality equation (3.2) leads to the following:

$$v^\beta(i') = \frac{1}{i(1-\beta)}; \quad v^\beta(i) = \min\{\beta v^\beta(i); \quad -1 + \beta v^\beta(i')\}$$
$$\Longrightarrow v^\beta(i') = \frac{1}{i(1-\beta)}; \quad v^\beta(i) = \min\left\{0; \quad -1 + \frac{\beta}{i(1-\beta)}\right\}.$$

This bounded solution coincides with the Bellman function $v_x^{*,\beta}$ and $v_x^{*,\beta} = v_x^{\varphi^\beta,\beta}$, where the uniformly optimal stationary selector φ^β is given by

$$\varphi^\beta(i) = \varphi^\beta(i') = \begin{cases} 1, & \text{if } i \le \frac{\beta}{1-\beta}; \\ 2, & \text{otherwise.} \end{cases}$$

See Theorem 3.1(a). When ignoring the actions in states i', this stationary selector is the unique (up to the case when $i = \frac{\beta}{1-\beta}$) uniformly optimal strategy.

The stationary selector $\varphi^1(x) \equiv 1$ is 'good' (in the sense of Definition 3.10) because, for each $i \in \{1, 2, \ldots\}$, it is optimal if $\beta \geq \frac{i}{i+1} =: \beta_0(i)$. It is certainly Maitra optimal. By the way, a Maitra optimal strategy exists due to Theorem 3.5. At the same time, in this MDP there are no Blackwell optimal strategies because, as β approaches 1, the uniformly optimal strategy φ^β essentially changes for large values of i. □

3.3.23 *Maitra optimal strategy which is not "good" and not Blackwell optimal; limiting stationary selectors*

Suppose the discounted MDP is finite. Then the set Π^{SN} of all stationary selectors is finite, and we fix the discrete topology in this set. Consider uniformly optimal stationary selectors φ^β (if the discount factor is $\beta \in (0,1)$) and assume that $\varphi^\beta \to \varphi^*$ as $\beta \to 1-$. Then, for some $\beta_0 \in (0,1)$, $\varphi^* = \varphi^\beta$ for all $\beta \in [\beta_0, 1)$, and φ^* is Blackwell optimal. Recall that Blackwell optimal stationary selector exists in finite models [Blackwell (1962), Thm. 5], [Puterman (1994), Thm. 10.1.4].

If the state space $\mathbf{X}$ is countable and the action space is topological Borel, then, similarly, we fix the component-wise convergence in the space of stationary selectors: $\varphi^\beta(x) \to \varphi^*(x)$, $x \in \mathbf{X}$.

The presented below MDP (with finite $\mathbf{X}$ and topological Borel space $\mathbf{A}$), published in [Hordijk and Spieksma (1989)], is such that

- for $\beta < 1$ close to 1, there is a unique (in some sense) uniformly optimal stationary selector φ^β and $\lim_{\beta \to 1-} \varphi^\beta = \varphi^*$;
- Blackwell optimal strategy and "good" strategy do not exist;
- there exists a Maitra optimal stationary selector;
- the stationary selector φ^* is not Maitra optimal.

Note that, in [Hordijk and Spieksma (1989)], the maximization version was studied. The authors use the rigorous Laurent series approach, the deviation matrix etc which is rather cumbersome. Therefore, below we provide a little less rigorous investigation, highlighting the key aspects. Another example showing that the Blackwell optimality is stronger than the Maitra optimality in non-finite models is presented in [Hordijk and Yushkevich (2002), Conterexample 1].

Example. Let $\mathbf{X} = \{\Delta, 1, 2, 3, 4\}$; $\mathbf{A} = [0, \frac{1}{2}(\sqrt{3}-1)] \cup \{2\}$ with the Euclidean topology; $p(\Delta|\Delta, a) \equiv 1$, $p(\Delta|1, a) = p(\Delta|3, a) \equiv 1$, for $a < 2$ $p(1|2, a) = \frac{1}{4} + a^2$, $p(2|2, a) = \frac{1}{2} - a^2 - a$, $p(3|2, a) = \frac{1}{4} + a$, $p(4|2, 2) = 1$, $p(2|4, a) \equiv \frac{3-\sqrt{5}}{13+\sqrt{5}}$, $p(\Delta|4, a) \equiv \frac{10+2\sqrt{5}}{13+\sqrt{5}}$, with all other transition probabilities being zero; $c(\Delta, a) \equiv 0$, $c(1, a) \equiv -1$, $c(2, a) \equiv 0$, $c(3, a) \equiv -2$, $c(4, a) \equiv -\frac{10+6\sqrt{5}}{13+\sqrt{5}}$. See Fig. 3.29. Clearly, only actions in state 2 are important.

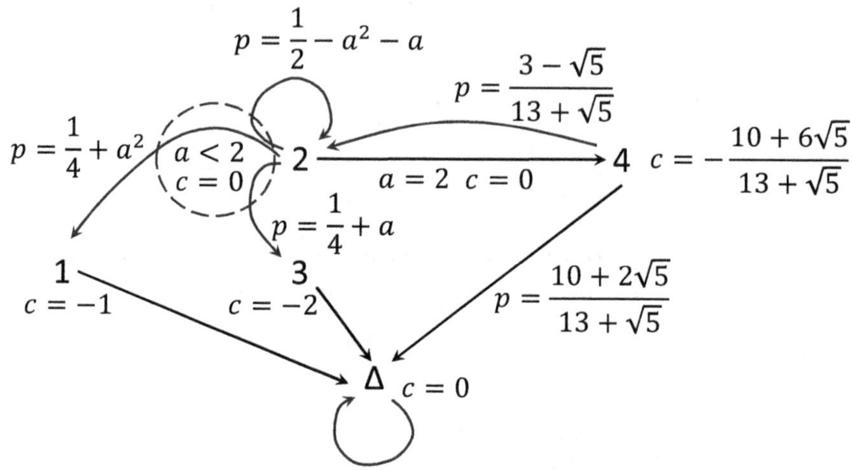

Fig. 3.29 Description of the MDP.

For a fixed $\beta \in (0, 1]$, the optimality equation (3.2) leads to the following:

$$\begin{aligned} v^\beta(\Delta) &= 0, \quad v^\beta(1) = -1, \quad v^\beta(3) = -2, \\ v^\beta(2) &= \beta \min\left\{ \inf_{a\in[0,\frac{1}{2}(\sqrt{3}-1)]} \left\{ -\frac{1}{4} - a^2 + \left(\frac{1}{2} - a^2 - a\right) v^\beta(2) - \frac{1}{2} - 2a \right\}, \right. \\ &\qquad \left. v^\beta(4) \right\}, \\ v^\beta(4) &= -\frac{10+6\sqrt{5}}{13+\sqrt{5}} + \beta\frac{3-\sqrt{5}}{13+\sqrt{5}} v^\beta(2). \end{aligned}$$

Let us show that, for all $\beta < 1$ close to 1,

$$a(\beta) := \frac{1}{\beta} - \frac{5}{4} + \frac{1}{2\beta}\sqrt{4 - 2\beta - \frac{3}{4}\beta^2}$$

is the unique action providing the infimum at state 2. First of all, when $\beta = 1$ we have $a(1) = \frac{\sqrt{5}-1}{4} < \frac{1}{2}(\sqrt{3} - 1)$, so that $a(\beta) \in \mathbf{A}$ for all β close to 1 because $a(\cdot)$ is a continuous function.

Assume that

$$v^\beta(2) = \beta \inf_{a\in[0,\frac{1}{2}(\sqrt{3}-1)]} \left\{ -\frac{1}{4} - a^2 + \left(\frac{1}{2} - a^2 - a\right) v^\beta(2) - \frac{1}{2} - 2a \right\}, \tag{3.21}$$

that is, $v^\beta(2) = \inf_{a\in[0,\frac{1}{2}(\sqrt{3}-1)]} f^\beta(a)$, where

$$f^\beta(a) := -\beta\frac{\frac{3}{4} + a^2 + 2a}{1 - \beta(1/2 - a^2 - a)}.$$

For all β close to 1, the only analytical minimum is attained at $a(\beta)$, function $f^\beta(\cdot)$ increases in a when $a > a(\beta)$, and, if we check $a = 0$, we obtain $f^\beta(0) = -\frac{3\beta}{4-2\beta}$. It remains to compare $f^1(a(1)) = -\frac{5+3\sqrt{5}}{5+\sqrt{5}}$ and $f^1(0) = \frac{-3}{4-2} = -\frac{3}{2}$:

$$\frac{-3}{2} + \frac{5+3\sqrt{5}}{5+\sqrt{5}} = \frac{\sqrt{5}-2}{2} > 0.$$

$f^1(a(1)) < f^1(0)$, and $f^\beta(a(\beta)) < f^\beta(0)$ for all β close to 1, because all those functions are continuous. Therefore, if $v^\beta(2)$ comes from (3.21), then the infimum is provided by $a(\beta)$.

Finally, let us show that, if we substitute the obtained solution $v^\beta(2) := f^\beta(a(\beta))$ to the original optimality equation, then $v^\beta(2) \le \beta v^\beta(4)$. ($v^\beta(4)$ is calculated using the obtained function $v^\beta(2)$.) The rigorous (rather long) proof can be found in [Hordijk and Spieksma (1989)]. Here, we only provide the results of computer simulation: see the graphs of $v^\beta(2)$ and $\beta v^\beta(4)$ in Fig. 3.30. Actually, $v^1(2) = f^1(a(1)) = v^1(4) = -\frac{5+3\sqrt{5}}{5+\sqrt{5}}$ and $v^\beta(2) < \beta v^\beta(4)$ for all $\beta < 1$ close to 1.

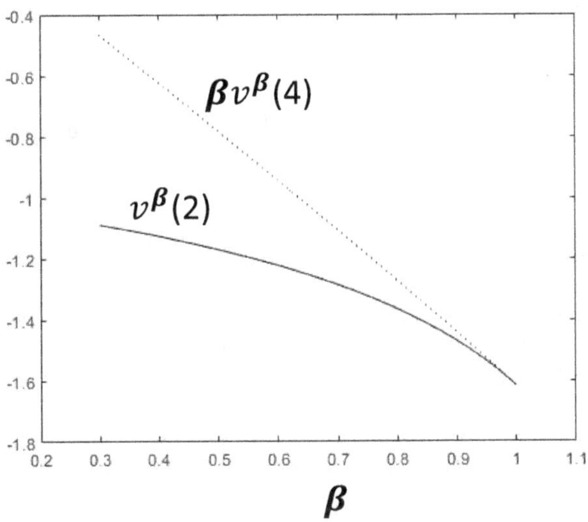

Fig. 3.30 Functions $v^\beta(2)$ and $\beta v^\beta(4)$.

We have found the maximal non-positive bounded solution to the optimality equation at all $\beta \leq 1$ close to 1:

$$v^\beta(\Delta) = 0, \quad v^\beta(1) = -1, \quad v^\beta(2) = f^\beta(a(\beta)), \quad v^\beta(3) = -2,$$
$$v^\beta(4) = -\frac{10 + 6\sqrt{5}}{13 + \sqrt{5}} + \beta \frac{3 - \sqrt{5}}{13 + \sqrt{5}} v^\beta(2),$$

which coincides with the Bellman function $v_x^{*,\beta}$: see [Hernández-Lerma and Lasserre (1999), Chapter 8]) and the paragraph below Theorem 2.3.

The infimum in state 2 is provided

- by $a(\beta)$ only, if $\beta < 1$;
- by both $a(1)$ and $a = 2$, if $\beta = 1$.

Since only actions at state 2 are important, we, less rigorously, say that the following proposition is valid.

Proposition 3.2. *Only the stationary selector $\varphi^\beta(x) \equiv a(\beta)$ is uniformly optimal if $\beta < 1$ (see Theorem 3.1(b)), and only the both stationary selectors*

$$\varphi^*(x) := \varphi^1(x) \equiv a(1) = \frac{\sqrt{5} - 1}{4} \quad \textit{and } \hat{\varphi}(x) \equiv 2$$

are equally uniformly optimal if $\beta = 1$.

For justification of the latter statement, see [Puterman (1994), Thm. 7.1.7(b)] or [Hernández-Lerma and Lasserre (1999), Thm. 9.6.10(b)]. Only the selectors φ^* and $\hat{\varphi}$ are conserving. They are equalizing because

$$\lim_{t\to\infty} \mathsf{P}_x^{\varphi^*}(X_t \neq \Delta) = \lim_{t\to\infty} \mathsf{P}_x^{\hat{\varphi}}(X_t \neq \Delta) = 0.$$

Note, the MDP is transient in the space $\mathbf{X} \setminus \{\Delta\}$, in the sense of Definition 9.6.1 of [Hernández-Lerma and Lasserre (1999)].

Clearly, $\varphi^\beta \to \varphi^*$ as $\beta \to 1-$; φ^* is the limiting stationary selector.

Blackwell optimal strategy does not exist. Indeed, if, for some $\beta_0 < 1$ close to 1, $v_x^{\pi^*,\beta_0} = v_x^{*,\beta_0}$ for all $x \in \mathbf{X}$, then π^* necessarily coincides with the uniformly optimal stationary selector $\varphi^{\beta_0}(x) \equiv a(\beta_0)$. But there are $\beta \in (\beta_0, 1)$ with $v_2^{\varphi^{\beta_0},\beta} > v_2^{\varphi^\beta,\beta} = v_2^{*,\beta}$ because only $\varphi^\beta(x) \equiv a(\beta)$ provides $\inf_\pi v_2^{\pi,\beta} = v_2^{*,\beta}$, and $a(\beta)$ is not constant. For the same reason, there are no 'good' strategies (in the sense of Definition 3.10) in this MDP.

Let us show that the stationary selector $\hat{\varphi}(x) \equiv 2$ is Maitra optimal.

Consider any strategy π which applies an action different from $a(1) = \frac{\sqrt{5}-1}{4}$ and $a = 2$ at state 2 with positive probability. This strategy results in $v_2^{\pi,1} > v_2^{\varphi^*,1} = v_2^{\hat{\varphi},1}$. Since the functions $v_2^{\pi,\beta}$ and $v_2^{\hat{\varphi},\beta}$ are continuous w.r.t. β, we have inequality $v_2^{\pi,\beta} \geq v_2^{\hat{\varphi},\beta}$ on some interval $[\beta_0(2,\pi), 1)$. This calculation is illustrated on Figures 3.31 and 3.32.

It remains to compare $v_2^{\varphi^*,\beta}$ and $v_2^{\hat{\varphi},\beta}$.

- Directly from the description of the MDP (see Fig. 3.29) we have for $a(1) = \frac{\sqrt{5}-1}{4}$

$$v_2^{\varphi^*,\beta} = \beta\left\{\left(\frac{1}{2} - [a(1)]^2 - a(1)\right) v_2^{\varphi^*,\beta} + \left(\frac{1}{4} + [a(1)]^2\right)\cdot(-1) + \left(\frac{1}{4} + a(1)\right)\cdot(-2)\right\},$$

so that

$$v_2^{\varphi^*,\beta} = -\frac{\beta(5+3\sqrt{5})}{8-\beta(3-\sqrt{5})}.$$

- Similarly, for $a = 2$,

$$v_2^{\hat{\varphi},\beta} = \beta\{v_4^{\hat{\varphi},\beta}\} = \beta\left\{-\frac{10+6\sqrt{5}}{13+\sqrt{5}} + \beta\frac{3-\sqrt{5}}{13+\sqrt{5}}\cdot v_2^{\hat{\varphi},\beta}\right\},$$

so that

$$v_2^{\hat{\varphi},\beta} = -\frac{\beta(10+6\sqrt{5})}{13+\sqrt{5}-\beta^2(3-\sqrt{5})}.$$

As expected, when $\beta = 1$, the values $v_2^{\varphi^*,1} = v_2^{\hat{\varphi},1} = -\frac{5+3\sqrt{5}}{5+\sqrt{5}}$ are equal. Moreover, direct calculations show that the derivatives

$$\left.\frac{dv_2^{\varphi^*,\beta}}{d\beta}\right|_{\beta=1} = \left.\frac{dv_2^{\hat{\varphi},\beta}}{d\beta}\right|_{\beta=1} = -\frac{4\sqrt{5}}{5}$$

also coincide when $\beta = 1$. But the second derivatives are such that

$$\left.\frac{d^2v_2^{\varphi^*,\beta}}{d\beta^2}\right|_{\beta=1} = \frac{16-8\sqrt{5}}{5} > \left.\frac{d^2v_2^{\hat{\varphi},\beta}}{d\beta^2}\right|_{\beta=1} = \frac{37-19\sqrt{5}}{10}.$$

Therefore, for $\beta < 1$ close to 1, $v_2^{\varphi^*,\beta} > v_2^{\hat{\varphi},\beta}$, and, in case a strategy π applies only actions $a(1) = \frac{\sqrt{5}-1}{4}$ and $a = 2$ at state 2 with $\mathsf{P}_2^\pi(A_t = a(1)) > 0$ at some $t = 1, 2, \ldots$, then again $v_2^{\pi,\beta} > v_2^{\hat{\varphi},\beta}$ in the neighbourhood of 1: the stationary selector $\hat{\varphi}$ is the unique uniformly optimal one if we consider the action space $\{a(1), 2\}$.

All the presented calculations are illustrated on Figures 3.31 and 3.32. The required $\beta_0(2,\pi) \in (0,1)$ exists for each strategy π, and the stationary selector $\hat{\varphi}(x) \equiv 2$ is Maitra optimal.

Absolutely the same reasoning holds for state 4:

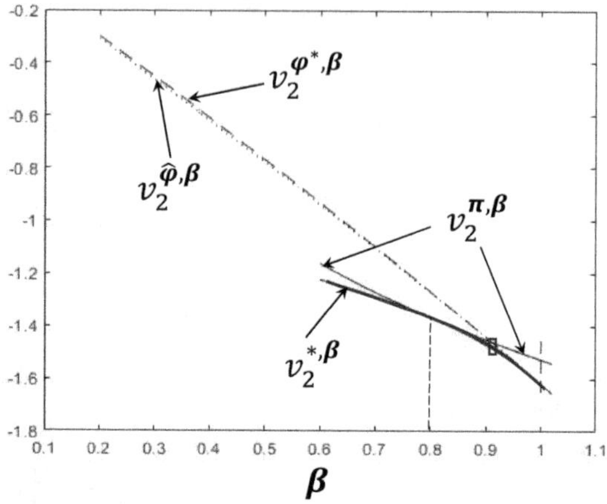

Fig. 3.31 Thick line is the graph of $v^\beta(2) = v_2^{*,\beta} = v_2^{\varphi^\beta,\beta}$; thin solid line, tangent to the previous one at $\beta = 0.8$, is the graph of $v_2^{\pi,\beta}$ for π being the stationary selector $\varphi(x) \equiv a(0.8)$; dashed line is the graph of $v_2^{\varphi^*,\beta}$, and dotted line, slightly below the previous one, is the graph of $v_2^{\hat\varphi,\beta}$.

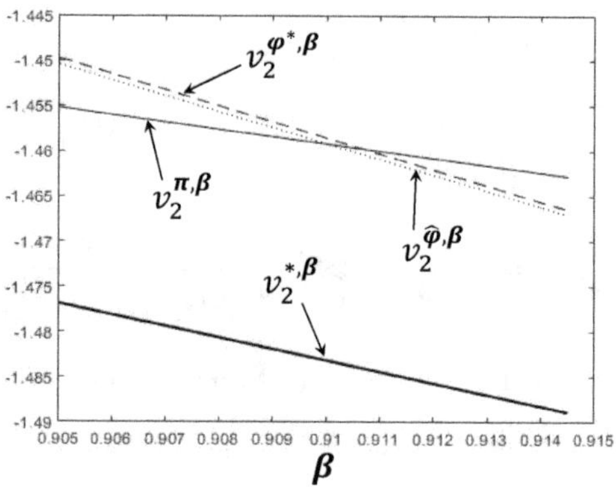

Fig. 3.32 Enlarged squared fragment from Fig. 3.31.

$$v_4^{\varphi,\beta} = -\frac{10+6\sqrt{5}}{13+\sqrt{5}} + \beta\frac{3-\sqrt{5}}{13+\sqrt{5}}v_2^{\varphi,\beta}$$

for each stationary selector φ, so the functions $v_4^{*,\beta}, v_4^{\hat\varphi,\beta}$ and $v_4^{\varphi^*,\beta}$ exhibit the same properties as the corresponding functions for state 2. See also Proposition 3.2. The values $v_\Delta^{\pi,\beta} = 0$, $v_1^{\pi,\beta} = -1$, $v_3^{\pi,\beta} = -2$ are constant, independent on π and β. Therefore, the stationary selector $\hat\varphi(x) \equiv 2$ is Maitra optimal, and the limiting stationary selector $\varphi^*(x) \equiv a(1)$ is not. Recall that we actually consider the equivalent classes of strategies: the actions in the states $\Delta, 1, 3, 4$ may be arbitrary, they play no role. Let us also emphasize that, in this example, a Maitra optimal strategy exists according to Theorem 3.5: the limiting stochastic matrix $Q(\varphi)$ is such that $Q_{\Delta|x} \equiv 1$ for all $x \in \mathbf{X}$ and all stationary selectors φ; other conditions of that theorem are obviously satisfied. □

3.3.24 *Vanishing discount factor and MDPs with the average cost*

Very often, solutions of MDPs with discounted cost help to solve problems with the expected *average* cost, that is, with the performance functional

$$\limsup_{T\to\infty} \frac{1}{T} \mathsf{E}^{\pi}_{\mathrm{P}_0} \left[\sum_{t=1}^{T} c(X_{t-1}, A_t) \right] \to \inf_{\pi} : \tag{3.22}$$

see (4.1).

We need the following:

Condition 3.1. The state space $\mathbf{X}$ is discrete, the action space $\mathbf{A}$ is finite, and $\sup_{(x,a)\in\mathbf{X}\times\mathbf{A}} |c(x,a)| < \infty$.

Theorem 3.6. [Ross (1983), Chapter V, Thm. 2.1, 2.2]. *Let Condition 3.1 be satisfied.*

(a) Suppose, for some $N < \infty$,

$$|v_x^{*,\beta} - v_y^{*,\beta}| < N \quad \text{for all } x, y \in \mathbf{X}, \ \beta \in (0,1). \tag{3.23}$$

Then there exist a bounded function $h(\cdot)$ on $\mathbf{X}$ and a constant ρ satisfying equation

$$\rho + h(x) = \min_{a\in\mathbf{A}} \left\{ c(x,a) + \sum_{y\in\mathbf{X}} p(y|x,a) h(y) \right\}. \tag{3.24}$$

If $z \in \mathbf{X}$ is an arbitrarily fixed state, then one can take

$$h(x) = \lim_{n\to\infty} (v_x^{*,\beta_n} - v_z^{*,\beta_n})$$

for an appropriately chosen sequence $\beta_n \to 1-$, and $\rho = \lim_{\beta\to 1-}(1-\beta) v_z^{*,\beta}$.

(b) If a bounded function $h(\cdot)$ on $\mathbf{X}$ and a constant ρ satisfy equation (3.24) and the map $\varphi^* : \mathbf{X} \to \mathbf{A}$ provides there the minimum, then the stationary selector φ^* is the uniformly optimal strategy in problem (3.22) and $\rho = \inf_\pi \limsup_{T\to\infty} \frac{1}{T} \mathsf{E}^{\pi}_{x_0} \left[\sum_{t=1}^{T} c(X_{t-1}, A_t) \right]$ for all $x_0 \in \mathbf{X}$.

Under additional conditions, this theorem was generalized for arbitrary Borel spaces $\mathbf{X}$ and $\mathbf{A}$ in [Hernández-Lerma and Lasserre (1996a), Thm. 5.5.4]. Note, $\langle \rho, h, \varphi^* \rangle$ as in Theorem 3.6 is a canonical triplet (see Theorem 4.1), and the statement (b) of Theorem 3.6 follows from Theorem 4.3(a).

Remark 3.3. If there is a Blackwell optimal stationary selector φ^*, then, under the assumptions of Theorem 3.6, it is uniformly optimal in problem (3.22). Indeed that selector provides the minimum in the equation

$$\begin{aligned}(1-\beta)v_z^{*,\beta} + h^\beta(x) &= \min_{a\in\mathbf{A}}\left\{c(x,a) + \beta\sum_{y\in\mathbf{X}} p(y|x,a)h^\beta(y)\right\} \\ &= c(x,\varphi^*(x)) + \beta\sum_{y\in\mathbf{X}} p(y|x,\varphi^*(x))h^\beta(y)\end{aligned}$$

for all β sufficiently close to 1. Here $h^\beta(x) := v_x^{*,\beta} - v_z^{*,\beta}$ at an arbitrarily fixed $z \in \mathbf{X}$. As the result, for $h(\cdot)$ and ρ as in Theorem 3.6, $\langle \rho, h, \varphi^* \rangle$ is a canonical triplet and φ^* is uniformly optimal in (3.22) by Theorem 4.3(a).

The fist example, similar to that described in Proposition 7.1 of [Guo *et al.* (2024)], shows that requirement (3.23) is crucial for the equality

$$\inf_\pi \limsup_{T\to\infty} \frac{1}{T}\mathsf{E}^\pi_{x_0}\left[\sum_{t=1}^T c(X_{t-1}, A_t)\right] = \lim_{\beta\to 1-}(1-\beta)v_{x_0}^{*,\beta}. \tag{3.25}$$

Example. Let $\mathbf{X} = \{0,1,2,\ldots\}$, $\mathbf{A} = \{0\}$ (a dummy action, so action $a = 0$ and (the unique) strategy π are omitted below); $p(i+1|i) = 1$ for all $i \in \mathbf{X}$ with all other transition probabilities being zero; $\{c(i) := 1 - u(i)\}_{i=0}^\infty$ is the non-negative bounded sequence, where $u(0), u(1), \ldots$ are as in Example 2 of [Bishop *et al.* (2014)]:

$$c(i) = \begin{cases} 1, & \text{if } k! \le i < 2k!, \quad k = 1, 2\ldots; \\ 0, & \text{otherwise.}\end{cases}$$

See Fig. 3.33.

According to [Bishop *et al.* (2014)],

$$\rho := \limsup_{n\to\infty} \frac{1}{n}\sum_{i=0}^{n-1} c(i) = \frac{1}{2} > \limsup_{\beta\to 1-}(1-\beta)\sum_{i=0}^\infty \beta^i c(i) = \frac{1}{4};$$

in particular, $\lim_{\beta\to 1-}\sum_{i=0}^\infty \beta^i c(i) = \infty$.

Clearly,

$$\limsup_{T\to\infty}\frac{1}{T}\mathsf{E}_0\left[\sum_{t=1}^T c(X_{t-1})\right] = \rho = \frac{1}{2}, \quad v_i^{*,\beta} = \sum_{l=0}^\infty \beta^l c(i+l) \;\; \forall i \in \mathbf{X},$$

and equality (3.25) is not valid at $x_0 = 0$.

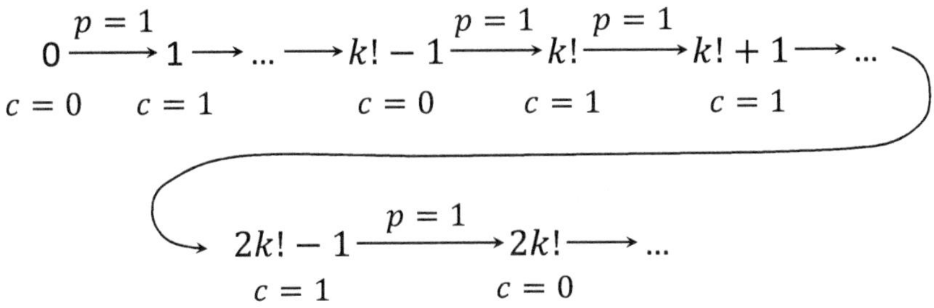

Fig. 3.33 Equality (3.25) is not valid.

In what follows, we show that condition (3.23) is not satisfied and the statements of Theorem 3.6 are not valid. Let $N > 0$ be arbitrarily fixed. Since $\lim_{\beta\to 1-}\sum_{i=0}^{\infty}\beta^i c(i) = \infty$, one can fix $\beta \in (0,1)$ such that

$$v_0^{*,\beta} = \sum_{i=0}^{\infty}\beta^i c(i) > N+1.$$

Next, let $k \geq 1$ be such that

$$\sum_{l=(k+1)!-2k!}^{\infty}\beta^l = \frac{\beta^{(k+1)!-2k!}}{1-\beta} < 1$$

and consider the state $i = 2k!$:

$$v_i^{*,\beta} = \sum_{l=0}^{\infty}\beta^l c(i+l) \leq \sum_{l=0}^{(k+1)!-1-2k!}\beta^l c(i+l) + \sum_{l=(k+1)!-2k!}^{\infty}\beta^l < 1$$

because $c(2k!) = c(2k!+1) = \ldots = c((k+1)!-1) = 0$ and $0 \leq c(l) \leq 1$ for all $l = 0,1,\ldots$. As the result, $v_0^{*,\beta} - v_{i=2k!}^{*,\beta} > N$, and condition (3.23) is not satisfied.

Equation (3.24) cannot have a bounded solution $h(\cdot)$. Indeed, suppose that ρ' and $h'(\cdot)$ satisfy that equation. Then, for each $k = 1,2,\ldots$

$$h'(k!+l) = \rho' + h'(k!) - l, \quad l = 1,2,\ldots,2k!-1$$

and

$$h'(k!+2k!-1) - h'(k!) = \rho' - (2k!-1):$$

function $h'(\cdot)$ exhibits arbitrarily large fluctuations because k is an arbitrary number. So, function $h'(\cdot)$ cannot be bounded.

By the way, Condition 4.2(b) also fails to hold because, otherwise formula

(4.10) implies that

$$v_0^{*,\beta} - v_{i=2k!}^{*,\beta} = v_0^{*,\beta} - v_z^{*,\beta} - (v_{i=2k!}^{*,\beta} - v_z^{*,\beta}) \le b(0) + M,$$

contradicting the unboundedness of the difference $v_0^{*,\beta} - v_{i=2k!}^{*,\beta}$. Theorem 4.4 is not applicable. □

The following simple example shows that condition (3.23) is important even in finite models (see [Hernández-Lerma and Lasserre (1996b), Ex. 6.1]). See also Subsections 4.3.11 and 4.3.12.

Example. Let $\mathbf{X} = \{1, 2, 3\}$, $\mathbf{A} = \{0\}$ (a dummy action), $p(1|1, a) = 1$, $p(2|2, a) = 1$, $p(1|3, a) = \alpha \in (0, 1)$, $p(2|3, a) = 1 - \alpha$, with all other transition probabilities being zero; $c(x, a) = x$ (see Fig. 3.34).

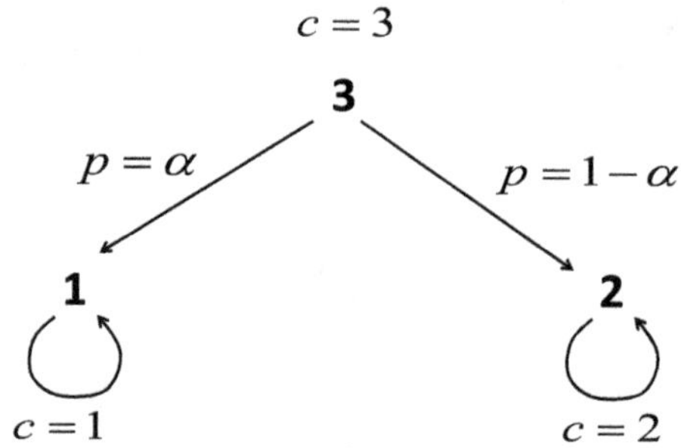

Fig. 3.34 No solutions to equation (3.24).

Obviously,

$$v_1^{*,\beta} = \frac{1}{1-\beta}, \quad v_2^{*,\beta} = \frac{2}{1-\beta}, \quad v_3^{*,\beta} = 3 + \beta[\alpha v_1^{*,\beta} + (1-\alpha)v_2^{*,\beta}] = \frac{3-\beta-\beta\alpha}{1-\beta}.$$

Condition (3.23) is violated and the (minimal) value of the expected average cost (3.22) depends on the initial state x:

$$\rho(1) = 1, \quad \rho(2) = 2, \quad \rho(3) = 2 - \alpha.$$

Equation (3.24) has no bounded solutions, because otherwise the value of ρ would have coincided with the state-independent (minimal) value of the performance functional (3.22) [Ross (1983), Chapter V, Thm. 2.1]; see also Theorems 4.1 and 4.3. □

Another theorem on the approach to MDPs with the expected average cost (3.22) via the vanishing discount factor is as follows [Ross (1968), Thm. 3.3].

Theorem 3.7. *Let Condition 3.1 be satisfied. Suppose, for some sequence $\beta_k \to 1-$, there exists a constant $N < \infty$ such that*

$$|v_x^{*,\beta_k} - v_y^{*,\beta_k}| < N \tag{3.26}$$

for all $k = 1, 2, \ldots$ and all $x, y \in \mathbf{X}$. Let $\varphi^{\beta_k}(x)$ be the uniformly optimal stationary selector in the corresponding discounted MDP. Then there exists a stationary selector solving problem (3.22) which is a limit point of φ^{β_k}. Moreover, for any $\varepsilon > 0$, for large k, the selectors φ^{β_k} are ε-optimal in the sense that

$$\limsup_{T\to\infty} \frac{1}{T} \mathsf{E}_{x_0}^{\varphi^{\beta_k}} \left[\sum_{t=1}^{T} c(X_{t-1}, A_t) \right]$$

$$\leq \inf_{\pi} \left\{ \limsup_{T\to\infty} \frac{1}{T} \mathsf{E}_{x_0}^{\pi} \left[\sum_{t=1}^{T} c(X_{t-1}, A_t) \right] \right\} + \varepsilon, \qquad x_0 \in \mathbf{X}.$$

The existence of uniformly optimal stationary selectors φ^{β_k} under Condition 3.1 follows from Corollary 9.17.1 in [Bertsekas and Shreve (1978)]. Similarly to Remark 3.3, if there is a Blackwell optimal strategy then, under the assumptions of Theorem 3.7, it is uniformly optimal in problem (3.22).

The following example, based on [Ross (1968), p. 417], shows that condition (3.26) is important. See also Subsections 4.3.11 and 4.3.12.

Example. Let $\mathbf{X} = \{(i,j) : \; i = 1, 2, \ldots, \;\; j = 0, 1, \ldots, i\} \cup \{\Delta\}$, $\mathbf{A} = \{1, 2\}$,

$$p((k,j)|(i,0),a) = \begin{cases} 1, & \text{if } a = 1, \; k = i+1, \; j = 0 \text{ or} \\ & \text{if } a = 2, \; k = i, \; j = 1; \\ 0, & \text{otherwise,} \end{cases}$$

$$p((i,j+1)|(i,j),a) \equiv 1 \text{ if } 0 < j < i,$$

$$p(\Delta|(i,i),a) = p(\Delta|\Delta,a) \equiv 1 \;\; \text{for all } i \geq 1,,$$

with all other transition probabilities being zero; $c((i,0),a) \equiv 1$, $c(\Delta,a) \equiv 2$, with other one-step costs being zero (see Fig. 3.35).

The optimality equation (3.2) is given by

$$\begin{aligned} v^\beta(\Delta) &= 2 + \beta v^\beta(\Delta), \\ v^\beta((i,i)) &= \beta v^\beta(\Delta), \; i = 1, 2, \ldots, \\ v^\beta((i,j)) &= \beta v^\beta((i,j+1)), \; i = 1, 2, \ldots, \;\; j = 1, 2, \ldots, i-1, \\ v^\beta((i,0)) &= \min\{1 + \beta v^\beta((i+1,0)); \;\; 1 + \beta v^\beta((i,1))\}, \;\; i = 1, 2, \ldots, \end{aligned}$$

so that

$$v^\beta(\Delta) = \frac{2}{1-\beta}; \quad v^\beta((i,j)) = \frac{2\beta^{i+1-j}}{1-\beta}, \; i = 1, 2, \ldots, \;\; j = 1, 2, \ldots, i,$$

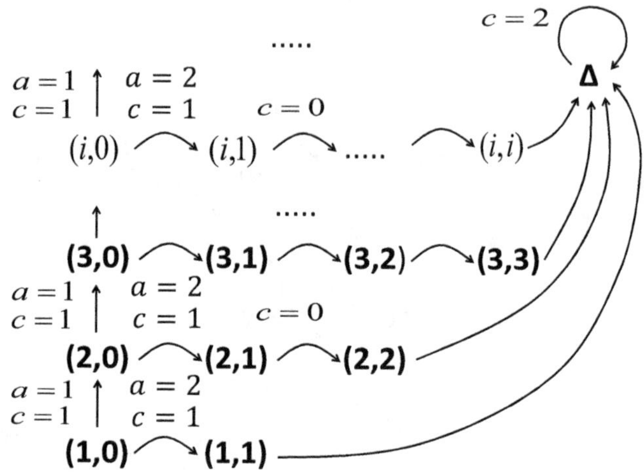

Fig. 3.35 Discount-optimal strategies are not ε-optimal in the problem with average cost.

and the last equation takes the form

$$v^{\beta}((i,0)) = \min\left\{1 + \beta v^{\beta}((i+1,0));\quad 1 + \frac{2\beta^{i+1}}{1-\beta}\right\}. \tag{3.27}$$

Lemma 3.2. *Let* $n(\beta) := \min\{k \in \{1,2,\ldots\}: \quad \beta^k(1+\beta) \leq \frac{1}{2}\}$. *Then*

$$v^{\beta}((i,0)) = \begin{cases} \dfrac{1 - \beta^{n(\beta)-i+1} + 2\beta^{2n(\beta)-i+1}}{1-\beta}, & \text{if } i < n(\beta); \\[2ex] 1 + \dfrac{2\beta^{i+1}}{1-\beta}, & \text{if } i \geq n(\beta), \end{cases}$$

and the stationary selector $\varphi^{\beta}((i,0)) = \begin{cases} 1, & \text{if } i < n(\beta); \\ 2, & \text{if } i \geq n(\beta) \end{cases}$ *is uniformly optimal. (Actions in other states obviously play no role.) This is the unique uniformly optimal strategy (up to the case when* $i = n(\beta)$*).*

The proof is presented in Appendix A. $v^{\beta}(x) = v^{\infty}(x)$ is the limiting function in value iteration (3.3). The Bellman function $v_x^{*,\beta}$ coincides with $v^{\beta}(x)$ [Bertsekas and Shreve (1978), Prop. 9.14].

It is easy to show that condition (3.26) is violated. Let $0 \leq i_1 < i_2$ and put

$\delta = i_2 - i_1$. Then, for large enough $n(\beta)$ (i.e., when β is close to 1),

$$\begin{aligned} v^{*,\beta}_{(i_1,0)} - v^{*,\beta}_{(i_2,0)} &= \frac{\beta^{n(\beta)+1-i_2}}{1-\beta}\{2\beta^{n(\beta)+\delta} - \beta^\delta + 1 - 2\beta^{n(\beta)}\} \\ &= \beta^{n(\beta)+1-i_2}\sum_{k=0}^{\delta-1}\beta^k(1 - 2\beta^{n(\beta)}). \end{aligned}$$

We know that $\beta^{n(\beta)} \leq \frac{1}{2(1+\beta)}$ and $\beta^{n(\beta)-1} > \frac{1}{2(1+\beta)}$. Thus

$$v^{*,\beta}_{(i_1,0)} - v^{*,\beta}_{(i_2,0)} > \frac{\beta^{2-i_2}}{2(1+\beta)}\sum_{k=0}^{\delta-1}\beta^k \cdot \frac{\beta}{1+\beta},$$

so that

$$\lim_{\beta\to 1-} |v^{*,\beta}_{(i_1,0)} - v^{*,\beta}_{(i_2,0)}| \geq \frac{\delta}{8},$$

and there does not exist a constant N for which (3.26) holds for any $\delta \in \{1, 2, \ldots\}$, that is, for any $x = (i_1, 0)$, $y = (i_2, 0)$.

When $\beta \to 1-$, $n(\beta)$ approaches infinity. The selector $\varphi(x) \equiv 1$ is the limit point of $\varphi^\beta(x)$ and is uniformly optimal in problem (3.22):

$$\limsup_{T\to\infty} \frac{1}{T}\mathsf{E}^{\varphi}_{x_0}\left[\sum_{t=1}^{T} c(X_{t-1}, A_t)\right] = 1, \quad x_0 \in \mathbf{X}.$$

But, for any $\beta \in (0, 1)$, $x_0 \in \mathbf{X}$

$$\limsup_{T\to\infty} \frac{1}{T}\mathsf{E}^{\varphi^\beta}_{x_0}\left[\sum_{t=1}^{T} c(X_{t-1}, A_t)\right] = 2,$$

because the chain gets absorbed at state Δ with $c(\Delta, a) \equiv 2$. Thus, discount-optimal selectors are not ε-optimal in problem (3.22) for $\varepsilon > 1$.

In this example, there are no Blackwell optimal strategies because, as β approaches 1, the uniformly optimal strategy φ^β essentially changes for $(i, 0)$ with large values of i.

There are no "good" strategies (in the sense of Definition 3.10) for the same reason as above: under each initial state $(i, 0) \in \mathbf{X}$ the optimal strategy does not stop to change when β approaches 1. Note also that any Blackwell optimal strategy is automatically a "good" one. □

Chapter 4

Homogeneous Infinite-Horizon Models: Average Cost and Other Related Criteria

The main performance functional in this chapter is the expected average cost per time unit. But there exist many strongly connected definitions of optimality like Blackwell, Maitra, overtaking etc optimality. The corresponding definitions are given as soon as they are needed in particular subsections. After the general knowledge of such type MDPs in Sections 4.1 and 4.2, examples and counterexamples are presented in Section 4.3.

4.1 Preliminaries

This chapter is mainly about the following problem

$$v^\pi_{\mathsf{P}_0} = \limsup_{T\to\infty} \frac{1}{T}\mathsf{E}^\pi_{\mathsf{P}_0}\left[\sum_{t=1}^{T} c(X_{t-1}, A_t)\right] \to \inf_\pi . \tag{4.1}$$

As usual, $v^\pi_{\mathsf{P}_0}$ is called the *performance functional.* Under rather general conditions, problem (4.1) is well defined, e.g., if the cost function $c(\cdot,\cdot)$ is bounded below. That is the case in all the examples in Section 4.3. As previously, we write P^π_x, E^π_x and v^π_x, if the initial distribution is concentrated at a single point $x \in \mathbf{X}$. In this connection,

$$v^*_x := \inf_\pi v^\pi_x$$

is the *Bellman function* (v^π_x is defined similarly to (4.1)). A strategy π^* is uniformly optimal if, for all $x \in \mathbf{X}$, $v^{\pi^*}_x = v^*_x$. In this context, such strategies will be called *AC-optimal*, i.e., average-cost-optimal [Hernández-Lerma and Lasserre (1999), Section 10.1]. A strategy π is called *AC-ε-optimal* if $v^\pi_x \le v^*_x + \varepsilon$ for all $x \in \mathbf{X}$, assuming $|v^*_x| < \infty$. If the model is finite then there exists a stationary AC-optimal selector [Puterman (1994), Thm. 9.1.8]. The situation becomes more complicated if either space $\mathbf{X}$ or $\mathbf{A}$ is not finite. The dynamic programming approach leads to the following concepts.

Definition 4.1. Let $\rho(\cdot)$ and $h(\cdot)$ be real-valued measurable functions on $\mathbf{X}$, and φ^* is a given stationary selector. Then $\langle\rho, h, \varphi^*\rangle$ is said to be a *canonical triplet* if, for all $x \in \mathbf{X}$ and $T = 0, 1, 2, \ldots$,

$$\mathsf{E}_x^{\varphi^*}\left[\sum_{t=1}^{T} c(X_{t-1}, A_t) + h(X_T)\right] = \inf_{\pi} \mathsf{E}_x^{\pi}\left[\sum_{t=1}^{T} c(X_{t-1}, A_t) + h(X_T)\right]$$
$$= T\rho(x) + h(x).$$

Theorem 4.1. **[Arapostatis** *et al.* **(1993), Thm. 6.2].** *Suppose the cost function $c(\cdot,\cdot)$ is bounded. Then the bounded measurable functions $\rho(\cdot)$ and $h(\cdot)$ and the stationary selector φ^* form a canonical triplet if and only if the following canonical equations are satisfied:*

$$\begin{aligned}\rho(x) &= \inf_{a \in \mathbf{A}} \int_{\mathbf{X}} \rho(y) p(dy|x,a) = \int_{\mathbf{X}} \rho(y) p(dy|x, \varphi^*(x));\\ \rho(x) + h(x) &= \inf_{a \in \mathbf{A}} \left\{ c(x,a) + \int_{\mathbf{X}} h(y) p(dy|x,a) \right\} \\ &= c(x, \varphi^*(x)) + \int_{\mathbf{X}} h(y) p(dy|x, \varphi^*(x)), \quad x \in \mathbf{X}.\end{aligned} \tag{4.2}$$

Remark 4.1. If the triplet $\langle \rho, h, \varphi^* \rangle$ solves equations (4.2), then so does the triplet $\langle \rho, h + const, \varphi^* \rangle$ for any value of *const*. Thus one can put $h(\hat{x}) = 0$ for an arbitrarily chosen state $\hat{x}$.

In the case when a stationary selector φ^* is an element of a canonical triplet, it is called *canonical*. Canonical triplets exist if the model is finite [Puterman (1994), Thm. 9.1.4]; the corresponding canonical selector is AC-optimal [Puterman (1994), Thm. 9.1.7]. If the cost function $c(\cdot,\cdot)$ is bounded below and $\langle \rho, h, \varphi^* \rangle$ is a canonical triplet wih bounded function $h(\cdot)$, then the stationary selector φ^* is AC-optimal: see Theorem 4.3(a).

Remark 4.2. In case $c(\cdot,\cdot) \geq -K > -\infty$ is bounded below, one can pass to the positive model by adding K to the cost function $c(\cdot,\cdot)$. The performance functional (4.1) will simply increase by K.

The similar reasoning applies to the case when the cost function is bounded from above.

Definition 4.2. Suppose the state space is finite or denumerable.

(a) The model is called (aperiodic) *unichain* if, for every stationary selector, the controlled process is a unichain Markov process with a single (aperiodic) positive-recurrent class plus a possibly empty set of transient states; absorption into the positive-recurrent class takes place in a finite expected time.

(b) The model is called *communicating* if, for any two states x, y, there is a stationary selector such that y is reachable from x.

(c) Suppose $\mathbf{R}$ is the set of all states which are recurrent under some stationary selector, and any two states $x, y \in \mathbf{R}$ communicate in the sense of (b). Then the model is called *weakly communicating* [Puterman (1994), Section 9.5].

(d) The model is called *recurrent* if, under any stationary strategy π^s, the controlled process X_t is positive recurrent (without transient states).

If the MDP is finite and unichain (or communicating), then one can put $\rho(x) \equiv \rho$ in Definition 4.1 and in equations (4.2): see [Puterman (1994), Thm. 8.4.3] and [Kallenberg (2010), Subsection 6.3.1].

For general MDPs, sufficient conditions for the existence of canonical triplets (including the case $\rho(x) = const$) can be found in [Hernández-Lerma and Lasserre (1996a), Section 5.5], [Hernández-Lerma and Lasserre (1999), Section 10.3] and [Puterman (1994), Sections 8.4, 9.1].

Note also Remark 2.1 about Markov and semi-Markov strategies which concerns the (uniform) AC-optimality.

Definition 4.3. A 0-discount optimal strategy (see Definition 3.9) is called *bias optimal* [Puterman (1994), Section 5.4.3].

The following condition plays a big role in MDP with average cost.

Condition 4.1.

(a) The spaces $\mathbf{X}$ and $\mathbf{A}$ are topological Borel and $\mathbf{A}$ is compact.

(b) $v_{\hat{x}}^{\hat{\pi}} < \infty$ for some strategy $\hat{\pi}$ and some initial state $\hat{x}$.

(c) The cost function $c(\cdot,\cdot)$ is non-negative and lower semi-continuous; moreover, it is *inf-compact*; that is, the set $\{(x,a) \in \mathbf{X} \times \mathbf{A} : \ c(x,a) \le r\}$ is compact for every number $r \in \mathbb{R}$.

(d) The transition probability p is a weakly continuous stochastic kernel.

(e) $1 + \int_{\mathbf{X}} \min_{a \in \mathbf{A}} c(y,a) p(dy|x,a) \le k \cdot [1 + c(x,a)]$ for some constant k, for all $(x,a) \in \mathbf{X} \times \mathbf{A}$.

Note that, if function $c(\cdot,\cdot)$ is inf-compact, then it is *strictly unbounded*, that is, there is a nondecreasing sequence of compact sets $B_n \subset \mathbf{X} \times \mathbf{A}$ such that $B_n \uparrow \mathbf{X} \times \mathbf{A}$ and $\liminf_{n\to\infty}\{c(x,a) : \ (x,a) \notin B_n\} = \infty$: see Remark 11.4.2 of [Hernández-Lerma and Lasserre (1999)].

Suppose the model is finite. It is known that a bias optimal strategy does exist. Clearly, any Blackwell optimal strategy is also bias optimal. Any bias optimal strategy is -1-discount optimal, and -1-discount optimality is equivalent to the AC-optimality [Kallenberg (2010), Thm. 7.1]. But there can exist AC-optimal strategies which are not bias optimal (see Subsection 4.3.14).

Remark 4.3. Suppose the state space $\mathbf{X}$ is denumerable (or finite), π^s is a stationary strategy such that the controlled process X_t under π^s is positive-recurrent (i.e., $\mathbf{X}$ is the positive-recurrent closed communicating class). Then, for any fixed initial distribution P_0, the limit $\lim_{T\to\infty}\frac{1}{T}\sum_{t=1}^{T}\mathsf{P}_{\mathsf{P}_0}^{\pi^s}(X_{t-1}=j)$ exists and equals $\hat{\eta}^{\pi^s}(j)$, the stationary probability of the state j [Kemeny*et al.* (1976), Prop. 6.5 and 6.25]. Moreover, $\hat{\eta}^{\pi^s}(j)=\frac{1}{M_{jj}}$, where $M_{jj}>1$ is the mean recurrence time from state j to j. Recall that the strictly positive row vector $\{\hat{\eta}^{\pi^s}(j)\}_{j\in\mathbf{X}}$ is the unique solution to equations $\hat{\eta}^{\pi^s}\mathsf{P}=\hat{\eta}^{\pi^s}$, $\sum_{j\in\mathbf{X}}\hat{\eta}^{\pi^s}(j)=1$, where P is the transition probability of the Markov chain X_t: $\mathsf{P}(i,j)=\mathsf{P}_{\mathsf{P}_0}^{\pi^s}(X_t=j|X_{t-1}=i)=\int_{\mathbf{A}}p(j|i,a)\pi^s(da|i)$. If the cost function $c(\cdot,\cdot)$ is bounded, then, by the Dominated Convergence Theorem,

$$\begin{aligned}
&\sum_{j\in\mathbf{X}}\int_{\mathbf{A}}c(j,a)\pi^s(da|j)\hat{\eta}^{\pi^s}(j)\\
&=\lim_{T\to\infty}\frac{1}{T}\sum_{t=1}^{T}\sum_{j\in\mathbf{X}}\int_{\mathbf{A}}c(j,a)\pi^s(da|j)\mathsf{P}_{\mathsf{P}_0}^{\pi^s}(X_{t-1}=j)\\
&=\lim_{T\to\infty}\frac{1}{T}\sum_{t=1}^{T}\mathsf{E}_{\mathsf{P}_0}^{\pi^s}[c(X_{t-1},A_t)]=v_{\mathsf{P}_0}^{\pi^s}.
\end{aligned}\tag{4.3}$$

One can easily generalize this equation to bounded from below cost functions $c(\cdot,\cdot)$: introduce $c^N(x,a):=\min\{c(x,a),N\}$ and pass to the limit as $N\to\infty$ in equation (4.3) for $c^N(\cdot,\cdot)$. Similar reasoning applies to bounded from above cost functions $c(\cdot,\cdot)$.

This observation leads to the so called *convex analytic approach*, when the following *Primal (Linear) Program* in the space of finite measures η on $\mathcal{B}(\mathbf{X}\times\mathbf{A})$ comes to the stage (cf (2.12)):

$$\begin{aligned}
&\int_{\mathbf{X}\times\mathbf{A}}c(x,a)d\eta(x,a)\to\inf,\\
\eta(\Gamma\times\mathbf{A})&=\int_{\mathbf{X}\times\mathbf{A}}p(\Gamma|x,a)d\eta(x,a)\text{ for all }\Gamma\in\mathcal{B}(\mathbf{X}),\\
\eta(\mathbf{X}\times\mathbf{A})&=1.
\end{aligned}\tag{4.4}$$

The dual linear program in the space of bounded functions $h(\cdot)$ on $\mathbf{X}$ looks as follows

$$\rho\to\sup_{\rho,h}\tag{4.5}$$

$$\rho+h(x)\le c(x,a)+\int_{\mathbf{X}}h(y)p(dy|x,a),\quad x\in\mathbf{X},\ a\in\mathbf{A}$$

(see [Hernández-Lerma and Lasserre (1999), Section 12.3]; compare with (4.2)). If the space $\mathbf{X}$ is discrete, then without loss of generality one can put $h(\hat{x}) = 0$ for some $\hat{x} \in \mathbf{X}$. If Condition 4.1 is satisfied then the duality gap is absent [Hernández-Lerma and Lasserre (1999), Thm. 12.3.4].

Note that linear program (4.4) is called dual in some books [Kallenberg (1983); Puterman (1994)]. In the case of a discrete model, we have

$$\begin{aligned} \sum_{i\in\mathbf{X}}\sum_{a\in\mathbf{A}} c(i,a)\eta(i,a) \to \inf_{\eta}&, \\ \sum_{a\in\mathbf{A}} \eta(i,a) &= \sum_{j\in\mathbf{X}}\sum_{a\in\mathbf{A}} p(i|j,a)\eta(j,a), \\ \sum_{i\in\mathbf{X}}\sum_{a\in\mathbf{A}} \eta(i,a) &= 1, \qquad \eta(i,a) \geq 0. \end{aligned} \tag{4.6}$$

Note that here the first set of equations is linearly dependent:

$$\sum_{i\in\mathbf{X}}\sum_{a\in\mathbf{A}} \eta(i,a) = \sum_{i\in\mathbf{X}}\sum_{j\in\mathbf{X}}\sum_{a\in\mathbf{A}} p(i|j,a)\eta(j,a) = \sum_{j\in\mathbf{X}}\sum_{a\in\mathbf{A}} \eta(j,a).$$

One can call variables $\eta(\cdot,\cdot)$ "limiting state-action frequencies": see Theorem 4.11. They play the same role as occupation measures in Chapters 2,3.

Theorem 4.2. *Suppose the finite model is recurrent, that is,* $\mathbf{X}$ *is the single positive-recurrent class under each stationary strategy. Let* η^* *be the optimal basic feasible solution to (4.6). Then, for each* $i \in \mathbf{X}$*, there exists exactly one value* $a^* =: \varphi^*(i) \in \mathbf{A}$ *with* $\eta^*(i,a^*) > 0$*, and the stationary selector* φ^* *is AC-optimal.*

For the proof see [Kallenberg (1983), Thm. 6.4], [Puterman (1994), Cor. 8.8.3 and 8.8.4]. Existence of feasible solutions to the linear program (4.6) for a finite recurrent MDP (hence, the solvability of (4.6)) was proved, e.g., in [Puterman (1994), Thm. 8.8.2]. The meaning of a feasible solution η to the linear program (4.6) is as follows: if $\pi^{\mathrm{s}}(a|i)$ is the stationary strategy defined by

$$\pi^{\mathrm{s}}(a|i) := \frac{\eta(i,a)}{\sum_{b\in\mathbf{A}} \eta(i,b)}, \tag{4.7}$$

then

$$\eta(i,a) = \lim_{T\to\infty} \frac{1}{T} \sum_{t=1}^{T} \mathsf{P}_{\mathsf{P}_0}^{\pi^{\mathrm{s}}}(X_{t-1} = i, A_t = a) \tag{4.8}$$

for each initial distribution P_0: see Remark 4.3. Inequality $\sum_{b\in\mathbf{A}} \eta(i,b) > 0$ for all $i \in \mathbf{X}$, if η is feasible, was proved in [Puterman (1994), Thm. 8.8.2(b)]. Equations (4.7),(4.8) define a 1-1 correspondence between feasible solutions η to

(4.6) and stationary strategies π^s [Puterman (1994), Thm. 8.8.2]. Moreover, those equations define a 1-1 correspondence between *basic* feasible solutions η to (4.6) and stationary *selectors* [Puterman (1994), Cor. 8.8.3].

Optimal solution η^* to linear program (4.6) leads to an AC-optimal stationary strategy also in case the finite model is unichain: one has to put

$$\pi^s(a|i) := \frac{\eta^*(i,a)}{\sum_{b\in\mathbf{A}}\eta^*(i,b)}, \tag{4.9}$$

whenever the denominator is positive. In all the other states the actions can be arbitrary. Moreover, if $\sum_{b\in\mathbf{A}}\eta^*(i,b) > 0$, one can fix $\pi^s(a|i) = \delta_a(i)$ for an arbitrary a with $\eta^*(i,a) > 0$ leading to an AC-optimal stationary selector [Kallenberg (2010), Thm. 6.9].

4.2 Several Important Statements

Theorem 4.3. [Hernández-Lerma and Lasserre (1996a), Thm. 5.2.4]. *Suppose the cost function $c(\cdot,\cdot)$ is bounded below, and $\langle\rho, h, \varphi^*\rangle$ is a canonical triplet.*

(a) If, for any π and any $x \in \mathbf{X}$,

$$\lim_{T\to\infty} \mathsf{E}_x^\pi[h(X_T)/T] = 0,$$

then φ^* is an AC-optimal strategy and

$$v_x^* = \rho(x) = v_x^{\varphi^*} = \lim_{T\to\infty}\frac{1}{T}\mathsf{E}_x^{\varphi^*}\left[\sum_{t=1}^T c(X_{t-1},A_t)\right], \quad x\in\mathbf{X}$$

(note the standard limit).

(b) If, additionally, for each $x \in \mathbf{X}$

$$\lim_{T\to\infty}\sup_\pi \mathsf{E}_x^\pi[h(X_T)/T] = 0$$

then, for all π, $x \in \mathbf{X}$,

$$v_x^{\varphi^*} \le \liminf_{T\to\infty}\frac{1}{T}\mathsf{E}_x^\pi\left[\sum_{t=1}^T c(X_{t-1},A_t)\right]$$

and

$$\lim_{T\to\infty}\left\{\mathsf{E}_x^{\varphi^*}\left[\sum_{t=1}^T c(X_{t-1},A_t)\right] - \inf_\pi \mathsf{E}_x^\pi\left[\sum_{t=1}^T c(X_{t-1},A_t)\right]\right\}/T = 0.$$

In case (b) the stationary selector φ^* is obviously AC-optimal and, if the cost function $c(\cdot,\cdot)$ is bounded, then it is optimal in problem (4.1) at any initial distribution P_0.

Recall also Theorems 3.6 and 3.7 which are relevant for the average cost problem (4.1).

Definition 4.4. We say that the model is *semi-continuous* if

(a) the state and action spaces $\mathbf{X}, \mathbf{A}$ are topological Borel spaces and $\mathbf{A}$ is compact;

(b) the transition probability $p(dy|x,a)$ is *strongly continuous*, i.e. integral $\int_{\mathbf{X}} u(y)p(dy|x,a)$ is a continuous function on $\mathbf{X}\times\mathbf{A}$ for any measurable bounded function $u(\cdot)$;

(c) the cost function $c(\cdot,\cdot)$ is bounded below and lower semi-continuous.

Note that this definition is slightly different from those introduced in Chapters 1 and 2. To avoid misunderstanding, we always give a reference to a specific condition when talking about semi-continuous models.

If the model is semi-continuous in the sense of Definition 4.4, then, for a fixed $x \in \mathbf{X}$, the function $c(x,\cdot)$ is lower semi-continuous by Lemma 7.13 of [Bertsekas and Shreve (1978)]; hence the set $\{a \in \mathbf{A} : \ c(x,a) \le k\}$ is compact for each $k \in \mathbb{R}$.

Condition 4.2.

(a) The Bellman function for the discounted problem (3.1) $v_x^{*,\beta}$ is finite and, for all β close to 1, the product $(1-\beta)v_z^{*,\beta} \le M < \infty$ is uniformly bounded from above for a particular state $z \in \mathbf{X}$.

(b) There is a function $b(x)$ such that the inequality

$$-M \le h^\beta(x) := v_x^{*,\beta} - v_z^{*,\beta} \le b(x) < \infty \qquad (4.10)$$

holds for all $x \in \mathbf{X}$ and all β close to 1.

Condition 4.3. Condition 4.2 holds and, additionally,

(a) For each $x \in \mathbf{X}$ there is $a \in \mathbf{A}$ such that $\sum_{y\in\mathbf{X}} p(y|x,a)b(y) < \infty$.

(b) For all $x \in \mathbf{X}$ and $a \in \mathbf{A}$ $\sum_{y\in\mathbf{X}} p(y|x,a)b(y) < \infty$.

Theorem 4.4. *Suppose Condition 4.2 is satisfied, the model is positive, the state space $\mathbf{X}$ is denumerable (or finite) and the action space $\mathbf{A}$ is finite.*

(a) *[Cavazos-Cadena (1991), Thm. 2.1] Under Condition 4.3(a), there exists a triplet $\langle \rho, h, \varphi^* \rangle$ with ρ being constant, such that*

$$\rho + h(x) \geq c(x, \varphi^*(x)) + \sum_{y \in \mathbf{X}} h(y) p(y|x, \varphi^*(x)), \quad x \in \mathbf{X}. \tag{4.11}$$

See also [Sennott (1989), p. 628].

(b) *[Hernández-Lerma and Lasserre (1996a), Thm. 5.5.4] Under Condition 4.3(b), there exists a solution to the canonical equations (4.2) with $\rho(x) \equiv \rho$.*

In the both cases, the corresponding stationary selector φ^* is AC-optimal in problem (4.1): $v^*_x = v^{\varphi^*}_x = \rho$, and $\rho = \lim_{\beta \to 1-} (1 - \beta) v^{*,\beta}_z$; for each $x \in \mathbf{X}$, $h(x)$ is a limiting point for $h^\beta(x)$ as $\beta \to 1-$.

Theorem 4.4 can be also applied to the models with bounded from below cost function $c(\cdot, \cdot)$: see Remark 4.2.

Theorem 4.5. [Hernández-Lerma and Lasserre (1996a), Thm. 5.4.3]. *Suppose the model is semi-continuous (see Definition 4.4), and Condition 4.2 is satisfied. Then there exists a triplet $\langle \rho, h, \varphi^* \rangle$ with ρ being constant such that*

$$\rho + h(x) \geq \min_{a \in \mathbf{A}} \left\{ c(x, a) + \int_{y \in \mathbf{X}} h(y) p(dy|x, a) \right\} \quad \textit{and}$$
$$\rho + h(x) \geq c(x, \varphi^*(x)) + \int_{y \in \mathbf{X}} h(y) p(dy|x, \varphi^*(x)), \quad x \in \mathbf{X}.$$

The stationary selector φ^* is AC-optimal and $v^*_x = v^{\varphi^*}_x = \rho$ for all $x \in \mathbf{X}$.

Theorem 4.6. [Fainberg (1977), Thm. 3]. *Suppose the state space $\mathbf{X}$ is finite, the model is semi-continuous (see Definition 4.4), and there is a strategy optimal in problem (4.1) under each initial state. Then there exists an AC-optimal stationary selector.*

Theorem 4.7. *Suppose Condition 4.1 is satisfied. Then the following statements are valid.*

(a) *[Hernández-Lerma and Lasserre (1999), Thm. 12.3.3(a)]. There exists an optimal solution η^* to the linear program (4.4) and its minimal value $\int_{\mathbf{X} \times \mathbf{A}} c(x, a) d\eta^*(y, a)$ coincides with $\inf_{\mathsf{P}_0, \pi} v^\pi_{\mathsf{P}_0}$.*

(b) *[Hernández-Lerma and Lasserre (1999), Lemma 11.4.5(a)]. There is a stationary strategy π^s and a corresponding invariant probability measure $\hat{\eta}$ on $\mathbf{X}$ such that*

$$\hat{\eta}(\Gamma) = \int_{\mathbf{X}} \int_{\mathbf{A}} p(\Gamma|x, a) \pi^s(da|x) \hat{\eta}(dx) \textit{ for all } \Gamma \in \mathcal{B}(\mathbf{X}),$$

the pair $(\hat{\eta}, \pi^s)$ *solves problem*

$$v^{\pi}_{P_0} = \limsup_{T\to\infty} \frac{1}{T} \mathsf{E}^{\pi}_{P_0} \left[\sum_{t=1}^{T} c(X_{t-1}, A_t) \right] \to \inf_{P_0,\pi}, \tag{4.12}$$

and the measure

$$\eta^*(\Gamma^{\mathbf{X}} \times \Gamma^{\mathbf{A}}) := \int_{\Gamma^{\mathbf{X}}} \pi^s(\Gamma^{\mathbf{A}}|x)\hat{\eta}(dx)$$

on $\mathbf{X}\times\mathbf{A}$ *solves the linear program (4.4). Note that* $\eta^*(\Gamma^{\mathbf{X}}\times\mathbf{A}) = \hat{\eta}(\Gamma^{\mathbf{X}})$.

The very last statement follows from the following reasoning. The measure η^* is feasible in (4.4). If $P_0(dx) = \hat{\eta}(dx) = \eta^*(dx \times \mathbf{A})$, then, under the strategy π^s, we have

$$\mathsf{E}^{\pi^s}_{\hat{\eta}}[c(X_{t-1}, A_t)] = \int_{\mathbf{X}\times\mathbf{A}} c(x, a) d\eta^*(x, a), \quad t = 1, 2, \ldots$$

because $\hat{\eta}$ is the invariant measure for the controlled process X_t under the strategy π^s. Therefore,

$$v^{\pi^s}_{\hat{\eta}} = \int_{\mathbf{X}\times\mathbf{A}} c(x, a) d\eta^*(x, a) = \inf_{P_0,\pi} v^{\pi}_{P_0};$$

hence η^* is optimal in (4.4) according to (a).

Let us emphasize the following ergodicity property.

Theorem 4.8. *Suppose* π^s *is a stationary strategy such that the controlled process* X_t *is positive Harris recurrent, that is,*

- *there exists a non-zero* σ*-finite measure* μ *on* $\mathcal{B}(\mathbf{X})$ *such that* $P^{\pi^s}_x(\inf\{t \geq 1 : X_t \in B\} < \infty) = 1$ *for all* $x \in \mathbf{X}$ *whenever* $\mu(B) > 0$,
- *and there exists the unique invariant probability measure (i.e., the stationary probability)* $\hat{\eta}^{\pi^s}$ *on* $\mathcal{B}(\mathbf{X})$ *as in Theorem 4.7(b).*

Then, for each integrable w.r.t. $\hat{\eta}^{\pi^s}$ *function* $u(\cdot)$ *on* $\mathbf{X}$ *and for each initial distribution* P_0,

$$\lim_{T\to\infty} \frac{1}{T} \sum_{t=1}^{T} u(X_{t-1}) = \int_{\mathbf{X}} u(x)\hat{\eta}^{\pi^s}(dx) \quad P^{\pi^s}_{P_0}\text{-a.s.}$$

Moreover,

$$\int_{\mathbf{X}} u(x)\hat{\eta}^{\pi^s}(dx) = \lim_{T\to\infty} \frac{1}{T} \mathsf{E}^{\pi^s}_x \left[\sum_{t=1}^{T} u(X_{t-1}) \right] \text{ for } \mu\text{-almost all } x \in \mathbf{X}.$$

For the proof see [Hernández-Lerma and Lasserre (1999), Thm. 11.2.1].

4.3 Examples

In this section, examples and counterexamples illustrate the theory which is well developed in [Altman (1999); Dynkin and Yushkevich (1979); Hernández-Lerma and Lasserre (1996a, 1999); Kallenberg (2010); Piunovskiy (1997); Puterman (1994)]. Theorems 4.1 and 4.3 describe the situations when the canonical stationary selector is AC-optimal; linear programming approach is often successful by Theorem 4.7; strategy and value iterations can provide the AC-optimal solution, etc. Most of the examples in this section demonstrate that all the conditions in the aforementioned and other theorems are important.

In Subsection 4.3.1 the performance functional (4.1) is discussed. Subsections 4.3.2–4.3.5 are about canonical triplets and canonical equations. Those equations can have no solutions (even if the model is semi-continuous) or multiple solutions, not all canonical selectors are AC-optimal, a canonical triplet can fail to satisfy the canonical equations, a non-canonical strategy can be AC-optimal and so on.

If the model is semi-continuous and/or unichain and/or communicating etc, then one can expect that an AC-optimal or AC-ε-optimal stationary (or Markov) strategy/selector exists. Counterexamples in Subsections 4.3.6–4.3.13 show that, if this or that mild condition is not satisfied, then these conjectures may be false.

The common numerical methods for solving problem (4.1) like strategy and value iterations and linear programming are discussed in Subsections 4.3.14–4.3.20. Again, the corresponding counterexamples show that, if some conditions are violated, then those methods may be unsuccessful.

A number of different definitions of optimality, strongly connected with the expected average cost (4.1), are considered in Subsections 4.3.21–4.3.27. Here the target is to show that all those types of the performance functional are different and can lead to different optimal solutions.

More specific examples are given in Subsections 4.3.28–4.3.33. Examples on the discounted MDPs from Subsection 3.3.24 are also relevant for the average cost problem (4.1).

4.3.1 *Why* lim sup*?*

As mentioned in [Puterman (1994), Section 8.1], one can also consider the following expected average cost criterion:

$$v^{\pi}_{\mathsf{P}_0} = \liminf_{T\to\infty} \frac{1}{T} \mathsf{E}^{\pi}_{\mathsf{P}_0}\left[\sum_{t=1}^{T} c(X_{t-1}, A_t)\right] \to \inf_{\pi}. \tag{4.13}$$

Formula (4.1) corresponds to comparing strategies in terms of the worst-case limiting performance, while (4.13) corresponds to comparison in terms of the best-case performance.

From the formal viewpoint, the both these critera define the maps $v_{\mathsf{P}_0} : \mathcal{P}_{\mathsf{P}_0} \to \mathbb{R} \cup \{+\infty\} \cup \{-\infty\}$ from the strategic measures space $\mathcal{P}_{\mathsf{P}_0}$ to (extended) real numbers. When using the conventions about infinity described in

Sections 1.2 and 2.1, the mathematical expectation is a convex (in fact, affine) functional on the strategic measures space. Indeed, for $\mathsf{P}^1, \mathsf{P}^2 \in \mathcal{P}_{\mathsf{P}_0}$, for each $\lambda \in [0,1]$, for each measurable function $W: \mathbf{H} \to \mathbb{R} \cup \{+\infty\} \cup \{-\infty\}$

$$\int_{\mathbf{H}} W^{\pm}(h)[(1-\lambda)\mathsf{P}^1 + \lambda\mathsf{P}^2](dh)$$
$$= (1-\lambda)\int_{\mathbf{H}} W^{\pm}(h)\mathsf{P}^1(dh) + \lambda \int_{\mathbf{H}} W^{\pm}(h)\mathsf{P}^2(dh);$$

hence, in any case (even if we face expressions "$+\infty$" + "$-\infty$"),

$$\int_{\mathbf{H}} W(h)[(1-\lambda)\mathsf{P}^1 + \lambda\mathsf{P}^2](dh) = (1-\lambda)\int_{\mathbf{H}} W(h)\mathsf{P}^1(dh) + \lambda \int_{\mathbf{H}} W(h)\mathsf{P}^2(dh).$$

Recall that $(1-\lambda)\mathsf{P}^1+\lambda\mathsf{P}^2 \in \mathcal{P}_{\mathsf{P}_0}$: see Proposition 2.1. Therefore, all the problems discussed in chapters 1,2 and 3 were convex. In this connection, the performance functional (4.1) is also convex because for $\mathsf{P}^1, \mathsf{P}^2 \in \mathcal{P}_{\mathsf{P}_0}$, for each $\lambda \in [0,1]$, for functions $W_T(h) := \frac{1}{T}\sum_{t=1}^{T} c(x_{t-1}, a_t)$

$$\limsup_{T\to\infty} \int_{\mathbf{H}} W_T(h)[(1-\lambda)\mathsf{P}^1 + \lambda\mathsf{P}^2](dh)$$
$$\leq (1-\lambda)\limsup_{T\to\infty}\int_{\mathbf{H}} W_T(h)\mathsf{P}^1(dh) + \lambda \limsup_{T\to\infty}\int_{\mathbf{H}} W_T(h)\mathsf{P}^2(dh).$$

As is shown in Subsection 4.3.13, the functional (4.1) may be not affine. At the same time, similarly, formula (4.13) defines the functional on $\mathcal{P}_{\mathsf{P}_0}$ which is concave and not necessarily convex, as the presented below example confirms. More about the convexity of performance functionals can be found in [Piunovskiy (1997), Section 2.1.2], see also Remark 4.5. Note that the theory of mathematical programming is better developed for the minimization of convex (rather than concave) functionals over convex sets (see, e.g., [Rockafellar (1987)]).

Below we illustrate that the lower limit leads to the degeneration of many concrete problems.

Example. Let $\mathbf{X} = \{0\}$ (in fact, the controlled process is absent); $\mathbf{A} = \{0, 1\}$. Suppose we are interested in minimizing two performance functionals, with one-step costs ${}^1c(x,a) = a$ and ${}^2c(x,a) = -a$. The objectives are contradictory because when the first objective decreases, the second one is expected to increase; the decision maker is interested in the trade-off between these objectives. This trade-off appears consistent with intuition if we accept formula (4.1). If we use formula (4.13), it is possible to make the both objectives minimal as is sown below.

Let

$$\varphi_t(x) = \begin{cases} 0, & \text{if } 2^{2^m} \leq t < 2^{2^{m+1}}, \quad m = 0, 2, 4, \ldots, \text{ or if } t = 1; \\ 1, & \text{otherwise.} \end{cases}$$

Then, for any $N \geq 1$ and $\varepsilon > 0$, one can find $T = 2^{2^{m+1}} - 1 > N$ with an even value of m, such that

$$\frac{1}{T}\sum_{t=1}^{T} {}^1c(x, \varphi_t(x)) = \frac{1}{2^{2^{m+1}} - 1}\sum_{t=1}^{2^{2^m}-1} {}^1c(x, \varphi_t(x)) \leq \frac{2^{2^m} - 1}{2^{2^{m+1}} - 1} < \varepsilon,$$

because $\lim_{m\to\infty} \frac{2^{2^m} - 1}{2^{2^{m+1}} - 1} = 0$. Therefore,

$$\inf_{T>N} \frac{1}{T}\mathsf{E}_0^{\varphi}\left[\sum_{t=1}^{T} {}^1c(X_{t-1}, A_t)\right] = \inf_{T>N}\left\{\frac{1}{T}\sum_{t=1}^{T} {}^1c(x, \varphi_t(x))\right\} = 0.$$

Similarly, when taking $T = 2^{2^{m+1}} - 1 > N$ with a sufficiently large odd value of m, we obtain

$$\frac{1}{T}\sum_{t=1}^{T} {}^2c(x, \varphi_t(x)) = -1 + \frac{1}{T}\sum_{t=1}^{T}[1 + {}^2c(x, \varphi_t(x))] \leq -1 + \frac{2^{2^m} - 1}{2^{2^{m+1}} - 1} < -1 + \varepsilon,$$

so that

$$\inf_{T>N} \frac{1}{T}\mathsf{E}_0^{\varphi}\left[\sum_{t=1}^{T} {}^2c(X_{t-1}, A_t)\right] = \inf_{T>N}\left\{\frac{1}{T}\sum_{t=1}^{T} {}^2c(x, \varphi_t(x))\right\} = -1.$$

Therefore, if we use formula (4.13), then selector φ provides the minimal possible values for the both objectives:

$${}^1v_0^{\varphi} = 0, \qquad {}^2v_0^{\varphi} = -1.$$

The values of the objectives calculated using formula (4.1) give ${}^1v_0^{\varphi} = 1$ and ${}^2v_0^{\varphi} = 0$. Note also that, for any stationary strategy π^s, we have ${}^1v_0^{\pi^s} + {}^2v_0^{\pi^s} = 0$ in the both cases when we use formula (4.1) or (4.13).

A similar example can be found in [Altman and Shwartz (1991b), Counterexample 2.7].

Now consider the single cost function $c(x, a) = a$ omitting the index 1, the same selector φ and the complimentary selector $\bar{\varphi}_t(x) := 1 - \varphi_t(x)$. Let us accept formula (4.13). The presented above and similar calculations result in $v_0^{\varphi} = v_0^{\bar{\varphi}} = 0$. The strategic measure P_0^{φ} is concentrated on the single trajectory

$$h = (0, \varphi_1(0) = 0, 0, \varphi_2(0) = 0, 0, \varphi_3(0) = 0, 0, \varphi_4(0) = 1, 0, \varphi_5(0) = 1, \ldots),$$

and $\mathsf{P}_0^{\bar\varphi}$ is concentrated on the single trajectory

$$h = (0, \bar\varphi_1(0) = 1, 0, \bar\varphi_2(0) = 1, 0, \bar\varphi_3(0) = 1, 0, \bar\varphi_4(0) = 0, 0, \bar\varphi_5(0) = 0, \ldots).$$

For $\lambda = \frac{1}{2}$, the strategic measure $\mathsf{P} := (1-\lambda)\mathsf{P}_0^{\varphi} + \lambda\mathsf{P}_0^{\bar\varphi} = \frac{1}{2}[\mathsf{P}_0^{\varphi} + \mathsf{P}_0^{\bar\varphi}]$ is such that $\mathsf{P}(X_{t-1} = 0) = 1$ and

$$\mathsf{P}(A_t = 0) = \mathsf{P}(A_t = 1) = \frac{1}{2}, \quad t = 1, 2, \ldots.$$

It is generated by some strategy π because the strategic measures space $\mathcal{P}_0$ is convex. Thus,

$$\begin{aligned} v_0^{\pi} &= \liminf_{T\to\infty} \frac{1}{T}\mathsf{E}_0^{\pi}\left[\sum_{t=1}^{T} c(X_{t-1}, A_t)\right] = \liminf_{T\to\infty} \frac{1}{T}\frac{T}{2} = \frac{1}{2} \\ &> 0 = (1-\lambda)v_0^{\varphi} + \lambda v_0^{\bar\varphi}; \end{aligned}$$

so the mapping $v_0 : \mathcal{P}_0 \to \mathbb{R}$ is not convex. □

4.3.2 *Canonical triplets and canonical equations*

The first example below shows that there may be many different canonical triplets and that not all canonical selectors are optimal.

Example. Let $\mathbf{X} = \{\Delta, 0, 1, 2, \ldots\}$, $\mathbf{A} = \{1, 2\}$, $p(\Delta|\Delta, a) \equiv 1$, $p(\Delta|0, 1) = 1$, $p(1|0, 2) = 1$, $p(i+1|i, a) = 1$ for all $i \geq 1$; other transition probabilities are zero. Let $c(\Delta, a) = c(0, a) = 0$, $c(i, a) = 1$ for all $i \geq 1$ (see Fig. 4.1).

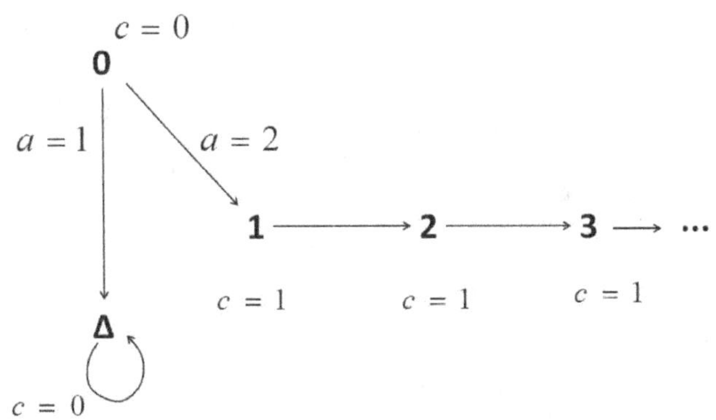

Fig. 4.1 Multiple canonical triplets.

Obviously, $\varphi^*(x) \equiv 1$ is the AC-optimal strategy and action $a = 2$ is not optimal if $x_0 = 0$.

Suppose

$$\left.\begin{aligned} \rho(0) &= \rho(\Delta) = 0, \quad \rho(i) \equiv \hat{\rho} \geq 0 \text{ (any number for } i = 1, 2, \ldots); \\ h(0) &= h(\Delta) = 0, \quad h(1) = \hat{h} \geq 0 \text{ (any number)}; \\ h(i+1) &= \hat{h} - (1 - \hat{\rho})i, \quad i = 1, 2, \ldots; \\ \varphi^*(x) &= 1. \end{aligned}\right\} \quad (4.14)$$

Then $\langle \rho, h, \varphi^* \rangle$ is a canonical triplet. The canonical equations (4.2) are also satisfied. The most natural values, leading to the bounded function $h(\cdot)$, are $\hat{\rho} = 1$, $\hat{h} = 0$. (Note that, if $\hat{\rho} = \hat{h} = 0$, then one can also take $\varphi^*(x) = 2$.)

On the other hand, if we put $\rho(\Delta) = 0$, $h(\Delta) = 0$, $\rho(0) = \rho(1) = \rho(2) = \ldots = \hat{\rho} \leq 0$ (any number), $h(1) = \hat{h} < 0$ (any number), $h(0) = \hat{h} - \hat{\rho}$, $h(i+1) = \hat{h} - (1 - \hat{\rho})i$ for $i = 1, 2, \ldots$, $\varphi^*(x) \equiv 2$, then we again obtain a canonical triplet satisfying equations (4.2).

Theorem 4.3 is applicable to the constructed canonical triplets only if $\lim_{T\to\infty} E_x^\pi[h(X_T)/T] = 0$ for all π and $x \in \mathbf{X}$, that is, when the function $h(\cdot)$ is bounded: $v_x^* = \rho(x) = v_x^{\varphi^*}$, where $\langle \rho, h, \varphi^* \rangle$ is the canonical triplet (4.14) with $\hat{\rho} = 1$, $\hat{h} = 0$.

Theorem 10.3.7 in [Hernández-Lerma and Lasserre (1999)], concerning the uniqueness of the solution to the canonical equations, is not applicable because Assumption 10.2.2 in [Hernández-Lerma and Lasserre (1999)] on the ergodicity of the controlled process is not fulfilled. □

Another example, which confirms that equations (4.2) can hold for ρ, h and φ^*, the stationary selector φ^* being not AC-optimal, can be found in [Robinson (1976), p. 161].

The second example shows that boundedness of function $h(\cdot)$ is important in Theorem 4.1: the canonical triplet can fail to satisfy equations (4.2).

Example. Let $\mathbf{X} = \{s_0, \Delta, 0, 1, 2, \ldots\}$, $\mathbf{A} = \{1, 2, \}$, $p(\Delta|\Delta, a) = p(0|0, a) = p(0|i, a) \equiv 1$ for $i = 1, 2, \ldots$, $p(\Delta|s_0, 1) = 1$, $p(i|s_0, 2) = \left(\frac{1}{2}\right)^i$ for $i = 1, 2, \ldots$, with the other transition probabilities being zero. Let $c(s_0, a) = c(0, a) \equiv 0$, $c(\Delta, a) \equiv 1$, $c(i, a) = 2^i$ for $i = 1, 2, \ldots$ (see Fig. 4.2).

Let $\rho(x) = \begin{cases} 1, & \text{if } x = s_0 \text{ or } x = \Delta; \\ 0, & \text{otherwise,} \end{cases}$

$$h(x) = \begin{cases} 0, & \text{if } x = \Delta \text{ or } x = 0; \\ -1, & \text{if } x = s_0; \\ 2^i, & \text{if } x = i \in \{1, 2, \ldots\}, \end{cases} \qquad \varphi^*(x) \equiv 1.$$

It is easy to check that $\langle \rho, h, \varphi^* \rangle$ is a canonical triplet and φ^* is the AC-optimal strategy. But equations (4.2) are not satisfied: for $x = s_0$,

$$\min\{\rho(\Delta), \ \sum_{i=1}^{\infty} p(i|s_0, 2)\rho(i)\} = \min\{1, \ 0\} = 0 \neq \rho(s_0) = 1.$$

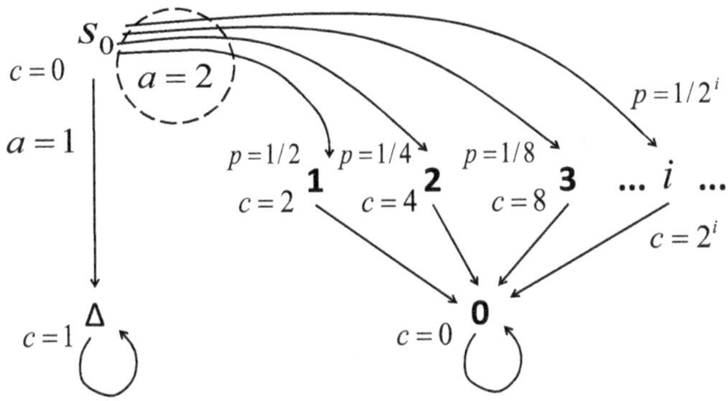

Fig. 4.2 Canonical equations have no solutions.

Moreover, equations (4.2) have no finite solutions at all. Indeed, from the second equation under $x = 0, \Delta$ or $i \in \{1, 2, \dots\}$, we deduce that $h(0)$ and $h(\Delta)$ are arbitrary numbers, $\rho(0) = 0$, $\rho(\Delta) = 1$ and $\rho(i) + h(i) = 2^i + h(0)$. The first equation at $x = i \in \{1, 2, \dots\}$ shows that $\rho(i) = \rho(0) = 0$. Now $h(i) = h(0) + 2^i$, and from the second equation (4.2) under $x = s_0$ we see that $\varphi^*(s_0) = 1$ and $\rho(s_0) + h(s_0) = h(\Delta)$, because $\sum_{i=1}^{\infty} h(i)p(i|s_0, 2) = \infty$. But the first equation implies that $\rho(s_0) = 0$ and $\varphi^*(s_0) = 2$. The resulting contradiction confirms that equations (4.2) have no finite solutions. □

4.3.3 *AC-optimal non-canonical stationary selectors*

Example. Let $\mathbf{X} = \{1, 2\}$, $\mathbf{A} = \{1, 2\}$, $p(1|1, 1) = p(2|1, 1) = 1/2$, $p(2|1, 2) = 1$, $p(2|2, a) \equiv 1$, with all other transition probabilities being zero. Let $c(1, 1) = -5$, $c(1, 2) = -10$, $c(2, a) \equiv 1$ (see Fig. 4.3). A similar example was presented in [Puterman (1994), Ex. 8.4.3]. This model is finite and unichain. (See Definition 4.2.) Thus, one can put $\rho(x) \equiv \rho$ in Definition 4.1 and in equations (4.2) [Puterman (1994), Thm. 8.4.3]:

$$\rho + h(1) = \min\left\{-5 + \frac{1}{2}h(1) + \frac{1}{2}h(2), \quad -10 + h(2)\right\};$$
$$\rho + h(2) = 1 + h(2).$$

We see that $\rho = 1$, and one can put $h(1) = 0$ as usual. (See Remark 4.1.) Now it is easy to see that $h(2) = 12$ and $\varphi^*(1) = 1$. The actions $\varphi^*(2)$ in state $x = 2$ play no role. The triplet $\langle \rho, h, \varphi^* \rangle$ is canonical according to Theorem 4.1. Thus, the stationary selector $\varphi^*(x) \equiv 1$ is canonical and hence AC-optimal, according to Theorem 4.3; the value of the infimum in (4.1) equals $\rho = 1$.

On the other hand, the stationary selector $\varphi^2(x) \equiv 2$ (as well as any other strategy) is also AC-optimal because, for any initial distribution P_0, $v_{\mathsf{P}_0}^{\varphi^2} = 1$ (the process will be ultimately absorbed at state 2). But this selector is not canonical. □

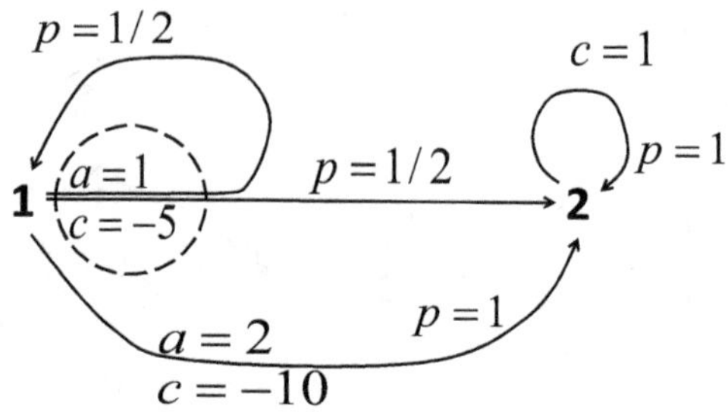

Fig. 4.3 AC-optimal non-canonical selector.

Example. If the model is not finite, then equations (4.2) may have no solutions. As an example, let $\mathbf{X} = \{1, 2\}$, $\mathbf{A} = [0, 1]$, $p(2|1, a) = 1 - p(1|1, a) = a^2$, $p(2|2, a) \equiv 1$, $c(1, a) = -a$, $c(2, a) \equiv 1$ (see Fig. 4.4). This model is semi-continuous in the sense of Definition 4.4. (We fix the discrete topology in $\mathbf{X}$ and the Euclidean topology in $\mathbf{A}$.)

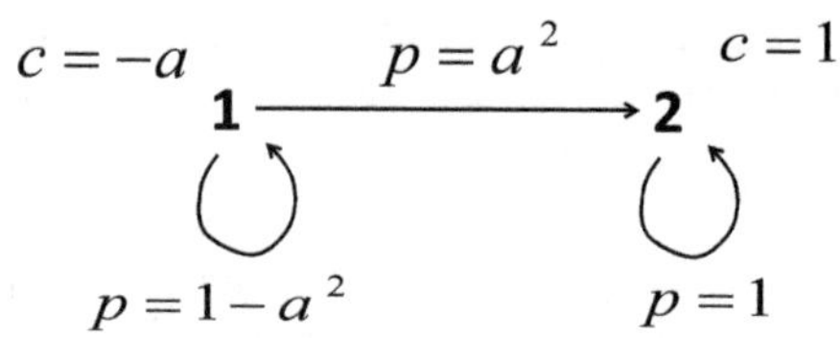

Fig. 4.4 No canonical selectors in a semi-continuous model.

Let us rewrite equations (4.2) as follows:

$$\begin{aligned}\rho(1) &= \inf_{a \in \mathbf{A}} \{a^2 \rho(2) + (1 - a^2)\rho(1)\};\\ \rho(2) &= \rho(2);\\ \rho(1) + h(1) &= \inf_{a \in \mathbf{A}} \{-a + a^2 h(2) + (1 - a^2) h(1)\};\\ \rho(2) + h(2) &= 1 + h(2)\end{aligned}$$

and show that canonical selector does not exist.

First of all, it follows from the fourth equation that $\rho(2) = 1$ and, as usual, we put $h(1) = 0$ without loss of generality. From the first equation, which has the form $\inf_{a \in [0,1]}\{a^2[1 - \rho(1)]\} = 0$, we conclude that either $\rho(1) = 1$ or $\rho(1) < 1$ (and the minimum is provided by $a^* = 0$).

The third equation has the form $\rho(1) = \inf_{a\in[0,1]}\{-a + a^2 h(2)\}$, and one can fix the value of $h(2)$ arbitrarily leading to the following.

- If $\frac{1}{2h(2)} \in [0,1]$, then $\rho(1) = -\frac{1}{4h(2)}$ (and the minimum is provided by $a^* = \frac{1}{2h(2)}$).
- if $\frac{1}{2h(2)} \notin [0,1]$ (or $h(2) = 0$), then $\rho(1) = -1 + h(2)$ (and the minimum is provided by $a^* = 1$).

In any case, $\rho(1) < 1$ for all finite values of $h(2)$. Although the presented equations have finite solutions, the common stationary selector φ^* for all of them does not exist. According to Theorem 4.1, there are no canonical triplets in this model. Nevertheless, below we show that the stationary selector $\varphi^*(x) \equiv 0$ is AC-optimal with $v_1^* = v_1^{\varphi^*} = 0$, $v_2^* = v_2^{\varphi^*} = 1$.

It is reasonable to switch to the negative model by putting $\tilde{c}(2,a) \equiv 0 < c(2,a) \equiv 1$; $\tilde{c}(1,a) := c(1,a) = -a$. As the result, we obtain exactly the blackmailer's dilemma investigated in Subsection 2.3.16 under the total cost criterion (see Fig. 2.21). Denote the average cost criterion (4.1) for the modified MDP with the initial state $x_0 = 1$ as V_1^π. Since $v_1^\pi \geq V_1^\pi$ for all strategies π, $v_1^* \geq \inf_\pi V_1^\pi$. Moreover,

$$\inf_\pi V_1^\pi = \inf_\pi \limsup_{T\to\infty} \frac{1}{T} \mathsf{E}_1^\pi \left[\sum_{t=1}^T \tilde{c}(X_{t-1}, A_t)\right]$$

$$\geq \limsup_{T\to\infty} \frac{1}{T} \inf_\pi \mathsf{E}_1^\pi \left[\sum_{t=1}^T \tilde{c}(X_{t-1}, A_t)\right].$$

The negative values

$$\inf_\pi \mathsf{E}_1^\pi \left[\sum_{t=1}^T \tilde{c}(X_{t-1}, A_t)\right] = v^T(1), \quad T = 1, 2, \ldots$$

were calculated in Subsection 2.3.16 when applying the value iteration algorithm:

$$v^1(1) = -1; \quad v^{T+1}(1) = v^T(1) + \frac{1}{4v^T(1)}, \quad T = 1, 2, \ldots.$$

Denote

$$y_T := \frac{2[v^T(1)]^2}{T+1}, \quad T = 1, 2, \ldots.$$

Now

$$y_1 = 1;$$

$$y_{T+1} = \frac{2}{T+2}\left\{[v^T(1)]^2 + \frac{1}{2} + \frac{1}{16[v^T(1)]^2}\right\}$$

$$= y_T \frac{T+1}{T+2} + \frac{1}{T+2} + \frac{1}{4y_T(T+1)(T+2)}, \quad T = 1, 2, \ldots,$$

and we shall prove that $1 \leq y_T \leq \frac{9}{8}$, $T = 1, 2, \ldots$.

These inequalities hold for $T = 1$. Assume that they are valid for some $T \geq 1$ and consider $T+1$:

$$y_{T+1} \geq \frac{T+1}{T+2} + \frac{1}{T+2} + \frac{1}{\frac{9}{2}(T+1)(T+2)} = 1 + \frac{2}{9(T+1)(T+2)} > 1;$$

$$y_{T+1} \leq \frac{9}{8} \cdot \frac{T+1}{T+2} + \frac{1}{T+2} + \frac{1}{4(T+1)(T+2)}$$
$$\leq \frac{1}{T+2}\left\{\frac{9}{8}(T+1) + 1 + \frac{1}{8}\right\} = \frac{9}{8}.$$

(The inequality in the second formula holds because $T \geq 1$.) Now

$$-\sqrt{\frac{y_T(T+1)}{2}} = v^T(1) \in \left[-\frac{3}{4}\sqrt{T+1},\ -\frac{1}{\sqrt{2}}\sqrt{T+1}\right]$$

and

$$\limsup_{T\to\infty} \frac{1}{T} v^T(1) = \lim_{T\to\infty} \frac{1}{T} v^T(1) = 0.$$

Therefore,

$$v_1^* \geq \inf_\pi V_1^\pi \geq \limsup_{T\to\infty} \frac{1}{T} v^T(1) = 0.$$

Since $v_1^{\varphi^*} = 0$, we conclude that $v_1^* = v_1^{\varphi^*} = 0$. Equalities $v_2^* = v_2^{\varphi^*} = 1$ are obvious.

All the presented reasoning also hold for any other positive value of the constant $c(2,a)$; in this case $v_2^* = v_2^{\varphi^*} = c(2,a)$. In the framework of the blackmailer's dilemma, one can say that $c(2,a) > 0$ is the cost of being in prison for one time interval (e.g. a day), after the victim refuses to yield to the blackmailer's demand and takes him to the police. In such a case, the optimal behaviour is not to blackmail at all.

Consider the discounted version of the same model with the discount factor β close to 1. Standard calculations lead to the following expressions for the Bellman function:

$$v_1^{*,\beta} = \frac{\beta - \sqrt{\beta}}{2\beta(1-\beta)}; \quad v_2^{*,\beta} = \frac{1}{1-\beta},$$

and the optimal action in state 1 $\varphi^{*,\beta}(1) = \frac{1-\beta}{\beta+\sqrt{\beta}}$ is positive and approaches zero as $\beta \to 1-$. The actions in state 2 are immaterial, and we put $\varphi^{*,\beta}(2) \equiv 0$. Note that Blackwell optimal strategy does not exist. We also see that Condition 4.2(b) is violated:

$$v_1^{*,\beta} - v_2^{*,\beta} = -\frac{\beta + \sqrt{\beta}}{2\beta(1-\beta)} \to -\infty \text{ as } \beta \to 1-,$$

so that Theorem 4.5 is not applicable. Finally, note that $\lim_{\beta\to 1-} \varphi^{*,\beta}(x) = \varphi^*(x) \equiv 0$ and

$$\lim_{\beta\to 1-} (1-\beta) v_1^{*,\beta} = v_1^* = 0, \qquad \lim_{\beta\to 1-} (1-\beta) v_2^{*,\beta} = v_2^* = 1.$$

Such equalities hold for many MDP models: see, e.g., Theorems 4.4, 3.7 and [Puterman (1994), Cor. 8.2.5]. □

4.3.4 *Multiple solutions to the canonical equations in finite models*

If the model is finite and, additionally, unichain or weakly communicating (see Definition 4.2), then in equations (4.2) (and in Definition 4.1) one should put $\rho(x) \equiv \rho$; the remainder equation

$$\begin{aligned}\rho + h(x) &= \inf_{a\in\mathbf{A}}\left\{c(x,a) + \sum_{y\in\mathbf{X}} h(y)p(y|x,a)\right\} \\ &= c(x,\varphi^*(x)) + \sum_{y\in\mathbf{X}} h(y)p(y|x,\varphi^*(x)).\end{aligned} \tag{4.15}$$

is solvable according to [Puterman (1994), Section 8.4.2] and [Scweitzer (1987), Thm. 1]. The triplet $\langle\rho, h, \varphi^*\rangle$ is canonical by Theorem 4.1 and the stationary selector φ^* is AC-optimal by Theorem 4.3. Moreover, the value of ρ is unique and, in the unichain case, if $h^1(\cdot)$ and $h^2(\cdot)$ are two solutions, then $h^1(x) - h^2(x) = const.$ At the same time, there can be several different canonical stationary selectors: see Subsection 4.3.14. The following example, first presented in [Scweitzer (1987), Ex. 3] shows that the unichain condition is important.

Example. Let $\mathbf{X} = \{1,2\}$, $\mathbf{A} = \{1,2\}$, $p(1|1,1) = p(2|2,1) = 1$, $p(1|1,2) = p(2|1,2) = p(1|2,2) = p(2|2,2) = 1/2$, $c(1,1) = c(2,1) = 0$, $c(1,2) = 1$, $c(2,2) = 2$ (see Fig. 4.5).

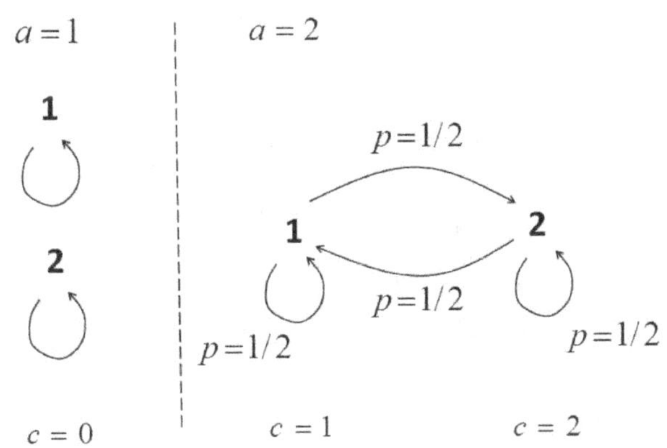

Fig. 4.5 Communicating model which is not unichain.

Equation (4.15) takes the form

$$\rho + h(1) = \min\{h(1),\ 1 + 0.5(h(1) + h(2))\},$$
$$\rho + h(2) = \min\{h(2),\ 2 + 0.5(h(1) + h(2))\},$$

and, without loss of generality, we can put $h(1) = 0$. (See Remark 4.1.)

Assume that $h(2) < -2$. Then from the first equation $\rho = 1 + h(2)/2$ and, from the second equation we see that $h(2) = 1$, which is a contradiction. Similarly, $h(2)$ cannot be greater than 4. But any value $h(2) \in [-2, 4]$ together with $\rho = 0$ solves the presented equations. All the corresponding triplets $\langle \rho = 0, h, \varphi^* \rangle$ with $\varphi^*(x) \equiv 1$ are canonical and the stationary selector φ^* is AC-optimal. □

If the state space is not finite then, even in communicating models, it can happen that equation (4.15) is solvable, but the corresponding stationary selector φ^* is not AC-optimal (see Subsection 4.3.11, Remark 4.4).

4.3.5 *Canonical equations have no solutions: the finite action space*

The following unichain model, based on [Cavazos-Cadena (1991), Section 3] shows that, under conditions 4.2 and 4.3(a), it can happen that the canonical equations (4.2) and (4.15) have no finite solution. The latter assertion was already illustrated in Subsection 4.3.7.

Example. Let $\mathbf{X} = \{0, 1, 2, \ldots\}$, $\mathbf{A} = \{1, 2\}$, $p(x-1|x,a) \equiv 1$ for all $x \geq 1$, $p(1|0,2) = 1$, $p(y|0,1) = q_y$, an arbitrary probability distribution on $\{1, 2, \ldots\}$. Other transition probabilities are zero. Finally, we put $c(x,a) = \begin{cases} 0, & \text{if } x = 0,\ a = 1; \\ 1, & \text{otherwise.} \end{cases}$ See Fig. 4.6.

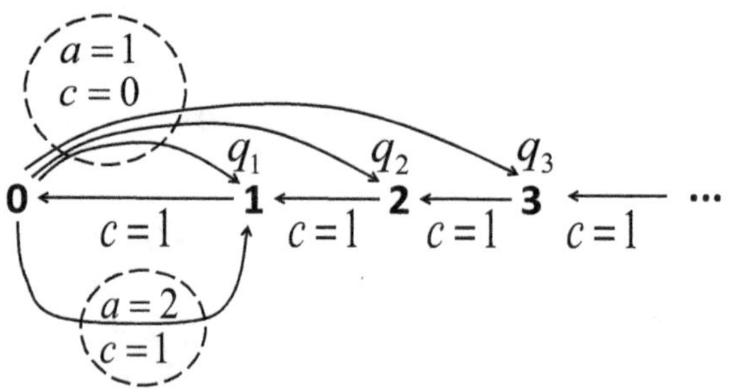

Fig. 4.6 Canonical equations (4.2) have no finite solutions if $\sum_{j=1}^{\infty} jq_j = \infty$.

First, we check conditions 4.2 and 4.3. The discounted optimality equation (3.2) can be explicitly solved: action $\varphi^*(0) = 1$ is optimal for any $\beta \in (0, 1)$ (so that the stationary selector $\varphi^*(x) \equiv 1$ is Blackwell optimal),

$$v_i^{*,\beta} = v_i^{\varphi^*,\beta} = \begin{cases} \dfrac{\beta}{1-\beta} \cdot \dfrac{1 - \sum_{j=1}^{\infty} \beta^j q_j}{1 - \beta \sum_{j=1}^{\infty} \beta^j q_j}, & \text{if } i = 0; \\ \\ \dfrac{1-\beta^i}{1-\beta} + \dfrac{\beta^{i+1}}{1-\beta} \cdot \dfrac{1 - \sum_{j=1}^{\infty} \beta^j q_j}{1 - \beta \sum_{j=1}^{\infty} \beta^j q_j}, & \text{if } i \geq 1, \end{cases}$$

and one can put $z = 0$. Now $(1-\beta)v_0^{*,\beta} \leq 2$ and Condition 4.2(a) is satisfied.

For Condition 4.2(b), note that

$$\begin{aligned} h^\beta(i) & \\ := v_i^{*,\beta} - v_0^{*,\beta} &= \frac{1-\beta^i}{1-\beta} + \frac{\beta^{i+1}-\beta}{1-\beta} \cdot \frac{1 - \sum_{j=1}^{\infty} \beta^j q_j}{1 - \beta \sum_{j=1}^{\infty} \beta^j q_j} \\ &= \frac{1-\beta^i}{1 - \beta \sum_{j=1}^{\infty} \beta^j q_j}. \end{aligned}$$

Applying L'Hôpital's rule at $i \geq 1$, we obtain

$$\lim_{\beta \to 1-} h^\beta(i) = \lim_{\beta \to 1-} \frac{i\beta^{i-1}}{\sum_{j=1}^{\infty} (1+j)\beta^j q_j} = \frac{i}{1 + \sum_{j=1}^{\infty} j q_j} < i < \infty,$$

and inequality (4.10) holds, e.g., for $M = 2$ and $b(i) = i$. Condition 4.2(b) is satisfied.

Condition 4.3(a) is also satisfied: one can take $a = 2$.

Now by Theorem 4.4(a) there exists a triplet $\langle \rho, h, \varphi^* \rangle$ with

$$\begin{aligned} \rho = \lim_{\beta \to 1-} (1-\beta) v_0^{*,\beta} &= \lim_{\beta \to 1-} \frac{\beta - \beta \sum_{j=1}^{\infty} \beta^j q_j}{1 - \beta \sum_{j=1}^{\infty} \beta^j q_j} = \text{(L'Hôpital's rule)} \\ = \lim_{\beta \to 1-} \frac{1 - \sum_{j=1}^{\infty} (j+1)\beta^j q_j}{-\sum_{j=1}^{\infty} (j+1)\beta^j q_j} &= \frac{\sum_{j=1}^{\infty} j q_j}{1 + \sum_{j=1}^{\infty} j q_j} \quad (\text{here } \tfrac{\infty}{\infty} := 1), \end{aligned}$$

$h(i) = \lim_{\beta \to 1-} h^\beta(i) = \frac{i}{1+\sum_{j=1}^{\infty} j q_j}$ and AC-optimal stationary selector $\varphi^*(x)$, satisfying inequality (4.11). Let us check explicitly that inequality and confirm that $\varphi^*(x) \equiv 1$.

$$\begin{aligned} \rho + h(i) &= \frac{\sum_{j=1}^{\infty} j q_j}{1 + \sum_{j=1}^{\infty} j q_j} + \frac{i}{1 + \sum_{j=1}^{\infty} j q_j} = 1 + \frac{i-1}{1 + \sum_{j=1}^{\infty} j q_j} \\ &= c(i, \varphi^*(i)) + \sum_{j=0}^{\infty} h(j) p(j|i, \varphi^*(i)), \quad i \geq 1. \end{aligned}$$

For $i = 0$, on the right-hand side we have

$$c(0, \varphi^*(0)) + \sum_{j=0}^{\infty} h(j)p(j|0, \varphi^*(0)) = \sum_{j=1}^{\infty} h(j)q_j = \frac{\sum_{j=1}^{\infty} jq_j}{1 + \sum_{j=1}^{\infty} jq_j} = \rho + h(0)$$

if $\sum_{j=1}^{\infty} jq_j < \infty$, and $\sum_{j=1}^{\infty} h(j)q_j = 0 < \rho + h(0) = 1$ if $\sum_{j=1}^{\infty} jq_j = \infty$. Actions in sates $i \geq 1$ are immaterial, and if we check $a = 2$ at state 0, we see that

$$\begin{aligned} & c(0,2) + \sum_{j=0}^{\infty} h(j)p(j|0,2) \\ &= 1 + h(1) \\ &= \frac{2 + \sum_{j=1}^{\infty} jq_j}{1 + \sum_{j=1}^{\infty} jq_j} \begin{cases} > \rho = \frac{\sum_{j=1}^{\infty} jq_j}{1 + \sum_{j=1}^{\infty} jq_j}, & \text{if } \sum_{j=1}^{\infty} jq_j < \infty; \\ = 1 = \rho > 0 = \sum_{j=1}^{\infty} h(j)q_j, & \text{if } \sum_{j=1}^{\infty} jq_j = \infty. \end{cases} \end{aligned}$$

Suppose $\sum_{j=1}^{\infty} jq_j < \infty$. Then Condition 4.3(b) is also satisfied because, as was shown above, one can fix $b(i) = i$:

$$\sum_{j=0}^{\infty} p(j|i,a)b(j) = b(i-1) = i - 1 \quad \text{for } i \geq 1;$$

$$\sum_{j=0}^{\infty} p(j|0,1)b(j) = \sum_{j=1}^{\infty} jq_j < \infty; \quad \sum_{j=0}^{\infty} p(j|0,2)b(j) = b(1) = 1.$$

Theorem 4.4(b) implies that the triplet $\langle \rho, h, \varphi^* \rangle$ satisfies canonical equations (4.2). By the way, the latter assertion was explicitly checked above.

Consider now the case $\sum_{j=1}^{\infty} jq_j = \infty$: here $h(x) \equiv 0$, $\rho = 1$, and inequality (4.11) holds strictly at $x = 0$ with $\varphi^*(0) = 1$. Therefore, the triplet $\langle \rho, h, \varphi^* \rangle$ does not satisfy canonical equations (4.2). One can put $\tilde{\varphi}(0) = 2$: as a result we obtain another AC-optimal strategy $\tilde{\varphi}(x) \equiv 2$ for which (4.11) becomes the equality. But $\min_{a \in \{1,2\}} \{c(0,a) + \sum_{j=0}^{\infty} h(j)p(j|0,a)\} = 0$ is provided by $a = 1$ and is smaller than $\rho + h(0) = 1$. Thus, the triplet $\langle \rho, h, \tilde{\varphi} \rangle$ does not satisfy the canonical equations (4.2) either.

Proposition 4.1. *Suppose $\sum_{j=1}^{\infty} jq_j = \infty$. Then*

(a) the canonical equations (4.2) have no finite solution,

(b) Condition 4.3(b) is not satisfied,

(c) under the control strategy $\varphi^*(x) \equiv 1$, there is no stationary distribution.

If $\sum_{j=1}^{\infty} jq_j < \infty$ then there exists a stationary distribution $\eta(x)$ under the control strategy $\varphi^*(x) \equiv 1$, and

$$\sum_{x \in \mathbf{X}} \eta(x)c(x, \varphi^*(x)) = 1 - \eta(0) = \frac{\sum_{j=1}^{\infty} jq_j}{1 + \sum_{j=1}^{\infty} jq_j} = \rho.$$

The proof is given in Appendix A. If $\sum_{j=1}^{\infty} jq_j = \infty$, then Theorem 4.4(b) is not applicable since Condition 4.3(b) is violated. □

4.3.6 *No AC-optimal strategies*

Example. Let $\mathbf{X} = \{1, 1', 2, 2', \ldots\}$, $\mathbf{A} = \{1, 2\}$; for all $i \geq 1$ we put $p(i'|i'a) \equiv 1$, $p(i+1|i,1) \equiv 1$, $p(i'|i,2) \equiv 1$, with all other transition probabilities being zero; $c(i,a) \equiv 1$, $c(i',a) = \frac{1}{i}$ (see Fig. 4.7). A similar example was presented in [Derman (1966), p.1547] and in [Sennott (2002), Ex. 5.1].

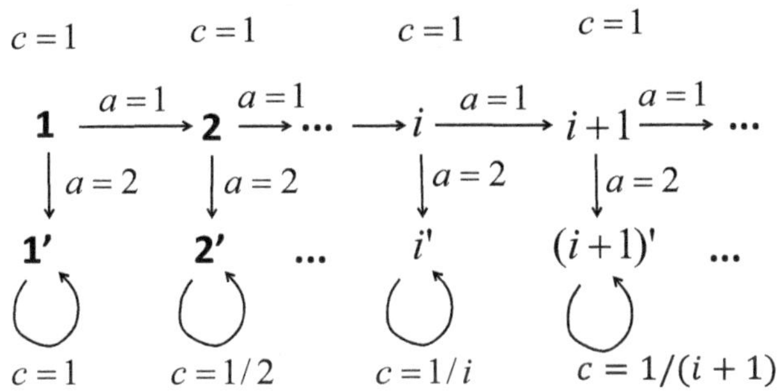

Fig. 4.7 No AC-optimal strategies.

The performance functional (4.1) is non-negative, and, for any $\varepsilon > 0$, the stationary selector $\varphi^\varepsilon(x) = \begin{cases} 1, & \text{if } x = i < \frac{1}{\varepsilon}; \\ 2, & \text{otherwise} \end{cases}$ gives $v_i^{\varphi^\varepsilon} \leq \varepsilon$, so that $v_i^* \equiv 0$. (Obviously, $v_{i'}^* = \frac{1}{i}$.) On the other hand, if $x = i \in \mathbf{X}$ is fixed then, for any control strategy π, $v_i^\pi > 0$ because, if $\mathsf{P}_i^\pi\{\forall t \geq 1\ A_t = 1\} = 1$, then $v_i^\pi = 1$; otherwise, if $\mathsf{P}_i^\pi\{\exists t \geq 1 :\ A_t = 2\} = p > 0$, then $v_i^\pi \geq \frac{p}{i+t-1} > 0$.

The canonical equations (4.2) have no solution with a bounded function $h(\cdot)$ and Theorem 4.1 is not applicable. At the same time, the functions $\rho(i) \equiv 0$, $\rho(i') = \frac{1}{i}$, $h(i) = 1 - i$, $h(i') \equiv 0$ and the stationary selector $\varphi^*(x) \equiv 1$ form a canonical triplet and satisfy equations (4.2). Note that $\mathsf{E}_i^{\varphi^*}[h(X_T)] = h(i+T) = 1 - (i+T)$ and Theorem 4.3 is not applicable either. □

4.3.7 *Only a randomized stationary strategy is AC-optimal*

The following example published in [Derman (1966), p.1547] shows that, even if the model is communicating, then it can happen that solution to equation (4.15) does not exist and only a randomized stationary strategy is AC-optimal.

Example. Let $\mathbf{X} = \{0, 1, 2, \ldots\}$, $\mathbf{A} = \{1, 2\}$, $p(0|0,a) \equiv 0$, $p(i|0,a) \equiv p_i > 0$ for $i \geq 1$ with $\sum_{i=1}^{\infty} p_i = 1$, $p(i|i,1) = p(0|i,2) \equiv 1$ for $i \geq 1$ with the other transition probabilities being zero. We introduce $p_0 = 0$ for convenience. Let $c(i,a) \equiv C_i > 0$, $i \in \mathbf{X}$, where $\{C_i\}_{i=0}^{\infty}$ is a decreasing to zero sequence. See Fig. 4.8. Obviously, this model is communicating (see Definition 4.2).

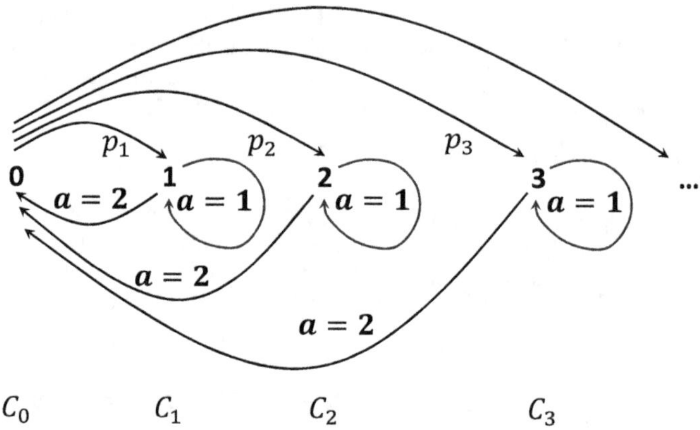

Fig. 4.8 No AC-optimal stationary selectors.

First of all, let us show that, for each stationary selector φ, $v_0^\varphi > 0$.

Denote $S := \{i > 0 : \varphi(i) = 1\}$ and consider the finite horizon T. The function

$$v_i^{\varphi,T} := \mathsf{E}_i^\varphi \left[\sum_{t=1}^{T} c(X_{t-1}, A_t) \right]$$

satisfies equations

$$v_i^{\varphi,0} \equiv 0, \qquad i \in \mathbf{X};$$
$$v_i^{\varphi,T} = \begin{cases} C_i + v_i^{\varphi,T-1}, & \text{if } i \in S; \\ C_i + v_0^{\varphi,T-1}, & \text{if } i \notin S \text{ and } i \neq 0; \end{cases}$$
$$v_0^{\varphi,T} = C_0 + \sum_{i=1}^{\infty} p_i v_i^{\varphi,T-1}, \qquad T \geq 1.$$

Therefore, $v_i^{\varphi,T} = TC_i$ for $i \in S$ and for $T \geq 2$ we have

$$\begin{aligned} v_0^{\varphi,T} &= C_0 + \sum_{i \in S} p_i (T-1) C_i + \sum_{i \notin S} p_i v_i^{\varphi,T-1} \\ &= C_0 + (T-1) \sum_{i \in S} p_i C_i + \sum_{i \notin S} p_i [C_i + v_0^{\varphi,T-2}] \\ &= v_0^{\varphi,T-2} \sum_{i \notin S} p_i + C_0 + (T-1) \sum_{i \in S} p_i C_i + \sum_{i \notin S} p_i C_i. \end{aligned}$$

If $S \neq \emptyset$, then $\sum_{i\in S} p_i > 0$ and

$$\limsup_{T\to\infty} \frac{v_0^{\varphi,T}}{T} \le \limsup_{T\to\infty} \frac{v_0^{\varphi,T-2}}{T-2} \sum_{i\notin S} p_i + \limsup_{T\to\infty} \left[-\frac{2v_0^{\varphi,T-2}}{T(T-2)} \sum_{i\notin S} p_i \right] + \sum_{i\in S} p_i C_i;$$

$$\liminf_{T\to\infty} \frac{v_0^{\varphi,T}}{T} \ge \liminf_{T\to\infty} \frac{v_0^{\varphi,T-2}}{T-2} \sum_{i\notin S} p_i + \liminf_{T\to\infty} \left[-\frac{2v_0^{\varphi,T-2}}{T(T-2)} \sum_{i\notin S} p_i \right] + \sum_{i\in S} p_i C_i.$$

Since the cost function $c(\cdot,\cdot)$ is bounded, the ratio $\frac{v_0^{\varphi,T-2}}{T}$ is bounded and

$$\limsup_{T\to\infty} \left[-\frac{2v_0^{\varphi,T-2}}{T(T-2)} \right] = \liminf_{T\to\infty} \left[-\frac{2v_0^{\varphi,T-2}}{T(T-2)} \right] = 0$$

leading to the standard limit

$$v_0^{\varphi} = \lim_{T\to\infty} \frac{v_0^{\varphi,T}}{T} = \frac{\sum_{i\in S} p_i C_i}{1 - \sum_{i\notin S} p_i} = \frac{\sum_{i\in S} p_i C_i}{\sum_{i\in S} p_i} > 0.$$

If $S = \emptyset$, then $\mathsf{P}_0^{\varphi}(X_0 = X_2 = X_4 = \ldots = 0) = 1$ and the expected cost over each pair of steps $t, t+1$ equals $C_0 + \sum_{i=1}^{\infty} p_i C_i$ leading to $v_0^{\varphi} = \frac{1}{2}\left(C_0 + \sum_{i=1}^{\infty} p_i C_i\right) > 0$.

Now consider the randomized stationary strategy $\pi^{\mathrm{s}}(2|i) = 1 - \pi^{\mathrm{s}}(1|i) = d_i \in (0,1)$ such that $\sum_{i=1}^{\infty} \frac{p_i}{d_i} = \infty$ (e.g., $d_i = p_i$ for $i \ge 1$; the value of d_0 is immaterial). The resulting Markov chain is communicating (hence, irreducible). $\mathsf{P}_i^{\pi^{\mathrm{s}}}(X_0 = X_1 = X_2 = \ldots = 0) = 0$ because $d_i > 0$ and the process will leave the initial state $x_0 = i$ sooner or later. Thus,

$$\mathsf{P}_0^{\pi^{\mathrm{s}}}(X_t \neq 0 \text{ for all } t > 0) = \sum_{i=1}^{\infty} p_i \mathsf{P}_i^{\pi^{\mathrm{s}}}(X_0 = X_1 = X_2 = \ldots = i) = 0,$$

and state 0 is recurrent as well as all the other states of this irreducible Markov chain. However, the mean recurrence time to 0 is $1 + \sum_{i=1}^{\infty} \frac{p_i}{d_i} = \infty$. ($\frac{1}{d_i}$ is the mean time the Markov chain spends in state $i \ge 1$: that random time is the geometric random variable with parameter d_i.) Therefore, the state 0 is null recurrent. Note that the recurrent Markov chain (under the strategy π^{s}) is null recurrent if one of the following equivalent statements holds for all states $i, j \in \mathbf{X}$ [Kemeny*et al.* (1976), Prop. 6.25]:

- the mean recurrence time to i (starting from i) is infinity;
- $\lim_{T\to\infty} \frac{1}{T}\sum_{t=0}^{T} \mathsf{P}_i^{\pi^s}(X_t = j) = 0,\ j \in \mathbf{X}.$

It also follows from Proposition 6.25 of [Kemenyet *al.* (1976)] that the null (or positive) recurrence is a class property, so that all states are null recurrent in this irreducible Markov chain.

Now, for an arbitrarily fixed $j \in \mathbf{X}$,

$$
\begin{aligned}
v_0^{\pi^s} &= \limsup_{T\to\infty} \frac{1}{T}\sum_{i=0}^{\infty}\sum_{t=1}^{T} C_i \mathsf{P}_0^{\pi^s}(X_{t-1} = i)\\
&= \limsup_{T\to\infty}\left[\frac{1}{T}\sum_{i=0}^{j-1}\sum_{t=1}^{T} C_i \mathsf{P}_0^{\pi^s}(X_{t-1} = i) + \frac{1}{T}\sum_{i=j}^{\infty}\sum_{t=1}^{T} C_i \mathsf{P}_0^{\pi^s}(X_{t-1} = i)\right]\\
&\le C_0 \sum_{i=0}^{j-1} \limsup_{T\to\infty}\frac{1}{T}\sum_{t=1}^{T} \mathsf{P}_0^{\pi^s}(X_{t-1} = i)\\
&\quad + C_j \limsup_{T\to\infty}\frac{1}{T}\sum_{t=1}^{T} \mathsf{P}_0^{\pi^s}(X_{t-1} \ge j) \le C_j
\end{aligned}
$$

because $\mathsf{P}_0^{\pi^s}(X_{t-1} \ge j) \le 1$. The change of the order of summation is legal because all the series converge absolutely. Recall also that the sequence $\{C_i\}_{i=0}^{\infty}$ decreases to zero. Since $v_0^{\pi^s} \le C_j$ for all $j = 0, 1, \ldots$, we conclude that $v_0^{\pi^s} = 0$.

For an arbitrarily fixed $j \ge 1$ we have

$$
0 = v_0^{\pi^s} = \limsup_{T\to\infty}\frac{1}{T}\left[C_0 + p_j v_j^{\pi^s,T-1} + \sum_{i=1, i\ne j}^{\infty} p_i v_i^{\pi^s,T-1}\right],
$$

where, as usual, $v_i^{\pi^s,T-1} := \mathsf{E}_i^{\pi^s}\left[\sum_{t=1}^{T-1} c(X_{t-1}, A_t)\right]$. Now, since $C_i > 0$ for all $i \in \mathbf{X}$ and $\sum_{i=1,i\ne j}^{\infty} p_i v_i^{\pi^s,T-1} > 0$ for $T > 1$,

$$
\begin{aligned}
\limsup_{T\to\infty}\frac{1}{T-1} p_j v_j^{\pi^s,T-1} &= \limsup_{T\to\infty}\frac{p_j}{T}\left[v_j^{\pi^s,T-1} + \frac{v_j^{\pi^s,T-1}}{T-1}\right]\\
&\le \limsup_{T\to\infty}\frac{1}{T}\Bigg[C_0 + p_j v_j^{\pi^s,T-1} + \sum_{i=1,i\ne j}^{\infty} p_i v_i^{\pi^s,T-1}\\
&\quad + \frac{p_j v_j^{\pi^s,T-1}}{T-1}\Bigg]\\
&\le v_0^{\pi^s} + \limsup_{T\to\infty}\frac{1}{T}\left[\frac{p_j v_j^{\pi^s,T-1}}{T-1}\right] = v_0^{\pi^s} = 0
\end{aligned}
$$

because the fraction $\frac{v_j^{\pi^s,T-1}}{T-1}$ is bounded. Therefore, $v_j^{\pi^s}=0$ because $p_j>0$ and obviously $v_j^{\pi^s}\geq 0$.

Since there are no AC-optimal stationary selectors, canonical triplets with bounded functions $\rho(\cdot)$ and $h(\cdot)$ do not exist (see Theorem 4.3(a)). The reasoning as at the beginning of Subsection 4.3.4 is not applicable because the model is not finite. Anyway, let us show that equation (4.15) has no finite solutions. If such a solution exists, we as usual can fix $h(0)=0$ (see Remark 4.1) :

$$\rho = C_0 + \sum_{i=1}^{\infty} p_i h(i);$$
$$\rho + h(i) = C_i + \min\{h(i),\ 0)\}, \qquad i \geq 1.$$

Suppose $\rho \neq C_i$ for all $i\geq 1$. Then from the second equation we have

$$h(i)\geq 0, \quad \rho + h(i) = C_i \Longrightarrow h(i) = C_i - \rho, \qquad i\geq 1,$$

and from the first equation

$$\rho = C_0 + \sum_{i=1}^{\infty} p_i C_i - \rho \Longrightarrow \rho = \frac{1}{2}\left[C_0 + \sum_{i=1}^{\infty} p_i C_i\right].$$

Now

$$\sum_{i=1}^{\infty} p_i h(i) \geq 0 \Longrightarrow \sum_{i=1}^{\infty} p_i C_i - \rho = \sum_{i=1}^{\infty} p_i C_i - \frac{C_0}{2} - \frac{1}{2}\sum_{i=1}^{\infty} p_i C_i \geq 0;$$

hence $\sum_{i=1}^{\infty} p_i C_i \geq C_0$ which contradicts the inequalities $C_0 > C_1 > C_2 > \ldots$.

Suppose $\rho = C_j$ for some $j\geq 1$. Then from the second equation we have

$$C_j + h(i) = C_i + \min\{h(i),\ 0\}, \qquad i\geq 1,\ i\neq j.$$

Since $C_i \neq C_j$, $h(i)\geq 0$ and

$$C_j + h(i) = C_i \Longrightarrow h(i) = C_i - C_j \quad \text{for all } i \geq 1,\ i\neq j.$$

We again obtain the contradiction because $C_i < C_j$ for $i>j$. □

4.3.8 *No AC-ε-optimal stationary strategies in a finite state model*

It is known that, in homogeneous discounted models with the bounded one-step cost there exists a stationary uniformly ε-optimal selector [Bertsekas and Shreve (1978), Prop. 9.19]. The situation is different in the case of average cost: it can happen that, for any stationary strategy, the controlled process gets absorbed in a 'bad' state; however one can make that absorption probability arbitrarily small using a non-stationary strategy. The following example illustrating this point is similar to that published in [Dynkin and Yushkevich (1979), Chapter 7, Section 8].

Example. Let $\mathbf{X} = \{0,1\}$; $\mathbf{A} = \{1,2,\ldots\}$; $p(0|0,a) \equiv 1$, $p(1|1,a) = 1 - p(0|1,a) = q_a$, where $\{q_a\}_{a=1}^{\infty}$ are given probabilities such that, for all $a \in \mathbf{A}$, $q_a \in (0,1)$ and $\lim_{a\to\infty} q_a = 1$. We put $c(0,a) \equiv 1$, $c(1,a) \equiv 0$ (see Fig. 4.9).

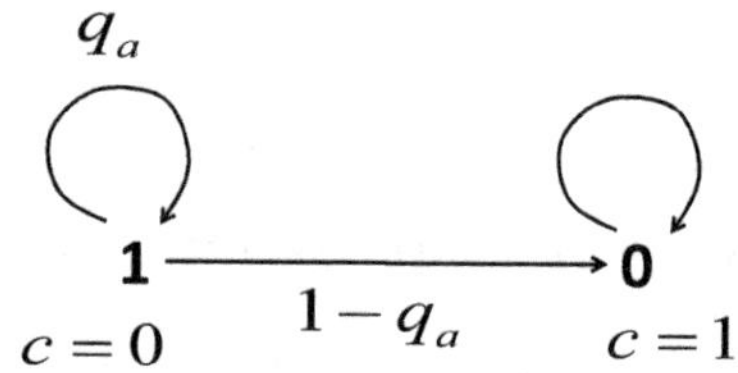

Fig. 4.9 No AC-ε-optimal stationary strategies in a finite state model.

For any stationary strategy π^s, $\sum_{a=1}^{\infty} \pi^s(a|1) q_a < 1$, so, starting from the initial state $x_0 = 1$, the controlled process will ultimately be absorbed at 0 with the absorbtion time having a finite expectation. Thus $v_1^{\pi^s} = 1$. On the other hand, for any number $Q \in (0,1)$, there is a sequence $a_t \to \infty$ for which $\prod_{t=1}^{\infty} q_{a_t} \geq Q$: it is sufficient to take a_t such that $q_{a_t} \geq Q^{\frac{1}{2^t}}$. Now, if $\varphi_t(1) = a_t$, then the controlled process X_t starting from $x_0 = 1$ will never be absorbed at 0 with probability $\prod_{t=1}^{\infty} q_{a_t} \geq Q$, and

$$v_1^{\varphi} \leq \limsup_{T\to\infty} \frac{1}{T} \mathsf{E}_1^{\varphi}\left[\sum_{T=1}^{T} c(X_{t-1}, A_t) \middle| B\right] \mathsf{P}_1^{\varphi}(B)$$
$$+ \limsup_{T\to\infty} \frac{1}{T} \mathsf{E}_1^{\varphi}\left[\sum_{T=1}^{T} c(X_{t-1}, A_t) \middle| B^c\right] \mathsf{P}_1^{\varphi}(B^c) \leq 1 - Q,$$

where $B := \mathbb{I}\{x_0 = x_1 = \cdots = 1\}$. Therefore, $v_1^* = \inf_\pi v_1^\pi = 0$ and no-one stationary strategy is AC-ε-optimal if $\varepsilon < 1$.

It can easily be shown that, for any $\beta \in (0,1)$, $v_0^{*,\beta} = \frac{1}{1-\beta}$ and $v_1^{*,\beta} = 0$, so that inequality (4.10) is violated and Theorem 4.4 is not applicable.

If we add point ∞ to the action space $\mathbf{A}$ (one-point compactification, i.e. $\lim_{a\to\infty} a := \infty$) and put $p(0|0,\infty) = p(1|1,\infty) = 1$, $c(0,\infty) = 1$, $c(1,\infty) = 0$, then the stationary selector $\varphi^*(x) \equiv \infty$ is AC-optimal. In this connection, it is interesting to note that the sequence of stationary selectors $\varphi^i(x) := i \in \{1,2,\ldots\}$, as functions of $x \in \mathbf{X}$, converges to $\varphi^*(x)$. But $v_x^{\varphi^i} \equiv 1$ and $v_x^{\varphi^*} = 1 - x = v_x^*$. Convergence of strategies therefore does not imply the convergence of performance functionals. Note that the model under consideration is semi-continuous. □

4.3.9 No AC-optimal strategies in a finite-state semi-continuous model

The following example, published in [Dynkin and Yushkevich (1979), Chapter 7, Section 8] shows that Condition 4.2 in Theorem 4.5 cannot be omitted.

Example. Let $\mathbf{X} = \{0, 1, \Delta\}$; $\mathbf{A} = [0, \frac{1}{2}]$; $p(0|0,a) = p(\Delta|\Delta,a) \equiv 1$, $p(0|1,a) = a$, $p(\Delta|1,a) = a^2$, $p(1|1,a) = 1 - a - a^2$; $c(0,a) \equiv 0$, $c(1,a) = c(\Delta,a) \equiv 1$ (see Fig. 4.10).

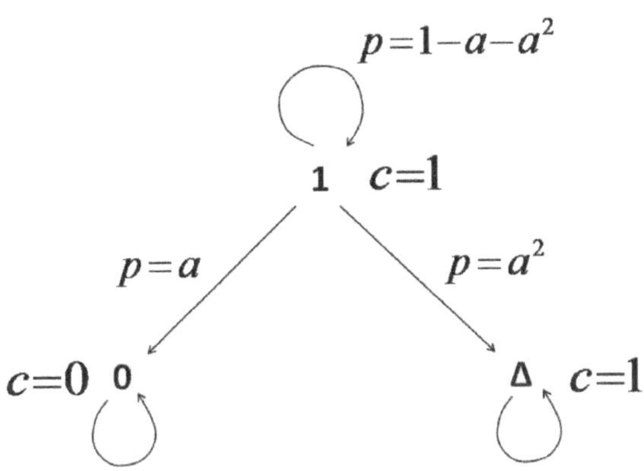

Fig. 4.10 Semi-continuous model.

Suppose $x_0 = 1$. For any control strategy π, either $A_t \equiv 0$ and the controlled process never leaves state 1, or at the first moment τ when $X_\tau \neq 1$, there is a strictly positive chance $A_\tau^2 > 0$ of absorption at state Δ. In all cases, $v_1^\pi > 0$. At the same time, for any stationary selector φ with $\varphi(1) = a$, we have

$$P^\varphi(\tau < \infty,\ X_\tau = 0) = \sum_{i=1}^{\infty}(1 - a - a^2)^{i-1}a = \begin{cases} 0, & \text{if } a = 0; \\ \frac{1}{1+a}, & \text{if } a > 0 \end{cases}$$

and

$$\inf_\pi v_1^\pi = \inf_\varphi[1 - P^\varphi(\tau < \infty,\ X_\tau = 0)] = 0.$$

If we consider the discounted version then the problem is solvable. One can check that, for $\beta \in \left(\frac{4}{5}, 1\right)$, the stationary selector

$$\varphi^*(x) = \frac{5\beta - 4}{2[\beta - 2\sqrt{\beta(1-\beta)}]} - \frac{1}{2}$$

is optimal, and

$$(1-\beta)v_x^{*,\beta} = \begin{cases} 0, & \text{if } c=0; \\ 1, & \text{if } x=\Delta; \\ \frac{4(1-\beta)-2\sqrt{\beta(1-\beta)}}{4-5\beta}, & \text{if } x=1. \end{cases}$$

The statement of Theorem 4.5 is false because Condition 4.2(b) is not satisfied: the functions

$$\frac{1}{1-\beta}\left[\frac{4(1-\beta)-2\sqrt{\beta(1-\beta)}}{4-5\beta}\right] \text{ and } \frac{1}{1-\beta}\left[\frac{4(1-\beta)-2\sqrt{\beta(1-\beta)}}{4-5\beta}-1\right]$$

are not bounded when $\beta \to 1-$, and inequality (4.10) cannot hold for $z = 0, 1$, or Δ.

We now show that canonical equations (4.2) have no finite solution. From the second equation at $x = 0, \Delta$ and 1, we deduce that $\rho(0) = 0$, $\rho(\Delta) = 1$ and

$$\rho(1) = \inf_{a \in \mathbf{A}} \{1 + a^2 h(\Delta) - (a + a^2)h(1)\} \tag{4.16}$$

correspondingly. (We have put $h(0) = 0$ following Remark 4.1.) Keep in mind that the infima in the both equations (4.2) must be provided by the same actions $\varphi^*(x)$. Now the first equation in (4.2) at $x = 1$ gives

$$\inf_{a \in \mathbf{A}} \{a^2 - (a + a^2)\rho(1)\} = 0,$$

meaning that $\rho(1) \leq 0$, because otherwise the function in the parentheses decreases in the neighbourhood of $a = 0$. Hence either $\varphi^*(1) = 0$ and, according to (4.16), $\rho(1) = 1$; or $\varphi^*(1) = \frac{\rho(1)}{1-\rho(1)} < 0$. Both these cases lead to contradictions. □

4.3.10 *Only a non-stationary selector is AC-optimal*

The following example, based on [Fainberg (1977)], shows that the statement of Theorem 4.6 may be false if the model (even being recurrent) is not semi-continuous.

Example. Let $\mathbf{X} = \{1, 2\}$, $\mathbf{A} = \{1, 2, \ldots\}$, $p(2|1, a) = 1 - p(1|1, a) = 1/a$, $p(1|2, a) \equiv 1$, $p(2|2, a) \equiv 0$, $c(1, a) \equiv 0$, $c(2, a) \equiv 1$ (see Fig. 4.11). Below, we fix an arbitrary initial distribution P_0.

For the stationary selector $\varphi^n(x) \equiv n$, the stationary probability of state 2 equals $\hat{\eta}(2) = \frac{1}{n+1}$, so that $v_{\mathsf{P}_0}^{\varphi^n} = \frac{1}{n+1}$: see Remark 4.3. Thus, no-one stationary selector is AC-optimal. Note, the actions in state 2 are immaterial.

On the other hand, the non-stationary Markov selector $\varphi_t^*(x) = t$ is AC-optimal because $v_{\mathsf{P}_0}^{\varphi^*} = 0$ for each initial distribution P_0. To show this, fix an arbitrary $\varepsilon > 0$ and ignore the first several decision epochs which do not influence

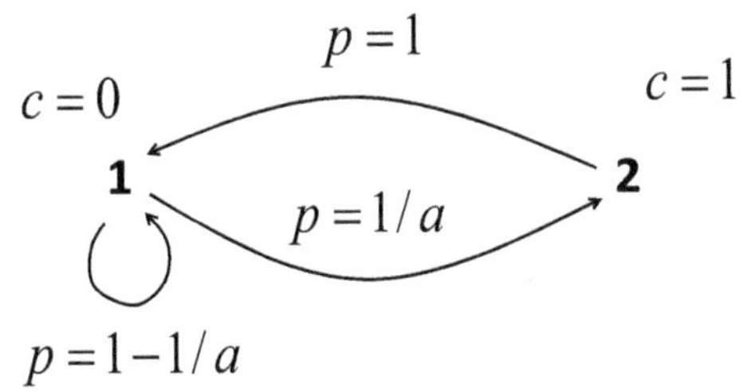

Fig. 4.11 No AC-optimal stationary selectors.

the value of $v^{\varphi^*}_{\mathsf{P}_0}$: without loss of generality, we accept that $\varphi^*_t(x) \geq 1/\varepsilon$ for all $t \geq 1$. Now

$$\mathsf{P}^{\varphi^*}_{\mathsf{P}_0}(X_t = 2) = \mathsf{P}^{\varphi^*}_{\mathsf{P}_0}(X_{t-1} = 1)/\varphi^*_t(1) \leq \varepsilon, \quad t \geq 1,$$

so that $v^{\varphi^*}_{\mathsf{P}_0} \leq \varepsilon$ and, since ε is arbitrarily positive, $v^{\varphi^*}_{\mathsf{P}_0} = 0$.

In the above example, the action space **A** was not compact. The same example can illustrate that requirement (c) in Definition 4.4 is also important. For instance, add action "∞" to **A** (one-point compactification) and put $p(2|1,\infty) = 1 - p(1|1,\infty) = 0$; $c(1,\infty) = c(2,\infty) = 1$. The transition probability is strongly continuous, but the cost function $c(\cdot,\cdot)$ is not lower semi-continuous. The same non-stationary selector φ^* is optimal for all initial distributions P_0, but $v^{\varphi}_{\mathsf{P}_0} > 0$ for any stationary selector φ, as before. (Clearly, $v^{\varphi^\infty}_{\mathsf{P}_0} = 1$ for $\varphi^\infty(x) \equiv \infty$.) In case we fix $c(1,\infty) = 0$, we obtain the semi-continuous model (see Definition 4.4), and, in accordance with Theorem 4.6, the stationary selector $\varphi^\infty(x) \equiv \infty$ is AC-optimal. □

4.3.11 *No AC-optimal stationary strategies in a recurrent model with a finite action space*

This example, first published in [Fisher and Ross (1968)], illustrates that the requirement $-M \leq h^\beta(x)$ in Condition 4.2 is important. Moreover, this model is semi-continuous, recurrent, and aperiodic unichain (see Definition 4.2).

Example. Let $\mathbf{X} = \{0, 1, 1', 2, 2', \ldots\}$, $\mathbf{A} = \{1, 2\}$; for all $i > 0$, $p(i|0,a) = p(i'|0,a) \equiv \frac{3}{2} \cdot \left(\frac{1}{4}\right)^i$; $p(0|i,1) = 1 - p(i'|i,1) = \left(\frac{1}{2}\right)^i$; $p(0|i,2) = p(i+1|i,2) = \frac{1}{2}$; $p(0|i',a) = 1 - p(i'|i',a) \equiv \left(\frac{1}{2}\right)^i$. Other transition probabilities are zero. We put $c(0,a) \equiv 1$, and all the other costs are zero. See Fig. 4.12.

Clearly, for each stationary strategy the controlled process X_t is recurrent.

Proposition 4.2.

(a) For the stationary selectors of the form

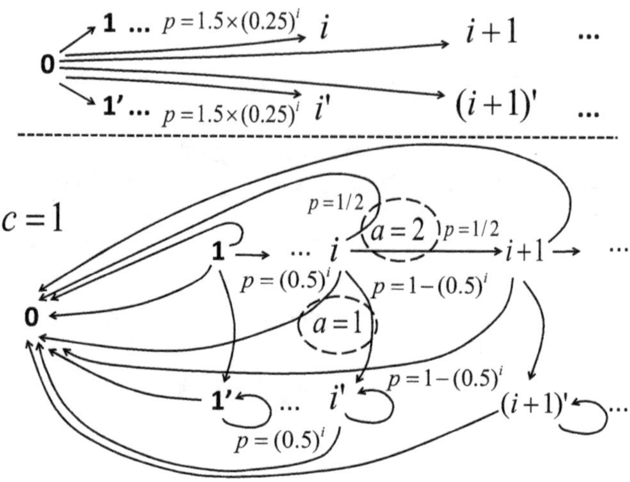

Fig. 4.12 No AC-optimal stationary strategies in a recurrent model. The transition probabilities are shown only to and from states i, i'.

$$\varphi^n(x) = \begin{cases} 2, & \text{if } x = i < n; \\ 1, & \text{otherwise} \end{cases} \qquad n = 1, 2, \ldots, \infty \tag{4.17}$$

the mean recurrence time from state 0 to 0 $M_{00}(\varphi^n) < 5$ *for* $n < \infty$ *and* $M_{00}(\varphi^\infty) = \frac{7}{2}$. *Thus, for any initial distribution* P_0, $v^{\varphi^n}_{\mathsf{P}_0} > \frac{1}{5}$ *for* $n < \infty$ *and* $v^{\varphi^\infty}_{\mathsf{P}_0} = \frac{2}{7}$. *Moreover,* $M_{00}(\varphi^n) \uparrow 5$ *and* $v^{\varphi^n}_{\mathsf{P}_0} \downarrow \frac{1}{5}$ *as* $n \to \infty$.

(b) *For any stationary strategy* π^s *the controlled process is positive recurrent and for each initial distribution* P_0 $v^{\pi^s}_{\mathsf{P}_0} > \frac{1}{5}$.

(c) *For an arbitrary control strategy* π, *for any initial distribution* P_0, *the inequality* $v^\pi_{\mathsf{P}_0} \geq \frac{1}{5}$ *holds. In particular,* $v^\pi_x \geq \frac{1}{5}$ *for all* $x \in \mathbf{X}$.

(d) *There exists a non-stationary selector* $\varphi^*(x)$ *such that* $v^{\varphi^*}_{\mathsf{P}_0} = \frac{1}{5}$ *for an arbitrary initial distribution* P_0. *(Hence selector* φ^* *is AC-optimal.)*

The proof, presented in Appendix A, is based on the following idea. Since $v^{\varphi^n}_{\mathsf{P}_0} \downarrow \frac{1}{5}$ as $n \to \infty$, the selector φ^* is constructed as a combination of the stationary selectors φ^n: $\varphi^*_t(x) = \varphi^{n_k}(x)\mathbb{I}\{T_{k-1} < t \leq T_k\}$. Here the random intervals $(T_0 = 0, T_1]$, $(T_1, T_2], \ldots$ become longer and longer in such a way that only the limiting recurrence time from 0 to 0, equal 5, affects the objective making $v^{\varphi^*}_{\mathsf{P}_0} = \frac{1}{5}$.

Consider the discounted version of this example with $\beta \in (0, 1)$. Obviously, $v^{*,\beta}_0 \leq \frac{1}{1-\beta}$ because $|c(x, a)| \leq 1$. During the proof of Proposition 4.2(c), we established that $v_0^{*,\beta} \geq \frac{1}{1-\beta^5}$ and, for any $i \geq 1$,

$$v_i^{*,\beta}, \quad v_{i'}^{*,\beta} \leq \frac{\beta\left(\frac{1}{2}\right)^i v_0^{*,\beta}}{1-\beta\left[1-\left(\frac{1}{2}\right)^i\right]}.$$

Now it is obvious that for all $x \in \mathbf{X}$ $(1-\beta)v_x^{*,\beta} \leq 1$ because the function $\frac{\beta\left(\frac{1}{2}\right)^i}{1-\beta\left[1-\left(\frac{1}{2}\right)^i\right]}$ decreases with i. Thus, Condition 4.2(a) is satisfied.

We now show that Condition 4.2(b) is violated. Clearly, if we take $z = 0$ then $h^\beta(x) < 0$ (see the proof of Proposition 4.2(c): $v_x^{*,\beta} \leq v_0^{*,\beta}$), but, for $x = i$ or i' with $i \geq 1$,

$$\begin{aligned} h^\beta(x) &\leq \frac{\beta\left(\frac{1}{2}\right)^i v_0^{*,\beta}}{1-\beta\left[1-\left(\frac{1}{2}\right)^i\right]} - \frac{1}{1-\beta^5} \\ &\leq \frac{1}{1-\beta}\left\{\frac{\beta\left(\frac{1}{2}\right)^i}{1-\beta\left[1-\left(\frac{1}{2}\right)^i\right]} - \frac{1}{1+\beta+\cdots+\beta^4}\right\} < 0. \end{aligned} \tag{4.18}$$

For an arbitrary $M > 0$, fix $\beta > 1 - \frac{1}{12M}$ such that $\frac{1}{1+\beta+\cdots+\beta^4} > \frac{1}{6}$ and i such that $\frac{\beta\left(\frac{1}{2}\right)^i}{1-\beta\left[1-\left(\frac{1}{2}\right)^i\right]} < \frac{1}{12}$. Now we have

$$h^\beta(i) < 12M\left\{\frac{1}{12} - \frac{1}{6}\right\} = -M.$$

The left-hand inequality (4.10) is also violated for any other value of z. In such cases (if $z = j$ or j' with $j \geq 1$), $v_z^{*,\beta} \geq \beta^{2+2j}\frac{1}{1-\beta^5}$: see the proof of Proposition 4.2(c). Hence, in a similar manner to (4.18),

$$h^\beta(x) \leq \frac{1}{1-\beta}\left\{\frac{\beta\left(\frac{1}{2}\right)^i}{1-\beta\left[1-\left(\frac{1}{2}\right)^i\right]} - \frac{\beta^{2+2j}}{1+\beta+\cdots+\beta^4}\right\},$$

and the above reasoning shows that a finite value of M, for which $-M \leq h^\beta(x)$, does not exist. Theorems 4.4 and 4.5 are not applicable.

Note that Theorems 3.6 and 3.7 are not applicable here, because inequalities (3.23) and (3.26) fail to hold. □

Simpler examples, showing that stationary selectors are not sufficient if the state space is not finite, are given in [Ross (1970), Section 6.6] and in [Puterman (1994), Ex. 8.10.2]; see also [Sennott (2002), Ex. 5.2]. But those models are not unichain. All of them are very similar to the following MDP:

Example. $\mathbf{X} = \{1, 2, \ldots\}$, $\mathbf{A} = \{1, 2\}$, $p(i+1|i, 1) = p(i|i, 2) \equiv 1$, with all other transition probabilities being zero; $c(i, 1) = 1$, $c(i, 2) = 1/i$ (see Fig. 4.13).

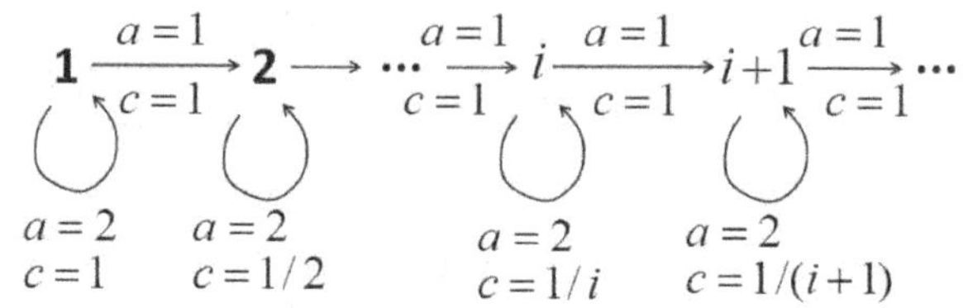

Fig. 4.13 Only a non-stationary selector is AC-optimal.

For the stationary selector $\varphi^1(x) \equiv 1$, $v_{\mathsf{P}_0}^{\varphi^1} = 1$ for each initial distribution P_0. If a stationary selector φ chooses an action $a = 2$ in state $i > x$, then $v_x^{\varphi} = 1/i$. In all cases $v_x^{\varphi} > 0$ and $v_x^{\pi} \geq 0$ for all strategies π. But $v_x^{\varphi^*} = 0$ for the following non-stationary AC-optimal selector: when the process X_t enters state i, φ^* applies action 2 i consecutive times, and then applies action 1. More details in the given above references.

Remark 4.4. One can slightly modify the model and make it communicating: introduce the third action $a = 3$ and put $p(1|1, 3) = 1$, $p(i-1|i, 3) = 1$ for $i \geq 2$; $c(i, 3) = 0$. Now equation (4.15) has a solution $\rho = 0$, $h(i) = 1 - i$, $\varphi^1(i) \equiv 1$, but the stationary selector φ^1 is not AC-optimal. Another solution to equation (4.15) is given by the canonical triplet $\langle \rho \equiv 0, h \equiv 0, \varphi^3 \rangle$ with $\varphi^3(i) \equiv 3$, and the stationary selector φ^3 is obviously AC-optimal.

□

4.3.12 *No AC-ε-optimal stationary strategies in a finite action model*

This example is based on [Ross (1971)], [Ross (1983), Ch. 5, §1]; the model is semi-continuous.

Example. Let $\mathbf{X} = \{0, 1, 1', 2, 2', \ldots\}$, $\mathbf{A} = \{1, 2\}$; for $x = i \geq 1$, $p(i+1|i, 1) = 1$, $p(i'|i, 2) = 1 - p(0|i, 2) = q_i$, where $\{q_i\}_{i=1}^{\infty}$ are given probabilities such that $q_i \in (0, 1)$, $q_i < q_{i+1}$ for all $i \geq 1$, and $\prod_{j=1}^{\infty} q_j = Q > 0$. We put $p((i-1)'|i', a) \equiv 1$ for all $i > 1$, $p(1|1', a) \equiv 1$, and $p(0|0, a) \equiv 1$. Other transition probabilities are zero. Finally, $c(0, a) \equiv 2$, $c(i', a) \equiv 0$ and $c(i, a) \equiv 2$ for all $i \geq 1$. See Fig. 4.14.

Starting from any $x_0 = x \neq 0$, under an arbitrarily fixed stationary strategy π^{s}, the controlled process will $\mathsf{P}_x^{\pi^{\mathrm{s}}}$-a.s. be either absorbed at 0, or it will go to infinity along the trajectory $1 \to 2 \to 3 \to \cdots$ (e.g, if $A_t \equiv 1$). Indeed, the third possibility is to visit state x (which below is denoted as either i or i') infinitely

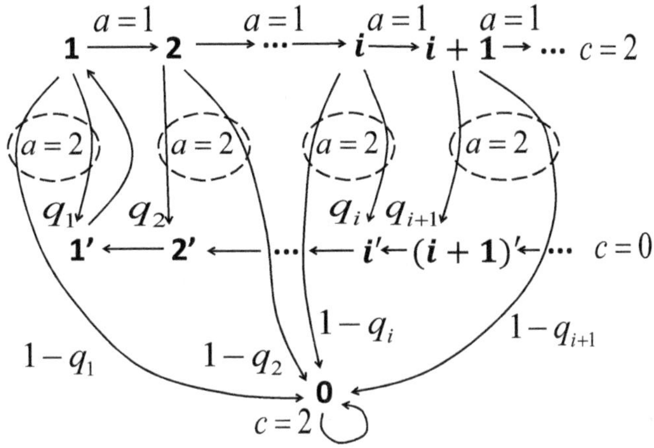

Fig. 4.14 No AC-ε-optimal stationary strategies in a finite action model.

many times through the loops

$$i \to i+1 \to \ldots \to j \to j' \to (j-1)' \to \ldots \to 1' \to 1 \to 2 \to \ldots \to i$$

(if $x = i$) or

$$i' \to (i-1)' \to \ldots \to 1' \to 1 \to 2 \to \ldots \to j \geq i \to j' \to (j-1)' \to \ldots \to i'$$

(if $x = i'$) which is impossible (i.e., occurs with $\mathsf{P}_x^{\pi^s}$-probability zero) because of the following. Leaving aside the trivial case $\pi^s(2|k) = 0$ for all $k = 1, 2, \ldots$, suppose $k \geq 1$ is the minimal number with $\pi^s(2|k) = p > 0$. Then in the above loops $j \geq k$ $\mathsf{P}_x^{\pi^s}$-a.s., and the state k is visited infinitely many times which is impossible: the transition $k \to k+1$ cannot occur infinitely many times because $\pi^s(1|k) = 1 - p < 1$ and the transition $k \to k'$ cannot occur infinitely many times because $q_k < 1$.

Thus, $\lim_{T\to\infty} \frac{1}{T} \sum_{t=1}^{T} c(X_{t-1}, A_t) = 2$ $\mathsf{P}_x^{\pi^s}$-a.s. and, by the Dominated Convergence Theorem,

$$\mathsf{E}_x^{\pi^s}\left[\lim_{T\to\infty} \frac{1}{T} \sum_{t=1}^{T} c(X_{t-1}, A_t)\right] = \lim_{T\to\infty} \frac{1}{T} \mathsf{E}_x^{\pi^s}\left[\sum_{t=1}^{T} c(X_{t-1}, A_t)\right] = v_x^{\pi^s} \equiv 2.$$

On the other hand, consider a control strategy π which initially chooses action 2, and then on the nth return to state 1 (if any), chooses action 1 n times and then chooses action 2. The process starting from $x_0 = i'$ (from $x_0 = i > 0$) will be never absorbed at 0 with probability $\prod_{j=2}^{\infty} q_j$ ($q_i \prod_{j=2}^{\infty} q_j$). In this case, after the initial return to state 1 (in i or $i+1$ steps), the trajectory and associated costs are as follows:

X_t	1	2	2'	1'	1	2	3	3'	2'	1'	1	2	3	...
$c(X_t, A_{t+1})$	2	2	0	0	2	2	2	0	0	0	2	2	2	...

This shows that the average cost equals 1. The complementary probability corresponds to absorption at 0 at a finite time moment, with the average cost being 2. Therefore, for example, if $x_0 = 1$ then, by the Dominated Convergence Theorem,

$$\mathsf{E}_1^\pi\left[\lim_{T\to\infty}\frac{1}{T}\sum_{t=1}^{T} c(X_{t-1}, A_t)\right] = Q + 2(1-Q) = 2 - Q$$

$$= \lim_{T\to\infty}\frac{1}{T}\mathsf{E}_1^\pi\left[\sum_{t=1}^{T} c(X_{t-1}, A_t)\right] = v_1^\pi$$

and no-one stationary strategy is AC-ε-optimal if $\varepsilon < Q$.

Consider the discounted version of this example with discount factor $\beta \in (0,1)$. Clearly, $v_0^{*,\beta} = \frac{2}{1-\beta}$ and, from the discounted optimality equation (3.2), we obtain:

$v_{i'}^{*,\beta} = \beta^i \cdot v_1^{*,\beta}$ for all $i \geq 1$;

$v_i^{*,\beta} \leq 2 + \beta(1-q_i)\frac{2}{1-\beta} + \beta q_i \beta^i v_1^{*,\beta}$ for all $i \geq 1$.

Hence

$$\begin{aligned} v_1^{*,\beta} &\leq 2 + \beta[q_1 v_{1'}^{*,\beta} + (1-q_1)v_0^{*,\beta}] \\ &= 2 + q_1\beta^2 v_1^{*,\beta} + (1-q_1)\frac{2\beta}{1-\beta} \\ \Longrightarrow \quad v_1^{*,\beta} &\leq \frac{2-2\beta q_1}{(1-\beta)(1-q_1\beta^2)} \end{aligned}$$

and

$$v_i^{*,\beta} - v_0^{*,\beta} < 2\left[\frac{q_i\beta^{i+1}(1-\beta q_1)}{(1-\beta)(1-q_1\beta^2)} - \frac{\beta q_i}{1-\beta}\right].$$

Now, for any fixed $M > 0$, one can take $\beta \in (0,1)$ such that $\frac{\beta q_1}{1-\beta} > M \Longrightarrow \frac{\beta q_i}{1-\beta} > M$ for all $i \geq 1$ and afterwards take i such that $\frac{q_i\beta^{i+1}(1-\beta q_1)}{(1-\beta)(1-q_1\beta^2)} < \frac{M}{2}$. Then $v_i^{*,\beta} - v_0^{*,\beta} < -M$. We see that requirements (3.23) and (3.26) are not fulfilled, and Theorems 3.6 and 3.7 are not applicable. Since there are no AC-optimal stationary selectors, conditions in Theorem 4.4 are not satisfied either. □

4.3.13 *No AC-ε-optimal Markov strategies*

According to Remark 2.1, Markov strategies are sufficient for solving almost all optimization problems if the initial distribution is fixed, but an AC-ε-optimal strategy π must satisfy the inequality $v_x^\pi \leq v_x^* + \varepsilon$ for all $x \in \mathbf{X}$ simultaneously. If the state space $\mathbf{X}$ is finite and the cost function $c(\cdot,\cdot)$ is bounded below then, for any $\varepsilon > 0$, there is an AC-ε-optimal Markov selector [Fainberg (1980), Thm. 1]. The following example, based on [Fainberg (1980), Section 5], shows that, if $\mathbf{X}$ is not finite, it can happen that only a semi-Markov strategy is AC-optimal, and no-one Markov strategy is AC-ε-optimal if $\varepsilon < 1/2$, even if the cost function is bounded.

Example. Let $\mathbf{X} = \{0, 1, 1', 2, 2', \ldots\}$, $\mathbf{A} = \{1, 2\}$, $p(3|0,1) = p(3'|0,2) = 1$, $p(0|1,a) = p(2|1,a) = p(0|1',a) = p(2'|1',a) \equiv 1/2$; for $j \geq 2$ $p(j+1|j,a) = p((j+1)'|j',a) \equiv 1$, with all other transition probabilities zero. We put $c(0,a) = c(1,a) = c(1',a) = c(2,a) = c(2',a) = 0$ and

$$c(j,a) = q_j = \begin{cases} +1, & \text{if } 2^{2^m} < j \leq 2^{2^{m+1}}, \quad m = 0, 2, 4, \ldots; \\ -1, & \text{for other } j > 2, \end{cases}$$

$c(j',a) = q_{j'} = -q_j = -c(j,a)$ for all $j > 2$ (see Fig. 4.15).

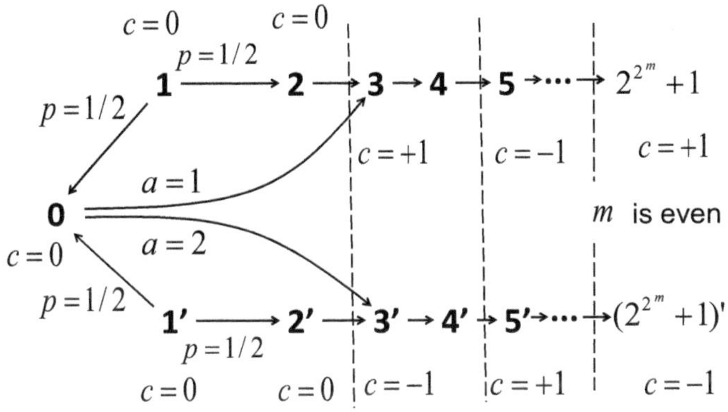

Fig. 4.15 Only a semi-Markov strategy is AC-optimal.

For arbitrary $N \geq 1$ and $\varepsilon > 0$, one can find $T = 2^{2^{m+1}} > N$ with an even value of $m \geq 0$, such that $\frac{2 \cdot 2^{2^m}}{2^{2^{m+1}}} < \varepsilon$. Now

$$\frac{1}{T} \sum_{j=3}^{T} q_j \geq \frac{1}{2^{2^{m+1}}} \left[2^{2^{m+1}} - 2 \cdot 2^{2^m} \right] > 1 - \varepsilon.$$

(We estimated the first 2^{2^m} terms from below: $q_j \geq -1$ for all $j \leq 2^{2^m}$. Other values of q_j, $j = 2^{2^m}+1, \ldots, 2^{2^{m+1}}$ equal 1.) Therefore, $\limsup_{T\to\infty} \frac{1}{T} \sum_{j=3}^{T} q_j = 1$. Similarly, one can show that $\limsup_{T\to\infty} \frac{1}{T} \sum_{j=3}^{T} q_{j'} = 1$.

Now it is clear that, for any strategy π, $v_3^\pi = \limsup_{T\to\infty} \frac{1}{T} \sum_{j=3}^{T+2} q_j = 1 = v_3^*$ and $v_{3'}^\pi = \limsup_{T\to\infty} \frac{1}{T} \sum_{j=3}^{T+2} q_{j'} = 1 = v_{3'}^*$; the same equalities hold for all initial states i, i' with $i \geq 2$.

When starting from $x_0 = 0$, the next states X_t for $t = 1, 2, \ldots$ can be either of the form $\{(2+t)\}_{t=1}^\infty$ or of the form $\{(2+t)'\}_{t=1}^\infty$. If such events occur equiprobably (i.e. $\pi_1(1|0) = \pi_1(2|0) = 1/2$), then $\mathsf{E}_0^\pi\left[c(X_{t-1}, A_t)\right] = \frac{1}{2}q_{t+1} + \frac{1}{2}q_{(t+1)'} = 0$ for all $t \geq 2$, and $v_0^\pi = 0$. Otherwise, $v_0^\pi > 0$: for instance, if $\pi_1(1|0) = \alpha > 1/2$ then

$$\mathsf{E}_0^\pi\left[c(X_{t-1}, A_t)\right] = \alpha q_{t+1} + (1-\alpha)q_{(t+1)'} = (2\alpha - 1)q_{t+1} \quad (\text{for } t \geq 2),$$

and we know that $\limsup_{T\to\infty} \frac{1}{T}\sum_{t=2}^T q_{t+1} = 1$.

Remark 4.5. As was explained in Subsection 4.3.1, the performance functional (4.1) is convex in the space of strategic measures. In the current example it is not affine: for the (1/2, 1/2)-mixture $\hat{\pi}$ of the stationary selectors $\varphi^1(x) \equiv 1$ and $\varphi^2(x) \equiv 2$ (meaning that $\mathsf{P}_0^{\hat{\pi}} = \frac{1}{2}\mathsf{P}_0^{\varphi^1} + \frac{1}{2}\mathsf{P}_0^{\varphi^2}$), we have $v_0^{\hat{\pi}} = 0$ while $v_0^{\varphi^1} = v_0^{\varphi^2} = 1$.

Suppose that $x_0 = 1$ or $1'$. The next states X_t for $t = 2, 3, \ldots$ can be either of the form $\{(1+t)\}_{t=2}^\infty$ or of the form $\{(1+t)'\}_{t=2}^\infty$.. The above reasoning implies that, for any strategy π, $v_1^\pi, v_{1'}^\pi \geq 0$, and, for the semi-Markov strategy $\tilde{\pi}$ satisfying

$$\tilde{\pi}_2(2|x_0 = 1) = \tilde{\pi}(1|x_0 = 1') = 1, \quad \tilde{\pi}_1(1|x_0 = 0) = \tilde{\pi}_1(2|x_0 = 0) = 1/2,$$

the sequences $\{X_t\}_{t=2}^\infty$ of the forms $\{(1+t)\}_{t=2}^\infty$ and $\{(1+t)\}_{t=2}^\infty$ appear equiprobably leading to

$$v_1^{\tilde{\pi}} = v_{1'}^{\tilde{\pi}} = 0 = v_1^* = v_{1'}^*.$$

Equality

$$v_0^{\tilde{\pi}} = 0 = v_0^*$$

has also been justified. Since the values of v_j^π and $v_{j'}^\pi$ at $j \geq 2$ do not depend on the strategy π (and, as was shown, equal 1), the strategy $\tilde{\pi}$ is (uniformly) AC-optimal.

On the other hand, consider an arbitrary Markov strategy π^{m} with $\pi_2^{\mathrm{m}}(1|0) = \alpha \in [0, 1]$:

$$\begin{aligned} v_1^{\pi^{\mathrm{m}}} &= \limsup_{T\to\infty} \frac{1}{T} E_1^{\pi^{\mathrm{m}}}\left[\sum_{t=1}^T c(X_{t-1}, A_t)\right] \\ &= \limsup_{T\to\infty} \frac{1}{T}\sum_{i=3}^T \left[\left(\frac{1}{2} + \frac{1}{2}\alpha\right) q_i + \frac{1}{2}(1-\alpha)q_{i'}\right] \\ &= \limsup_{T\to\infty} \frac{1}{T}\sum_{i=3}^T \alpha q_i = \alpha \end{aligned}$$

because for all $i \geq 3$ $q_i + q_{i'} \equiv 0$ and $q_i - q_{i'} \equiv 2q_i$. Similarly, $v_{1'}^{\pi^m} = (1-\alpha)$. It is clear that one cannot have $v_1^{\pi^{\mathrm{m}}} = v_{1'}^{\pi^{\mathrm{m}}} = 0$; the strategy π^{m} is not (uniformly) AC-ε-optimal if $\varepsilon < 1/2$. But there certainly exists an optimal Markov strategy for a fixed initial distribution P_0: that may be the strategy π^{m} corresponding, according to Lemma 2.1, to the AC-optimal semi-Markov strategy $\tilde{\pi}$

constructed above. Note that different initial distributions can lead to different optimal Markov strategies π^{m}. □

4.3.14 *Strategy iteration in a unichain model*

The basic *strategy iteration algorithm* can be written as follows [Puterman (1994), Section 8.6.1].

1. Set $n = 0$ and select a stationary selector φ^0 arbitrarily enough.
2. Obtain a scalar ρ_n and a bounded function $h_n(\cdot)$ on $\mathbf{X}$ by solving equation
$$\rho_n + h_n(x) = c(x, \varphi^n(x)) + \int_{\mathbf{X}} h_n(y) p(dy|x, \varphi^n(x)).$$
(Clearly, $h_n(\cdot) + const$ is also a solution for any value of $const$; we leave aside the question of the measurability of the function $h_n(\cdot)$.)
3. Choose $\varphi^{n+1} : \mathbf{X} \to \mathbf{A}$ such that
$$c(x, \varphi^{n+1}(x)) + \int_{\mathbf{X}} h_n(y) p(dy|x, \varphi^{n+1}(x))$$
$$= \inf_{a \in \mathbf{A}} \left\{ c(x, a) + \int_{\mathbf{X}} h_n(y) p(dy|x, a) \right\},$$
setting $\varphi^{n+1}(x) = \varphi^n(x)$ whenever possible.
4. If $\varphi^{n+1} = \varphi^n$, then stop and set $\varphi^* = \varphi^n$; $\rho = \rho_n$; $h(\cdot) = h_n(\cdot)$. Otherwise increment n by 1 and return to step 2.

It is known that, in a finite unichain model, this algorithm converges in a finite number of iterations to a solution of the canonical equations (4.2); furthermore, $\langle \rho, h, \varphi^* \rangle$ is a canonical triplet and the stationary selector φ^* is AC-optimal [Puterman (1994), Thm. 8.6.6].

The following example shows that the stationary selector returned by the strategy iteration needs not be bias optimal.

Example. Let $\mathbf{X} = \{1, 2\}$, $\mathbf{A} = \{1, 2\}$, $p(1|x, 1) \equiv 1$, $p(2|1, 2) = p(1|2, 2) = 1$, with all other transition probabilities being zero; $c(1, 1) = -4$, $c(1, 2) = 0$, $c(2, a) \equiv -8$ (see Fig. 4.16).

Denote $\hat{\varphi}(x) \equiv 1$ and $\hat{\hat{\varphi}}(x) \equiv 2$. For the discounted version of this example, under all values $\beta \in (0, 1)$ of the discount factor, we have
$$v_1^{*,\beta} = v_1^{\hat{\varphi},\beta} = \frac{-4}{1-\beta}; \quad v_2^{*,\beta} = v_2^{\hat{\varphi},\beta} = \frac{4\beta - 8}{1-\beta}:$$
it is straightforward to check that these values satisfy optimality equation (3.2).
$$v(1) = \min\{-4 + \beta v(1); \quad 0 + \beta v(2)\};$$
$$v(2) = \min\{-8 + \beta v(1); \quad 0 + \beta v(1)\},$$

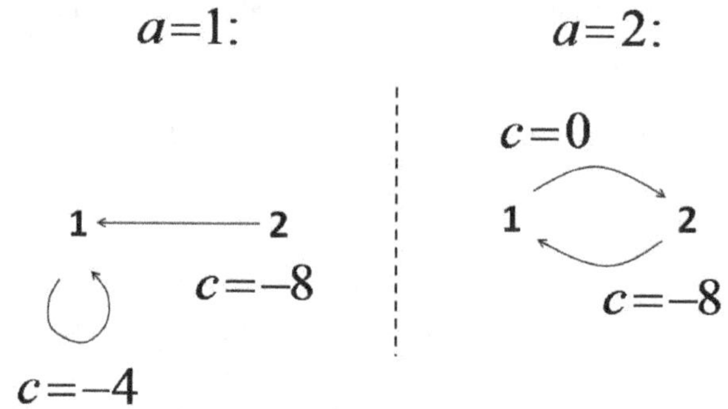

Fig. 4.16 Strategy iteration does not return a bias optimal stationary selector.

and the minima are provided only by action $a = 1$. Therefore, only the stationary selector $\hat{\varphi}$ is Blackwell optimal. It is bias optimal and AC-optimal with $v^*_{1,2} = -4$.

The stationary selector $\hat{\hat{\varphi}}$ obviously is also AC-optimal with $v^{\hat{\hat{\varphi}}}_{1,2} = -4$, but it is not bias optimal (and not Blackwell optimal):

$$v_x^{\hat{\hat{\varphi}},\beta} - v_x^{\hat{\varphi},\beta} = \begin{cases} \dfrac{-8\beta}{1-\beta^2} - \dfrac{-4}{1-\beta} = \dfrac{4}{1+\beta} \geq 2 > 0, & \text{if } x = 1; \\[2ex] \dfrac{-8}{1-\beta^2} - \dfrac{4\beta-8}{1-\beta} = \dfrac{4\beta}{1+\beta} \geq 2 > 0, & \text{if } x = 2. \end{cases}$$

The values of $v_x^{\hat{\hat{\varphi}},\beta}$ come from equations

$$v(1) = \beta v(2) = \beta[-8 + \beta v(1)].$$

At the same time, the strategy iteration presented above, starting from $\varphi^0 = \hat{\hat{\varphi}}$, gives the following:

$$\rho_0 = -4, \quad h_0(x) = \begin{cases} 0, & \text{if } x = 1; \\ -4, & \text{if } x = 2, \end{cases} \quad \varphi^1 = \hat{\hat{\varphi}},$$

and we terminate the algorithm concluding that the stationary selector $\hat{\hat{\varphi}}$ is AC-optimal and $\langle \rho_0, h_0, \varphi^0 = \hat{\hat{\varphi}} \rangle$ is the associated canonical triplet. Note that $\langle \rho_0, h_0, \hat{\varphi} \rangle$ is another canonical triplet. A similar example was considered in [Puterman (1994), Ex. 8.6.2]. □

In finite unichain models, if $\mathbf{X} = \mathbf{X}^r \cup \mathbf{X}^t$, where, under any stationary selector, $\mathbf{X}^r$ is the (same) set of recurrent states, $\mathbf{X}^t$ being the transient subset, then one can apply the strategy iteration algorithm to the recurrent subset $\mathbf{X}^r$. The transient states can be ignored.

There was a conjecture [Hordijk and Puterman (1987), Thm. 4.2] that if, in a finite unichain model, for some constant ρ, function $h(\cdot)$ and stationary selector φ the following equations hold

$$\begin{aligned} & c(x,\varphi(x)) + \sum_{y\in\mathbf{X}} h(y)p(y|x,\varphi(x)) \\ & = \inf_{a\in\mathbf{A}}\left\{c(x,a)+\sum_{y\in\mathbf{X}} h(y)p(y|x,a)\right\} \text{ for all } x\in\mathbf{X}; \\ \rho + h(x) & = \inf_{a\in\mathbf{A}}\left\{c(x,a)+\sum_{y\in\mathbf{X}} h(y)p(y|x,a)\right\} \text{ for all } x\in\mathbf{X}^{r,\varphi}, \end{aligned} \tag{4.19}$$

then the stationary selector φ is AC-optimal. Here $\mathbf{X}^{r,\varphi}$ is the set of recurrent states under the selector φ.

The following example, based on [Golubin (2003), Ex. 1], shows that this statement may be incorrect.

Example. We consider the same model as before (Fig. 4.16), but with the different cost function: $c(1,1)=1$, $c(2,1)=3$, $c(x,2)\equiv 0$ (see Fig. 4.17).

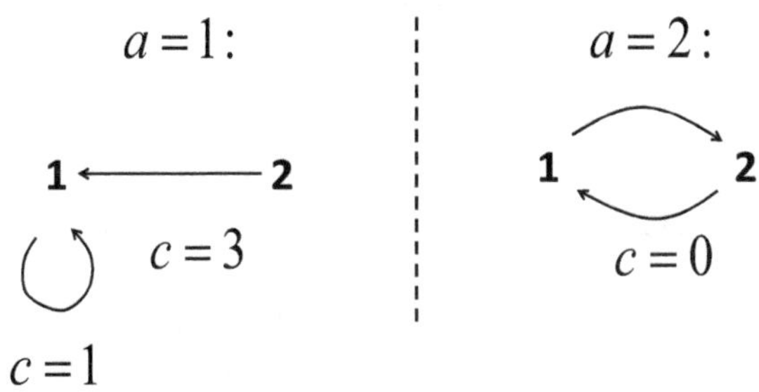

Fig. 4.17 Different recurrent subsets.

Take $\varphi(x)=x$, $\rho=1$ and $h(x)=\begin{cases}0, & \text{if } x=1;\\ 2, & \text{if } x=2.\end{cases}$ Then $\mathbf{X}^{r,\varphi}=\{1\}$ and equations (4.19) are satisfied, but the stationary selector φ is obviously not AC-optimal: $v_x^*=v_x^{\varphi^*}\equiv 0$ for $\varphi^*(x)\equiv 2$ and $v_1^\varphi=v_2^\varphi=1$.

The same example shows that it is insufficient to apply step 3 of the algorithm only to the subset $\mathbf{X}^{r,\varphi^n}$. Indeed, if we take $\varphi^0(x)\equiv 1$, then direct calculations show that $\rho_0 = 1$ and $h_0(x) = \begin{cases} 0, & \text{if } x = 1; \\ 2, & \text{if } x = 2. \end{cases}$ Equation

$$c(x, \varphi^0(x)) + \sum_{y \in \mathbf{X}} h_0(y) p(y|x, \varphi^0(x)) = \inf_{a \in \mathbf{A}} \left\{ c(x, a) + \sum_{y \in \mathbf{X}} h_0(y) p(y|x, a) \right\}$$

holds for all $x \in \mathbf{X}^{r,\varphi^0}$, i.e. for $x = 1$. But the iterations are unfinished because, after further steps, we will obtain $\varphi^1(x) = x$ and $\varphi^2(x) = \varphi^*(x) \equiv 2$ resulting in the final answer $\langle \rho = 0, h(x) \equiv 0, \varphi^*(x) \equiv 2 \rangle$. □

According to [Puterman (1994), Section 9.5.1], if a finite model is communicating, then there exists an AC-optimal stationary selector such that the controlled process has a single communicating class, i.e. it is a unichain Markov process. Therefore, one might conjecture that we can solve such a problem using the unichain strategy iteration algorithm described at the beginning of the current subsection. The following example, based on the same idea as [Puterman (1994), Ex. 9.5.1] shows that this needs not happen.

Example. Let $\mathbf{X} = \{1, 2\}$, $\mathbf{A} = \{1, 2\}$, $p(a|x, a) \equiv 1$, with all other transition probabilities being zero; $c(1, a) = a$, $c(2, 1) = 2$, $c(2, 2) = 0$ (see Fig. 4.18).

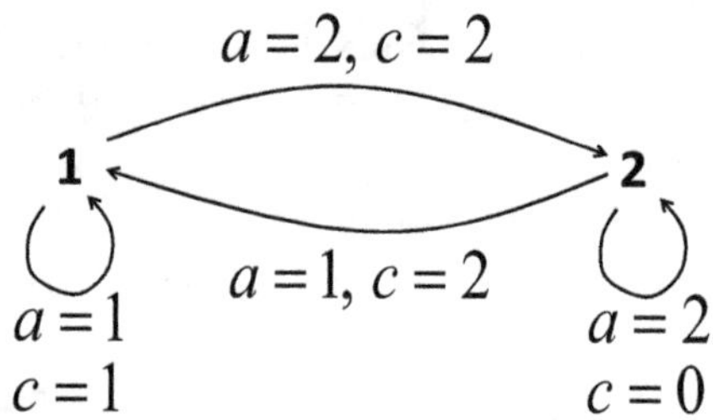

Fig. 4.18 Unichain strategy iteration algorithm is not applicable for communicating models.

We try to apply the unichain strategy iteration algorithm starting from $\varphi^0(x) \equiv 1$. The controlled process under the stationary selector φ^0 is unichain, and one can find $\rho_0 = 1$, $h_0(x) = x - 1$. At step 3, we obtain the improved stationary selector $\varphi^1(x) = x$. On the next iteration, step 2, we have the equations:

$$\rho_1 + h_1(1) = 1 + h_1(1); \qquad \rho_1 + h_1(2) = 0 + h_1(2),$$

which have no solutions. By the way, the stationary selector $\varphi^*(x) \equiv 2$ is AC-optimal and $\langle \rho = 0, h, \varphi^* \rangle$ is the canonical triplet, where $h(x) = \begin{cases} 0, & \text{if } x = 1; \\ -2, & \text{if } x = 2. \end{cases}$

□

There exist special strategy iteration algorithms applicable to communicating and general finite models [Puterman (1994), Sections 9.5 and 9.2], [Kallenberg (2010), Algorithm 5.6]; see also Section 4.3.15.

4.3.15 *General strategy iteration and semi-continuous model*

If the model is multichain (i.e. the requirement in Definition 4.2(a) fails to hold), then the strategy iteration algorithm, in the case of finite or denumerable *ordered* state space $\mathbf{X}$, can be written as follows [Puterman (1994), Section 9.2.1].

1. Set $n = 0$ and select a stationary selector φ^0 arbitrarily enough.
2. Obtain bounded functions ρ_n and h_n on $\mathbf{X}$ by solving the equations

$$\rho_n(x) = \sum_{y \in \mathbf{X}} \rho_n(y)p(y|x, \varphi^n(x));$$

$$\rho_n(x) + h_n(x) = c(x, \varphi^n(x)) + \sum_{y \in \mathbf{X}} h_n(y)p(y|x, \varphi^n(x)).$$

   At this step, one must determine the structure of the controlled process under the selector φ^n, denote its recurrent classes by $R_1, R_2, \ldots$, and put $h_n(x_i) = 0$, where x_i is the minimal state in R_i.
3. Choose $\varphi^{n+1} : \mathbf{X} \to \mathbf{A}$ such that

$$\sum_{y \in \mathbf{X}} \rho_n(y)p(y|x, \varphi^{n+1}(x)) = \inf_{a \in \mathbf{A}} \left\{ \sum_{y \in \mathbf{X}} \rho_n(y)p(y|x, a) \right\},$$

   setting $\varphi^{n+1}(x) = \varphi^n(x)$ whenever possible. If $\varphi^{n+1} = \varphi^n$, go to step 4; otherwise, increment n by 1 and return to step 2.
4. Choose $\varphi^{n+1} : \mathbf{X} \to \mathbf{A}$ such that

$$c(x, \varphi^{n+1}(x)) + \sum_{y \in \mathbf{X}} h_n(y)p(y|x, \varphi^{n+1}(x))$$

$$= \inf_{a \in \mathbf{A}} \left\{ c(x, a) + \sum_{y \in \mathbf{X}} h_n(y)p(y|x, a) \right\},$$

   setting $\varphi^{n+1}(x) = \varphi^n(x)$ whenever possible. If $\varphi^{n+1} = \varphi^n$, stop and set $\varphi^* = \varphi^n$, $\rho = \rho_n$, $h = h_n$. Otherwise increment n by 1 and return to step 2.

It is known that, in a finite model, this algorithm converges in a finite number of iterations to a solution of the canonical equations (4.2); $\langle \rho, h, \varphi^* \rangle$ is a canonical triplet and the stationary selector φ^* is AC-optimal [Puterman (1994), Cor. 9.2.7].

The following example, based on [Dekker (1987), Thm. 1], shows that, if the action space is not finite, then this algorithm may fail to converge, even in a semi-continuous model (see Definition 4.4).

Example. Let $\mathbf{X} = \{1,2,3,4\}$, $\mathbf{A} = \{\hat{0}\} \cup \{\alpha : 0 \le \alpha \le 1/3\}$. The maximal possible value of α is $\frac{\sqrt{3}-1}{2}$, and it is convenient to fix it at the level of 1/3.

$$p(1|1,a) = p(2|2,a) = p(3|4,a) \equiv 1, \quad p(4|3,\hat{0}) = 1,$$

$$p(y|3,\alpha) = \begin{cases} \frac{1}{4}+\alpha^2, & \text{if } y=1; \\ \frac{1}{4}+\alpha, & \text{if } y=2; \\ \frac{1}{2}-\alpha-\alpha^2, & \text{if } y=3, \end{cases}$$ with all other transition probabilities being zero; $c(1,a) \equiv 5$, $c(2,a) \equiv 4$, $c(4,a) \equiv 6$, $c(3,\hat{0}) = 0$, $c(3,\alpha) = 6$ (see Fig. 4.19). The topology in $\mathbf{X}$ is discrete, $\hat{0}$ is the isolated point in $\mathbf{A}$, and the topology in $[0,1/3]$ is the natural Euclidean one.

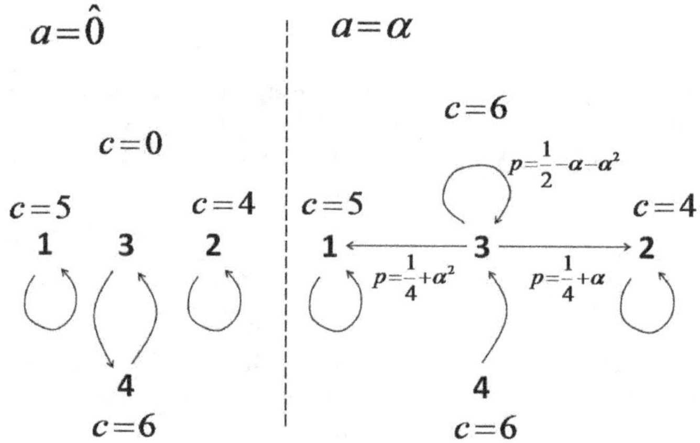

Fig. 4.19 Strategy iteration algorithm does not converge.

The canonical equations (4.2) have the following solution:

$$\rho(x) = \begin{cases} 5, & \text{if } x=1; \\ 4, & \text{if } x=2; \\ 3, & \text{if } x=3 \text{ or } 4 \end{cases} \qquad h(x) = \begin{cases} 3, & \text{if } x=4; \\ 0, & \text{otherwise} \end{cases} \qquad \varphi^*(x) \equiv \hat{0}.$$

Thus $\langle \rho, h, \varphi^* \rangle$ is a canonical triplet by Theorem 4.1 and the canonical stationary selector φ^* is AC-optimal by Theorem 4.3. (The actions in the uncontrolled states $x = 1,2,4$ may be ignored).

The multichain strategy iteration presented above, starting from $\varphi^0(x) = \tilde{\alpha} \in [0, 1/3]$, where $\tilde{\alpha}$ is a rational number, results in the following.

For all $n = 0, 1, 2, \dots$,

$$\rho_n(1) \equiv 5, \quad \rho_n(2) \equiv 4, \quad \rho_n(4) = \rho_n(3), \quad h_n(1) \equiv 0, \quad h_n(2) \equiv 0.$$

Lemma 4.1. *Suppose $\varphi^n(3) \in [0, 1/3]$ is a rational number for some $n = 0, 1, \dots$ and $\varphi^n(1, 2, 4) = \tilde{\alpha}$. Then on step 3 of the algorithm $\varphi^{n+1}(3) \in [0, 1/3]$ is again rational, $\varphi^{n+1}(3) \neq \varphi^n(3)$, and $\varphi^{n+1}(1, 2, 4) = \tilde{\alpha}$.*

The proof is given in Appendix A.

According to Lemma 4.1, the presented at the beginning strategy iterations never terminate. □

If a semi-continuous model is unichain then, again, it can happen that the (unichain) strategy iteration as in Subsection 4.3.14 never terminates if the action space is not finite: see Theorem 3 in [Dekker (1987)].

4.3.16 *When value iteration is not successful*

The basic value iteration algorithm can be written as follows [Puterman (1994), Subsection 8.5.1].

1. Set $n = 0$, specify a small enough $\varepsilon > 0$, and select a bounded measurable function $v^0(\cdot) \in \mathbf{B}(\mathbf{X})$.
2. Compute

$$v^{n+1}(x) = \inf_{a \in \mathbf{A}} \left\{ c(x, a) + \int_{\mathbf{X}} v^n(y) p(dy|x, a) \right\}, \quad x \in \mathbf{X} \tag{4.20}$$

(we leave aside the question of the measurability of $v^{n+1}(\cdot)$).

3. If

$$\sup_{x \in \mathbf{X}} [v^{n+1}(x) - v^n(x)] - \inf_{x \in \mathbf{X}} [v^{n+1}(x) - v^n(x)] < \varepsilon,$$

then stop and choose $\varphi^*(x)$ providing the infimum in (4.20). Otherwise increment n by 1 and return to step 2.

For a finite model, sufficient conditions for the successful termination of this algorithm and for the AC-ε-optimality of the stationary selector φ^* are given in [Puterman (1994), Subsections 8.5.2 and 8.5.3]; see also Condition 4.4.

Theorem 4.9. [Puterman (1994), Thm. 8.5.6]. *Suppose the model is finite and unichain. Then, if the value iteration terminates, then the stationary selector φ^* is AC-ε-optimal.*

In what follows, for $v(\cdot) \in \mathbf{B}(\mathbf{X})$,

$$sp(v) := \sup_{x \in \mathbf{X}} [v(x)] - \inf_{x \in \mathbf{X}} [v(x)]$$

is the so-called *span* of the (bounded) function $v(\cdot)$. It exhibits all the properties of a seminorm and is convenient for the comparison of classes of equivalence when we do not distinguish between two bounded functions $v_1(\cdot)$ and $v_2(\cdot)$ if $v_2(x) \equiv v_1(x) + const$. See Remark 4.1 in this connection. It can easily happen that $\sup_{x\in\mathbf{X}} |v^{n+1}(x) - v^n(x)|$ does not approach zero, but $\lim_{n\to\infty} sp(v^{n+1} - v^n) = 0$, and the value iteration returns an AC-ε-optimal selector in a finite number of steps.

Example. The MDP described in Subsection 4.3.3 (Fig. 4.3) confirms this statement. Value iteration starting from $v^0(x) \equiv 0$ results in the following.

Obviously, $v^{n+1}(2) = 1 + v^n(2) \Longrightarrow v^n(2) = n$, and $v^1(1) = -10$, $v^2(1) = -9.5 = -12 + 2 + (0.5)^1$.

If, for some $n \geq 2$, $v^n(1) = -12 + n + (0.5)^{n-1}$, then

$$v^{n+1}(1) = \min\{-5 + \frac{1}{2}[-12 + n + (0.5)^{n-1}] + \frac{1}{2}n; \quad -10 + n\},$$

and the minimum is given by the first expression, equal to $-12 + (n+1) + (0.5)^n$, because

$$-10 + n + 5 + 6 - n - (0.5)^n = 1 - (0.5)^n > 0.$$

Therefore, for all $n \geq 2$,

$$v^n(1) = -12 + n + (0.5)^{n-1}$$

and the minimum is provided by $a = 1$; the actions in state 2 are immaterial.

We see that $\sup_{x\in\mathbf{X}} |v^{n+1}(x) - v^n(x)| = 1$ and $sp(v^{n+1} - v^n) = 0.5^n$. The stationary selector $\varphi^*(x) \equiv 1$ is AC-optimal, just like any other control strategy, because the process will ultimately be absorbed at state 2;

$\left\langle \rho \equiv 1, h(x) = \begin{cases} 0, & \text{if } x = 1; \\ 12, & \text{if } x = 2, \end{cases} \varphi^* \right\rangle$ is the canonical triplet. □

Condition 4.4. The model is finite and either

(a) $\max_{x\in\mathbf{X}, a\in\mathbf{A}, y\in\mathbf{X}, b\in\mathbf{A}} [1 - \sum_{z\in\mathbf{X}} \min\{p(z|x,a), p(z|y,b)] < 1$, or

(b) there exists a state $\hat{x} \in \mathbf{X}$ and an integer K such that, for any stationary selector φ, for all $x \in \mathbf{X}$, the K-step transition probability satisfies $p^K(\hat{x}|x, \varphi(x)) > 0$, or

(c) the model is unichain and $p(x|x,a) > 0$ for all $x \in \mathbf{X}$, $a \in \mathbf{A}$.

It is known that, under Condition 4.4, value iteration achieves the stopping criterion for any $\varepsilon > 0$ [Puterman (1994), Thm. 8.5.3]. In the above example, all the Conditions 4.4(a,b,c) are satisfied.

In the next example [Puterman (1994), Ex. 8.5.1], Conditions 4.4(a,b,c) are obviously violated, and the value iteration never terminates.

Example. Let $\mathbf{X} = \{1, 2\}$, $\mathbf{A} = \{0\}$ (dummy action); $p(2|1,0) = p(1|2,0) = 1$, with all other transition probabilities being zero. We put $c(1,0) = c(2,0) = 0$. See Fig. 4.20. If $v^0(1) = r^1$ and $v^0(2) = r^2$ then $v^1(1) = r^2$, $v^1(2) = r^1$,

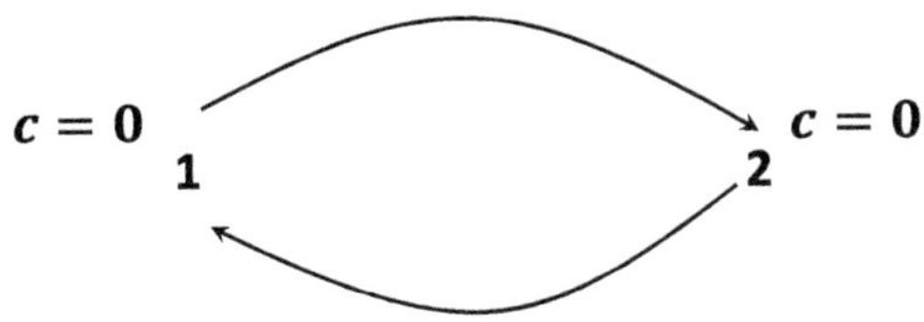

Fig. 4.20 Unsuccessful value iteration.

$v^2(1) = r^1$, $v^2(2) = r^2$, and so on: the function $v^n(\cdot)$ oscillates with period 2, $\sup_{x\in\mathbf{X}} |v^{n+1}(x) - v^n(x)| = |r^1 - r^2|$, and $sp(v^{n+1} - v^n) = 2|r^1 - r^2|$. Value iteration is unsuccessful except for the case $r^1 = r^2$. □

Note that iterations (4.20) are the same as (1.16), and sufficient conditions for the convergence $\lim_{n\to\infty} sp(v^{n+1} - v^n) = 0$ in finite models are also given in Lemma 1.2. As was shown in Subsection 1.4.6, the average cost version of the example presented on Fig. 1.8 is such that the values of $v^n(\cdot)$ (denoted as $W^n(\cdot)$ in Subsection 1.4.6) oscillate, $sp(v^{n+1} - v^n)$ remains separated from zero, and the actions providing the infimum in (4.20) also oscillate. Condition 4.4 is here violated. At the same time, in this example $|sp(v^{n+1} - v^n) - sp(v^{n+2} - v^{n+1})| \to 0$ as $n \to \infty$ because the sequence $\{w_k\}_{k=1}^{\infty}$, as in Lemma 1.3, converges. Recall also that the both stationary selectors φ^o and φ^e providing the minima in the value iteration are not AC-optimal. This observation confirms that the following conjecture is not valid in general.

- If value iteration achieves the step when $|sp(v^{n+1} - v^n) - sp(v^{n+2} - v^{n+1})| < \varepsilon$, then the last mapping providing the infimum in (4.20) is AC-ε-optimal [Puterman (1994), p. 477].

The following example, similar to that presented in Theorem 5 of [Della Vecchia *et al.* (2012)], also confirms that this conjecture is not valid in general and, moreover, may be misleading.

Example. Let $\mathbf{X} = \{1, 2\}$, $\mathbf{A} = \{1, 2\}$; $p(1|1, 1) = p(2|1, 2) = 1$, $p(2|2, a) \equiv 1$ with all other transition probabilities begin zero. Put $c(1, 1) = -10$, $c(1, 2) = -1$ and $c(2, a) \equiv -11$: see Fig. 4.21. Obviously, only actions in state 1 are important and the stationary selector $\varphi^*(x) \equiv 2$ is AC-optimal; $v^*_x = v^{\varphi^*}_x \equiv -11$. Value iteration starting from $v^0(x) \equiv 0$ results in the following.

Clearly, $v^n(2) = -11n$.

For $n = 0, 1, \ldots, 8$, one can easily check by induction that only action $a = 1$

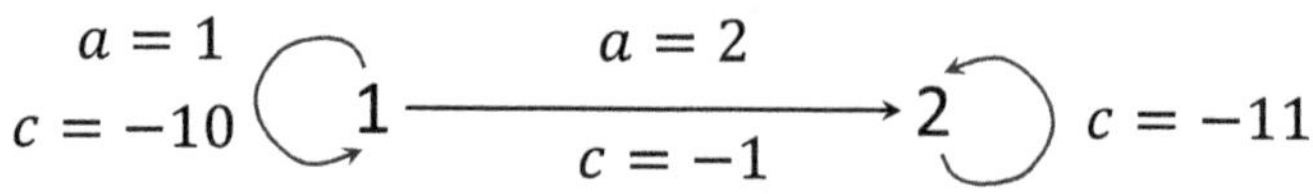

Fig. 4.21 Misleading termination criterion in value iteration.

provides the minimum in (4.20) at $x = 1$ and

$$\begin{aligned} v^n(1) &= -10n: \\ v^{n+1}(1) &= \min\{-10 + v^n(1) = -10(n+1); \quad -1 + v^n(2) = -1 - 11n\} \\ &= -10(n+1). \end{aligned}$$

For $n = 9$, the both actions $a = 1$ and $a = 2$ provide the minimum:

$$v^{10}(1) = \min\{-10 - 10 \cdot 9; \quad -1 - 11 \cdot 9\} = -100 = -11 \cdot 10 + 10.$$

For $n > 9$, again by induction, only action $a = 2$ provides the minimum in (4.20) at $x = 1$ and

$$\begin{aligned} v^n(1) &= -11n + 10: \\ v^{n+1}(1) &= \min\{-10 + v^n(1) = -11n; \quad -1 + v^2(n) = -1 - 11n\} \\ &= -11(n+1) + 10. \end{aligned}$$

Let us fix an arbitrary $\varepsilon \in (0, 1)$. According to the presented calculations,

$$sp(v^{n+1} - v^n) = \begin{cases} sp(-10, \ -11) = 1, & \text{if } 0 \le n \le 9; \\ sp(-11, \ -11) = 0, & \text{if } n > 9. \end{cases}$$

Thus, the standard value iteration, as at the beginning of the current subsection, terminates at $n = 10$ and returns the AC-ε-optimal (in fact, AC-optimal) stationary selector $\varphi^*(x) = 2$. But if one relies an the above presented conjecture, then the value iteration should be stopped at the second step (when $n = 1$) with $|sp(v^1 - v^0) - sp(v^2 - v^1)| = |1 - 1| = 0$. The last mapping providing the minimum in (4.20) is $\varphi^1(x) \equiv 1$ (recall that the actions in state 2 are immaterial), which is not AC-ε-optimal: $v_1^{\varphi^1} = -10 > v_1^{\varphi^*} = -11$. □

4.3.17 *Finite-horizon approximation does not work*

One might think that, by solving the finite-horizon problem (1.1) with the final cost $C(x) \equiv 0$ for a large enough value of T, then an AC-optimal control strategy will be approximated in some sense and $V_x^T/T := \inf_\pi E_x^\pi \left[\sum_{t=1}^T c(X_{t-1}, A_t)\right]/T$ will converge to v_x^* as $T \to \infty$. The following example, based on [Flynn (1980), Ex. 1], shows that this conjecture is false in general.

Example. Let $\mathbf{X} = \{0, 0', 1, 1', \ldots\}$, $\mathbf{A} = \{0, 1, 2\}$, $p(0|0, a) = p(0'|0', a) \equiv 1$, $p((i-1)'|i', a) \equiv 1$ for all $i \geq 1$, $p(0|i, 0) = p(i+1|i, 1) = p(i'|i, 2) = 1$ for all $i \geq 1$, all the other transition probabilities being zero; $c(0, a) \equiv 0$, $c(0', a) \equiv 1$, $c(i, a) \equiv 1$, $c(i', a) \equiv -3$ (see Fig. 4.22).

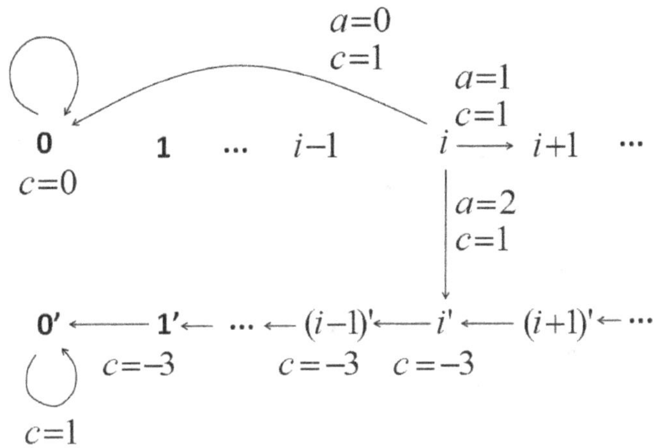

Fig. 4.22 Optimal strategy in a finite-horizon model is not AC-optimal.

If the initial state x_0 is 0, or $0'$, or i' ($i \geq 1$), then in fact the process is uncontrolled and, for any $T < \infty$, the values

$$\frac{V_0^T}{T} = 0, \quad \frac{V_{0'}^T}{T} = 1, \quad \frac{V_{i'}^T}{T} = \begin{cases} -3, & \text{if } T \leq i; \\ 1 - 4i/T, & \text{if } T > i \end{cases}$$

indeed approach the corresponding long-run average costs $v_{x_0}^* = \begin{cases} 0, & \text{if } x_0 = 0; \\ 1, & \text{if } x_0 = i',\ i \geq 0. \end{cases}$

Below we assume that the initial state $x_0 = i \geq 1$ is fixed.

Consider the finite-horizon version of this model:

$$V_i^{\pi,T} := \mathsf{E}_i^\pi \left[\sum_{t=1}^T c(X_{t-1}, A_t) \right] \to \inf_\pi .$$

If $0 < T \leq i+1$, then, obviously, it is optimal to apply action $A_1 = 2$ in order to obtain the minimal possible total cost, equal to $1 - 3(T-1) = V_i^T := \inf_\pi V_i^{\pi,T}$.

Suppose $T \geq i+1$ and prove by induction w.r.t. $s := \left\lfloor \frac{T-(i+1)}{2} \right\rfloor = 0, 1, 2, \ldots$

($\lfloor\cdot\rfloor$ is the integer part) that

$$V_i^T = \begin{cases} -3i - 2s + 1, & \text{if } T - (i+1) = 2s; \\ -3i - 2s + 2, & \text{if } T - (i+1) = 2s + 1 \end{cases} \tag{4.21}$$

for each $i \geq 1$.

If $s = 0$, then there are two cases.
(i) $T = i + 1$. Now $V_i^T = 1 - 3(T-1) = -3i + 1$.
(ii) $T = i + 2$. Now

$$\begin{aligned} V_i^T &= \min\{1 + 0,\ 1 + V_{i+1}^{T-1},\ 1 + V_{i'}^{T-1}\} \\ &= \min\{1,\ 1 + 1 - 3(T-2),\ 1 + (T - 1 - 4i)\} \\ &= \min\{1,\ 2 - 3i,\ 2 - 3i\} = -3i + 2. \end{aligned}$$

Note, in this situation it is suggested to accept that the minimum is provided by action $a = 2$, which is as optimal as action $a = 1$.

We proved that equality (4.21) holds for each $i \geq 1$ if $s = 0$. Suppose it holds for each $i \geq 1$ and some $s \geq 0$ and consider $s + 1$.

Again there are two cases.
(i) $T = (i+1) + 2s + 2$. Now

$$\begin{aligned} V_i^T &= \min\{1 + 0,\ 1 + V_{i+1}^{T-1},\ 1 + V_{i'}^{T-1}\} \\ &= \min\{1,\ 1 + [-3(i+1) - 2s + 1],\ 1 + (T - 1 - 4i)\}. \end{aligned}$$

Here $(T-1) - (i+2) = 2s$ and $V_{i+1}^{T-1} = -3(i+1) - 2s + 1$ by the inductive supposition. Thus,

$$\begin{aligned} V_i^T &= \min\{1,\ -1 - 3i - 2s,\ 1 + i + 2s + 2 - 4i = 3 - 3i + 2s\} \\ &= -1 - 3i - 2s = -3i - 2(s+1) + 1. \end{aligned}$$

(ii) $T = (i+1) + 2s + 3$. Now

$$\begin{aligned} V_i^T &= \min\{1 + 0,\ 1 + V_{i+1}^{T-1},\ 1 + V_{i'}^{T-1}\} \\ &= \min\{1,\ 1 + [-3(i+1) - 2s + 2],\ 1 + (T - 1 - 4i)\}. \end{aligned}$$

Here $(T-1) - (i+2) = 2s + 1$ and $V_{i+1}^{T-1} = -3(i+1) - 2s + 2$ by the inductive supposition. Thus,

$$\begin{aligned} V_i^T &= \min\{1,\ -3i - 2s,\ 1 + i + 2s + 3 - 4i = 4 - 3i + 2s\} \\ &= -3i - 2s = -3i - 2(s+1) + 2. \end{aligned}$$

The proof is completed.

It was also shown that in case $T > i + 1$ one has to apply action 1, that is, move to the right, up to (and excluding) the moment t when $(T - t) \leq X_t + 2 = i + t + 2$, and after that apply action $A_t = 2$. The presented inequality for

t can be rewritten as $t \geq \frac{T-(i+1)}{2} - \frac{1}{2}$, i.e., one can calculate $s := \left\lfloor \frac{T-(i+1)}{2} \right\rfloor$ and describe the optimal selector as follows. Apply action $a = 1$ in the states $X_t = 1, 2, \ldots, i, i+1, \ldots, i+s-1$ and action $a = 2$ at $X_t \geq i+s$. This optimal stationary selector depending on T and initial state i is denoted as $\varphi^{T,i}(x)$. As was shown, $V_i^{\varphi^{T,i},T} = V_i^T$ and $V_i^{\varphi^{T,i},T} \to -\infty$ as $T \to \infty$, that is, as $s \to \infty$: see (4.21). If time horizon T is large, it is reasonable to move from initial state $x_0 = i$ far enough to the right (with one step costs 1) and after that apply action $a = 2$ and move back along the trajectory $j' \to (j-1)' \to 0'$ with one step costs -3 leading to the large negative total cost.

In order to calculate the AC-optimal strategy for this long-run average MDP with initial state $x_0 \in \mathbf{X}$, one can argue as follows, leaving aside the trivial uncontrolled cases $x_0 = 0$ and $x_0 = i'$, $i = 0, 1, 2, \ldots$. Starting from each $x_0 = i > 0$, under an arbitrarily fixed control strategy π, there are only three types of trajectories with possibly positive P_i^π-probability:

$$
\begin{gathered}
i \to i+1 \to \ldots \text{ with} \\
\lim_{T\to\infty} \frac{1}{T} \sum_{t=1}^{T} c(X_{t-1}, A_t) = 1;
\end{gathered}
$$

$$
\begin{aligned}
& i \to i+1 \to \ldots \to j \to 0 \to 0 \ldots \text{ with} \\
& \lim_{T\to\infty} \frac{1}{T} \sum_{t=1}^{T} c(X_{t-1}, A_t) = 0; \\
& i \to i+1 \to \ldots \to j \to j' \to (j-1)' \to \ldots \to 0' \to 0' \ldots \text{ with} \\
& \lim_{T\to\infty} \frac{1}{T} \sum_{t=1}^{T} c(X_{t-1}, A_t) = 1.
\end{aligned}
$$

Thus, by the Dominated Convergence Theorem,

$$
\mathsf{E}_i^\pi \left[\lim_{T\to\infty} \frac{1}{T} \sum_{t=1}^{T} c(X_{t-1}, A_t) \right] = \lim_{T\to\infty} \frac{1}{T} \mathsf{E}_i^\pi \left[\sum_{t=1}^{T} c(X_{t-1}, A_t) \right] = v_i^\pi \geq 0,
$$

and $v_i^* = 0 = v_i^{\varphi^*}$, where $\varphi^*(x) \equiv 0$ is an AC-optimal stationary selector.

The finite-horizon optimal strategy $\varphi^{T,i}$ has nothing in common with φ^*: it never uses action $a = 0$. For a fixed initial state $x_0 = i > 0$, when the time horizon T goes to infinity, the difference between the performance functionals in the finite-horizon model under φ^*, that is $V_i^{\varphi^*,T} = 1$, and under the optimal control strategy $\varphi^{T,i}$ described above, goes to infinity, meaning that the AC-optimal selector φ^* becomes progressively less desirable as T increases. □

4.3.18 Linear programming approach to finite models

If a finite MDP is not recurrent, then it can still be reasonable to solve the linear program (4.6). According to Theorem 4.7, if the cost function $c(\cdot,\cdot)$ is non-negative, then there exists such an optimal solution η^* to (4.6) and a stationary strategy π^{s} that
– $\eta^*(i,a) = \pi^{\mathrm{s}}(a|i)\sum_{b\in\mathbf{A}}\eta^*(i,b)$, $i \in \mathbf{X}$, $a \in \mathbf{A}$;
– $\hat{\eta}^*(i) := \sum_{b\in\mathbf{A}}\eta^*(i,b)$ is the invariant probability on $\mathcal{B}(\mathbf{X})$ of the controlled process X_t under the control strategy π^{s}:

$$\hat{\eta}^*(j) = \sum_{i\in\mathbf{X}}\sum_{a\in\mathbf{A}} p(j|i,a)\pi^{\mathrm{s}}(a|i)\hat{\eta}^*(i) = \sum_{i\in\mathbf{X}}\sum_{a\in\mathbf{A}} p(j|i,a)\eta^*(i,a),\ j \in \mathbf{X};$$

– the pair $(\hat{\eta}^*, \pi^{\mathrm{s}})$ solves problem (4.12).

Consider the following example based on [Puterman (1994), Ex. 9.3.2].

Example. $\mathbf{X} = \{1,2,3,4\}$, $\mathbf{A} = \{1,2,3\}$, $p(2|1,2) \equiv 1$, $p(1|2,1) = 1$, $p(3|2,2) = 1$, $p(4|2,3) = 1$, $p(4|4,a) \equiv 1$, $c(1,a) \equiv 1$, $c(2,1) = 4$, $c(2,2) = 3$, $c(2,3) = 0$, $c(4,1) = 3$, $c(4,2) = c(4,3) = 4$. See Fig. 4.23. (All the transitions are deterministic).

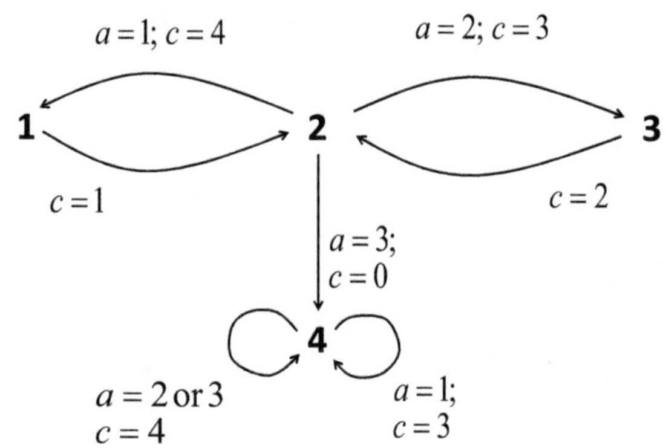

Fig. 4.23 Statement of Theorem 4.2 fails to hold; the basic optimal solution to the primal linear program gives a randomized strategy.

The linear program (4.6) can be written in the following optimal basic form:

$$\text{Objective to be minimized: } \frac{5}{2} + \frac{1}{2}\eta^*(4,1) + \frac{3}{2}\eta^*(4,2) + \frac{3}{2}\eta^*(4,3);$$

$$\eta^*(1,1) + \eta^*(1,2) + \eta^*(1,3) + \eta^*(2,2) + \frac{1}{2}\sum_{a=1}^{3}\eta^*(4,a) = \frac{1}{2};$$

$$\eta^*(2,1) + \eta^*(2,2) + \frac{1}{2}\sum_{a=1}^{3}\eta^*(4,a) = \frac{1}{2};$$

$$\eta^*(2,3) = 0;$$

$$\eta^*(3,1) + \eta^*(3,2) + \eta^*(3,3) - \eta^*(2,2) = 0.$$

The variables $\eta^*(1,1) = 1/2, \eta^*(2,1) = 1/2, \eta^*(2,3) = 0$, and $\eta^*(3,1) = 0$ are basic. All the other variables $\eta^*(1,2), \eta^*(1,3), \eta^*(2,2), \ldots$ equal zero. Now, the stationary selector satisfying $\varphi^*(1) = \varphi^*(2) = 1$ and the corresponding invariant probability

$$\hat{\eta}^*(1) = \sum_{a=1}^{3}\eta^*(1,a) = \frac{1}{2}, \quad \hat{\eta}^*(2) = \sum_{a=1}^{3}\eta^*(2,a) = \frac{1}{2},$$

$$\hat{\eta}^*(3) = \sum_{a=1}^{3}\eta^*(3,a) = 0, \quad \hat{\eta}^*(4) = \sum_{a=1}^{3}\eta^*(4,a) = 0$$

solve problem (4.12) in accordance with the first paragraph of the current subsection. Note that the actions in states 3 and 4 can be arbitrary, as those states will never be visited under strategy φ^* with $\mathsf{P}_0(i) = \hat{\eta}^*(i)$.

The model is not recurrent, and the statement of Theorem 4.2 fails to hold since $\eta^*(3,a) \equiv 0$ and $\eta^*(4,a) \equiv 0$.

The dual program (4.5) in the space of bounded functions $h(\cdot)$ on $\mathbf{X}$ looks as follows. (We put $h(4) = 0$ without loss generality.)

$$\rho \to \sup_{\rho,h}$$

$$\rho + h(1) \le 1 + h(2); \quad \rho + h(2) \le \begin{cases} 4 + h(1); \\ 3 + h(3); \\ 0 + h(4) = 0; \end{cases}$$

$$\rho + h(3) \le 2 + h(2); \quad \rho + h(4) = \rho \le 3.$$

From the second inequality, we have $h(2) - h(1) \le 4 - \rho$, and the first inequality leads to $\rho \le 1 + h(2) - h(1) \le 5 - \rho \Longrightarrow \rho \le \frac{5}{2}$. One of the optimal solutions is presented as follows:

$$\rho^* = \frac{5}{2}, \quad h^*(1) = -5, \quad h^*(2) = -\frac{7}{2}, \quad h^*(3) = -4, \quad h^*(4) = 0.$$

Note that

$$\frac{5}{2} = \rho + h^*(4) < \min_{a\in\mathbf{A}}\left\{c(4,a) + \sum_{j\in\mathbf{X}} h^*(j)p(j|4,a)\right\} = 3,$$

and there are no canonical triplets with $\rho \equiv const$ because otherwise $\rho = v^*_x$ by Theorem 4.3, but the Bellman function v^*_x really depends on the initial state $x \in \mathbf{X}$ as is shown below.

It is well known that canonical triplets exist in finite models [Puterman (1994), Thm. 9.1.4]. In the considered example, that is

$$\begin{aligned} &\rho^*(1) = \rho^*(2) = \rho(3) = \frac{5}{2}, \quad \rho^*(4) = 3, \\ &h^*(1) = -5, \quad h^*(2) = -\frac{7}{2}, \quad h^*(3) = -4, \quad h^*(4) = 0, \\ &\varphi^*(x) \equiv 1, \end{aligned} \tag{4.22}$$

and $v^*_x = \rho(x)$ by Theorem 4.3. □

Note that problem (4.12) is different from (4.1), where the initial distribution P_0 is fixed. In finite models, problem (4.1) can also be solved using the linear programming approach. In general case, one needs to investigate linear program

$$\sum_{x\in\mathbf{X}}\sum_{a\in\mathbf{A}} c(x,a)\eta(x,a) \to \inf_{\eta,\tilde{\eta}} \tag{4.23}$$

$$\sum_{a\in\mathbf{A}} \eta(x,a) = \sum_{y\in\mathbf{X}}\sum_{a\in\mathbf{A}} p(x|y,a)\eta(y,a), \quad x \in \mathbf{X}; \tag{4.24}$$

$$\eta(x,a) \geq 0;$$

$$\sum_{a\in\mathbf{A}} \eta(x,a) + \sum_{a\in\mathbf{A}} \tilde{\eta}(x,a) - \sum_{y\in\mathbf{X}}\sum_{a\in\mathbf{A}} p(x|y,a)\tilde{\eta}(y,a) = \alpha(x), \tag{4.25}$$

$$x \in \mathbf{X}; \quad \tilde{\eta}(x,a) \geq 0,$$

where $\alpha(x) > 0$ are arbitrarily fixed numbers such that $\sum_{x\in\mathbf{X}} \alpha(x) = 1$. Like in (4.6), at least one of equations in (4.24) is redundant.

If $(\eta^*, \tilde{\eta}^*)$ is an optimal solution to (4.23),(4.24),(4.25), then the following stationary strategy π^s is AC-optimal [Kallenberg (2010), Thm. 5.18(2)]:

$$\pi^s(a|x) = \begin{cases} \eta^*(x,a) \big/ \sum_{b\in\mathbf{A}} \eta^*(x,b), & \text{if } \sum_{b\in\mathbf{A}} \eta^*(x,b) > 0; \\ \tilde{\eta}^*(x,a) \big/ \sum_{b\in\mathbf{A}} \tilde{\eta}^*(x,b), & \text{if } \sum_{b\in\mathbf{A}} \eta^*(x,b) = 0. \end{cases} \tag{4.26}$$

For any feasible solution $(\eta, \tilde{\eta})$ one can also construct the strategy π^s using (4.26), and π^s is well defined because $\sum_{b\in\mathbf{A}} \eta(x,b) + \sum_{b\in\mathbf{A}} \tilde{\eta}(x,b) \geq \alpha(x) > 0$ for all $x \in \mathbf{X}$ in accordance with (4.25). We say that the strategy π^s in (4.26) is *induced* by the feasible solution $(\eta, \tilde{\eta})$.

Having in hand an optimal solution $(\eta^*, \tilde{\eta}^*)$ to linear program (4.23),(4.24),(4.25), one also can construct an AC-optimal stationary selector φ^* by putting

$$\varphi^*(x) = \begin{cases} a, & \text{if } \eta^*(x,a) > 0; \\ b, & \text{if } \sum_{a\in\mathbf{A}} \eta^*(x,a) = 0 \text{ and } \tilde{\eta}^*(x,b) > 0: \end{cases}$$

see [Kallenberg (2010), Thm. 5.16], [Puterman (1994), Thm. 9.3.8(b)].

It is known that, for any stationary strategy π^s, one can construct a feasible solution $(\eta, \tilde{\eta})$ to program (4.23),(4.24),(4.25), such that π^s is induced by $(\eta, \tilde{\eta})$ [Kallenberg (2010), Thm. 5.17], [Puterman (1994), Thm. 9.3.4]. Moreover, if the strategy $\pi^s(a|x) = \mathbb{I}\{a = \varphi(x)\}$ is actually a stationary selector, then that construction provides $(\eta, \tilde{\eta})$ being a *basic* feasible solution (see [Puterman (1994), Thm. 9.3.5]). In other words, some basic feasible solutions induce all the stationary selectors. In what follows, it will be shown that in the presented above example (see Fig. 4.23) a basic feasible solution can induce a randomized strategy π^s.

Example Fig. 4.23 continued. The linear program (4.23),(4.24),(4.25) with $\alpha(1) = 1/6$, $\alpha(2) = 1/3$, $\alpha(3) = 1/6$, $\alpha(4) = 1/3$ can be written in the following optimal basic form:

$$\text{Objective to be minimized: } \frac{8}{3} + \eta^*(4,2) + \eta^*(4,3) + \frac{1}{2}\tilde{\eta}^*(2,3);$$

$$\eta^*(1,1) + \eta^*(1,2) + \eta^*(1,3) - \sum_{a=1}^{3} \tilde{\eta}^*(3,a) + \tilde{\eta}^*(2,2) + \frac{1}{2}\tilde{\eta}^*(2,3) = \frac{1}{6};$$
$$\eta^*(2,1) - \sum_{a=1}^{3} \tilde{\eta}^*(3,a) + \tilde{\eta}^*(2,2) + \frac{1}{2}\tilde{\eta}^*(2,3) = \frac{1}{6};$$
$$\eta^*(2,2) + \sum_{a=1}^{3} \tilde{\eta}^*(3,a) - \tilde{\eta}^*(2,2) = \frac{1}{6};$$
$$\eta^*(2,3) = 0;$$

$$\eta^*(3,1) + \eta^*(3,2) + \eta^*(3,3) + \sum_{a=1}^{3} \tilde{\eta}^*(3,a) - \tilde{\eta}^*(2,2) = \frac{1}{6};$$
$$\eta^*(4,1) + \eta^*(4,2) + \eta^*(4,3) - \tilde{\eta}^*(2,3) = \frac{1}{3};$$
$$\tilde{\eta}^*(2,1) + \tilde{\eta}^*(2,2) + \frac{1}{2}\tilde{\eta}^*(2,3) - \sum_{a=1}^{3} [\tilde{\eta}^*(1,a) + \tilde{\eta}^*(3,a)] = 0.$$

The variables $\eta^*(1,1) = 1/6, \eta^*(2,1) = 1/6, \eta^*(2,2) = 1/6, \eta^*(2,3) = 0, \eta^*(3,1) = 1/6, \eta^*(4,1) = 1/3$, and $\tilde{\eta}^*(2,1) = 0$ are basic. Now, according to (4.26), $\pi^s(1|1) = \pi^s(1|3) = \pi^s(1|4) = 1$, but $\pi^s(1|2) = \pi^s(2|2) = 1/2$. This stationary strategy is AC-optimal.

One can make $\tilde{\eta}^*(2,2)$ basic, at level 0, instead of $\tilde{\eta}^*(2,1)$. That new basic solution will still be optimal, inducing the same randomized strategy π^s. Of course, there exist many other optimal basic solutions leading to AC-optimal stationary selectors $\varphi(1) = 1$, or 2, or 3, $\varphi(2) = 1$, or 2, $\varphi(3) = 1$, or 2, or 3, $\varphi(4) = 1$. For example, one can make $\tilde{\eta}^*(3,1)$ basic instead of $\eta^*(2,2)$ using the third constraint equality: simply substitute expression $\tilde{\eta}^*(3,1) =$

$1/6 - \eta^*(2,2) - \tilde{\eta}^*(3,2) - \tilde{\eta}^*(3,3) + \tilde{\eta}^*(2,2)$ in all the other formulae. We finish with $\sum_{a=1}^{3} \eta^*(3,1) = 0$ and, according to (4.26), $\pi^s(1|x) \equiv 1$. The new optimal basic solution induces the AC-optimal stationary selector $\varphi^*(x) \equiv 1$.

If one takes another distribution $\alpha(x)$, the linear program will produce AC-optimal strategies, but the value of the infimum in (4.23) will change. In fact, that infimum coincides with the long-run expected average cost (4.1) under an AC-optimal strategy if $\mathsf{P}_0(x) = \alpha(x)$ [Kallenberg (2010), Cor. 5.5]. In the considered example with $\alpha(1) = 1/6$, $\alpha(2) = 1/3$, $\alpha(3) = 1/6$, $\alpha(4) = 1/3$, the infimum equals 8/3. If the initial state x_0 equals 1,2, or 3, then the minimal long-run expected cost is $v^*_{x_0} = 5/2$. In case $x_0 = 4$, $v^*_4 = 3$. Thus, the minimal long-run expected cost under $\mathsf{P}_0(\cdot) = \alpha(\cdot)$ equals $2/3 \cdot 5/2 + 1/3 \cdot 3 = 8/3$.

As was mentioned above, solving linear program (4.6) for the considered MDP provides optimal actions in states 1 and 2; actions in states 3 and 4 can be arbitrary. (And a stationary selector φ with $\varphi(4) = 2$ or 3 is obviously not optimal in problem (4.1) if $\mathsf{P}_0(4) > 0$.) Conversely, linear program (4.23),(4.24),(4.25) provides all the optimal actions: it solves problem (4.1) for all initial states.

The dual linear program to (4.23),(4.24),(4.25) (in the space of bounded functions on $\mathbf{X}$) can be written as follows:

$$\sum_{x\in\mathbf{X}} \alpha(x)\rho(x) \to \sup_{\rho,h}$$

$$\rho(x) \le \sum_{y\in\mathbf{X}} \rho(y)p(y|x,a), \qquad x \in \mathbf{X}, a \in \mathbf{A}$$

$$\rho(x) + h(x) \le c(x,a) + \sum_{y\in\mathbf{X}} h(y)p(y|x,a), \qquad x \in \mathbf{X}, a \in \mathbf{A}.$$

(Compare with canonical equations (4.2)). In the above example, the optimal solution is given by formula (4.22), and $\langle \rho^*, h^*, \varphi^*(x) \equiv 1\rangle$ is a canonical triplet in the considered MDP. □

4.3.19 *Linear programming in infinite models*

First of all, we discuss Condition 4.1 and Theorem 4.7. In discrete spaces the discrete topology is fixed (unless specified differently), and in the subsets of $\mathbb{R}$ the topology is Euclidean.

Example. Consider Example 4.3.8, Fig. 4.9. Condition 4.1 is satisfied apart from items (a,c). The space $\mathbf{A}$ is not compact and the set $\{(x,a) \in \mathbf{X} \times \mathbf{A} : c(x,a) \le 1\} = \mathbf{X} \times \mathbf{A}$ is not compact. If we introduce another topology in $\mathbf{A}$, say, $\lim_{a\to\infty} a := 1$, then the transition probability p becomes not (weakly) continuous. The only feasible solutions to (4.4) are $\eta(1,a) \equiv 0$, $\eta(0,\mathbf{A}) = 1$ with $\eta(0,a)$ being arbitrary, so that $\sum_{x\in\mathbf{X}} \sum_{a\in\mathbf{A}} c(x,a)\eta(x,a) = 1$. At the same time, we know that $\inf_\pi v^\pi_1 = 0$ and $\inf_{\mathsf{P}_0,\pi} v^\pi_{\mathsf{P}_0} = 0$. The statements of Theorem 4.7 fail to hold. □

Example. In Example 4.3.9 (see Fig. 4.10) Condition 4.1 is satisfied and Theorem 4.7 holds: in (a) η^* is an arbitrary probability measure with $\eta^*(0,\mathbf{A}) = 1$;

$\sum_{x\in\mathbf{X}} c(x,a)\eta^*(x,a) = 0 = \inf_{\mathsf{P}_0,\pi} v^\pi_{\mathsf{P}_0}$. Recall that Theorem 4.7 deals with problem (4.12) which is different from problem (4.1): the latter concerns a specified initial distribution P_0. As for Item (b) of Theorem 4.7, an optimal pair in problem (4.12) is given by $(\eta^*(i,\mathbf{A}) = \delta_0(i), \pi^s)$ with an arbitrary stationary control strategy π^s. □

Example. Example 4.3.10 (see Fig. 4.11) can be investigated in a similar way. Feasible measures in (4.4) are those for which $\eta(2,\mathbf{A}) = \sum_{a=1}^\infty \frac{1}{a}\eta(1,a)$ and $\eta(1,\mathbf{A}) + \eta(2,\mathbf{A}) = 1$. Thus, for each feasible measure η, $\eta(2,\mathbf{A}) > 0$, $\int_{\mathbf{X}\times\mathbf{A}} c(x,a)d\eta(x,a) > 0$, and $\inf_{\mathsf{P}_0,\pi} v^\pi_{\mathsf{P}_0} = 0$ is not achieved. In Subsection 4.3.10 it was shown that the selector $\varphi^*_t(x) = t$ provides $v^{\varphi^*}_{\mathsf{P}_0} = 0$ for each initial distribution P_0. □

In the first example of Subsection 4.3.11 (Fig. 4.12) and in the example presented in Subsection 4.3.12 (Fig. 4.14) the cost function $c(\cdot,\cdot)$ is not inf-compact because, e.g., for $r = 3$ the set $\{(x,a)\in\mathbf{X}\times\mathbf{A}: c(x,a)\le r\} = \mathbf{X}\times\mathbf{A}$ is not compact. All the other conditions 4.1(a,b,d,e) are satisfied. The statement (b) of Theorem 4.7 fails to hold: it was shown in those subsections that, for all initial distributions P_0 and for all stationary strategies π^s, $v^{\pi^s}_{\mathsf{P}_0} > \inf_{\mathsf{P}_0,\pi} v^\pi_{\mathsf{P}_0}$.

The linear programming approach to infinite models was also developed in [Altman and Shwartz (1991b)] and [Altman (1999), Ch. 11] for the original problem (4.1). The corresponding results can be formulated as Theorem 4.10 below. In what follows in the current subsection, the spaces $\mathbf{X}$ and $\mathbf{A}$ are assumed to be denumerable (or finite) topological Borel.

Definition 4.5. The *expected frequencies* under a control strategy π and initial distribution P_0 are defined as

$$\bar f^T_{\pi,\mathsf{P}_0}(y,a) = \frac{1}{T}\sum_{t=1}^T \mathsf{P}^\pi_{\mathsf{P}_0}(X_{t-1}=y, A_t=a), \quad T=1,2,\ldots.$$

As usual, we write $\bar f^T_{\pi,x}$ if the initial distribution is concentrated at the singleton $x\in\mathbf{X}$.

Condition 4.5. For any control strategy π, for each $x\in\mathbf{X}$, the set of expected frequencies $\{\bar f^T_{\pi,x}\}_{T\ge1}$ is tight, that is, for each $\varepsilon>0$ there are T_0 and compact $\mathbf{K}\subset\mathbf{X}\times\mathbf{A}$ such that $\sum_{(y,a)\in\mathbf{K}} \bar f^T_{\pi,x}(y,a) > 1-\varepsilon$ for all $T > T_0$.

We emphasize that Condition 4.5 only holds if the action space $\mathbf{A}$ is compact (see [Altman and Shwartz (1991b), p. 799 and Counterexample 3.5]). Indeed, let $\{a_t\}_{t=1}^\infty$ be a sequence in $\mathbf{A}$ having no convergent sub-sequences, and consider the selector $\varphi_t(y)\equiv a_t$. Then, for any compact set $\mathbf{K}\subset\mathbf{X}\times\mathbf{A}$, only a finite number of values a_t can appear as the second components of elements $k=(y,a)\in\mathbf{K}$. Thus, starting from some τ, $P^\varphi_x(X_{t-1}=y, A_t=a) = 0$ for $(y,a)\in\mathbf{K}$ and $\bar f^{n\tau}_{\varphi,x}(\mathbf{K}) \le 1/n$.

Theorem 4.10. [Altman and Shwartz (1991b), Cor. 5.4, Thm. 7.1]. *Suppose the cost function $c(\cdot,\cdot)$ is bounded below and the model is recurrent (see*

Definition 4.2(d)). Then, under Condition 4.5, there exists an AC-optimal stationary strategy.

Moreover, if a stationary strategy π^s *is AC-optimal and* $\hat{\eta}^{\pi^s}$ *is the corresponding invariant probability measure on* $\mathbf{X}$*, then the measure* $\eta^*(x,a) := \pi^s(a|x)\hat{\eta}^{\pi^s}(x)$ *solves the linear program (4.4).*

Conversely, if η^* *solves that linear program, then the stationary strategy*

$$\pi^s(a|x) := \frac{\eta^*(x,a)}{\sum_{a\in\mathbf{A}}\eta^*(x,a)} \tag{4.27}$$

is AC-optimal. (If the denominator is zero, the probability distribution $\pi^s(\cdot|x)$ *is chosen arbitrarily.)*

In the first example of Subsection 4.3.11 (Fig. 4.12), as was mentioned in [Altman and Shwartz (1991b), p.797], Condition 4.5 is violated: for $x = 0$ and the constructed non-stationary selector φ^*, the set of expected frequencies $\{f^T_{\varphi^*,0}\}_{T\geq 1}$ is not tight. All the other conditions of Theorem 4.10 are satisfied, but the statement of that theorem is not valid: as was shown, no-one stationary strategy is AC-optimal.

Consider now the following simple example.

Example. $\mathbf{X} = \{0,1,2,\ldots\}$, $\mathbf{A} = \{0,1\}$, $p(0|0,a) \equiv 1$, $p(0|i,0) = p(i+1|i,1) = 1$ for all $i > 0$, with all the other transition probabilities being zero; $c(0,a) \equiv 1$, $c(i,a) \equiv 0$ for $i > 0$ (see Fig. 4.24).

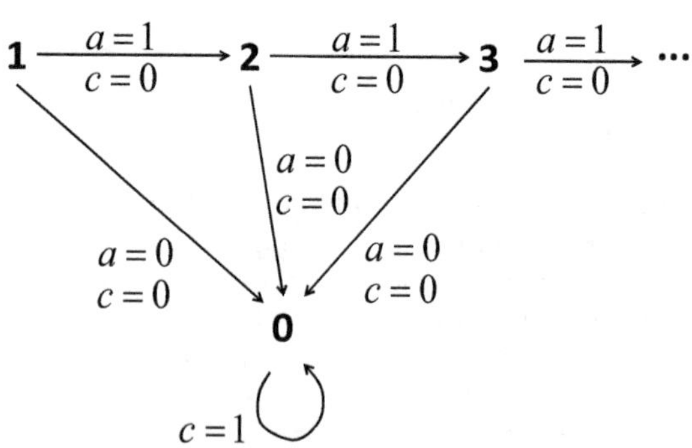

Fig. 4.24 Linear programming approach is not applicable.

The linear program (4.4) can be written as follows

$$\begin{aligned}\eta(0,\mathbf{A}) &\to \inf\\ \eta(0,\mathbf{A}) &= \eta(0,\mathbf{A}) + \sum_{i=1}^{\infty}\eta(i,0);\\ \eta(1,\mathbf{A}) &= 0;\\ \eta(i,\mathbf{A}) &= \eta(i-1,1) \text{ for all } i>1;\\ \eta(\mathbf{X}\times\mathbf{A}) &= 1.\end{aligned}$$

The feasible solutions satisfy equations $\eta^*(0,\mathbf{A}) = 1$, $\eta^*(i,\mathbf{A}) = 0$ for all $i > 0$, so that $\sum_{i\in\mathbf{X}}\sum_{a\in\mathbf{A}} c(i,a)\eta^*(i,a) = \eta^*(0,\mathbf{A}) = 1$. But $\inf_{\mathsf{P}_0,\pi} v^{\pi}_{\mathsf{P}_0} = 0$ and is provided by ($\mathsf{P}_0(i) = \delta_1(i)$, $\varphi^*(i) \equiv 1$). The statement of Theorem 4.7 fails to hold because Condition 4.1 is not satisfied: the cost function $c(\cdot,\cdot)$ is not inf-compact as the set $\{(i,a)\in\mathbf{X}\times\mathbf{A} : c(i,a)\le 1\} = \mathbf{X}\times\mathbf{A}$ is not compact.

In the current example, the dual program (4.5) takes the form:

$$\begin{aligned}\rho &\to \sup_{\rho,h}\\ \rho + h(0) &\le 1 + h(0);\\ \rho + h(i) &\le \min\{h(i+1),\ h(0)\} \text{ for all } i>0.\end{aligned}$$

Since $h(i+1) \ge h(i) + \rho \ge \cdots \ge h(1) + i\cdot\rho$ and $\sup_{i\in\mathbf{X}}|h(i)| < \infty$, we conclude that $\rho \le 0$. The optimal solution is given by $\rho^* = 0$ and $h^*(i) \equiv 0$, so that the duality gap equals 1.

Note that canonical equations (4.2) have solution $\rho(0) = 1$, $\rho(i) \equiv 0$ for all $i > 0$, and $h(i) \equiv 0$; $\varphi^*(i) \equiv 1$. Thus, $\langle\rho, h, \varphi^*\rangle$ is a canonical triplet by Theorem 4.1, and the stationary selector φ^* is AC-optimal (see Theorem 4.3).

In this example (Fig. 4.24) the statement of Theorem 4.10 also fails to hold: the model is not recurrent and Condition 4.5 is not satisfied as the set of expected frequencies $\{\bar f^T_{\varphi^*,1}\}_{T\ge 1}$ is not tight: for each compact (hence, finite) set $\mathbf{K}\subset\mathbf{X}\times\mathbf{A}$ $\lim_{T\to\infty}\sum_{(y,a)\in\mathbf{K}} \bar f^T_{\varphi^*,1}(y,a) = 0$.

One can slightly modify the model under study in such a way that Condition 4.1 is satisfied. Namely, add the state "∞" making the state space $\mathbf{X}$ compact: $\lim_{i\to\infty} i := \infty$. To make the transition probability p weakly continuous, we have to put $p(0|\infty,0) = 1$ and $p(\infty|\infty,1) = 1$; we also put $c(\infty,a) \equiv 0$. Now all the statements of Theorem 4.7 hold true.

The optimal solution to the linear program (4.4) looks as follows:

$$\begin{aligned}&\eta^*(i,0)\equiv 0, \quad i = 1,2,\ldots,\infty;\\ &\eta^*(i,1)\equiv 0, \quad i = 1,2,\ldots, \quad \text{and } \eta^*(\infty,1) = 1;\\ &\eta^*(0,\mathbf{A}) = 0.\end{aligned}$$

The stationary selector $\varphi^*(i)\equiv 1$ is AC-optimal and the pair $(\hat\eta,\varphi^*)$ with $\hat\eta(i) := \delta_\infty(i)$ solves problem (4.12).

The optimal solution to the dual program (4.5) is given by $\rho^* = 0$ and $h^*(i) \equiv 0$, and there is no duality gap. Like previously, $\langle \rho, h, \varphi^* \rangle$ is a canonical triplet, where $\rho(0) = 1$, $\rho(i) \equiv 0$, $h(i) \equiv 0$ for all $i = 1, 2, \ldots, \infty$. (See Theorem 4.1.) The stationary selector φ^* is AC-optimal by Theorem 4.3.

Theorem 4.10 is not applicable because the model is not recurrent. By the way, the last statement of Theorem 4.10 is not valid: for the provided above solution to the linear program (4.4) $\eta^*(i, \mathbf{A}) = \sum_{a \in \mathbf{A}} \eta^*(i, a) = 0$ for $i = 1, 2, \ldots$, but $\pi^s(a|i)$ cannot be arbitrary. It must equal $\delta_1(a)$ in order to be AC-optimal.

If we modify the basic model (Fig. 4.24) by omitting state 0 and action 0, then the linear program (4.4) will have no feasible solutions. The dual program (4.5) still has the same optimal solution $\rho = 0$, $h(i) \equiv 0$, and $\langle \rho, h, \varphi^*(i) \equiv 1 \rangle$ is the same canonical triplet. This trivial MDP appeared in [Altman and Shwartz (1991b), Counterexample 2.1]. Theorem 4.10 is not applicable because the model is not recurrent and Condition 4.5 is not satisfied. □

4.3.20 *Expected frequencies in finite models*

The expected frequencies $\bar{f}^T_{\pi, \mathsf{P}_0}$ were introduced in Definition 4.5. Let F_{π, P_0} be the set of all accumulation (or limit) points of the vectors $\bar{f}^1_{\pi, \mathsf{P}_0}, \bar{f}^2_{\pi, \mathsf{P}_0}, \ldots$. As usual, we write $F_{\pi, x}$ if $\mathsf{P}_0(x) = 1$.

Theorem 4.11. *Suppose the spaces* $\mathbf{X}$ *and* $\mathbf{A}$ *are finite.*

(a) *[Altman (1999), Thm. 4.2]. If the model is aperiodic unichain, then the sets* $\bigcup_{\pi \in \Pi^S} F_{\pi, \mathsf{P}_0} = \bigcup_{\pi \in \Pi^{\mathrm{All}}} F_{\pi, \mathsf{P}_0}$ *coincide, do not depend on the initial distribution* P_0, *and coincide with the closed convex hull of the set* $\bigcup_{\pi \in \Pi^{SN}} F_{\pi, \mathsf{P}_0}$ *and with the collection of all feasible solutions to the linear program (4.6).*

(b) *[Kallenberg (2010), Thm. 9.4]. The set* $\bigcup_{\pi \in \Pi^{\mathrm{All}}} F_{\pi, \mathsf{P}_0}$ *coincides with the collection*

$$\{\eta : \ (\eta, \tilde{\eta}) \text{ is feasible in the linear program (4.23), (4.24), (4.25)}$$
$$\text{with } \alpha(x) = \mathsf{P}_0(x)\}.$$

The following simple example illustrates that, in the multichain case, the inclusion $\bigcup_{\pi \in \Pi^S} F_{\pi, \mathsf{P}_0} \subset \bigcup_{\pi \in \Pi^{\mathrm{All}}} F_{\pi, \mathsf{P}_0}$, can be strict, so that statement (a) of Theorem 4.11 fails to hold.

Example. Let $\mathbf{X} = \{1, 2\}$, $\mathbf{A} = \{1, 2\}$, $\mathsf{P}_0(1) = 1$; $p(1|1, 1) = p(2|1, 2) = 1$, $p(2|2, a) \equiv 1$, with all the other transition probabilities being zero. The cost function $c(\cdot, \cdot)$ is immaterial. (See Fig. 4.39). For any stationary strategy π^s, if $f_{\pi^s,1} \in F_{\pi^s,1}$ then $f_{\pi^s,1}(1, 1)$ equals either 1 (if $\pi^s(1|1) = 1$) or 0 (otherwise). At the same time, there exists a non-stationary strategy π, in the form of the (1/2,1/2)-mixture of two stationary selectors $\varphi^1(x) \equiv 1$ and $\varphi^2(x) \equiv 2$, such that $f_{\pi,1} \in F_{\pi,1}$ and $f_{\pi,1}(1, 1) = 1/2$. More details are given in Subsection 4.3.29. The sets $\bigcup_{\pi \in \Pi^S} F_{\pi, \mathsf{P}_0}$ and $\bigcup_{\pi \in \Pi^{\mathrm{All}}} F_{\pi, \mathsf{P}_0}$ clearly depend on the initial distribution: $f_{\pi,2}(1, a) \equiv 0$ if $x_0 = 2$. □

The next example, based on [Derman (1964), Section 4], shows that statement (a) of Theorem 4.11 can fail to hold even if the model is communicating.

Example. Let $\mathbf{X} = \{1, 2, 3\}$, $\mathbf{A} = \{1, 2\}$, $p(2|1, a) \equiv 1$, $p(2|2, 1) = 1$, $p(3|2, 2) = 1$, $p(3|3, 1) = 1$, $p(1|3, 2) = 1$ with all the other transition probabilities being zero: the transitions are deterministic. The cost function $c(\cdot, \cdot)$ is immaterial. See Fig. 4.25.

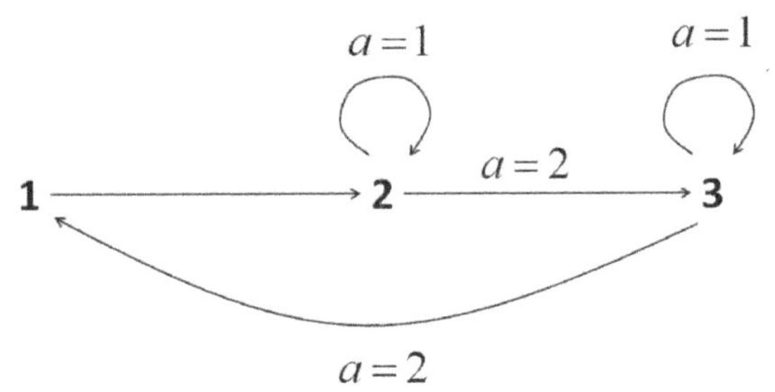

Fig. 4.25 Expected frequencies which are not generated by a stationary strategy.

Suppose that $\mathsf{P}_0(1) = 1$, and π^{s} is a stationary strategy. Below we use the notation

$$\hat{f}_{\pi,1} := \left(\sum_{a \in \mathbf{A}} f_{\pi,1}(1, a), \sum_{a \in \mathbf{A}} f_{\pi,1}(2, a), \sum_{a \in \mathbf{A}} f_{\pi,1}(3, a) \right)$$

for the vectors $f_{\pi,1} \in F_{\pi,1}$. The vector $\hat{f}_{\pi^{\mathrm{s}},1}$ coincides with the stationary distribution of the controlled process X_t governed by the control strategy π^{s} and has the form

$$\hat{f}_{\pi^{\mathrm{s}},1} = \begin{cases} (0, 1, 0), & \text{if } \pi^{\mathrm{s}}(1|2) = 1; \\ (0, 0, 1), & \text{if } \pi^{\mathrm{s}}(1|2) < 1 \text{ and } \pi^{\mathrm{s}}(1|3) = 1; \\ \left[1 + \frac{1}{p_2} + \frac{1}{p_3}\right]^{-1} \left(1, \frac{1}{p_2}, \frac{1}{p_3}\right), & \text{if } \pi^{\mathrm{s}}(2|2) = p_2 > 0 \\ & \text{and } \pi^{\mathrm{s}}(2|3) = p_3 > 0. \end{cases}$$

Actually, $f_{\pi^{\mathrm{s}},1}(x, a) = \hat{f}_{\pi^{\mathrm{s}},1}(x)\pi^{\mathrm{s}}(a|x)$. No-one vector $\hat{f}_{\pi^{\mathrm{s}},1}$ has components $\hat{f}_{\pi^{\mathrm{s}},1}(1) = 0$ and $\hat{f}_{\pi^{\mathrm{s}},1}(2) \in (0, 1)$ simultaneously.

On the other hand, consider the following Markov strategy:

$$\pi_t^{\mathrm{m}}(1|2) = 1 - \pi_t^{\mathrm{m}}(2|2) = \begin{cases} p \in (0, 1), & \text{if } t = 2; \\ 1, & \text{otherwise}, \end{cases} \qquad \pi_t^{\mathrm{m}}(1|3) \equiv 1,$$

where $\pi_t^{\rm m}(a|1)$ can be arbitrary as state 1 is uncontrolled. The (marginal) expected frequencies are

$$\hat{f}^1_{\pi^{\rm m},1}(2) := \sum_{a\in\mathbf{A}} \bar{f}^1_{\pi^{\rm m},1}(2,a) = 0, \qquad \hat{f}^2_{\pi^{\rm m},1}(2) := \sum_{a\in\mathbf{A}} \bar{f}^2_{\pi^{\rm m},1}(2,a) = 1/2,$$

$$\hat{f}^3_{\pi^{\rm m},1}(2) := \sum_{a\in\mathbf{A}} \bar{f}^3_{\pi^{\rm m},1}(2,a) = \frac{1}{3}(1+p),$$

$$\hat{f}^4_{\pi^{\rm m},1}(2) := \sum_{a\in\mathbf{A}} \bar{f}^4_{\pi^{\rm m},1}(2,a) = \frac{1}{4}(1+2p), \dots ,$$

$$\hat{f}^T_{\pi^{\rm m},1}(2) := \sum_{a\in\mathbf{A}} \bar{f}^T_{\pi^{\rm m},1}(2,a) = \frac{1}{4}(1+(T-2)p),$$

so that the only limit point equals

$$\hat{f}_{\pi^{\rm m},1}(2) = \lim_{T\to\infty} \hat{f}^T_{\pi^{\rm m},1}(2) = p.$$

Obviously, $\hat{f}_{\pi^{\rm m},1}(1) = 0$ $\hat{f}_{\pi^{\rm m},1}(3) = 1-p$ and, as was explained above, $f_{\pi^{\rm m},1} \notin \bigcup_{\pi\in\Pi^{\rm S}} F_{\pi,1}$.

Incidentally, $F_{\pi^{\rm m},1}$ contains a single point $f_{\pi^{\rm m},1}$: $f_{\pi^{\rm m},1}(1,a) \equiv 0$, $f_{\pi^{\rm m},1}(2,1) = p$, $f_{\pi^{\rm m},1}(2,2) = 0$, $f_{\pi^{\rm m},1}(3,1) = 1-p$, $f_{\pi^{\rm m},1}(3,2) = 0$ coincident with the feasible solution

$$\eta(1,a) \equiv 0, \quad \eta(2,1) = p, \quad \eta(2,2) = 0, \quad \eta(3,1) = 1-p, \quad \eta(3,2) = 0$$

to the linear program (4.6). □

4.3.21 *Blackwell optimal strategies and AC-optimality*

It is known that in a finite model a Blackwell optimal strategy exists and is AC-optimal [Bertsekas (2001), Prop. 4.2.2]; [Kallenberg (2010), Thm. 4.1, Cor. 5.3].

The first example shows that, even in finite models, an AC-optimal strategy may be not Blackwell optimal. The same occurs in the last example of Subsection 4.3.27.

Example. Let $\mathbf{X} = \{0,1\}$, $\mathbf{A} = \{0,1\}$, $p(0|0,a) \equiv 1$, $p(0|1,0) = 1$, $p(1|1,1) = 1$, with all other transition probabilities being zero; $c(0,a) \equiv 0$, $c(1,1) = 0$, $c(1,0) = 1$ (see Fig. 4.26).

Clearly, $v_x^{*,\beta} \equiv 0$ and the stationary selector $\varphi(x) \equiv x$ is Blackwell optimal (and also AC-optimal). At the same time, for any control strategy π we have $v_x^\pi = v_x^* = 0$, because only trajectories $(0,0,\dots)$, $(1,1,\dots)$, $(1,1,\dots,0,0,\dots)$ can be realized. Thus, all strategies are AC-optimal, but the stationary selector $\varphi(x) \equiv 0$ is not Blackwell optimal: it is not optimal for any value $\beta \in (0,1)$, if $x_0 = 1$. □

The second example, based on [Flynn (1974)], shows that, if the state space $\mathbf{X}$ is not finite, then a Blackwell optimal strategy may not be AC-optimal.

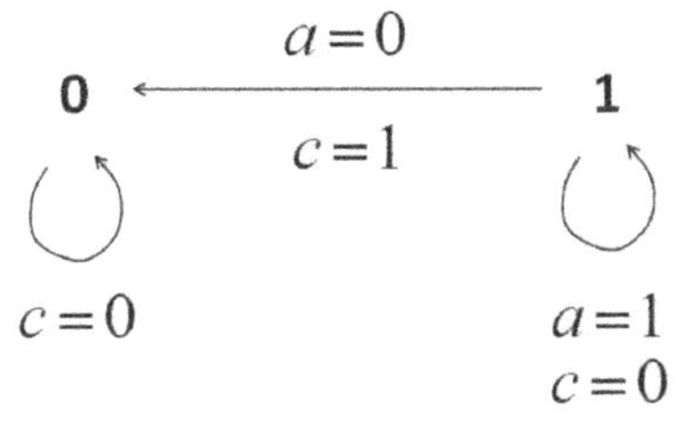

Fig. 4.26 AC-optimal stationary selector $\varphi(x) \equiv 0$ is not Blackwell optimal.

Example. Let $C_1, C_2, \ldots$ be a non-negative bounded sequence such that

$$C^* := \limsup_{n\to\infty} \frac{1}{n}\sum_{i=1}^{n} C_i > \limsup_{\beta\to 1-}(1-\beta)\sum_{i=1}^{\infty}\beta^{i-1}C_i := C_*.$$

Such a sequence exists: one can take $C_i := 1 - u_{i-1}$, where $\{u_i\}_{i=0}^{\infty}$ are as in Example 2 of [Bishop *et al.* (2014)]. Suppose $\mathbf{X} = \{1, 2, \ldots\}$; $\mathbf{A} = \{0, 1\}$; $p(1|1,0) = 1$, $p(2|1,1) = 1$, $p(i+1|i,a) \equiv 1$ for all $i > 1$; all other transition probabilities are zero. We put $c(1,0) := (C_* + C^*)/2$, $c(1,1) = C_1$, $c(i,a) \equiv C_i$ for all $i > 1$ (see Fig. 4.27).

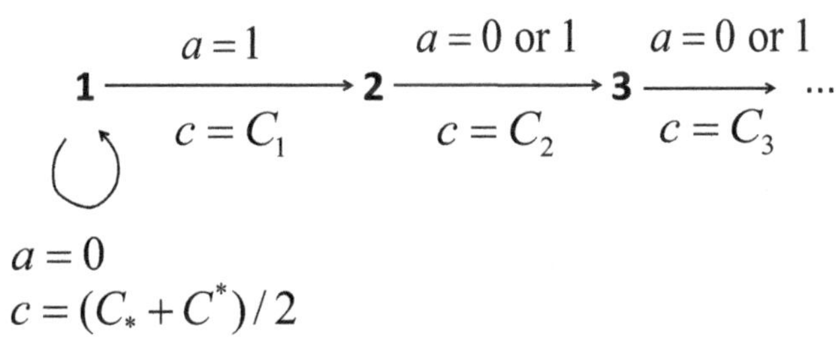

Fig. 4.27 Blackwell optimal strategy $\varphi^1(x) \equiv 1$ is not AC-optimal.

The process is uncontrolled if the initial state x_0 is $2, 3, \ldots$, and the actions in states $2, 3, \ldots$ are immaterial. Thus, we concentrate on $x_0 = 1$ and do not pay any attention to the actions in states $2, 3, \ldots$.

For the discounted version of this example with discount factor $\beta \in (0, 1)$

we have

$$v_2^{\pi,\beta} = v_2^{*,\beta} = \sum_{i=2}^{\infty} \beta^{i-2} C_i \quad \text{for all } \pi;$$

$$v_1^{*,\beta} = \min \left\{ (C_* + C^*)/2 + \beta v_1^{*,\beta};\ \sum_{i=1}^{\infty} \beta^{i-1} C_i \right\}.$$

To solve the second equation, let us compare $\frac{(C_*+C^*)/2}{1-\beta}$ and $\sum_{i=1}^{\infty} \beta^{i-1} C_i$. For all β sufficiently close to 1,

$$(1-\beta) \sum_{i=1}^{\infty} \beta^{i-1} C_i < C_* + \varepsilon$$

with $\varepsilon = (C_* + C^*)/2 - C_* > 0$, so that

$$(1-\beta) \sum_{i=1}^{\infty} \beta^{i-1} C_i < (C_* + C^*)/2$$

leading to the desired inequality

$$\sum_{i=1}^{\infty} \beta^{i-1} C_i < \frac{(C_* + C^*)/2}{1-\beta}$$

and meaning that the only Blackwell optimal strategy is the stationary selector $\varphi^1(x) \equiv 1$. (We ignore the actions in states $2, 3, \ldots$.)

Under the stationary selector φ^1, starting from the initial state $x_0 = 1$, only one trajectory $(1, 2, \ldots)$ can be realized ($\mathsf{P}_1^{\varphi^1}$-a.s.) with

$$\mathsf{E}_1^{\varphi^1} \left[\frac{1}{T} \sum_{t=1}^{T} c(X_{t-1}, A_t) \right] = \frac{1}{T} \sum_{i=1}^{T} C_i \quad \text{for all } T = 1, 2, \ldots.$$

Thus,

$$v_1^{\varphi^1} = \limsup_{T\to\infty} \frac{1}{T} \mathsf{E}_1^{\varphi^1} \left[\sum_{t=1}^{T} c(X_{t-1}, A_t) \right] = \limsup_{T\to\infty} \frac{1}{T} \sum_{i=1}^{T} C_i = C^*.$$

But for $\varphi^0(x) \equiv 0$

$$v_1^{\varphi^0} = (C_* + C^*)/2 < C^* = v_1^{\varphi^1},$$

so that the Blackwell optimal strategy φ^1 is not AC-optimal. □

4.3.22 *Overtaking optimal strategies*

Definition 4.6. [Puterman (1994), Section 5.4.2] A strategy π^* is called *overtaking optimal* if, for each strategy π,

$$\limsup_{T\to\infty}\left\{\mathsf{E}_x^{\pi^*}\left[\sum_{t=1}^{T} c(X_{t-1},A_t)\right]-\mathsf{E}_x^{\pi}\left[\sum_{t=1}^{T} c(X_{t-1},A_t)\right]\right\}\le 0,\quad x\in\mathbf{X}.$$

Any overtaking optimal strategy π^* is also AC-optimal, because for all π and $x\in\mathbf{X}$

$$\begin{aligned}
&\limsup_{T\to\infty}\frac{1}{T}\mathsf{E}_x^{\pi^*}\left[\sum_{t=1}^{T} c(X_{t-1},A_t)\right]\\
&\le \limsup_{T\to\infty}\frac{1}{T}\left\{\mathsf{E}_x^{\pi^*}\left[\sum_{t=1}^{T} c(X_{t-1},A_t)\right]-\mathsf{E}_x^{\pi}\left[\sum_{t=1}^{T} c(X_{t-1},A_t)\right]\right\}\\
&+\limsup_{T\to\infty}\frac{1}{T}\mathsf{E}_x^{\pi}\left[\sum_{t=1}^{T} c(X_{t-1},A_t)\right]\le \limsup_{T\to\infty}\frac{1}{T}\mathsf{E}_x^{\pi}\left[\sum_{t=1}^{T} c(X_{t-1},A_t)\right].
\end{aligned}$$

Similar reasoning confirms that any overtaking optimal strategy minimizes the opportunity loss; that is, it solves problem (3.20).

$$\begin{aligned}
&\forall\pi\ \ \forall x\in\mathbf{X}\ \ \limsup_{T\to\infty}\left\{\mathsf{E}_x^{\pi^*}\left[\sum_{t=1}^{T} c(X_{t-1},A_t)\right]-V_x^T\right\}\\
&\le \limsup_{T\to\infty}\left\{\mathsf{E}_x^{\pi^*}\left[\sum_{t=1}^{T} c(X_{t-1},A_t)\right]-\mathsf{E}_x^{\pi}\left[\sum_{t=1}^{T} c(X_{t-1},A_t)\right]\right\}\\
&+\limsup_{T\to\infty}\left\{\mathsf{E}_x^{\pi}\left[\sum_{t=1}^{T} c(X_{t-1},A_t)\right]-V_x^T\right\}\\
&\le \limsup_{T\to\infty}\left\{\mathsf{E}_x^{\pi}\left[\sum_{t=1}^{T} c(X_{t-1},A_t)\right]-V_x^T\right\}.
\end{aligned}$$

See also [Flynn (1980), Corollary 1], where this assertion is formulated for discrete models.

The following example, based on [Denardo and Miller (1968), p. 1221], shows that an AC-optimal, or Blackwell optimal, or bias optimal, or opportunity-loss optimal strategy may be not overtaking optimal, moreover an overtaking optimal strategy may not exist even in the simplest finite models.

Example. Let $\mathbf{X}=\{1,2,3\}$, $\mathbf{A}=\{1,2\}$, $p(2|1,1)=p(3|1,2)=1$, $p(3|2,a)\equiv p(2|3,a)\equiv 1$, with all other transition probabilities being zero. We put $c(1,1)=1$, $c(1,2)=0$, $c(2,a)\equiv 0$, $c(3,a)\equiv 2$ (see Fig. 4.28).

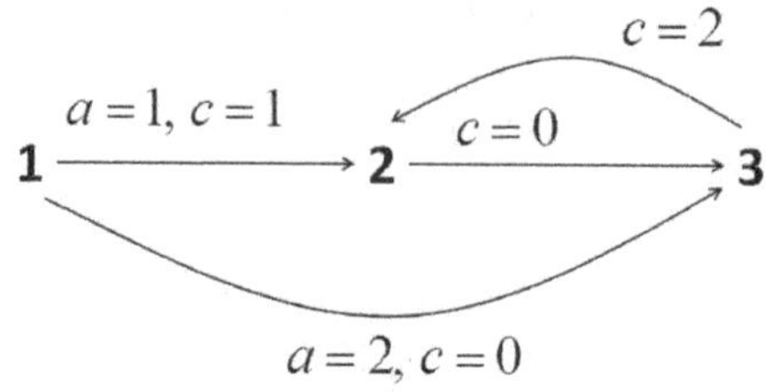

Fig. 4.28 No overtaking optimal strategies.

In fact, there are only two essentially different stationary selectors $\varphi^1(x) \equiv 1$ and $\varphi^2(x) \equiv 2$ (actions in states 2 and 3 play no role). Suppose $x_0 = 1$. Then the values of $\mathsf{E}_1^{\varphi^{1,2}}\left[\sum_{t=1}^T c(X_{t-1}, A_t)\right]$ for different values of T are given in the following table:

	$T = 1$	2	3	4	5	6	7	8	...
φ^1	1	1	3	3	5	5	7	7	...
φ^2	0	2	2	4	4	6	6	8	...

Therefore, for any strategy π^* there exists a strategy π such that

$$\limsup_{T\to\infty}\left\{\mathsf{E}_1^{\pi^*}\left[\sum_{t=1}^T c(X_{t-1}, A_t)\right] - \mathsf{E}_1^{\pi}\left[\sum_{t=1}^T c(X_{t-1}, A_t)\right]\right\} > 0,$$

and no-one strategy is overtaking optimal. At the same time, all strategies are equally AC-optimal with $v_x^* \equiv 1$. The canonical triplets are given by $\rho = 1$, $h(1) = h(2) = 0$, $h(3) = 1$ with φ^* being either φ^1 or φ^2: the both stationary selectors are canonical.

Investigation of the discounted version of this MDP results in

$$v_1^{*,\beta} = v_2^{*,\beta} = \frac{2\beta}{1-\beta^2}; \quad v_3^{*,\beta} = \frac{2}{1-\beta^2},$$

and only the stationary selector φ^2 is optimal for all $\beta \in (0, 1)$, hence Blackwell optimal. (Like previously, we ignore the actions in states 2 and 3). Like any Blackwell optimal strategy, φ^2 is bias optimal. Incidentally, the stationary selector φ^1 is also bias optimal:

$$v_1^{\varphi^1,\beta} - v_1^{\varphi^2,\beta} = \frac{1+\beta^2}{1-\beta^2} - \frac{2\beta}{1-\beta^2} \to 0 \text{ as } \beta \to 1-.$$

In the states $x = 2, 3$ the values $v_x^{\varphi^1,\beta}$ and $v_x^{\varphi^2,\beta}$ are obviously equal.

Statements on the existence of overtaking optimal strategies like [Denardo and Rothblum (1979), Cor. 2] and [Hernández-Lerma and Lasserre (1999), Thm. 10.3.11] strongly rely on the assumption that the controlled process X_t is aperiodic under each stationary selector which is not the case in the considered example.

The $(1/2, 1/2)$-mixture π^* of selectors φ^1 and φ^2 is an opportunity-loss optimal strategy: it solves problem (3.20). Indeed, one has to look only at state 1. According to the presented above table, $V_1^T = T - 1$, $\mathsf{E}_1^{\pi^*}\left[\sum_{t=1}^T c(X_{t-1}, A_t)\right] = T - 1/2$, and, for any other control strategy (i.e., for any other value of $\pi_1(1|1) \neq 1/2$), again according to that table, the left-hand side of (3.20) is larger than 1/2. Exactly the same reasoning confirms that π^* is *D-optimal* [Hernández-Lerma and Lasserre (1999), p. 119], i.e. it provides the minimum to

$$D(\pi, x) := \limsup_{T\to\infty}\left\{E_x^\pi\left[\sum_{t=1}^T c(X_{t-1}, A_t)\right] - Tv_x^*\right\}$$

for all $x \in \mathbf{X}$. Recall that $v_x^* \equiv 1$ in the current example.

Theorem 10.3.11 in [Hernández-Lerma and Lasserre (1999)] says that, under appropriate conditions, many types of optimality are equivalent. For instance, any stationary selector φ^*, D-optimal *among stationary selectors*, is also *weakly* overtaking optimal* among stationary selectors; that is, for each $\varepsilon > 0$ and any stationary selector φ,

$$\mathsf{E}_x^{\varphi^*}\left[\sum_{t=1}^T c(X_{t-1}, A_t)\right] \leq \mathsf{E}_x^{\varphi}\left[\sum_{t=1}^T c(X_{t-1}, A_t)\right] + \varepsilon$$

as soon as $T \geq N(\varphi^*, \varphi, x, \varepsilon)$. In the above example, φ^1 is D-optimal among stationary selectors, as well as φ^2: $D(\varphi^{1,2}, 1) \equiv 0$, $D(\varphi^{1,2}, 2) \equiv 0$, $D(\varphi^{1,2}, 3) \equiv 1$. But

$$\begin{aligned}
&\limsup_{T\to\infty}\left\{\mathsf{E}_1^{\varphi^1}\left[\sum_{t=1}^T c(X_{t-1}, A_t)\right] - \mathsf{E}_1^{\varphi^2}\left[\sum_{t=1}^T c(X_{t-1}, A_t)\right]\right\} \\
&= \limsup_{T\to\infty}\left\{\mathsf{E}_1^{\varphi^2}\left[\sum_{t=1}^T c(X_{t-1}, A_t)\right] - \mathsf{E}_1^{\varphi^1}\left[\sum_{t=1}^T c(X_{t-1}, A_t)\right]\right\} \\
&= 1 > 0.
\end{aligned}$$

Theorem 10.3.11 from [Hernández-Lerma and Lasserre (1999)] is not applicable here, because the controlled process X_t is not geometric ergodic under the strategies φ^1 and φ^2, since it is not aperiodic. □

Along with Definition 4.6, one can define the *weakly overtaking optimal* strategies π^* for which

$$\liminf_{T\to\infty}\left\{\mathsf{E}_x^{\pi^*}\left[\sum_{t=1}^{T} c(X_{t-1},A_t)\right]-\mathsf{E}_x^{\pi}\left[\sum_{t=1}^{T} c(X_{t-1},A_t)\right]\right\}\le 0,\quad x\in\mathbf{X}$$

for all strategies π: see, e.g., [Schochetman and Smith (2005)]. In the following model, similar to Example 3.15 of [Schochetman and Smith (2005)], all strategies are weakly overtaking optimal, but only one (in some sense) of them is overtaking optimal and AC-optimal.

Example. Let $\mathbf{X}=\{(i,0),\ i=0,1,2,\ldots\}\cup\{(i,1),\ i=1,2,\ldots\}$, $\mathbf{A}=\{0,1\}$, $p((1,0)|(0,0),0)=p((1,1)|(0,0),1)=p((i+1,k)|(i,k),a)\equiv 1$ for all $i=1,2,\ldots$, $k=0,1$ with all other transition probabilities being zero; $c((0,0),0)=1$, $c((i,0),a)\equiv 1$ for all $i=1,2,\ldots$; $c((0,0),1)=2$, $c((i,1),a)\equiv \begin{cases}2^m, & \text{if } i=2^m-2 \text{ for some } m=2,3,\ldots;\\ 0, & \text{otherwise}\end{cases}$ for all $i=1,2,\ldots$. See Fig. 4.29. We assume that the initial state is $x_0=(0,0)$.

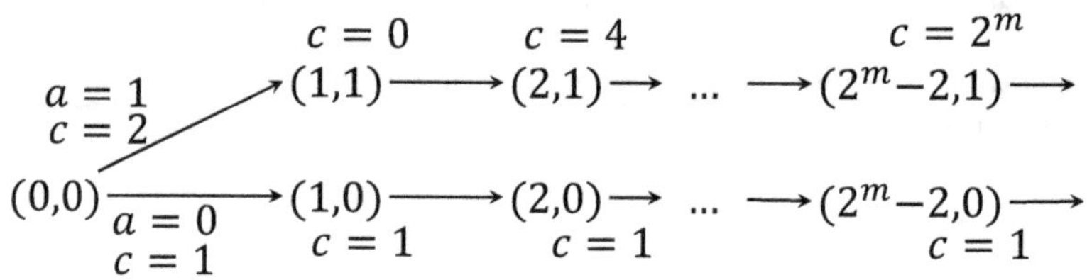

Fig. 4.29 Weakly overtaking selector is not AC-optimal.

Actions in all states, apart from $(0,0)$, are immaterial, so actually there are only two different stationary selectors $\varphi^0(x)\equiv 0$ and $\varphi^1(x)\equiv 1$, and any control strategy π with $\pi_1(0|(0,0))=\alpha$ leads to the same total expected cost on each finite interval $\{1,2,\ldots,T\}$ as the $(\alpha,1-\alpha)$-mixture of selectors φ^0 and φ^1. Note that, for each T, $\mathsf{E}_{(0,0)}^{\varphi^0}\left[\sum_{t=1}^{T} c(X_{t-1},A_t)\right]=T$.

If $T=2^m-2$ for some $m\ge 2$, then

$$\mathsf{E}_{(0,0)}^{\varphi^1}\left[\sum_{t=1}^{T} c(X_{t-1},A_t)\right]=\sum_{i=1}^{m-1}2^i=2^m-2=\mathsf{E}_{(0,0)}^{\varphi^0}\left[\sum_{t=1}^{T} c(X_{t-1},A_t)\right],$$

and, for any strategy π (equivalent to the mixture of selectors φ^0 and φ^1), the value of the total expected cost is the same.

Suppose $2^m - 2 < T \le 2^{m+1} - 3$ for some $m \ge 2$. Then

$$\mathsf{E}^{\varphi^1}_{(0,0)}\left[\sum_{t=1}^{T} c(X_{t-1}, A_t)\right] = 2^{m+1} - 2 > T = \mathsf{E}^{\varphi^0}_{(0,0)}\left[\sum_{t=1}^{T} c(X_{t-1}, A_t)\right],$$

and, for any strategy π, the value of the total expected cost is not smaller than T. To summarize, for all T, for all strategies π,

$$\mathsf{E}^{\varphi^0}_{(0,0)}\left[\sum_{t=1}^{T} c(X_{t-1}, A_t)\right] = T \le \mathsf{E}^{\pi}_{(0,0)}\left[\sum_{t=1}^{T} c(X_{t-1}, A_t)\right]$$

meaning that the stationary selector φ^0 is overtaking optimal and hence weakly overtaking optimal and AC-optimal.

Suppose $T = 2^m - 2$. As was shown above, for all strategies π^1 and π^2,

$$\mathsf{E}^{\pi^1}_{(0,0)}\left[\sum_{t=1}^{T} c(X_{t-1}, A_t)\right] - \mathsf{E}^{\pi^2}_{(0,0)}\left[\sum_{t=1}^{T} c(X_{t-1}, A_t)\right] = 0,$$

so that anyone strategy is weakly overtaking optimal.

Finally, let us show that no-one strategy different from φ^0 is overtaking optimal or AC-optimal. Recall that we ignore the actions in all states apart from $(0,0)$. As was mentioned above, we need to consider only an $(\alpha, 1-\alpha)$-mixture π of selectors φ^0 and φ^1 with $\alpha < 1$.

For the sequence $T_m := 2^{m+1} - 3 \to \infty$ (as $m \to \infty$), we have

$$\begin{aligned}
&\mathsf{E}^{\pi}_{(0,0)}\left[\sum_{t=1}^{T_m} c(X_{t-1}, A_t)\right] - \mathsf{E}^{\varphi^0}_{(0,0)}\left[\sum_{t=1}^{T_m} c(X_{t-1}, A_t)\right]\\
&= \alpha(2^{m+1} - 3) + (1-\alpha)(2^{m+1} - 2) - (2^{m+1} - 3) = 1 - \alpha > 0\\
&\Longrightarrow \limsup_{T\to\infty}\left\{\mathsf{E}^{\pi}_{(0,0)}\left[\sum_{t=1}^{T} c(X_{t-1}, A_t)\right] - \mathsf{E}^{\varphi^0}_{(0,0)}\left[\sum_{t=1}^{T} c(X_{t-1}, A_t)\right]\right\} > 0.
\end{aligned}$$

Hence, the strategy π does not overtake the selectorφ^0.

For the sequence $T_m := 2^m - 1 \to \infty$ (as $m \to \infty$), we have

$$\begin{aligned}
&\frac{1}{T_m}\mathsf{E}^{\pi}_{(0,0)}\left[\sum_{t=1}^{T_m} c(X_{t-1}, A_t)\right] = \frac{\alpha(2^m - 1) + (1-\alpha)(2^{m+1} - 2)}{2^m - 1}\\
&\to \alpha + (1-\alpha)\cdot 2 = 2 - \alpha > 1 \text{ as } m \to \infty.
\end{aligned}$$

Therefore, $\limsup_{T\to\infty} \frac{1}{T}\mathsf{E}^{\pi}_{(0,0)}\left[\sum_{t=1}^{T} c(X_{t-1}, A_t)\right] > 1$, while $\limsup_{T\to\infty} \frac{1}{T}\mathsf{E}^{\varphi^0}_{(0,0)}\left[\sum_{t=1}^{T} c(X_{t-1}, A_t)\right] = v^*_{(0,0)} = 1$.

The proof is completed. □

4.3.23 AC-optimal and average-overtaking optimal strategies

The standard average cost (4.1) is under-selective because it does not take into account the finite-horizon accumulated cost $\mathsf{E}_x^\pi\left[\sum_{t=1}^T c(X_{t-1}, A_t)\right]$.

Example. Consider the following MDP: $\mathbf{X} = \{\Delta, 1\}$, $\mathbf{A} = \{1, 2\}$, $p(\Delta|\Delta, a) \equiv 1$, $p(\Delta|1, a) \equiv 1$, $c(1, 1) = 0$, $c(1, 2) = 1$, $c(\Delta, a) \equiv 0$ (see Fig. 4.30).

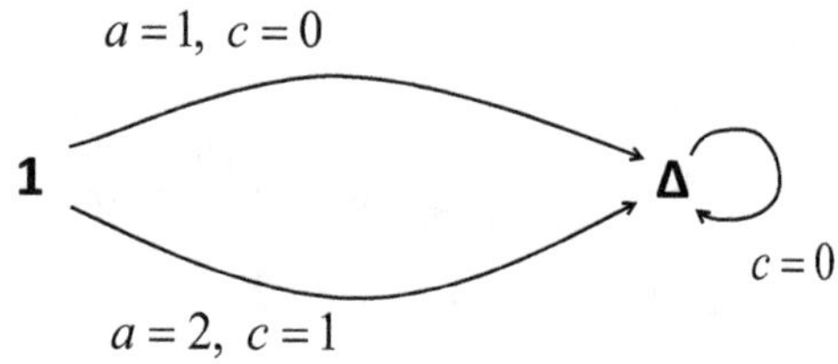

Fig. 4.30 AC-optimal selector φ^2 is not average-overtaking optimal.

In fact, there are only two essentially different strategies (stationary selectors) $\varphi^1(x) \equiv 1$ and $\varphi^2(x) \equiv 2$ (the actions in state Δ play no role). Suppose $x_0 = 1$. Then, for any finite $T \geq 1$,

$$\mathsf{E}_1^{\varphi^1}\left[\sum_{t=1}^T c(X_{t-1}, A_t)\right] = 0 \text{ and } \mathsf{E}_1^{\varphi^2}\left[\sum_{t=1}^T c(X_{t-1}, A_t)\right] = 1,$$

so that it is natural to say that selector φ^1 is better than φ^2. But formula (4.1) gives $v_x^{\varphi^1} \equiv v_x^{\varphi^2} \equiv 0$, meaning that all strategies are equally AC-optimal. On the other hand, as Subsection 4.3.22 shows, Definition 4.6 gives an over-selective notion of optimality. That is why the following definition is sometimes used.

Definition 4.7. [Puterman (1994), Section 5.4.2]. A strategy π^* is called *average-overtaking optimal* if, for each strategy π, for all $x \in \mathbf{X}$,

$$\limsup_{T\to\infty} \frac{1}{T}\sum_{t=1}^T \left\{\mathsf{E}_x^{\pi^*}\left[\sum_{\tau=1}^t c(X_{\tau-1}, A_\tau)\right] - \mathsf{E}_x^{\pi}\left[\sum_{\tau=1}^t c(X_{\tau-1}, A_\tau)\right]\right\} \leq 0.$$

In the above example (Fig. 4.30) the selector φ^1 is average-overtaking optimal, while φ^2 is not:

$$\limsup_{T\to\infty} \frac{1}{T}\sum_{t=1}^T \left\{\mathsf{E}_1^{\varphi^2}\left[\sum_{\tau=1}^t c(X_{\tau-1}, A_\tau)\right] - \mathsf{E}_1^{\varphi^1}\left[\sum_{\tau=1}^t c(X_{\tau-1}, A_\tau)\right]\right\} = 1.$$

□

The following example shows that an average-overtaking optimal strategy may be not AC-optimal (and hence not overtaking optimal).

Example. Let $\mathbf{X} = \{\Delta, 0, 1, 2, \ldots\}$, $\mathbf{A} = \{1, 2\}$, $p(1|0,1) = p(\Delta|0,2) = 1$, $p(i+1|i,a) \equiv 1$ for all $i \geq 1$, with all other transition probabilities being zero; $c(\Delta, a) \equiv 0$, $c(0,1) = -1$, $c(0,2) = 0$. To describe the cost function $c(i,a)$ for $i \geq 1$, we introduce the following increasing sequence $\{m_j\}_{j=0}^{\infty}$. $m_0 := 0$; for each $j \geq 0$, let n_j be the first natural number satisfying inequality $0.9n_j > 1.1 + 0.1m_j$, and put $m_{j+1} := m_j + 2n_j$. Now, for $i \geq 1$,

$$c(i,a) \equiv \begin{cases} 2n_j - 1, & \text{if } i = m_j + n_j \text{ for some } j \geq 0; \\ -1, & \text{otherwise} \end{cases}$$

Note that $\{n_j\}_{j=0}^{\infty}$ is a non-decreasing sequence. See Fig. 4.31.

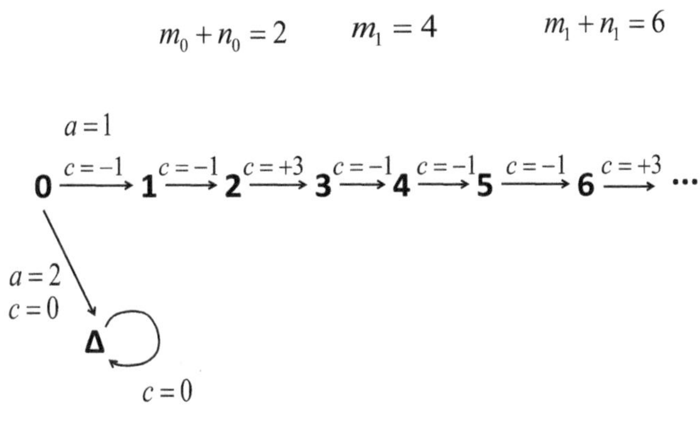

Fig. 4.31 Average-overtaking optimal strategy is not AC-optimal.

The first several values of $m_j, n_j, i = m_j + n_j$ and $c(i,a)$ are given in the following table

j	0	1	2	3	4	5	...
m_j	0	4	8	14	20	28	...
n_j	2	2	3	3	4	5	...
$i = m_j + n_j$	2	6	11	17	24	33	...
$c(i,a)$	3	3	5	5	7	9	...

In fact, there are only two essentially different strategies (stationary selectors) $\varphi^1(x) \equiv 1$ and $\varphi^2(x) \equiv 2$ (the actions in states $\Delta, 1, 2, \ldots$ play no role). Suppose $x_0 = 0$. All the other initial states are of no interest because, starting from them, the process X_t is uncontrolled. Clearly, $\mathsf{E}_0^{\varphi^2}\left[\sum_{t=1}^{T} c(X_{t-1}, A_t)\right] = 0$ for all $T = 1, 2, \ldots$.

For the strategy φ^1, $\mathsf{E}_0^{\varphi^1}\left[\sum_{t=1}^{m_j} c(X_{t-1},A_t)\right] = 0$ for all $j \geq 0$. To see this, notice that this equality holds at $j=0$. If it holds for some $j \geq 0$, then

$$\mathsf{E}_0^{\varphi^1}\left[\sum_{t=1}^{m_{j+1}} c(X_{t-1},A_t)\right] = -n_j + (2n_j - 1) - [m_{j+1} - (m_j + n_j + 1)] = 0.$$

Here the fist term corresponds to $t = m_j+1, m_j+2, \dots, m_j+n_j$ and the last term corresponds to $t = m_j + n_j + 2, m_j + n_j + 3, \dots, m_{j+1}$ when $c(X_{t-1},A_t) = -1$. Recall that $m_{j+1} = m_j + 2n_j$.

As a consequence,

$$E_0^{\varphi^1}\left[\sum_{t=1}^{m_j+n_j+1} c(X_{t-1},A_t)\right] = -n_j + (2n_j - 1) = n_j - 1$$

and, since function $\frac{y-a}{y+b} = 1 - \frac{a+b}{y+b}$ increases in y for $a + b > 0$,

$$\frac{1}{m_j+n_j+1} E_0^{\varphi^1}\left[\sum_{t=1}^{m_j+n_j+1} c(X_{t-1},A_t)\right] = \frac{0.9(n_j - 1)}{0.9(m_j + n_j + 1)}$$

$$> \frac{(1.1 + 0.1m_j) - 0.9}{0.9m_j + (1.1 + 0.1m_j) + 0.9} = 0.1,$$

so that the strategy φ^1 is not AC-optimal: recall that $v_0^{\varphi^2} = 0$.

On the other hand, we will show that, for each $j = 0, 1, \dots$,

$$\sum_{t=1}^{m_j} \mathsf{E}_0^{\varphi^1}\left[\sum_{\tau=1}^{t} c(X_{\tau-1},A_\tau)\right] = -\sum_{i=0}^{j-1} n_i \leq 0$$

and, for all $T \in [m_{j-1}, m_j]$, $\sum_{t=1}^{T} \mathsf{E}_0^{\varphi^1}\left[\sum_{\tau=1}^{t} c(X_{\tau-1},A_\tau)\right] \leq 0$. (When $j = 0$, we put $m_{j-1} = m_j = 0$.) Moreover, the latter inequality is strict if $T > 0$.

This assertion obviously holds for $j = 0$. Suppose it holds for some $j \geq 0$. Since $\mathsf{E}_0^{\varphi^1}\left[\sum_{\tau=1}^{m_j} c(X_{\tau-1},A_\tau)\right] = 0$ and there is only one value of $\tau \in [m_j, m_{j+1} = m_j + 2n_j]$ when $\tau = m_j + n_j + 1$ (hence $X_{\tau-1} = m_j + n_j$), the values of $\mathsf{E}_0^{\varphi^1}\left[\sum_{\tau=1}^{t} c(X_{\tau-1},A_\tau)\right]$ equal

$$0,\ -1,\ -2, \dots, -n_j,\ -n_j + 2n_j - 1 = n_j - 1,\ n_j - 2, \dots, n_j - n_j = 0$$

when

$$t = m_j, m_j + 1, m_j + 2, \dots, m_j + n_j, m_j + n_j + 1, m_j + n_j + 2, \dots, m_{j+1}.$$

Thus, denoting $K := -\sum_{i=1}^{j-1} n_i$, the values of $F(T) := \sum_{t=1}^{T} \mathsf{E}_0^{\varphi^1}\left[\sum_{\tau=1}^{t} c(X_{\tau-1}, A_\tau)\right]$, by the inductive supposition, equal

$$K \leq 0,\ K-1,\ K-1-2, \dots, K - \frac{n_j(n_j+1)}{2},$$

$$K - \frac{n_j(n_j+1)}{2} + n_j - 1, K - \frac{n_j(n_j+1)}{2} + n_j - 1 + n_j - 2,$$

$$\dots, K - \frac{n_j(n_j+1)}{2} + n_j \cdot n_j - \frac{n_j(n_j+1)}{2} = K - n_j = -\sum_{i=1}^{j} n_i$$

when, correspondingly,

$$T = m_j\ m_j+1,\ m_j+2, \dots, m_j+n_j,$$

$$m_j+n_j+1,\ m_j+n_j+2,$$

$$\dots, m_j + 2n_j = m_{j+1}.$$

Clearly, the values of $F(T)$ initially decrease from $-\sum_{i=1}^{j-1} n_i \leq 0$, equal $-\sum_{i=1}^{j-1} n_i - \frac{n_j(n_j+1)}{2} + n_j - 1 < 0$ when $T = m_j + n_j + 1$, and increase thereafter up to $-\sum_{i=1}^{j} n_i$ when $T = m_{j+1}$. The proof by induction is completed. The very last statement on the strict inequality is now obvious.

As the conclusion,

$$\sum_{t=1}^{T} \mathsf{E}_0^{\varphi^1}\left[\sum_{\tau=1}^{t} c(X_{\tau-1}, A_\tau)\right] < 0 \quad \text{for all } T > 0,$$

and the stationary selector φ^1 is average-overtaking optimal: one has to compare φ^1 with φ^2 only. □

4.3.24 *Blackwell optimal, bias optimal, average-overtaking optimal and AC-optimal strategies*

Theorem 4.12. (Tauberian Theorem) [Bishop *et al.* **(2014)].** *If a sequence $\{u_n\}_{n=1}^{\infty}$ is bounded above or below, then*

$$\liminf_{n\to\infty} \frac{1}{n}\sum_{i=1}^{n} u_i \leq \liminf_{\beta\to 1-}(1-\beta)\sum_{i=1}^{\infty} \beta^{i-1} u_i$$
$$\leq \limsup_{\beta\to 1-}(1-\beta)\sum_{i=1}^{\infty} \beta^{i-1} u_i \leq \limsup_{n\to\infty} \frac{1}{n}\sum_{i=1}^{n} u_i.$$

Lemma 4.2. *There exist bounded sequences $\{z_j\}_{j=1}^{\infty}$ such that*

(a)

$$\liminf_{n\to\infty} \frac{1}{n}\sum_{i=1}^{n}\sum_{j=1}^{i} z_j = -1$$

and

$$\lim_{\beta\to 1-}\sum_{j=1}^{\infty}\beta^{j-1}z_j = \limsup_{n\to\infty}\frac{1}{n}\sum_{i=1}^{n}\sum_{j=1}^{i} z_j = 0;$$

(b)

$$\lim_{n\to\infty}\frac{1}{n}\sum_{i=1}^{n}\sum_{j=1}^{i} z_j = \infty \quad \textit{and} \quad \liminf_{n\to\infty}\frac{1}{n}\sum_{j=1}^{n} z_j < 0.$$

For the proof see Lemmas 5 and 4 of [Flynn (1976)] correspondingly.

The following example, based on [Flynn (1976), Ex. 1], shows that a Blackwell and bias optimal strategy may be not average-overtaking optimal.

Example. Let $\mathbf{X} = \{\Delta, 0, 1, 2, \ldots\}$, $\mathbf{A} = \{1, 2\}$, $p(\Delta|\Delta, a) \equiv 1$, $p(1|0,1) = 1$, $p(\Delta|0,2) = 1$, $p(i+1|i,a) \equiv 1$ for all $i \geq 1$. Let $\{C_j\}_{j=1}^{\infty}$ be a bounded sequence such that

$$\limsup_{n\to\infty}\frac{1}{n}\sum_{i=1}^{n}\sum_{j=1}^{i} C_j = 1 \text{ and } \lim_{\beta\to 1-}\sum_{j=1}^{\infty}\beta^{j-1}C_j = \liminf_{n\to\infty}\frac{1}{n}\sum_{i=1}^{n}\sum_{j=1}^{i} C_j = 0$$

One can take $C_j := -z_j$, where the sequence $\{z_j\}_{j=1}^{\infty}$ is as in Lemma 4.2(a). We put $c(0,1) = C_1$, $c(i,a) \equiv C_{i+1}$ for all $i \geq 1$, $c(0,2) = 1/4$, $c(\Delta, a) \equiv 0$ (see Fig. 4.32).

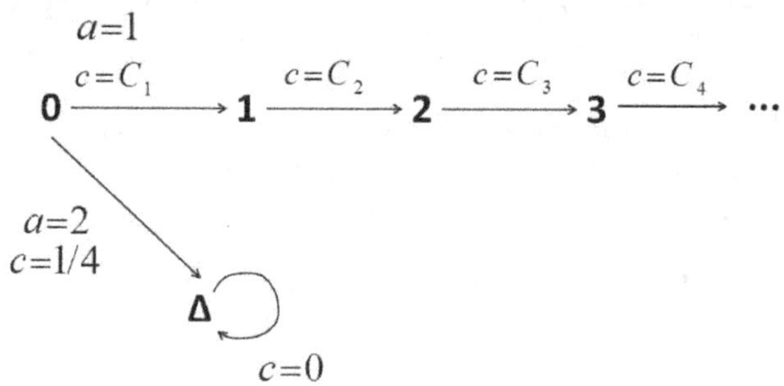

Fig. 4.32 Blackwell optimal strategy is not average-overtaking optimal.

In fact, there are only two essentially different strategies (stationary selectors) $\varphi^1(x) \equiv 1$ and $\varphi^2(x) \equiv 2$ (actions in states $\Delta, 1, 2, \ldots$ play no role). The selector

φ^1 is Blackwell optimal because

$$\lim_{\beta\to 1-} v_0^{\varphi^1,\beta} = \lim_{\beta\to 1-} \sum_{j=1}^{\infty} \beta^{j-1} C_j = 0 \text{ and } v_0^{\varphi^2,\beta} = 1/4.$$

It is also 0-discount optimal (i.e., bias optimal). At the same time,

$$\limsup_{T\to\infty} \frac{1}{T} \sum_{t=1}^{T} \left\{ \mathsf{E}_0^{\varphi^1} \left[\sum_{\tau=1}^{t} c(X_{\tau-1}, A_\tau) \right] - \mathsf{E}_0^{\varphi^2} \left[\sum_{\tau=1}^{t} c(X_{\tau-1}, A_\tau) \right] \right\}$$
$$= \limsup_{T\to\infty} \frac{1}{T} \sum_{t=1}^{T} \sum_{\tau=1}^{t} C_\tau - 1/4 = 3/4,$$

and hence φ^1 is not average-overtaking optimal. Note that selector φ^2 is not average-overtaking either, because

$$\limsup_{T\to\infty} \frac{1}{T} \sum_{t=1}^{T} \left\{ \mathsf{E}_0^{\varphi^2} \left[\sum_{\tau=1}^{t} c(X_{\tau-1}, A_\tau) \right] - \mathsf{E}_0^{\varphi^1} \left[\sum_{\tau=1}^{t} c(X_{\tau-1}, A_\tau) \right] \right\}$$
$$= 1/4 - \liminf_{T\to\infty} \frac{1}{T} \sum_{t=1}^{T} \sum_{\tau=1}^{t} C_\tau = 1/4.$$

Now, in the same example (Fig. 4.32), we can put $c(0,2) = 0$, so that $v_0^{\varphi^2,\beta} = 0$. The stationary selector φ^1 is again 0-discount optimal, but still not average-overtaking optimal, while φ^2 is both 0-discount and average-overtaking optimal. A similar example was presented in [Flynn (1976), Ex. 2].

Theorem 1 in [Lippman (1969)] states that, in finite models, any strategy is 0-discount optimal if and only if it is average-overtaking optimal. See also [Kallenberg (2010), Thm. 7.2]. The current example shows that, if the model is not finite, then a 0-discount optimal strategy can be not average-overtaking optimal.

Now consider the MDP with the same state and action spaces and the same transition probabilities (see Fig. 4.32), but with the different cost function $c(\cdot,\cdot)$. Namely, let $\{C_j\}_{j=1}^{\infty}$ be a bounded sequence such that

$$\lim_{n\to\infty} \frac{1}{n} \sum_{i=1}^{n} \sum_{j=1}^{i} C_j = -\infty \text{ and } \limsup_{n\to\infty} \frac{1}{n} \sum_{j=1}^{n} C_j > 0$$

One can take $C_j := -z_j$, where the sequence $\{z_j\}_{j=1}^{\infty}$ is as in Lemma 4.2(b). We put $c(0,1) = C_1$, $c(0,2) = 0$, $c(i,a) \equiv C_{i+1}$ for all $i \geq 1$, and $c(\Delta, a) \equiv 0$. Like previously, it is sufficient to consider only two strategies $\varphi^1(x) \equiv 1$ and $\varphi^2(x) \equiv 2$ and the initial state 0.

The stationary selector φ^1 is average-overtaking optimal, because

$$\limsup_{T\to\infty} \frac{1}{T}\sum_{t=1}^{T}\left\{\mathsf{E}_0^{\varphi^1}\left[\sum_{\tau=1}^{t} c(X_{\tau-1},A_\tau)\right] - \mathsf{E}_0^{\varphi^2}\left[\sum_{\tau=1}^{t} c(X_{\tau-1},A_\tau)\right]\right\}$$
$$= \limsup_{T\to\infty} \frac{1}{T}\sum_{t=1}^{T}\sum_{\tau=1}^{t} C_\tau = -\infty.$$

On the other hand, it is not AC-optimal, because

$$v_0^{\varphi^1} = \limsup_{n\to\infty} \frac{1}{n}\sum_{j=1}^{n} C_j > 0 \text{ and } v_0^{\varphi^2} = 0. \tag{4.28}$$

For finite models, the following statement holds [Kallenberg (2010), Lemma 7.3]: if a stationary selector is average-overtaking optimal, then it is AC-optimal. In the current example, the state space $\mathbf{X}$ is not finite.

The selector φ^2 is AC-optimal by (4.28), but not average-overtaking optimal:

$$\limsup_{T\to\infty} \frac{1}{T}\sum_{t=1}^{T}\left\{\mathsf{E}_0^{\varphi^2}\left[\sum_{\tau=1}^{t} c(X_{\tau-1},A_\tau)\right] - \mathsf{E}_0^{\varphi^1}\left[\sum_{\tau=1}^{t} c(X_{\tau-1},A_\tau)\right]\right\}$$
$$= -\liminf_{T\to\infty} \frac{1}{T}\sum_{t=1}^{T}\sum_{\tau=1}^{t} C_\tau = +\infty.$$

□

A similar example was presented in [Flynn (1976), Ex. 3].

4.3.25 *Nearly optimal and average-overtaking optimal strategies*

The following example, based on [Flynn (1976), Ex. 7], illustrates that an average-overtaking optimal strategy may be not nearly optimal (see Definition 3.4).

Example. Let $\mathbf{X} = \{0, 1, (1,1), (1,2), 2, (2,1), \dots, (2,4), 3, \dots, k, (k,1), \dots, (k,2k), k+1, \dots\}$, $\mathbf{A} = \{0,1\}$, $p(0|0,a) \equiv 1$, $p(k+1|k,0) = p((k,1)|k,1) \equiv 1$, $p((k,i+1)|(k,i),a) \equiv 1$ for all $i < 2k$; $p(0|(k,2k),a) \equiv 1$, with all other transition probabilities being zero. For all $k \ge 1$, we put

$$c(k,a) \equiv 0, \quad c((k,i),a) = \begin{cases} -1, & \text{if } 1 \le i \le k; \\ 2, & \text{if } k+1 \le i \le 2k. \end{cases}$$

Finally, $c(0,a) \equiv 0$: see Fig. 4.33.

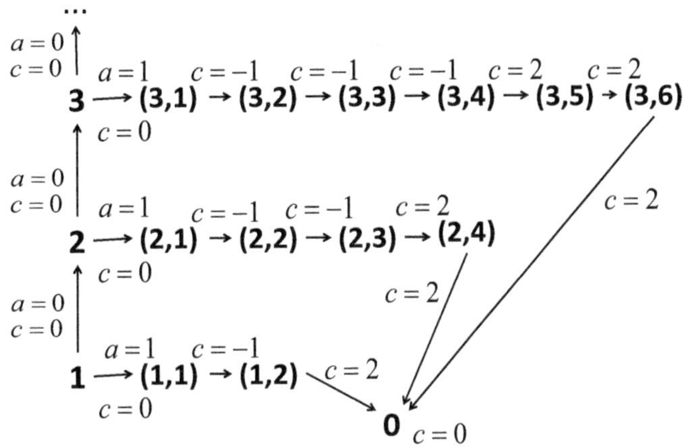

Fig. 4.33 Average-overtaking optimal strategy is not nearly optimal.

Proposition 4.3.

(a) For any $k \geq 1$, for any control strategy π,

$$\lim_{T\to\infty}\inf \frac{1}{T}\sum_{t=1}^{T} \mathsf{E}_k^\pi\left[\sum_{\tau=1}^{t} c(X_{\tau-1}, A_\tau)\right] \geq 0.$$

(b) There exists $\varepsilon > 0$ such that, for all $\beta \in (0,1)$ sufficiently close to 1, $v_1^{*,\beta} < -\varepsilon$.

The proof is given in Appendix A.

Consider the stationary selector $\varphi^0(x) \equiv 0$ with $c(X_{\tau-1}, A_\tau) \equiv 0$ $\mathsf{P}_k^{\varphi^0}$-a.s. for all $k \geq 1$. Proposition 4.3(a) implies that φ^0 is average-overtaking optimal. Note that it is sufficient to consider only the initial states $k \geq 1$:

$$\lim_{T\to\infty}\sup \frac{1}{T}\sum_{t=1}^{T}\left\{0 - \mathsf{E}_k^\pi\left[\sum_{\tau=1}^{t} c(X_{\tau-1}, A_\tau)\right]\right\}$$
$$= -\lim_{T\to\infty}\inf \frac{1}{T}\sum_{t=1}^{T} \mathsf{E}_k^\pi\left[\sum_{\tau=1}^{t} c(X_{\tau-1}, A_\tau)\right] \leq 0.$$

At the same time, according to Proposition 4.3(b),

$$\liminf_{\beta\to 1-}[v_1^{\varphi^0,\beta} - v_1^{*,\beta}] = \liminf_{\beta\to 1-}\left(-v_1^{*,\beta}\right) \geq \varepsilon > 0,$$

so that φ^0 is not nearly optimal. □

4.3.26 Strong-overtaking/average optimal, overtaking optimal, AC-optimal strategies and minimal opportunity loss

Definition 4.8. [Flynn (1980), Equation (7)]. A strategy π^* is called *strong-overtaking optimal* if, for all $x \in \mathbf{X}$,

$$\lim_{T\to\infty}\left\{\mathsf{E}_x^{\pi^*}\left[\sum_{\tau=1}^{T} c(X_{\tau-1},A_\tau)\right] - V_x^T\right\} = 0$$

(recall that $V_x^T = \inf_\pi E_x^\pi\left[\sum_{\tau=1}^T c(X_{\tau-1},A_\tau)\right]$). Such a strategy provides the minimal possible value to the (limiting) opportunity loss (3.20).

Definition 4.9. [Flynn (1980), Equation (5)]. A strategy π^* is called *strong-average optimal* if, for all $x \in \mathbf{X}$,

$$\lim_{T\to\infty}\frac{1}{T}\left\{\mathsf{E}_x^{\pi^*}\left[\sum_{\tau=1}^{T} c(X_{\tau-1},A_\tau)\right] - V_x^T\right\} = 0.$$

Any strong-overtaking optimal strategy is overtaking optimal and hence AC-optimal and opportunity-loss optimal (see Subsection 4.3.22). Any strong-average optimal strategy is AC-optimal [Hernández-Lerma and Vega-Amaya (1998), Remark 3.3]. The proof of the latter statement is similar to that given after Definition 4.6. If the model is finite, then the canonical stationary selector φ^*, the element of the canonical triplet $\langle\rho, h, \varphi^*\rangle$, is also strong-average optimal, because

$$\begin{aligned}
V_x^T &\geq \mathsf{E}_x^\pi\left[\sum_{t=1}^T c(X_{t-1},A_t) + h(X_T)\right] + \inf_\pi \mathsf{E}_x^\pi\left[-h(X_T)\right]\\
&\geq \inf_\pi \mathsf{E}_x^\pi\left[\sum_{t=1}^T c(X_{t-1},A_t) + h(X_T)\right] - \sup_\pi \mathsf{E}_x^\pi[h(X_T]\\
&\geq T\rho(x) + h(x) - \sup_{x\in\mathbf{X}}|h(x)|
\end{aligned}$$

and

$$\begin{aligned}
0 &\leq \mathsf{E}_x^{\varphi^*}\left[\sum_{\tau=1}^T c(X_{\tau-1},A_\tau)\right] - V_x^T = T\rho(x) + h(x) - \mathsf{E}_x^{\varphi^*}\left[h(X_T)\right] - V_x^T\\
&\leq 2\sup_{x\in\mathbf{X}}|h(x)|.
\end{aligned}$$

The following example, based on [Flynn (1980), Ex. 2], shows that a strong-overtaking optimal strategy may not exist, even in finite models, and even when an overtaking optimal strategy does exist.

Example. Let $\mathbf{X} = \mathbf{A} = \{0, 1\}$, $p(0|x, 0) \equiv 1$, $p(1|x, 1) \equiv 1$, with all other transition probabilities being zero; $c(0, 0) = 5$, $c(0, 1) = 10$, $c(1, 0) = -5$, $c(1, 1) = 0$ (see Fig. 4.34).

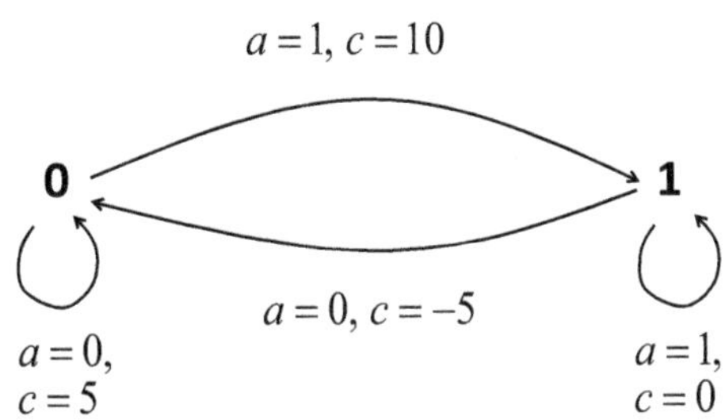

Fig. 4.34 Overtaking optimal strategy is not strong-overtaking optimal.

Equations (1.5) give the following:

$$\begin{aligned} V_0^1 &= 5, & V_1^1 &= -5, \\ V_0^2 &= 5, & V_1^2 &= -5, \ldots, \\ V_0^T &= 5, & V_1^T &= -5, \ldots. \end{aligned}$$

Starting from $x_0 = 0$ and from $x_0 = 1$, the trajectories $(x_0 = 0, a_1 = 1, x_1 = 1, a_2 = 1, \ldots)$ and $(x_0 = 1, a_1 = 1, x_1 = 1, a_2 = 1, \ldots)$ result in total costs 10 and 0 respectively, over any time interval $T \geq 1$. One can check that any other infinite trajectory results in greater (or equal) cost for all large enough values of T. Therefore, the stationary selector $\varphi^*(x) \equiv 1$ is overtaking optimal (and hence opportunity-loss optimal and AC-optimal: see Subsection 4.3.22). But the selector φ^* is not strong-overtaking optimal, because

$$E_x^{\varphi^*}\left[\sum_{\tau=1}^{T} c(X_{\tau-1}, A_\tau)\right] - V_x^T \equiv 5 > 0 \text{ for all } x \in \mathbf{X} \text{ and } T \geq 1.$$

At the same time, it is strong-average optimal. One can also check that $\rho = 0$, $h(x) = \begin{cases} 0, & \text{if } x = 0; \\ -10, & \text{if } x = 1, \end{cases}$ and $\varphi^*(x) \equiv 1$ form the canonical triplet: canonical equations (4.2) are satisfied. □

Example. Now consider the same MDP as in Subsection 2.3.4 (Fig. 2.4). One can check that $\rho(x) \equiv 0$, $h(x) = \begin{cases} 0, & \text{if } x = 0; \\ -1, & \text{if } x > 0, \end{cases}$ and $\varphi^*(x) \equiv 2$ form a canonical triplet $\langle \rho, h, \varphi^* \rangle$ (see Theorem 4.1). According to Theorem 4.3(a), the selector φ^* is AC-optimal and $v_x^* = \rho(x) \equiv 0$. In fact, all strategies in this MDP are equally AC-optimal and strong-average optimal. Note that the total expected cost on any time interval $1, 2, \ldots, T$ is non-positive and uniformly bounded from below by -1.

Straightforward calculations lead to the following expressions:

$$V_0^T = 0, \quad V_x^T = (x + T - 1)^{-1} - 1 \text{ for all } x > 0, \ T \geq 1.$$

No-one strategy is overtaking optimal. Indeed, for each strategy π, since $c(\cdot, \cdot) \leq 0$, there exists the following limit

$$\lim_{T\to\infty} E_1^\pi \left[\sum_{t=1}^{T} c(X_{t-1}, A_t) \right] =: F(\pi) > -1$$

(see Subsection 2.3.4). Therefore, for the stationary selector

$$\varphi(x) := \begin{cases} 2, & \text{if } x \leq \frac{1}{F(\pi)+1}; \\ 1, & \text{if } x > \frac{1}{F(\pi)+1} \end{cases}$$

we have

$$F(\varphi) = \frac{1}{\lfloor \frac{1}{F(\pi)+1} \rfloor + 1} - 1 < F(\pi),$$

and π is not overtaking optimal (see Definition 4.6; $\lfloor \cdot \rfloor$ is the integer part).

In this example, no-one strategy is strong-overtaking optimal, because $\lim_{T\to\infty} V_1^T = -1$ and $F(\pi) > -1$. No-one strategy is opportunity-loss optimal: for $x > 0$, $\lim_{T\to\infty} V_x^T \equiv -1$, $\lim_{T\to\infty} \mathsf{E}_x^\pi \left[\sum_{t=1}^{T} c(X_{t-1}, A_t) \right] > -1$ for all strategies π, and $\inf_\pi \lim_{T\to\infty} \mathsf{E}_x^\pi \left[\sum_{t=1}^{T} c(X_{t-1}, A_t) \right] \equiv -1$ is not achieved: as was explained in Subsection 2.3.4, there are no optimal strategies in the expected total cost version of this model. Similar reasoning shows that no-one strategy is D-optimal. Recall that $v_x^* \equiv 0$. □

This model was discussed in [Hernández-Lerma and Vega-Amaya (1998), Ex. 4.14] and in [Hernández-Lerma and Lasserre (1999), Section 10.9].

4.3.27 *Strong-overtaking optimal and strong*-overtaking optimal strategies*

In [Fernández-Gaucherand *et al.* (1994)] and in [Nowak and Vega-Amaya (1999)] a strategy π^* was called overtaking optimal if, for any strategy π, for each $x \in \mathbf{X}$, there is $N(\pi^*, \pi, x)$ such that

$$\mathsf{E}_x^{\pi^*} \left[\sum_{t=1}^{T} c(X_{t-1}, A_t) \right] \leq \mathsf{E}_x^\pi \left[\sum_{t=1}^{T} c(X_{t-1}, A_t) \right] \tag{4.29}$$

as soon as $T \geq N(\pi^*, \pi, x)$. (Compare this with weakly overtaking optimality, introduced at the end of Section 4.3.22.) This definition is stronger than Definition 4.6 of the overtaking optimal strategy. We shall call a strategy π^* satisfying (4.29) *strong*-overtaking optimal*, to distinguish from Definitions 4.6 and 4.8.

Remark 4.6. If inequality (4.29) holds for all strategies π from a specified class Π, then π^* is said to be strong*-overtaking optimal *in that class.* The same remark is valid for all other types of optimality.

The next two examples confirm that the notions of strong- and strong*-overtaking optimality are indeed different.

Example. Let $\mathbf{X} = \{0, (1,1), (1,2), \ldots, (2,1), (2,2), \ldots\}$, $\mathbf{A} = \{1, 2, \ldots\}$; $p((a,1)|0,a) \equiv 1$, for all $i,j \geq 1$ $p((i,j+1)|(i,j),a) \equiv 1$, with all other transition probabilities being zero; $c(0,a) = \left(\frac{1}{2}\right)^{a-1}$, $c((i,j),a) \equiv -\left(\frac{1}{2}\right)^{i+j-1}$ (see Fig. 4.35).

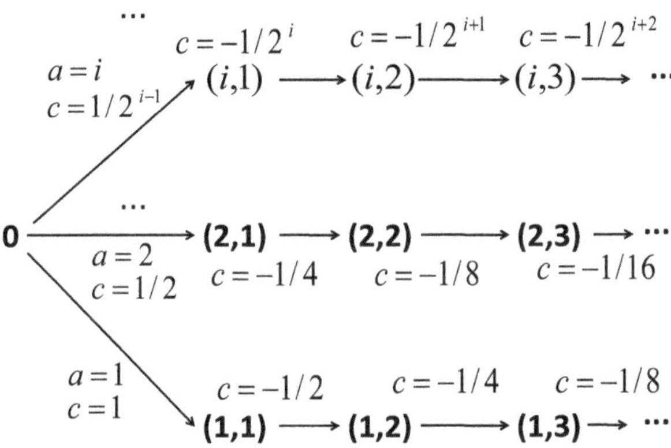

Fig. 4.35 Strategy, that is strong-overtaking optimal, is not strong*-overtaking optimal.

Obviously, one needs to consider only the initial state $x_0 = 0$.

For any finite $T \geq 1$, $V_0^T := \inf_\pi \left\{ \mathsf{E}_0^\pi \left[\sum_{t=1}^T c(X_{t-1}, A_t) \right] \right\} = 0$ because $\mathsf{E}_0^\pi \left[\sum_{t=1}^T c(X_{t-1}, A_t) \right] > 0$ for each π and $\mathsf{E}_0^\varphi \left[\sum_{t=1}^T c(X_{t-1}, A_t) \right] < \frac{1}{2^{i-1}}$ for $\varphi(x) \equiv i$. For any strategy π^*, $\lim_{T\to\infty} \mathsf{E}_0^{\pi^*} \left[\sum_{t=1}^T c(X_{t-1}, A_t) \right] = 0$. Therefore, all strategies are equally strong-overtaking optimal.

On the other hand, for an arbitrarily fixed strategy π^*, let $a^* := \min\{a \in \mathbf{A} : \pi_1^*(a|0) = q > 0\}$ and fix $I > 0$ such that $\frac{1}{2^I} < q$. Then

$$\mathsf{E}_0^{\pi^*} \left[\sum_{t=1}^T c(X_{t-1}, A_t) \right] > \frac{1}{2^I} \left[\frac{1}{2^{a^*-1}} - \frac{1}{2^{a^*}} - \ldots - \frac{1}{2^{a^*+T-2}} \right], \quad T = 1, 2, \ldots$$

and

$$\mathsf{E}_0^{\varphi}\left[\sum_{t=1}^{T} c(X_{t-1}, A_t)\right] = \frac{1}{2^{I+a^*-1}} - \frac{1}{2^{I+a^*}} - \cdots - \frac{1}{2^{I+a^*+T-2}}$$
$$< \mathsf{E}_0^{\pi^*}\left[\sum_{t=1}^{T} c(X_{t-1}, A_t)\right], \quad T = 1, 2, \ldots$$

for $\varphi(x) \equiv I + a^*$. Therefore, no-one strategy is strong*-overtaking optimal. □

Example. Consider now the following model: $\mathbf{X} = \{0, 1, 2, \ldots\}$, $\mathbf{A} = \{1, 2\}$, $p(0|0, a) \equiv 1$, for all $i \geq 1$ $p(i+1|i, 1) \equiv 1$, $p(0|i, 2) \equiv 1$, with all other transition probabilities being zero; $c(0, a) \equiv 1$, for all $i \geq 1$ $c(i, 1) \equiv 0$ and $c(i, 2) \equiv -1$ (see Fig. 4.36).

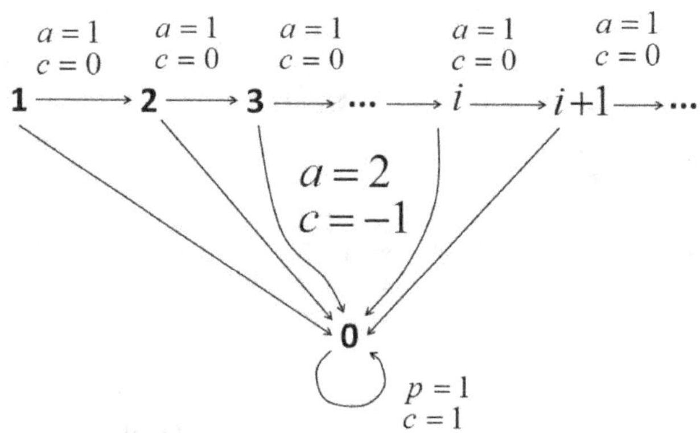

Fig. 4.36 Strategy, that is strong*-overtaking optimal, is not strong-overtaking optimal.

For any initial state $x_0 = x \geq 1$, for each $T \geq 1$, $V_x^T = -1$: one applies action $a = 2$ only at the last step: $A_T = 2$. But, for any strategy π, for each initial state $x_0 = x \geq 1$, we have

$$\lim_{T\to\infty} \mathsf{E}_x^{\pi}\left[\sum_{t=1}^{T} c(X_{t-1}, A_t)\right] = \begin{cases} 0, & \text{if } \mathsf{P}_x^{\pi}(A_t = 1 \text{ for all } t \geq 1) = 1; \\ \infty, & \text{otherwise,} \end{cases}$$

meaning that $\lim_{T\to\infty}\left\{\mathsf{E}_x^{\pi}\left[\sum_{t=1}^{T} c(X_{t-1}, A_t)\right] - V_x^T\right\} > 0$. Thus, no-one strategy is strong-overtaking optimal. At the same time, the stationary selector $\varphi(x) \equiv 1$ is strong*-overtaking optimal because, for any other strategy π, for each initial state $x_0 = x \geq 1$, either $\mathsf{E}_x^{\pi}\left[\sum_{t=1}^{T} c(X_{t-1}, A_t)\right] = \mathsf{E}_x^{\varphi}\left[\sum_{t=1}^{T} c(X_{t-1}, A_t)\right] = 0$ for all $T \geq 1$, or $\lim_{T\to\infty} \mathsf{E}_x^{\pi}\left[\sum_{t=1}^{T} c(X_{t-1}, A_t)\right] = \infty$ after absorption at 0.

The case of $x_0 = 0$ is trivial. □

There was an attempt to prove that, under appropriate conditions, there exists a strong*-overtaking stationary selector (in the class of AC-optimal stationary selectors): see Lemma 6.2 and Theorems 6.1 and 6.2 in [Fernández-Gaucherand *et al.* (1994)]. In fact, the stationary selector described in Theorem 6.2 is indeed strong*-overtaking optimal if it is unique. But the following example, published in [Nowak and Vega-Amaya (1999)], shows that it is possible for *two* stationary selectors to be equal candidates, but neither of them overtakes the other. As a result, a strong*-overtaking optimal stationary selector does not exist. Note that the controlled process under consideration below is irreducible and aperiodic under any stationary strategy, and the model is finite.

Example. Let $\mathbf{X} = \{1,2,3\}$, $\mathbf{A} = \{1,2\}$, $p(1|1,1) = p(3|1,2) = 0.7$, $p(3|1,1) = p(1|1,2) = 0.1$, $p(2|1,a) \equiv 0.2$, $p(1|2,a) = p(3|2,a) = p(1|3,a) = p(2|3,a) \equiv 0.5$, with all other transition probabilities being zero. We put $c(1,1) = 1.4$, $c(1,2) = 0.2$, $c(2,a) \equiv -9$, $c(3,a) \equiv 6$ (see Fig. 4.37).

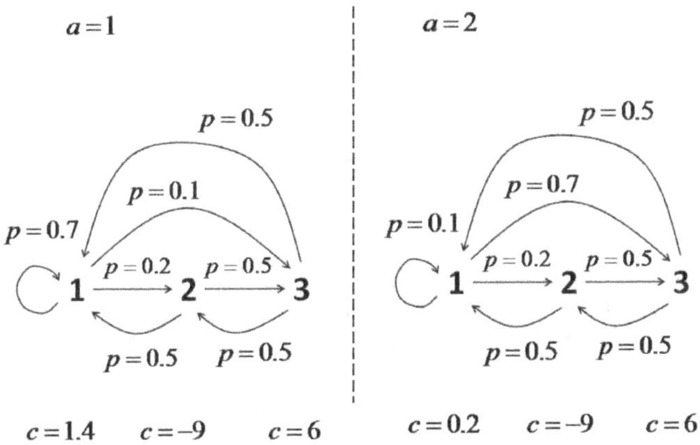

Fig. 4.37 No strong*-overtaking optimal stationary selectors.

Only actions in state 1 are important; so we study only the stationary selectors $\varphi^1(x) \equiv 1$ and $\varphi^2(x) \equiv 2$.

Proposition 4.4.

(a) $\rho^*(x) \equiv 0$ *and* $h(x) = \begin{cases} 8, & \textit{if } x = 1; \\ 0, & \textit{if } x = 2; \\ 10, & \textit{if } x = 3 \end{cases}$ *are the unique functions satisfying equations (4.2). (As usual, function $h(\cdot)$ can be modified by adding an arbitrary constant.) The both stationary selectors $\varphi^1(x) \equiv 1$ and $\varphi^2(x) \equiv 2$ provide the minimums in (4.2).*

(b) *The stationary distributions of the controlled process under the stationary selectors φ^1 and φ^2 are as follows:*

$$\hat{\eta}^{\varphi^1}(1) = 15/24, \quad \hat{\eta}^{\varphi^1}(2) = 5/24, \quad \hat{\eta}^{\varphi^1}(3) = 4/24;$$

$$\hat{\eta}^{\varphi^2}(1) = 15/42, \quad \hat{\eta}^{\varphi^2}(2) = 11/42, \quad \hat{\eta}^{\varphi^2}(3) = 16/42.$$

Thus,

$$\sum_{x\in\mathbf{X}} \hat{\eta}^{\varphi^1}(x)h(x) = \sum_{x\in\mathbf{X}} \hat{\eta}^{\varphi^2}(x)h(x) = \frac{20}{3}.$$

(c) *For all $T = 1, 2, \ldots$*

$$\mathsf{E}_x^{\varphi^1}\left[\sum_{t=1}^{T} c(X_{t-1}, A_t)\right] = \begin{cases} \frac{4}{3} + \frac{1}{21}\left[-10\left(-\frac{1}{2}\right)^T - 18\left(\frac{1}{5}\right)^T\right], & \textit{if } x = 1; \\ -\frac{20}{3} + \frac{1}{21}\left[110\left(-\frac{1}{2}\right)^T + 30\left(\frac{1}{5}\right)^T\right], & \textit{if } x = 2; \\ \frac{10}{3} + \frac{1}{21}\left[-100\left(-\frac{1}{2}\right)^T + 30\left(\frac{1}{5}\right)^T\right], & \textit{if } x = 3, \end{cases}$$

$$\mathsf{E}_x^{\varphi^2}\left[\sum_{t=1}^{T} c(X_{t-1}, A_t)\right] = \begin{cases} \frac{4}{3} + \frac{1}{3}\left[50\left(-\frac{1}{2}\right)^T - 54\left(\frac{-2}{5}\right)^T\right], & \textit{if } x = 1; \\ -\frac{20}{3} + \frac{1}{3}\left[-10\left(-\frac{1}{2}\right)^T + 30\left(\frac{-2}{5}\right)^T\right], & \textit{if } x = 2; \\ \frac{10}{3} + \frac{1}{3}\left[-40\left(-\frac{1}{2}\right)^T + 30\left(\frac{-2}{5}\right)^T\right], & \textit{if } x = 3. \end{cases}$$

The proof is given in Appendix A.

According to Proposition 4.4(a), $\langle \rho^*, h, \varphi^1\rangle$ and $\langle \rho^*, h, \varphi^2\rangle$ are canonical triplets and the both stationary selectors φ^1 and φ^2 are AC-optimal: see Theorems 4.1 and 4.3. It also follows from Proposition 4.4(a,b) that the conditions in Theorem 6.2 of [Fernández-Gaucherand *et al.* (1994)] are satisfied for the both selectors φ^1 and φ^2.

At the same time, according to Proposition 4.4(c),

$$21 \cdot 2^T \left[E_x^{\varphi^1}\left[\sum_{t=1}^{T} c(X_{t-1}, A_t)\right] - E_x^{\varphi^2}\left[\sum_{t=1}^{T} c(X_{t-1}, A_t)\right]\right]$$
$$= \begin{cases} -360(-1)^T + o(1), & \text{if } x = 1; \\ 180(-1)^T + o(1), & \text{if } x = 2; \\ 180(-1)^T + o(1), & \text{if } x = 3, \end{cases}$$

where $\lim_{T\to\infty} o(1) = 0$. Hence, inequality (4.29) holds neither for selector φ^1 nor for φ^2. Thus, neither φ^1 nor φ^2 is strong*-overtaking optimal: there are no strong*-overtaking optimal stationary selectors in this model.

Proposition 4.5.

(a) *There exists an increasing sequence $T_1, T_2, \ldots$ such that the selectors φ^1 and φ^2 do not overtake (in the sense of Definition 4.6) the non-stationary* *selector*

$$\varphi_t(x) \equiv \begin{cases} 2, & \text{if } t = T_1, T_2, \dots \\ 1, & \text{otherwise.} \end{cases}$$

(b) *The stationary selector φ^2 is Blackwell optimal, hence n-discount optimal for any $n \geq -1$ [Puterman (1994), Thm. 10.1.5]. The both selectors φ^1 and φ^2 are bias optimal (hence, AC-optimal); the selector φ^1 is not 1-discount optimal.*

The proof is given in Appendix A.

Since any strong-overtaking optimal strategy is also overtaking optimal, according to Proposition 4.5(a), neither φ^1, nor φ^2 is strong-overtaking optimal.

It is also interesting to look at the optimality equation (2.5) which after the trivial rearrangement, similar to that which was done for equations (A.13), takes the form

$$v(1) = \min\{v(1),\ v(1)\}; \quad v(2) = -8 + v(1); \quad v(3) = 2 + v(1).$$

Clearly, there are no other solutions except for

$$v(x) = r + \begin{cases} \frac{4}{3}, & \text{if } x = 1; \\ -\frac{20}{3}, & \text{if } x = 2; \\ \frac{10}{3}, & \text{if } x = 3, \end{cases}$$

where $r \in \mathbb{R}$ is an arbitrary constant. The both selectors φ^1 and φ^2 provide the minimum (and also the maximum).

Note also that, for each control strategy π, for any initial distribution P_0,

$$\mathsf{E}^\pi_{\mathsf{P}_0}\left[\sum_{t=1}^\infty c^+(X_{t-1}, A_t)\right] = +\infty, \quad \mathsf{E}^\pi_{\mathsf{P}_0}\left[\sum_{t=1}^\infty c^-(X_{t-1}, A_t)\right] = -\infty,$$

so that Condition 2.1 is violated. If convention (2.3) is accepted, then the total-cost performance functional (2.1) equals $+\infty$. But one certainly can define the total-cost performance functional as either

$$\tilde{v}^\pi_{\mathsf{P}_0} := \limsup_{T\to\infty} \mathsf{E}^\pi_{\mathsf{P}_0}\left[\sum_{t=1}^T c(X_{t-1}, A_t)\right]$$

$$\text{or} \quad \tilde{\tilde{v}}^\pi_{\mathsf{P}_0} := \limsup_{\beta\to 1-} \mathsf{E}^\pi_{\mathsf{P}_0}\left[\sum_{t=1}^\infty \beta^{t-1} c(X_{t-1}, A_t)\right].$$

In the both cases, according to Proposition 4.4(c) and to the proof of Proposition 4.5(b),

$$\tilde{v}^{\varphi^1,\varphi^2}_{\mathsf{P}_0} = \tilde{\tilde{v}}^{\varphi^1,\varphi^2}_{\mathsf{P}_0} = \frac{4}{3}\mathsf{P}_0(1) - \frac{20}{3}\mathsf{P}_0(2) + \frac{10}{3}\mathsf{P}_0(3).$$

Moreover, since the selector φ^2 is Blackwell optimal, $\tilde{\tilde{v}}^{\pi}_{\mathsf{P}_0} \geq \tilde{\tilde{v}}^{\varphi^1,\varphi^2}_{\mathsf{P}_0}$ for all strategies π. □

4.3.28 *Singular perturbation of an MDP*

Suppose the state space is finite (or denumerable). We say that an MDP is *perturbed* if the transition probabilities (and possibly the cost function) change slightly according to $p(y|x,a) + \varepsilon d(y|x,a)$, where ε is a small parameter and $\sum_{y\in\mathbf{X}} d(y|x,a) \equiv 1$. A perturbation is *singular* if it changes the ergodic structure of the underlying Markov chain. In such cases it can happen that one stationary selector φ^ε is AC-optimal for all small enough $\varepsilon > 0$, but an absolutely different selector φ^* is AC-optimal for $\varepsilon = 0$. The following confirming this example is based on [Avrachenkov *et al.* (2002), Ex. 2.1].

Example. Let $\mathbf{X} = \{1, 2\}$, $\mathbf{A} = \{1, 2\}$, $p(1|1,a) \equiv p(2|2,a) \equiv 1$, with all other transition probabilities being zero. Let $d(1|1,2) = -1$, $d(2|1,2) = 1$, $d(1|2,a) \equiv 1$, $d(2|2,a) \equiv -1$, with other values of function $d(\cdot)$ being zero. We put $c(1,1) = 1$, $c(1,2) = 1.5$, $c(2,a) \equiv 0$ (see Fig. 4.38).

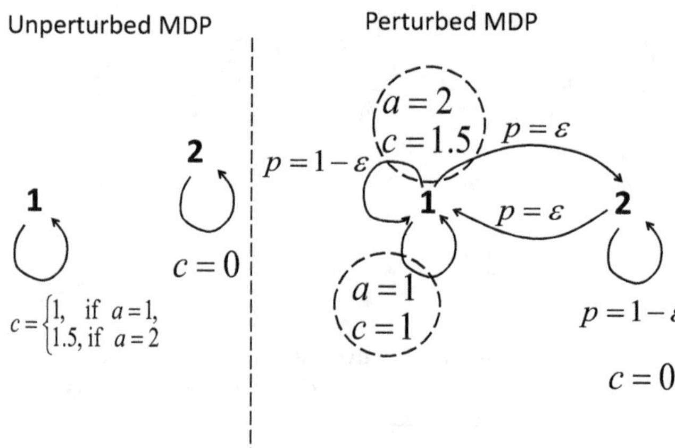

Fig. 4.38 Singularly perturbed MDP.

The solution of the unperturbed MDP (when $\varepsilon = 0$) is obvious: the stationary selector $\varphi^*(x) \equiv 1$ is optimal; $\rho(1) = 1$, $\rho(2) = 0$.

When $\varepsilon \in (0, 1]$, the model is communicating and one can put $\rho(x) \equiv \rho^\varepsilon$ as a constant in equations (4.2): see [Puterman (1994), Thm. 8.4.3] and [Kallenberg (2010), Subsection 6.3.1]. After we fix $h^\varepsilon(1) = 0$ (see Remark 4.1), the canonical equations (4.2) take the form

$$\rho^\varepsilon = \min\{1,\ 1.5 + \varepsilon h^\varepsilon(2)\};$$
$$\rho^\varepsilon + h^\varepsilon(2) = (1-\varepsilon)h^\varepsilon(2)$$

and have the single solution $\rho^\varepsilon = 0.75$, $h^\varepsilon(2) = -0.75/\varepsilon$ leading to the canonical triplet $\langle \rho^\varepsilon, h^\varepsilon, \varphi^\varepsilon\rangle$, where $\varphi^\varepsilon(x) \equiv 2$ is the AC-optimal canonical stationary selector for all $\varepsilon \in (0,1)$ by Theorem 4.1 and [Puterman (1994), Thm. 9.1.7]. There are no other AC-optimal stationary selectors since one can ignore any actions at the uncontrolled state 2, and another stationary selector $\varphi^1(x) \equiv 1$ leads to $v_x^{\varphi^1} = 1 > 0.75$ because the controlled process will be absorbed at state 1 with $c(1,1) = 1$. One can also see that the limit $\lim_{\varepsilon\to 0}\rho^\varepsilon = 0.75$ is different from $\rho(1)$ and $\rho(2)$. □

4.3.29 *Constrained optimization*

Suppose we have two cost functions ${}^1c(x,a)$ and ${}^2c(x,a)$. Then every control strategy π results in two performance functionals ${}^1v^\pi_{\mathsf{P}_0}$ and ${}^2v^\pi_{\mathsf{P}_0}$ defined according to (4.1). The *constrained* problem can be expressed as

$$ {}^1v^\pi_{\mathsf{P}_0} \to \inf_\pi; \qquad {}^2v^\pi_{\mathsf{P}_0} \le d, \tag{4.30}$$

where d is a chosen number. Strategies satisfying the above inequality are called *admissible*.

If the finite model is aperiodic unichain and there is at least one admissible strategy, then there exists a stationary strategy solving problem (4.30): see [Altman (1999), Thm. 4.3]. One should complement the linear program (4.6) with the obvious inequality

$$\sum_{i\in\mathbf{X}}\sum_{a\in\mathbf{A}} {}^2c(i,a)\eta(i,a) \le d,$$

write down the objective as

$$\sum_{i\in\mathbf{X}}\sum_{a\in\mathbf{A}} {}^1c(i,a)\eta(i,a) \to \inf_\eta$$

and build the optimal stationary strategy using formula (4.9) for the optimal solution η^* of the obtained linear program.

The following example from [Piunovskiy (1997), p. 149] shows that the unichain condition is important.

Example. Let $\mathbf{X} = \{1,2\}$; $\mathbf{A} = \{1,2\}$; $p(1|1,1) = 1$, $p(2|1,2) = 1$, $p(2|2,a) \equiv 1$, with all other transition probabilities being zero; ${}^1c(x,a) = \mathbb{I}\{x=2\}$, ${}^2c(x,a) = \mathbb{I}\{x=1\}$ (see Fig. 4.39).

Suppose $d = \frac{1}{2}$ and $\mathsf{P}_0(1) = 1$. For a stationary strategy with $\pi^s(1|1) = 1$ we have ${}^1v_1^{\pi^s} = 0$, ${}^2v_1^{\pi^s} = 1$, so that such strategies are not admissible for (4.30). If $\pi^s(1|1) < 1$, then the process will be absorbed at state 2, leading to ${}^1v_1^{\pi^s} = 1$,

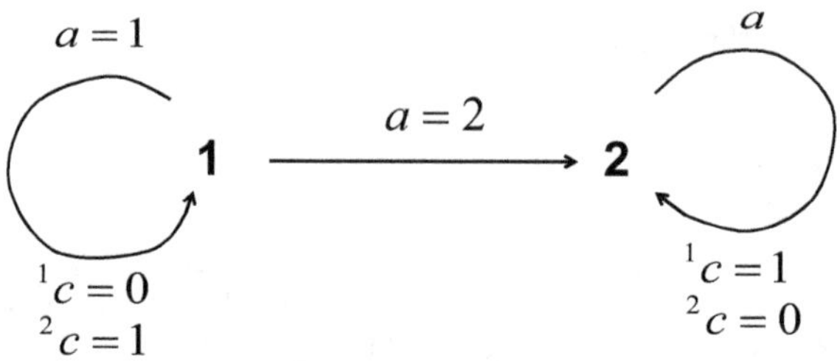

Fig. 4.39 Stationary strategies are not sufficient in constrained MDP.

${}^2v_1^{\pi^s} = 0$. At the same time, as will be shown, the solution to problem (4.30) is given by a strategy π^* with ${}^1v_1^{\pi^*} = 1/2$, ${}^2v_1^{\pi^*} = 1/2$.

Definition 4.10. For $\alpha \in [0,1]$ the $(\alpha, 1-\alpha)$-*mixture* of two strategies π^1 and π^2 is a strategy π such that the corresponding strategic measure satisfies equality $\mathsf{P}^\pi_{\mathsf{P}_0} = \alpha \mathsf{P}^{\pi^1}_{\mathsf{P}_0} + (1-\alpha)\mathsf{P}^{\pi^2}_{\mathsf{P}_0}$.

The existence of a mixture follows from the convexity of the space of all strategic measures $\mathcal{P}_{\mathsf{P}_0}$, [Piunovskiy (1997), Thm. 8]. One can say that the decision maker tosses a coin at the very beginning and applies strategies π^1 or π^2 with probabilities α and $(1-\alpha)$ correspondingly. In finite models, one can replace lim sup in formula (4.1) with the standard limit, if π^s is a stationary strategy: the limit $\left\{\lim_{T\to\infty} \frac{1}{T}\mathsf{E}^{\pi^s}_{\mathsf{P}_0}[\mathbb{I}\{X_t = i\}]\right\}_{i\in\mathbf{X}}$ equals the vector of stationary distribution of the controlled process X_t under the control strategy π^s. Therefore, the functional $v^\pi_{\mathsf{P}_0}$ is linear w.r.t. $\mathsf{P}^\pi_{\mathsf{P}_0}$ if these strategic measures correspond to finite mixtures of stationary strategies.

If $\pi^{s,1}$ and $\pi^{s,2}$ are stationary strategies then the $(\alpha, 1-\alpha)$-mixture π is usually non-stationary, but there always exists an equivalent Markov strategy π^m with $v^{\pi^m}_{\mathsf{P}_0} = v^\pi_{\mathsf{P}_0}$ (the one-step cost $c(\cdot,\cdot)$ can be arbitrary): see Lemma 2.1.

In the considered example π^* can be taken as the $(1/2, 1/2)$-mixture of the stationary selectors $\varphi^1(x) \equiv 1$ and $\varphi^2(x) \equiv 2$. Below, we show that this mixture solves problem (4.30).

Lemma 4.3. **[Piunovskiy (1997), Thm. 13].** *Let the functionals ${}^1v^\pi_{\mathsf{P}_0}$ and ${}^2v^\pi_{\mathsf{P}_0}$ be finite for any control strategy π, ${}^1v^\pi_{\mathsf{P}_0} \ge K > -\infty$, and let the Slater condition be satisfied, i.e. the inequality in (4.30) is strict for at least one strategy π. A strategy π^* solves the constrained problem (4.30) if and only if there is a Lagrange multiplier $\lambda^* \ge 0$ such that*

$$ {}^1v^{\pi^*}_{\mathsf{P}_0} + \lambda^*({}^2v^{\pi^*}_{\mathsf{P}_0} - d) = \min_\pi\{{}^1v^\pi_{\mathsf{P}_0} + \lambda^*({}^2v^\pi_{\mathsf{P}_0} - d)\} $$

and $\lambda^*({}^2v^{\pi^*}_{\mathsf{P}_0} - d) = 0$, ${}^2v^{\pi^*}_{\mathsf{P}_0} \le d$.

Note that this lemma holds for any performance functionals ${}^1v^\pi_{\mathsf{P}_0}$ and ${}^2v^\pi_{\mathsf{P}_0}$ which can be expressed as *convex* functionals on the space of strategic measures $\mathcal{P}_{\mathsf{P}_0}$ [Piunovskiy and Zhang (2020), Prop. B.2.6]. The functional ${}^2v^\pi_{\mathsf{P}_0}$ can be multi-dimensional (column vector as well as d), in which case λ^* is a row vector.

In the example under consideration, $\lambda^* = 1$,

$$\begin{aligned} {}^1v_1^{\pi^*} &= \limsup_{T\to\infty} \frac{1}{T}\left\{\frac{1}{2}\mathsf{E}_1^{\varphi^1}\left[\sum_{t=1}^T {}^1c(X_{t-1},A_t)\right] + \frac{1}{2}\mathsf{E}_1^{\varphi^2}\left[\sum_{t=1}^T {}^1c(X_{t-1},A_t)\right]\right\} \\ &= \frac{1}{2}\lim_{T\to\infty}\frac{1}{T}\mathsf{E}_1^{\varphi^1}\left[\sum_{t=1}^T {}^1c(X_{t-1},A_t)\right] + \frac{1}{2}\lim_{T\to\infty}\frac{1}{T}\mathsf{E}_1^{\varphi^2}\left[\sum_{t=1}^T {}^1c(X_{t-1},A_t)\right] \\ &= 0 + 1/2 = 1/2 \end{aligned}$$

(note the standard limits). Similarly, ${}^2v_1^{\pi^*} = 1/2$. Now, for any strategy π,

$$\begin{aligned} &{}^1v_1^\pi + \lambda^*({}^2v_1^\pi - 1/2) = {}^1v_1^\pi + ({}^2v_1^\pi - 1/2) \\ &\geq \limsup_{T\to\infty}\frac{1}{T}\mathsf{E}_1^\pi\left[\sum_{t=1}^T({}^1c(X_{t-1},A_t) + {}^2c(X_{t-1},A_t))\right] - 1/2 \\ &\equiv 1/2 = {}^1v_1^{\pi^*} + \lambda^*({}^2v_1^{\pi^*} - 1/2). \end{aligned}$$

Therefore, the $(1/2, 1/2)$-mixture of stationary selectors φ^1 and φ^2 really solves problem (4.30) by Lemma 4.3.

A solution to a *finite unichain* model can be found in the form of a *time-sharing* strategy [Altman and Shwartz (1993)]. Such a strategy switches between several stationary selectors in such a way that the expected frequencies (see Definition 4.5) converge, as $T \to \infty$, to the values of $\eta^*(y,a)$ solving the corresponding linear program. If, for example, $\eta^*(x,1) = \eta^*(x,2) = 1/2$ for a particular recurrent state x, then one applies actions $\varphi^1(x) = 1$ and $\varphi^2(x) = 2$ in turn, every time the controlled process visits state x.

The current example (Fig. 4.39) shows that a time-sharing strategy π cannot solve the constrained problem if the model is not unichain. If action $a = 2$ is applied (in state $x_0 = 1$) at least once, then ${}^1v_1^\pi = 1$. Otherwise ${}^2v_1^\pi = 1$, meaning that π is not admissible.

An algorithm for solving constrained problems for general finite models, based on the linear programming approach, is presented in [Kallenberg (2010), Alg. 9.1]. It leads to an optimal (non-stationary) Markov strategy if there is at least one admissible strategy.

It is interesting to consider the constrained MDP with discounted costs and to see what happens when the discount factor β approaches 1. According to [Kallenberg (2010), Thm. 7.1], in finite unconstrained models the AC-optimality is equivalent to the (-1)-discount optimality (see Definition 3.9). Therefore, we normalize the expected discounted costs, multiplying them by $(1-\beta)$:

$$(1-\beta)\ {}^1v^{\pi,\beta}_{\mathsf{P}_0} \to \inf_\pi; \qquad (1-\beta)\ {}^2v^{\pi,\beta}_{\mathsf{P}_0} \leq d. \tag{4.31}$$

It is known that stationary strategies are sufficient for solving the constrained discounted MDP of the form (4.31); see [Piunovskiy (1997), Section 3.2.3.2]. In the current example (Fig. 4.39), any such strategy π^s is characterized by the single number $\pi^s(1|1) = p$ because $p(2|2,a) \equiv 1$. One can compute

$$(1-\beta)\ {}^1v_1^{\pi^s,\beta} = \frac{\beta(1-p)}{1-\beta p}; \qquad (1-\beta)\ {}^2v_1^{\pi^s,\beta} = \frac{1-\beta}{1-\beta p},$$

so that $(1-\beta)[\ {}^1v_1^{\pi^s,\beta} + {}^2v_1^{\pi^s,\beta}] = 1$ and, for $\beta \in [1/2, 1)$, the strategy π^{s*} with $p^* = 2 - \frac{1}{\beta}$ solves problem (4.31). If β approaches 1, the optimal strategy π^{s*} does not stop to change: we have nothing similar to the Blackwell optimality. Moreover, $\lim_{\beta\to 1-} p^* = 1$, but we already know that the stationary selector φ^1 is not admissible in problem (4.30). Note that

$$\lim_{\beta\to 1-} (1-\beta)\ {}^1v_1^{\pi^{s*},\beta} = \lim_{\beta\to 1-} \frac{\beta(1-2+1/\beta)}{1-\beta(2-1/\beta)} = \frac{1}{2} \neq {}^1v_1^{\varphi^1} = 0,$$

$$\lim_{\beta\to 1-} (1-\beta)\ {}^2v_1^{\pi^{s*},\beta} = \lim_{\beta\to 1-} \frac{1-\beta}{1-\beta(2-1/\beta)} = \frac{1}{2} \neq {}^2v_1^{\varphi^1} = 1.$$

One can conclude that investigation of the discounted version does not help to solve the constrained average cost problem. Recall that in the unconstrained finite model there is a Blackwell optimal strategy which is also AC-optimal [Bertsekas (2001), Prop. 4.2.2], [Kallenberg (2010), Thm. 4.1, Cor. 5.3]. □

Similar analysis was performed in [Altman *et al.* (2002), Ex. 1].

4.3.30 *Weighted MDP*

If the model is finite then, as mentioned in Section 3.1 (Theorem 3.2(b)) and in Section 4.1, there exist uniformly optimal stationary selectors both in the discounted model and in problem (4.1). The following example, published in [Krass *et al.* (1992), Ex. 2], confirms that, if one considers the combination of such criteria, then it can happen that, for a fixed initial state, no-one strategy is optimal.

Example. Let $\mathbf{X} = \{1,2\}$, $\mathbf{A} = \{1,2\}$, $p(1|1,1) = p(2|1,2) = p(1|2,a) \equiv 1$ (all the transitions are deterministic), $c(1,1) = 0$, $c(1,2) = 10$, $c(2,a) \equiv -12$. Only actions in state 1 are meaningful: see Fig. 4.40. Let the discount factor β be equal $\frac{1}{2}$.

As mentioned in Section 3.1, the Bellman function in the discounted problem (3.1) is the unique bounded solution to the optimality equation (3.2):

$$v(1) = \min\{\beta v(1); \qquad 10 + \beta v(2)\};$$
$$v(2) = -12 + \beta v(1).$$

Therefore, $v_1^{*,\beta} = 0$; $v_2^{*,\beta} = -12$. The stationary selector $\varphi^1(x) \equiv 1$ is uniformly optimal by Theorem 3.1(b).

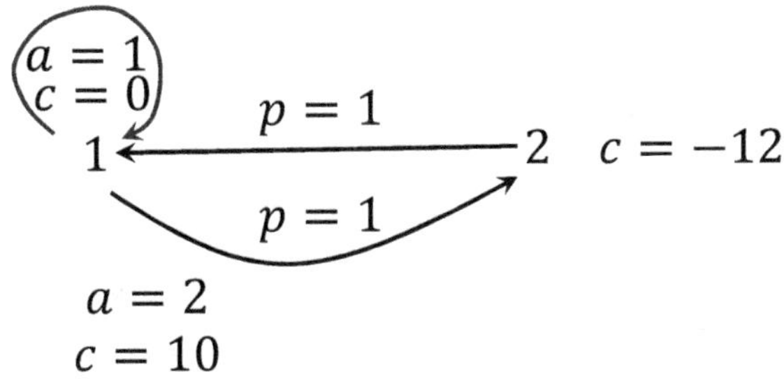

Fig. 4.40 No optimal strategies in the finite weighted MDP.

Since this MDP is communicating (all the states communicate under the stationary selector $\varphi^2(x) \equiv 2$), $\rho(x) \equiv const =: \rho$ [Kallenberg (2010), Subsection 6.3.1], and the canonical equations (4.2) take the form

$$\rho + h(1) = \min\{h(1); \quad 10\};$$
$$\rho = -12 + h(1).$$

(We put $h(2) = 0$ without loss of generality.) Therefore, $h(1) = 11$ and $\rho = -1$. The canonical stationary selector $\varphi^2(x) \equiv 2$ is AC-optimal by Theorem 4.3(a), providing the minimum in (4.1) equal to $\rho = -1$ for all initial distributions.

Consider now the combined criterion

$$V^\pi_{\mathsf{P}_0} := \frac{1}{4}\mathsf{E}^\pi_{\mathsf{P}_0}\left[\sum_{t=1}^\infty \beta^{t-1}c(X_{t-1}, A_t)\right] + \frac{3}{4}\limsup_{T\to\infty}\frac{1}{T}\mathsf{E}^\pi_{\mathsf{P}_0}\left[\sum_{t=1}^T c(X_{t-1}, A_t)\right]$$
$$\to \inf_\pi. \tag{4.32}$$

Suppose $\mathsf{P}_0(dx) = \delta_1(dx)$. Then the above infimum is not smaller than $\frac{1}{4}v_1^{*,\beta} + \frac{3}{4}\rho = -\frac{3}{4}$. For an arbitrarily fixed $\varepsilon > 0$, consider the following selector: $\varphi^\varepsilon_t(x) \equiv \begin{cases} 1, & \text{if } t < N; \\ 2, & \text{if } t \geq N, \end{cases}$ where the fixed $N \geq 1$ is such that $\beta^{N-1} \leq \frac{\varepsilon}{12}$. Then

$$\begin{aligned}\mathsf{E}_1^{\varphi^\varepsilon}\left[\sum_{t=1}^\infty \beta^{t-1}c(X_{t-1}, A_t)\right] &\leq \mathsf{E}_1^{\varphi^1}\left[\sum_{t=1}^\infty \beta^{t-1}c(X_{t-1}, A_t)\right] \\ &\quad + 2\sum_{t=N}^\infty \beta^{t-1}\max_{(x,a)\in\mathbf{X}\times\mathbf{A}}|c(x,a)| \\ &= 0 + 24\cdot\frac{\beta^{N-1}}{1-\beta} = 4\varepsilon.\end{aligned}$$

When calculating the criterion (4.1), one can obviously ignore the first N steps leading to

$$v_1^{\varphi^\varepsilon} = \limsup_{T\to\infty} \frac{1}{T} \mathsf{E}_1^{\varphi^\varepsilon}\left[\sum_{t=1}^{T} c(X_{t-1}, A_t)\right] = \rho = -1.$$

Therefore, the left-hand side of (4.32) under the selector φ^ε is not bigger than $\varepsilon - \frac{3}{4}$, and $\inf_\pi V_1^\pi = -\frac{3}{4}$.

Let us show now that there are no optimal strategies in problem (4.32). If such a strategy π^* exists, then it is different from selector $\varphi^1(x) \equiv 1$ because

$$\limsup_{T\to\infty} \frac{1}{T} \mathsf{E}_1^{\varphi^1}\left[\sum_{t=1}^{T} c(X_{t-1}, A_t)\right] = \mathsf{E}_1^{\varphi^1}\left[\sum_{t=1}^{\infty} \beta^{t-1} c(X_{t-1}, A_t)\right] = 0$$

and $V_1^{\pi^*} = 0$. Therefore, at some finite (minimal) value of $N \geq 1$, $\pi_N^*(2|1) = p > 0$. (Recall that the actions in state 2 are immaterial.) Now, even if $\limsup_{T\to\infty} \frac{1}{T}\mathsf{E}_1^{\pi^*}\left[\sum_{t=1}^{T} c(X_{t-1}, A_t)\right] = \rho = -1$, the value of $\mathsf{E}_1^{\pi^*}\left[\sum_{t=1}^{\infty} \beta^{t-1} c(X_{t-1}, A_t)\right]$ is not smaller than $\beta^{N-1} p[10 + \beta v_2^{*,\beta}] = p\beta^{N-1}\cdot 4 > 0$, and $V_1^{\pi^*} \geq p\beta^{N-1} - \frac{3}{4} > -\frac{3}{4}$. The strategy π^* is not optimal.

Similar reasoning applies to the initial state $x_0 = 2$. □

4.3.31 *Sample-path optimality*

"The expected average criterion (4.1) is quite appropriate if the controller repeats the underlying random dynamical experiment many times under similar conditions, but not for a single trial. In this latter case, it is interesting to study the average reward from a *sample-path* point of view." [Cavazos-Cadena *et al.* (2014)].

Definition 4.11. A strategy π^* is called *sample-path average optimal* with the optimal value ρ if

- for each state $x \in \mathbf{X}$ $\lim_{T\to\infty} \frac{1}{T}\sum_{t=1}^{T} c(X_{t-1}, A_t) = \rho$ $\mathsf{P}_x^{\pi^*}$-a.s. and
- for every strategy π and $x \in \mathbf{X}$ $\liminf_{T\to\infty} \frac{1}{T}\sum_{t=1}^{T} c(X_{t-1}, A_t) \geq \rho$ P_x^{π}-a.s.

Remark 4.7. Suppose the state space $\mathbf{X}$ is denumerable (or finite), the cost function $c(\cdot,\cdot)$ is bounded, and the model is recurrent. Then, under each stationary strategy, the controlled process X_t is positive Harris recurrent with $\mu(x) \equiv 1$ and, in view of Theorem 4.8, if π^{s*} is an AC-optimal stationary strategy, then it is sample-path average optimal within the class of stationary strategies, that is, in the last line of Definition 4.11 we have $\lim_{T\to\infty} \frac{1}{T}\sum_{t=1}^{T} c(X_{t-1}, A_t) \geq \rho$ $\mathsf{P}_x^{\pi^s}$-a.s. for all $x \in \mathbf{X}$ and for all stationary strategies π^s.

In what follows, we need to modify Definition 2.7 in the following way. The fixed state (say, 0) is not necessarily a cemetery and the requirement (i) of Definition 2.7 is replaced with the following one:

$$1 + |c(x,a)| + \sum_{y \in \mathbf{X} \setminus \{0\}} p(y|x,a)\mu(y) \le \mu(x), \quad x \in \mathbf{X}, \ a \in \mathbf{A}.$$

(Recall that the state space $\mathbf{X}$ is assumed to be denumerable or finite.) Such a function $\mu(\cdot)$ will be called *general* Lyapunov function.

Theorem 4.13. [Cavazos-Cadena and Montes-de-Oca (2012), Thm. 3.4.1]. *Suppose the space* $\mathbf{X}$ *is denumerable (or finite) and the following conditions are satisfied:*

- *there exits a general Lyapunov function* $\mu(\cdot)$*;*
- *there also exists a general Lyapunov function if we replace the cost function with* $\tilde{c}(x,a) \equiv \mu^\gamma(x)$*, where* $\gamma > 2$*;*
- *the space* $\mathbf{A}$ *is compact; the mappings* $a \to c(x,a)$ *and* $a \to p(y|x,a)$ *are continuous for all* $x, y \in \mathbf{X}$*.*

Then there exist a number ρ *and a function* $h(\cdot)$ *on* $\mathbf{X}$ *such that*

$$\rho + h(x) = \inf_{a \in \mathbf{A}} \left\{ c(x,a) + \sum_{y \in \mathbf{X}} h(y) p(y|x,a) \right\}, \quad x \in \mathbf{X} \tag{4.33}$$

and, if a measurable mapping $\varphi^*(\cdot) : \mathbf{X} \to \mathbf{A}$ *provides the infimum in (4.33), then this stationary selector* φ^* *is sample-path average optimal with* $\lim_{T\to\infty} \frac{1}{T} \sum_{t=1}^{T} c(X_{t-1}, A_t) = \rho$ $\mathsf{P}_x^{\varphi^*}$*-a.s. for all* $x \in \mathbf{X}$*.*

The presented below example, published in [Cavazos-Cadena *et al.* (2014)], confirms that the requirement regarding the general Lyapunov function for the cost function $\tilde{c}(x,a) \equiv \mu^\gamma(x)$ is important.

Example. Let $\mathbf{X} = \{0, 1, 2 \ldots\}$, $\mathbf{A} = \{0, 1, 1/2, 1/3, \ldots\}$. The topology in $\mathbf{X}$ is discrete, and the topology in $\mathbf{A}$ is the trace (restriction) of the Euclidean topology.

$$p(0|x,a) \equiv 1 \text{ for all } x = 1, 2, \ldots; \qquad p(0|0,0) = 1;$$
$$p(x|0,1/x) = 1 - p(0|0,1/x) = \frac{1}{x \ln(12+x)} \text{ for all } x = 1, 2, \ldots$$

with all the other transition probabilities being zero. Finally, $c(x,a) \equiv -x$. (See Fig. 4.41.)

First of all, note that the space $\mathbf{A}$ is compact and the mappings $a \to c(x,a)$ and $a \to p(y|x,a)$ are continuous for all $x, y \in \mathbf{A}$.

Let us show that the function $\mu(\cdot)$ on $\mathbf{X}$, given by

$$\mu(0) := 1 + \frac{2}{\ln(13)}; \qquad \mu(x) := x + 1 \text{ for } x = 1, 2, \ldots, \tag{4.34}$$

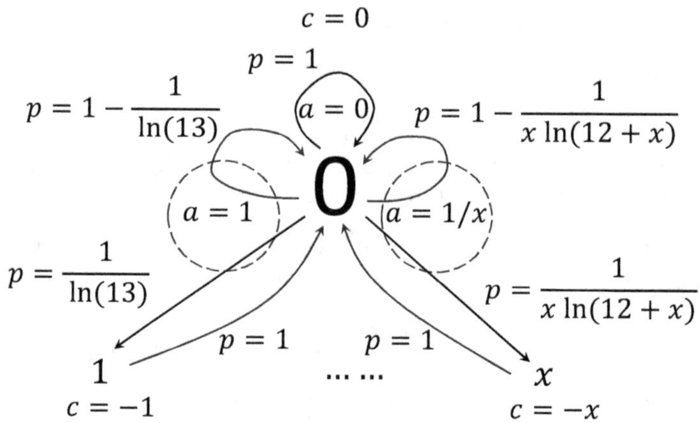

Fig. 4.41 No sample-path average optimal stationary policies.

is a general Lyapunov function. Formulae

$$1 + |c(x,a)| + \sum_{y\in\mathbf{X}\setminus\{0\}} p(y|x,a)\mu(y) = 1 + x = \mu(x), \quad x = 1,2,\ldots, \quad a \in \mathbf{A};$$

$$1 + |c(0,0)| + \sum_{y\in\mathbf{X}\setminus\{0\}} p(y|0,0)\mu(y) = 1 < \mu(0);$$

$$1 + |c(0,1/k)| + \sum_{y\in\mathbf{X}\setminus\{0\}} p(y|0,1/k)\mu(y) = 1 + \frac{k+1}{k\ln(12+k)}$$
$$\le 1 + \frac{2}{\ln(13)} = \mu(0), \quad k = 1,2,\ldots$$

are obvious. (In the last line, the inequality is valid because function $\frac{y+1}{y\ln(12+y)}$ decreases for $y > 0$.) For Item (ii) of Definition 2.7, note that $\sum_{y\in\mathbf{X}\setminus\{0\}} p(y|x,a)\mu(y) \equiv 0$ if $x \ne 0$ and

$$\lim_{a\to 0} \sum_{y\in\mathbf{X}\setminus\{0\}} p(y|0,a)\mu(y) = \lim_{k\to\infty} \frac{k+1}{k\ln(12+k)} = 0 = \sum_{y\in\mathbf{X}\setminus\{0\}} p(y|0,0)\mu(y).$$

For Item (iii) of Definition 2.7, it is sufficient to notice that $T_0 := \min\{t > 0 : X_t = 0\} \le 2$ P_x^π-a.s. Hence $\mathsf{E}_x^\pi[\mu(X_t)\mathbb{I}\{T > t\}] = 0$ for all $t \ge 2$.

Let us show that

$$\rho := -\frac{1}{\ln(16) + 1/4} \approx -0.330842 \quad \text{and} \quad h(0) := 0; \quad h(x) := -\rho - x, \quad x > 0$$

satisfy equation (4.33) and the stationary selector $\varphi^*(y) \equiv 1/4$ provides the infimum therein. Suppose $x = 0$. (The case of $x > 0$ is trivial.) In the parentheses of (4.33) we have expression

$$g(y) := \frac{-\rho - y}{y \ln(12 + y)} \text{ if } y > 0, \text{ i.e. } a = 1/y > 0, \text{ or } 0, \text{ if } a = 0.$$

After we compute the derivative

$$\frac{dg(y)}{dy} = \frac{\frac{y^2+\rho y}{12+y} + \rho \ln(12+y)}{[y \ln(12+y)]^2},$$

we can see that function $\frac{y^2+\rho y}{12+y} + \rho \ln(12+y)$ increases to ∞ starting from $\frac{1+\rho}{13} + \rho \ln(13) \approx -0.797$ at $y = 1$ because its derivative, equal to $\frac{y^2+(24+\rho)y+24\rho}{(12+y)^2}$, is positive at $y \geq 1$. Therefore, there is the single value of $y > 1$ with $\frac{dg(y)}{dy} = 0$ providing the single analytical minimum in $[1, \infty)$ to function $g(\cdot)$. Since y takes only integer values, it is sufficient to estimate

$$g(y) = \frac{-\rho - y}{y \ln(12 + y)} \approx \begin{cases} -0.3286, & \text{if } y = 3; \\ -0.3308, & \text{if } y = 4; \\ -0.3295, & \text{if } y = 5 \end{cases}$$

to ensure that

$$\inf_{y=1,2,\ldots} g(y) = g(4) = \frac{-\rho - 4}{4 \ln(16)} = \rho.$$

Thus, equation (4.33) is satisfied at $x = 0$ and the minimum on the right-hand side is provided by $a = 1/4$.

The stationary selector $\varphi^*(x) \equiv 1/4$ is AC-optimal among stationary selectors. This can be proved by calculating the invariant probability for $\varphi(x) \equiv 1/y$, $y = 1, 2, \ldots$:

$$\begin{aligned} \hat{\eta}^\varphi(0) &= \hat{\eta}^\varphi(0)\left[1 - \frac{1}{y \ln(12+y)}\right] + (1 - \hat{\eta}^\varphi(0)); \\ \hat{\eta}^\varphi(y) &= 1 - \hat{\eta}^\varphi(0) = \frac{1}{1 + y \ln(12+y)}, \quad \hat{\eta}^\varphi(x) = 0 \text{ for } x \neq 0, y; \\ \Longrightarrow v^\varphi_{\mathsf{P}_0} &= \frac{-y}{1 + y \ln(12+y)} \text{ for all } \mathsf{P}_0, \end{aligned}$$

and $\min_\varphi v^\varphi_{\mathsf{P}_0} = v^{\varphi^*}_{\mathsf{P}_0} = \rho < 0 = v^{\varphi^0}_{\mathsf{P}_0}$ for $\varphi^0(x) \equiv 0$. Moreover, the stationary selector φ^* is AC-optimal (among all strategies) by Proposition 4.1(b) of [Cavazos-Cadena *et al.* (2014)].

Arguing like in Remark 4.7, we conclude that φ^* is sample-path average optimal in the class of stationary selectors:

- For each initial state $x \in \mathbf{X}$, starting from $t = 1$, the controlled process X_t under φ^* takes only values 0 and 4, and

$$\lim_{T\to\infty} \frac{1}{T} \mathsf{E}_x^{\varphi^*}\left[\sum_{t=1}^{T} c(X_{t-1}, A_t)\right] = \lim_{T\to\infty} \frac{1}{T} \sum_{t=1}^{T} c(X_{t-1}, A_t) = \rho \quad \mathsf{P}_x^{\varphi^*}\text{-a.s.}$$

- For every stationary selector φ and $x \in \mathbf{X}$, similarly,

$$\lim_{T\to\infty} \frac{1}{T} \mathsf{E}_x^{\varphi}\left[\sum_{t=1}^{T} c(X_{t-1}, A_t)\right] = \lim_{T\to\infty} \frac{1}{T} \sum_{t=1}^{T} c(X_{t-1}, A_t) \geq \rho \quad \mathsf{P}_x^{\varphi}\text{-a.s.}$$

In what follows, we show that there exists a selector $\hat{\varphi}$ such that for all $x \in \mathbf{X}$

$$\liminf_{T\to\infty} \frac{1}{T} \sum_{t=1}^{T} c(X_{t-1}, A_t) \leq -\frac{1}{2} < \rho \quad \mathsf{P}_x^{\hat{\varphi}}\text{-a.s.} \tag{4.35}$$

Let N_T be the (random) number of visits to state 0 up to time T, i.e.,

$$N_T(\omega) = N_T(H_T(\omega)) = N_T(X_0, A_1, X_1, \ldots, A_T, X_T) := \sum_{t=1}^{T+1} \mathbb{I}\{X_{t-1} = 0\}.$$

Obviously, for each $t \geq 0$, the case $X_t \neq 0$ and $X_{t+1} \neq 0$ is excluded (that is, has P_x^{π}-probability zero for all π, $x \in \mathbf{X}$). Thus, for all π, $x \in \mathbf{X}$

$$\mathsf{P}_x^{\pi}\left(N_T \geq \frac{T}{2}\right) = 1 : \tag{4.36}$$

the minimal value of N_T corresponds to the case $X_0 \neq 0$, $X_1 = 0$, $X_2 \neq 0$, $X_3 = 0, \ldots$. The non-stationary selector $\hat{\varphi}$ of our interest is defined as follows:

$$\hat{\varphi}_t(h_{t-1}) := \frac{1}{n_{t-1}(h_{t-1}) + 1}.$$

(According to the conventional notations, $n_T(h_T) := \sum_{t=1}^{T+1} \mathbb{I}\{x_{t-1} = 0\}$ is just function of h_T; capital letters are for the random variables, i.e., functions of $\omega \in \Omega = (\mathbf{X} \times \mathbf{A})^\infty$.) In what follows, the initial state $x \in \mathbf{X}$ is arbitrarily fixed.

Let us show that for each $m = 0, 1, \ldots$ $\mathsf{P}_x^{\hat{\varphi}}\left(\bigcup_{t=m}^{\infty}\{X_t \neq 0\}\right) \equiv 1$: in words, on each time interval $\{m, m+1, \ldots\}$ sooner or later a non-zero value of the process X_t will appear. The complement event reads as $\bigcap_{t=m}^{\infty}\{X_t = 0\}$ and

$$\mathsf{P}_x^{\hat{\varphi}}\left(\bigcap_{t=m}^{\infty}\{X_t = 0\}\right) = \mathsf{E}_x^{\hat{\varphi}}\left[\mathsf{E}_x^{\hat{\varphi}}\left[\mathbb{I}\left\{\bigcap_{t=m}^{\infty}\{X_t = 0\}\right\} | X_0, A_1, \ldots, X_m, A_{m+1}\right]\right].$$

Further,

$$\begin{aligned}
&\mathsf{E}_x^{\varphi}\left[\mathbb{I}\left\{\bigcap_{t=m}^{\infty}\{X_t=0\}\right\}|X_0,A_1,\ldots,X_m,A_{m+1}\right]\\
&=\mathsf{P}_x^{\varphi}\left(\bigcap_{t=m}^{\infty}\{X_t=0\}|X_0,A_1,\ldots,X_m,A_{m+1}\right)\\
&=\begin{cases}0, & \text{if } X_m\neq 0;\\ \prod_{k=0}^{\infty}p\left(0|0,\frac{1}{N_m+1+k}\right)=\prod_{k=0}^{\infty}\left[1-\frac{1}{(N_m+1+k)\ln(12+N_m+1+k)}\right] & \\ \qquad =0, \quad \text{if } X_m=0.\end{cases}
\end{aligned}$$

Here and below, all the equalities of random variables and conditional probabilites hold $\mathsf{P}_x^{\hat{\varphi}}$-a.s. The very last equality holds because

$$\begin{aligned}
&\sum_{k=0}^{\infty}\ln\left[1-\frac{1}{(N_m+1+k)\ln(12+N_m+1+k)}\right]\\
&<-\sum_{k=0}^{\infty}\frac{1}{(N_m+1+k)\ln(12+N_m+1+k)},
\end{aligned}$$

and the series $\sum_{i=1}^{\infty}\frac{1}{i\ \ln(12+i)}=\infty$ diverges. Therefore, $\mathsf{P}_x^{\hat{\varphi}}\left(\bigcap_{t=m}^{\infty}\{X_t=0\}\right)\equiv 0$ and $\mathsf{P}_x^{\hat{\varphi}}\left(\bigcup_{t=m}^{\infty}\{X_t\neq 0\}\right)\equiv 1$.

Since for all $t=0,1,\ldots$ $\mathsf{P}_x^{\hat{\varphi}}(X_t\neq 0,\ X_{t+1}\neq 0)=0$, we have for all $m=0,1,\ldots$

$$\begin{aligned}
1&=\mathsf{P}_x^{\hat{\varphi}}\left(\bigcup_{t=m}^{\infty}\{X_{t+1}\neq 0\}\right)\\
&=\mathsf{P}_x^{\hat{\varphi}}\left(\bigcup_{t=m}^{\infty}\{X_t=0,\ X_{t+1}\neq 0\}\cup\{X_t\neq 0,\ X_{t+1}\neq 0\}\right)\\
&=\mathsf{P}_x^{\hat{\varphi}}\left(\bigcup_{t=m}^{\infty}\{X_t=0,\ X_{t+1}\neq 0\}\right)=\mathsf{P}_x^{\hat{\varphi}}\left(\bigcup_{t=m}^{\infty}\{X_t=0,\ X_{t+1}=N_t+1\}\right).
\end{aligned}$$

The last equality is due to the definition of $\hat{\varphi}_{t+1}(H_t)=\frac{1}{N_t+1}$ and of the transition probability: if $X_{t+1}\neq 0$, then $X_{t+1}=\frac{1}{\hat{\varphi}_{t+1}(H_t)}$. Since $c(\cdot,\cdot)\leq 0$, for all $t=0,1,\ldots$ we have inclusions

$$\begin{aligned}
&\{X_t=0,\ X_{t+1}=N_t+1\}\subseteq\{c(X_{t+1},A_{t+2})=-(N_t+1)\}\\
&\subseteq\left\{\sum_{i=1}^{t+2}c(X_{i-1},A_i)\leq-(N_t+1)\right\}=\left\{\frac{1}{t+2}\sum_{i=1}^{t+2}c(X_{i-1},A_i)\leq-\frac{N_t+1}{t+2}\right\}
\end{aligned}$$

and, according to (4.36), we obtain

$$\begin{aligned}
&\mathsf{P}_x^{\hat{\varphi}}(X_t = 0,\ X_{t+1} = N_t + 1) \\
&\le \mathsf{P}_x^{\hat{\varphi}}\left(\left\{\frac{1}{t+2}\sum_{i=1}^{t+2} c(X_{i-1}, A_i) \le -\frac{N_t+1}{t+2}\right\} \cap \left\{N_t \ge \frac{t}{2}\right\}\right) \\
&\le \mathsf{P}_x^{\hat{\varphi}}\left(\frac{1}{t+2}\sum_{i=1}^{t+2} c(X_{i-1}, A_i) \le -\frac{t/2+1}{t+2} = -\frac{1}{2}\right).
\end{aligned}$$

Therefore, for all $m = 0, 1, 2, \ldots$

$$\begin{aligned}
&\mathsf{P}_x^{\hat{\varphi}}\left(\bigcup_{t=m}^{\infty}\left\{\frac{1}{t+2}\sum_{i=1}^{t+2} c(X_{i-1}, A_i) \le -\frac{1}{2}\right\}\right) \\
&\ge \mathsf{P}_x^{\hat{\varphi}}\left(\bigcup_{t=m}^{\infty}\{X_t = 0,\ X_{t+1} = N_t + 1\}\right) = 1,
\end{aligned}$$

or, equivalently, for all $m = 2, 3, \ldots$

$$\mathsf{P}_x^{\hat{\varphi}}\left(\bigcup_{t=m}^{\infty}\left\{\frac{1}{t}\sum_{i=1}^{t} c(X_{i-1}, A_i) \le -\frac{1}{2}\right\}\right) = 1.$$

The last equality implies (4.35) because

$$\begin{aligned}
&\forall m = 1, 2, \ldots \quad \inf_{T>m} \frac{1}{T}\sum_{t=1}^{T} c(X_{t-1}, A_t) \le -\frac{1}{2} \quad \mathsf{P}_x^{\hat{\varphi}}\text{-a.s.} \\
&\iff \liminf_{T\to\infty} \frac{1}{T}\sum_{t=1}^{T} c(X_{t-1}, A_t) \le -\frac{1}{2} \quad \mathsf{P}_x^{\hat{\varphi}}\text{-a.s.}
\end{aligned}$$

To summarize, in the considered MDP there are no sample-path average optimal stationary selectors because, for each stationary selector φ, for all $x \in \mathbf{X}$,

$$\lim_{T\to\infty} \frac{1}{T}\mathsf{E}_x^{\varphi}\left[\sum_{t=1}^{T} c(X_{t-1}, A_t)\right] = \lim_{T\to\infty} \frac{1}{T}\sum_{t=1}^{T} c(X_{t-1}, A_t) \ge \rho > -\frac{1}{2} \quad \mathsf{P}_x^{\varphi}\text{-a.s.}$$

The statement of Theorem 4.13 is false: the stationary selector $\varphi^*(x) \equiv 1/4$ provides the infimum in (4.33), but is not sample-path average optimal. As was mentioned, it is sample-path average optimal only within the class of stationary selectors. Theorem 4.13 is not applicable because there is no general Lyapunov function for the cost $\tilde{c}(x, a) \equiv \mu^{\gamma}(x)$ with $\gamma > 2$, where $\mu(\cdot)$ was calculated earlier: see (4.34). This general Lyapunov function $\tilde{\mu}(\cdot) : \mathbf{X} \to [1, \infty)$ (if exists) must satisfy inequalities

$$1 + (x+1)^{\gamma} \le \tilde{\mu}(x) \Longrightarrow \tilde{\mu}(x) > x^{\gamma}, \text{ if } x > 0,$$

and

$$1+\left(1+\frac{2}{\ln(13)}\right)^{\gamma}+\frac{\tilde{\mu}(y)}{y\ln(12+y)}\le\tilde{\mu}(0) \text{ for all } a=1/y,\ y=1,2,\dots,\ \text{if } x=0,$$

which is impossible because $\frac{\tilde{\mu}(y)}{y\ln(12+y)} > \frac{y^{\gamma-1}}{\ln(12+y)} \to \infty$ as $y\to\infty$ for $\gamma>2$. □

4.3.32 *Model with partial information*

A series of classical models with incomplete information are called bandit problems.

> "The game is as follows: Imagine we have two slot machines with unknown success probabilities θ_1 and θ_2... At each stage we have to choose one of the arms. We receive one Euro if the arm wins, else no cash flow appears. The aim is to maximize the expected total reward." [Bäuerle and Rieder (2011), §5.5].

The rigorous mathematical statement and solution for the discounted version can be found in [Bäuerle and Rieder (2011), §7.6.4].

In the current subsection, we study the following modification similar to that described in [Platzman (1980), Ex. 5]. Suppose either $\theta_1=1$ and $\theta_2=0$, or $\theta_1=0$ and $\theta_2=1$. Clearly, after the first trial, after winning or loosing, the situation becomes totally clear if the result of this trial is seen. We assume that the reward (0 or 1 unit) is not seen, but the total value is discovered only after the game is finished. Assume also that, apart from the actions 1 and 2 (meaning the chosen arm), there is action 3 ('approach the oracle') which leads to no income, but results in the independent answers 1 (or 2) with probability $p\in(1/2,1)$ if $\theta_1=1$ (if $\theta_2=1$); $1-p<1/2$ is the probability of the prediction error. Below we study the minimization version of the problem with the average cost.

Example. The action space is $\mathbf{A}=\{1,2,3\}$ and $a=1$ or 2 means trying arm a, while $a=3$ means approaching the oracle. The state space is $\mathbf{X}=(0,1)\times\{1,2\}$. State $x_t=(q_t,i_t)\in\mathbf{X}$ means the *a priori* (if $t=0$) or *a posteriori* (if $t>0$) probability of $(\theta_1=1,\theta_2=0)$ equals q_t and the last return on action 3 was i_t. When $t=0$, the value i_0 plays no role and can be chosen arbitrarily. As usual, the sufficient statistic q_t appears as the component of the state [Bertsekas and Shreve (1978), Chapter 10].

The transition probability is $p(dy|x,a)=\delta_x(dy)$ if $a=1$ or 2, and in this case, for $x=(q,i)$, the expected cost equals minus expectation of the coincidence $\theta_a=1$: $c(x,a)=\begin{cases}-q, & \text{if } a=1;\\ -(1-q), & \text{if } a=2.\end{cases}$ In case $a=3$, $c(x,3)\equiv 0$, and the transition probability is calculated as follows. The value of the component j of the new state $y=(\tilde{q},j)$ equals 1 or 2 with probability $qp+(1-q)(1-p)$ or $q(1-p)+(1-q)p$ correspondingly. In the first case, $\tilde{q}=\frac{qp}{qp+(1-q)(1-p)}$, and in the second case $\tilde{q}=\frac{q(1-p)}{q(1-p)+(1-q)p}$. See Fig. 4.42. The initial state is $x_0=(q_0,i_0)$, where, as mentioned above, i_0 can be fixed arbitrarily.

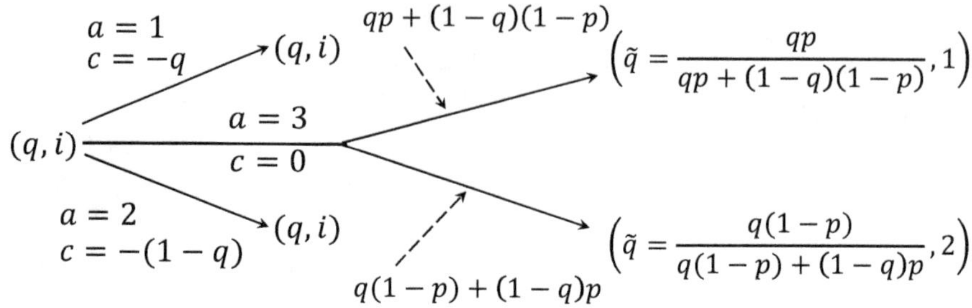

Fig. 4.42 Model with partial information.

First of all, we show that

$$v^*_{(q_0,i_0)} := \inf_\pi v^\pi_{(q_0,i_0)} = \inf_\pi \limsup_{T\to\infty} \frac{1}{T}\mathsf{E}^\pi_{(q_0,i_0)}\left[\sum_{t=1}^T c(X_{t-1},A_t)\right] = -1. \qquad (4.37)$$

To do so, for an arbitrarily fixed $\varepsilon \in (0,1)$ take $N > \sqrt{\frac{p-p^2}{(p-0.5)^2\varepsilon}}$ and consider the following non-stationary, past dependent selector φ^N:

$$\varphi^N_t(h_{t-1}) = 3 \text{ at } t = 1,2,\ldots,N \text{ and}$$
$$\varphi^N_t(h_{t-1}) = \begin{cases} 1, & \text{if } z(h_N) < 1.5; \\ 2, & \text{if } z(h_N) \geq 1.5 \end{cases} \quad \text{at } t > N,$$

where, for $h_N = ((q_0,i_0),3,(q_1,i_1),3,(q_2,i_2),\ldots,3,(q_N,i_N))$, $z(h_N) := \frac{1}{N}\sum_{\tau=1}^N i_\tau$. Histories h_N of another form have $\mathsf{P}^{\varphi^N}_{(q_0,i_0)}$-measure zero. In words, at the first steps we estimate the value of θ_1, and after that apply action 1 (or 2) if the estimation is in favor of $\theta_1 = 1$ (in favor of $\theta_1 = 0 \iff \theta_2 = 1$).

If $\theta_1 = 1$, then $\{I_\tau\}_{\tau=1}^N$, the random components of H_N, are the sequence of independent Bernoulli random variables with values 1 or 2 having probabilities $\mathsf{P}(I_\tau = 1) = p$ and $\mathsf{P}(I_\tau = 2) = 1-p$. Thus, $\mathsf{E}[I_\tau] = 2-p$; $Var[I_\tau] = p-p^2$; for the random variable $Z = z(H_N)$, $\mathsf{E}[Z] = 2-p$; $Var[Z] = \frac{p-p^2}{N^2}$, and, by the Chebyshev inequality,

$$\mathsf{P}^{\varphi^N}_{(q_0,i_0)}(|Z-(2-p)| \geq p-0.5|\theta_1 = 1) \leq \frac{p-p^2}{N^2(p-0.5)^2} < \varepsilon.$$

Therefore,

$$\begin{aligned}
& \mathsf{P}^{\varphi^N}_{(q_0,i_0)}(Z \geq 1.5)|\theta_1 = 1) \\
\leq\ & \mathsf{P}^{\varphi^N}_{(q_0,i_0)}(Z - 2 + p \geq p - 0.5 \text{ or } Z - 2 + p \leq -p + 0.5|\theta_1 = 1) \\
=\ & \mathsf{P}^{\varphi^N}_{(q_0,i_0)}(|Z - (2-p)| \geq p - 0.5|\theta_1 = 1) \leq \frac{p - p^2}{N^2(p-0.5)^2} < \varepsilon \\
\Longrightarrow\ & \mathsf{P}^{\varphi^N}_{(q_0,i_0)}(Z < 1.5|\theta_1 = 1) > 1 - \varepsilon.
\end{aligned}$$

Similarly, if $\theta_2 = 1$, then $\mathsf{E}[Z] = 1 + p$, $Var[Z] = \frac{p-p^2}{N^2}$, and

$$\begin{aligned}
& \mathsf{P}^{\varphi^N}_{(q_0,i_0)}(Z < 1.5)|\theta_2 = 1) \leq \mathsf{P}^{\varphi^N}_{(q_0,i_0)}(|Z - (1+p)| > p - 0.5|\theta_2 = 1) \\
\leq\ & \frac{p - p^2}{N^2(p-0.5)^2} < \varepsilon \Longrightarrow \mathsf{P}^{\varphi^N}_{(q_0,i_0)}(Z \geq 1.5|\theta_2 = 1) > 1 - \varepsilon
\end{aligned}$$

and

$$\mathsf{P}^{\varphi^N}_{(q_0,i_0)}(\theta_1 = 1 \ \&\ Z < 1.5) + \mathsf{P}^{\varphi^N}_{(q_0,i_0)}(\theta_2 = 1 \ \&\ Z \geq 1.5) > 1 - \varepsilon. \tag{4.38}$$

The state $X_t = (Q_t, I_t) = (Q_N, I_N)$ does not change at $t > N$ and

$$\begin{aligned}
& \mathsf{E}^{\varphi^N}_{(q_0,i_0)}[c(X_{t-1}, A_t)] \\
=\ & -\mathsf{E}^{\varphi^N}_{(q_0,i_0)}[Q_N|Z < 1.5]\mathsf{P}^{\varphi^N}_{(q_0,i_0)}(Z < 1.5) \\
& -\mathsf{E}^{\varphi^N}_{(q_0,i_0)}[(1 - Q_N)|Z \geq 1.5]\mathsf{P}^{\varphi^N}_{(q_0,i_0)}(Z \geq 1.5), \quad \forall t > N.
\end{aligned}$$

Recall that $Q_N = \mathsf{P}^{\varphi^N}_{(q_0,i_0)}(\theta_1 = 1|H_N)$ is the *a posteriori* probability that $\theta_1 = 1$ and

$$\begin{aligned}
\mathsf{E}^{\varphi^N}_{(q_0,i_0)}[Q_N|Z < 1.5] &= \mathsf{E}^{\varphi^N}_{(q_0,i_0)}[\mathsf{E}^{\varphi^N}_{(q_0,i_0)}[\mathbb{I}\{\theta_1 = 1\}|H_N]|Z < 1.5] \\
&= \mathsf{E}^{\varphi^N}_{(q_0,i_0)}[\mathbb{I}\{\theta_1 = 1\}|Z < 1.5] = \mathsf{P}^{\varphi^N}_{(q_0,i_0)}(\theta_1 = 1|Z < 1.5).
\end{aligned}$$

Similarly,

$$\mathsf{E}^{\varphi^N}_{(q_0,i_0)}[(1 - Q_N)|Z \geq 1.5] = \mathsf{P}^{\varphi^N}_{(q_0,i_0)}(\theta_2 = 1|Z \geq 1.5).$$

Thus, by (4.38), for all $t > N$,

$$\begin{aligned}
& \mathsf{E}^{\varphi^N}_{(q_0,i_0)}[c(X_{t-1}, A_t)] \\
=\ & -\left\{\mathsf{P}^{\varphi^N}_{(q_0,i_0)}(\theta_1 = 1 \ \&\ Z < 1.5) + \mathsf{P}^{\varphi^N}_{(q_0,i_0)}(\theta_2 = 1 \ \&\ Z \geq 1.5)\right\} < -1 + \varepsilon.
\end{aligned}$$

The cost function $c(\cdot,\cdot) \geq -1$ is bounded; thus, we conclude that

$$v^{\varphi^N}_{(q_0,i_0)} = \limsup_{T\to\infty} \frac{1}{T} \mathsf{E}^{\varphi^N}_{(q_0,i_0)} \left[\sum_{t=1}^{N} c(X_{t-1},A_t) + \sum_{t=N+1}^{T} c(X_{t-1},A_t)\right]$$
$$< \lim_{T\to\infty} \frac{T-N}{T}[-1+\varepsilon] = -1+\varepsilon.$$

Since $\varepsilon \in (0,1)$ was fixed arbitrarily, formula (4.37) follows.

Canonical equations (4.2) have no solutions with bounded function $h(\cdot)$. Hence, there are no canonical triplets with bounded function $h(\cdot)$ by Theorem 4.1. These assertions hold because, otherwise, the canonical stationary selector φ^*, existing because the space $\mathbf{A}$ is finite, would have been AC-optimal according to Theorem 4.3. The latter statement cannot be valid:

- if $\varphi^*((q_0,i_0)) = 1$ (or 2) for some $q_0 \in (0,1)$, then $v^{\varphi^*}_{(q_0,i_0)} = -q_0 > -1$ (or $v^{\varphi^*}_{(q_0,i_0)} = -(1-q_0) > -1$) because $X_t \equiv (q_0,i_0)$ for all $t = 0,1,2,\ldots$;
- if $\varphi^*((q_0,i_0)) = 3$ for all $q_0 \in (0,1)$, then $v^{\varphi^*}_{(q_0,i_0)} \equiv 0 > -1$ because $Q_t \in (0,1)$ for all $t = 0,1,2,\ldots$.

One of the AC-optimal selectors looks as follows. For $t = N^2$ $(N = 1,2,\ldots)$, put $\varphi^*_t(h_{t-1}) \equiv 3$ and compute $z(h_t) := \frac{1}{N}\sum_{n=1}^{N} i_{n^2}$. Like previously, here

$$h_t = ((q_0,i_0), a_1 = 3, (q_1,i_1), a_2, (q_2,i_2), \ldots, a_t = 3, (q_t,i_t)).$$

For $t \in \{N^2+1, N^2+2, \ldots, (N+1)^2-1\}$ put $\varphi^*_t(h_{t-1}) = \begin{cases} 1, & \text{if } z(h_{N^2}) < 1.5; \\ 2, & \text{if } z(h_{N^2}) \geq 1.5. \end{cases}$

Let us estimate $v^{\varphi^*}_{(q_0,i_0)}$. For an arbitrary $\varepsilon \in (0,1)$ take $N > \sqrt{\frac{p-p^2}{(p-0.5)\varepsilon}}$. Similarly to the above presented reasoning, for all $t > N^2$,

$$\mathsf{E}^{\varphi^*}_{(q_0,i_0)}[c(X_{t-1},A_t)] \quad \begin{cases} < -1+\varepsilon, & \text{if } t \neq n^2 \ (n = 1,2,\ldots); \\ = 0, & \text{if } t = n^2 \text{ for some } n = 1,2,\ldots. \end{cases}$$

Thus,

$$v^{\varphi^*}_{(q_0,i_0)} = \limsup_{T\to\infty} \frac{1}{T} \mathsf{E}^{\varphi^*}_{(q_0,i_0)} \left[\sum_{t=1}^{N^2} c(X_{t-1},A_t) + \sum_{t=N^2+1}^{T} c(X_{t-1},A_t)\right]$$
$$< \lim_{T\to\infty} \frac{1}{T}[(T - N^2 - \lfloor\sqrt{T}\rfloor)(-1+\varepsilon)] = -1+\varepsilon.$$

$\lfloor\cdot\rfloor$ is the integer part, and $\lfloor\sqrt{T}\rfloor$ is the estimation of the number of decision epochs with $A_t = 3$ and zero cost. Since this inequality holds for all $\varepsilon \in (0,1)$, $v^{\varphi^*}_{(q_0,i_0)} = -1 = v^*_{(q_0,i_0)}$. □

4.3.33 *Parrondo's paradox*

This paradox can be described as follows [Parrondo and Dinis (2004), p. 147]: "Two losing gambling games, when alternated in a periodic or random fashion, can produce a winning game." There exist many examples to illustrate this; we present the simplest one.

Example. Let $\mathbf{X} = \{1,2,3\}$, $\mathbf{A} = \{1,2\}$, $p(2|1,1) = p(3|2,1) = p(1|3,1) = 0.49$, $p(3|1,1) = p(1|2,1) = p(2|3,1) = 0.51$, $p(2|1,2) = 1 - p(3|1,2) = 0.09$, $p(3|2,2) = 1-p(1|2,2) = p(1|3,2) = 1-p(2|3,2) = 0.74$, with all other transition probabilities being zero. We put $c(1,a) = p(3|1,a)-p(2|1,a)$, $c(2,a) = p(1|2,a)-p(3|2,a)$, $c(3,a) = p(2|3,a) - p(1|3,a)$ (see Fig. 4.43).

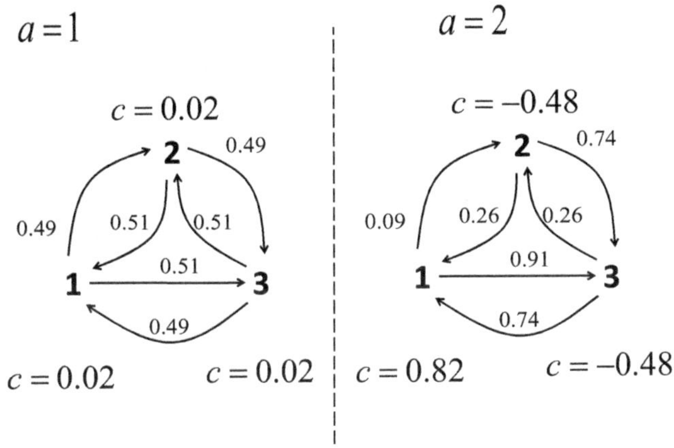

Fig. 4.43 Parrondo's paradox. The arrows are marked with their corresponding transition probabilities.

One can say that the process moves clockwise or anticlockwise, with the probabilities depending on the actions. The gambler gains one pound for each clockwise step of the walk, and loses one pound for each anticlockwise step. The objective is to minimize the expected average cost per time unit.

After we as usual put $h(1) = 0$ and keeping in mind that the model is unichain (hence $\rho(x) \equiv \rho$), the canonical equations (4.2) take the form:

$$\begin{aligned}
\rho &= \min\{0.02 + 0.49\ h(2) + 0.51\ h(3); \quad 0.82 + 0.09\ h(2) + 0.91\ h(3)\};\\
\rho + h(2) &= \min\{0.02 + 0.49\ h(3); \quad -0.48 + 0.74\ h(3)\};\\
\rho + h(3) &= \min\{0.02 + 0.51\ h(2); \quad -0.48 + 0.26\ h(2)\}.
\end{aligned}$$

One can check that the solution is given by $h(2) = -\frac{5075}{13433} \approx -0.3778$; $h(3) = -\frac{525}{1919} \approx -0.2736$; $\rho = 0.02 + 0.49\ h(2) + 0.51\ h(3) = -\frac{204617}{671650} \approx -0.305$; the stationary selector $\varphi^*(x) = \begin{cases} 1, & \text{if } x = 1;\\ 2, & \text{if } x = 2 \text{ or } 3 \end{cases}$

is AC-optimal according to Theorems 4.1 and 4.3.

Consider the stationary selector $\varphi^1(x) \equiv 1$. Analysis of the canonical equations (4.2) gives the following values: $h^1(x) \equiv 0$, $\rho^1 = 0.02$. Similarly, for the stationary selector $\varphi^2(x) \equiv 2$ we obtain: $h^2(1) = 0$, $h^2(2) = -\frac{13195}{12313} \approx -1.0716$, $h^2(3) = -\frac{1365}{1759} \approx -0.7760$, $\rho^2 = 0.82 + 0.09\ h^2(2) + 0.91\ h^2(3) = \frac{10703}{615650} \approx 0.017$. This means that the outcomes for the both pure games, where action 1 or action 2 is always chosen, are unfavourable: the expected average cost per time unit is positive.

The optimal strategy φ^* results in a winning game, but more excitingly, a random choice of actions 1 and 2 at each step also results in a winning game. To be more precise, we analyze the stationary randomized strategy $\pi(1|x) = \pi(2|x) \equiv 0.5$. The canonical equations are as follows:

$$\begin{aligned}
\tilde{\rho} + \tilde{h}(1) &= 0.42 + 0.29\ \tilde{h}(2) + 0.71\ \tilde{h}(3);\\
\tilde{\rho} + \tilde{h}(2) &= -0.23 + 0.615\ \tilde{h}(3) + 0.385\ \tilde{h}(1);\\
\tilde{\rho} + \tilde{h}(3) &= -0.23 + 0.615\ \tilde{h}(1) + 0.385\ \tilde{h}(2),
\end{aligned}$$

and the solution is given by $\tilde{h}(1) = 0$, $\tilde{h}(2) = -\frac{11631230}{24541369} \approx -0.4739$, $\tilde{h}(3) = -\frac{36010}{88597} \approx -0.4064$, $\tilde{\rho} = 0.42 + 0.29\ \tilde{h}(2) + 0.71\ \tilde{h}(3) = -\frac{7388421}{1227068450} \approx -0.006$. Since $\tilde{\rho} < 0$, this random choice of losing games (i.e. actions 1 and 2) results in a winning game. $\square$

Chapter 5

Continuous-Time Markov Decision Processes

In this chapter we deal with the homogeneous infinite-horizon continuous-time version of MDP abbreviated as CTMDP. After the brief presentation of the necessary material in Sections 5.1 and 5.2, a number of examples are given in Section 5.3. Let us also underline that a very big part of Chapter 5 is based on the recent monograph [Piunovskiy and Zhang (2020)].

5.1 Preliminaries

Like previously, $\mathbf{X}$ and $\mathbf{A}$ are the Borel state and action spaces. The dynamics is described by the *transition rate* $q(dy|x,a)$, the signed kernel on $\mathcal{B}(\mathbf{X})$ given $(x,a) \in \mathbf{X} \times \mathbf{A}$ such that $q(\Gamma|x,a) \geq 0$ for $\Gamma \not\ni \{x\}$ and $q(\mathbf{X}|x,a) = 0$: $q(\{x\}|x,a) = -q(\mathbf{X} \setminus \{x\}|x,a)$ (that is, the transition rate is *conservative*). At any time moment, if $x \in \mathbf{X}$ and $a \in \mathbf{A}$ are the current state and action, then the *cost rate* is $c(x,a)$.

In what follows, $q_x(a) := -q(\{x\}|x,a) \geq 0$ is the *jump intensity* and $\tilde{q}(\Gamma|x,a) := q(\Gamma \setminus \{x\}|x,a)$. We always assume that the transition rate is *stable*: $\bar{q}_x := \sup_{a\in\mathbf{A}} q_x(a) < \infty$ for all $x \in \mathbf{X}$.

5.1.1 *Informal description*

The trajectory of the controlled process is piecewise constant. The (random) initial state $X_0 \in \mathbf{X}$ has the distribution P_0. After the value x_0 of X_0 is realized, the decision maker can choose an action $A(0)$, possibly using a *randomized* control $\varpi_1(da|x_0)$, a stochastic kernel on $\mathcal{B}(\mathbf{A})$ given $x_0 \in \mathbf{X}$. Otherwise, (s)he chooses a so called *relaxed* control, a stochastic kernel $\pi_1(da|x_0,s)$ on $\mathcal{B}(\mathbf{A})$ given $(x_0,s) \in \mathbf{X} \times \mathbb{R}_+$. As a result, the sojourn time Θ_1 and the new state X_1 are realized with the following joint distribution. (Here we assume for simplicity that $0 < \delta \leq q_x(a) \leq K < \infty$ for some constants δ, K; this is not needed in the formal construction in the next subsection.)

- In the first case, under the randomized control ϖ_1,

$$\begin{aligned}&\mathsf{P}(\Theta_1 \le t, X_1 \in \Gamma_X | X_0 = x_0)\\&= \int_{(0,t]} \int_{\mathbf{A}} \tilde{q}(\Gamma_X | x_0, a) e^{-q_{x_0}(a)s} \varpi_1(da|x_0) ds, \quad t \ge 0,\ \Gamma_X \in \mathcal{B}(\mathbf{X}).\end{aligned}$$

- In the second case, under the relaxed control π_1, for $t \ge 0,\ \Gamma_X \in \mathcal{B}(\mathbf{X})$ we have

$$\begin{aligned}&\mathsf{P}(\Theta_1 \le t, X_1 \in \Gamma_X | X_0 = x_0)\\&= \int_{(0,t]} \int_{\mathbf{A}} \tilde{q}(\Gamma_X | x_0, a) \pi_1(da|x_0, s) e^{-\int_{(0,s]} \int_{\mathbf{A}} q_{x_0}(a) \pi_1(da|x_0,u) du} ds.\end{aligned}$$

Even if $\pi_1(da|x_0)$ does not depend on time s, the controls ϖ_1 and π_1 are essentially different: in the first case, the sojourn time Θ_1 is a convex combination of exponential random variables with parameters $q_{x_0}(a)$, and in the second case that is just the exponential random variable with parameter $\int_{\mathbf{A}} q_{x_0}(a)\pi_1(da|x_0)$. The distributions of X_1 are also different. If $\varpi_1(da|x_0) = \delta_{a^*}(da)$ or $\pi_1(da|x_0, s) = \delta_{a^*}(da)$, then the sojourn time Θ_1 is exponential with parameter $q_{X_0}(a^*)$ (assuming $q_{X_0}(a^*) > 0$), and the distribution of X_1 is given by $\frac{\tilde{q}(dy|X_0,a^*)}{q_{X_0}(a^*)}$.

After the piece of trajectory $h_1 = (x_0, \theta_1, x_1)$ is realized, the process develops in the similar manner. The kernels ϖ_2 or π_2 may depend on h_1. And so on.

Let us underline that usually, when talking about continuous-time MDPs, only the relaxed controls are considered [Guo and Hernández-Lerma (2009); Prieto-Rumeau and Hernández-Lerma (2012)]. In case only randomized controls are in use, we obtain the so called *'exponential semi-Markov decision process'* [Feinberg (2004)], [Ross (1970), Ch. 7]. Combined randomized-relaxed controls (and more general control strategies) appeared in [Piunovskiy and Zhang (2020)].

5.1.2 *Strategies, strategic measures and optimal control problems*

The canonical (Borel) *sample space* Ω contains sequences (*trajectories*) of the following types:

$$\omega = (x_0, \theta_1, x_1, \ldots) \text{ and } \omega = (x_0, \theta_1, x_1, \ldots, \theta_{m-1}, x_{m-1}, \infty, x_\infty, \ldots),$$

where $\theta_1, \theta_2, \ldots \in (0, \infty)$, $x_1, x_2, \ldots \in \mathbf{X}$ and $x_\infty \notin \mathbf{X}$ is the isolated artificial point which comes to the stage if there were only finite number of jumps on the infinite horizon $[0, \infty)$. x_0 is the initial state and $\theta_1, \theta_2, \ldots \in (0, \infty)$ are the *sojourn times* with the after-jump states $x_1, x_2, \ldots \in \mathbf{X}$. As usual, the capital letters like X_i and Θ_i denote the random variables, i.e., the projection mappings on Ω. In what follows, $T_n := \sum_{i=1}^n \Theta_i$ are the *jump epochs*, $T_\infty := \lim_{n\to\infty} T_n$,

$\mathbf{H}_n := \{(x_0, \theta_1, x_1, \ldots, \theta_n, x_n) : (x_0, \theta_1, x_1, \ldots) \in \Omega\}$ is the space of n-term *histories*.

Definition 5.1. A *control strategy* is a sequence $S = \{S_n\}_{n=1}^{\infty}$, where, for each fixed $n = 1, 2, \ldots$, S_n is either a stochastic kernel $\varpi_n(da|h_{n-1})$ on $\mathcal{B}(\mathbf{A})$ given $h_{n-1} \in \mathbf{H}_{n-1}$, or a stochastic kernel $\pi_n(da|h_{n-1}, s)$ on $\mathcal{B}(\mathbf{A})$ given $(h_{n-1}, s) \in \mathbf{H}_{n-1} \times (0, \infty)$. If $S_n = \varpi_n$ for all $n = 1, 2, \ldots$, then the strategy is purely *randomized*; it is denoted as $\varpi = \{\varpi_n\}_{n=1}^{\infty}$. Similarly, a purely *relaxed* strategy with $S_n = \pi_n$ for all $n = 1, 2, \ldots$ will be denoted as $\pi = \{\pi_n\}_{n=1}^{\infty}$ and will be also called a "*π-strategy*". Much more general control strategies were introduced in [Piunovskiy and Zhang (2020), Ch. 6].

If $S_n = \pi_n$ is a relaxed control, we introduce the following notations for each $\Gamma_X \in \mathcal{B}(\mathbf{X})$ and $s \in (0, \infty)$:

$$
\begin{aligned}
q(\Gamma_X|x_{n-1}, \pi_n, s) &:= \int_{\mathbf{A}} q(\Gamma_X|x_{n-1}, a)\pi_n(da|h_{n-1}, s) \\
\tilde{q}(\Gamma_X|x_{n-1}, \pi_n, s) &:= q(\Gamma_X \setminus \{x_{n-1}\}|x_{n-1}, \pi_n, s)
\end{aligned}
\tag{5.1}
$$

and

$$
q_{x_{n-1}}(\pi_n, s) := -q(\{x_{n-1}\}|x_{n-1}, \pi_n, s) = \int_{\mathbf{A}} q_{x_{n-1}}(a)\pi_n(da|h_{n-1}, s). \tag{5.2}
$$

If the initial distribution P_0 and a strategy S are fixed, then one can define the strategic probability measure $\mathsf{P}^S_{\mathsf{P}_0}$ on $(\Omega, \mathcal{B}(\Omega))$ using the Ionescu Tulcea Theorem [Bertsekas and Shreve (1978), Prop. 7.28], [Hernández-Lerma and Lasserre (1996a), Prop. C10], [Piunovskiy (1997), Thm. A1.11]. That measure will exhibit the natural properties informally described in Subsection 5.1.1. $\mathsf{E}^S_{\mathsf{P}_0}$ is the mathematical expectation w.r.t. $\mathsf{P}^S_{\mathsf{P}_0}$. More about strategies and strategic measures in [Piunovskiy and Zhang (2020)]. The *controlled* process is defined by

$$
X(t) := \sum_{n=0}^{\infty} \mathbb{I}\{T_n \le t < T_{n+1}\}X_n + \mathbb{I}\{T_\infty \le t\}x_\infty, \qquad t \in [0, \infty).
$$

The controlled process is not investigated after the accumulation of jumps (after the time moment $T_\infty < \infty$): we accept that $X(t)$ is absorbed at x_∞ after T_∞ and $c(x_\infty, a) := 0$ for all $a \in \mathbf{A}$.

Definition 5.2.

(a) A π-strategy $\pi = \{\pi_n\}_{n=1}^{\infty}$ is called *natural Markov* and denoted as $\breve{\pi}$ if there exists a stochastic kernel $\breve{\pi}$ on $\mathcal{B}(\mathbf{A})$ given $(x, t) \in \mathbf{X} \times (0, \infty)$ such that, for each $n \ge 1$, it holds that

$$
\pi_n(da|x_0, \theta_1, \ldots, \theta_{n-1}, x_{n-1}, s) = \breve{\pi}(da|x_{n-1}, t_{n-1} + s).
$$

with $t_{n-1} = \sum_{i=1}^{n-1} \theta_i$.

(b) A π-strategy is called *Markov* and denoted as $\pi^{\rm m}$, if

$$\pi_n(da|x_0,\theta_1,x_1,\theta_2,\ldots,x_{n-1},s)=\pi^{\rm m}_n(da|x_{n-1},s)$$

for all $n=1,2,\ldots$.

(c) A randomized strategy is called *Markov* and denoted as $\varpi^{\rm m}$, if

$$\varpi_n(da|x_0,\theta_1,x_1,\theta_2,\ldots,x_{n-1})=\varpi^{\rm m}_n(da|x_{n-1})$$

for all $n=1,2,\ldots$.

(d) A Markov π-strategy (respectively, randomized strategy) is called *stationary* and denoted as $\pi^{\rm s}$ (respectively, $\varpi^{\rm s}$), if $\pi^{\rm m}_n(da|x,s)=\pi^{\rm s}(da|x)$ (respectively, $\varpi^{\rm m}_n(da|x)=\varpi^{\rm s}(da|x)$) for all $n=1,2,\ldots$.

(e) If all the kernels π_n (respectively, ϖ_n) are degenerate, concentrated at points $\varphi_n(h_{n-1},s)\in\mathbf{A}$ (respectively, $\varphi_n(h_{n-1})\in\mathbf{A}$), with φ_n being a measurable $\mathbf{A}$-valued mapping, then the π-strategy (respectively, randomized strategy) is called *deterministic* and denoted as $\varphi=\{\varphi_n\}_{n=1}^\infty$.

(f) A *deterministic stationary* strategy is defined by the measurable mapping $\varphi:\ \mathbf{X}\to\mathbf{A}$ and denoted as φ. It can be equally called a π-strategy or a randomized strategy in the sense that the corresponding strategic measure $\mathsf{P}^\pi_{\mathsf{P}_0}=\mathsf{P}^\varpi_{\mathsf{P}_0}$ is the same.

Note that, in the most works on CTMDPs, only natural Markov strategies are considered: [Guo and Hernández-Lerma (2009); Hernández-Lerma *et al.* (2023); Prieto-Rumeau and Hernández-Lerma (2012)].

Remark 5.1.

(a) Assume for simplicity that $T_\infty=\infty$. If the strategy ϖ is randomized then one can introduce the *control* process $\{A(t)\}_{t\in(0,\infty)}$: on each interval $(T_{n-1},T_n]$ the constant value of $A(t)$ is the random variable with distribution $\varpi_n(da|H_{n-1})$. The control process is also well defined if the π-strategy is deterministic.

(b) Suppose $\mathbf{X}=\{0\}$, $\mathbf{A}=\{1,2\}$ and consider the relaxed strategy $\pi(1|x_0=0)=\pi(2|x_0=0)=\frac{1}{2}$: for each finite set of distinct points $t_1,t_2,\ldots,t_n$ $\{A(t_1),A(t_2),\ldots,A(t_n)\}$ is the sequence of independent Bernoulli random variables. According to Kolmogorov's Extension Theorem [Oksendal (2013), Thm. 2.1.5], there exists a probability space $(\tilde\Omega,\tilde{\mathcal{F}},\tilde{\mathsf{P}})$ and stochastic process thereon (i.e., the family of random variables $\{A(t)\}_{t\in(0,\infty)}$) exhibiting the above formulated property. But this random process cannot be jointly measurable w.r.t. $(\tilde\omega,t)$: see Example 1.2.5 of [Kallianpur (1980)]. One can say that the control process is not realizable. This is the typical situation if a π-strategy is not deterministic. More about realizable/non-realizable strategies in [Piunovskiy (2018)] and [Piunovskiy and Zhang (2020), §1.1.4].

Remark 5.2. Note that, under a Markov π-strategy, the controlled process $X(\cdot)$ is not necessarily Markov because the transition rate on $(T_{n-1}, T_n]$, given by $\int_{\mathbf{A}} q(dy|X_{n-1}, a)\pi_n^M(da|X_{n-1}, t - T_{n-1})$, can depend on T_{n-1}. As was shown in [Piunovskiy and Zhang (2020), Ch. 2], under a natural Markov strategy, the controlled process $X(\cdot)$ is Markov. Note also that any stationary π-strategy is natural Markov.

Definition 5.3. The *marginal distribution* of the CTMDP with the initial distribution P_0 under a strategy $S = \{S_n\}_{n=1}^{\infty}$ is a substochastic kernel on $\mathcal{B}(\mathbf{X} \times \mathbf{A})$ given $t \in (0, \infty)$ defined by

$$\begin{aligned}
&\bar{\mathsf{P}}_{\mathsf{P}_0}^S(t, dy \times da) \\
&:= \sum_{n=0}^{\infty} \Bigg(\mathbb{I}\{S_{n+1} = \varpi_{n+1}\} \mathsf{E}_{\mathsf{P}_0}^S \Big[\mathbb{I}\{X(t-) = X_n \in dy\} \varpi_{n+1}(da|H_n) \\
&\quad \times \mathbb{I}\{t \in (T_n, T_{n+1}]\} \Big] \\
&\quad + \mathbb{I}\{S_{n+1} = \pi_{n+1}\} \mathsf{E}_{\mathsf{P}_0}^S \Big[\mathbb{I}\{X(t-) = X_n \in dy\} \pi_{n+1}(da|H_n, t - T_n) \\
&\quad \times \mathbb{I}\{t \in (T_n, T_{n+1}]\} \Big] \Bigg).
\end{aligned}$$

We also put

$$\bar{\mathsf{P}}_{\mathsf{P}_0}^S(t, dy \times \mathbf{A}) =: \bar{\mathsf{P}}_{\mathsf{P}_0}^S(t, dy)$$

and $\bar{\mathsf{P}}_{\mathsf{P}_0}^S(0, dy \times \mathbf{A}) := \mathsf{P}_0(dy)$ with a slight abuse of notations.

The *total expected α-discounted cost*, i.e., the *performance functional* is defined by

$$W^\alpha(S, \mathsf{P}_0) := \int_{(0,\infty)} e^{-\alpha t} \int_{\mathbf{X}\times\mathbf{A}} c(x, a) \bar{\mathsf{P}}_{\mathsf{P}_0}^S(t, dx \times da) dt, \quad \alpha \in [0, \infty), \tag{5.3}$$

where the positive and negative parts of the given cost rate $c(\cdot, \cdot) : \mathbf{X} \times \mathbf{A} \to \mathbb{R}$ are integrated separately with the convention $\infty - \infty := +\infty$. $\alpha \in [0, \infty)$ is the *discount factor*. If $\alpha = 0$, then the CTMDP is undiscounted.

Another common performance functional looks as follows:

$$W^A(S, \mathsf{P}_0) := \limsup_{T\to\infty} \frac{1}{T} \int_{(0,\min\{T,T_\infty\}]} \int_{\mathbf{X}\times\mathbf{A}} c(x, a) \bar{\mathsf{P}}_{\mathsf{P}_0}^S(t, dx \times da) dt \tag{5.4}$$

If $S = \breve{\pi}$ is a natural Markov strategy, then $\bar{\mathsf{P}}_{\mathsf{P}_0}^{\breve{\pi}}(t, dy \times da) = \mathsf{E}_{\mathsf{P}_0}^{\breve{\pi}} [\mathbb{I}\{X(t-) \in dy\} \times \breve{\pi}(da|X(t-), t)]$ and performance functionals (5.3) and (5.4) take the well-known form

$$W^\alpha(\breve{\pi}, \mathsf{P}_0) = \mathsf{E}^{\breve{\pi}}_{\mathsf{P}_0}\left[\int_{(0,\infty)} e^{-\alpha t} \int_{\mathbf{A}} c(X(t-), a)\breve{\pi}(da|X(t-), t)dt\right]; \tag{5.5}$$

$$W^A(\breve{\pi}, \mathsf{P}_0) = \limsup_{T\to\infty} \frac{1}{T}\mathsf{E}^{\breve{\pi}}_{\mathsf{P}_0}\left[\int_{(0,\min\{T,T_\infty\}]} \int_{\mathbf{A}} c(X(t-), a)\breve{\pi}(da|X(t-), t)dt\right]. \tag{5.6}$$

[Guo and Hernández-Lerma (2009); Prieto-Rumeau and Hernández-Lerma (2012)]. One can obviously replace here $X(t-)$ by $X(t)$.

The goal is to find an *optimal* control strategy S^* solving the problem

$$W^\alpha(S, \mathsf{P}_0) \longrightarrow \inf_S \tag{5.7}$$

or solving the problem

$$W^A(S, \mathsf{P}_0) \longrightarrow \inf_S. \tag{5.8}$$

These problems are below called *total cost (discounted) problem* and *average cost problem* correspondingly.

In what follows, $W^{*\alpha}(\mathsf{P}_0) := \inf_S W^\alpha(S, \mathsf{P}_0)$; $W^{*A}(\mathsf{P}_0) := \inf_S W^A(S, \mathsf{P}_0)$. If the initial distribution $\mathsf{P}_0(dx) = \delta_{x_0}(dx)$ is degenerate, we use notations $\mathsf{P}^S_{x_0}, \bar{\mathsf{P}}^S_{x_0}, W^\alpha(S, x_0)$ etc. $W^{*\alpha}(x_0)$ is the *Bellman function* for the total cost problem.

Definition 5.4. A strategy S^* is called *uniformly optimal* for the total cost problem (5.7) if

$$W^\alpha(S^*, x) = \inf_S W^\alpha(S, x) =: W^{*\alpha}(x), \ \forall\, x \in \mathbf{X}.$$

It is called average *uniformly optimal* if

$$W^A(S^*, x) = \inf_S W^A(S, x) =: W^{*A}(x), \ \forall\, x \in \mathbf{X}.$$

The continuous-time Markov decision process (*CTMDP model*) is defined by the collection $\{\mathbf{X}, \mathbf{A}, q, c\}$.

One can represent the performance functional (5.3) as the sum over the intervals (T_{n-1}, T_n) where the value of the controlled process is constant:

$$W^\alpha(S, \mathsf{P}_0) = \sum_{n=1}^\infty \int_{\mathbf{X}\times\mathbf{A}} c(x, a) m^{S,\alpha}_{\mathsf{P}_0,n}(dx \times da) \longrightarrow \inf_S. \tag{5.9}$$

The jump epochs T_n have the Lebesgue measure zero and are excluded from integration. Here the *detailed occupation measures* $m^{S,\alpha}_{\mathsf{P}_0,n}$ are given by

$$\begin{aligned}
&m^{S,\alpha}_{\mathsf{P}_0,n}(\Gamma_X \times \Gamma_A) \\
&:= \mathsf{E}^S_{\mathsf{P}_0}\left[\int_{(T_{n-1},T_n)} e^{-\alpha t}\mathbb{I}\{X_{n-1} \in \Gamma_X\}\pi_n(\Gamma_A|H_{n-1}, t - T_{n-1})dt\right] \\
&= \mathsf{E}^S_{\mathsf{P}_0}\left[\mathbb{I}\{X_{n-1} \in \Gamma_X\}e^{-\alpha T_{n-1}}\int_{(0,\infty)} e^{-\alpha\theta - \int_{(0,\theta]} q_{X_{n-1}}(\pi_n,s)ds}\right. \\
&\quad \left.\times\, \pi_n(\Gamma_A|H_{n-1},\theta)d\theta\right],
\end{aligned}$$

if $S_n = \pi_n$ ($q_{x_{n-1}}(\pi_n, s)$ was defined in (5.2)), and by

$$\begin{aligned}
&m^{S,\alpha}_{\mathsf{P}_0,n}(\Gamma_X \times \Gamma_A) \\
&:= \mathsf{E}^S_{\mathsf{P}_0}\left[\mathbb{I}\{X_{n-1} \in \Gamma_X\}e^{-\alpha T_{n-1}}\int_{\Gamma_A}\int_{(0,\infty)} e^{-\alpha\theta - q_{X_{n-1}}(a)\theta}d\theta\varpi_n(da|H_{n-1})\right] \\
&= \mathsf{E}^S_{\mathsf{P}_0}\left[\mathbb{I}\{X_{n-1} \in \Gamma_X\}e^{-\alpha T_{n-1}}\int_{\Gamma_A}\frac{1}{\alpha + q_{X_{n-1}}(a)}\varpi_n(da|H_{n-1})\right],
\end{aligned}$$

if $S_n = \varpi_n$. Here $\Gamma_X \in \mathcal{B}(\mathbf{X})$, $\Gamma_A \in \mathcal{B}(\mathbf{A})$ and $0 \times \infty := 0$. Particular points $0 = T_0, T_1, \ldots$ do not influence the values of the integrals. One can also define the (total) occupation measure

$$\mathbb{M}^{S,\alpha}_{\mathsf{P}_0} := \sum_{n=1}^{\infty} m^{S,\alpha}_{\mathsf{P}_0,n}$$

which has the same meaning as the occupation measure $\eta^{\pi}_{\mathsf{P}_0}$ in Definition 2.1. More about occupation measures in [Piunovskiy and Zhang (2020), Section 1.3].

Recall that by a *cemetery* we mean an absorbing state Δ (sometimes denoted as 0) in the state space with the zero cost rate: for all $a \in \mathbf{A}$ $q_\Delta(a) = c(\Delta, a) \equiv 0$. In the case when there is a measurable subset $\tilde{\mathbf{X}} \subset \mathbf{X}$ such that $q(\mathbf{X} \setminus \tilde{\mathbf{X}}|x, a) \equiv 0$ and $c(x, a) \equiv 0$ for all $x \in \tilde{\mathbf{X}}$, we merge together all the states from $\tilde{\mathbf{X}}$ in one cemetery Δ with the obvious adjustment of the transition rate q.

One can also consider the following situation. Suppose the model is positive (i.e., $c(\cdot,\cdot) \geq 0$), there is a measurable subset $\tilde{\mathbf{X}} \in \mathcal{B}(\mathbf{X})$ such that $q(\mathbf{X} \setminus \tilde{\mathbf{X}}|x, a) \equiv 0$ for all $x \in \tilde{\mathbf{X}}$, and, for a measurable mapping $a^* : \tilde{\mathbf{X}} \to \mathbf{A}$, $c(x, a^*(x)) = 0$ for all $x \in \tilde{\mathbf{X}}$. Since we deal with minimization problems, actions $a^*(x)$ are optimal for $x \in \tilde{\mathbf{X}}$, and we again merge together all the states from $\tilde{\mathbf{X}}$ in one cemetery Δ.

In this framework, in case there is a cemetery state $\Delta \in \mathbf{X}$, we introduce the space $\mathbf{X}^\Delta := \mathbf{X} \setminus \{\Delta\}$. Clearly, the values of the detailed occupation measures on $\{\Delta\} \times \mathbf{A}$ play no role when solving the optimal control problems (5.7) and (5.8).

Thus, without more explanations, we sometimes consider the measures $m^{S,\alpha}_{\mathsf{P}_0,n}$ and $\mathbb{M}^{S,\alpha}_{\mathsf{P}_0}$ on $\mathcal{B}(\mathbf{X}^\Delta \times \mathbf{A})$.

Theorem 5.1. *Let $S = \{S_n\}_{n=1}^\infty$ be an arbitrary strategy. Then the following assertions are valid.*

(a) *[Piunovskiy and Zhang (2020), Thm. 1.1.1]. There exists a π-strategy $\pi = \{\pi_n\}_{n=1}^\infty$ such that on $\mathcal{B}(\mathbf{X} \times \mathbf{A})$,*

$$\bar{\mathsf{P}}^S_{\mathsf{P}_0}(t, dy \times da) = \bar{\mathsf{P}}^\pi_{\mathsf{P}_0}(t, dy \times da), \ t \in [0, \infty)$$

for all initial distributions P_0. For each $n \in \{1, 2, \dots\}$, if $S_n = \tilde{\pi}_n$, then put $\pi_n := \tilde{\pi}_n$; if $S_n = \varpi_n$, then put

$$\pi_n(da|h_{n-1}, s) := \frac{e^{-q_{x_{n-1}}(a)s}\varpi_n(da|h_{n-1})}{\int_{\mathbf{A}} e^{-q_{x_{n-1}}(a)s}\varpi_n(da|h_{n-1})}$$

on $\mathcal{B}(\mathbf{A})$ for each $s > 0$.

(b) *[Piunovskiy and Zhang (2020), Thm. 1.3.1]. For each initial distribution P_0 and discount factor $\alpha \ge 0$ there exists a Markov π-strategy π^{m} such that $\{m^{\pi^{\mathrm{m}},\alpha}_{\mathsf{P}_0,n}\}_{n=1}^\infty = \{m^{S,\alpha}_{\mathsf{P}_0,n}\}_{n=1}^\infty$.*

According to this theorem, one can restrict to (relaxed) Markov π-strategies when solving problems (5.7) and (5.8). One can also restrict to natural Markov strategies if there is no explosion. (See Remark 5.4.) But, in view of Remark 5.1, it is desirable to try to find the optimal control strategy which is randomized or is a deterministic π-strategy.

5.1.3 *Dynkin's formula and optimality equation*

To give the idea of the dynamic programming approach, consider the undiscounted model ($\alpha = 0$) and assume for simplicity that there is a costless cemetery Δ with $q_\Delta(a) = c(\Delta, a) \equiv 0$, the initial distribution P_0 is concentrated on $\mathbf{X}^\Delta := \mathbf{X} \setminus \{\Delta\}$, and, for some $0 < \varepsilon < K < \infty$, for all $x \in \mathbf{X}^\Delta$, $a \in \mathbf{A}$, $q_x(a) \in [\varepsilon, K]$, $q(\{\Delta\}|x, a) \ge \varepsilon$.

Lemma 5.1. *In the described above model, the following assertions are valid.*

(a) *If the cost rate $c(\cdot,\cdot)$ is uniformly bounded, then $|W^0(S, \mathsf{P}_0)| \le L$ for some constant $L < \infty$.*

(b) *If a measurable function $V(\cdot) : \mathbf{X} \to \mathbb{R}$ is bounded and $V(\Delta) = 0$, then, for each control strategy S and initial distribution P_0, $\lim_{t\to\infty} \mathsf{E}^S_{\mathsf{P}_0}[V(X(t))] = 0$.*

The proof is given in Appendix A.

The Dynkin formula for a relaxed strategy π and a function $V(\cdot): \mathbf{X} \to \mathbb{R}$ looks as follows

$$\begin{aligned} &\mathsf{E}_x^\pi[V(X(t))] - V(x) \\ &= \mathsf{E}_x^\pi \left[\int_{(0,t]} \int_{\mathbf{X}} \int_{\mathbf{A}} \sum_{n=0}^{\infty} \mathbb{I}\{T_n < v \le T_{n+1}\} \right. \\ &\left. \times\ \pi_{n+1}(da|H_n, v - T_n) q(dy|X_n, a) V(y) dv \right] \quad t \ge 0,\ x \in \mathbf{X}. \end{aligned} \tag{5.10}$$

Condition 5.1.

(a) There exist a measurable function $w(\cdot): \mathbf{X} \to [0, \infty)$ and constants $\rho \in \mathbb{R}$, $L, b \in [0, \infty)$ such that the following assertions hold.

- $\bar{q}_x := \sup_{\mathbf{A}} q_x(a) \le Lw(x)$ for each $x \in \mathbf{X}$ and $a \in \mathbf{A}$.
- For each $x \in \mathbf{X}$ and $a \in \mathbf{A}$,

$$\int_{\mathbf{X}} w(y) q(dy|x, a) \le \rho w(x) + b.$$

(b) There exist an $[1, \infty)$-valued measurable function $w'(\cdot)$ on $\mathbf{X}$ and constants $L', b' \ge 0$ and $\rho' \in \mathbb{R}$ such that the following assertions hold:

- $(\bar{q}_x + 1) w'(x) \le L' w(x)$ for each $x \in \mathbf{X}$.
- $\int_{\mathbf{X}} q(dy|x, a) w'(y) \le \rho' w'(x) + b'$ for each $x \in \mathbf{X}$ and $a \in \mathbf{A}$.

Theorem 5.2. [Piunovskiy and Zhang (2020), Thm. 2.4.4]. *If Condition 5.1 is satisfied, then Dynkin's formula (5.10) is valid for each relaxed strategy π and each measurable function $V(\cdot)$ such that $\sup_{x \in \mathbf{X}} \frac{|V(x)|}{w'(x)} < \infty$. All the expressions in (5.10) are finite.*

Condition 5.1 is satisfied for the model described at the beginning of the current subsection: one can put $w(x) = w'(x) \equiv 1$, $\rho = b = \rho' = b' = 0$, $L' = \sup_{x \in \mathbf{X}} \bar{q}_x + 1$. Below, the cost rate $c(\cdot, \cdot)$ is assumed to be uniformly bounded and a relaxed strategy π is fixed.

Let us add expression

$$\mathsf{E}_x^\pi \left[\int_{(0,t]} \int_{\mathbf{A}} \sum_{n=0}^{\infty} \mathbb{I}\{T_n < v \le T_{n+1}\} c(X_n, a) \pi_{n+1}(da|H_n, v - T_n) dv \right]$$

to the Dynkin formula (5.10):

$$\mathsf{E}_x^\pi\left[\int_{(0,t]}\int_{\mathbf{A}}\sum_{n=0}^{\infty}\mathbb{I}\{T_n<v\le T_{n+1}\}c(X_n,a)\pi_{n+1}(da|H_n,v-T_n)dv\right]$$
$$=V(x)+\mathsf{E}_x^\pi\left[\int_{(0,t]}\sum_{n=0}^{\infty}\mathbb{I}\{T_n<v\le T_{n+1}\}\int_{\mathbf{A}}\left[c(X_n,a)+\right.\right. \tag{5.11}$$
$$\left.\left.+\int_{\mathbf{X}}V(y)q(dy|X_n,a)\right]\pi_{n+1}(da|H_n,v-T_n)dv\right]-\mathsf{E}_x^\pi(V(X(t))].$$

Here and below, the interchange of the order of integrals, summations and mathematical expectation is legal by the Dominated Convergence Theorem.

Suppose a measurable bounded function $V(\cdot)$ is such that $V(\Delta)=0$ and

$$\inf_{a\in\mathbf{A}}\left\{c(x,a)+\int_{\mathbf{X}}V(y)q(dy|x,a)\right\}=0,\quad \forall x\in\mathbf{X}. \tag{5.12}$$

Then, passing to the limit as $t\to\infty$ in the equality (5.11), we obtain by Lemma 5.1:

$$W^0(\pi,x)=V(x)+\lim_{t\to\infty}\mathsf{E}_x^\pi\left[\int_{(0,t]}\sum_{n=0}^{\infty}\mathbb{I}\{T_n<v\le T_{n+1}\}\int_{\mathbf{A}}\left[c(X_n,a)+\right.\right.$$
$$\left.\left.+\int_{\mathbf{X}}V(y)q(dy|X_n,a)\right]\pi_{n+1}(da|H_n,v-T_n)dv\right].$$

Obviously, for each relaxed strategy π and all $x\in\mathbf{X}$, $W^0(\pi,x)\ge V(x)$ according to (5.12) and, if a deterministic stationary strategy $\varphi^*:\ \mathbf{X}\to\mathbf{A}$ is such that

$$c(x,\varphi^*(x))+\int_{\mathbf{X}}V(y)q(dy|x,\varphi^*(x))=0,\quad \forall x\in\mathbf{X},$$

then

$$\inf_\pi W^0(\pi,x)=W^0(\varphi^*,x)=V(x).$$

According to Theorem 5.1, φ^* is uniformly optimal and $V(x)=W^{*0}(x)$ is the Bellman function. Equation (5.12) is the optimality equation for the considered model.

It is well known that the positive discount factor $\alpha>0$ means that the life time of the controlled process is exponential with parameter α. One can also say that, with intensity α, the process $X(t)$ jumps to the specific isolated state $\square\notin\mathbf{X}$, similar to the costless cemetery Δ. The coefficient $e^{-\alpha t}$ in formula (5.3) equals the probability that $X(t)\in\mathbf{X}$. With this interpretation in mind, one can add α to the jump intensity $q_x(a)$, so that the optimality equation (5.12) takes the form

$$\alpha V(x)=\inf_{a\in\mathbf{A}}\left\{c(x,a)+\int_{\mathbf{X}}V(y)q(dy|x,a)\right\},\quad x\in\mathbf{X}. \tag{5.13}$$

The term $\alpha V(\square) = 0$ on the right is omitted. This equation for the models with the discrete state space appeared in many articles and in the monographs [Guo and Hernández-Lerma (2009); Prieto-Rumeau and Hernández-Lerma (2012)]: see, correspondingly, formulas (6.1) and (4.10).

5.2 Several Important Statements

5.2.1 *Explosion*

Many statements hold in case there is no *explosion*, i.e., for all strategies S, for each initial distribution P_0 $T_\infty = \infty$ $\mathsf{P}^S_{\mathsf{P}_0}$-a.s.; equivalently, $\bar{\mathsf{P}}^S_{\mathsf{P}_0}(t, \mathbf{X}) = 1$ for all $t \in [0, \infty)$. One can also say that, under initial state x_0 and control strategy S, the controlled process $X(\cdot)$ is *non-explosive* if $T_\infty = \infty$ $\mathsf{P}^S_{x_0}$-a.s.; equivalently, $\bar{\mathsf{P}}^S_{x_0}(t, \mathbf{X}) = 1$ for all $t \in [0, \infty)$. Otherwise, the controlled process $X(\cdot)$ is called *explosive*. Note also that if, under a strategy S the controlled process $X(\cdot)$ is non-explosive for each initial state x_0, then it is non-explosive under each initial distribution P_0.

Condition 5.2. There exist a monotone nondecreasing sequence of measurable subsets $\{V_l\}_{l=1}^\infty$ of $\mathbf{X}$, a measurable function $w(\cdot) : \mathbf{X} \to [0, \infty)$ and constants $\rho \in \mathbb{R}$, $b \in [0, \infty)$ such that the following assertions hold.

(a) $\mathbf{X} = \bigcup_{l=1}^\infty V_l, \quad \sup_{x \in V_l} w(x) < \infty.$

(b) For every $l = 1, 2, \ldots$, $\sup_{x \in V_l,\ a \in \mathbf{A}} q_x(a) < \infty$.

(c) $\lim_{l \to \infty} \inf_{x \in \mathbf{X} \setminus V_l} w(x) = \infty$.

(d) For each $x \in \mathbf{X}$ and $a \in \mathbf{A}$,

$$\int_{\mathbf{X}} w(y) q(dy|x, a) \leq \rho w(x) + b.$$

Evidently, if $\sup_{x \in \mathbf{X}, a \in \mathbf{A}} q_x(a) < \infty$, then Condition 5.2 is satisfied. Note also that Condition 5.1(a) implies Condition 5.2.

Theorem 5.3. [Piunovskiy and Zhang (2020), Thm. 2.2.5]. *If Condition 5.2 is satisfied, then there is no explosion.*

Remark 5.3. Suppose there is no explosion and consider the discounted total cost problem (5.7) with $\alpha > 0$ or the average cost problem (5.8). If the cost rate $c(\cdot, \cdot) \geq -K$ is bounded from below, then the solutions to the modified problems with $\tilde{c}(\cdot, \cdot) := c(\cdot, \cdot) + K \geq 0$ obviously provide the solutions to the original problems, since the performance functionals only increase by a constant.

Definition 5.5. (See Lemma 1 of [Feinberg *et al.* (2021)].) Suppose the initial distribution P_0 is fixed. The natural Markov strategy $\breve{\pi}$ is said to be *induced* by a strategy S if

$$\bar{\mathsf{P}}^S_{\mathsf{P}_0}(t, dy \times da) = \bar{\mathsf{P}}^S_{\mathsf{P}_0}(t, dy) \breve{\pi}(da|y, t), \ \forall\, t \in (0, \infty)$$

on $\mathcal{B}(\mathbf{X} \times \mathbf{A})$.

In Definition 2.2.2 of [Piunovskiy and Zhang (2020)] induced strategies were introduced for π-strategies. But, in view of Theorem 5.1(a), one can consider all strategies S.

Theorem 5.4. [Feinberg *et al.* (2021), Thm.1], [Piunovskiy and Zhang (2020), Cor. 2.2.5]. *Let the initial distribution* P_0 *and a strategy* S *be fixed. Then for the induced natural Markov strategy* $\breve{\pi}$

$$\bar{\mathsf{P}}^{\breve{\pi}}_{\mathsf{P}_0}(t, dy \times da) \le \bar{\mathsf{P}}^{S}_{\mathsf{P}_0}(t, dy \times da), \quad \forall t \in [0, \infty).$$

If the controlled process $X(\cdot)$ *is non-explosive under the strategy* $\breve{\pi}$*, then it is non-explosive under the strategy* S *and*

$$\bar{\mathsf{P}}^{\breve{\pi}}_{\mathsf{P}_0}(t, dy \times da) = \bar{\mathsf{P}}^{S}_{\mathsf{P}_0}(t, dy \times da), \quad \forall t \in [0, \infty).$$

Corollary 5.1. *If the controlled process* $X(\cdot)$ *is non-explosive under each natural Markov strategy* $\breve{\pi}$ *and each initial state* x_0*, then there is no explosion at all (under all strategies and all initial distributions).*

Remark 5.4. If there is no explosion, then, according to Theorem 5.4, one can restrict to natural Markov strategies in problems (5.7) and (5.8).

For a fixed natural Markov strategy $\breve{\pi}$ (stationary π-strategy π^{s}) below we use notations

$$q(dy|x,t) := \int_{\mathbf{A}} q(dy|x,a)\breve{\pi}(da|x,t) \quad \left(q(dy|x) := \int_{\mathbf{A}} q(dy|x,a)\pi^{\mathrm{s}}(da|x)\right);$$

$$\tilde{q}(dy|x,t) := \int_{\mathbf{A}} \tilde{q}(dy|x,a)\breve{\pi}(da|x,t) \quad \left(\tilde{q}(dy|x) := \int_{\mathbf{A}} \tilde{q}(dy|x,a)\pi^{\mathrm{s}}(da|x)\right);$$

$$q_x(t) := \int_{\mathbf{A}} q_x(a)\breve{\pi}(da|x,t) \quad \left(q_x := \int_{\mathbf{A}} q_x(a)\pi^{\mathrm{s}}(da|x)\right)$$

for brevity.

Theorem 5.5. [Piunovskiy and Zhang (2020), Thms. 2.2.3, 2.2.4(c,d), Cor. 2.2.3].

(a) *The controlled process* $X(\cdot)$ *is non-explosive under a fixed natural Markov strategy* $\breve{\pi}$ *and all initial states* x_0 *if and only if, for any* $\alpha > 0$*, the equation*

$$u(v,s) = \int_{(0,\infty)} \int_{\mathbf{X}} u(v+t,y)\tilde{q}(dy|x,v+t)e^{-\alpha t - \int_{(0,t]} q_x(s+v)ds}dt,$$
$$\forall\, x \in \mathbf{X},\; v \in [0,\infty) \tag{5.14}$$

has no other nonnegative bounded solutions except for $u(v,x) \equiv 0$.
In case of a stationary π*-strategy* π^{s}, *this equation takes the form*

$$\alpha u(x) = \int_{\mathbf{X}} u(y) q(dy|x), \ \forall\, x \in \mathbf{X}, \tag{5.15}$$

and the process $X(\cdot)$ *is non-explosive for all initial states* x_0 *if and only if, for any* $\alpha > 0$, *there are no nontrivial* $[0,1]$*-valued measurable solutions.*

(b) The controlled process $X(\cdot)$ *is explosive under a fixed natural Markov strategy* $\breve{\pi}$ *and some state* x_0 *if and only if, for some* $\alpha > 0$, *there is a nontrivial bounded nonnegative measurable function* $u(\cdot,\cdot)$ *on* $[0,\infty) \times \mathbf{X}$ *satisfying the following inequality:*

$$u(v,x) \le \int_{(0,\infty)} \int_{\mathbf{X}} u(v+t,y) \tilde{q}(dy|x, v+t) e^{-\alpha t - \int_{(0,t]} q_x(s+v)ds} dt,$$
$$\forall\, x \in \mathbf{X},\ v \in [0,\infty). \tag{5.16}$$

In case of a stationary π*-strategy* π^{s}, *this inequality takes the form*

$$\alpha u(x) \le \int_{\mathbf{X}} u(y) q(dy|x), \ \forall\, x \in \mathbf{X}. \tag{5.17}$$

Obviously, in Theorem 5.5 one can always restrict to $[0,1]$-valued functions $u(v,x)$ and $u(x)$ by passing to the normalized functions $\tilde{u}(v,x) := \frac{u(v,x)}{\sup_{(v,x)\in[0,\infty)\times\mathbf{X}} |u(v,x)|}$ and $\tilde{u}(x) := \frac{u(x)}{\sup_{x\in\mathbf{X}} |u(x)|}$. The following lemma is less obvious.

Lemma 5.2. *Suppose for some* $\alpha > 0$ *there is a nontrivial function* $u^0(v,x) : [0,\infty)\times\mathbf{X} \to [0,1]$ *(*$u^0(x) : \mathbf{X} \to [0,1]$*) satisfying inequality (5.16) (inequality (5.17)). Then there is a nontrivial function* $u(v,x) : [0,\infty)\times\mathbf{X} \to [0,1]$ *(*$u(x) : \mathbf{X} \to [0,1]$ *) satisfying equation (5.14) (equation (5.15)).*

The proof is given in Appendix A.

5.2.2 *Total cost problem*

Condition 5.3.

(a) $q_x(a) + \alpha > 0$ for all $(x,a) \in \mathbf{X}^{\Delta} \times \mathbf{A}$.

(b) For each $x \in \mathbf{X}^{\Delta}$, there exists some $\varepsilon = \varepsilon(x) > 0$ such that $\inf_{a\in\mathbf{A}} q_x(a) + \alpha \ge \varepsilon$.

Theorem 5.6. [Piunovskiy and Zhang (2020), Thm. 1.3.2].

(a) Suppose Condition 5.3(a) is satisfied. Then, for any strategy S, *for each fixed* $\alpha \in [0,\infty)$ *and initial distribution* P_0, *there is a randomized Markov*

strategy $\varpi^{\rm m}$ *such that*

$$m^{\varpi^{\rm m},\alpha}_{\mathsf{P}_0,n} \geq m^{S,\alpha}_{\mathsf{P}_0,n} \tag{5.18}$$

on $\mathbf{X}^\Delta \times \mathbf{A}$ *for all* $n = 1, 2, \dots$. *Hence, randomized Markov strategies are sufficient for solving problem (5.9) with nonpositive cost rate* $c(\cdot,\cdot)$.

(b) *If Condition 5.3(b) is satisfied, then the inequality in (5.18) can be taken as equality.*

(c) *In the both cases(a,b), according to Theorem 5.1(b), without loss of generality, one can take* $S = \pi^{\rm m}$ *as a relaxed Markov strategy. In this case the desired randomized Markov strategy (for* $x_{n-1} \neq \Delta$*) has the form*

$$\varpi^{\rm m}_n(\Gamma_A|x_{n-1}) = \frac{\int_{(0,\infty)} \int_{\Gamma_A} (q_{x_{n-1}}(a)+\alpha)\pi^{\rm m}_n(da|x_{n-1},\theta) e^{-\int_{(0,\theta]}(q_{x_{n-1}}(\pi^{\rm m}_n,u)+\alpha)du}\, d\theta}{1 - e^{-\int_{(0,\infty)}(q_{x_{n-1}}(\pi^{\rm m}_n,u)+\alpha)du}}.$$

Recall that $q_x(\pi_n, s)$ *was defined in (5.2). In particular, if* $\pi^{\rm m} = \pi^{\rm s}$ *is a relaxed stationary strategy, then one obtains the randomized stationary strategy* $\varpi^{\rm s}$.

According to Theorem 5.6(b), if the CTMDP is discounted (i.e., $\alpha > 0$), then randomized Markov strategies are sufficient for solving the total cost problems (5.7) and (5.9).

Remark 5.5. According to Theorems 5.1 and 5.6, under Condition 5.3(b), there are many different control strategies with the same marginal distribution $\bar{\mathsf{P}}_{\mathsf{P}_0}$ or with the same detailed occupation measures $m_{\mathsf{P}_0,n}$, $n = 1, 2, \dots$ under an arbitrarily fixed P_0 (below called *equivalent* for brevity):

- For each control strategy S, there is an equivalent relaxed and an equivalent relaxed Markov strategy by Theorem 5.1.
- For each relaxed Markov strategy, there is an equivalent randomized Markov strategy by Theorem 5.6(c).

Note also that, if $\pi^{\rm s}$ is a relaxed stationary strategy, then the equivalent randomized Markov strategy $\varpi^{\rm s}$ as in Theorem 5.6(c) will be stationary. But the equivalent with $\varpi^{\rm s}$ relaxed strategy π as in Theorem 5.1(a) will be in general non-stationary (s-dependent).

Therefore, usually the optimal strategy in problem (5.7) and in its constrained version (5.46) is not unique.

Remark 5.6. Sometimes, it is natural to introduce (measurable) *lump sum costs* $C(x,y)$ incurred at the jump epochs T_n if $X(T_n-) = x$ and $X(T_n) = y$, and consider the following performance functional:

$$W^\alpha(S,\mathsf{P}_0) = \int_{(0,\infty)} e^{-\alpha t}\int_{\mathbf{X}\times\mathbf{A}} c(x,a)\bar{\mathsf{P}}^S_{\mathsf{P}_0}(t,dx\times da)dt \qquad (5.19)$$

$$+\mathsf{E}^S_{\mathsf{P}_0}\left[\sum_{n=1}^{\infty} e^{-\alpha T_n} C(X_{T_{n-1}},X_{T_n})\right] \longrightarrow \inf_S .$$

As was shown in Subsection 1.1.5 of [Piunovskiy and Zhang (2020)], in such situation, one should simply switch to the new cost rate $\tilde{c}(x,a) := c(x,a) + \int_{\mathbf{X}} C(x,y)\tilde{q}(dy|x,a)$ and investigate the total cost problem (5.7): for each control strategy S and initial distribution P_0

$$W^\alpha(S,\mathsf{P}_0) = \tilde{W}^\alpha(S,\mathsf{P}_0) := \int_{(0,\infty)} e^{-\alpha t}\int_{\mathbf{X}\times\mathbf{A}} \tilde{c}(x,a)\bar{\mathsf{P}}^S_{\mathsf{P}_0}(t,dx\times da)dt. \quad (5.20)$$

Consider the following homogeneous discrete-time MDP (DTMDP) $\{\mathbf{X}_\Box, \mathbf{A}, p_\alpha, l\}$, called *induced*, where

- $\mathbf{X}_\Box = \mathbf{X}\cup\{\Box\}$, where $\Box$ is the cemetery state;
- $\mathbf{A}$ is the action space;
- transition probability is given for each $\Gamma\in\mathcal{B}(\mathbf{X}_\Box)$ by

$$p_\alpha(\Gamma|x,a) = \begin{cases} \dfrac{\tilde{q}(\Gamma\cap\mathbf{X}|x,a)+\alpha\,\mathbb{I}\{\Gamma\ni\Box\}}{\alpha+q_x(a)}, & \text{if } x\in\mathbf{X}\ , a\in\mathbf{A};\\ \mathbb{I}\{\Box\in\Gamma\}, & \text{if } x=\Box; \end{cases}$$

  here $\frac{0}{0} := 0$ and $p_\alpha(\{x\}|x,a) := 1 - p_\alpha(\mathbf{X}_\Box\setminus\{x\}|x,a) = 1$ only if $\alpha+q_x(a)=0$;
- the cost function is $l(x,a) = \frac{c(x,a)}{q_x(a)+\alpha}$ for each $(x,a)\in\mathbf{X}\times\mathbf{A}$, $\frac{0}{0}:=0$, and $l(\Box,a)\equiv 0$.

The strategies in this DTMDP will be denoted as σ and the strategic measures – as $\mathbb{P}^\sigma_{\mathsf{P}_0}$, just to distinguish from the relaxed strategies and strategic measures in CTMDP. Similarly, the time parameter in the DTMDP will be denoted as n, not t. Further, accept that the performance functional in the induced DTMDP is calculated using formula (2.4). Recall that, under Condition 2.1, expressions (2.3) and (2.4) coincide. In what follows, for $\Gamma_X\in\mathcal{B}(\mathbf{X}_\Box)$, $\Gamma_A\in\mathcal{B}(\mathbf{A})$,

$$\eta^{\sigma,\alpha}_{\mathsf{P}_0,n}(\Gamma_X\times\Gamma_A) := \mathbb{P}^\sigma_{\mathsf{P}_0}(X_{n-1}\in\Gamma_X, A_n\in\Gamma_A), \quad n=1,2,\ldots,$$

so that $\eta^{\sigma,\alpha}_{\mathsf{P}_0}(\Gamma_X\times\Gamma_A) = \sum_{n=1}^\infty \eta^{\sigma,\alpha}_{\mathsf{P}_0,n}(\Gamma_X\times\Gamma_A)$ is the occupation measure in the induced DTMDP: see Definition 2.1.

Theorem 5.7. [Piunovskiy and Zhang (2020), Thm. 4.2.3]. *Suppose either $\alpha>0$ or (if $\alpha=0$), for each $x\in\mathbf{X}^\Delta$, there exists some $\varepsilon(x)>0$ such that*

$\inf_{a\in\mathbf{A}} q_x(a) \geq \varepsilon(x)$. *Suppose also that* P_0 *is the common initial distribution in the CTMDP* $\{\mathbf{X}, \mathbf{A}, q, c\}$ *and in the induced DTMDP* $\{\mathbf{X}_\square, \mathbf{A}, p_\alpha, l\}$. *Consider a randomized Markov strategy* $\varpi^{\mathrm{m}} = \{\varpi_n^{\mathrm{m}}(da|x)\}_{n=1}^\infty$ *in the CTMDP* $\{\mathbf{X}, \mathbf{A}, q, c\}$ *and the Markov strategy* $\sigma^{\mathrm{m}} = \{\sigma_n^{\mathrm{m}}(da|x)\}_{n=1}^\infty = \{\varpi_n^{\mathrm{m}}(da|x)\}_{n=1}^\infty$ *in the DTMDP* $\{\mathbf{X}_\square, \mathbf{A}, p_\alpha, l\}$. *Then, for each* $n \in \{1, 2, \ldots\}$,

$$m_{\mathsf{P}_0,n}^{\varpi^{\mathrm{m}},\alpha}(dx \times da) = \frac{1}{q_x(a) + \alpha}\eta_{\mathsf{P}_0,n}^{\sigma^{\mathrm{m}},\alpha}(dx \times da)$$

on $\mathcal{B}(\mathbf{X}^\Delta \times \mathbf{A})$. *Hence*

$$\int_{\mathbf{X}^\Delta\times\mathbf{A}} c(x,a) m_{\mathsf{P}_0,n}^{\varpi^{\mathrm{m}},\alpha}(dx \times da) = \int_{\mathbf{X}^\Delta\times\mathbf{A}} l(x,a)\eta_{\mathsf{P}_0,n}^{\sigma^{\mathrm{m}},\alpha}(dx \times da).$$

This theorem also holds if Condition 5.3(a) is satisfied. For the proof, it is sufficient to consider the case $\alpha = 0$, omit $\square$ and ignore the costless cemetery Δ. Here, for each randomized Markov strategy ϖ^{m}, we actually deal with a discrete-time MDP: at each decision epoch T_n with the current state $X_n \in \mathbf{X}^\Delta$, one has to choose action $a \in \mathbf{A}$ leading to the new state X_{n+1} at $T_{n+1} = T_n + \Theta_{n+1}$ distributed according to $\frac{\tilde{q}(dy|X_n,a)}{q_{X_n}(a)}$, exactly like in the induced DTMDP. Therefore, if $\varpi^{\mathrm{m}} = \sigma^{\mathrm{m}}$, then

$$\begin{aligned} &\mathsf{P}_{\mathsf{P}_0}^{\varpi^{\mathrm{m}}}(X_{n-1} \in \Gamma_X, \varpi_n^{\mathrm{m}}(\Gamma_A|X_{n-1})) \\ &= \mathbb{P}_{\mathsf{P}_0}^{\sigma^{\mathrm{m}}}(X_{n-1} \in \Gamma_X, A_n \in \Gamma_A) \\ &= \eta_{\mathsf{P}_0,n}^{\sigma^{\mathrm{m}},0}(\Gamma_X \times \Gamma_A), \quad \Gamma_X \in \mathcal{B}(\mathbf{X}^\Delta),\ \Gamma_A \in \mathcal{B}(\mathbf{A}) \end{aligned} \tag{5.21}$$

for all $n = 1, 2, \ldots$. According to the definition, to obtain the measure $m_{\mathsf{P}_0,n}^{\varpi^{\mathrm{m}},0}$, one has to integrate function $\frac{1}{q_x(a)}$ w.r.t. the presented measure:

$$m_{\mathsf{P}_0,n}^{\varpi^{\mathrm{m}},0}(\Gamma_X \times \Gamma_A) = \int_{\Gamma_X\times\Gamma_A} \frac{1}{q_x(a)}\eta_{\mathsf{P}_0,n}^{\sigma^{\mathrm{m}},0}(dx \times da), \quad \Gamma_X \in \mathcal{B}(\mathbf{X}^\Delta),\ \Gamma_A \in \mathcal{B}(\mathbf{A})$$

$n = 1, 2, \ldots$.

For the case of $\alpha > 0$, the similar theorem was presented in [Feinberg (2010)]: Theorem 5.4.1.

According to Theorem 5.6, if Condition 5.3(a) is satisfied and the cost rate $c(\cdot,\cdot)$ is nonpositive, or Condition 5.3(b) is satisfied, then randomized Markov strategies are sufficient for solving the total cost problem (5.7), and Markov strategies in any DTMDP are also sufficient by Lemma 2.1. Therefore, due to Theorem 5.7, optimization problem (5.7) is reduced to discrete-time MDP and all the corresponding theory is applicable: if a Markov strategy $\sigma^{*\mathrm{m}}$ is the (uniformly) optimal strategy in the induced DTMDP (homogeneous model and total expected cost: see Chapter 2), then the randomized Markov strategy $\varpi^{*\mathrm{m}} = \sigma^{*\mathrm{m}}$

is (uniformly) optimal in the underlying CTMDP and *vice versa*. Note that this assertion holds also in multi-objective problems with a series of cost functions $\{c_j(\cdot,\cdot)\}_{j=0}^{J}$ when one is looking for Pareto-optimal solutions or considers constrained optimization.

Condition 5.4.

(a) The (Borel) action space $\mathbf{A}$ is compact and $\mathbf{X}$ is a topological Borel space.

(b) For each bounded continuous function $f(\cdot)$ on $\mathbf{X}$, the mapping $(x,a) \in \mathbf{X}\times\mathbf{A} \to \int_{\mathbf{X}} f(y)\tilde{q}(dy|x,a)$ is bounded and continuous.

(c) The cost rate $c(\cdot,\cdot)$ is lower semicontinuous.

Condition 5.5.

(a) The (Borel) action space $\mathbf{A}$ is compact and $\mathbf{X}$ is a topological Borel space.

(b) For each bounded measurable function $f(\cdot)$ on $\mathbf{X}$ and $x \in \mathbf{X}$, the integral $\int_{\mathbf{X}} f(y)\tilde{q}(dy|x,a)$ is continuous in $a \in \mathbf{A}$.

(c) The cost rate $c(\cdot,\cdot)$ is lower semicontinuous.

Theorem 5.8. [Piunovskiy and Zhang (2020), Thm. 4.2.4]. *Suppose either $\alpha > 0$ or (if $\alpha = 0$), for each $x \in \mathbf{X}^{\Delta}$, there exists some $\varepsilon(x) > 0$ such that $\inf_{a\in\mathbf{A}} q_x(a) \geq \varepsilon(x)$. Assume additionally that the cost rate $c(\cdot,\cdot)$ is nonnegative. Then the following assertions are valid.*

(a) The Bellman function $W^{*\alpha}(x)$ coincides with the infimum of $W^{\alpha}(\pi^{\mathrm{m}},x)$ (or with the infimum of $W^{\alpha}(\varpi^{\mathrm{m}},x)$) over Markov π-strategies (or over randomized Markov strategies respectively) and coincides with the minimal nonnegative lower semianalytic solution to optimality equation:

$$V(x) = \inf_{a\in\mathbf{A}}\left\{\frac{c(x,a)}{\alpha+q_x(a)} + \int_{\mathbf{X}} V(y)\frac{\tilde{q}(dy|x,a)}{\alpha+q_x(a)}\right\}, \quad x \in \mathbf{X}. \tag{5.22}$$

(b) A deterministic stationary strategy φ^* is uniformly optimal for problem (5.7) if and only if

$$\begin{aligned}&\inf_{a\in\mathbf{A}}\left\{\frac{c(x,a)}{\alpha+q_x(a)} + \int_{\mathbf{X}} W^{*\alpha}(y)\frac{\tilde{q}(dy|x,a)}{\alpha+q_x(a)}\right\}\\ &= \frac{c(x,\varphi^*(x))}{\alpha+q_x(\varphi^*(x))} + \int_{\mathbf{X}} W^{*\alpha}(y)\frac{\tilde{q}(dy|x,\varphi^*(x))}{\alpha+q_x(\varphi^*(x))}\end{aligned}$$

for each $x \in \mathbf{X}$.

(c) *Suppose Condition 5.4 (or Condition 5.5) is satisfied. Then there exists a uniformly optimal deterministic stationary strategy φ^* for the CTMDP problem (5.7). The Bellman function $W^{*\alpha}$ is the minimal nonnegative lower semicontinuous (respectively, measurable) solution to (5.22).*

If there is a costless cemetery Δ, then $V(\Delta) := 0$ and one can consider only $x \in \mathbf{X}^\Delta$ in the presented theorem. Note also that, if $\alpha + q_x(a) \geq \varepsilon(x) > 0$ for some function $\varepsilon(\cdot) > 0$ on $\mathbf{X}^\Delta$, then equations (5.22) and (5.13) are equivalent in the space of finite-valued functions $V(\cdot)$ on $\mathbf{X}^\Delta$.

5.2.3 *Average cost problem*

Firstly, assume that the state and action spaces $\mathbf{X}$ and $\mathbf{A}$ are finite.

Proposition 5.1. *Consider the following linear program*

$$\sum_{x\in\mathbf{X}}\sum_{a\in\mathbf{A}} c(x,a)\eta(x,a) \to \inf_{\eta,\tilde{\eta}} \tag{5.23}$$

$$\sum_{y\in\mathbf{X}}\sum_{a\in\mathbf{A}} q(x|y,a)\eta(y,a) = 0, \quad x \in \mathbf{X}; \tag{5.24}$$

$$\eta(x,a) \geq 0;$$

$$\sum_{a\in\mathbf{A}}\eta(x,a) - \sum_{y\in\mathbf{X}}\sum_{a\in\mathbf{A}} q(x|y,a)\tilde{\eta}(y,a) = \alpha(x), \quad x \in \mathbf{X}; \tag{5.25}$$

$$\tilde{\eta}(x,a) \geq 0,$$

where $\alpha(x) > 0$ are arbitrarily fixed numbers such that $\sum_{x\in\mathbf{X}}\alpha(x) = 1$. At least one of equations in (5.24) is redundant.

If $(\eta^*, \tilde{\eta}^*)$ is an optimal solution to (5.23), (5.24), (5.25), then the following stationary π-strategy π^{s} is average uniformly optimal:

$$\pi^{\mathrm{s}}(a|x) = \begin{cases} \eta^*(x,a)\,/\sum_{b\in\mathbf{A}}\eta^*(x,b)\,, & \text{if } \sum_{b\in\mathbf{A}}\eta^*(x,b) > 0; \\ \tilde{\eta}^*(x,a)\,/\sum_{b\in\mathbf{A}}\tilde{\eta}^*(x,b)\,, & \text{if } \sum_{b\in\mathbf{A}}\eta^*(x,b) = 0. \end{cases} \tag{5.26}$$

For justification see Theorem 3.25 of [Guo and Hernández-Lerma (2009)]. In that book, only natural Markov strategies were considered, but in finite models they are sufficient for solving problems (5.7) and (5.8): see Remark 5.4.

For any feasible solution $(\eta, \tilde{\eta})$ one can also construct the strategy π^{s} using (5.26), and π^{s} is well defined because, if $\sum_{b\in\mathbf{A}}\eta^*(x,b) = 0$, then $\sum_{b\in\mathbf{A}}\tilde{\eta}^*(x,b) > 0$: otherwise, if $\sum_{b\in\mathbf{A}}\tilde{\eta}^*(x,b) = 0$, too, then (5.25) implies that $0 = \alpha(x) + \sum_{y\neq x} q(x|y,a)\tilde{\eta}(y,a)$ which is a contradiction. We say that the strategy π^{s} in (5.26) is *induced* by a feasible solution $(\eta, \tilde{\eta})$. The induced by an optimal solution $(\eta^*, \tilde{\eta}^*)$ relaxed stationary strategy may be not deterministic, but one can always find another optimal solution to the linear program (5.23), (5.24), (5.25) leading to an average uniformly optimal deterministic stationary strategy [Guo and Hernández-Lerma (2009), Rem. 3.26].

One can show that linear program (5.23), (5.24), (5.25) is equivalent to linear program (4.23), (4.24), (4.25) for the discrete-time MDP with the same state and action spaces $\mathbf{X}$ and $\mathbf{A}$, with the cost function $\bar{c}(x,a) := \frac{c(x,a)}{1+\bar{q}}$ and with the transition probabilities $\bar{p}(y|x,a) := \frac{q(y|x,a)}{1+\bar{q}} + \mathbb{I}\{y = x\}$, where $\bar{q} := \sup_{x\in\mathbf{X}} \bar{q}_x = \sup_{(x,a)\in\mathbf{X}\times\mathbf{A}} q_x(a)$.

- Constant $1 + \bar{q}$ does not affect the objective, so (5.23) can be written as (4.23).
- After dividing (5.24) by $1+\bar{q}$ and adding $\sum_{a\in\mathbf{A}} \eta(x,a)$ to the both sides, we obtain (4.24):

$$\sum_{a\in\mathbf{A}} \eta(x,a) = \frac{1}{1+\bar{q}} \sum_{y\in\mathbf{X}} \sum_{a\in\mathbf{A}} q(x|y,a)\eta(y,a) + \sum_{a\in\mathbf{A}} \eta(x,a)$$
$$= \sum_{y\in\mathbf{X}} \sum_{a\in\mathbf{A}} \bar{p}(x|y,a)\eta(y,a).$$

- Equality (5.25) is equivalent to

$$\sum_{a\in\mathbf{A}} \eta(x,a) + \sum_{a\in\mathbf{A}} (1+\bar{q})\tilde{\eta}(x,a) - \sum_{y\in\mathbf{X}} \sum_{a\in\mathbf{A}} q(x|y,a)\tilde{\eta}(y,a)$$
$$- \sum_{a\in\mathbf{A}} (1+\bar{q})\tilde{\eta}(x,a) = \alpha(x)$$
$$\Longleftrightarrow \sum_{a\in\mathbf{A}} \eta(x,a) + \sum_{a\in\mathbf{A}} (1+\bar{q})\tilde{\eta}(x,a)$$
$$- \sum_{y\in\mathbf{X}} \sum_{a\in\mathbf{A}} \bar{p}(x|y,a)(1+\bar{q})\tilde{\eta}(y,a) = \alpha(x).$$

  The last equality coincides with (4.25) with $\tilde{\eta}(\cdot,\cdot)$ replaced by $(1 + \bar{q})\tilde{\eta}(\cdot,\cdot)$.

Therefore, AC-optimal strategies in the constructed DTMDP are average uniformly optimal in the underlying CTMDP. This transformation to the equivalent DTMDP is called the *uniformization technique*: see [Guo and Hernández-Lerma (2009), p. 51] and [Puterman (1994), Section 11.5].

If under each deterministic stationary strategy the controlled process $X(\cdot)$ is *ergodic* (i.e., all states communicate), then there is no need in the variables $\tilde{\eta}(\cdot,\cdot)$, that is, put $\tilde{\eta}(x,a) \equiv 0$. Such models will be called *ergodic*. Equality (5.25) should be omitted, and the linear program (5.23), (5.24) should be considered in the space of probability distributions $\eta(\cdot,\cdot)$ on $\mathbf{X}\times\mathbf{A}$: see also the linear program (5.39), (5.40) with $w(x) \equiv 1$ and (4.6) for the equivalent DTMDP. In this case, and also if the model is unichain, the optimal value of problem (5.8) does not depend on the initial distribution P_0 and equals the optimal value of the linear program (5.23), (5.24): see the end of Subsection 4.1 for the equivalent DTMDP. Recall that the model is *unichain* if under each deterministic stationary strategy the Markov process $X(\cdot)$ is unichain, with a single positive-recurrent class plus a possibly empty set of transient states.

Now consider the general Borel model and introduce the so called vanishing discount factor approach. Let

$$m_\alpha := \inf_{x\in\mathbf{X}} W^{*\alpha}(x),$$
$$h_\alpha(x) := W^{*\alpha}(x) - m_\alpha \geq 0, \ \forall\, \alpha > 0,$$

where the convention $\infty - \infty := \infty$ is in use. $W^{*\alpha}(\cdot)$ is the Bellman function for the discounted total cost problem. Let

$$g := \limsup_{\alpha\downarrow 0} \alpha m_\alpha \in [0,\infty] \tag{5.27}$$

and

$$h(x) := \liminf_{\alpha\downarrow 0,\ y\to x} h_\alpha(y) := \sup_{\delta>0,\ \epsilon>0}\left\{\inf_{0<\alpha\leq\delta,\ \rho(x,y)\leq\epsilon} h_\alpha(y)\right\} \in [0,\infty], \tag{5.28}$$

where ρ denotes the predetermined metric on $\mathbf{X}$ consistent with its topology.

Theorem 5.9. [Piunovskiy and Zhang (2020), Thm. 5.1.2]. *Suppose Condition 5.4 is satisfied, the cost rate $c(\cdot,\cdot)$ is nonnegative, and $h(x) < \infty$ for all $x \in \mathbf{X}$. Then there exists a deterministic stationary average uniformly optimal strategy.*

If additionally $\inf_{x\in\mathbf{X}} W^{*A}(x) < \infty$, then any deterministic stationary strategy φ^* satisfying

$$\begin{aligned} g &\geq \inf_{a\in\mathbf{A}}\left\{c(x,a) + \int_{\mathbf{X}} h(y)q(dy|x,a)\right\} \\ &= c(x,\varphi^*(x)) + \int_{\mathbf{X}} h(y)q(dy|x,\varphi^*(x)), \quad \forall\, x\in\mathbf{X} \end{aligned} \tag{5.29}$$

is average uniformly optimal. The constant g and function $h(\cdot)$ are defined by (5.27) and (5.28), respectively. In this case,

$$g = \inf_{x\in\mathbf{X}} \limsup_{\alpha\downarrow 0} \alpha W^{*\alpha}(x) = W^{*A}(x) \in [0,\infty),\ \forall\, x\in\mathbf{X}.$$

Formula in [Piunovskiy and Zhang (2020), Thm. 5.1.2.] looks a little different, but is equivalent to (5.29) if $h(\cdot) < \infty$. In case $c(\cdot,\cdot) \geq -K > -\infty$ and there is no explosion, one can immediately pass to the positive model according to Remark 5.3. Statements similar to Theorem 5.9 appeared in many articles and in monographs [Guo and Hernández-Lerma (2009); Prieto-Rumeau and Hernández-Lerma (2012)]: see, correspondingly, Theorems 5.7 and 3.20.

Note that in general $W^{*A}(\cdot)$ is not a constant, e.g., if there are several ergodic classes in the Markov process $X(\cdot)$ under some deterministic stationary strategies. In such cases Theorem 5.9 is not applicable. As a simplest example, consider the following uncontrolled model with $\mathbf{A} = \{a_0\}$ being a singleton: $\mathbf{X} = \{1,2\}$, $q_x \equiv 0$

and $c(i) = i$. Here $\{1\}$ and $\{2\}$ are ergodic classes and, obviously, $W^{*A}(i) = i$. In this model $W^{*\alpha}(i) = \frac{i}{\alpha}$, $m_\alpha = \frac{1}{\alpha}$, $h_\alpha(1) = 0$, $h_\alpha(2) = \frac{1}{\alpha}$ and $h(1) = 0$, but $h(2) = \infty$. Condition $h(\cdot) < \infty$ is violated.

5.3 Examples

In this section, examples and counterexamples illustrate the theory which is well developed in [Guo and Hernández-Lerma (2009); Piunovskiy and Zhang (2020); Prieto-Rumeau and Hernández-Lerma (2012)]. In general, randomized strategies and relaxed stationary strategies are not sufficient in problem (5.7): examples in Subsection 5.3.1 illustrate this point. Strategic measures are discussed in Subsection 5.3.2. Different approaches to the lump sum costs are presented in Subsection 5.3.3. Conditions for explosion/no explosion are illustrated in Subsection 5.3.4. Example 5.3.5 illustrates the cases when it is possible (impossible) to pass to the equivalent DTMDP. The validity of Dynkin's formula is discussed in Subsection 5.3.6. Example 5.3.7 shows that optimality equations (5.13) and (5.22) are not equivalent. Subsections 5.3.8–5.3.12 are about the average cost problem (5.8). In particular, the constrained version is discussed in subsections 5.3.10–5.3.12, and the applicability of the linear programming approach is discussed in subsections 5.3.9 and 5.3.10. Example 5.3.11 shows that the average cost problem sometimes cannot be approximated by the discounted problem as $\alpha \downarrow 0$. Subsection 5.3.12 confirms that restricting to mixed strategies is not always legal. Another version of the constrained average cost problem, popular in the field of semi-Markov decision processes, is discussed in Subsection 5.3.13. Subsection 5.3.14 is about the long-run average variance. The so called fluid approximation to birth-and-death processes is described in Subsections 5.3.15 and 5.3.16. Finally, some unexpected results from the field of queueing theory are presented in subsections 5.3.17 and 5.3.18.

According to Theorems 5.6(b) and 5.7, under Condition 5.3(b) the optimization problem (5.7) in the space of all strategies is equivalent to the expected total cost DTMDP, and examples from Chapter 2 are relevant for the continuous-time MDP. Similarly, if the model is finite, then using the uniformization technique, one can pass to DTMDP, and the corresponding examples from Chapter 4 are relevant for CTMDP.

5.3.1 *When randomized and relaxed stationary strategies are not sufficient*

In this subsection we show that the conditions in Theorem 5.6 are important. The presented below examples were published in [Piunovskiy and Zhang (2020), §1.3.4].

The first simple example shows that in general randomized strategies are not sufficient for solving optimization problems if the performance functional is undiscounted. Neither item of Condition 5.3 is satisfied.

Example. Suppose $\mathbf{X} = \{0\}$, $\mathbf{A} = (0, 1]$, $\mathsf{P}_0(0) = 1$, $\alpha = 0$, and $q(0|0, a) \equiv 0$:

the single state 0 is absorbing, but we do not call it "cemetery" because $c(0,a) := a$. For an arbitrary randomized strategy ϖ we have

$$m_{0,1}^{\varpi,0}(\{0\}\times\Gamma_A) = \varpi_1(\Gamma_A|0)\times\infty, \text{ where } 0\times\infty = 0$$

and $W^0(\varpi,0) = \int_{(0,1]} a\; m_{0,1}^{\varpi,0}(\{0\}\times da) = \infty$: the cost rate is constant and positive on the infinite interval $(0,\infty)$.

At the same time, e.g., for the deterministic π-strategy φ given by $\varphi_1(0,s) = e^{-\kappa s}$ with $\kappa > 0$, we have

$$m_{0,1}^{\varphi,0}(\{0\}\times\Gamma_A) = \int_{(0,\infty)} \delta_{e^{-\kappa s}}(\Gamma_A)ds,$$

so that, e.g., $m_{0,1}^{\varphi,0}(\{0\}\times[\frac{1}{2},1]) = -\ln\left(\frac{1}{2}\right)/\kappa$ is different from $0, \infty$, the measures $m_{0,1}^{\varphi,0} \neq m_{0,1}^{\varpi,0}$ are different, and, for the same cost rate $c(0,a) = a$,

$$\begin{aligned} W^0(\varphi,0) &= \int_{(0,\infty)}\int_{(0,1]} a\delta_{e^{-\kappa s}}(da)ds = \int_{(0,\infty)} e^{-\kappa s}ds = \frac{1}{\kappa} \\ &\Longrightarrow \inf_{S\in\mathbf{S}} W_0^0(S,0) = 0. \end{aligned}$$

In this example, statement (a) of Theorem 5.6 still holds true, although Condition 5.3(a) is not satisfied. For a strategy S, take the Markov π-strategy π^{m} equivalent to S in the sense of Theorem 5.1(b). Next, introduce the following randomized Markov strategy ϖ^{m}:

$$\varpi_1^{\mathrm{m}}(\Gamma_A|0) = \sum_{i=1}^{\infty}\left(\frac{1}{2}\right)^i \int_{(i-1,i]} \pi_1^{\mathrm{m}}(\Gamma_A|0,s)ds, \qquad \Gamma_A\in\mathcal{B}(\mathbf{A}).$$

Now $m_{0,1}^{\pi^{\mathrm{m}},0}(\{0\}\times\Gamma_A) > 0$ if and only if $\varpi_1^{\mathrm{m}}(\Gamma_A|0) > 0$, i.e., $m_{0,1}^{\varpi^{\mathrm{m}},0}(\{0\}\times\Gamma_A) = \infty$, so that $m_{0,1}^{\varpi^{\mathrm{m}},0} \geq m_{0,1}^{\pi^{\mathrm{m}},0}$. □

Let us show that the statement of Theorem 5.6(a) can fail to hold if the transition rate takes zero values and $\alpha = 0$.

Example. Suppose $\mathbf{X} = \{1,\Delta\}$, where Δ is the cemetery with $q_\Delta(a)\equiv 0$. We put $\mathbf{A} = \{1,2\}$, $q_1(1) = 0$, $q_1(2) = q(\Delta|1,2) = 1$, $\alpha = 0$, $\mathsf{P}_0(1) = 1$. Only the first component of the strategy (π_1 or ϖ_1) plays role because either $\Theta_1 = \infty$ or $X_1 = \Delta$. See Fig. 5.1. The cost rate $c(\cdot,\cdot)$ is immaterial.

Consider the deterministic π-strategy φ given by

$$\varphi_1(1,s) = \mathbb{I}\{s\in(0,1]\}\cdot 1 + \mathbb{I}\{s\in(1,\infty)\}\cdot 2.$$

The controlled process spends 1 time unit in the initial state 1; after that the jump intensity to the cemetery Δ equals 1. Then

$$m_{\mathsf{P}_0,1}^{\varphi,0}(\{1\}\times\{1\}) = 1$$

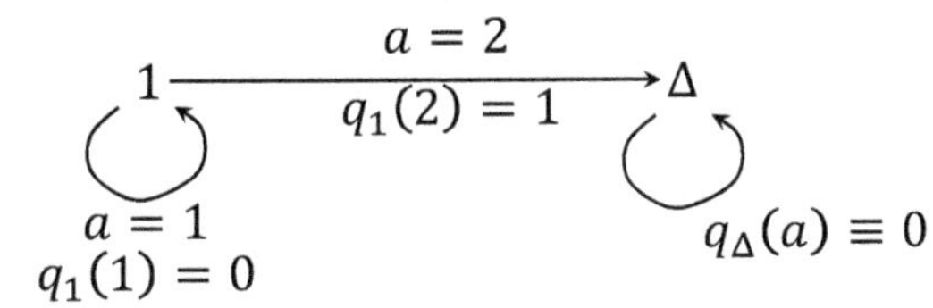

Fig. 5.1 No-one randomized Markov strategy satisfies inequality (5.18).

and

$$m^{\varphi,0}_{\mathsf{P}_0,1}(\{1\}\times\{2\}) = \mathsf{E}^{\varphi}_{\mathsf{P}_0}\left[\int_{(1,T_1)} dt\right] = \mathsf{E}^{\varphi}_{\mathsf{P}_0}[(T_1-1)] = 1.$$

On the other hand, for an arbitrary randomized Markov strategy ϖ^{m} with

$$\varpi^{\mathrm{m}}_1(2|1) = 1 - \varpi^{\mathrm{m}}_1(1|1) = \beta \in [0,1],$$

we have

$$m^{\varpi^{\mathrm{m}},0}_{\mathsf{P}_0,1}(\{1\}\times\{1\}) = \begin{cases} 0, & \text{if } \beta = 1; \\ \infty, & \text{if } \beta < 1, \end{cases} \quad \text{and } m^{\varpi^{\mathrm{m}},0}_{\mathsf{P}_0,1}(\{1\}\times\{2\}) = \beta. \tag{5.30}$$

Hence there is no such $\beta \in [0,1]$ that $m^{\varpi^{\mathrm{m}},0}_{\mathsf{P}_0,1} \geq m^{\varphi,0}_{\mathsf{P}_0,1}$.

It is interesting to look what happens if we slightly increase $q_1(1)$ and take $q_1(1) = q(\Delta|1,1) = \varepsilon > 0$. Now, for the same deterministic π-strategy φ, we have

$$m^{\varphi,0}_{\mathsf{P}_0,1}(\{1\}\times\{1\}) = \mathsf{E}^{\varphi}_{\mathsf{P}_0}\left[\int_{(0,\Theta_1\wedge 1)} 1\, dt\right] = e^{-\varepsilon} + \int_{(0,1)} \varepsilon e^{-\varepsilon u} u\, du.$$

The first (second) term corresponds to the case $\Theta_1 > 1$ ($\Theta_1 \leq 1$). Thus

$$m^{\varphi,0}_{\mathsf{P}_0,1}(\{1\}\times\{1\}) = \frac{1}{\varepsilon}(1-e^{-\varepsilon}) \text{ and } m^{\varphi,0}_{\mathsf{P}_0,1}(\{1\}\times\{2\}) = e^{-\varepsilon}.$$

Consider the following randomized Markov strategy ϖ^{m}:

$$\varpi^{\mathrm{m}}_1(1|1) = 1 - e^{-\varepsilon}; \quad \varpi^{\mathrm{m}}_1(2|1) = e^{-\varepsilon}.$$

Now

$$m^{\varpi^{\mathrm{m}},0}_{\mathsf{P}_0,1}(\{1\}\times\{1\}) = \frac{1}{\varepsilon}(1-e^{-\varepsilon}); \qquad m^{\varpi^{\mathrm{m}},0}_{\mathsf{P}_0,1}(\{1\}\times\{2\}) = e^{-\varepsilon} :$$

Θ_1 is exponential with parameter ε or 1 with probabilities $1-e^{-\varepsilon}$ and $e^{-\varepsilon}$ correspondingly. Therefore, $m^{\varpi^{\mathrm{m}},0}_{\mathsf{P}_0,1} = m^{\varphi,0}_{\mathsf{P}_0,1}$ on $\mathbf{X}^{\Delta}\times\mathbf{A}$ in accordance with Theorem 5.6(b).

When ε approaches zero, we have

$$\lim_{\varepsilon\to 0} m^{\varpi^{m},0}_{\mathsf{P}_0,1}(\{1\}\times\{1\}) = \lim_{\varepsilon\to 0} m^{\varphi,0}_{\mathsf{P}_0,1}(\{1\}\times\{1\}) = 1;$$

$$\lim_{\varepsilon\to 0} m^{\varpi^{m},0}_{\mathsf{P}_0,1}(\{1\}\times\{2\}) = \lim_{\varepsilon\to 0} m^{\varphi,0}_{\mathsf{P}_0,1}(\{1\}\times\{2\}) = 1,$$

but $m^{\varpi^{m},0}_{\mathsf{P}_0,1}(\{1\}\times\{1\})$ as a function of ε is not continuous at zero: $m^{\varpi^{m},0}_{\mathsf{P}_0,1}(\{1\}\times\{1\}) = 0$ when $\varepsilon = 0$ by (5.30) at $\beta = 1$, and $m^{\varphi,0}_{\mathsf{P}_0,1}(\{1\}\times\{1\}) = 1$ when $\varepsilon = 0$.

Below, for the same unperturbed example ($\varepsilon = 0$), we show that, if π^{s} is a stationary π-strategy, then the detailed occupation measure $m^{\pi^{s},0}_{\mathsf{P}_0,1}$ cannot be generated by a stationary randomized strategy and *vice versa*: for each stationary randomized strategy ϖ^{s} there are no stationary π-strategies having the same measure $m^{\varpi^{s},0}_{\mathsf{P}_0,1}$. For an arbitrary stationary π-strategy π^{s} we have,

$$\begin{aligned}
\text{either } m^{\pi^{s},0}_{\mathsf{P}_0,1}(\{1\}\times\{1\}) \in (0,\infty) \ \text{ and } \ & m^{\pi^{s},0}_{\mathsf{P}_0,1}(\{1\}\times\{2\}) \in (0,\infty)\\
& (\text{if } \pi^{s}(1|1)\in(0,1)),\\
\text{or } m^{\pi^{s},0}_{\mathsf{P}_0,1}(\{1\}\times\{2\}) = 0 \ \text{ and } \ & m^{\pi^{s},0}_{\mathsf{P}_0,1}(\{1\}\times\{1\}) = \infty\\
& (\text{if } \pi^{s}(2|1) = 0),\\
\text{or } m^{\pi^{s},0}_{\mathsf{P}_0,1}(\{1\}\times\{2\}) = 1 \ \text{ and } \ & m^{\pi^{s},0}_{\mathsf{P}_0,1}(\{1\}\times\{1\}) = 0\\
& (\text{if } \pi^{s}(2|1) = 1).
\end{aligned}$$

Note that here Θ_1 is the exponential random variable with parameter $\pi^{s}(2|1)$.

For an arbitrary stationary randomized strategy ϖ^{s} we have

$$\begin{aligned}
\text{either } m^{\varpi^{s},0}_{\mathsf{P}_0,1}(\{1\}\times\{2\}) = \varpi^{s}(2|1) \in (0,1) \ \text{ and } \ & m^{\varpi^{s},0}_{\mathsf{P}_0,1}(\{1\}\times\{1\}) = \infty\\
& (\text{if } \varpi^{s}(1|1)\in(0,1)),\\
\text{or } m^{\varpi^{s},0}_{\mathsf{P}_0,1}(\{1\}\times\{2\}) = 0 \ \text{ and } \ & m^{\varpi^{s},0}_{\mathsf{P}_0,1}(\{1\}\times\{1\}) = \infty\\
& (\text{if } \varpi^{s}(2|1) = 0),\\
\text{or } m^{\varpi^{s},0}_{\mathsf{P}_0,1}(\{1\}\times\{2\}) = 1 \ \text{ and } \ & m^{\varpi^{s},0}_{\mathsf{P}_0,1}(\{1\}\times\{1\}) = 0\\
& (\text{if } \varpi^{s}(2|1) = 1).
\end{aligned}$$

Here Θ_1 is the exponential random variable with parameter 1 with probability $\varpi^{s}(2|1)$ and $\Theta_1 = \infty$ with probability $\varpi^{s}(1|1)$.

If, for a stationary π-strategy π^{s}, $\pi^{s}(1|1)\in(0,1)$ then $m^{\pi^{s},0}_{\mathsf{P}_0,1}$ cannot be generated by a stationary randomized strategy. If, for a stationary randomized strategy ϖ^{s}, $\varpi^{s}(1|1)\in(0,1)$ then $m^{\varpi^{s},0}_{\mathsf{P}_0,1}$ cannot be generated by a stationary π-strategy. At the same time, there exists a Markov π-strategy π^{m} such that $m^{\pi^{m},0}_{\mathsf{P}_0,1} = m^{\varpi,0}_{\mathsf{P}_0,1}$ for each randomized strategy ϖ by Theorem 5.1(b). □

Now we show that even if the transition rate $q_x(a)$ is strictly positive, but not separated from zero on $\mathbf{X}^{\Delta}$, then the statement (b) of Theorem 5.6 can fail to hold, and randomized strategies, as well as stationary π-strategies, may be not sufficient for solving undiscounted optimization problems with positive cost rate.

Example. Let $\mathbf{X} = \{1, \Delta\}$, where Δ is the costless cemetery with $q_\Delta(a) \equiv 0$. We put $\mathbf{A} = (0, 1]$, $q_1(a) = q(\Delta|1, a) = a$, $c(1, a) = a$, $c(\Delta, a) \equiv 0$, $\mathsf{P}_0(1) = 1$. Similarly to the second example in this subsection, only the first component of the strategy (π_1 or ϖ_1) plays role. See Fig. 5.2. Assume that $\alpha = 0$ and consider the total cost problem (5.7).

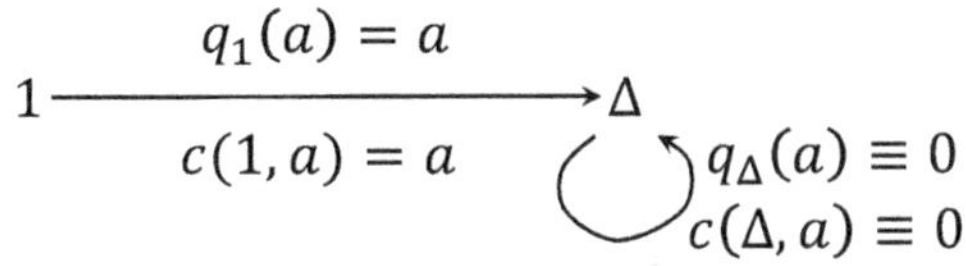

Fig. 5.2 Randomized strategies are not sufficient for solving the total cost problem (5.7).

Clearly, for any randomized strategy ϖ in this model,

$$m^{\varpi,0}_{\mathsf{P}_0,1}(\{1\} \times \Gamma_A) = \mathsf{E}^{\varpi}_{\mathsf{P}_0}\left[\int_{\Gamma_A}\int_{(0,\infty)} e^{-as} ds\, \varpi_1(da|X_0)\right] = \int_{\Gamma_A} \frac{1}{a}\varpi_1(da|1).$$

Therefore,

$$W^0(\varpi, \mathsf{P}_0) = \int_{\mathbf{A}} c(1, a) m^{\varpi,0}_{\mathsf{P}_0,1}(\{1\} \times da) = \varpi_1(\mathbf{A}|1) = 1.$$

Similarly, for an arbitrary stationary π-strategy π^s, we have

$$m^{\pi^s,0}_{\mathsf{P}_0,1}(\{1\} \times \Gamma_A) = \mathsf{E}^{\pi^s}_{\mathsf{P}_0}\left[\int_{(0,\infty)} e^{-q_1(\pi^s)u}\pi^s(\Gamma_A|X_0)du\right] = \pi^s(\Gamma_A|1)/q_1(\pi^s),$$

where $q_1(\pi^s) := \int_{\mathbf{A}} q_1(a)\pi^s(da|1) = \int_{\mathbf{A}} a\pi^s(da|1)$. Again

$$W^0(\pi^s, \mathsf{P}_0) = \int_{\mathbf{A}} c(1, a) m^{\pi^s,0}_{\mathsf{P}_0,1}(\{1\} \times da) = \int_{\mathbf{A}} a\pi^s(da|1)/q_1(\pi^s) = 1.$$

Now consider the deterministic π-strategy φ given by $\varphi_1(1, s) = e^{-\kappa s}$, where $\kappa > 0$ is an arbitrarily fixed constant. The sojourn time Θ_1 has the distribution function given by $1 - e^{\frac{-1+e^{-\kappa\theta}}{\kappa}}$ and, for an arbitrarily fixed $u \in (0,1]$, we have

$$\begin{aligned}
m^{\varphi,0}_{\mathsf{P}_0,1}(\{1\} \times (u,1]) &= \mathsf{E}^{\varphi}_{\mathsf{P}_0}\left[\int_{(0,\infty)} e^{-\int_{(0,s]} e^{-\kappa v} dv} \mathbb{I}\{e^{-\kappa s} \in (u,1]\} ds\right] \\
&= \int_{(0,\infty)} e^{\frac{-1+e^{-\kappa s}}{\kappa}} \mathbb{I}\{e^{-\kappa s} \in (u,1]\} ds \\
&= \int_{(0,1)} e^{\frac{-1+y}{\kappa}} \mathbb{I}\{y \in (u,1]\} \frac{dy}{\kappa y} = \int_{(u,1)} \frac{e^{\frac{-1+y}{\kappa}}}{\kappa y} dy.
\end{aligned}$$

This detailed occupation measure has density $\frac{e^{\frac{-1+a}{\kappa}}}{\kappa a}$ on $\mathbf{A}$. The total expected time spent by the controlled process in state 1 under the strategy φ equals $m^{\varphi,0}_{\mathsf{P}_0,1}(\{1\} \times \mathbf{A}) = \int_{(0,1)} \frac{e^{\frac{-1+y}{\kappa}}}{\kappa y} dy = \infty$: the jump intensity $q_1(e^{-\kappa t}) = e^{-\kappa t}$, although positive, rapidly decreases to zero. The measure $m^{\varphi,0}_{\mathsf{P}_0,1}$ is different from the measures $m^{\varpi,0}_{\mathsf{P}_0,1}$ and $m^{\pi^s,0}_{\mathsf{P}_0,1}$ presented above because, for the same cost rate $c(\cdot,\cdot)$, we have

$$W^0(\varphi, \mathsf{P}_0) = \int_{\mathbf{A}} c(1,a) m^{\varphi,0}_{\mathsf{P}_0,1}(\{1\} \times da) = \int_{(0,1]} \frac{1}{\kappa} e^{\frac{-1+a}{\kappa}} da = 1 - e^{-\frac{1}{\kappa}}.$$

Therefore (as $\kappa \to \infty$) $\inf_S W^0(S, \mathsf{P}_0) = 0$, but, for each randomized strategy ϖ, $W^0(\varpi, \mathsf{P}_0) = 1$, and, for each stationary π-strategy π^s, $W^0(\pi^s, \mathsf{P}_0) = 1$, as was shown above. □

5.3.2 *Strategic measures and performance functional*

For a fixed initial distribution P_0 and strategy S, the strategic measure $\mathsf{P}^S_{\mathsf{P}_0}$ on $\Omega = \{(x_0, \theta_1, x_1, \ldots)\}$ only defines the controlled process $X(\cdot)$. Since it does not refer explicitly to the actions, it can easily happen that $\mathsf{P}^{S^1}_{\mathsf{P}_0} = \mathsf{P}^{S^2}_{\mathsf{P}_0}$, but $W^\alpha(S^1, \mathsf{P}_0) \neq W^\alpha(S^2, \mathsf{P}_0)$ as the following elementary examples shows.

Example. Let $\mathbf{X} = \{1, \Delta\}$, where Δ is the (costless) cemetery, $\mathsf{P}_0(1) = 1$, $\mathbf{A} = \{1,2\}$ and $q(\Delta|1,1) = q(\Delta|1,2) = \lambda > 0$. Consider the deterministic stationary strategies $\varphi^1(x) \equiv 1$ and $\varphi^2(x) \equiv 2$. Clearly, $\mathsf{P}^{\varphi^1}_1 = \mathsf{P}^{\varphi^2}_1$ because the jump rate is the same under actions 1 and 2. If $c_0(1,1) = 0$ and $c_0(1,2) = 1$, then

$$W^0(\varphi^1, \mathsf{P}_0) = 0; \quad W^0(\varphi^2, \mathsf{P}_0) = \frac{1}{\lambda}.$$

See Fig. 5.3.

Note also that $\frac{1}{2}\mathsf{P}^{\varphi^1}_1 + \frac{1}{2}\mathsf{P}^{\varphi^2}_1 = \mathsf{P}^{\varphi^1}_1$ because $\mathsf{P}^{\varphi^1}_1 = \mathsf{P}^{\varphi^2}_1$. But

$$\frac{1}{2}W^0(\varphi^1, \mathsf{P}_0) + \frac{1}{2}W^0(\varphi^2, \mathsf{P}_0) = \frac{1}{2\lambda} \neq W^0(\varphi^1, \mathsf{P}_0) = 0:$$

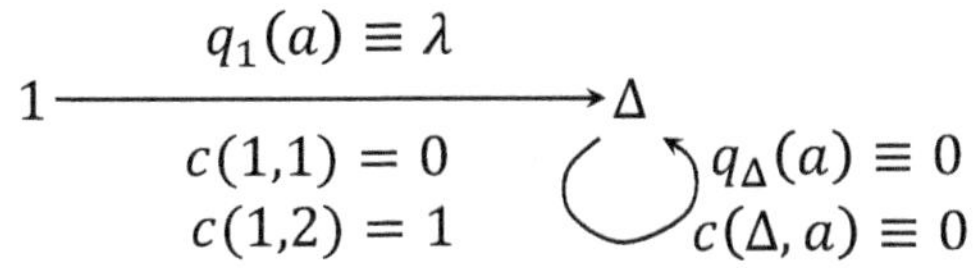

Fig. 5.3 Same strategic measures, but different performance functionals.

the convex combination of strategic measures does not lead to the convex combination of performance functionals. □

The second example shows that the space of strategic measures generated by relaxed strategies may be not convex.

Example. Let $\mathbf{X} = \{0, 1, 2\}$, where 1 and 2 are absorbing states, $\mathsf{P}_0(0) = 1$, $\mathbf{A} = \{1, 2\}$, $q(1|0,1) = -q(0|0,1) = \lambda_1$, $q(2|0,2) = -q(0|0,2) = \lambda_2$, $0 < \lambda_1 < \lambda_2$, $\lambda_1 + \lambda_2 = 2$, and other transition rates are zero. The cost rate $c(\cdot,\cdot)$ is of no importance. See Fig. 5.4.

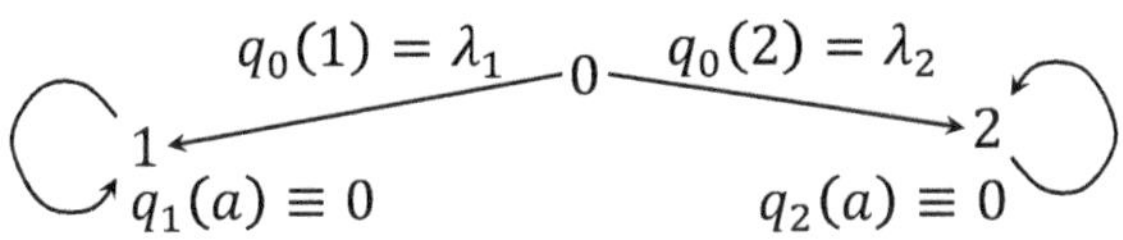

Fig. 5.4 Non-convex strategic space generated by π-strategies.

Consider the deterministic stationary strategies $\varphi^1(x) \equiv 1$ and $\varphi^2(x) \equiv 2$. Then

$$\mathsf{P}_0^{\varphi^1}(\Theta_1 \in dt,\ X_1 = 1) = \lambda_1 e^{-\lambda_1 t} dt,\quad \mathsf{P}_0^{\varphi^1}(\Theta_1 \in dt,\ X_1 = 2) = 0;$$
$$\mathsf{P}_0^{\varphi^2}(\Theta_1 \in dt,\ X_1 = 2) = \lambda_2 e^{-\lambda_2 t} dt,\quad \mathsf{P}_0^{\varphi^2}(\Theta_1 \in dt,\ X_1 = 1) = 0.$$

For the convex combination $\mathsf{P} := \frac{1}{2}\mathsf{P}_0^{\varphi^1} + \frac{1}{2}\mathsf{P}_0^{\varphi^2}$ we have

$$\mathsf{P}(\Theta_1 \le s) = 1 - \frac{1}{2}\left(e^{-\lambda_1 s} + e^{-\lambda_2 s}\right);\quad \mathsf{P}(X_1 = 1) = \mathsf{P}(X_1 = 2) = \frac{1}{2}. \tag{5.31}$$

Suppose that $\mathsf{P} = \mathsf{P}_0^\pi$ for some relaxed strategy π. In order to have

$$F(s) := \mathsf{P}_0^\pi(\Theta_1 \le s) = 1 - \frac{1}{2}\left(e^{-\lambda_1 s} + e^{-\lambda_2 s}\right),$$

it is necessary and sufficient that the hazard rate of the distribution characterized by F equals $\lambda_1\pi_1(1|0,s)+\lambda_2\pi_1(2|0,s)$, that is,

$$\lambda_1\pi_1(1|0,s)+\lambda_2[1-\pi_1(1|0,s)]=\frac{\lambda_1 e^{-\lambda_1 s}+\lambda_2 e^{-\lambda_2 s}}{e^{-\lambda_1 s}+e^{-\lambda_2 s}},$$

which leads to

$$\pi_1(1|0,s)=\frac{e^{-\lambda_1 s}}{e^{-\lambda_1 s}+e^{-\lambda_2 s}};\ \pi_1(2|0,s)=\frac{e^{-\lambda_2 s}}{e^{-\lambda_1 s}+e^{-\lambda_2 s}}.$$

Now

$$\begin{aligned}
&\mathsf{P}_0^\pi(X_1=1)\\
&=\int_{(0,\infty)}\frac{e^{-\lambda_1 t}}{e^{-\lambda_1 t}+e^{-\lambda_2 t}}e^{-\int_{(0,t]}\frac{\lambda_1 e^{-\lambda_1 s}+\lambda_2 e^{-\lambda_2 s}}{e^{-\lambda_1 s}+e^{-\lambda_2 s}}ds}dt\\
&=\left.\frac{-e^{-\lambda_1 t}}{\lambda_1 e^{-\lambda_1 t}+\lambda_2 e^{-\lambda_2 t}}e^{-\int_{(0,t)}\frac{\lambda_1 e^{-\lambda_1 s}+\lambda_2 e^{-\lambda_2 s}}{e^{-\lambda_1 s}+e^{-\lambda_2 s}}ds}\right|_0^\infty\\
&+\int_{(0,\infty)}\frac{(\lambda_2-\lambda_1)\lambda_2 e^{-\lambda_1 t-\lambda_2 t}}{[\lambda_1 e^{-\lambda_1 t}+\lambda_2 e^{-\lambda_2 t}]^2}e^{-\int_{(0,t]}\frac{\lambda_1 e^{-\lambda_1 s}+\lambda_2 e^{-\lambda_2 s}}{e^{-\lambda_1 s}+e^{-\lambda_2 s}}ds}dt\\
&=\frac{1}{2}+\int_{(0,\infty)}\frac{(\lambda_2-\lambda_1)\lambda_2 e^{-\lambda_1 t-\lambda_2 t}}{[\lambda_1 e^{-\lambda_1 t}+\lambda_2 e^{-\lambda_2 t}]^2}e^{-\int_{(0,t]}\frac{\lambda_1 e^{-\lambda_1 s}+\lambda_2 e^{-\lambda_2 s}}{e^{-\lambda_1 s}+e^{-\lambda_2 s}}ds}dt>\frac{1}{2},
\end{aligned}$$

where the second equality is by integrating by parts. Thus, we see that in this example it is impossible to construct a relaxed strategy π such that the strategic measure P_0^π satisfies equalities (5.31) and equals $\mathsf{P}=\frac{1}{2}\mathsf{P}_0^{\varphi^1}+\frac{1}{2}\mathsf{P}_0^{\varphi^2}$. The similar example was presented in [Piunovskiy and Zhang (2020), Ex. 6.4.2].

At the same time, the space of strategic measures coming from randomized strategies ϖ is convex. Only trajectories of the form

$$(0,\theta_1,x,\infty,x_\infty,\ldots),\quad x\in\{1,2\}$$

can be realized with

$$\begin{aligned}
&\mathsf{P}_0^\varpi(\Theta_1\in dt,\ X_1=x)\\
&=\varpi_1(1|0)\lambda_1 e^{-\lambda_1 t}\mathbb{I}\{x=1\}+\varpi_1(2|0)\lambda_2 e^{-\lambda_2 t}dt\mathbb{I}\{x=2\},
\end{aligned}$$

so that, for $\beta\in[0,1]$, the measure $\beta\mathsf{P}_0^{\varpi^1}+(1-\beta)\mathsf{P}_0^{\varpi^2}$ is generated by the randomized strategy $\beta\varpi^1+(1-\beta)\varpi^2$. □

5.3.3 *Lump sum costs*

Suppose the discount factor $\alpha\in[0,\infty)$ is arbitrarily fixed.

According to Remark 5.6, in the presence of the lump sum cost $C(\cdot,\cdot)$ one can introduce the modified cost rate $\tilde{c}(x,a):=c(x,a)+\int_{\mathbf{X}}C(x,y)\tilde{q}(dy|x,a)$ such that for the performance functional $W^\alpha(S, \mathsf{P}_0)$ defined in (5.19) equals $\tilde{W}^\alpha(S, \mathsf{P}_0)$ for all strategies S and initial distributions P_0. Further, by Theorem 5.4, if there is no explosion, then, for a fixed control strategy S and initial distribution P_0, $\tilde{W}^\alpha(S, \mathsf{P}_0) = \tilde{W}^\alpha(\breve{\pi}, \mathsf{P}_0)$, where the natural Markov strategy $\breve{\pi}$ is induced by S. The same reasoning as in Remark 5.6 applied to $\breve{\pi}$ leads to the equality $W^\alpha(\breve{\pi}, \mathsf{P}_0) = \tilde{W}^\alpha(\breve{\pi}, \mathsf{P}_0)$, so that

$$W^\alpha(S, \mathsf{P}_0) = W^\alpha(\breve{\pi}, \mathsf{P}_0) \text{ for all strategies } S \text{ and arbitrarily fixed } \mathsf{P}_0. \tag{5.32}$$

For simplicity below we assume that $c(x, a) \equiv 0$ and concentrate on the lump sum costs at the jump epochs. We also assume that there is no explosion and the initial distribution P_0 is fixed.

Suppose now that the lump sum cost $C(x, a, y)$ depends also on the action at the jump epoch. If a π-strategy is deterministic, denoted as φ, one can again define the performance functional similarly to (5.19):

$$W^\alpha(\varphi, \mathsf{P}_0) = \mathsf{E}^\varphi_{\mathsf{P}_0}\left[\sum_{n=1}^\infty e^{-\alpha T_n} C(X(T_{n-1}), A(T_n), X(T_n))\right],$$

where $A(T_n) = \varphi_n(H_{n-1}, T_n - T_{n-1})$. As was shown in Corollary 4.4 of [Feinberg (2004)], now equality (5.20) holds for

$$\tilde{c}(x, a) := \int_{\mathbf{X}} C(x, a, y)\tilde{q}(dy|x, a), \tag{5.33}$$

and equality (5.32) is valid, as was explained above. It looks that the similar result holds also for any randomized strategy and, a bit more generally, if the control process $A(t)$ is well defined.

The situation becomes less clear for general relaxed strategies π, since the control process $A(t)$ may be not measurable. (See Remark 5.1(b); the random process $A(t)$ exists by the Kolmogorov Extension/Consistency Theorem.) The natural decision is to switch again to the cost rate (5.33). But the justification is less clear because it is not obvious how to construct accurately the analogue of formula (5.19).

One of the ways applicable for any strategy S is as follows. (For simplicity here and below we accept that $q_x(a) > 0$.) Let the action at the jump epoch T_n be $A(T_n)$. Then the expected lump sum cost is $Y_n(A(T_n)) := \int_{\mathbf{X}} C(X(T_{n-1}), A(T_n), y)\frac{\tilde{q}(dy|X(T_{n-1}), A(T_n))}{q_{X(T_{n-1})}(A(T_n))}$; so

$$W^\alpha(S, \mathsf{P}_0) = \mathsf{E}^S_{\mathsf{P}_0}\left[\sum_{n=1}^\infty e^{-\alpha T_n} Y_n(A(T_n))\right].$$

Since the jump intensity is $q_{X(T_{n-1})}(A(T_n))$, the cost rate at any time moment t

equals

$$e^{-\alpha t}\int_{\mathbf{X}} C(X(t),A(t),y)\frac{\tilde{q}(dy|X(t),A(t))}{q_{X(t)}(A(t))}\cdot q_{X(t)}(A(t)).$$

It remains to integrate w.r.t. the marginal distribution $\bar{\mathsf{P}}^S_{\mathsf{P}_0}$ leading to formula (5.20) with the cost rate $\tilde{c}(\cdot,\cdot)$ as in (5.33). Now all the reasoning as at the beginning of the current subsection applies resulting in equality (5.32).

Another way was suggested in [Feinberg *et al.* (2021), p. 1284] for relaxed strategies π. The (random) jump rate at any time moment $t>0$ is

$$J(\omega,t):=\int_{\mathbf{A}} q_{X(t)}(a)\Pi(\omega,t,da),$$

where

$$\Pi(\omega,t,\Gamma_A):=\sum_{n=1}^{\infty}\mathbb{I}\{T_{n-1}<t\le T_n\}\pi_n(\Gamma_A|H_{n-1},t-T_{n-1})$$

is the (predictable) $\mathbf{P}(\mathbf{A})$-valued random process. At any jump from the current state $X(t)$ the random lump sum cost can be written as

$$L(\omega,t):=\int_{\mathbf{A}}\int_{\mathbf{X}} C(X(t),a,y)\frac{\tilde{q}(dy|X(t),a)}{q_{X(t)}(a)}\Pi(\omega,t,da).$$

After that we got rid of the parameter $a\in\mathbf{A}$ and the performance functional can be written as

$$W^\alpha(\pi,\mathsf{P}_0)=\mathsf{E}^\pi_{\mathsf{P}_0}\left[\int_{(0,\infty)} e^{-\alpha t}J(\omega,t)L(\omega,t)dt\right]. \tag{5.34}$$

Example 1 of [Feinberg *et al.* (2021)] presented below shows that this way of reasoning can disrupt equality (5.32). Consequently, there is no such a cost rate $\tilde{c}(\cdot,\cdot)$ that

$$W^\alpha(\pi,\mathsf{P}_0)=\tilde{W}^\alpha(\pi,\mathsf{P}_0)=\int_{(0,\infty)} e^{-\alpha t}\int_{\mathbf{X}\times\mathbf{A}}\tilde{c}(x,a)\bar{\mathsf{P}}^\pi_{\mathsf{P}_0}(t,dx\times da)$$

for all relaxed strategies π and initial distributions P_0 because, otherwise, as at the beginning of the current subsection, equality (5.32) would have been valid.

Example. Let $\mathbf{X}=\{1,2\}$; $\mathbf{A}=\{a_1,a_2\}$; $q_1(a)=\tilde{q}(2|1,a)\equiv 1$; $q_2(a_1)=\tilde{q}(1|2,a_1)=2$; $q_2(a_2)=\tilde{q}(1|2,a_2)=1$. We put $c(x,a)\equiv 0$ and $C(1,a,2)\equiv 0$, $C(2,a_1,1)=1$, $C(2,a_2,1)=2$. Suppose $\mathsf{P}_0(2)=1$ and $\alpha>0$ is arbitrarily fixed. See Fig. 5.5. Clearly, there is no explosion in this model.

If a π-strategy φ is deterministic, then, as explained before, one can use equality (5.20) with function $\tilde{c}(1,a)\equiv 0$ and $\tilde{c}(2,a_1)=C(2,a_1,1)\tilde{q}(1|2,a_1)=2$, $\tilde{c}(2,a_2)=C(2,a_2,1)\tilde{q}(1|2,a_2)=2$: see (5.33). Thus,

$$W^\alpha(\varphi,\mathsf{P}_0)=2\int_{(0,\infty)} e^{-\alpha t}\bar{\mathsf{P}}^\varphi_{\mathsf{P}_0}(t,\{2\})dt. \tag{5.35}$$

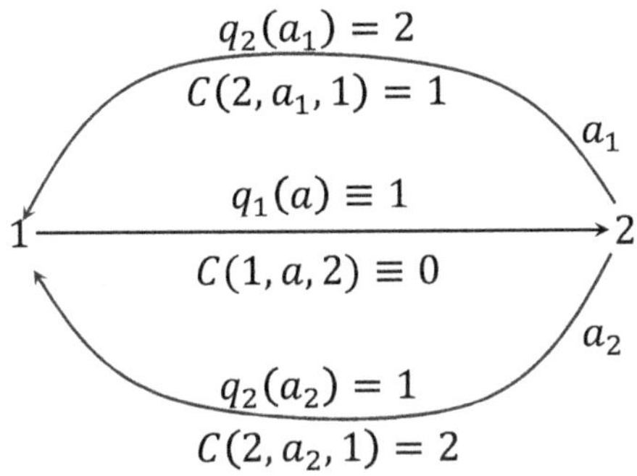

Fig. 5.5 Action-dependent lump sum cost: equality (5.32) is not valid.

Recall that $\bar{\mathsf{P}}^S_{\mathsf{P}_0}(t, dy) := \bar{\mathsf{P}}^S_{\mathsf{P}_0}(t, dy \times \mathbf{A})$.

Obviously, starting from $x_0 = 2$, only the following trajectories can be realized:

$$\omega = (2, \theta_1, 1, \theta_2, 2, \theta_3, 1 \ldots).$$

The actions in state 1 are of no importance.

Consider the following past-dependent deterministic π-strategy φ:

$$\begin{aligned}&\varphi_{4n+1}(h_{4n}, s) \equiv a_1, \quad \varphi_{4n+3}(h_{4n+2}, s) \equiv a_2,\\ &\varphi_{4n+2}(h_{4n+1}, s) = \varphi_{4n+4}(h_{4n+3}, s) \equiv a_1, \quad n = 0, 1, 2, \ldots.\end{aligned}$$

In words, after even (odd) visits to state 2, the action at $X(t-) = 2$ is a_1 (a_2). Then the performance functional $W^\alpha(\varphi, \mathsf{P}_0)$ is given by (5.35).

At each time moment $t > 0$, if $X(t) = 2$, there are positive probabilities that $X(t) = X_{4n}$ (even visits to state 2) and $X(t) = X_{4n+2}$ (odd visits to state 2) for some $n = 0, 1, 2, \ldots$. Therefore, for the marginal distribution we have

$$\bar{\mathsf{P}}^\varphi_{\mathsf{P}_0}(t, \{2\} \times \{a_1\}) > 0 \text{ and } \bar{\mathsf{P}}^\varphi_{\mathsf{P}_0}(t, \{2\} \times \{a_2\}) > 0, \quad t > 0.$$

As the result, the induced by φ natural Markov strategy $\breve{\pi}$ is a non-deterministic relaxed strategy with

$$\breve{\pi}(\{a_1\}|1, t) \equiv 1, \quad \breve{\pi}(\{a_1\}|2, t) > 0, \quad \breve{\pi}(\{a_2\}|2, t) > 0 \text{ at } t > 0.$$

Let us compute $W^\alpha(\breve{\pi}, \mathsf{P}_0)$ using expression (5.34). Since $C(1, a, 2) \equiv 0$, we need to estimate only $J(\omega, t)\mathbb{I}\{X(t) = 2\}$ and $L(\omega, t)\mathbb{I}\{X(t) = 2\}$. If $X(t) = 2$, then

$$\Pi(\omega, t+, \{a_1\}) = \breve{\pi}(\{a_1\}|2, t) > 0 \text{ and } \Pi(\omega, t+, \{a_2\}) = \breve{\pi}(\{a_2\}|2, t) > 0.$$

Therefore (up to the Lebesgue measure zero on $(0,\infty)$),

$$J(\omega,t)\mathbb{I}\{X(t)=2\} = q_2(a_1)\breve{\pi}(\{a_1\}|2,t) + q_2(a_2)\breve{\pi}(\{a_2\}|2,t)$$

and

$$L(\omega,t)\mathbb{I}\{X(t)=2\} = C(2,a_1,1)\breve{\pi}(\{a_1\}|2,t) + C(2,a_2,1)\breve{\pi}(\{a_2\}|2,t)$$

resulting in

$$\begin{aligned}
W^\alpha(\breve{\pi},\mathsf{P}_0) &= \mathsf{E}^{\breve{\pi}}_{\mathsf{P}_0}\left[\int_{(0,\infty)} \mathbb{I}\{X(t)=2\}e^{-\alpha t}[2\breve{\pi}(\{a_1\}|2,t)+\breve{\pi}(\{a_2\}|2,t)]\right.\\
&\qquad\left.\times\ [\breve{\pi}(\{a_1\}|2,t)+2\breve{\pi}(\{a_2\}|2,t)]dt\right]\\
&= \int_{(0,\infty)} \bar{\mathsf{P}}^{\breve{\pi}}_{\mathsf{P}_0}(t,\{2\})e^{-\alpha t}[1+\breve{\pi}(\{a_1\}|2,t)]\cdot[1+\breve{\pi}(\{a_2\}|2,t)]dt\\
&= 2\int_{(0,\infty)} e^{-\alpha t}\bar{\mathsf{P}}^{\breve{\pi}}_{\mathsf{P}_0}(t,\{2\})dt\\
&\quad + \int_{(0,\infty)} e^{-\alpha t}\bar{\mathsf{P}}^{\breve{\pi}}_{\mathsf{P}_0}(t,\{2\})\breve{\pi}(\{a_1\}|2,t)\breve{\pi}(\{a_2\}|2,t)dt\\
&> W^\alpha(\varphi,\mathsf{P}_0).
\end{aligned}$$

Recall that $\bar{\mathsf{P}}^{\varphi}_{\mathsf{P}_0} = \bar{\mathsf{P}}^{\breve{\pi}}_{\mathsf{P}_0}$ by Theorem 5.4. Equality (5.32) fails to hold.

If we accept formula (5.20):

$$W^\alpha(S,\mathsf{P}_0) = \tilde{W}^\alpha(S,\mathsf{P}_0) := \int_{(0,\infty)} e^{-\alpha t}\int_{\mathbf{X}\times\mathbf{A}} \tilde{c}(x,a)\bar{\mathsf{P}}^{S}_{\mathsf{P}_0}(t,dx\times da)dt$$

with $\tilde{c}(\cdot,\cdot)$ as in (5.33): $\tilde{c}(1,a)\equiv 0$, $\tilde{c}(2,a)\equiv 2$, then

$$\begin{aligned}
W^\alpha(\breve{\pi},\mathsf{P}_0) &= \int_{(0,\infty)} e^{-\alpha t}[\tilde{c}(2,a_1)\bar{\mathsf{P}}^{\breve{\pi}}_{\mathsf{P}_0}(t,\{2\}\times\{a_1\})\\
&\qquad + \tilde{c}(2,a_2)\bar{\mathsf{P}}^{\breve{\pi}}_{\mathsf{P}_0}(t,\{2\}\times\{a_2\})]dt\\
&= 2\int_{(0,\infty)} e^{-\alpha t}\bar{\mathsf{P}}^{\breve{\pi}}_{\mathsf{P}_0}(t,\{2\})dt = W^\alpha(\varphi,\mathsf{P}_0)
\end{aligned}$$

as expected. □

To summarize, it looks more reasonable to accept formula (5.20) with $\tilde{c}(\cdot,\cdot)$ given by (5.33).

5.3.4 *Conditions for no explosion*

Examples in this subsection are taken from [Piunovskiy and Zhang (2020), §2.3.1].

Example. Suppose $\mathbf{X} = \{0, 1, 2, \ldots\}$ and the Borel action space $\mathbf{A}$ is arbitrary. Let $q(j|i,a) = q_i(a)H(j-i|i,a)$ for $j \neq i$, where, for each fixed $i \in \mathbf{X}$ and $a \in \mathbf{A}$, $q_i(a) \geq 0$ and $H(\cdot|i,a)$ is the probability distribution on the set $\{-i, -i+1, \ldots, -1, +1, +2, \ldots\}$ of all the possible increments of the process $X(\cdot)$ after any one jump from the current state $i \in \mathbf{X}$. The cost rate $c(\cdot,\cdot)$ is of no importance. See Fig. 5.6.

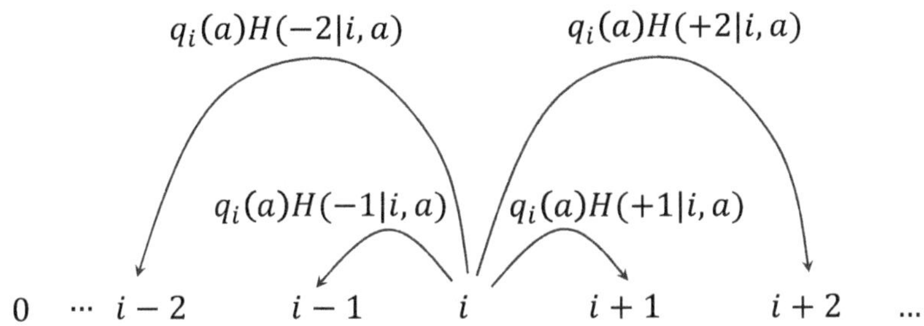

Fig. 5.6 No explosion in a discrete-space model.

Suppose, for a fixed constant $m \in \{1, 2, \ldots\}$, the series

$$\sum_{k\neq 0} |k|^m H(k|i,a) = \tilde{H}(i,a)$$

converges, the function $\tilde{H}(\cdot,\cdot)$ is uniformly bounded: $\tilde{H}(i,a) \leq \bar{H} < \infty$ for all $(i,a) \in \mathbf{X}\times\mathbf{A}$, and $\sup_{i\in\mathbf{X},a\in\mathbf{A}} \frac{q_i(a)}{i+1} = \bar{h} < \infty$. Then one can show that Condition 5.2 is satisfied for

$$V_n = \{0, 1, 2, \ldots, n\},\ w(i) = (i+1)^m,\ \rho = (2^m - 1)\bar{h}\bar{H},\ b = 0.$$

Conditions 5.2(a,b,c) are trivially satisfied. For the case (d), the elementary calculations imply

$$\begin{aligned}
&\sum_{j\in\mathbf{X}} q(j|i,a)(j+1)^m \\
&= q_i(a)\left\{\sum_{j\neq i} H(j-i|i,a)(j+1)^m - (i+1)^m\right\} \\
&= q_i(a)\left\{\sum_{k\neq 0} H(k|i,a)(k+i+1)^m - (i+1)^m\right\}
\end{aligned}$$

$$= q_i(a)\left\{\sum_{k\neq 0} H(k|i,a)\left[\sum_{l=0}^{m-1}\frac{m!}{l!(m-l)!}(i+1)^l k^{m-l}\right]\right\}$$
$$\le (i+1)\bar{h}\left[\sum_{l=0}^{m-1}\frac{m!}{l!(m-l)!}(i+1)^{m-1}\right]\sum_{k\neq 0}|k|^m H(k|i,a)$$
$$\le \bar{h}(i+1)^m[2^m-1]\bar{H} = \rho(i+1)^m.$$

The proof is completed.

Therefore, under the imposed conditions, there is no explosion (for each initial distribution P_0 under each control strategy S) by Theorem 5.3.

Clearly, this model covers all birth-and-death processes with (for $j \neq i$)

$$q(j|i,a) = \lambda_i(a)\mathbb{I}\{j=i+1\} + \mu_i(a)\mathbb{I}\{j=i-1\}$$

when $\sup_{i\in\mathbf{X},a\in\mathbf{A}} \frac{\lambda_i(a)+\mu_i(a)}{i+1} < \infty$. □

Example: Birth-and-death model. Suppose $\mathbf{X} = \{0,1,2,\ldots\}$ and the Borel action space $\mathbf{A}$ is arbitrary. Let

$$q(j|i,a) = \lambda_i(a)\mathbb{I}\{j=i+1\} + \mu_i(a)\mathbb{I}\{j=i-1\} \quad \text{for } j\neq i.$$

Here $\mu_0(a) \equiv 0$. We assume that $\sup_{a\in\mathbf{A}}\{\lambda_i(a)+\mu_i(a)\} < \infty$ for each $i \in \mathbf{X}$. There are no conditions on the behavior of the jump intensity $\lambda_i(a)+\mu_i(a)$ as $i \to \infty$, but the dynamics is supposed to have a negative trend: for some $I \in \{0,1,\ldots\}$

$$\inf_{i\ge I, a\in\mathbf{A}} \frac{\mu_i(a)+1}{\lambda_i(a)+1} = h > 1.$$

Then Condition 5.2 is satisfied for

$$V_n = \{0,1,2,\ldots,n\},\ w(i) = h^{i+1},\ \rho = 0,\ b = \max\{0, b_0, b_1, \ldots, b_{I-1}\},$$

where

$$b_i = \sup_{a\in\mathbf{A}} h^i\{\lambda_i(a)h^2 + \mu_i(a) - [\lambda_i(a)+\mu_i(a)]h\}.$$

Conditions 5.2(a,b,c) are trivially satisfied. For part (d), the elementary calculations imply

$$\sum_{j\in\mathbf{X}} q(j|i,a)h^{j+1} = h^i[\lambda_i(a)h^2 + \mu_i(a) - \{\lambda_i(a)+\mu_i(a)\}h]$$
$$\le h^i[(\lambda_i(a)+1)h^2 - \{\lambda_i(a)+\mu_i(a)+2\}h + (\mu_i(a)+1)]$$
$$= h^i(\lambda_i(a)+1)\left[h^2 - \left(\frac{\mu_i(a)+1}{\lambda_i(a)+1}+1\right)h + \frac{\mu_i(a)+1}{\lambda_i(a)+1}\right].$$

Since $h > 1$, the last square bracket decreases as $\frac{\mu_i(a)+1}{\lambda_i(a)+1}$ increases. Therefore, for $i \geq I$,

$$\sum_{j \in \mathbf{X}} q(j|i,a)h^{j+1} \leq h^i(\lambda_i(a)+1)[h^2 - (h+1)h + h] = 0,$$

and the proof follows.

Therefore, under the imposed conditions, there is no explosion (for each initial distribution P_0 under each control strategy S) by Theorem 5.3. $\square$

Example. Let $\mathbf{X} = \{0, 1, 2, \ldots\}$, $\mathbf{A} = (0, \infty)$, $q(j|i,a) = \begin{cases} i^a, & \text{if } j = i+1; \\ -i^a, & \text{if } j = i; \\ 0, & \text{otherwise.} \end{cases}$ The cost rate $c(\cdot,\cdot)$ is of no importance.

Consider a deterministic stationary strategy $\varphi^m(i) \equiv m \in \mathbf{A}$, that is, the controlled process $X(\cdot)$ is just the homogeneous pure birth process with the transition rate $q(j|i) = \mathbb{I}\{j = i+1\}i^m - \mathbb{I}\{j = i\}i^m$. See Fig 5.7.

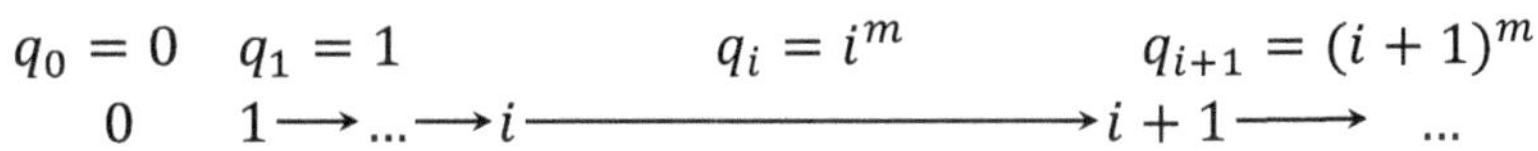

Fig. 5.7 Explosion in a pure birth process.

Equation (5.15) takes the form

$$\alpha u(i) = i^m u(i+1) - i^m u(i), \quad i = 0, 1, 2, \ldots$$

Obviously, for an arbitrarily fixed $\alpha > 0$, function

$$u(i) = \begin{cases} 0, & \text{if } i = 0, \\ k \prod_{j=1}^{i-1} \frac{\alpha + j^m}{j^m}, & \text{if } i \geq 1 \end{cases}$$

is a solution, where $u(1) = k \geq 0$ is arbitrarily fixed, and any nonnegative solution has this form.

Note that

$$K := \prod_{j=1}^{\infty} \frac{\alpha + j^m}{j^m} = \prod_{j=1}^{\infty} \left(1 + \frac{\alpha}{j^m}\right) \begin{cases} = \infty, & \text{if } m \leq 1, \\ \in (0, \infty), & \text{if } m > 1, \end{cases}$$

because the series $\sum_{j=1}^{\infty} \frac{\alpha}{j^m}$ diverges or converges for the corresponding values of m: $\lim_{j\to\infty} \ln\left(1 + \frac{\alpha}{j^m}\right) / \left(\frac{\alpha}{j^m}\right) = 1$ and, by the limit comparison test, $\prod_{j=1}^{\infty} \left(1 + \frac{\alpha}{j^m}\right) < \infty \Longleftrightarrow \sum_{j=1}^{\infty} \ln\left(1 + \frac{\alpha}{j^m}\right) < \infty \Longleftrightarrow \sum_{j=1}^{\infty} \frac{\alpha}{j^m} < \infty$.

Therefore, equation (5.15) has no bounded nontrivial solutions if and only if $m \le 1$. In case $m > 1$, one can always normalize the nonnegative function $u(\cdot)$ and take

$$u(i) = \begin{cases} 0, & \text{if } i = 0, \\ \frac{1}{K} \prod_{j=1}^{i-1} \frac{\alpha + j^m}{j^m}, & \text{if } i \ge 1. \end{cases}$$

According to Theorem 5.5(a), the controlled process $X(\cdot)$ is non-explosive for all initial states x_0 if and only if $m \le 1$.

Let us remind that the standard deterministic analogue of the considered birth process is given by the ordinary differential equation $\frac{dx}{dt} = x^m$, and, if $x(0) = x_0 > 0$ then, for $m > 0$, $m \ne 1$, $x(t) = \left[(1-m)t + x_0^{1-m}\right]^{\frac{1}{1-m}}$. For any $m < 1$, $x(t)$ is well defined for all $t \ge 0$, but in case $m > 1$, $\lim_{t \uparrow x_0^{1-m}/(m-1)} x(t) = \infty$, which means explosion. The special case of $m = 1$ is trivial: $x(t) = x_0 e^t$ is finite for all $t \ge 0$. □

5.3.5 *Reduction to discrete-time MDP: total cost problem*

Example. Consider again the last example from Subsection 5.3.1, see Fig. 5.2. It shows that Condition 5.3(b) is important for the justification of the reduction to DTMDP as in Theorem 5.7 and below it. Here $\mathbf{X} = \{1, \Delta\}$, $\mathbf{A} = (0, 1]$, $\alpha = 0$, $\mathsf{P}_0(1) = 1$, $q_x(a) = a > 0$ for $x \in \mathbf{X}^\Delta = \{1\}$, but this jump intensity is not separated from zero and the cost rate $c(1, a) = a$ is positive. Note that Condition 5.3(a) is satisfied and the statement of Theorem 5.7 is valid, but randomized Markov strategies are not sufficient in the total cost problem (5.7).

The cemetery state □ in the induced DTMDP can be omitted as it is never achieved in case $\alpha = 0$; $p_0(\Delta|1, a) = \frac{a}{a} = 1$; $l(1, a) = \frac{a}{a} = 1$; $p_0(\Delta|\Delta, a) = 1 - p_0(1|\Delta, a) = 1$; $l(\Delta, a) = 0$; $\mathsf{P}_0(1) = 1$. For each (Markov) strategy σ^{m} in this DTMDP, $X_1 = \Delta$ and $\eta_{\mathsf{P}_0,1}^{\sigma^{\mathrm{m}},0}(\{1\} \times da) = \sigma_1^{\mathrm{m}}(da|1)$. The total expected cost equals $\int_{\mathbf{A}} l(1, a) \eta_{\mathsf{P}_0,1}^{\sigma^{\mathrm{m}},0}(\{1\} \times da) = 1$ and coincides with the minimal total expected cost by Lemma 2.1: any strategy σ^{m} is optimal.

In accordance with Theorem 5.7,

$$m_{\mathsf{P}_0,1}^{\varpi^{\mathrm{m}},0}(\{1\} \times da) = \frac{1}{a} \varpi_1^{\mathrm{m}}(da|1) = \frac{1}{a} \eta_{\mathsf{P}_0,1}^{\sigma^{\mathrm{m}},0}(\{1\} \times da)$$

for a (Markov) randomized strategy $\varpi^{\mathrm{m}} = \sigma^{\mathrm{m}}$. For each ϖ^{m}, $W^0(\varpi^{\mathrm{m}}, \mathsf{P}_0) = 1$: the minimal value of performance (5.3) in the class of randomized Markov strategies coincides with the minimal total expected cost in the induced DTMDP. At the same time, as was shown in Subsection 5.3.1, in the underlying CTMDP $\inf_S W^0(S, \mathsf{P}_0) = 0$. Solution to the induced DTMDP does not provide solution to the CTMDP. Reduction to DTMDP is unsuccessful because Condition 5.3(b) is not satisfied and the cost rate $c(\cdot, \cdot)$ is positive. A similar example was published in [Guo and Zhang (2017), Ex. 3.1].

Suppose now that in this CTMDP $\alpha > 0$. In the induced DTMDP, $p_\alpha(\Delta|1,a) = \frac{a}{\alpha+a}$, $p_\alpha(\Box|1,a)\frac{\alpha}{\alpha+a}$, $p_\alpha(\Delta|\Delta,a) \equiv 0$, $p_\alpha(\Box|\Delta,a) \equiv 1$, $p_\alpha(\Box|\Box,a) \equiv 1$, $l(1,a) = \frac{a}{\alpha+a}$, $l(\Delta,a) = l(\Box,a) \equiv 0$, $\mathsf{P}_0(1) = 1$. For each (Markov) strategy σ^{m} in this DTMDP, $l(X_n,a) \equiv 0$ at $n \geq 1$ and $\eta^{\sigma^{\mathrm{m}},\alpha}_{\mathsf{P}_0,1}(\{1\}\times da) = \sigma^{\mathrm{m}}_1(da|1)$. The total expected cost equals

$$\int_{\mathbf{A}} l(1,a)\eta^{\sigma^{\mathrm{m}},\alpha}_{\mathsf{P}_0,1}(\{1\}\times da) = \int_{\mathbf{A}} \frac{a}{\alpha+a}\sigma^{\mathrm{m}}_1(da|1),$$

and the infimum equals zero: for $\sigma^{\mathrm{m},k}_1(da|1) = \delta_{\frac{1}{k}}(da)$, $k = 1,2,\ldots$, that cost equals $\frac{1}{\alpha k+1}$ approaching zero as $k \to \infty$.

In accordance with Theorem 5.7,

$$m^{\varpi^{\mathrm{m}},\alpha}_{\mathsf{P}_0,1}(\{1\}\times da) = \frac{1}{\alpha+a}\varpi^{\mathrm{m}}_1(da|1) = \frac{1}{\alpha+a}\eta^{\sigma^{\mathrm{m}},0}_{\mathsf{P}_0,1}(\{1\}\times da)$$

for a (Markov) randomized strategy $\varpi^{\mathrm{m}} = \sigma^{\mathrm{m}}$. For the sequence of strategies with $\varpi^{\mathrm{m},k}_1(da|1) = \sigma^{\mathrm{m},k}_1(da|1) = \delta_{\frac{1}{k}}(da)$ we have

$$\begin{aligned} W^\alpha(\varpi^{\mathrm{m},k},\mathsf{P}_0) &= \int_{\mathbf{A}} c(1,a)m^{\varpi^{\mathrm{m},k},\alpha}_{\mathsf{P}_0,1}(\{1\}\times da) = \int_{\mathbf{A}} a\cdot\frac{1}{\alpha+a}\delta_{\frac{1}{k}}(da) \\ &= \frac{1}{\alpha k+1} \longrightarrow 0 = \inf_S W^\alpha(S,\mathsf{P}_0) \quad \text{as } k \to \infty. \end{aligned}$$

Solution to the induced DTMDP provides the solution to the underlying CTMDP. Reduction is successful because Condition 5.3(b) is satisfied when $\alpha > 0$.

There is always a possibility to satisfy Condition 5.3(b) by splitting the states in which the jump intensity $q_x(a)$ is not separated from zero. This idea is strictly connected with introducing the so called Poisson-related strategies [Piunovskiy and Zhang (2020), Ch. 4] and was used for investigation of piece-wise deterministic Markov decision processes in [Guo and Zhang (2021)].

Suppose again that $\alpha = 0$ and replace state 1 with states 1 and 1' in the following way: $q(1'|1,a) = q(1|1',a) = 1$, $q(\Delta|1,a) = q(\Delta|1',a) = a$, $c(1,a) = c(1',a) = a$. The initial distribution P_0 is arbitrary, concentrated on $\{1,1'\}$. See Fig. 5.8. When the original chain is in state 1, one can mimic the transitions to 1' and back. As the result, the randomized Markov strategy in the modified model corresponds to the strategy in the original CTMDP which can change with time on the interval with constant $X(t) = 1$. Randomized Markov strategies are sufficient by Theorem 5.6, and the optimal strategy comes from the reduction to DTMDP.

The cemetery state $\Box$ in the induced DTMDP can be omitted as it is never achieved in case $\alpha = 0$; $p_0(1'|1,a) = p_0(1|1',a) = \frac{1}{1+a}$, $p_0(\Delta|1,a) = p_0(\Delta|1',a) = \frac{a}{1+a}$; $l(1,a) = l(1',a) = \frac{a}{1+a}$; $p_0(\Delta|\Delta,a) = 1 - p_0(\mathbf{X}^\Delta|\Delta,a) = 1$; $l(\Delta,a) = 0$. See Fig. 5.9.

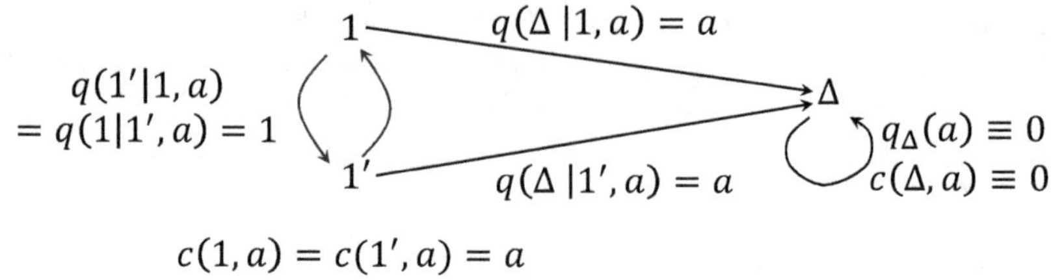

Fig. 5.8 Splitting a state.

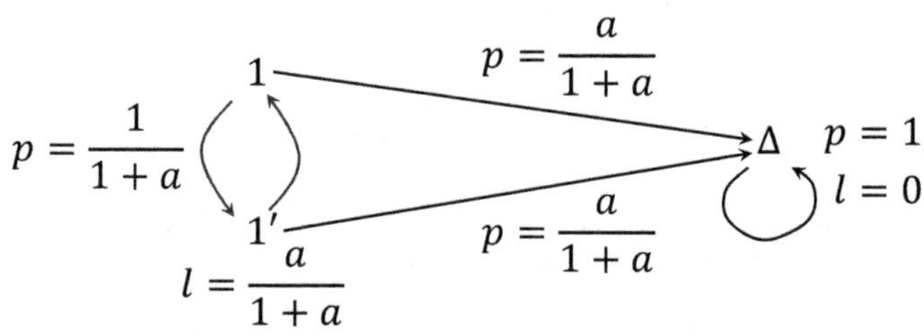

Fig. 5.9 Induced DTMDP.

The Bellman function v_x^* in the induced DTMDP is the minimal non-negative solution to the optimality equation

$$\begin{aligned} v_1^* &= \inf_{a\in(0,1]} \left\{ \frac{a}{1+a} + \frac{1}{1+a} v_{1'}^* + \frac{a}{1+a} v_\Delta^* \right\}; \\ v_{1'}^* &= \inf_{a\in(0,1]} \left\{ \frac{a}{1+a} + \frac{1}{1+a} v_1^* + \frac{a}{1+a} v_\Delta^* \right\}; \\ v_\Delta^* &= 0 + v_\Delta^* \end{aligned}$$

according to Theorem 2.3 and the paragraph below it. Thus $v_x^* \equiv 0$.

If σ^{s} is a stationary strategy in the induced DTMDP, then, for the performance functional (2.1), we have $v_1^{\sigma^{\text{s}}} = v_{1'}^{\sigma^{\text{s}}} = 1$. To show this, based on Remark 2.2, and putting $v_\Delta^{\sigma^{\text{s}}} = 0$, one can solve equations

$$\begin{aligned} v_1^{\sigma^{\text{s}}} &= \int_{\mathbf{A}} l(1,a)\sigma^{\text{s}}(da|1) + \int_{\mathbf{A}} p_O(1'|1,a)\sigma^{\text{s}}(da|1) v_{1'}^{\sigma^{\text{s}}}; \\ v_{1'}^{\sigma^{\text{s}}} &= \int_{\mathbf{A}} l(1',a)\sigma^{\text{s}}(da|1') + \int_{\mathbf{A}} p_O(1|1',a)\sigma^{\text{s}}(da|1') v_1^{\sigma^{\text{s}}} \end{aligned}$$

which have the form

$$v_1^{\sigma^{\text{s}}} = (1-k_1) + k_1 v_{1'}^{\sigma^{\text{s}}}; \quad v_{1'}^{\sigma^{\text{s}}} = (1-k_2) + k_2 v_1^{\sigma^{\text{s}}},$$

where

$$k_1 := \int_{\mathbf{A}} \frac{1}{1+a} \sigma^{\mathrm{s}}(da|1), \quad k_2 := \int_{\mathbf{A}} \frac{1}{1+a} \sigma^{\mathrm{s}}(da|1').$$

As the result, since $k_1 k_2 < 1$,

$$v_1^{\sigma^{\mathrm{s}}} = (1-k_1) + k_1 \left[(1-k_2) + k_2 v_1^{\sigma^{\mathrm{s}}}\right] \Longrightarrow v_1^{\sigma^{\mathrm{s}}} = 1$$

and similarly $v_{1'}^{\sigma^{\mathrm{s}}} = 1$. To summarize, there are no stationary ε-optimal strategies in the induced DTMDP if $\varepsilon < 1$.

Now consider the following non-stationary Markov selector φ^{m} in the induced DTMDP: $\varphi_n^{\mathrm{m}}(x) \equiv a_n$, where $\{a_n\}_{n=1}^{\infty}$ is some sequence in $\mathbf{A}$. The values of $\eta_{\mathsf{P}_0,n}^{\varphi^{\mathrm{m}},0}(\Delta \times da)$ are of no importance as $l(\Delta, a) \equiv 0$, $p_0(\Delta|\Delta, a) \equiv 1$, and the values of $\eta_{\mathsf{P}_0,n}^{\varphi^{\mathrm{m}},0}(\{1,1'\} \times da)$ can be calculated in the following way:

$$\eta_{\mathsf{P}_0,n}^{\varphi^{\mathrm{m}},0}(\{1,1'\} \times da) = \prod_{j=1}^{n-1} \frac{1}{1+a_j} \cdot \delta_{a_n}(da), \quad n = 1, 2, \ldots.$$

This expression is obviously valid when $n = 1$ because $\mathsf{P}_0(\{1,1'\}) = 1$. Suppose it holds for some n and consider the case of $n+1$:

$$\eta_{\mathsf{P}_0,n+1}^{\varphi^{\mathrm{m}},0}(\{1,1'\} \times da) = \eta_{\mathsf{P}_0,n+1}^{\varphi^{\mathrm{m}},0}(\{1,1'\} \times \mathbf{A}) \delta_{a_{n+1}}(da),$$

and

$$\begin{aligned} \eta_{\mathsf{P}_0,n+1}^{\varphi^{\mathrm{m}},0}(\{1,1'\} \times \mathbf{A}) &= \mathbb{P}_{\mathsf{P}_0}^{\varphi^{\mathrm{m}}}(X_{n-1} = 1) p_0(1'|1, a_n) \\ &\quad + \mathbb{P}_{\mathsf{P}_0}^{\varphi^{\mathrm{m}}}(X_{n-1} = 1') p_0(1|1', a_n) \\ &= \mathbb{P}_{\mathsf{P}_0}^{\varphi^{\mathrm{m}}}(X_{n-1} \in \{1,1'\}) \frac{1}{1+a_n} \\ &= \eta_{\mathsf{P}_0,n}^{\varphi^{\mathrm{m}},0}(\{1,1'\} \times \mathbf{A}) \frac{1}{1+a_n} = \prod_{j=1}^{n} \frac{1}{1+a_j}. \end{aligned}$$

The proof is completed. Therefore,

$$\begin{aligned} v_{\mathsf{P}_0}^{\varphi^{\mathrm{m}}} &= \int_{\mathbf{X} \times \mathbf{A}} l(x,a) \eta_{\mathsf{P}_0}^{\varphi^{\mathrm{m}},0}(dx \times da) = \sum_{n=1}^{\infty} \int_{\mathbf{A}} \frac{a}{1+a} \eta_{\mathsf{P}_0,n}^{\varphi^{\mathrm{m}},0}(\{1,1'\} \times da) \\ &= \sum_{n=1}^{\infty} \frac{a_n}{1+a_n} \prod_{j=1}^{n-1} \frac{1}{1+a_j} \leq \sum_{n=1}^{\infty} \frac{a_n}{1+a_n}. \end{aligned}$$

For an arbitrarily fixed $\varepsilon > 0$ take such a sequence $\{a_n\}_{n=1}^{\infty}$ that $\frac{a_n}{1+a_n} < \varepsilon \left(\frac{1}{2}\right)^n$. Then $v_{\mathsf{P}_0}^{\varphi^{\mathrm{m}}} < \varepsilon$, and the Markov selector φ^{m} is ε-optimal. The corresponding non-stationary Markov deterministic strategy φ^{m} in the underlying CTMDP with the split state 1 (Fig. 5.8) is also ε-optimal: reduction to DTMDP is successful.

For $\varepsilon < 1$ and the above described Markov selector $\varphi^{\rm m}$ in the induced DTMDP,

$$\eta_{\rm P_0}^{\varphi^{\rm m},0}(\{1,1'\}\times\mathbf{A}) = \sum_{n=1}^{\infty}\eta_{{\rm P_0},n}^{\varphi^{\rm m},0}(\{1,1'\}\times\mathbf{A}) > \sum_{n=1}^{\infty}\prod_{j=1}^{n-1}\left(1-\varepsilon\left(\frac{1}{2}\right)^j\right).$$

Let us show that $\prod_{j=1}^{n-1}\left(1-\varepsilon\left(\frac{1}{2}\right)^j\right) \geq \frac{\varepsilon}{n}$ for all $n = 1, 2, \ldots$. This inequality is obvious at $n = 1$. If it is valid for some $n = 1, 2, \ldots$, then

$$\prod_{j=1}^{n}\left(1-\varepsilon\left(\frac{1}{2}\right)^j\right) \geq \frac{\varepsilon}{n}\left(1-\varepsilon\left(\frac{1}{2}\right)^n\right) > \frac{\varepsilon}{n}\cdot\frac{n}{n+1}$$

because $\varepsilon\left(\frac{1}{2}\right)^n < \frac{1}{n+1}$ for $n \geq 1$ leading to $\left(1-\varepsilon\left(\frac{1}{2}\right)^n\right) > 1 - \frac{1}{n+1} = \frac{n}{n+1}$. Therefore, $\eta_{\rm P_0}^{\varphi^{\rm m},0}(\{1,1'\}\times\mathbf{A}) > \varepsilon\sum_{n=1}^{\infty}\frac{1}{n} = \infty$: the induced DTMDP is not absorbing (at Δ) and Lemma 2.4 about sufficiency of stationary strategies is not applicable. □

5.3.6 *When Dynkin's formula is not valid*

In this subsection, we assume that a stationary π-strategy $\pi^{\rm s}$ is fixed.

By formula (5.1), the signed kernel $q(dy|x,\pi^{\rm s}) := \int_{\mathbf{A}} q(dy|x,a)\pi^{\rm s}(da|x)$ does not depend on the time parameter s and is called (stationary) *Q-function* in what follows.

Definition 5.6.

(a) A $(0,\infty)$-valued measurable function $V(\cdot)$ on $\mathbf{X}$ is called *0-drift* function with respect to q if

$$\int_{\mathbf{X}} V(y)q(dy|x,\pi^{\rm s}) \leq 0, \quad x \in \mathbf{X}.$$

(b) Let $\mathbf{X}_\delta := \mathbf{X}\cup\{\delta\}$, where $\delta \notin \mathbf{X}$, $\delta \neq x_\infty$ is an isolated point. For a 0-drift function $V(\cdot)$ with respect to q, the *V-transformed* Q-function q^V on $\mathbf{X}_\delta$ is defined by

$$q^V(\Gamma|x,\pi^{\rm s}) = \begin{cases} \dfrac{\int_\Gamma V(y)q(dy|x,\pi^{\rm s})}{V(x)}, & \text{if } x\in\mathbf{X},\ \Gamma\in\mathcal{B}(\mathbf{X}); \\[2ex] -\dfrac{\int_{\mathbf{X}} V(y)q(dy|x,\pi^{\rm s})}{V(x)}, & \text{if } x\in\mathbf{X},\ \Gamma=\{\delta\}; \\[2ex] 0, & \text{if } x=\delta. \end{cases}$$

As usual, $q_x^V(\pi^s) := -q^V(\{x\}|x,\pi^s) = q_x(\pi^s)$.

For a 0-drift function $V(\cdot)$ w.r.t. q,

$$\mathsf{E}_x^{\pi^s}[V(X(t))] \le V(x), \quad x \in \mathbf{X},\ t \ge 0 \tag{5.36}$$

by Lemma 2.4.2 of [Piunovskiy and Zhang (2020)], and the Dynkin formula (5.10) takes the form

$$\mathsf{E}_x^{\pi^s}[V(X(t))] - V(x) = \int_{(0,t]} \mathsf{E}_x^{\pi^s}\left[\int_{\mathbf{X}} V(y)q(dy|X(v),\pi^s)\right] dv, \quad x \in \mathbf{X},\ t \ge 0. \tag{5.37}$$

If equality (5.37) is valid, then

$$\int_{(0,t]} \mathsf{E}_x^{\pi^s}\left[\left|\int_{\mathbf{X}} V(y)q(dy|X(v),\pi^s)\right|\right] dv < \infty, \quad x \in \mathbf{X},\ t \ge 0$$

by Lemma 2.4.7 of [Piunovskiy and Zhang (2020)], and the interchange of the integration w.r.t. v and the mathematical expectation is legal by the Fubini theorem. By the way, conditions

$$\left.\begin{array}{ll} \mathsf{E}_x^{\pi^s}[V(X(t))] < \infty, & \forall\, t \ge 0\ x \in \mathbf{X}; \\ \mathsf{E}_x^{\pi^s}\left[\int_{(0,t]} \left|\int_{\mathbf{X}} V(y)q(y|X(s))\right| ds\right] < \infty, & \forall\, t \ge 0,\ x \in \mathbf{X} \end{array}\right\} \tag{5.38}$$

are necessary for the Dynkin formula (5.37).

Theorem 5.10. [Piunovskiy and Zhang (2020), Thm. 2.4.2]. *For a 0-drift function $V(\cdot)$ w.r.t q the Dynkin formula (5.37) holds if and only if the jump Markov process in $\mathbf{X}_\delta$, governed by the V-transformed Q-function q^V, is non-explosive under all initial states $x \in \mathbf{X}_\delta$.*

The following example published in [Spieksma (2016)] shows that in general conditions (5.38) are not sufficient for (5.37), so that the statement C.3 of [Guo and Hernández-Lerma (2009)] and Theorem 2.2 in Ch. 9 of [Brémaud (2000)] are inaccurate.

Example. Let $\mathbf{X} = \{0, 1, 2, \dots\}$, $\mathbf{A} = \{0\}$ (the dummy action meaning that the process is not controlled). The component a can be everywhere omitted, and for the single stationary π-strategy $\pi^s(0|x) = 1$, the Q-function $q(y|x,\pi^s)$, where we again omit the component π^s, is given by

$$q(y|x) = \begin{cases} \frac{5}{12}2^x, \text{ if } x \neq 0, \ y = x+1; \\ \frac{7}{12}2^x, \text{ if } x \neq 0, \ y = x-1; \\ -2^x, \quad \text{if } x \neq 0, \ y = x; \\ 0, \qquad \text{otherwise.} \end{cases}$$

See Fig. 5.10.

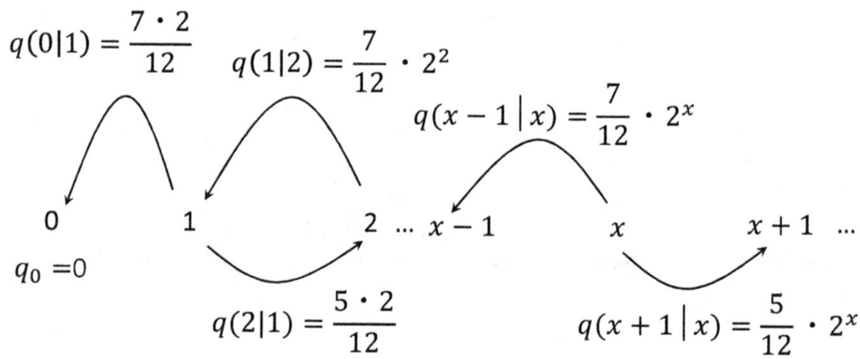

Fig. 5.10 Birth-and-death process.

Consider the function $V(\cdot)$ on $\mathbf{X}$ defined by

$$V(x) = \left(\frac{7}{5}\right)^x, \ \forall\, x \in \mathbf{X}.$$

Then one can verify that

$$\sum_{y \in \mathbf{X}} V(y) q(y|x) = 0, \ \forall\, x \in \mathbf{X},$$

i.e., $V(\cdot)$ is a 0-drift function w.r.t. q. It is clear that Condition 5.2 is satisfied with $V_l := \{0, 1, \ldots, l\}$, $w(\cdot)$ being replaced by $V(\cdot)$, and $\rho = b = 0$. From Theorem 5.3, we see that the Markov pure jump process corresponding to the Q-function $q(y|x)$ is non-explosive.

Now the V-transformed Q function q^V on $\mathbf{X}_\delta$ is defined by

$$q^V(y|x) = \begin{cases} \frac{7}{12}2^x, & \text{if } x \neq \delta,\ x \neq 0,\ y = x+1; \\ \frac{5}{12}2^x, & \text{if } x \neq \delta,\ x \neq 0,\ y = x-1; \\ -2^x, & \text{if } x \neq \delta,\ x \neq 0,\ y = x; \\ 0, & \text{otherwise.} \end{cases}$$

We shall verify that the jump Markov process in $\mathbf{X}_\delta$, governed by the V-transformed Q-function q^V is explosive under some initial states $x \in \mathbf{X}$ (in fact, under all $x > 0$). The isolated point δ is absorbing and is not accessible from any other states. Thus, we can ignore it and consider, without loss of generality, q^V as a Q-function on $\mathbf{X}$, still denoted as q^V. This Q-function q^V is *upwardly skip-free* in the sense that $q^V(x+1|x) > 0$ for each $x = 1, 2, \ldots$, and $q^V(y|x) = 0$ for each $x = 0, 1, 2, \ldots$, and $y > x+1$. For such Q-functions, the following criterion for the non-explosion is known [Chen *et al.* (2005), Thm. 2].

Proposition 5.2. *Let $R_0 = 1$, and for each $n = 1, 2, \ldots$ put*

$$R_n := \frac{1}{q^V(n+1|n)}\left(1 + \sum_{m=1}^{n}\sum_{k=0}^{m-1} q^V(k|n)R_{m-1}\right).$$

Then the jump Markov process governed by the Q-function q^V is non-explosive under all initial states $x \in \mathbf{X}$ if and only if

$$\sum_{n=1}^{\infty} R_n = \infty.$$

Note that in our case, we have

$$R_n = \frac{1}{q^V(n+1|n)}\left(1 + q^V(n-1|n)R_{n-1}\right) = \frac{12}{7}\frac{1}{2^n} + \frac{5}{7}R_{n-1},\ \forall\, n = 1, 2, \ldots.$$

Simple iterations result in

$$R_n = \sum_{i=0}^{n-1}\left(\frac{5}{7}\right)^i \frac{12}{7}\frac{1}{2^{n-i}} + \left(\frac{5}{7}\right)^n \leq \sum_{i=0}^{n}\left(\frac{5}{7}\right)^i \frac{12}{7}\frac{1}{2^{n-i}},\ \forall\, n = 1, 2, \ldots.$$

Thus,

$$\sum_{n=1}^{\infty} R_n \leq \sum_{n=0}^{\infty}\sum_{i=0}^{n}\left(\frac{5}{7}\right)^i \frac{12}{7}\frac{1}{2^{n-i}} = \frac{12}{7}\sum_{i=0}^{\infty}\sum_{n=i}^{\infty}\left(\frac{5}{7}\right)^i \frac{1}{2^{n-i}} = 12 < \infty.$$

Thus, by the criterion stated in the above, we see that the jump Markov process in $\mathbf{X}_\delta$, governed by the V-transformed Q-function q^V is explosive under some initial states $x \in \mathbf{X}$.

By Theorem 5.10, the Dynkin formula (5.37) does not hold for the 0-drift function $V(\cdot)$, even though conditions (5.38) are satisfied. (Recall that $\mathsf{E}_x^{\pi^s}[V(X(t))] \le V(x)$ by (5.36) and $\sum_{y\in\mathbf{X}} V(y)q(y|x) = 0$.)

According to Theorems 5.2 and 5.3, Condition 5.1 with $w'(\cdot)$ satisfying inequality $\inf_{x\in\mathbf{X}} \frac{w'(x)}{V(x)} = \inf_{x\in\mathbf{X}} w'(x)\left(\frac{5}{7}\right)^x > 0$ and Condition 5.2 for the Q-function q^V are not fulfilled. □

5.3.7 *Optimality equations for the undiscounted total cost problem*

The following simple example reported by Dr. Y.Zhang in a private conversation shows that, if the sum $\alpha + q_x(a)$ is positive but not separated from zero, then equations (5.13) and (5.22) are not equivalent in the space of finite-valued functions $V(\cdot)$.

Example. Let $\mathbf{X} = \{1, 2, \dots\}$ and $\mathbf{A} = (0, 1]$; $q(i+1|i, a) = a$, $q_i(a) \equiv a > 0$ for all $i \in \mathbf{X}$, with all the other transition rates being zero; $c(i, a) \equiv 0$, and put $\alpha = 0$. See Fig. 5.11.

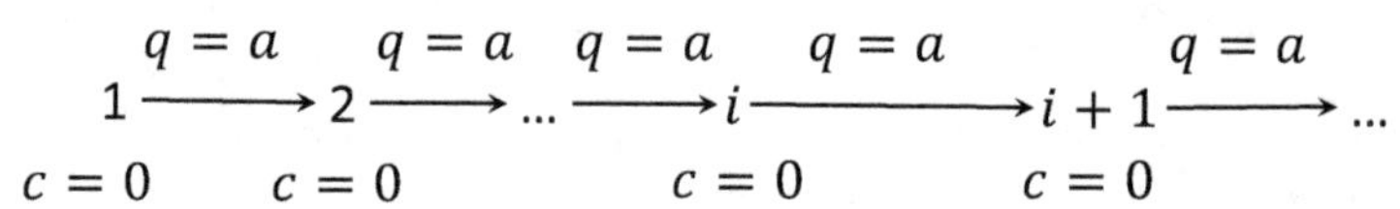

Fig. 5.11 Equations (5.22) and (5.13) are not equivalent.

Let $W(\cdot)$ be a bounded real-valued function on $\mathbf{X}$ such that $W(i)$ is strictly increasing in i. Then

$$\inf_{a\in\mathbf{A}}\left\{c(i,a) + \sum_{j\in\mathbf{X}} W(j)q(j|i,a)\right\} = \inf_{a\in(0,1]}\{a(W(i+1) - W(i))\} = 0$$

for all $i \in \mathbf{X}$. So $W(\cdot)$ satisfies equation (5.13).

On the other hand,

$$\inf_{a\in\mathbf{A}}\left\{\frac{c(i,a)}{q_i(a)} + \sum_{j\in\mathbf{X}} W(j)\frac{\tilde{q}(j|i,a)}{q_i(a)}\right\} = W(i+1) \neq W(i).$$

for all $i \in \mathbf{X}$. So $W(\cdot)$ does not satisfy equation (5.22).

Although Theorem 5.8 is not applicable, the Bellman function $W^{*0}(\cdot)$ obviously coincides with the minimal nonnegative solution to the equations (5.22) and (5.13): $W^{*0}(i) \equiv V(i) \equiv 0$. □

5.3.8 *Average uniformly optimal deterministic stationary strategies*

Example. Let $\mathbf{X} = \{0, 1, \dots\}$ and $\mathbf{A} = \{1, 2\}$. The transition and cost rates are defined by

$$\begin{aligned} &q_x(a) = -q(\{x\}|x, a) = 1, \ \forall\, x \in \mathbf{X},\ a \in \mathbf{A}, \\ &q(x-1|x, a) = 1,\ q(x|0, a) = p_x,\ \forall\, x \in \{1, 2, \dots\},\ a \in \mathbf{A}, \\ &c(x, a) = 1,\ \forall\, x \in \{1, 2, \dots\},\ a \in \mathbf{A}, \\ &c(0, a) = \mathbb{I}\{a = 2\},\ \forall\, a \in \mathbf{A}, \end{aligned}$$

where $0 \le p_x$ for all $x = 1, 2, \dots$, and $\sum_{y=1}^{\infty} p_y = 1$. Other transition rates are zero. See Fig. 5.12. The similar example was published in Subsubsection 5.1.1.4 of [Piunovskiy and Zhang (2020)], see also Example 5.2 of [Guo and Hernández-Lerma (2009)].

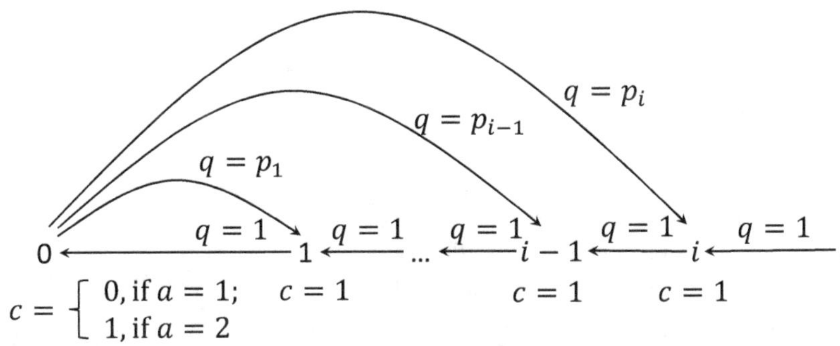

Fig. 5.12 Equality in the first line of formula (5.29) is not achieved.

Let $\alpha > 0$ be fixed for now and denote $\beta := \frac{1}{1+\alpha} < 1$. It is clear that a deterministic stationary (α-discounted) uniformly optimal strategy is given by $\varphi^*(x) \equiv 1$ because the actions affect only the cost rate $c(0, a)$.

Equation (5.13) takes the form

$$\begin{aligned} \alpha V(i) &= 1 + V(i-1) - V(i), \quad \text{if } i \ge 1; \\ \alpha V(0) &= 0 + \sum_{i=1}^{\infty} p_i V(i) - V(0) \end{aligned}$$

and has the following unique non-negative finite-valued solution

$$V(i) = \frac{\beta}{1-\beta}\left[1-\beta^i+\beta^{i+1}\cdot\frac{1-\sum_{j=1}^{\infty}p_j\beta^j}{1-\beta\sum_{j=1}^{\infty}p_j\beta^j}\right], \quad i\geq 1;$$
$$V(0) = \frac{\beta^2}{1-\beta}\cdot\frac{1-\sum_{j=1}^{\infty}p_j\beta^j}{1-\beta\sum_{j=1}^{\infty}p_j\beta^j}.$$

Since $\alpha > 0$, the presented function $V(\cdot)$ is also the unique non-negative finite-valued solution to equation (5.22) and coincides with the Bellman function $W^{*\alpha}(\cdot)$ by Theorem 5.8.

One can verify that $m_\alpha := \inf_{i\in\mathbf{X}} W^{*\alpha}(i) = W_0^{*\alpha}(0)$ and hence

$$\alpha m_\alpha = \beta\frac{1-\sum_{j=1}^{\infty}p_j\beta^j}{1-\beta\sum_{j=1}^{\infty}p_j\beta^j},$$
$$h_\alpha(i) := W^{*\alpha}(i)-m_\alpha = \frac{\beta(1-\beta^i)}{1-\beta\sum_{j=1}^{\infty}p_j\beta^j} > 0, \quad i=1,2,\ldots,$$
$$h_\alpha(0) = 0.$$

Applying L'Hôpital's rule we obtain

$$g = \lim_{\alpha\downarrow 0}\alpha m_\alpha = \lim_{\beta\uparrow 1}\beta\frac{1-\sum_{j=1}^{\infty}p_j\beta^j}{1-\beta\sum_{j=1}^{\infty}p_j\beta^j} = \frac{\sum_{j=1}^{\infty}jp_j}{1+\sum_{j=1}^{\infty}jp_j}, \text{ where } \frac{\infty}{\infty}:=1;$$
$$h(i) = \lim_{\alpha\downarrow 0}h_\alpha(i) = \lim_{\beta\uparrow 1}\frac{\beta(1-\beta^i)}{1-\beta\sum_{j=1}^{\infty}p_j\beta^j} = \frac{i}{1+\sum_{j=1}^{\infty}jp_j}, \ \forall\, i\in\mathbf{X}.$$

All the conditions of Theorem 5.9 are satisfied.

If $\sum_{j=1}^{\infty}jp_j < \infty$, then there are equalities in formula (5.29), and $\varphi^*(i)\equiv 1$ provides the infimum and thus, by Theorem 5.9, defines the average uniformly optimal deterministic stationary strategy; $W^{*A}(i)\equiv g = \frac{\sum_{j=1}^{\infty}jp_j}{1+\sum_{j=1}^{\infty}jp_j}$.

Suppose $\sum_{j=1}^{\infty}jp_j = \infty$. Then $g=1$, $h(i)\equiv 0$. Equality in the first line of (5.29) holds for all $i\geq 1$. But, at $i=0$, $g=1>0$ and the inequality in (5.29) is strict. The deterministic stationary strategy $\varphi^*(i)\equiv 1$ is average uniformly optimal providing the infimum in (5.29). At the same time, the strategy $\varphi(i)\equiv 2$ is also average uniformly optimal with $W^A(\varphi,i) = W^{*A}(i)\equiv g = 1$, although it does not achieve the infimum in (5.29). Equality (5.29) is sufficient but not necessary for the optimality of deterministic stationary strategy φ^*.

Let us show that, if $\sum_{j=1}^{\infty}jp_j=\infty$, then equation

$$g = \min\left\{c(i,a)+\sum_{j=0}^{\infty}h(j)q(j|i,a)\right\}$$

does not have finite solutions. Assume for contradiction that g' and $h'(\cdot)$ satisfy this equation. Then

$$g' = 1 + h'(i-1) - h'(i) \Longrightarrow h'(i) = i(1-g') + h'(0) \quad \forall i \geq 1$$

and at $i = 0$

$$g' = \sum_{j=1}^{\infty} h'(j)p_j - h'(0) = (1-g')\sum_{j=1}^{\infty} jp_j.$$

The case $g' = 1$ leading to $1 = 0$ is excluded, as well as any other value providing $\pm\infty$ on the right side. □

By the way, the presented example shows that Theorem 5.1.4 in [Piunovskiy and Zhang (2020)] is inaccurate.

5.3.9 *Linear programming approach to the average cost problem*

Under appropriate conditions, including Condition 5.1(a) with $\rho < 0$, one can solve problem (5.8) using the linear programming approach (see (5.28) of [Piunovskiy and Zhang (2020)]):

$$\text{Minimize} \quad \text{over } \eta \in \mathbf{P}(\mathbf{X} \times \mathbf{A}): \int_{\mathbf{X}\times\mathbf{A}} c(x,a)\eta(dx,da) \tag{5.39}$$

$$\text{subject to} \int_{\mathbf{X}} w(x)\eta(dx \times \mathbf{A}) < \infty \text{ and}$$

$$\int_{\mathbf{X}} \left(\int_{\mathbf{A}} q(\Gamma|x,a)\pi^{\mathrm{s}}(da|x) \right) \eta(dx \times \mathbf{A}) = 0, \tag{5.40}$$

for all $\Gamma \in \mathcal{B}(\mathbf{X})$ satisfying $\sup_{x\in\Gamma} \int_{\mathbf{A}} q_x(a)\pi^{\mathrm{s}}(da|x) < \infty$.

Here $w(\cdot)$ is the function from Condition 5.1(a) and $\pi^{\mathrm{s}} = \pi_\eta$ is the stochastic kernel from the decomposition $\eta(dx \times da) = \eta(dx \times \mathbf{A})\pi_\eta(da|x)$. Equality (5.40) is the characteristic equation of the *invariant* probability measure $\eta(dx \times \mathbf{A})$ of the controlled process $X(\cdot)$ under the relaxed stationary strategy $\pi^{\mathrm{s}}(da|x)$, e.g., in case $X(\cdot)$ is ergodic.

If η^* is the optimal solution to this linear program, then $\pi^{\mathrm{s}}(da|x) := \pi_{\eta^*}(da|x)$ is a uniformly average optimal stationary π-strategy. Precise conditions and rigorous formulations of the relevant statements can be found in Subsection 5.2.1 of [Piunovskiy and Zhang (2020)].

If the state and action spaces $\mathbf{X}$ and $\mathbf{A}$ are finite and, under each stationary π-strategy π^{s}, the controlled process $X(\cdot)$ is ergodic (i.e., all states communicate), then the similar linear program, called dual, is presented in Subsection 3.6.1 of [Guo and Hernández-Lerma (2009)]. In this special case all the conditions imposed in Subsection 5.2.1 of [Piunovskiy and Zhang (2020)] are satisfied with $w(x) \equiv 1$, and equalities $\int_{\mathbf{X}} w(x)\eta(dx \times \mathbf{A}) = 1 < \infty$ and $\sup_{x\in\Gamma} \int_{\mathbf{A}} q_x(a)\pi_\eta(da|x) < \infty$ for all $\Gamma \in \mathcal{B}(\mathbf{X})$ trivially hold because

$\sup_{(x,a)\in\mathbf{X}\times\mathbf{A}} q_x(a) < \infty$. Note also that, in the finite ergodic model, problem (5.39), (5.40) transforms to (5.23), (5.24), (5.25), with $\tilde{\eta} \equiv 0$, and Proposition 5.1 is applicable.

The following example published in [Piunovskiy and Zhang (2020), Ex. 5.2.1] illustrates that Condition 5.1(a) is important for the validity of the described linear programming approach.

Example. Let $\mathbf{X} = \{0, \pm 1, \pm 2, \dots\}$, $\mathbf{A} = \{a_1, a_2\}$ and let $0 < \mu < \lambda < 2\mu$ be fixed constants such that $\lambda + \mu = 1$. The transition rate is given by

$$
\begin{aligned}
q_0(a_1) &= q(\{1\}|0, a_1) = \lambda = q_0(a_2) = q(\{-1\}|0, a_2);\\
q_x(a) &= q(\{x-1\}|x, a) = 1, \ \forall\, x \in \{-1, -2, \dots\},\ a \in \mathbf{A};\\
q_x(a) &= 2^x, \quad q(\{x+1\}|x, a) = \lambda 2^x,\ q(\{x-1\}|x, a) = \mu 2^x,\\
&\qquad \forall\, x \in \{1, 2, \dots\},\ a \in \mathbf{A}.
\end{aligned}
$$

All the other transition rates are zero.

$$
\begin{aligned}
c(x, a) &= 0, \forall\, x \in \{0, 1, \dots\},\ a \in \mathbf{A};\\
c(x, a) &= -1, \forall\, x \in \{-1, -2, \dots\},\ a \in \mathbf{A},
\end{aligned}
$$

Note that the process is essentially controlled only at the state 0. See Fig. 5.13.

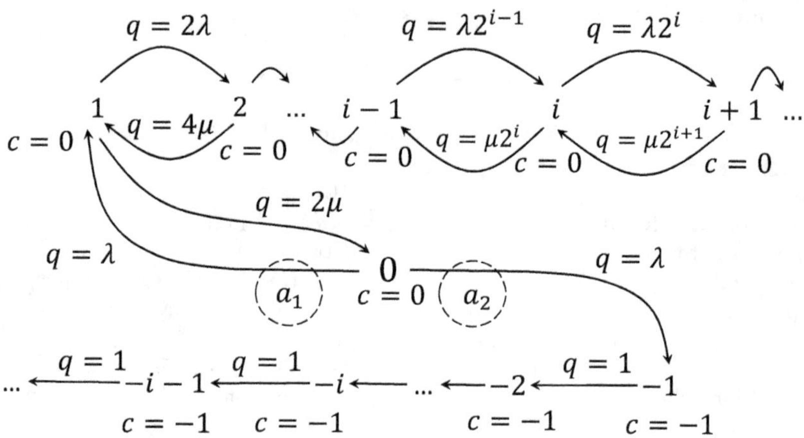

Fig. 5.13 Linear programming approach is not valid.

Let us show that Condition 5.1(a) is not satisfied and the linear program (5.39), (5.40) has no connection to problem (5.8).

In order to solve the linear program (5.39), (5.40), we first consider $\Gamma = \{i\}$ as singletons. Then equality (5.40) takes the form

$$\begin{aligned}
\lambda 2^{i-1}\eta(\{i-1\}\times\mathbf{A}) + \mu 2^{i+1}\eta(\{i+1\}\times\mathbf{A}) - 2^i\eta(\{i\}\times\mathbf{A}) &= 0, \quad i \geq 2;\\
\lambda\eta(\{0\}\times\{a_1\}) + 4\mu\eta(\{2\}\times\mathbf{A}) - 2\eta(\{1\}\times\mathbf{A}) &= 0, \quad i = 1;\\
2\mu\eta(\{1\}\times\mathbf{A}) - \lambda\eta(\{0\}\times\mathbf{A}) &= 0, \quad i = 0;\\
\lambda\eta(\{0\}\times\{a_2\}) - \eta(\{-1\}\times\mathbf{A}) &= 0, \quad i = -1;\\
\eta(\{i+1\}\times\mathbf{A}) - \eta(\{i\}\times\mathbf{A}) &= 0, \quad i \leq -2.
\end{aligned}$$

If $\eta(\{0\}\times\{a_2\}) > 0$, then $\eta(\{-1\}\times\mathbf{A}) = \eta(\{-2\}\times\mathbf{A}) = \ldots > 0$, and η cannot be a probability measure. Thus, $\eta(\{0\}\times\{a_2\}) = 0$; $\eta(\{i\}\times\mathbf{A}) = 0$ for all $i < 0$ and $\eta(\{0\}\times\mathbf{A}) = \eta(\{0\}\times\{a_1\})$. If $\eta(\{0\}\times\{a_1\}) = 0$, then $\eta(\{1\}\times\mathbf{A}) = 0 \Longrightarrow \eta(\{2\}\times\mathbf{A}) = 0 \Longrightarrow \ldots$ and again η cannot be a probability measure. The only chance is to put $\eta(\{0\}\times\{a_1\}) > 0$. Hence, the unique solution to equation (5.40) is given by the values on singletons $\eta(i,a) = \eta(i\times\mathbf{A})\pi_\eta(a|i)$, where

$$\begin{aligned}
&\eta(\{i\}\times\mathbf{A}) = 0 \quad \forall i < 0;\\
&\eta(\{i\}\times\mathbf{A}) = \left(1 - \frac{\lambda}{2\mu}\right)\left(\frac{\lambda}{2\mu}\right)^i \quad \forall i \geq 0
\end{aligned}$$

and the stochastic kernels $\pi_\eta(a|i)$ are arbitrary for all $i \neq 0$, and $\pi_\eta(a_1|0) = 1$. The sets $\Gamma \in \mathcal{B}(\mathbf{X})$ satisfying $\sup_{i\in\Gamma}\sum_{a\in\mathbf{A}} q_i(a)\pi_\eta(a|i) < \infty$ are those which have finite number of positive points $i \in \mathbf{X}$, and for all such sets equality (5.40) is valid. We ignore inequality $\sum_{i\in\mathbf{X}} w(i)\eta(\{i\}\times\mathbf{A}) < \infty$, as Condition 5.1(a) is not satisfied (this will be proved below) and function $w(\cdot)$ is undefined.

To summarize, the optimal solution η^* to the linear program (5.39), (5.40) is such that $\pi_{\eta^*}(a_1|0) = 1$. But such a stationary π-strategy $\pi^s := \pi_{\eta^*}$ is not uniformly average optimal: if $\mathsf{P}_0(0) = 1$, then

$$W^A(\pi^s, 0) = 0 = \sum_{(i,a)\in\mathbf{X}\times\mathbf{A}} c(i,a)\eta^*(\{i\}\times\{a\}),$$

and $W^A(\varphi^*, 0) = -1$ for the deterministic stationary strategy $\varphi^*(x) \equiv a_2$.

It remains to show that Condition 5.1(a) is not satisfied. Under the stationary π-strategy $\pi^s := \pi_{\eta^*}$, starting from $i = x_0 \geq 0$, the controlled process $X(\cdot)$ is a birth-and-death process in the state subspace $\{j : j \geq 0\} \subset \mathbf{X}$. For such processes, the expected time to explosion is given by

$$\mathsf{E}_i^{\pi^s}[T_\infty] = \sum_{j=i}^{\infty} \frac{1}{q(\{j+1\}|j,a)\gamma(j)} \sum_{k=0}^{j} \gamma(k), \quad i \geq 0,$$

where $\gamma(k) := \prod_{l=1}^{k} \frac{q(\{l\}|l-1,a)}{q(\{l-1\}|l,a)}$: see Proposition 2 of [Pollett (2003)]. In the current example,

$$\gamma(k) = \prod_{l=1}^{k} \frac{\lambda 2^{l-1}}{\mu 2^l} = \left(\frac{\lambda}{2\mu}\right)^k, \quad k \geq 0;$$

$$\mathsf{E}_i^{\pi^s}[T_\infty] = \sum_{j=i}^{\infty} \frac{(2\mu)^j}{\lambda 2^j \lambda^j} \sum_{k=0}^{j} \left(\frac{\lambda}{2\mu}\right)^k$$

$$= \sum_{j=i}^{\infty} \frac{\mu^j}{\lambda^{j+1}} \frac{1-\left(\frac{\lambda}{2\mu}\right)^{j+1}}{1-\frac{\lambda}{2\mu}} < \frac{2\mu}{2\mu-\lambda} \cdot \frac{1}{\lambda} \sum_{j=i}^{\infty} \left(\frac{\mu}{\lambda}\right)^j < \infty, \quad i \geq 0.$$

Thus, the controlled process $X(\cdot)$ is explosive and Condition 5.2 is not satisfied according to Theorem 5.3. Since Condition 5.1(a) implies Condition 5.2, we conclude that Condition 5.1(a) is not satisfied either. See also Remark 5.7. □

5.3.10 *Constrained average cost problem: linear programming approach*

Suppose there are several cost rates $c_0(\cdot,\cdot), c_1(\cdot,\cdot), \ldots c_J(\cdot,\cdot)$, $J \geq 1$. Then the *constrained* average cost problem (for a fixed initial distribution P_0) is

$$\left.\begin{array}{l} W_0^A(S,\mathsf{P}_0) \to \inf_S \\ W_j^A(S,\mathsf{P}_0) \leq d_j, \quad j=1,2,\ldots,J, \end{array}\right\} \tag{5.41}$$

where $\{d_j\}_{j=1}^J$ are predetermined constants and the average costs $W_j^A(S,\mathsf{P}_0)$ are given by formula (5.4), respectively with cost rates $c_j(\cdot,\cdot)$. Strategies satisfying all the constraints-inequalities will be called *feasible*. In problem (5.41) one can restrict to relaxed strategies or to natural Markov strategies: see Theorem 5.1 and Remark 5.4. In what follows, it is assumed that there is no explosion and there exists a strategy S such that $W_j^A(S,\mathsf{P}_0) \leq d_j$, $j=1,2,\ldots,J$ and $W_0^A(S,\mathsf{P}_0) < \infty$.

Definition 5.7. [Kaspi and Mandelbaum (1994), Thm. 1]. The controlled process $X(\cdot)$, under a relaxed stationary strategy π^s, is called *Harris recurrent* if there exists a nonzero σ-finite measure μ on $\mathcal{B}(\mathbf{X})$ such that

$$\mu(\Gamma) > 0 \Longrightarrow \mathsf{P}_x^{\pi^s}(X(t) \in \Gamma \text{ for some } t > 0) = 1 \quad \forall x \in \mathbf{X}.$$

If the space $\mathbf{X}$ is finite, then Harris recurrence is equivalent to the requirement that the model is unichain (with a single communicating class under each relaxed stationary strategy).

Theorem 5.11. *Suppose that, under each relaxed stationary strategy π^s, the controlled process $X(\cdot)$ is Harris recurrent and has a unique finite invariant measure satisfying equation (5.40) (i.e, the process $X(\cdot)$ is positive Harris recurrent). Assume that the following conditions are satisfied:*

- *there exists a feasible strategy in problem (5.41);*
- $\sup_{x\in\mathbf{X}} \bar{q}_x < \infty$ *(recall that* $\bar{q}_x := \sup_{a\in\mathbf{A}} q_x(a)$*);*
- *the spaces* $\mathbf{X}$ *and* $\mathbf{A}$ *are compact;*
- *the cost rates* $c_j(\cdot,\cdot)$, $j = 0, 1, 2, \ldots, J$ *are bounded and continuous;*
- *for each bounded continuous function* $f(\cdot)$ *on* $\mathbf{X}$*, the mapping* $(x, a) \in \mathbf{X} \times \mathbf{A} \to \int_{\mathbf{X}} f(y)q(dy|x, a)$ *is bounded and continuous.*

Then there exists an optimal solution η^* *to Linear Program*

$$\text{Minimize} \quad \text{over } \eta \in \mathbf{P}(\mathbf{X}\times\mathbf{A})\text{:} \int_{\mathbf{X}\times\mathbf{A}} c_0(x,a)\eta(dx\times da) \tag{5.42}$$

$$\text{subject to} \int_{\mathbf{X}\times\mathbf{A}} q(\Gamma|x,a)\eta(dx\times da) = 0, \quad \forall\Gamma\in\mathcal{B}(\mathbf{X}) \quad \text{and}$$
$$\int_{\mathbf{X}\times\mathbf{A}} c_j(x,a)\eta(dx\times da) \le d_j, \quad j = 1,2,\ldots,J,$$

and the relaxed stationary strategy $\pi^{*s} = \pi_{\eta^*}$ *coming from the decomposition* $\eta^*(dx \times da) = \eta^*(dx \times \mathbf{A})\pi_{\eta^*}(da|x)$ *solves problem (5.41) for any fixed initial distribution* P_0*. The optimal value of problem (5.41) does not depend on the initial distribution* P_0 *and equals the optimal value of linear program (5.42).*

For the proof see Theorem 4.1 of [Guo *et al.* (2012)], where the more general case, with weaker conditions, was investigated, but the Harris recurrence is essential.

In [Guo (2007)] the following modification was studied. Suppose the set

$$\begin{aligned}\hat{\Gamma} := \{(\breve{\pi}, \tilde{\mathsf{P}}_0) \in \breve{\Pi}\times\mathbf{P}(\mathbf{X}) : \ & W_0^A(\breve{\pi}, \tilde{\mathsf{P}}_0) < \infty \\ & \text{and } W_j^A(\breve{\pi}, \tilde{\mathsf{P}}_0) \le d_j, \ j = 1,2,\ldots,J\}\end{aligned} \tag{5.43}$$

is not empty, where $\breve{\Pi}$ is the collection of all natural Markov strategies. Then the modified problem looks as follows:

$$W_0^A(\breve{\pi}, \tilde{\mathsf{P}}_0) \to \inf_{(\breve{\pi}, \tilde{\mathsf{P}}_0)\in\hat{\Gamma}}. \tag{5.44}$$

Theorem 5.12. [Guo (2007), Thm. 3.1, Remark 3.2]. *Assume that the following conditions are satisfied:*

- *the spaces* $\mathbf{X}$ *and* $\mathbf{A}$ *are topological Borel;*
- *the cost rate* $c_0(\cdot,\cdot)$ *is inf-compact (see Condition 4.1(c));*
- *all the cost rates* $c_j(\cdot,\cdot)$, $j = 1, 2, \ldots, J$ *are lower semicontinuous;*
- *for each bounded continuous function* $f(\cdot)$ *on* $\mathbf{X}$*, the mapping* $(x, a) \in \mathbf{X} \times \mathbf{A} \to \int_{\mathbf{X}} f(y)q(dy|x, a)$ *is bounded and continuous;*
- $\sup_{x\in\mathbf{X}} \bar{q}_x < \infty$*. (Recall that* $\bar{q}_x := \sup_{a\in\mathbf{A}} q_x(a)$*.)*

Then there exists an optimal solution η^* to linear program (5.42) and the pair $\tilde{\mathsf{P}}_0^*(dx) = \eta^*(dx \times \mathbf{A})$ and $\pi^{*s} = \pi_{\eta^*}$ solves problem (5.44). Like previously, π_{η^*} comes from the decomposition $\eta^*(dx \times da) = \eta^*(dx \times \mathbf{A})\pi_{\eta^*}(da|x)$. The optimal values of linear program (5.42) and of problem (5.44) are equal.

Remark 5.7. One can certainly use theorems 5.11 and 5.12 for the investigation of the unconstrained problem (5.8). Already the example in Subsection 5.3.9 shows that the Harris recurrence is important in Theorem 5.11. Theorem 5.12 is not applicable in that example either, because the cost rate is not inf-compact in the discrete topology in $\mathbf{X}$.

Now consider the finite model assuming that the set $\hat{\Gamma}$ defined in (5.43) is not empty. All the conditions of Theorem 5.12 are satisfied and linear program (5.42) takes the form

$$\begin{aligned} \text{Minimize } & \sum_{x\in\mathbf{X}}\sum_{a\in\mathbf{A}} c_0(x,a)\eta(x,a) \qquad (5.45)\\ \text{subject to } & \eta(x,a) \ge 0, \quad \sum_{x\in\mathbf{X}}\sum_{a\in\mathbf{A}} \eta(x,a) = 1,\\ & \sum_{x\in\mathbf{X}}\sum_{a\in\mathbf{A}} q(y|x,a)\eta(x,a) = 0 \quad \forall y \in \mathbf{X} \text{ and}\\ & \sum_{x\in\mathbf{X}}\sum_{a\in\mathbf{A}} c_j(x,a)\eta(x,a) \le d_j, \quad j = 1,2,\ldots,J. \end{aligned}$$

Suppose the finite model is unichain: under each relaxed stationary strategy (equivalently, under each deterministic stationary strategy), the embedded chain $X_0, X_1, \ldots$ is unichain. Assuming that problem (5.41) is *feasible*, i.e., there is at least one strategy satisfying the constraints, all the conditions in Theorem 5.11 are satisfied and relaxed stationary strategies are sufficient in problem (5.41). In (5.45) one can recognize the linear program (5.23), (5.24), (5.25), supplemented by the constraints-inequalities.

The following example shows that an optimal relaxed stationary strategy in problem (5.41) may not exist if the finite model is not unichain.

Example. Let $\mathbf{X} = \{1,2\}$, $\mathbf{A} = \{a_1, a_2\}$, $J = 1$ and $\mathsf{P}_0(1) = 1$. The transition rate is given by

$$q(2|1,a_1) = \lambda > 0, \quad q(1|1,a_1) = -\lambda, \quad q(2|1,a_2) = q(1|1,a_2) = 0;$$

$$q(1|2,a_1) = q(1|2,a_2) = q(2|2,a_1) = q(2|2,a_2) = 0,$$

and the cost rates are

$$c_0(1,a_1) = c_0(2,a_1) = c_0(2,a_2) = 0, \quad c_0(1,a_2) = 2;$$

$$c_1(1,a_1) = c_1(2,a_1) = c_1(2,a_2) = 2, \quad c_1(1,a_2) = 0.$$

Finally, let $d_1 = 1$. See Fig. 5.14.

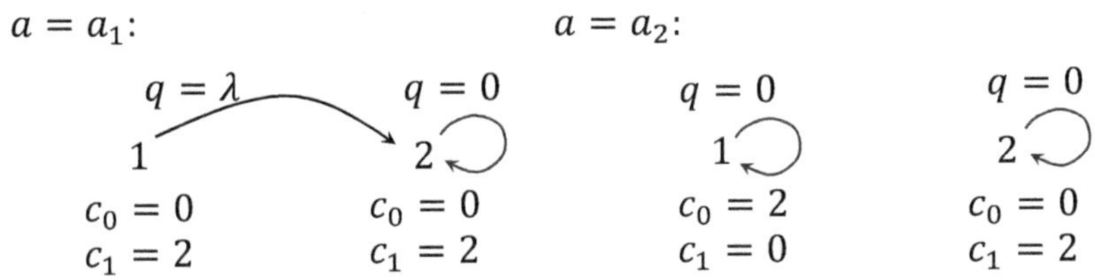

Fig. 5.14 Multichain CTMDP.

There are only two essentially different relaxed stationary strategies: $\pi^{1s}(a_1|1) > 0$ and $\pi^{2s}(a_1|1) = 0$. The actions in state 2 are immaterial. $W_0^A(\pi^{1s}, \mathsf{P}_0) = 0$ and $W_1^A(\pi^{1s}, \mathsf{P}_0) = 2 > 1 = d_1$ because the process $X(\cdot)$ is ultimately absorbed in state 2 under π^{1s}. $W_0^A(\pi^{2s}, \mathsf{P}_0) = 2$ and $W_1^A(\pi^{2s}, \mathsf{P}_0) = 0 < 1 = d_1$ because now state 1 is absorbing. Thus, π^{2s} is the only feasible relaxed stationary strategy with $W_0^A(\pi^{2s}, \mathsf{P}_0) = 2$.

Obviously, the optimal value of problem (5.41) cannot be smaller than the optimal value of problem (5.44) because one can restrict to natural Markov strategies in (5.41) and

$$\{(\breve{\pi}, \mathsf{P}_0) : \ \breve{\pi} \in \breve{\Pi} \text{ and } W_1^A(\breve{\pi}, \mathsf{P}_0) \le d_1\} \subset \hat{\Gamma}.$$

To solve problem (5.44), we use Theorem 5.12 in which all the conditions are satisfied. Linear program (5.45) takes the form

$$\begin{aligned} 2\eta(1,a_2) &\longrightarrow \min \\ \eta(1,a_1)+\eta(1,a_2)+\eta(2,a_1)+\eta(2,a_2) &= 1, \quad \eta(x,a) \ge 0, \\ -\lambda\eta(1,a_1) &= 0 \text{ (when } y=1), \\ \lambda\eta(1,a_1) &= 0 \text{ (when } y=2), \\ 2\eta(1,a_1)+2\eta(2,a_1)+2\eta(2,a_2) &\le 1. \end{aligned}$$

The optimal solution is $\eta^*(1,a_1) = 0$, $\eta^*(1,a_2) = \frac{1}{2}$, $\eta^*(2,a_1)+\eta^*(2,a_2) = \frac{1}{2}$. The pair $\tilde{\mathsf{P}}_0(1) = \tilde{\mathsf{P}}_0(2) = \frac{1}{2}$, $\pi^{*s}(a_2|1) = 1$ solves problem (5.44). ($\pi^{*s}(a_1|2) \in [0,1]$ can be fixed arbitrarily.) The optimal value of linear program (5.45) and of problem (5.44) is 1.

Consider the following randomized stationary strategy: $\varpi^{*s}(a_1|x) = \varpi^{*s}(a_2|x) = \frac{1}{2}$. Starting from $X_0 = 1$, the controlled process $X(\cdot)$ is absorbed at 1 and 2 with equal probabilities and $W_0^A(\varpi^{*s}, \mathsf{P}_0) = W_1^A(\varpi^{*s}, \mathsf{P}_0) = 1$. This strategy is feasible in problem (5.41) and, as explained above, is optimal.

According to Theorem 5.1, there exists a relaxed Markov strategy π^{*m} equivalent to ϖ^{*s}:

$$\pi^{*m}(a_1|1,s) = \frac{e^{-\lambda s}}{1+e^{-\lambda s}}, \quad \pi^{*m}(a_2|1,s) = \frac{1}{1+e^{-\lambda s}},$$
$$\pi^{*m}(a_1|2,s) = \frac{1}{2}, \quad \pi^{*m}(a_2|2,s) = \frac{1}{2}.$$

Let us explicitly show that $W_0^A(\pi^{*m}, \mathsf{P}_0) = W_1^A(\pi^{*m}, \mathsf{P}_0) = 1$. Recall that $\mathsf{P}_0(1) = 1$.

$$\begin{aligned}
\mathsf{P}_{\mathsf{P}_0}^{\pi^{*m}}(X(t)=1) &= e^{-\int_0^t \lambda\pi^{*m}(a_1|1,s)ds} = e^{-\int_0^t \frac{\lambda e^{-\lambda s}}{1+e^{-\lambda s}}ds} = \frac{1+e^{-\lambda t}}{2};\\
\mathsf{P}_{\mathsf{P}_0}^{\pi^{*m}}(X(t)=2) &= \frac{1-e^{-\lambda t}}{2};\\
\bar{\mathsf{P}}_{\mathsf{P}_0}^{\pi^{*m}}(t, 1\times a_1) &= \frac{1+e^{-\lambda t}}{2}\cdot\frac{e^{-\lambda t}}{1+e^{-\lambda t}} = \frac{e^{-\lambda t}}{2};\\
\bar{\mathsf{P}}_{\mathsf{P}_0}^{\pi^{*m}}(t, 1\times a_2) &= \frac{1+e^{-\lambda t}}{2}\cdot\frac{1}{1+e^{-\lambda t}} = \frac{1}{2}\\
\bar{\mathsf{P}}_{\mathsf{P}_0}^{\pi^{*m}}(t, 2\times a_1) &= \frac{1-e^{-\lambda t}}{2}\cdot\frac{1}{2};\\
\bar{\mathsf{P}}_{\mathsf{P}_0}^{\pi^{*m}}(t, 2\times a_2) &= \frac{1-e^{-\lambda t}}{2}\cdot\frac{1}{2}.
\end{aligned}$$

Note, there is only one jump $1 \to 2$ in this example.

Therefore,

$$\begin{aligned}
&W_0^A(\pi^{*m}, \mathsf{P}_0)\\
&= \lim_{T\to\infty}\frac{1}{T}\int_0^T 2\bar{\mathsf{P}}_{\mathsf{P}_0}^{\pi^{*m}}(t, 1\times a_2)dt = \lim_{T\to\infty}\frac{1}{T}\int_0^T 2\cdot\frac{1}{2}dt = 1;\\
&W_1^A(\pi^{*m}, \mathsf{P}_0)\\
&= \lim_{T\to\infty}\frac{1}{T}\int_0^T 2[\bar{\mathsf{P}}_{\mathsf{P}_0}^{\pi^{*m}}(t, 1\times a_1) + \bar{\mathsf{P}}_{\mathsf{P}_0}^{\pi^{*m}}(t, 2\times a_1) + \bar{\mathsf{P}}_{\mathsf{P}_0}^{\pi^{*m}}(t, 2\times a_2)]dt\\
&= \lim_{T\to\infty}\frac{1}{T}\int_0^T 2\left[\frac{e^{-\lambda t}}{2} + \frac{1-e^{-\lambda t}}{4} + \frac{1-e^{-\lambda t}}{4}\right]dt = 1.
\end{aligned}$$

The proof is completed. □

The following example published in [Feinberg (2002a), Ex. 3.1] shows that in a finite unichain model there may be no optimal randomized stationary strategies in problem (5.41).

Example. Let $\mathbf{X} = \{1, 2\}$, $\mathbf{A} = \{a_1, a_2\}$, $J = 1$ and $\mathsf{P}_0(1) = 1$. The transition rate is given by

$$q(2|1,a_1) = \lambda > 0, \quad q(1|1,a_1) = -\lambda, \quad q(1|2,a) \equiv \lambda, \quad q(2|2,a) \equiv -\lambda,$$
$$q(2|1,a_2) = q(1|1,a_2) = 0,$$

and the cost rates are

$$c_0(1, a_1) = c_0(2, a) \equiv 0, \quad c_0(1, a_2) = 2;$$

$$c_1(1, a_1) = c_1(2, a) \equiv 2 \quad c_1(1, a_2) = 0.$$

Finally, let $d_1 = 1$. See Fig. 5.15.

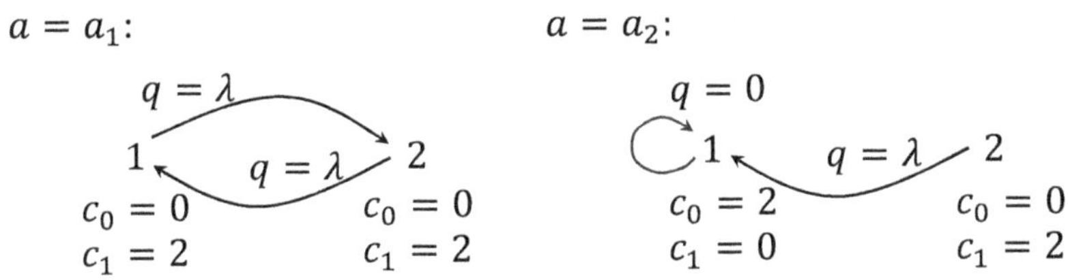

Fig. 5.15 Unichain CTMDP with absorbing state.

There are only two essentially different randomized stationary strategies: $\varpi^{1s}(a_2|1) = 0$ and $\varpi^{2s}(a_2|1) > 0$. The actions in state 2 are immaterial. $W_0^A(\varpi^{1s}, \mathsf{P}_0) = 0$ and $W_1^A(\varpi^{1s}, \mathsf{P}_0) = 2 > 1 = d_1$; $W_0^A(\varpi^{2s}, \mathsf{P}_0) = 2$ and $W_1^A(\varpi^{2s}, \mathsf{P}_0) = 0 < 1 = d_1$ because under the strategy ϖ^{2s} the controlled process $X(\cdot)$ will be absorbed at state 1. Thus, ϖ^{2s} is the only feasible randomized stationary strategy in problem (5.41) with $W_0^A(\varpi^{2s}, \mathsf{P}_0) = 2$.

At the same time, the model is unichain and one can build the optimal strategy using Theorem 5.11. Linear program (5.45) takes the form

$$\begin{aligned} 2\eta(1, a_2) &\longrightarrow \min \\ \eta(1, a_1) + \eta(1, a_2) + \eta(2, a_1) + \eta(2, a_2) &= 1, \quad \eta(x, a) \geq 0, \\ \lambda\eta(2, a_1) + \lambda\eta(2, a_2) - \lambda\eta(1, a_1) &= 0 \ \text{(when } y = 1), \\ \lambda\eta(1, a_1) - \lambda\eta(2, a_1) - \lambda\eta(2, a_2)) &= 0 \ \text{(when } y = 2), \\ 2\eta(1, a_1) + 2\eta(2, a_1) + 2\eta(2, a_2) &\leq 1. \end{aligned}$$

The optimal solution is $\eta^*(1, a_1) = \frac{1}{4}$, $\eta^*(1, a_2) = \frac{1}{2}$, $\eta^*(2, a_1) + \eta^*(2, a_2) = \frac{1}{4}$ and the optimal strategy in problem (5.41) is $\pi^{*s}(a_1|1) = \frac{1}{3}$, $\pi^{*s}(a_2|1) = \frac{2}{3}$; $\pi^{*s}(a_1|2) \in [0, 1]$ can be arbitrary. The optimal value of problem (5.41) and of the linear program equals 1. The optimal initial distribution as in Theorem 5.12 is $\tilde{\mathsf{P}}_0^*(1) = \frac{3}{4}$, $\tilde{\mathsf{P}}_0^*(2) = \frac{1}{4}$.

The obtained optimal stationary π-strategy π^{*s} with $\pi^{*s}(a_1|2) = 1$ is 2-mixed over deterministic stationary stable strategies $\varphi^1(x) \equiv a_1$ and $\varphi^2(x) \equiv a_2$ with weights $\alpha_1 = \alpha_2 = \frac{1}{2}$: see Remark 5.8. □

Assuming that there exists a feasible strategy, an optimal solution to problem (5.41) in the form of a randomized stationary strategy exists in a finite unichain model if $\lambda_x(a) > 0$ for all $x \in \mathbf{X}$, $a \in \mathbf{A}$: see Theorem 2.1 of [Feinberg (2002a)]. It has the form

$$\varpi^{*s}(a|x) = \frac{q_x(a)\eta^*(x,a)}{\sum_{b\in\mathbf{A}} q_x(b)\eta^*(x,b)}$$

if $\sum_{b\in\mathbf{A}} q_x(b)\eta^*(x,b) > 0$, and can be fixed arbitrarily otherwise. Here η^* again is the optimal solution to linear program (5.45). In the above example this theorem is not applicable because $q_1(a_2) = 0$.

5.3.11 *Constrained discounted and average cost problems*

Along with the constrained average cost problem (5.41), one can consider the constrained discounted cost problem with $\alpha > 0$:

$$\left.\begin{array}{l} W_0^\alpha(S, \mathsf{P}_0) \to \inf_S \\ W_j^\alpha(S, \mathsf{P}_0) \le d_j, \quad j = 1, 2, \ldots, J, \end{array}\right\} \tag{5.46}$$

where again $\{d_j\}_{j=1}^J$ are predetermined constants and the α-discounted costs $W_j^\alpha(S, \mathsf{P}_0)$ are given by formula (5.3), respectively with cost rates $c_j(\cdot,\cdot)$.

Suppose the model is finite and problem (5.46) is *feasible*, that is, all the inequalities are satisfied for some strategy S. Then the following linear program has an optimal solution η^*:

$$\text{Minimize} \quad \frac{1}{\alpha}\sum_{x\in\mathbf{X}}\sum_{a\in\mathbf{A}} c_0(x,a)\eta(x,a) \tag{5.47}$$

$$\text{subject to} \quad \eta(x,a) \ge 0, \quad \sum_{x\in\mathbf{X}}\sum_{a\in\mathbf{A}} \eta(x,a) = 1,$$

$$\hat{\eta}(x) := \sum_{b\in\mathbf{A}} \eta(x,b) = \mathsf{P}_0(x) + \frac{1}{\alpha}\sum_{y\in\mathbf{X}}\sum_{a\in\mathbf{A}} q(x|y,a)\eta(y,a) \quad \forall x \in \mathbf{X} \tag{5.48}$$

$$\text{and} \quad \frac{1}{\alpha}\sum_{x\in\mathbf{X}}\sum_{a\in\mathbf{A}} c_j(x,a)\eta(x,a) \le d_j, \quad j = 1, 2, \ldots, J.$$

At least one of equations in (5.48) is redundant. The relaxed stationary strategy $\pi^{*\alpha s} = \pi_{\eta^*}$ coming from the decomposition $\eta^*(x,a) = \hat{\eta}^*(x)\pi_{\eta^*}(a|x)$ is optimal in problem (5.46). For justification see [Piunovskiy and Zhang (2020), Section 3.2].

Proposition 5.3. [Prieto-Rumeau and Hernández-Lerma (2012), Prop. 8.18]. *Suppose the model is finite and ergodic (i.e., all states communicate under each deterministic stationary strategy), $J = 1$, $c_1(\cdot,\cdot) \ge 0$ and $i_0 \in \mathbf{X}$ is an arbitrarily fixed initial state: $\mathsf{P}_0(i_0) = 1$. Assume that d_1^α depends on the discount factor $\alpha > 0$, for each (small) $\alpha > 0$, $W_1^\alpha(S, i_0) < d_1^\alpha$ for some strategy S, and* *put*

$$d_1^{\min} := \inf_S W_1^A(S, i_0), \qquad d_1^{\max} := \sup_S W_1^A(S, i_0).$$

(In finite ergodic models these numbers do not depend on i_0 [Prieto-Rumeau and Hernández-Lerma (2012), p. 175]. See also the linear programming approach at the beginning of Subsection 5.2.3.) If there exists $\lim_{\alpha\downarrow 0} \alpha d_1^\alpha =: d_1^0 \in (d_1^{\min}, d_1^{\max}]$, then the following assertions hold:

(a) Any limiting as $\alpha \downarrow 0$ relaxed stationary strategy π^{*s} of $\{\pi^{*\alpha s}\}_{\alpha>0}$ (the strategies $\pi^{*\alpha s}$ are optimal in problem (5.46)) is optimal in the constrained average cost problem (5.41) with $d_1 = d_1^0$.

(b) $\lim_{\alpha\downarrow 0} \alpha W_0^\alpha(\pi^{*\alpha s}, i_0)$ equals the optimal value of problem (5.41) with $d_1 = d_1^0$, which is i_0-independent by Theorem 5.11.

The following example published in [Prieto-Rumeau and Hernández-Lerma (2012), Ex. 8.19] shows that the strict inequality $d_1^0 > d_1^{\min}$ in Proposition 5.3 is important.

Example. Let $\mathbf{X} = \{0, 1\}$, $\mathbf{A} = \{a_1, a_2, a_3\}$, $J = 1$,

$$\begin{gathered}
q(1|0,a) = -q(0|0,a) \equiv 1, \quad q(0|1,a_1) = -q(1|1,a_1) = 2, \\
q(0|1,a_2) = -q(1|1,a_2) = q(0|1,a_3) = -q(1|1,a_3) = 4; \\
c_0(0,a) \equiv -1, \quad c_0(1,a) \equiv 0, \\
c_1(0,a) \equiv 0, \ c_1(1,a_1) = 3, \quad c_1(1,a_2) = 5, \ c_1(1,a_3) = 10.
\end{gathered}$$

The actions in state 0 are immaterial. See Fig. 5.16. Put $i_0 = 0$ and $d_1^\alpha = \frac{4}{\alpha(\alpha+4)}$.

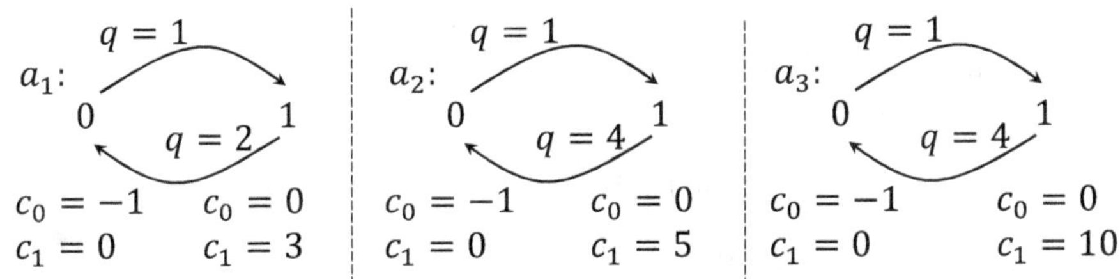

Fig. 5.16 Limiting strategy of discounted optimal strategies (as $\alpha \downarrow 0$) is not average optimal.

Firstly, compute $\inf_S W_1^\alpha(S, 0)$ using Theorem 5.8. Equation (5.22) takes the form

$$\begin{aligned}
V(0) &= 0 + \frac{V(1)}{\alpha+1}; \\
V(1) &= \min\left\{\frac{3}{\alpha+2} + \frac{2V(0)}{\alpha+2}, \ \frac{5}{\alpha+4} + \frac{4V(0)}{\alpha+4}, \ \frac{10}{\alpha+4} + \frac{4V(0)}{\alpha+4}\right\}
\end{aligned}$$

and has solution

$$V(0) = W_1^{*\alpha}(0) = \frac{3}{\alpha(\alpha+3)}; \quad V(1) = W_1^{*\alpha}(1) = \frac{3(\alpha+1)}{\alpha(\alpha+3)},$$

so that

$$\inf_S W_1^\alpha(S,0) = W_1^{*\alpha}(0) = \frac{3}{\alpha(\alpha+3)} < \frac{4}{\alpha(\alpha+4)} = d_1^\alpha,$$

and condition $W_1^\alpha(S,0) < d_1^\alpha$ for some strategy S is fulfilled.

To compute the optimal strategies $\pi^{*\alpha s}$ in problem (5.46) for $\mathsf{P}_0(0) = 1$, we solve linear program (5.47), (5.48) which takes the from

$$\begin{aligned}
\text{Minimize} \quad & \frac{1}{\alpha}[-\hat\eta(0)] \\
\text{subject to} \quad & \hat\eta(0),\ \eta(1,a_1),\ \eta(1,a_2),\ \eta(1,a_3) \ge 0, \\
& \hat\eta(0) + \eta(1,a_1) + \eta(1,a_2) + \eta(1,a_3) = 1, \\
& \hat\eta(0) = 1 + \frac{1}{\alpha}[-\hat\eta(0) + 2\eta(1,a_1) + 4\eta(1,a_2) + 4\eta(1,a_3)] \\
\text{and} \quad & \frac{1}{\alpha}[3\eta(1,a_1) + 5\eta(1,a_2) + 10\eta(1,a_3)] \le \frac{4}{\alpha(\alpha+4)}.
\end{aligned}$$

Here $\hat\eta(x) := \eta(x,a_1) + \eta(x,a_2) + \eta(x,a_3)$. The unique optimal solution is given by

$$\hat\eta^*(0) = \frac{\alpha+3}{\alpha+4}, \quad \eta^*(1,a_1) = \eta^*(1,a_2) = \frac{1}{2(\alpha+4)}, \quad \eta^*(1,a_3) = 0,$$

and the unique optimal relaxed stationary strategy (up to the immaterial values of $\pi^{*\alpha s}(a|0)$) is $\pi^{*\alpha s}(a_1|1) = \pi^{*\alpha s}(a_2|1) = \frac{1}{2}$ with $W_0^\alpha(\pi^{*\alpha s},0) = -\frac{\alpha+3}{\alpha(\alpha+4)}$, $W_1^\alpha(\pi^{*\alpha s},0) = d_1^\alpha = \frac{4}{\alpha(\alpha+4)}$.

To compute $d_1^{\min}$, we solve linear program (5.23),(5.24), (5.25) for $c(\cdot,\cdot) = c_1(\cdot,\cdot)$, which takes the form

$$\begin{aligned}
3\eta(1,a_1) + 5\eta(1,a_2) + 10\eta(1,a_3) &\to \min \\
-\hat\eta(0) + 2\eta(1,a_1) + 4\eta(1,a_2) + 4\eta(1,a_3) &= 0, \\
\hat\eta(0) + \eta(1,a_1) + \eta(1,a_2) + \eta(1,a_3) &= 1, \\
\hat\eta(0),\ \eta(1,a_1),\ \eta(1,a_2),\ \eta(1,a_3) &\ge 0.
\end{aligned}$$

Since the model is ergodic, $\tilde\eta(x,a) \equiv 0$; here $\hat\eta(0) := \eta(0,a_1) + \eta(0,a_2) + \eta(0,a_3)$. The optimal solution is given by $\eta^*(1,a_3) = 0$, $\eta^*(1,a_2) \in \left[0,\frac{1}{5}\right]$ is arbitrary and $\hat\eta^*(0) = \frac{2+2\eta^*(1,a_2)}{3}$, $\eta^*(1,a_1) = \frac{1-5\eta^*(1,a_2)}{3}$. The optimal value of this program, equal to $d_1^{\min}$, is 1 and, e.g., $W_1^A(\varphi^2,0) = 1$ for the deterministic stationary strategy $\varphi^2(1) := a_2$, which corresponds to the case $\eta^*(1,a_2) = \frac{1}{5}$. The actions in state 0 are immaterial. The value $d_1^{\max} = 2$ can be calculated in the similar way.

Finally, we solve the constrained average cost problem (5.41) with $d_1 = d_1^0 = \lim_{\alpha\downarrow 0} \alpha d_1^\alpha = 1$. This problem is feasible because, as was shown above,

$W_1^A(\varphi^2, 0) = 1$. Thus, one can use Theorem 5.11. Linear program (5.42) takes the form

$$\begin{aligned}
-\hat{\eta}(0) &\to \min \\
-\hat{\eta}(0) + 2\eta(1, a_1) + 4\eta(1, a_2) + 4\eta(1, a_3) &= 0, \\
\hat{\eta}(0) + \eta(1, a_1) + \eta(1, a_2) + \eta(1, a_3) &= 1, \\
\hat{\eta}(0),\ \eta(1, a_1),\ \eta(1, a_2),\ \eta(1, a_3) &\ge 0, \\
3\eta(1, a_1) + 5\eta(1, a_2) + 10\eta(1, a_3) &\le 1.
\end{aligned}$$

Here $\hat{\eta}(0) := \eta(0, a_1) + \eta(0, a_2) + \eta(0, a_3)$. The unique optimal solution is given by

$$\hat{\eta}^*(0) = \frac{4}{5}, \quad \eta^*(1, a_1) = 0, \quad \eta^*(1, a_2) = \frac{1}{5}, \quad \eta^*(1, a_3) = 0,$$

and the unique optimal relaxed stationary strategy (up to the immaterial values of $\pi^{*s}(a|0)$) is $\pi^{*s}(a_2|1) = 1$ with $W_0^A(\pi^{*s}, 0) = -\frac{4}{5} = W_0^A(\pi^{*s}, \mathsf{P}_0)$ for any P_0; $W_1^A(\pi^{*s}, 0) = 1 = W_1^A(\pi^{*s}, \mathsf{P}_0)$ for any P_0.

All the conditions in Proposition 5.3 are satisfied apart from $d_1^0 \in (d_1^{\min}, d_1^{\max}]$: the values $d_1^0 = 1$, $d_1^{\min} = 1$ and $d_1^{\max} = 2$ have been calculated above. The assertions of Proposition 5.3 fail to hold:

- The unique limiting as $\alpha \downarrow 0$ relaxed stationary strategy of $\{\pi^{*\alpha s}\}_{\alpha>0}$ is $\pi^s(a_1|1) = \pi^s(a_2|1) = \frac{1}{2}$ (up to the immaterial values of $\pi^s(a|0)$). As was shown, this strategy is not optimal in the constrained average cost problem (5.41) with $d_1 = d_1^0 = 1$.
- The limit $\lim_{\alpha\downarrow 0} \alpha W_0^\alpha(\pi^{*\alpha s}, 0) = -\frac{3}{4}$ is different from the optimal value of problem (5.41) with $d_1 = d_1^0 = 1$, which equals $-\frac{4}{5}$ as calculated above.

□

5.3.12 *Constrained average cost problem: mixed strategies*

Below, several definitions and statements from [Guo and Zhang (2016)], regarding the constrained average cost problem (5.41), are provided. As was mentioned at the beginning of Subsection 5.3.10, it is legal to restrict to natural Markov strategies provided there is no explosion: see Remark 5.4.

Definition 5.8. Suppose Condition 5.2 is satisfied with $\rho > 0$ and $\sup_{x\in\mathbf{X}} \frac{\bar{q}_x}{w(x)} < \infty$. A probability measure η on $\mathcal{B}(\mathbf{X} \times \mathbf{A})$ is called *stable* if

$$\int_{\mathbf{X}} w(x)\eta(dx \times \mathbf{A}) \le 1 + \frac{b}{\rho} \quad \text{and}$$

$$\int_{\mathbf{X}} \left(\int_{\mathbf{A}} q(\Gamma|x, a)\pi_\eta(da|x) \right) \eta(dx \times \mathbf{A}) = 0 \quad \forall \Gamma \in \mathcal{B}(\mathbf{X}).$$

The underlying stationary π-strategy $\pi^s = \pi_\eta$ is said to be *stable*, too. Here, as usual, π_η comes from the decomposition $\eta(dx \times da) = \eta(dx \times \mathbf{A})\pi_\eta(da|x)$.

Condition 5.6. Suppose Conditions 5.2 with $\rho > 0$ and 5.4(a,b) are satisfied, along with the following conditions.

(a) There exist a continuous *moment* function $w'(\cdot) \geq 1$ on $\mathbf{X}$ (i.e., there is an increasing sequence of compact sets $\mathbf{X}'_n \uparrow \mathbf{X}$ such that $\lim_{n\to\infty} \inf_{x\in\mathbf{X}\setminus\mathbf{X}'_n} w'(x) = +\infty$) and a constant $M' \geq 0$ satisfying inequalities $\bar{q}_x \leq M'w'(x)$ and $\sup_{a\in\mathbf{A}} |c_j(x,a)| \leq M'w'(x)$ for all $x \in \mathbf{X}$ and $j = 0, 1, \ldots, J$.

(b) There exists an increasing sequence of compact sets $\mathbf{X}_m \uparrow \mathbf{X}$ such that $\lim_{m\to\infty} \inf_{x\in\mathbf{X}\setminus\mathbf{X}_m} \frac{w(x)}{w'(x)} = +\infty$. Function $w(\cdot)$ is as in Condition 5.2.

(c) $\int_{\mathbf{X}} w(y)\mathsf{P}_0(dy) < \infty$.

(d) The functions $c_j(\cdot,\cdot)$, $j = 0, 1, \ldots, J$ are continuous.

(e) There exist constants $\bar{M} \in \mathbb{R}$ and $\rho' > 0$, $b' \geq 0$ such that, for all $x \in \mathbf{X}$, $a \in \mathbf{A}$,

$$(\bar{q}_x + 1)w'(x) \leq \bar{M}w(x); \qquad \int_{\mathbf{X}} w'(y)q(dy|x,a) \leq -\rho' w'(x) + b'.$$

Proposition 5.4. [Guo and Zhang (2016), Prop. 1]. *If Condition 5.6(a-c) is satisfied, then the space of stable measures is nonempty, convex and compact in the space of probability measures μ on $\mathcal{B}(\mathbf{X} \times \mathbf{A})$ equipped with the w'-weak topology (the weakest topology such that, for each continuous function $u(\cdot,\cdot)$ on $\mathbf{X} \times \mathbf{A}$ with $\sup_{x\in\mathbf{X}} \frac{\sup_{a\in\mathbf{A}} |u(x,a)|}{w'(x)} < \infty$, the mapping $\mu \to \int_{\mathbf{X}\times\mathbf{A}} u(x,a)\mu(dx \times da)$ is continuous).*

Definition 5.9. A stable π-strategy, corresponding to a stable measure η, is called *L-mixed* (over deterministic stationary stable strategies φ^l, $l = 1, 2, \ldots, L$, corresponding to stable measures η_l) if $\eta = \sum_{l=1}^{L} \alpha_l \eta_l$ with $\alpha_l > 0$, $\sum_{l=1}^{L} \alpha_l = 1$.

Theorem 5.13. [Guo and Zhang (2016), Thm. 2]. *Suppose the state space $\mathbf{X}$ is countable, the model is unichain, Condition 5.6 is satisfied, and there is at least one feasible strategy in problem (5.41).*

Then there exists a $(J+1)$-mixed (stable) π-strategy solving problem (5.41), and the corresponding stable measure is optimal in linear program (5.42).

In particular, all the conditions in Theorem 5.13 are satisfied if the model is finite and unichain. In Conditions 5.2 and 5.6, one can put $w(x) = w'(x) \equiv 1$ for all $x \in \mathbf{X}$.

The presented below example is the simplified version of Example 2 of [Guo and Zhang (2016)]. It illustrates that, if the model is unichain, but the action space is not compact, then the statement of Theorem 5.13 may fail to hold.

Example. Let $\mathbf{X} = \{1, 2\}$, $\mathbf{A} = \{1, 2, \ldots\}$, $J = 0$, the initial distribution P_0 is arbitrary, $q(2|1, a) = q(1|2, a) \equiv 1$, $q(1|1, a) = q(2|2, a) \equiv -1$, $c_0(x, a) = -2^a$ for all $x \in \mathbf{X}$. See Fig. 5.17.

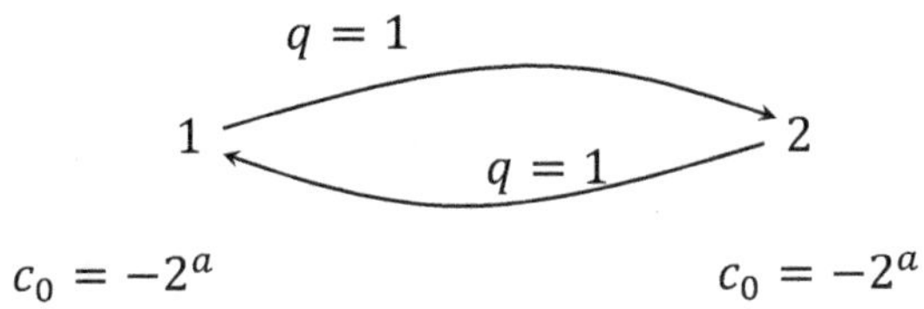

Fig. 5.17 The optimal stationary π-strategy is not finitely mixed.

Evidently, this model is unichain and Condition 5.6 is satisfied apart from the compactness of $\mathbf{A}$ and boundedness of the cost rate $c_0(\cdot, \cdot)$. In Conditions 5.2 and 5.6 one can put $w(x) = w'(x) \equiv 1$ for all $x \in \mathbf{X}$.

All stationary π-strategies are stable:

$$\sum_{x \in \mathbf{X}} \left(\sum_{a \in \mathbf{A}} q(y|x, a)\pi_\eta(a|x) \right) \eta(x \times \mathbf{A}) = \eta((3-y) \times \mathbf{A}) - \eta(y \times \mathbf{A}) = 0, \quad y \in \mathbf{X};$$

$\pi_\eta(a|x)$ can be arbitrary and $\eta(1 \times \mathbf{A}) = \eta(2 \times \mathbf{A}) = \frac{1}{2}$. Under any deterministic stationary strategy φ, the function $c_0(x, \varphi(x)) : \mathbf{X} \to \mathbb{R}$ is bounded, so that $-\infty < W_0^A(\varphi, \mathsf{P}_0) < 0$. On the other hand, for the stationary π-strategy $\pi^{*s}(a|x) := \left(\frac{1}{2}\right)^a$, $\sum_{a \in \mathbf{A}} c_0(x, a)\pi^{*s}(a|x) = -\infty$ and $W_0^A(\pi^{*s}, \mathsf{P}_0) = -\infty$. Thus, π^{*s} is average uniformly optimal and optimal in the unconstrained version of problem (5.41). Furthermore, no-one optimal stationary π-strategy is L-mixed for finite L. □

Remark 5.8. CTMDP from the last example in Subsection 5.3.10 satisfies all the conditions of Theorem 5.13. The 2-mixed π-strategy solving problem (5.41) is as follows:

$$\pi^{*s}(a_1|1) = \frac{1}{3}, \quad \pi^{*s}(a_2|1) = \frac{2}{3}, \quad \pi^{*s}(a_1|2) = 1.$$

The corresponding stable measure is given by

$$\eta(1, a_1) = \frac{1}{4}, \quad \eta(1, a_2) = \frac{1}{2}, \quad \eta(2, a_1) = \frac{1}{4}, \quad \eta(2, a_2) = 0.$$

It has the form $\eta = \frac{1}{2}\eta_1 + \frac{1}{2}\eta_2$, where the stable measures η_1 and η_2 are associated with deterministic stationary strategies $\varphi^1(x) \equiv a_1$ and $\varphi^2(x) \equiv a_2$ correspondingly:

$$\eta_1(1, a_1) = \frac{1}{2}, \quad \eta_1(2, a_1) = \frac{1}{2}, \quad \eta_1(1, a_2) = \eta_1(2, a_2) = 0;$$

$$\eta_2(1,a_2)=1,\quad \eta_2(1,a_1)=\eta_2(2,a_1)=\eta_2(2,a_2)=0.$$

5.3.13 *Ratio-average cost*

The current subsection was motivated by [Feinberg (1994)]. In that article, only randomized strategies were considered which is usual for semi-Markov decision processes.

Assume that there is not explosion. Along with problem (5.41), one can consider the following constrained problem:

$$\left.\begin{array}{l} G_0^A(S,\mathsf{P}_0)\to\inf_S \\ G_j^A(S,\mathsf{P}_0)\le d_j,\quad j=1,2,\dots,J, \end{array}\right\} \tag{5.49}$$

where the *ratio-average* costs are defined by

$$G_j^A(S,\mathsf{P}_0)=\limsup_{N\to\infty}\frac{\mathsf{E}^S_{\mathsf{P}_0}\left[\int_{(0,T_N]} c_j(x,a)\bar{\mathsf{P}}^S_{\mathsf{P}_0}(t,dx\times da)dt\right]}{\mathsf{E}^S_{\mathsf{P}_0}[T_n]}. \tag{5.50}$$

Below, we consider the case of a finite unichain model with $q_x(a)>0$ for all $x\in\mathbf{X}$, $a\in\mathbf{A}$ and compare problems (5.41) and (5.49) assuming that there exists a feasible strategy in each of them. Note that these problems are feasible or unfeasible simultaneously [Feinberg (1994), Cor. 8.2(ii)]. It is natural to call problem (5.41) *time-average* constrained problem.

According to Theorem 5.6(b), for each strategy S, there is an (equivalent) randomized Markov strategy ϖ^{m} such that $m^{\varpi^{\mathrm{m}},0}_{\mathsf{P}_0,n}=m^{S,0}_{\mathsf{P}_0,n}$, $n=1,2,\dots$. Therefore, randomized Markov strategies are sufficient in problem (5.49). Moreover, in this problem one can restrict to randomized stationary strategies ϖ^{s} by [Feinberg (1994), Thm. 6.4(ii)].

Suppose a randomized stationary strategy ϖ^{s} is fixed. One can obviously represent the nominator and the denominator in (5.50) using the detailed occupation measures $m^{\varpi^{\mathrm{s}},0}_{\mathsf{P}_0,n}$:

$$\mathsf{E}^{\varpi^{\mathrm{s}}}_{\mathsf{P}_0}\left[\int_{(0,T_N]} c_j(x,a)\bar{\mathsf{P}}^{\varpi^{\mathrm{s}}}_{\mathsf{P}_0}(t,dx\times da)dt\right]=\sum_{n=1}^N\sum_{(x,a)\in\mathbf{X}\times\mathbf{A}} c_j(x,a)m^{\varpi^{\mathrm{s}},0}_{\mathsf{P}_0,n}(x,a);$$

$$\mathsf{E}^{\varpi^{\mathrm{s}}}_{\mathsf{P}_0}[T_n]=\sum_{n=1}^N m^{\varpi^{\mathrm{s}},0}_{\mathsf{P}_0,n}(\mathbf{X}\times\mathbf{A}).$$

According to Theorem 5.7, one can pass to the induced DTMDP with $\sigma^{\mathrm{s}} = \varpi^{\mathrm{s}}$:

$$\begin{aligned}\sum_{(x,a)\in\mathbf{X}\times\mathbf{A}} c_j(x,a)m^{\varpi^{\mathrm{s}},0}_{\mathsf{P}_0,n}(x,a) &= \sum_{(x,a)\in\mathbf{X}\times\mathbf{A}} \left(\frac{c_j(x,a)}{q_x(a)}\right)\sigma^{\mathrm{s}}(a|x)\mathbb{P}^{\sigma^{\mathrm{s}}}_{\mathsf{P}_0}(X_{n-1}=x)\\ &= \mathbb{E}^{\sigma^{\mathrm{s}}}_{\mathsf{P}_0}\left[\sum_{a\in\mathbf{A}}\left(\frac{c_j(X_{n-1},a)}{q_{X_{n-1}}(a)}\right)\sigma^{\mathrm{s}}(a|X_{n-1})\right];\\ m^{\varpi^{\mathrm{s}},0}_{\mathsf{P}_0,n}(\mathbf{X}\times\mathbf{A}) &= \sum_{(x,a)\in\mathbf{X}\times\mathbf{A}}\left(\frac{1}{q_x(a)}\right)\sigma^{\mathrm{s}}(a|x)\mathbb{P}^{\sigma^{\mathrm{s}}}_{\mathsf{P}_0}(X_{n-1}=x)\\ &= \mathbb{E}^{\sigma^{\mathrm{s}}}_{\mathsf{P}_0}\left[\sum_{a\in\mathbf{A}}\left(\frac{1}{q_{X_{n-1}}(a)}\right)\sigma^{\mathrm{s}}(a|X_{n-1})\right],\end{aligned}$$

where $\mathbb{P}^{\sigma^{\mathrm{s}}}_{\mathsf{P}_0}$ is the strategic measure in the induced DTMDP and $\mathbb{E}^{\sigma^{\mathrm{s}}}_{\mathsf{P}_0}$ is the corresponding mathematical expectation. The discrete-time controlled Markov process X_n, $n=0,1,\ldots$ under the strategy σ^{s} is unichain. Therefore, there exist standard limits

$$\begin{aligned}&\lim_{N\to\infty}\frac{1}{N}\mathbb{E}^{\sigma^{\mathrm{s}}}_{\mathsf{P}_0}\left[\sum_{n=1}^{N}\sum_{a\in\mathbf{A}}\left(\frac{c_j(X_{n-1},a)}{q_{X_{n-1}}(a)}\right)\sigma^{\mathrm{s}}(a|X_{n-1})\right]\\ &= \sum_{x\in\mathbf{X}}\left(\sum_{a\in\mathbf{A}}\left(\frac{c_j(x,a)}{q_x(a)}\right)\sigma^{\mathrm{s}}(a|x)\right)\hat{\eta}^{\sigma^{\mathrm{s}}}(x);\\ &\lim_{N\to\infty}\frac{1}{N}\mathbb{E}^{\sigma^{\mathrm{s}}}_{\mathsf{P}_0}\left[\sum_{n=1}^{N}\sum_{a\in\mathbf{A}}\left(\frac{1}{q_{X_{n-1}}(a)}\right)\sigma^{\mathrm{s}}(a|X_{n-1})\right]\\ &= \sum_{x\in\mathbf{X}}\left(\sum_{a\in\mathbf{A}}\left(\frac{1}{q_x(a)}\right)\sigma^{\mathrm{s}}(a|x)\right)\hat{\eta}^{\sigma^{\mathrm{s}}}(x)\end{aligned}$$

by Theorem 4.8, where $\hat{\eta}^{\sigma^{\mathrm{s}}}$ is the invariant probability measure on $\mathcal{B}(\mathbf{X})$, that is the unique normalized solution to equation

$$\hat{\eta}^{\sigma^{\mathrm{s}}}(x) = \sum_{y\in\mathbf{X},y\neq x}\left(\sum_{a\in\mathbf{A}}\frac{\tilde{q}(x|y,a)}{q_y(a)}\sigma^{\mathrm{s}}(a|y)\right)\hat{\eta}^{\sigma^{\mathrm{s}}}(y),\quad x\in\mathbf{X}. \tag{5.51}$$

Therefore,

$$G^A_j(\varpi^{\mathrm{s}},\mathsf{P}_0) = \frac{\sum_{x\in\mathbf{X}}\left(\sum_{a\in\mathbf{A}}\left(\frac{c_j(x,a)}{q_x(a)}\right)\sigma^{\mathrm{s}}(a|x)\right)\hat{\eta}^{\sigma^{\mathrm{s}}}(x)}{\sum_{x\in\mathbf{X}}\left(\sum_{a\in\mathbf{A}}\left(\frac{1}{q_x(a)}\right)\sigma^{\mathrm{s}}(a|x)\right)\hat{\eta}^{\sigma^{\mathrm{s}}}(x)}, \tag{5.52}$$

see also [Feinberg (1994), (3.3)]. In particular, $G^A_j(\varpi^{\mathrm{s}},\mathsf{P}_0)$ does not depend on P_0.

Suppose $\{\varphi^l\}_{l=1}^L$ is a collection of stationary selectors in the induced DT-MDP with the corresponding invariant probability measures $\hat{\eta}^{\varphi^l}$. The stationary strategy $\sigma^s(a|y)$, such that

$$\sigma^s(a|x)\sum_{l=1}^L \beta_l \hat{\eta}^{\varphi^l}(x) = \sum_{l=1}^L \beta_l \mathbb{I}\{a = \varphi^l(x)\}\hat{\eta}^{\varphi^l}(x), \quad x \in \mathbf{X}$$

is called *L-mixed*. Here $\beta_l \geq 0$, $\sum_{l=1}^L \beta_l = 1$. According to [Feinberg (1994), Thm. 6.5(iv)], there exists a randomized stationary strategy $\varpi^s = \sigma^s$ which is $(J+1)$-mixed and optimal in problem (5.49). Therefore, the optimal value of problem (5.49) equals

$$\begin{array}{rcl} \text{optimal value of (5.49)} & & \text{optimal value of (5.49)} \\ \text{in the space of randomized} & = & \text{in the space of randomized} \\ \text{stationary strategies} & & (J+1)\text{-mixed strategies.} \end{array}$$

According to Theorem 5.13, there exists an optimal in problem (5.41) relaxed $(J+1)$-mixed strategy, and the optimal value of problem (5.41) coincides with the optimal value $\mathbb{C}$ of linear program (5.42) restricted to stable probability measures of the form $\eta = \sum_{l=1}^L \alpha_l \eta_l$ with η_l corresponding to deterministic stationary strategies φ^l. ($\alpha_l \geq 0$, $\sum_{l=1}^L \alpha_l = 1$.) In its turn, $\mathbb{C}$ equals the optimal value of problem (5.49) in the space of randomized $(J+1)$-mixed strategies by the proof of Theorem 8.4 of [Feinberg (1994)]: see the linear programs (6.8)-(6.11) and (8.1)-(8.4) therein. Finally, the optimal values of problems (5.41) and (5.49) in the space of randomized stationary strategies coincide by [Feinberg (1994), Cor. 8.2(ii)].

To summarize, the optimal value of problem (5.41) equals

$$\begin{array}{rcl} \text{optimal value of (5.41)} & = & \text{optimal value of (5.49)} \\ \text{in the space of relaxed} & & \text{in the space of randomized} \\ (J+1)\text{-mixed strategies} & & (J+1)\text{-mixed strategies} \end{array}$$

$$\begin{array}{rrcl} = & \text{optimal value of (5.49)} & = & \text{optimal value of (5.41)} \\ & \text{in the space of randomized} & & \text{in the space of randomized} \\ & \text{stationary strategies} & & \text{stationary strategies} \end{array}$$

equals the optimal value of problem (5.49).

Recall that all the mentioned above problems are solvable by Theorem 5.13 and by Corollary 8.2 and Theorem 6.5 of [Feinberg (1994)], that is, all the mentioned optimal $(J+1)$-mixed strategies and randomized stationary strategies do exist, provided the constrained problems are feasible.

If problems (5.41) and (5.49) are unconstrained (i.e., $J = 0$), then the optimal solutions are given by randomized 1-mixed strategies, that is just by deterministic stationary strategies φ. As was shown in [Feinberg (1994), p. 262], in this case $G_0^A(\varphi, \mathsf{P}_0) = W_0^A(\varphi, \mathsf{P}_0)$ for each φ, and these functions do not depend on P_0, as was already mentioned above.

In (feasible) constrained problems (5.41) and (5.49) with $J > 0$, there exist optimal randomized stationary strategies, and a randomized stationary strategy, optimal in one problem, is also optimal in another problem [Feinberg (1994), Cor. 8.2(ii)]. On the contrary, optimal solutions to problems (5.41) and (5.49) are usually provided by *different* $(J+1)$-mixed strategies, as the following example illustrates.

Example. Let $\mathbf{X} = \{1, 2\}$, $\mathbf{A} = \{a_1, a_2\}$, $J = 1$; the initial distribution P_0 is immaterial. The transition rate is given by

$$q(2|1, a_1) = 1, \quad q(1|1, a_1) = -1, \quad q(2|1, a_2) = 2, \quad q(1|1, a_2) = -2,$$

$$q(1|2, a) \equiv 1, \quad q(2|2, a) \equiv -1,$$

and the cost rates are

$$c_0(1, a_1) = 6, \quad c_0(1, a_2) = 0, \quad c_0(2, a) \equiv 6;$$

$$c_1(1, a_1) = 0, \quad c_1(1, a_2) = 6, \quad c_1(2, a) \equiv 0.$$

Finally, let $d_1 = 1$. See Fig. 5.18.

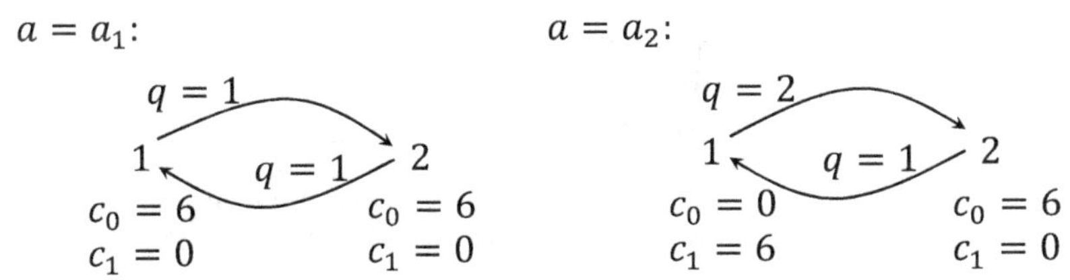

Fig. 5.18 Time-average and ratio-average constrained problems are indeed different.

The actions in state 2 are immaterial, and in this model there are only two essentially different deterministic stationary strategies: $\varphi^1(1) = a_1$ and $\varphi^2(1) = a_2$.

Let us calculate the optimal (relaxed) 2-mixed strategy as in Definition 5.9 for problem (5.41). The case of φ^1 is trivial: $W_0^A(\varphi^1, \mathsf{P}_0) = 6$ and $W_1^A(\varphi^1, \mathsf{P}_0) = 0$. Calculation for φ^2 can be based on the ergodic property of the process $X(\cdot)$ under φ^2 as in Proposition 5.1 and in Theorem 5.11. The invariant probability measure $\hat{\eta}_2$ comes from the balance equations:

$$-2\hat{\eta}_2(1) + \hat{\eta}_2(2) = 0, \quad \hat{\eta}_2(1) + \hat{\eta}_2(2) = 1, \quad \Longrightarrow \hat{\eta}_2(1) = \frac{1}{3}, \ \hat{\eta}_2(2) = \frac{2}{3}.$$

Thus,

$$W_0^A(\varphi^2, \mathsf{P}_0) = 6\hat{\eta}_2(2) = 4, \quad W_1^A(\varphi^2, \mathsf{P}_0) = 6\hat{\eta}_2(1) = 2.$$

The stable measures as in Definition 5.8, corresponding to the (stable) deterministic stationary strategies φ^1 and φ^2 are as follows

$$\eta_1(1, a_1) = \hat{\eta}_1(1) = \frac{1}{2}, \quad \eta_1(1, a_2) = 0, \quad \hat{\eta}_1(2) = \eta_1(2, a_1) + \eta_1(2, a_2) = \frac{1}{2};$$

$$\eta_2(1, a_1) = 0, \quad \eta_2(1, a_2) = \hat{\eta}_2(1) = \frac{1}{3}, \quad \hat{\eta}_2(2) = \eta_2(2, a_1) + \eta_2(2, a_2) = \frac{2}{3}.$$

The relaxed 2-mixed strategy π corresponds to the stable measure $\eta = \alpha\eta_1 + (1-\alpha)\eta_2$, $\alpha \in [0, 1]$; see Definition 5.9. For this strategy π, we have

$$\begin{aligned} W_0^A(\pi, \mathsf{P}_0) &= \alpha W_0^A(\varphi^1, \mathsf{P}_0) + (1-\alpha) W_0^A(\varphi^2, \mathsf{P}_0) = 6\alpha + 4(1-\alpha); \\ W_1^A(\pi, \mathsf{P}_0) &= \alpha W_1^A(\varphi^1, \mathsf{P}_0) + (1-\alpha) W_1^A(\varphi^2, \mathsf{P}_0) = 2(1-\alpha), \end{aligned}$$

so that $W_0^A(\pi, \mathsf{P}_0) + W_1^A(\pi, \mathsf{P}_0) = 6$, and the unique optimal value of α is $\alpha^* = \frac{1}{2}$ leading to $W_0^A(\pi^*, \mathsf{P}_0) = 5$ and $W_1^A(\pi^*, \mathsf{P}_0) = 1$. The optimal relaxed 2-mixed strategy $\pi^* = \pi_{\eta^*}$ comes from the decomposition of the measure $\eta^* := \alpha^*\eta_1 + (1-\alpha^*)\eta_2$:

$$\begin{aligned} &\pi^*(a_1|1)\{\alpha^*[\eta_1(1, a_1) + \eta_1(1, a_2)] + (1-\alpha^*)[\eta_2(1, a_1) + \eta_2(1, a_2)]\} \\ &= \alpha^*\eta_1(1, a_1) + (1-\alpha^*)\eta_2(1, a_1) : \end{aligned}$$

$$\pi^*(a_1|1) = \frac{\frac{1}{4}}{\frac{1}{4} + \frac{1}{2} \cdot \frac{1}{3}} = \frac{3}{5}; \quad \pi^*(a_2|1) = \frac{2}{5}.$$

The actions in state 2 can be fixed arbitrarily. This relaxed 2-mixed strategy is stationary and optimal in problem (5.41).

Now calculate the optimal randomized 2-mixed strategy as on p.418 for problem (5.49). The invariant probability measures (in the induced DTMDP) $\hat{\eta}^{\varphi^1}$ and $\hat{\eta}^{\varphi^2}$ come from equation (5.51) and are the same:

$$\hat{\eta}^{\varphi^{1.2}}(1) = \hat{\eta}^{\varphi^{1.2}}(2), \quad \hat{\eta}^{\varphi^{1.2}}(1) + \hat{\eta}^{\varphi^{1.2}}(2) = 1 \implies \hat{\eta}^{\varphi^{1.2}}(1) = \hat{\eta}^{\varphi^{1.2}}(2) = \frac{1}{2}.$$

Actually, the invariant probability measure is the same for all stationary strategies in the induced DTMDP:

$$\hat{\eta}^{\sigma^s}(1) = \hat{\eta}^{\sigma^s}(2) = \frac{1}{2} \quad \forall \sigma^s. \tag{5.53}$$

The 2-mixed randomized strategy with coefficients $\beta \in [0, 1]$ and $(1-\beta)$, correspondingly for the stationary selectors φ^1 and φ^2, is given by

$$\begin{aligned} \varpi^s(a_1|1) &= \sigma^s(a_1|1) = \beta\mathbb{I}\{a_1 = \varphi^1(x)\} + (1-\beta)\mathbb{I}\{a_1 = \varphi^2(x)\} = \beta, \\ \varpi^s(a_2|1) &= \sigma^s(a_2|1) = 1 - \beta. \end{aligned}$$

The actions in state 2 are immaterial as usual. Therefore, we obtain from (5.52) and (5.53):

$$G_0^A(\varpi^s, \mathsf{P}_0) = \frac{[6\beta + 6]\frac{1}{2}}{[\beta + \frac{1}{2}(1-\beta) + 1]\frac{1}{2}} = \frac{12\beta + 12}{3+\beta};$$
$$G_1^A(\varpi^s, \mathsf{P}_0) = \frac{[3(1-\beta)]\frac{1}{2}}{[\beta + \frac{1}{2}(1-\beta) + 1]\frac{1}{2}} = \frac{6-6\beta}{3+\beta}.$$

The value of $G_0^A(\varpi^s, \mathsf{P}_0)$ increases and the value of $G_1^A(\varpi^s, \mathsf{P}_0)$ decreases with β; their sum equals 6. Thus, the unique optimal value of β is $\beta^* = \frac{3}{7}$ providing $G_0^A(\varpi^{*s}, \mathsf{P}_0) = 5$ and $G_1^A(\varpi^{*s}, \mathsf{P}_0) = 1$; $\varpi^{*s}(a_1|1) = \beta^* = \frac{3}{7}$, $\varpi^{*s}(a_2|1) = 1-\beta^* = \frac{4}{7}$. We see that the optimal mixing coefficients $\alpha^* = \frac{1}{2}$ in problem (5.41) and $\beta^* = \frac{3}{7}$ in problem (5.49) are different. The obtained randomized stationary strategy ϖ^{*s} is optimal in problem (5.49).

Let us calculate the optimal randomized stationary strategy ϖ^{*s} in problem (5.41). The values of $\varpi^{*s}(a_1|2)$ and $\varpi^{*s}(a_2|2)$ are of no importance. Suppose the value of $\varpi^s(a_1|1)$ is fixed and $\varpi^s(a_2|1) = 1 - \varpi^s(a_1|1)$. We have the continuous-time Markov chain with states $(1, a_1)$, $(1, a_2)$ and $(2, .)$: the states $(2, a_1)$ and $(2, a_2)$ are merged together to the state $(2, .)$. The transition rates are as follows:

$$\begin{aligned} \lambda((2,.)|(1,a_1)) &= 1, & \lambda((1,a_1))|(1,a_1)) &= -1,\\ \lambda((2,.)|(1,a_2)) &= 2, & \lambda((1,a_2))|(1,a_2)) &= -2,\\ \lambda((1,a_1)|(2,.)) &= \varpi^s(a_1|1), & \lambda((1,a_2)|(2,.)) &= 1-\varpi^s(a_1|1),\\ & & \lambda((2,.)|(2,.)) &= -1. \end{aligned}$$

The transition rate to the original state 1 from 2 is 1; after that the state $(1, a_1)$ is realized with probability $\varpi^s(a_1|1)$ and remains constant up to the next jump. The cost rates are

$$c_0(1,a_1) = 6, \;\; c_0(1,a_2) = 0, \;\; c_0(2,.) = 6,$$
$$c_1(1,a_1) = 0, \;\; c_1(1,a_2) = 6, \;\; c_1(2,.) = 0.$$

See Fig. 5.19.

The stationary probabilities $\eta(1, a_1)$, $\eta(1, a_2)$ and $\hat{\eta}(2)$, correspondingly of the states $(1, a_1)$, $(1, a_2)$ and $(2, .)$, come from the balance equations

$$\begin{aligned} -\eta(1,a_1) + \varpi^s(a_1|1)\hat{\eta}(2) &= 0;\\ -2\eta(1,a_2) + (1-\varpi^s(a_1|1))\hat{\eta}(2) &= 0;\\ \eta(1,a_1) + \eta(1,a_2) + \hat{\eta}(2) &= 1: \end{aligned}$$

$$\eta(1,a_1) = \frac{2\varpi^s(a_1|1)}{3+\varpi^s(a_1|1)}, \;\; \eta(1,a_2) = \frac{1-\varpi^s(a_1|1)}{3+\varpi^s(a_1|1)}, \;\; \eta(2) = \frac{2}{3+\varpi^s(a_1|1)}.$$

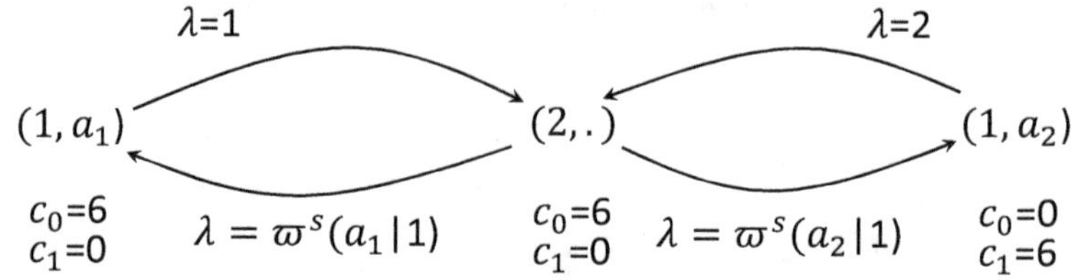

Fig. 5.19 Calculation of $W_j^A(\varpi^{\rm s}, \mathsf{P}_0)$.

The jump process under study is ergodic and

$$W_0^A(\varpi^{\rm s}, \mathsf{P}_0) = 6\eta(1,a_1) + 6\hat{\eta}(2) = \frac{12\varpi^{\rm s}(a_1|1) + 12}{3 + \varpi^{\rm s}(a_1|1)};$$

$$W_1^A(\varpi^{\rm s}, \mathsf{P}_0) = 6\eta(1,a_2) = \frac{6 - 6\varpi^{\rm s}(a_1|1)}{3 + \varpi^{\rm s}(a_1|1)}.$$

The value of $W_0^A(\varpi^{\rm s}, \mathsf{P}_0)$ increases and the value of $W_1^A(\varpi^{\rm s}, \mathsf{P}_0)$ decreases with $\varpi^{\rm s}(a_1|1)$; their sum equals 6. Thus, the optimal value of $\varpi^{\rm s}(a_1|1)$ in problem (5.41) equals $\varpi^{*\rm s}(a_1|1) = \frac{3}{7}$ providing $W_0^A(\varpi^{*\rm s}, \mathsf{P}_0) = 5$ and $W_1^A(\varpi^{*\rm s}, \mathsf{P}_0) = 1$.

One can also compute the relaxed stationary strategy optimal in problem (5.41) using Theorem 5.11. The linear program (5.45) has the form

$$\begin{aligned} 6\eta(1,a_1) + 6\hat{\eta}(2) &\to \min \\ \eta(1,a_1),\ \eta(1,a_2),\ \hat{\eta}(2) \geq 0, \qquad & \eta(1,a_1) + \eta(1,a_2) + \hat{\eta}(2) = 1, \\ -\eta(1,a_1) - 2\eta(1,a_2) + \hat{\eta}(2) &= 0 \quad (\text{for } y = 1), \\ 6\eta(1,a_2) &\leq 1. \end{aligned}$$

Here again we merge together the probabilities $\eta(2,a_1)$ and $\eta(2,a_2)$: $\hat{\eta}(2) := \eta(2,a_1) + \eta(2,a_2)$. The optimal solution is

$$\eta^*(1,a_1) = \frac{1}{4}, \quad \eta^*(1,a_2) = \frac{1}{6}, \quad \hat{\eta}^*(2) = \frac{7}{12},$$

and the optimal relaxed stationary strategy is given by

$$\pi^{*\rm s} = \pi_{\eta^*}: \qquad \pi^{*\rm s}(a_1|1) = \frac{3}{5}, \quad \pi^{*\rm s}(a_2|1) = \frac{2}{5};$$

the actions in state 2 can be fixed arbitrarily. As expected, this strategy $\pi^{*\rm s}$ coincides with the optimal in problem (5.41) (relaxed) 2-mixed strategy π^* calculated earlier, and the measure η^* coincides with the optimal stable measure η^* corresponding to π^* (up to the immaterial specific values $\eta^*(2,a_1)$ and $\eta^*(2,a_2)$).

The randomized stationary strategy, equivalent to π^{*s} in the sense of *equality* (5.18) comes from Theorem 5.6(c):

$$\varpi^{*s}(a_1|1) = \frac{\pi^{*s}(a_1|1)\int_0^\infty e^{-[\pi^{*s}(a_1|1)+2\pi^{*s}(a_2|1)]\theta}d\theta}{1-e^{-\int_0^\infty[\pi^{*s}(a_1|1)+2\pi^{*s}(a_2|1)]du}} = \frac{3}{7},\ \varpi^{*s}(a_2|1) = \frac{4}{7}.$$

No surprise, it coincides with the optimal randomized stationary strategy ϖ^{*s} calculated earlier. The actions in state 2 can be fixed arbitrarily.

Now compute the optimal randomized stationary strategy ϖ^{*s} in problem (5.49) straightforwardly using formulas (5.52) and (5.53). For $\varpi^s = \sigma^s$ we have

$$G_0^A(\varpi^s, \mathsf{P}_0) = \frac{6\sigma^s(a_1|1)+6}{\sigma^s(a_1|1)+\frac{1}{2}\sigma^s(a_2|1)+1} = \frac{12\sigma^s(a_1|1)+12}{3+\sigma^s(a_1|1)};$$
$$G_1^A(\varpi^s, \mathsf{P}_0) = \frac{\frac{6}{2}\sigma^s(a_2|1)}{\sigma^s(a_1|1)+\frac{1}{2}\sigma^s(a_2|1)+1} = \frac{6-6\sigma^s(a_1|1)}{3+\sigma^s(a_1|1)}.$$

We obtained $G_{0,1}^A(\varpi^s, \mathsf{P}_0) = W_{0,1}^A(\varpi^s, \mathsf{P}_0)$. Hence, the optimal randomized stationary strategy is $\varpi^{*s}(a_1|1) = \frac{3}{7}$ and $\varpi^{*s}(a_2|1) = \frac{4}{7}$ as expected, providing $G_0^A(\varpi^{*s}, \mathsf{P}_0) = 5$ and $G_1^A(\varpi^{*s}, \mathsf{P}_0) = 1$. The values of $\varpi^{*s}(a|2)$ play no role. □

5.3.14 *Average variance*

Suppose there is no explosion.

In case there are several deterministic stationary strategies φ providing the infimum in problem (5.8), it makes sense to choose the one providing the minimum to the *long-run average variance*

$$\sigma^2(\varphi, x) := \limsup_{T\to\infty} \frac{1}{T} Var(\varphi, x, T),$$

where

$$Var(\varphi, x, T) := \mathsf{E}_x^\varphi\left[\left(\int_{(0,T]} c(X(t-), \varphi(X(t-)))dt\right.\right.$$
$$\left.\left.-\mathsf{E}_x^\varphi[\int_{(0,T]} c(X(t-), \varphi(X(t-)))dt]\right)^2\right].$$

Below in this subsection, we consider the finite model (with finite state and action spaces $\mathbf{X}$ and $\mathbf{A}$) and a fixed deterministic stationary strategy φ. Thus, parameter φ is omitted for brevity (so that, e.g., $W^A(i) := W^A(\varphi, i)$). The matrix of transition rates $q(j|i, \varphi(i))$ in the canonical form is as follows:

$$Q = \begin{bmatrix} Q^1 & 0 & \dots & 0 & 0 \\ 0 & Q^2 & \dots & 0 & 0 \\ \dots & \dots & \dots & \dots & \dots \\ 0 & 0 & \dots & Q^m & 0 \\ W^1 & W^2 & \dots & W^m & W^{m+1} \end{bmatrix}.$$

m is the number of closed communicating (i.e., ergodic) classes and $W^1, \dots, W^m$ (W^{m+1}) are the matrices of transition rates from transient states to the corresponding ergodic classes (respectively, to the transient states). The matrix of limiting probabilities (see Definition 5.3) $\lim_{t\to\infty} \bar{\mathsf{P}}_i(t,j) =: \mathsf{P}^*_{ij}$ again has the block structure

$$\mathsf{P}^* = \begin{bmatrix} \mathsf{P}^{*1} & 0 & \dots & 0 & 0 \\ 0 & \mathsf{P}^{*2} & \dots & 0 & 0 \\ \dots & \dots & \dots & \dots & \dots \\ 0 & 0 & \dots & \mathsf{P}^{*m} & 0 \\ W^{*1} & W^{*2} & \dots & W^{*m} & 0 \end{bmatrix};$$

$\mathsf{P}^{*k} = \mathbf{1}^k p^k$, where $\mathbf{1}^k$ is the column vector of ones of the corresponding dimensionality and p^k, the row vector, is the invariant probability measure for the k-th ergodic class; $W^k \mathsf{P}^{*k} + W^{m+1} W^{*k} = 0$. Finally, denote

$$h_i := \int_0^\infty \sum_{j\in\mathbf{X}} [\bar{\mathsf{P}}_i(t,j) - \mathsf{P}^*_{ij}] c(j) dt, \quad i \in \mathbf{X}.$$

Note that $W^A(i) = \sum_{j\in\mathbf{X}} \mathsf{P}^*_{ij} c(j)$. All the introduced matrices and vectors are finite and well defined. For details see [Chung (1967)].

Theorem 5.14. [Guo and Song (2009), Lemma 5, Thm. 3].

(a)

$$\sigma^2(i) = \limsup_{T\to\infty} \frac{1}{T} \mathsf{E}_i \left[\left(\int_0^T [c(X(t)) - W^A(i)] dt \right)^2 \right] \quad \forall i \in \mathbf{X}.$$

(b) If i belongs to an ergodic class, then

$$\sigma^2(i) = 2 \sum_{j\in\mathbf{X}} \mathsf{P}^*_{ij} [c(j) - W^A(j)] h_j.$$

(c) If each transient state reaches only one ergodic class, then

$$\sigma^2(i) = 2 \sum_{j\in\mathbf{X}} \mathsf{P}^*_{ij} [c(j) - W^A(j)] h_j \quad \forall i \in \mathbf{X}.$$

The following example, published in [Guo and Song (2009), Ex. 4.1], shows that the condition in Theorem 5.14(c) is important.

Example. Let $\mathbf{X} = \{1, 2, 3, 4\}$, the matrix of transition rates is

$$Q = \begin{bmatrix} 0 & 0 & 0 & 0 \\ 0 & -0.7 & 0.7 & 0 \\ 0 & 0.7 & -0.7 & 0 \\ 0.8 & 0 & 0.2 & -1 \end{bmatrix},$$

$c(i) = i$. See Fig. 5.20.

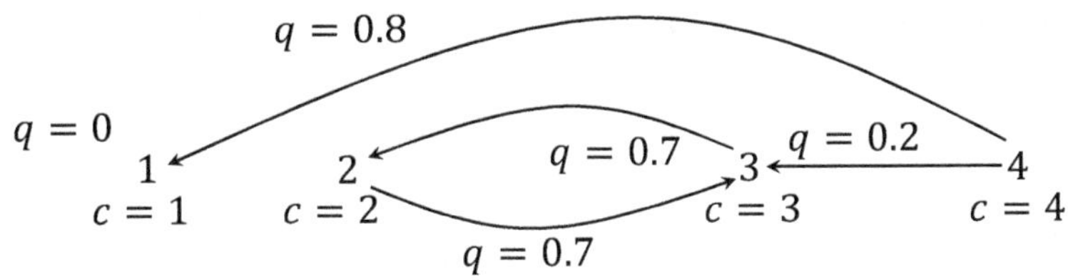

Fig. 5.20 Infinite average variance.

Straightforward calculations give

$$\mathsf{P}^* = \begin{bmatrix} 1 & 0 & 0 & 0 \\ 0 & 0.5 & 0.5 & 0 \\ 0 & 0.5 & 0.5 & 0 \\ 0.8 & 0.1 & 0.1 & 0 \end{bmatrix}; \quad W^A = \begin{bmatrix} 1 \\ 2.5 \\ 2.5 \\ 1.3 \end{bmatrix}.$$

There are two ergodic classes $\{1\}$ and $\{2, 3\}$. State 4 is transient, but can reach the both of them, and the statement (c) of Theorem 5.14 is not valid since $\sigma^2(4) = \infty$. To show this, note that $\bar{\mathsf{P}}_4(t_0, 1) > 0$ for some (and all) $t_0 > 0$ because $q(1|4) = 0.8 > 0$. According to Theorem 5.14(a),

$$\begin{aligned} \sigma^2(4) &= \limsup_{T\to\infty} \frac{1}{T} \mathsf{E}_4 \left[\mathsf{E}_4 \left[\left(\int_0^{t_0} [c(X(t)) - W^A(4)]dt \right.\right.\right. \\ &\quad \left.\left.\left. + \int_{t_0}^T [c(X(t)) - W^A(4)]dt \right)^2 \right| X(t_0) \right] \right] \\ &\geq \limsup_{T\to\infty} \frac{1}{T} \bar{\mathsf{P}}_4(t_0, 1) \left\{ \mathsf{E}_4 [K(T) | X(t_0) = 1] \right. \\ &\quad \left. + \mathsf{E}_4 \left[\left(\int_{t_0}^T [c(X(t)) - W^A(4)]dt \right)^2 \right| X(t_0) = 1 \right] \right\}, \end{aligned}$$

where

$$K(T) := 2 \int_0^{t_0} [c(X(t)) - W^A(4)]dt \times \int_{t_0}^T [c(X(t)) - W^A(4)]dt.$$

If $X(t_0) = 1$, then for all $t \geq t_0$ $c(X(t)) = c(1) = 1$ because state 1 is absorbing. Thus, for $T > t_0$ we have

$$\begin{aligned} \mathsf{E}_4[K(T)|X(t_0 = 1] &\geq 2t_0 \max_{i\in\mathbf{X}}[c(i) - W^A(4)] \cdot [c(1) - W^A(4)] \cdot (T - t_0) \\ &= 2t_0 \cdot 2.7 \cdot (-0.3)(T - t_0) \end{aligned}$$

resulting in the desired inequality:

$$\sigma^2(4) \geq \limsup_{T\to\infty} \frac{1}{T}\bar{\mathsf{P}}_4(t_0, 1)\left\{-1.62t_0(T - t_0) + [(-0.3)(T - t_0)]^2\right\} = \infty.$$

□

5.3.15 *When fluid approximation does not work*

Before describing the fluid approximation to birth-and-death processes, consider the following real-life example: see [Piunovskiy (2009b)] and references therein. When studying information flows (e.g. on the Internet) the nodes are often considered as $M/M/1$ queues. Small packets of information, size $\frac{1}{n}$ Mb (or Gb), arrive in a buffer according to a Poisson stream with intensity λ Mb/sec $= n\lambda$ packets/sec and go through according to the service intensity μ Mb/sec $= n\mu$ packets/sec. Intensities $\lambda = \lambda(y)$ and $\mu = \mu(y)$ can depend on the state of the buffer, i.e., the buffer filling size y Mb $= yn$ packets, in case there is a feedback from the buffer to the sources of information and to the server(processor). Holding y Mb in the buffer is associated with the reward/cost of $g(y)$ units/sec. All the presented quantities can also depend on the current action a. Suppose one is interested in the total undiscounted expected cost during the time interval leading to the empty buffer, in case there are i packets at the beginning.

The described above situation leads to the following CTMDP. $\mathbf{X} = \{0, 1, \ldots\}$, where 0 is the costless cemetery, $\mathbf{A}$ is an arbitrary Borel space. The state $i \in \mathbf{X}$ means there are i packets in the system. Suppose $\lambda(y, a), \mu(y, a) > 0$ are two bounded functions on $(0, \infty) \times \mathbf{A}$ and $g(y, a) : [0, \infty) \times \mathbf{A} \to [0, \infty)$ is a measurable function with $g(0, a) \equiv 0$. The transition rates are given by

$$q(j|i, a) = \begin{cases} n\lambda(i/n, a), & \text{if } i > 0, j = i + 1; \\ n\mu(i/n, a), & \text{if } i > 0, j = i - 1; \\ -n(\lambda(i/n, a) + \mu(i/n, a)), & \text{if } j = i > 0; \\ 0, & \text{otherwise,} \end{cases}$$

and the cost rate is $c(i, a) = g(i/n, a)$. See Fig. 5.21.

Condition 5.7. There exist constants $\underline{\mu} > 0$, $\tilde{\eta} > 1$ and $\eta \in (1, \tilde{\eta})$ such that

$$\inf_{a\in\mathbf{A}} \inf_{y>0} \mu(y, a) > \underline{\mu}, \quad \inf_{a\in\mathbf{A}} \inf_{y>0} \frac{\mu(y, a)}{\lambda(y, a)} > \tilde{\eta}; \quad \sup_{a\in\mathbf{A}} \sup_{y>0} \frac{|g(y, a)|}{\eta^y} < \infty.$$

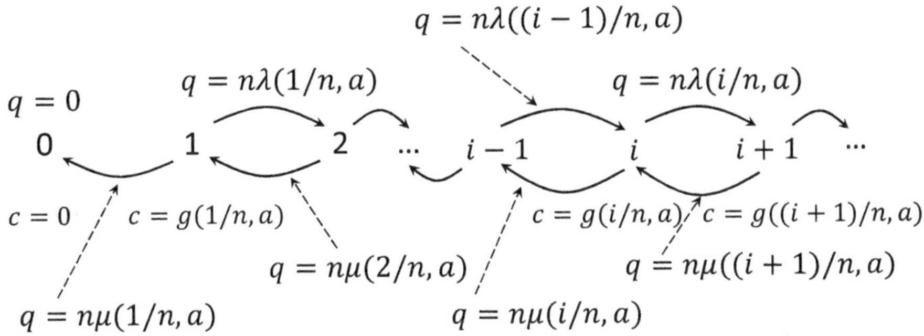

Fig. 5.21 Birth-and-death process suitable for fluid approximation.

Under Condition 5.7, there is no explosion: one can refer to the so called Reuter's condition, equivalent to the one in Proposition 5.2 in our case:

$$\sum_{i=1}^{\infty}\left[\frac{1}{n\lambda(i/n,a)}+\frac{\mu(i/n,a)}{n\lambda(i/n)\lambda((i-1)/n,a)}+\ldots+\frac{\mu(i/n,a)\ldots\mu(2/n,a)}{n\lambda(i/n)\ldots\lambda(1/n,a)}\right.$$
$$\left.+\frac{\mu(i/n,a)\ldots\mu(1/n,a)}{\lambda(i/n)\ldots\lambda(1/n,a)}\right]\geq\sum_{i=1}^{\infty}\tilde{\eta}^i=\infty.$$

If $\mathbf{A}$ is compact and the functions $\lambda(y,\cdot),\mu(y,\cdot)$ and $g(y,\cdot)$ are continuous in a for each $y\in(0,\infty)$, then, by Theorem 5.8, there exists a uniformly optimal deterministic stationary strategy φ^*. Thus, when solving the undiscounted problem (5.7) with $\alpha=0$, one can restrict to deterministic stationary strategies φ. The corresponding performance functional is denoted as $V^{(n)}(\varphi,i):=W^0(\varphi,i)$ to underline the dependence on parameter $n=1,2,\ldots$.

Fluid approximation means the following. Consider dynamical system

$$\frac{dy}{dt}=\begin{cases}\lambda(y,\psi(y))-\mu(y,\psi(y)), & \text{if } y>0;\\ 0, & \text{if } y=0\end{cases}\tag{5.54}$$

with the initial condition $y(0)=y_0>0$ and calculate

$$v^{\psi}(y_0):=\int_0^{\infty}g(y(t),\psi(y(t)))dt.$$

Here $\psi(\cdot):\ (0,\infty)\to\mathbf{A}$ is the (measurable) feedback control and, under Condition 5.7, the functions $y(\cdot)$ and $v^{\psi}(\cdot)$ are well defined.

Theorem 5.15. *Suppose Condition 5.7 is satisfied and $\psi(\cdot)$ is a measurable feedback control in dynamical system (5.54) such that all the functions*

$\lambda(y,\psi(y)), \mu(y,\psi(y))$ and $g(y,\psi(y))$ on $(0,\infty)$ are piece-wise continuously differentiable.

Then, for the deterministic stationary strategy $\varphi(i) := \psi(i/n)$, the values $V^{(n)}(\varphi,i)$ are such that

$$\lim_{n\to\infty} \sup_{0\le i\le n\tilde{y}} |V^{(n)}(\varphi,i) - v^{\psi}(i/n)| = 0.$$

Here $\tilde{y} \ge 0$ is an arbitrarily fixed number.

For the proof see Theorem 13.3.1 of [Piunovskiy and Zhang (2012)] and Lemma 4 and Theorem 1 of [Piunovskiy (2009b)].

If ψ^* is the optimal feedback control in system (5.54) providing $\inf_{\psi} v^{\psi}(y)$ for all $y \ge 0$, then one can expect that, for large enough n, the deterministic stationary strategy $\varphi^*(i) := \psi^*(i/n)$ is almost optimal for the undiscounted total cost problem (5.7) for all initial states $i \le n\tilde{y}$. For justification, note that $V^{(n)}(\varphi^*,i) \approx v^{\psi^*}(i/n)$ for $i \le n\tilde{y}$, and other strategies φ, associated with other feedback controls ψ in the sense that $\varphi(i) = \psi(i/n)$, are such that $V^{(n)}(\varphi,i) \approx v^{\psi}(i/n) \ge v^{\psi^*}(i/n)$. See also Theorem 2 of [Piunovskiy (2009b)].

Below we present Example 13.3.2 from [Piunovskiy and Zhang (2012)] showing that requirement $\sup_{y>0} \frac{|g(y,a)|}{\eta^y} < \infty$ in Condition 5.7 is important.

Example. Let $\mathbf{A} = [1,2]$, $\lambda(y,a) = ad^+$, $\mu(y,a) = ad^-$ for $y > 0$, where $d^- > d^+ > 0$ are fixed numbers. Put $g(y,a) = a^2\gamma^{y^2}$ for $y > 0$, where $\gamma > 1$ is a constant.

To construct the optimal feedback control ψ^* in system (5.54), note that

$$\frac{dy}{dt} = \psi(y)(d^+ - d^-) \le d^+ - d^- < 0.$$

Thus, function $y(\cdot)$ decreases and achieves the costless cemetery 0 in a finite time, so that

$$v^{\psi}(y_0) = \int_{y_0}^{0} \frac{g(y,\psi(y)}{\psi(y)(d^+ - d^-)}dy = \int_0^{y_0} \frac{g(y,\psi(y))}{\psi(y)(d^- - d^+)}dy.$$

Therefore, $\psi^*(y)$ must provide

$$\inf_{a\in\mathbf{A}} \frac{g(y,a)}{a(d^- - d^+)} = \inf_{a\in[1,2]} \frac{a\gamma^{y^2}}{(d^- - d^+)},$$

and $\psi^*(y) \equiv 1$ is the optimal feedback control with

$$v^{\psi^*}(y) = \int_0^y \frac{\gamma^{u^2}}{d^- - d^+}du < \infty.$$

Note also that, for any measurable feedback control ψ, the value of $v^{\psi}(y)$ is finite.

On the other hand, for any deterministic stationary strategy φ in the underlying CTMDP, $V^{(n)}(\varphi, i) = \infty$ for all $i > 0$, $n = 1, 2, \ldots$. Indeed, looking at the embedded chain $X_0 = i, X_1, \ldots$, starting from state $i > 0$, the probability of reaching state $i + k > i$ is not smaller than $\left(\frac{d^+}{d^+ + d^-}\right)^k$. Since $\mathsf{E}_i^\varphi[\Theta_{i+k+1}] \geq \frac{1}{2n(d^+ + d^-)}$,

$$V^{(n)}(\varphi, i) \geq \left(\frac{d^+}{d^+ + d^-}\right)^k \cdot \inf_{a \in \mathbf{A}} c(i + k, a) \cdot \frac{1}{2n(d^+ + d^-)}$$
$$= \frac{(d^+)^k}{2n(d^+ + d^-)^{k+1}} \cdot \gamma^{\left(\frac{i+k}{n}\right)^2}$$

for all $k = 1, 2, \ldots$. Hence

$$V^{(n)}(\varphi, i) \geq \lim_{k \to \infty} \frac{(d^+)^k}{2n(d^+ + d^-)^{k+1}} \cdot \gamma^{\left(\frac{i+k}{n}\right)^2} = \infty$$

for each $n = 1, 2, \ldots$ because $\gamma > 1$.

Condition 5.7 is not satisfied, and the statement of Theorem 5.15 is not valid. □

5.3.16 *Fluid approximation: refined model*

Consider the same situation as in Subsection 5.3.15 and assume that Condition 5.7 is satisfied, except for $\mu(y, a) > \underline{\mu} > 0$. Since $\mu(\cdot, \cdot)$ can approach zero and $0 < \lambda(y, a) < \mu(y, a)$ for $y > 0$, under fixed n and fixed deterministic stationary strategy φ, the controlled process $X(t)$ can spend too much time around the (nearly absorbing) state $i > 0$ for which $\mu(i/n, \varphi(i/n)) \approx \lambda(i/n, \varphi(i/n)) \approx 0$, so that the performance functional $V^{(n)}(\varphi, i)$ can become large and can even approach infinity as $n \to \infty$. The situation becomes good again if, instead of inequalities

$$\mu(y, a) > \underline{\mu} > 0 \qquad \sup_{a \in \mathbf{A}} \sup_{y > 0} \frac{|g(y, a)|}{\eta^y} < \infty,$$

we impose condition

$$\sup_{a \in \mathbf{A}} \sup_{y > 0} \frac{|g(y, a)|}{[\lambda(y, a) + \mu(y, a)]\eta^y} < \infty.$$

Suppose the parameter $n = 1, 2, \ldots$ and a measurable feedback control $\psi(\cdot)$ as in Theorem 5.15 are fixed and $\varphi(i) := \psi(i/n)$. Applying the (random) change of time when $X(t) > 0$

$$\frac{d\tau}{dt} = q_{X(t)}(\varphi(X(t))), \quad \text{where as usual } q_i(a) = -q(i|i, a),$$

we obtain the continuous-time Markov chain in the time scale τ with parameters $\hat{q}(j|i, \varphi(i)) = \frac{q(j|i, \varphi(i))}{q_i(\varphi(i))}$ for $i > 0$. Recall that $q_i(a) > 0$ for all $i > 0, a \in \mathbf{A}$. The performance functional (5.5) takes the form

$$V^{(n)}(\varphi,i) = \mathsf{E}_i^\varphi\left[\int_{(0,\infty)}\int_{\mathbf{A}} c(X(t),\varphi(X(t)))dt\right]$$
$$= \mathsf{E}_i^\varphi\left[\int_{(0,\infty)}\int_{\mathbf{A}} \frac{c(X(\tau),\varphi(X(\tau)))}{q_{X(\tau)}(\varphi(X(\tau)))}d\tau\right].$$

In other words, $V^{(n)}(\varphi,i)$ equals the performance functional $\hat{V}^{(n)}(\varphi,i)$ in the 'hat' CTMDP with parameters

$$\hat{\lambda}(y,a) = \frac{\lambda(y,a)}{\lambda(y,a)+\mu(y,a)}, \quad \hat{\mu}(y,a) = \frac{\mu(y,a)}{\lambda(y,a)+\mu(y,a)},$$

$$\hat{g}(y,a) = \frac{g(y,a)}{\lambda(y,a)+\mu(y,a)}.$$

Condition 5.8. There exist constants $\tilde{\eta} > 1$ and $\eta \in (1,\tilde{\eta})$ such that

$$\inf_{a\in\mathbf{A}}\inf_{y>0}\frac{\mu(y,a)}{\lambda(y,a)} > \tilde{\eta}; \quad \sup_{a\in\mathbf{A}}\sup_{y>0}\frac{|g(y,a)|}{[\lambda(y,a)+\mu(y,a)]\eta^y} < \infty.$$

Condition 5.8 implies Condition 5.7 for the 'hat' model: for $\underline{\mu} := 1-\frac{1}{1+\tilde{\eta}}$, $\hat{\mu}(y,a) = 1-\frac{\lambda(y,a)}{\lambda(y,a)+\mu(y,a)} > 1-\frac{\lambda(y,a)}{\lambda(y,a)+\lambda(y,a)\tilde{\eta}} = \underline{\mu}$. Therefore, using Theorem 5.15 for the 'hat' CTMDP, we obtain the following statement representing 'refined' fluid approximation.

Corollary 5.2 (from Theorem 5.15). *Suppose Condition 5.8 is satisfied and $\psi(\cdot): (0,\infty)\to\mathbf{A}$ is a measurable feedback control in the fluid model*

$$\frac{d\hat{y}}{d\tau} = \begin{cases} \frac{\lambda(\hat{y},\psi(\hat{y}))-\mu(\hat{y},\psi(\hat{y}))}{\lambda(\hat{y},\psi(\hat{y}))+\mu(\hat{y},\psi(\hat{y}))}, & \text{if } \hat{y}>0;\\ 0, & \text{if } \hat{y}=0\end{cases}$$

such that all the functions $\lambda(\hat{y},\psi(\hat{y})), \mu(\hat{y},\psi(\hat{y}))$ and $g(\hat{y},\psi(\hat{y}))$ on $(0,\infty)$ are piece-wise continuously differentiable. For the initial condition $\hat{y}(0)=\hat{y}_0>0$, put

$$\hat{v}^\psi(\hat{y}_0) := \int_0^\infty \frac{g(\hat{y}(\tau),\psi(\hat{y}(\tau)))}{\lambda(\hat{y}(\tau),\psi(\hat{y}(\tau)))+\mu(\hat{y}(\tau),\psi(\hat{y}(\tau)))}d\tau.$$

Then, for the deterministic stationary strategy $\varphi(i) := \psi(i/n)$, the values $V^{(n)}(\varphi,i)$ are such that

$$\lim_{n\to\infty}\sup_{0\le i\le n\tilde{y}}|V^{(n)}(\varphi,i)-\hat{v}^\psi(i/n)| = 0.$$

Here $\tilde{y}\ge 0$ is an arbitrarily fixed number.

Condition 5.7 implies Condition 5.8: since $[\lambda(y,a)+\mu(y,a)] > \underline{\mu}$ for $y > 0$,

$$\sup_{a\in\mathbf{A}}\sup_{y>0}\frac{|g(y,a)|}{[\lambda(y,a)+\mu(y,a)]\eta^y} < \frac{1}{\underline{\mu}}\sup_{a\in\mathbf{A}}\sup_{y>0}\frac{|g(y,a)|}{\eta^y} < \infty.$$

Therefore, refined fluid approximation is more general than that presented in Subsection 5.3.15. The following example published in [Piunovskiy and Zhang (2011), Ex. 1] confirms that, when Condition 5.8 is satisfied, the standard fluid approximation as in Subsection 5.3.15 may be very inaccurate.

Example. Let $\mathbf{A}$ be a singleton, that is, we consider an uncontrolled case. Therefore, argument a and the strategies ψ and φ are below omitted.

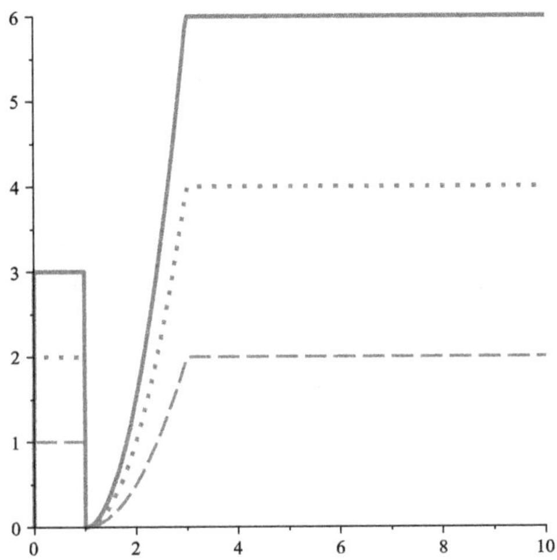

Fig. 5.22 Plots of $\lambda(y), \mu(y), g(y)$ against y. The solid line (resp. dotted line, dashed line) corresponds to $g(\cdot)$ (resp. $\mu(\cdot)$, $\lambda(\cdot)$).

Suppose

$$\begin{aligned}\lambda(y) &= \mathbb{I}\{y\in(0,1]\} + \frac{1}{2}(y-1)^2\mathbb{I}\{y\in(1,3]\} + 2\mathbb{I}\{y\in(3,\infty)\},\\ \mu(y) &= 2\mathbb{I}\{y\in(0,1]\} + (y-1)^2\mathbb{I}\{y\in(1,3]\} + 4\mathbb{I}\{y\in(3,\infty)\},\\ g(y) &= 3\mathbb{I}\{y\in(0,1]\} + \frac{3}{2}(y-1)^2\mathbb{I}\{y\in(1,3]\} + 6\mathbb{I}\{y\in(3,\infty)\}\end{aligned}$$

(see Figure 5.22). Then Condition 5.8 is satisfied and, in the framework of Corollary 5.2, $\hat{v}(\hat{y}_0) = 3\hat{y}_0$: $\frac{d\hat{y}}{d\tau} = -1/3$, the fluid variable $\hat{y}(\cdot)$ reaches the steady state 0 during $\tau = 3\hat{y}_0$ time units, and the cost rate $\frac{g(\hat{y})}{\lambda(\hat{y})+\mu(\hat{y})}$ equals 1.

Suppose for a fixed $n = 1, 2, \ldots$ the initial state in the birth-and-death Markov chain $X(t)$ as on Fig. 5.21 equals $X(0) = X_0 = 2n$. Then, by Corollary 5.2 at $\tilde{y} = 2$, $\lim_{n\to\infty} |V^{(n)}(2n) - \hat{v}(2)| = 0$, i.e., $\lim_{n\to\infty} V^{(n)}(2n) = 6$. The process $X(t)$ is absorbed at 0 because $\lambda(y) < \mu(y)$ and the functions $\lambda(\cdot)$ and $\mu(\cdot)$ are bounded. But as n increases, the moment of absorption is postponed till later and later becuase the process $X(t)$ spends more and more time in the right neighbourhood of the state n, where $q(i+1|i) = n\lambda(i/n) \approx 0$ and $q(i-1|i) = n\mu(i/n) \approx 0$. See Fig. 5.23 representing computer simulation.

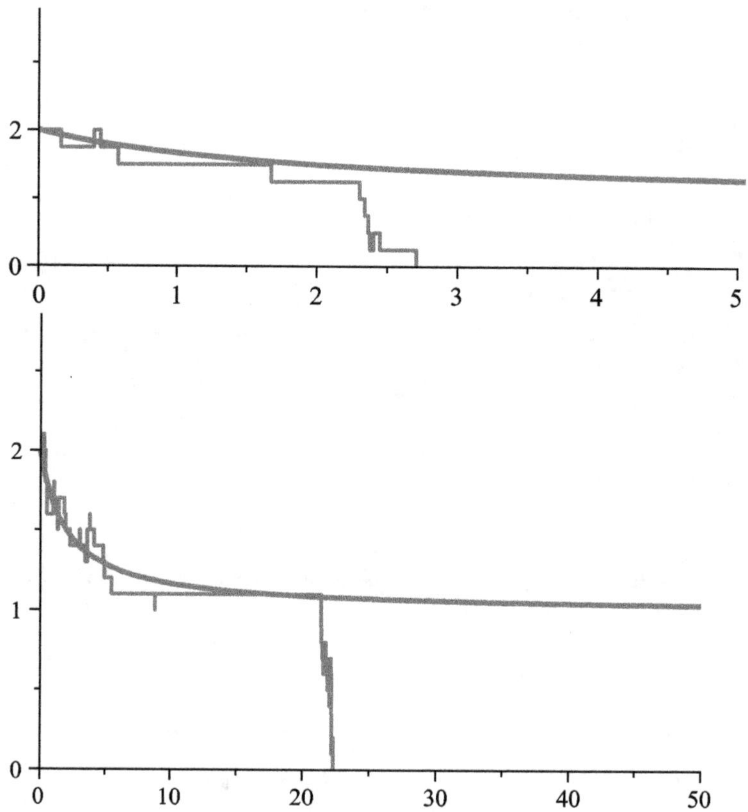

Fig. 5.23 Trajectories of the fluid model $y(t)$ with $y_0 = 2$ and of the birth-and-death chain $X(t)/n$ starting from $2n$ for $n = 4$ and $n = 10$.

The standard fluid approximation as in Subsection 5.3.15 is not applicable and the statement of Theorem 5.15 is not valid because Condition 5.7 is not satisfied: the function $\mu(\cdot)$ is not separated from zero. One can solve equation

(5.54) for $y_0 = 2$:

$$\frac{dy}{dt} = -\frac{1}{2}(y-1)^2 \implies y(t) = \frac{2}{t+2} + 1 \to 1 \text{ as } t \to \infty.$$

See Fig. 5.23. Thus, $v(2)$, as in Theorem 5.15, equals

$$v(2) = \int_0^\infty \frac{3}{2}(y(t)-1)^2 dt = \int_0^\infty \frac{3}{2}\left(\frac{2}{t+2}\right)^2 dt = 3$$

and $\lim_{n\to\infty} |V^{(n)}(2n) - v(2)| = 6 - 3 = 3$: the standard fluid approximation is wrong as it does not take into account the dynamics of $X(t)$ at the states $X(t) \le n$.

Calculations as in Theorem 5.15 for $\tilde{y} = 1$ and $X(0) = X_0 = n$ give the following:

$$\frac{dy}{dt} = -1, \quad y_0 = 1, \quad y(t) = \begin{cases} 1-t, & \text{if } t \le 1; \\ 0, & \text{if } t > 1, \end{cases} \quad v(1) = \int_0^1 3\, dt = 3$$

and

$$\lim_{n\to\infty} |V^{(n)}(n) - v(1)| = 0 \implies \lim_{n\to\infty} V^{(n)}(n) = 3$$

as expected. The total expected cost $V^{(n)}(2n) = 6$ is the sum of the cost when $X(t) > n$ (equal 3) and of the cost when $X(t) \le n$ (equal 3). □

5.3.17 *Optimal service strategy in a queueing system*

Consider a two-server queueing system in which the service time distributions of the two servers are different, namely the service time for server 1 is stochastically smaller than that for server 2. There is no space for waiting, so that the customers confronted by an occupied system are lost. If the system is empty, the arriving customer can be served by either of the servers. Intuitively, the best strategy is to send that customer to server 1, in order to minimize the average number of lost customers. The following example, published in [Seth (1977), Section 2], shows that this decision is not necessarily optimal.

Example. Assume that the input stream is Poisson with parameter $\lambda > 0$. Server 1 has a mixed service-time distribution: $\mathbf{T}_1 = 0$ with probability 1/2 and $\mathbf{T}_1 \sim exp(\mu/2)$ with probability 1/2. Server 2 also has a mixed service-time distribution: $\mathbf{T}_2 \sim exp(\mu)$ with probability 1/2 and $\mathbf{T}_2 \sim exp(\mu/2)$ with probability 1/2. The positive numbers λ and μ are fixed and the above mentioned exponential random variables are totally independent. The mode of the service for each server is realized independently of anything else at the moment when the service begins. It is easy to see that $\mathbf{T}_1$ is stochastically smaller than $\mathbf{T}_2$: $\mathsf{P}(\mathbf{T}_1 > z) < \mathsf{P}(\mathbf{T}_2 > z)$ for all $z \ge 0$. The state of the process is encoded as (i, j), where $i = 0$ if server 1 is free, and $i = 2$ if server 1 is performing the $exp(\mu/2)$ service; $j = 0$ if server 2 is free, $j = 1$ (or 2) if server 2 is performing the $exp(\mu)$ (or $exp(\mu/2)$) service. Thus, $\mathbf{X} = \{(0,0), (0,1), (0,2), (2,0), (2,1), (2,2)\}$. Action

$a \in \mathbf{A} = \{1, 2\}$ means that a new customer arriving at the free system is sent to server a.

According to the verbal description of the model,

$$q((2,0)|(0,0),1) = \lambda/2, \quad q((0,1)|(0,0),2) = \lambda/2, \quad q((0,2)|(0,0),2) = \lambda/2,$$

$$q((0,0)|(0,0),1) = -\lambda/2, \quad q((0,0)|(0,0),2) = -\lambda$$

and

$$q(y|(0,1),a) \equiv \begin{cases} -\lambda/2 - \mu, & \text{if } y = (0,1); \\ \lambda/2, & \text{if } y = (2,1); \\ \mu, & \text{if } y = (0,0), \end{cases}$$

$$q(y|(0,2),a) \equiv \begin{cases} -\lambda/2 - \mu/2, & \text{if } y = (0,2); \\ \lambda/2, & \text{if } y = (2,2); \\ \mu/2, & \text{if } y = (0,0), \end{cases}$$

$$q(y|(2,0),a) \equiv \begin{cases} -\lambda - \mu/2, & \text{if } y = (2,0); \\ \lambda/2, & \text{if } y = (2,1); \\ \lambda/2, & \text{if } y = (2,2); \\ \mu/2, & \text{if } y = (0,0), \end{cases}$$

$$q(y|(2,1),a) \equiv \begin{cases} -3\mu/2, & \text{if } y = (2,1); \\ \mu, & \text{if } y = (2,0); \\ \mu/2, & \text{if } y = (0,1), \end{cases}$$

$$q(y|(2,2),a) \equiv \begin{cases} -\mu, & \text{if } y = (2,2); \\ \mu/2, & \text{if } y = (0,2); \\ \mu/2, & \text{if } y = (2,0), \end{cases}$$

and all the other transition rates are zero.

$$c(x,a) \equiv \begin{cases} \lambda, & \text{if } x = (2,1); \text{ or } (2,2), \\ 0, & \text{otherwise} \end{cases}$$

(see Fig. 5.24). We are going to study the average cost problem (5.8).

Proposition 5.5. *If the spaces* $\mathbf{X}$ *and* $\mathbf{A}$ *are finite and, under each deterministic stationary strategy* φ*, the Markov chain* $X(t)$ *is unichain (i.e. the embedded discrete-time chain* $X_0 \to X_1 \to \ldots$ *is unichain), then the deterministic stationary strategy* φ^* *is average uniformly optimal if and only if it provides the minimum in the average optimality equation*

$$\begin{aligned} g &= \min_{a \in \mathbf{A}} \left\{ c(x,a) + \sum_{y \in \mathbf{X}} h(y) q(y|x,a) \right\} \\ &= c(x, \varphi^*(x)) + \sum_{y \in \mathbf{X}} h(y) q(y|x, \varphi^*(x)), \quad \forall\, x \in \mathbf{X}. \end{aligned} \tag{5.55}$$

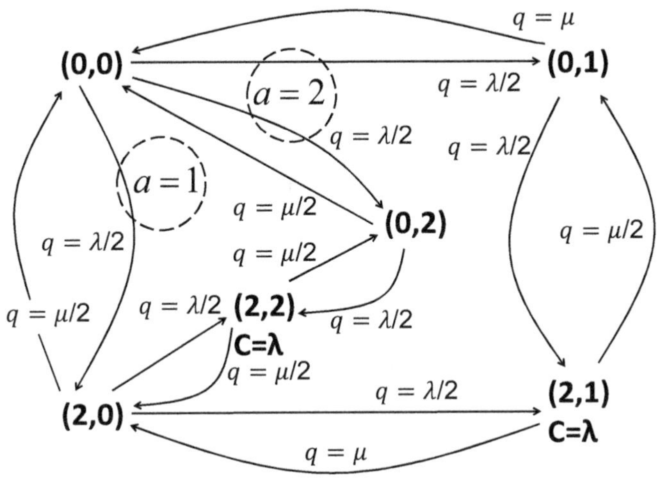

Fig. 5.24 Queueing system.

This equation has a solution in the space of finite numbers g and finite-valued functions $h(\cdot): \mathbf{X} \to (-\infty, +\infty)$*; the value of g is unique and* $W^{*A}(x) \equiv g$.

For the justification, see [Prieto-Rumeau and Hernández-Lerma (2012), Thm. 3.20] and p.318 of [Piunovskiy and Zhang (2020)]. Equation (5.55) is sometimes called *Poisson equation* [Hernández-Lerma *et al.* (2023)]. One can also call it canonical equation, similarly to (4.2).

Clearly, all the conditions of Proposition 5.5 are satisfied for the queueing system under study.

Since only actions in state $(0,0)$ are important, without loss of generality, we accept that only deterministic stationary strategies $\varphi^1(x) \equiv 1$ and $\varphi^2(x) \equiv 2$ are in use.

The canonical equations (5.55), respectively for states $(0,0)$, $(0,1)$, $(0,2)$, $(2,0)$, $(2,1)$ and $(2,2)$, have the form

$$g = \min\left\{-\frac{\lambda}{2}h(0,0) + \frac{\lambda}{2}h(2,0); \quad -\lambda h(0,0) + \frac{\lambda}{2}h(0,1) + \frac{\lambda}{2}h(0,2)\right\};$$

$$g = -(\mu + \lambda/2)h(0,1) + \mu h(0,0) + \frac{\lambda}{2}h(2,1);$$

$$g = -(\mu/2 + \lambda/2)h(0,2) + \frac{\mu}{2}h(0,0) + \frac{\lambda}{2}h(2,2);$$

$$g = -(\mu/2 + \lambda)h(2,0) + \frac{\mu}{2}h(0,0) + \frac{\lambda}{2}h(2,2) + \frac{\lambda}{2}h(2,1);$$

$$g = \lambda - \frac{3\mu}{2}h(2,1) + \mu h(2,0) + \frac{\mu}{2}h(0,1);$$

$$g = \lambda - \mu h(2,2) + \frac{\mu}{2}h(2,0) + \frac{\mu}{2}h(0,2).$$

Here we omit the double brackets for brevity and write $h(0,0)$, not $h((0,0))$ etc.

If we put $h(0,0) := 0$, as usual, and denote $\gamma := \lambda/\mu$ then, assuming that $\varphi^*(x) = \varphi^2(x) \equiv 2$, we have equation

$$g = \frac{\lambda}{2}h(0,1) + \frac{\lambda}{2}h(0,2).$$

After some trivial algebra, we obtain

$$g = \lambda\gamma^2 \frac{3\gamma^2 + 9\gamma + 4}{3\gamma^4 + 14\gamma^3 + 23\gamma^2 + 19\gamma + 6};$$

$$h(2,0) = \gamma^2 \frac{6\gamma^2 + 19\gamma + 10}{3\gamma^4 + 14\gamma^3 + 23\gamma^2 + 19\gamma + 6}.$$

Other values of the function $h(\cdot)$ are of no importance, but they can certainly be calculated, too. For example:

$$h(0,1) = \gamma^2 \frac{2\gamma^2 + 5\gamma + 2}{3\gamma^4 + 14\gamma^3 + 23\gamma^2 + 19\gamma + 6},$$

and so on. To complete the proof, it only remains to compare g with $\frac{\lambda}{2}h(2,0)$:

$$\begin{aligned}\frac{\lambda}{2}h(2,0) - g &= \lambda\gamma^2 \left[\frac{6\gamma^2 + 19\gamma + 10}{2(3\gamma^4 + 14\gamma^3 + 23\gamma^2 + 19\gamma + 6)}\right.\\ &\qquad \left. - \frac{3\gamma^2 + 9\gamma + 4}{3\gamma^4 + 14\gamma^3 + 23\gamma^2 + 19\gamma + 6}\right]\\ &= \lambda\gamma^2 \frac{\gamma + 2}{2(3\gamma^4 + 14\gamma^3 + 23\gamma^2 + 19\gamma + 6)} > 0\end{aligned}$$

for any values of $\gamma > 0$. Thus, the triplet $\langle g, h, \varphi^* \rangle$ satisfies equation (5.55), and the deterministic stationary strategy φ^* is average uniformly optimal. If the system is free then it is better to send the arriving customer to server 2 with a stochastically *larger* service time.

To understand this better, we present comments given in [Seth (1977), Table 1]. Suppose three customers arrive in a fixed time slot to the initially empty system and assume that the random modes of service for the first accepted customer ($\mathbf{T}_1 = 0$ or $\mathbf{T}_1 \sim exp(\mu/2)$ for the first server and $\mathbf{T}_2 \sim exp(\mu)$ or $\mathbf{T}_2 \sim exp(\mu/2)$ for the second server) have been realized. Then four different situations can exist with equal probabilities 1/4:

	Server 1	Server 2	Optimal decision about the first customer:
Situation 1	0	$exp(\mu)$	both equally good
Situation 2	0	$exp(\mu/2)$	both equally good
Situation 3	$exp(\mu/2)$	$exp(\mu/2)$	both equally good
Situation 4	$exp(\mu/2)$	$exp(\mu)$	send to server 2.

As for the first two lines, if one server has service time zero, then all three customers will be served, no matter which strategy is used.

> *"The result obtained may be intuitively explained as follows. Consider the case when λ is small. The probability of more than three customers arriving during one busy period is very small. Under either policy, if one or two customers arrive during a busy period, they have to be served. When three customers arrive, four different situations (which would be of importance for the service of the third customer), could exist with equal probability (see the table above). It is clear that server 2 should serve the third customer." [Seth (1977)].*

Perhaps, there is a typo in the last sentence above: server 2 should serve the FIRST customer, just to increase the chance that all three customers are served. □

5.3.18 *Repair shop*

Consider two identical devices: one works and the other is reserved, replacing the first one when it is broken, so that the state space is $\mathbf{X} = \{0, 1, 2\}$, where $x \in \mathbf{X}$ is the number of broken devices. At any time moment only one device can be served, and in case the both are broken, one device is just waiting for repair. The operational time of each device is exponential, independent of anything else, with parameter $\lambda > 0$. The service can be done in one of two modes, so the action space is $\mathbf{A} = \{1, 2\}$. If $a = 1$ or $a = 2$, then the service time is exponential with parameter $0 < \mu_1 < \mu_2$ correspondingly. The cost rate has two components:

$$c(x, a) = C(x) + R(a),$$

where $C(x)$ is the cost rate associated with x broken devices, and $R(a)$ is the cost rate associated with the a-th mode of the service; $0 = C(0) < C(1) < C(2)$; $0 < R(1) < R(2)$. Note, even in state 0 there is the cost rate $R(a)$. Anyway, one can pass to the case when $c(0, 1) = 0$ by subtracting the constant $R(1)$ from the cost rate. The optimal strategy will not change. According to the verbal description, the non-zero transition rates are

$$q(1|0, a) = q(2|1, a) \equiv \lambda, \quad q(0|1, a) = q(1|2, a) = \mu_a,$$

$$q(0|0, a) = -\lambda, \quad q(1|1, a) = -(\lambda + \mu_a), \quad q(2|2, a) = -\mu_a.$$

One can think that the optimal service rate should not decrease when the number of broken devices increases. The following example shows that sometimes that is not the case.

Example. Consider the specific model with $\lambda = 1$, $\mu_1 = 9$, $\mu_2 = 9.1$, $C(0) = 0$, $C(1) = 99$, $C(2) = 100$, $R(1) = 1$ and $R(2) = 2$ (see Fig. 5.25), and solve the average cost problem (5.8) using the linear programming approach. Note that the controlled process $X(\cdot)$ is ergodic under any determinstic stationary strategy.

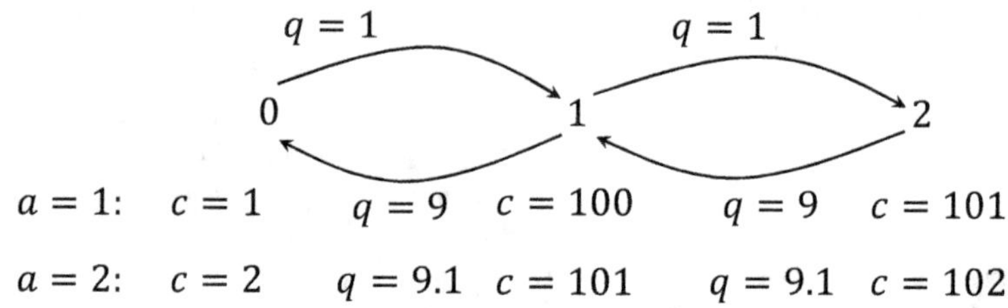

Fig. 5.25 Repair shop.

Linear program (5.39), (5.40) for $\Gamma = \{y\}$ being singletons (equivalently, (5.23), (5.24) and (5.42) with $J = 0$) looks as follows:

$$\begin{aligned}
&\eta(0,1) + 2\eta(0,2) + 100\eta(1,1) \\
&+101\eta(1,2) + 101\eta(2,1) + 102\eta(2,2) \longrightarrow \min_{\eta \in \mathbf{P}(\mathbf{X}\times\mathbf{A})} \\
&-\eta(0,1) - \eta(0,2) + 9\eta(1,1) + 9.1\eta(1,2) = 0; \\
&-9\eta(2,1) - 9.1\eta(2,2) + \eta(1,1) + \eta(1,2) = 0; \\
&\eta(0,1) + \eta(0,2) + \eta(1,1) + \eta(1,2) + \eta(2,1) + \eta(2,2) = 1.
\end{aligned}$$

Here the first two equations are for $y = 0, 2$, and the linearly dependent equation for $y = 1$ is omitted. Probabilities on $\mathbf{X} \times \mathbf{A}$ are defined by their values $\eta(\{x\} \times \{a\}) =: \eta(x, a)$ omitting the parentheses.

The optimal solution η^* provides the average uniformly optimal stationary π-strategy $\pi^* = \pi_{\eta^*}$: $\eta^*(x, a) = \pi_{\eta^*}(a|x) \sum_{b\in\mathbf{A}} \eta^*(x, b)$ according to Remark 5.7. Optimal basic solution, up to 10^{-4}, is as follows:

	const	$\eta(0,2)$	$\eta(1,1)$	$\eta(2,2)$
Objective =	11.8814	1.0000	0.08814	0.00979
$\eta(0,1) =$	0.89119	−1.0000	−0.01088	0.00990
$\eta(1,2) =$	0.09793	0	−0.99021	0.00109
$\eta(2,1) =$	0.01088	0	0.00109	−1.0110

Therefore, $\eta^*(0, 2) = \eta^*(1, 1) = \eta^*(2, 2) = 0$, and the optimal basic variables are $\eta^*(0, 1) = 0.89119$, $\eta^*(1, 2) = 0.09793$, $\eta^*(2, 1) = 0.01088$. The average uniformly optimal strategy is just the deterministic stationary strategy $\varphi^*(0) = 1$, $\varphi^*(1) = 2$, $\varphi^*(2) = 1$. The minimal long-run average cost rate equals $W^{*A}(\mathsf{P}_0) \equiv 11.8814$.

It looks surprising that the second mode with the higher service rate is optimal if only one device is broken and not optimal if the both devices are down. Perhaps, one can explain this like follows. Mode $a = 2$ is more expensive. Nevertheless it has an advantage to empty the system faster if only one device needs repair: the cost rate decreases almost by 100. If the both devices are down, then after one of them is repaired, the situation does not improve essentially: the cost rate decreases approximately by 1. Therefore, it makes sense to save a little by using the cheaper and less intensive service. □

This example was published in [Crabill, T.B. (1972)].

Appendix A

Proofs of Auxiliary Statements

Lemma A.1. *Let* $\mathbf{A}$ *and* $\mathbf{X}$ *be two Borel spaces,* $\mathsf{P}_0(dx)$ *be a non-atomic probability distribution on* $\mathbf{X}$*, and* $f(x)$ *be a real-valued measurable function on* $\mathbf{X}$*. Then, for any measurable stochastic kernel* $\pi(da|x)$ *on* $\mathbf{A}$ *given* $\mathbf{X}$*, there exists a measurable mapping (selector)* $\varphi: \mathbf{X} \to \mathbf{A}$ *such that, for any real measurable bounded function* $\rho(a)$ *on* $\mathbf{A}$*,*

$$\int_{\mathbf{X}} \int_{\mathbf{A}} \rho(a) f(x) \pi(da|x) \mathsf{P}_0(dx) = \int_{\mathbf{X}} \rho(\varphi(x)) f(x) \mathsf{P}_0(dx). \tag{A.1}$$

(We call π *and* φ *strongly equivalent w.r.t.* $f(\cdot)$*).*

Proof. Without loss of generality, we assume that $\mathbf{A} = \mathbf{X} = [0, 1]$, so that we deal with random variables and their distributions: see Proposition 7.16 and Corollary 7.16.1 of [Bertsekas and Shreve (1978)] or Appendix 1 of [Dynkin and Yushkevich (1979)]. (The case of discrete $\mathbf{A}$ is obviously a simplified version.)

Firstly, suppose that $f(x) \geq 0$ and $0 \neq \int_{[0,1]} f(x) \mathsf{P}_0(dx) < \infty$. Then

$$F_{\mathbf{X}}(x) := \int_0^x f(x) \mathsf{P}_0(dx) \Big/ \int_0^1 f(x) \mathsf{P}_0(dx)$$

is the cumulative distribution function (CDF) of a non-atomic probability measure on $\mathbf{X}$. (If $\int_0^1 f(x) \mathsf{P}_0(dx) = 0$ then the statement of the Lemma is trivial.) Let

$$F_{\mathbf{A}}(a) := \int_0^1 \left[\int_{[0,a]} \pi(da|x) \right] f(x) \mathsf{P}_0(dx) \Big/ \int_0^1 f(x) \mathsf{P}_0(dx)$$

be the cumulative distribution function associated with a probability measure on $\mathbf{A}$. Now put

$$\varphi(x) := \inf\{a : \ F_{\mathbf{A}}(a) \geq F_{\mathbf{X}}(x)\}.$$

See Fig. A.1 for illustration.

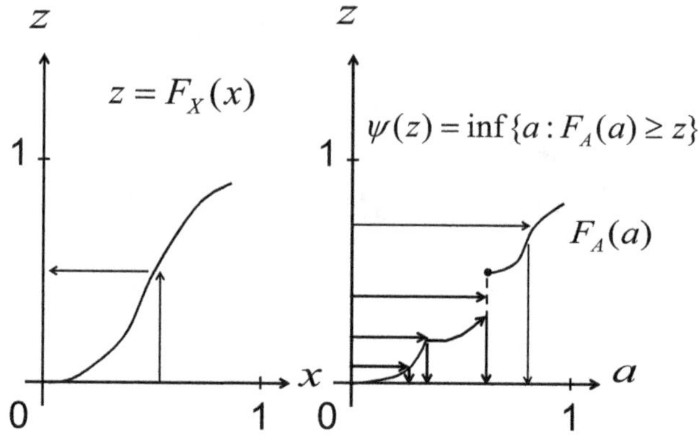

Fig. A.1 Construction of the selector φ.

We know that the image of measure $f(x)\mathsf{P}_0(dx)/\int_0^1 f(x)\mathsf{P}_0(dx)$ w.r.t. the map $z = F_{\mathbf{X}}(x)$ is uniform on $[0,1] \ni z$. Note, this is not the case if the measure $\mathsf{P}_0(dx)$ has atoms. We also know that the image of the uniform measure on $[0,1]$ w.r.t. the map $\psi(z) = \inf\{a : F_{\mathbf{A}}(a) \geq z\}$ coincides with the distribution on $\mathbf{A}$ defined by the CDF $F_{\mathbf{A}}(\cdot)$ [Piunovskiy and Zhang (2020), Prop.B.1.21] (see Fig. A.1). Therefore the image of the measure $f(x)\mathsf{P}_0(dx)/\int_0^1 f(x)\mathsf{P}_0(dx)$ w.r.t. $\varphi : \mathbf{X} \to \mathbf{A}$ coincides with the distribution on $\mathbf{A}$ defined by the CDF $F_{\mathbf{A}}(\cdot)$. Hence, for each measurable bounded function $\rho(a)$,

$$\frac{\int_{\mathbf{X}}\int_{\mathbf{A}} \rho(a) f(x)\pi(da|x)\mathsf{P}_0(dx)}{\int_0^1 f(x)\mathsf{P}_0(dx)} = \frac{\int_{\mathbf{X}} \rho(\varphi(x)) f(x)\mathsf{P}_0(dx)}{\int_0^1 f(x)\mathsf{P}_0(dx)},$$

and the proof is finished.

Now, if $f(x) \geq 0$ and $\int_{[0,1]} f(x)\mathsf{P}_0(dx) = \infty$, one should consider the subsets

$$\mathbf{X}_j := \{x \in \mathbf{X} : j-1 \leq f(x) < j\}, \quad j = 1, 2, \ldots,$$

and build the selector φ, strongly equivalent to π w.r.t. $f(\cdot)$, separately on each subset $\mathbf{X}_j$.

Finally, if the function $f(\cdot)$ is not non-negative, one should consider the subsets $\mathbf{X}_+ := \{x \in \mathbf{X} : f(x) \geq 0\}$ and $\mathbf{X}_- := \{x \in \mathbf{X} : f(x) < 0\}$ and build the selectors $\varphi_+(x)$ and $\varphi_-(x)$ on $\mathbf{X}_+$ and on $\mathbf{X}_-$, strongly equivalent to π w.r.t. $f(\cdot)$ and $-f(\cdot)$ correspondingly. The combined selector $\varphi(x) := \varphi_+(x)\mathbb{I}\{x \in \mathbf{X}_+\} + \varphi_-(x)\mathbb{I}\{x \in \mathbf{X}_-\}$ will satisfy equality (A.1). □

Remark A.1. If function $f(\cdot)$ is non-negative or non-positive, then equality (A.1) in Lemma A.1 holds for any measurable function $\rho(\cdot)$ (not necessarily bounded): for the proof one should consider $\min\{\rho^+(\cdot), K\}$ and $\min\{\rho^-(\cdot), K\}$ and pass to the limit as $K \to \infty$.

Theorem A.1. *Let Ω be the collection of all ordinals up to (and excluding) the first uncountable one, or, in other words, let Ω be the first uncountable ordinal. Let $h(\alpha)$ be a real-valued non-increasing function on Ω taking non-negative values and such that, in case $\inf_{\gamma<\alpha} h(\gamma) > 0$, the strict inequality $h(\alpha) < \inf_{\gamma<\alpha} h(\gamma)$ holds.*

Then $h(\alpha) = 0$ for some $\alpha \in \Omega$.

Proof. Suppose $h(\alpha) > 0$ for all $\alpha \in \Omega$. For each α, consider the open interval $(h(\alpha),\ \inf_{\gamma<\alpha} h(\gamma))$. Such intervals are non-empty and disjoint for different α. The total collection of such intervals is not more than countable, because each interval contains a rational number. However, Ω is not countable, so, by contradiction, $h(\alpha) = 0$ for some $\alpha \in \Omega$ (and $h(\gamma) = 0$ for all $\gamma > \alpha$ as well). □

Proof of Lemma 1.3. (a) This follows from the explicit formula $w_k = \left(\frac{h}{\varepsilon} - 1\right)\left[1 - (1-\varepsilon)^{k-1}\right]$.

(b) The proof for the cemetery Δ is trivial. For $k = 1, 2$ the assertions follow by the direct calculation. Suppose they hold for all $1 \le i \le k-1$ and consider the case of k.

If k is even, then, by induction,

$$W^{k+1}(0) = \min\{0 - 1,\quad h + \varepsilon w_k - (1-\varepsilon),\quad -1 + 2k\} = -1,$$

since $\varepsilon w_k + \varepsilon + h > \varepsilon w_k + \varepsilon - h > 0$ according to (a), and the minimum is provided by $a = \alpha$;

$$W^{k+1}(1) = \min\{0 + w_k,\quad h - \varepsilon + (1-\varepsilon)w_k,\quad 0 + 2k\} = w_{k+1},$$

since $w_{k+1} < w_k$ according to (a), and the minimum is provided by $a = \beta$.

If k is odd, then, by induction and similarly to above,

$$W^{k+1}(0) = \min\{0 + w_k,\quad h - \varepsilon + (1-\varepsilon)w_k,\quad -1 + 2k\} = w_{k+1},$$

and the minimum is provided by $a = \beta$;

$$W^{k+1}(1) = \min\{0 - 1,\quad h + \varepsilon w_k - (1-\varepsilon),\quad 0 + 2k\} = -1,$$

and the minimum is provided by $a = \alpha$. □

Proof of Corollary 2.1. Since $\eta(Y) \le K$ with $K > 0$, one can add an isolated point Δ to Y, put $\eta(\Delta) := K - \eta(Y)$, and normalize the obtained space of measures on $Y \cup \{\Delta\}$. As the result, we obtain a measurable subset $\tilde{\Xi}$ of the space of probability measures on $Y \cup \{\Delta\}$, equipped with the weak topology, so that $\tilde{\Xi}$ is a Borel space [Bertsekas and Shreve (1978), Cor.7.25.1, Prop.7.12].

The constructed 1-1 correspondence $\eta \leftrightarrow \tilde{\eta}$ between Ξ and $\tilde{\Xi}$ is continuous in the both directions, and $\eta(\cdot) = K\tilde{\eta}(\cdot)$ on Y. Consider the continuous stochastic kernel $q(dy|\tilde{\eta}) := \tilde{\eta}(dy)$ on Y given $\tilde{\Xi}$. Then, for each lower semi-continuous and bounded below function $c(\cdot) : Y \to \mathbb{R}^*$, for the function

$$f(\tilde{\eta}, y) := \begin{cases} Kc(y), & \text{if } y \in Y; \\ 0, & \text{if } y = \Delta \end{cases}$$

on $\tilde{\Xi} \times (Y \cup \{\Delta\})$ we know that the mapping

$$\tilde{g}(\tilde{\eta}) := \int_{Y \cup \{\Delta\}} f(\tilde{\eta}, y) q(dy|\tilde{\eta}) = \int_Y Kc(y)\tilde{\eta}(dy)$$

on $\tilde{\Xi}$ is lower semi-continuous and bounded below by Theorem 2.7. Thus, for the measures $\eta \leftrightarrow \tilde{\eta}$ in the correspondence,

$$g(\eta) := \tilde{g}(\tilde{\eta}) = \int_Y c(y)\eta(dy),$$

and $g(\cdot)$ is also a lower semi-continuous and bounded below function on Ξ. $\square$

Lemma A.2. *Suppose positive numbers λ_i, $i = 1, 2, \ldots$, and μ_i, $i = 2, 3, \ldots$, are such that $\lambda_1 \le 1$, $\lambda_i + \mu_i \le 1$ for $i \ge 2$, and $\sum_{j=2}^{\infty} \left(\frac{\mu_2\mu_3 \cdots \mu_j}{\lambda_2\lambda_3 \cdots \lambda_j} \right) < \infty$. Then equations*

$$\begin{aligned} \eta(1) &= 1 + \mu_2\eta(2); \\ \eta(i) &= \lambda_{i-1}\eta(i-1) + \mu_{i+1}\eta(i+1), \quad i = 2, 3, 4, \ldots \end{aligned}$$

have the minimal non-negative solution satisfying inequalities

$$\eta(1) \le \sum_{j=1}^{\infty} \left(\frac{\mu_2\mu_3 \cdots \mu_j}{\lambda_2\lambda_3 \cdots \lambda_j} \right) = 1 + \sum_{j=2}^{\infty} \left(\frac{\mu_2\mu_3 \cdots \mu_j}{\lambda_2\lambda_3 \cdots \lambda_j} \right);$$

$$\begin{aligned} \eta(i) &\le \sum_{j=i}^{\infty} \left(\frac{\mu_2\mu_3 \cdots \mu_j}{\lambda_2\lambda_3 \cdots \lambda_j} \right) \Big/ \left(\frac{\mu_2\mu_3 \cdots \mu_i}{\lambda_2\lambda_3 \cdots \lambda_{i-1}} \right) \\ &= \frac{1}{\lambda_i} \left[1 + \sum_{j=1}^{\infty} \frac{\mu_{i+1}\mu_{i+2} \cdots \mu_{i+j}}{\lambda_{i+1}\lambda_{i+2} \cdots \lambda_{i+j}} \right], \quad i = 2, 3, \ldots. \end{aligned}$$

Proof. The minimal non-negative solution can be built by successive approximations: $\eta(i) = \lim_{n\to\infty} \eta_n(i)$, where

$$\begin{aligned}
\eta_0(i) &\equiv 0;\\
\eta_{n+1}(1) &= 1 + \mu_2\eta_n(2);\\
\eta_{n+1}(i) &= \lambda_{i-1}\eta_n(i-1) + \mu_{i+1}\eta_n(i+1), \quad i = 2,3,4,\ldots;\\
n &= 0,1,2,\ldots.
\end{aligned}$$

For each $i \geq 1$, the value $\eta_n(i)$ increases with n, and we can prove inequalities

$$\begin{aligned}
\eta_n(1) &\leq \sum_{j=1}^{\infty}\left(\frac{\mu_2\mu_3\cdots\mu_j}{\lambda_2\lambda_3\cdots\lambda_j}\right) = 1 + \sum_{j=2}^{\infty}\left(\frac{\mu_2\mu_3\cdots\mu_j}{\lambda_2\lambda_3\cdots\lambda_j}\right);\\
\eta_n(i) &\leq \sum_{j=i}^{\infty}\left(\frac{\mu_2\mu_3\cdots\mu_j}{\lambda_2\lambda_3\cdots\lambda_j}\right) \Big/ \left(\frac{\mu_2\mu_3\cdots\mu_i}{\lambda_2\lambda_3\cdots\lambda_{i-1}}\right), \quad i = 2,3,\cdots
\end{aligned}$$

by induction w.r.t. n. These inequalities hold for $n = 0$. Suppose they are satisfied for some n. Then

$$\eta_{n+1}(1) \leq 1 + \mu_2 \sum_{j=2}^{\infty}\left(\frac{\mu_2\mu_3\cdots\mu_j}{\lambda_2\lambda_3\cdots\lambda_j}\right) \Big/ \mu_2.$$

For $i > 2$,

$$\begin{aligned}
\eta_{n+1}(i) \leq\ & \lambda_{i-1} \sum_{j=i-1}^{\infty}\left(\frac{\mu_2\mu_3\cdots\mu_j}{\lambda_2\lambda_3\cdots\lambda_j}\right) \Big/ \left(\frac{\mu_2\mu_3\cdots\mu_{i-1}}{\lambda_2\lambda_3\cdots\lambda_{i-2}}\right)\\
&+\mu_{i+1} \sum_{j=i+1}^{\infty}\left(\frac{\mu_2\mu_3\cdots\mu_j}{\lambda_2\lambda_3\cdots\lambda_j}\right) \Big/ \left(\frac{\mu_2\mu_3\cdots\mu_{i+1}}{\lambda_2\lambda_3\cdots\lambda_i}\right)\\
=\ & 1 + \left\{\sum_{j=i}^{\infty}\left(\frac{\mu_2\mu_3\cdots\mu_j}{\lambda_2\lambda_3\cdots\lambda_j}\right) \Big/ \left(\frac{\mu_2\mu_3\cdots\mu_{i-1}}{\lambda_2\lambda_3\cdots\lambda_{i-1}}\right)\right\}\\
&+ \left\{\sum_{j=i}^{\infty}\left(\frac{\mu_2\mu_3\cdots\mu_j}{\lambda_2\lambda_3\cdots\lambda_j}\right) \Big/ \left(\frac{\mu_2\mu_3\cdots\mu_i}{\lambda_2\lambda_3\cdots\lambda_i}\right)\right\} - 1.
\end{aligned}$$

In the last formula, expressions in the parentheses equal

$$\sum_{j=i}^{\infty}\left(\frac{\mu_2\mu_3\cdots\mu_j}{\lambda_2\lambda_3\cdots\lambda_j}\right) \Big/ \left(\frac{\mu_2\mu_3\cdots\mu_i}{\lambda_2\lambda_3\cdots\lambda_{i-1}}\right)$$

multiplied by μ_i and by λ_i correspondingly. Thus,

$$\eta_{n+1}(i) \le \left\{ \sum_{j=i}^{\infty} \left(\frac{\mu_2\mu_3\cdots\mu_j}{\lambda_2\lambda_3\cdots\lambda_j} \right) \Big/ \left(\frac{\mu_2\mu_3\cdots\mu_i}{\lambda_2\lambda_3\cdots\lambda_{i-1}} \right) \right\} (\mu_i + \lambda_i)$$
$$\le \sum_{j=i}^{\infty} \left(\frac{\mu_2\mu_3\cdots\mu_j}{\lambda_2\lambda_3\cdots\lambda_j} \right) \Big/ \left(\frac{\mu_2\mu_3\cdots\mu_i}{\lambda_2\lambda_3\cdots\lambda_{i-1}} \right).$$

For $i = 2$, similar calculations lead to inequalities

$$\eta_{n+1}(2) \le \lambda_1 + \left(\lambda_1 + \frac{\lambda_2}{\mu_2} \right) \sum_{j=2}^{\infty} \left(\frac{\mu_2\mu_3\cdots\mu_j}{\lambda_2\lambda_3\cdots\lambda_j} \right) - 1$$
$$\le \left\{ \sum_{j=2}^{\infty} \left(\frac{\mu_2\mu_3\cdots\mu_j}{\lambda_2\lambda_3\cdots\lambda_j} \right) \Big/ \mu_2 \right\} (\lambda_2 + \lambda_1\mu_2)$$
$$\le \sum_{j=2}^{\infty} \left(\frac{\mu_2\mu_3\cdots\mu_j}{\lambda_2\lambda_3\cdots\lambda_j} \right) \Big/ \mu_2.$$

□

Proof of Proposition 2.6. (a) The non-zero values of any occupation measure $\eta^{\pi}_{(0,0)}$ are given by

$$\eta^{\pi}_{(0,0)}((0,0)\times\Gamma^A) = \pi_1(\Gamma^A|(0,0));$$
$$\eta^{\pi}_{(0,0)}((1,\Gamma^X)\times\Gamma^A) = \int_{\Gamma^X} \pi_1(dx|(0,0))\pi_2(\Gamma^A|(0,0),x,(1,x)).$$

Thus, for each bounded continuous function $g(\cdot,\cdot)$ on $\{0,1\}\times[0,1]\times\mathbf{A}$ we have

$$\int_{\{0,1\}\times[0,1]\times\mathbf{A}} g((k,x),a)d\eta^{\pi^{\mathrm{m},i}}_{(0,0)}((k,x),a)$$
$$= \int_{\mathbf{A}} g((0,0),a)\pi_1^{\mathrm{m},i}(da|(0,0)) + \int_{\mathbf{A}}\int_{[0,1]} g((1,x),a)\pi_1^{\mathrm{m},i}(dx|(0,0))\pi_2^{\mathrm{m},i}(da|(1,x))$$
$$= i\int_0^{1/i} g((0,0),a)da + i\int_0^{1/i}\int_{\mathbf{A}} g((1,x),a)\mu(da)dx.$$

The last equality is by the Fubini Theorem. Note that $\int_{\mathbf{A}} g((1,x),a)\mu(da)$ is a bounded continuous function of x by the Dominated Convergence Theorem. When $i\to\infty$, according to L'Hopital's rule, we obtain in the limit

$$g((0,0),0) + \int_{\mathbf{A}} g((1,0),a)\mu(da) = \int_{\{0,1\}\times[0,1]\times\mathbf{A}} g((k,x),a)d\eta^{\pi^{\mathrm{m},\infty}}_{(0,0)}((k,x),a)$$

as required.

(b) Suppose for contradiction that the sequence $\{\eta^{\pi^{\mathrm{m},i}}_{(0,0)}\}_{i=1}^{\infty}$ has a convergent subsequence in $\mathbf{P}((\mathbf{X} \setminus \{\Delta\}) \times \mathbf{A})$ with respect to the ws-topology, say, $\{\eta^{\pi^{\mathrm{m},i_j}}_{(0,0)}\}_{j=1}^{\infty}$. Since the ws-toology is finer than the weak topology and $\{\eta^{\pi^{\mathrm{m},i}}_{(0,0)}\}_{i=1}^{\infty}$ converges to $\eta^{\pi^{\mathrm{m},\infty}}_{(0,0)}$ in the weak topology, we see that the limit $\lim_{j\to\infty} \eta^{\pi^{\mathrm{m},i_j}}_{(0,0)}$, assumed to exist, is necessarily $\eta^{\pi^{\mathrm{m},\infty}}_{(0,0)}$. Below we show that $\lim_{j\to\infty} \eta^{\pi^{\mathrm{m},i_j}}_{(0,0)} \neq \eta^{\pi^{\mathrm{m},\infty}}_{(0,0)}$ in the ws-topology.

Repeating the same calculations as above for the Carathéodory function $g((k,x),a) \equiv \mathbb{I}\{x > 0\}$, we obtain

$$\begin{aligned}
&\int_{\{0,1\}\times[0,1]\times\mathbf{A}} g((k,x),a)d\eta^{\pi^{\mathrm{m},i_j}}_{(0,0)}((k,x),a)\\
&= i_j \int_0^{1/i_j} g((0,0),a)da + i_j \int_0^{1/i_j} \int_{\mathbf{A}} g((1,x),a)\mu(da)dx\\
&= i_j \int_0^{1/i_j} \mathbb{I}\{x > 0\}dx = 1
\end{aligned}$$

for each $i_j = 1, 2, \ldots$. On the other hand, for the measure $\eta^{\pi^{\mathrm{m},\infty}}_{(0,0)}$ we have

$$\int_{\{0,1\}\times[0,1]\times\mathbf{A}} g((k,x),a)d\eta^{\pi^{\mathrm{m},\infty}}_{(0,0)}((k,x),a) = g((0,0),0) + \int_{\mathbf{A}} g((1,0),a)\mu(da) = 0.$$

Thus the limit $\lim_{j\to\infty} \eta^{\pi^{\mathrm{m},i_j}}_{(0,0)}$ in the ws-topology (if it exists) is different from $\eta^{\pi^{\mathrm{m},\infty}}_{(0,0)}$, and the sequence $\{\eta^{\pi^{\mathrm{m},i}}_{(0,0)}\}_{i=1}^{\infty}$ cannot have convergent subsequences in $\mathbf{P}((\mathbf{X} \setminus \{\Delta\}) \times \mathbf{A})$ with respect to the ws-topology. □

Proof of Lemma 2.6. The statement of this lemma will be proved after we prove by induction the following statements for $j \geq 1$.

(i) For even steps $i = 2j$,

$$v^*_{2j} = -2j + h,$$

and the minimum in (2.54) is provided by a_1 only.

(ii) For odd steps $i = 2j - 1$,

$$v^*_{2j-1} > \frac{-2\varepsilon(j-1) + (1+\varepsilon)h + 2}{\varepsilon}.$$

(iii) For odd steps $i = 2j + 1$,

$$v^*_{2j+1} = (1-\varepsilon)v^*_{2j-1} + (1+\varepsilon)h - 2\varepsilon(j-1),$$

and the minimum in (2.54) is provided by a_2 only.

When $j = 1$, Items (i) and (iii) are valid by the preliminary calculations, and Item (ii) comes from the following:

$$-2\varepsilon(j-1) + (1+\varepsilon)h + 2 - \varepsilon v^*_{2j-1} = (1+\varepsilon)h + 2 - \varepsilon h = h + 2 < 0.$$

Suppose statements (i), (ii) and (iii) hold for some $j \geq 1$ and consider $j+1$.
(i) For $i = 2(j+1)$, using the induction supposition, we estimate the difference

$$\begin{aligned} &-2 + v^*_{2j} - [h + (1-\varepsilon)v^*_{2j} + \varepsilon v^*_{2j-1}] \\ &< -2 - 2j + h - h - (1-\varepsilon)(-2j+h) - \varepsilon\frac{-2\varepsilon(j-1) + (1+\varepsilon)h + 2}{\varepsilon} \\ &= -4 - 2\varepsilon - 2h = 2[-2 - \varepsilon - h] < 0. \end{aligned}$$

The inequality is according to statement (ii) at j.

Thus, $v^*_{2(j+1)} = -2(j+1) + h$, and the minimum in (2.54) is provided only by a_1.

(ii)

$$\begin{aligned} v^*_{2j+1} &= (1-\varepsilon)v^*_{2j-1} + (1+\varepsilon)h - 2\varepsilon(j-1) \\ &> (1-\varepsilon)\frac{-2\varepsilon(j-1) + (1+\varepsilon)h + 2}{\varepsilon} + (1+\varepsilon)h - 2\varepsilon(j-1) \\ &= \frac{-2\varepsilon j + (1+\varepsilon)h + 2}{\varepsilon}, \end{aligned}$$

so statement (ii) is valid for $j+1$.

(iii) For $i = 2(j+1)+1$, using the induction supposition, we estimate the difference

$$\begin{aligned} &h + (1-\varepsilon)v^*_{2j+1} + \varepsilon v^*_{2j} - [-2 + v^*_{2j+1}] \\ &= h - \varepsilon v^*_{2j+1} + \varepsilon[-2j + h] + 2 \\ &< h(1+\varepsilon) - [-2\varepsilon j + h + 2 + \varepsilon h] - 2\varepsilon j + 2 = 0, \end{aligned}$$

where the inequality is by the proved above Item (ii) for $j+1$. Recall also that $v^*_{2j} = -2j + h$. Therefore,

$$v^*_{2(j+1)+1} = h + (1-\varepsilon)v^*_{2j+1} + \varepsilon[-2j + h).$$

We see that statement (iii) is valid for $j+1$ and the minimum in (2.54) is provided only by a_2. □

Proof of Lemma 3.1. Let the initial state $x_0 \in \mathbf{X}$ be arbitrarily fixed. The target is to show that, for any strategy π, $v^\pi_{x_0} \geq v^{\hat\pi}_{x_0}$. This inequality is obvious in the case $v^\pi_{x_0} = +\infty$, so below we assume that $v^\pi_{x_0} < \infty$. Equation (3.4) implies that, for each $T = 1, 2, \dots$,

$$\mathsf{E}^{\pi}_{x0}\left[\sum_{t=1}^{T}\beta^{t-1}\left\{c(X_{t-1}, A_t) - v^{\hat{\pi}}_{X_{t-1}} + \beta\int_{\mathbf{X}} v^{\hat{\pi}}_y p(dy|X_{t-1}, A_t)\right\}\right] \ge 0. \qquad \text{(A.2)}$$

On the left-hand side we have the following sum of finite expectations:

$$\begin{aligned}&\mathsf{E}^{\pi}_{x0}\left[\sum_{t=1}^{T}\beta^{t-1}c(X_{t-1}, A_t)\right] - \mathsf{E}^{\pi}_{x0}[v^{\hat{\pi}}_{X_0}] + \beta\mathsf{E}^{\pi}_{x0}\left[\int_{\mathbf{X}} v^{\hat{\pi}}_y p(dy|X_0, A_1)\right]\\ &-\beta\mathsf{E}^{\pi}_{x0}[v^{\hat{\pi}}_{X_1}] + \beta^2\mathsf{E}^{\pi}_{x0}\left[\int_{\mathbf{X}} v^{\hat{\pi}}_y p(dy|X_1, A_2)\right] - \dots\\ &+\beta^T\mathsf{E}^{\pi}_{x0}\left[\int_{\mathbf{X}} v^{\hat{\pi}}_y p(dy|X_{T-1}, A_T)\right].\end{aligned}$$

For each $t = 0, 1, \dots, T-1$

$$\mathsf{E}^{\pi}_{x0}\left[\int_{\mathbf{X}} v^{\hat{\pi}}_y p(dy|X_t, A_{t+1})\right] = \mathsf{E}^{\pi}_{x0}[v^{\hat{\pi}}_{X_{t+1}}] \in \mathbb{R},$$

and (A.2) takes the form

$$\mathsf{E}^{\pi}_{x0}\left[\sum_{t=1}^{T}\beta^{t-1}c(X_{t-1}, A_t)\right] - v^{\hat{\pi}}_{x0} + \beta^T\mathsf{E}^{\pi}_{x0}[v^{\hat{\pi}}_{X_T}] \ge 0.$$

The desired inequality $v^{\pi}_{x0} \ge v^{\hat{\pi}}_{x0}$ follows after passing to the limit as $T \to \infty$. □

Proof of Lemma 3.2. Since the cost function $c(\cdot,\cdot)$ is bounded, the optimality equation has a unique bounded solution, and we only need to check that the given function $v^{\beta}(\cdot)$ satisfies equation (3.27).

(a) Let $i < n(\beta)$ and introduce

$$\begin{aligned}\delta &:= \frac{1 - \beta^{n(\beta)-i+1} + 2\beta^{2n(\beta)-i+1}}{1-\beta} - \left[1 + \frac{2\beta^{i+1}}{1-\beta}\right]\\ &= \frac{\beta}{1-\beta}[2\beta^{2n(\beta)-i} - \beta^{n(\beta)-i} + 1 - 2\beta^i]\\ &= \frac{\beta^{1-i}}{1-\beta}[2\beta^{2n(\beta)} - \beta^{n(\beta)} + \beta^i - 2\beta^{2i}].\end{aligned}$$

For $i < n(\beta)$

$$\beta^{i+1} - 2\beta^{2(i+1)} - \beta^i + 2\beta^{2i} = \beta^i(\beta-1)[1 - 2\beta^i(1+\beta)] > 0,$$

so the function $\beta^i - 2\beta^{2i}$ increases with $i \in \{1, 2, \dots, n(\beta) - 1\}$, and the inequality

$$\begin{aligned}
&2\beta^{2n(\beta)} - \beta^{n(\beta)} + \beta^{n(\beta)-1} - 2\beta^{2(n(\beta)-1)} \\
&= \beta^{n(\beta)-1}[2\beta^{n(\beta)+1} - \beta + 1 - 2\beta^{n(\beta)-1}] \\
&= \beta^{n(\beta)-1}(1-\beta)[1 - 2\beta^{n(\beta)-1}(1+\beta)] < 0
\end{aligned}$$

implies that $\delta < 0$ for all $i < n(\beta)$.

For all $i < n(\beta) - 1$, the equality $v^\beta((i,0)) = 1 + \beta v^\beta((i+1,0))$ is obvious; it holds also for $i = n(\beta) - 1$:

$$1 + \beta v^\beta((n(\beta),0)) = 1 + \beta\left[1 + \frac{2\beta^{n(\beta)+1}}{1-\beta}\right] = \frac{1 - \beta^2 + 2\beta^{n(\beta)+2}}{1-\beta}.$$

Thus, equation (3.27) is satisfied.

(b) Let $i \geq n(\beta)$. Then

$$\delta := 1 + \frac{2\beta^{i+1}}{1-\beta} - [1 + \beta v^\beta((i+1,0))]$$

$$= \frac{2\beta^{i+1} - \beta(1-\beta) - 2\beta^{i+3}}{1-\beta} = 2\beta^{i+1}(1+\beta) - \beta \leq 0,$$

and equation (3.27) is satisfied, too.

The stationary selector φ^β is uniformly optimal because it is conserving: see Theorem 3.1(b). □

Proof of Proposition 4.1. (a) Suppose canonical equations (4.2) have a finite solution $\langle \rho, h, \varphi \rangle$. As usual, one can put $h(0) = 0$. (See Remark 4.1.) Then, for any $x \geq 1$, $\rho(x) = \rho(x-1)$, so that $\rho(x) \equiv \rho$. From the second equation (4.2) we obtain

$$\rho + h(x) = 1 + h(x-1), \qquad x \geq 1,$$

so that

$$h(x) = h(0) + (1-\rho)x = (1-\rho)x$$

and, for $x = 0$, we have

$$\rho + h(0) = \min\{0 + (1-\rho)\sum_{x\geq 1} x q_x; \quad 1 + 1 - \rho\}.$$

ρ cannot be greater than 1, because otherwise $\rho = -\infty$. If $\rho < 1$ then

$$\rho = 2 - \rho \Longrightarrow \rho = 1.$$

Therefore, $\rho = 1$ leading to $\rho = \min\{0, 1\} = 0$, which again is a contradiction.

(b) Condition 4.3(b) is certainly not satisfied, since otherwise there would have existed a triplet satisfying canonical equations (4.2). (See Theorem 4.4(b).) A more straightforward argument can be as follows. If, for $x = 0$, $a = 1$, the function $b(y)$ is summable w.r.t. the distribution $p(j|0,1) = q_j$, then, by the Lebesgue Dominated Convergence Theorem [Goffman and Pedrick (1983), Section 3.7],

$$\lim_{\beta\to 1-} \sum_{j=1}^{\infty} h^\beta(j) q_j = \sum_{j=1}^{\infty} \lim_{\beta\to 1-} h^\beta(j) q_j.$$

But on the left-hand side we have

$$\lim_{\beta\to 1-} \frac{1-\sum_{j=1}^\infty \beta^j q_j}{1-\beta\sum_{j=1}^\infty \beta^j q_j} = \lim_{\beta\to 1-} \frac{\sum_{j=1}^\infty j\beta^{j-1} q_j}{\sum_{j=1}^\infty \beta^j q_j + \sum_{j=1}^\infty j\beta^j q_j} = 1,$$

and on the right we have zero because

$$h(j) = \lim_{\beta\to 1-} h^\beta(j) = \frac{j}{1+\sum_{j=1}^\infty j q_j} = 0 \quad \text{for } j \geq 1.$$

Here L'Hôpital's rule was in use.

(c) If a stationary distribution $\hat{\eta}(x)$ on $\mathbf{X}$ exists, then it satisfies the equations

$$\hat{\eta}(0) = \hat{\eta}(1); \qquad \hat{\eta}(i) = \hat{\eta}(0) q_i + \hat{\eta}(i+1).$$

After we denote $\gamma := \hat{\eta}(0)$, the value of γ comes from the normalization condition:

$$\begin{aligned} &\gamma[1+1+(1-q_1)+(1-q_1-q_2)+(1-q_1-q_2-q_3)+\cdots] \\ &= \gamma\left[1+\lim_{n\to\infty}\{n-(n-1)q_1-(n-2)q_2-\cdots-q_{n-1}\}\right] \qquad \text{(A.3)} \\ &= \gamma\left[1+\lim_{n\to\infty}\{n-n\sum_{i=1}^{n-1} q_i+\sum_{i=1}^{n-1} i q_i\}\right] = 1. \end{aligned}$$

But $n - n\sum_{i=1}^{n-1} q_i \geq 0$ and $\lim_{n\to\infty}\sum_{i=1}^{n-1} i q_i = +\infty$, so that $\gamma\cdot\infty = 1$, which is a contradiction.

Let $\sum_{j=1}^\infty j q_j < \infty$. Then equation (A.3) implies that $\gamma = \hat{\eta}(0) = \dfrac{1}{1+\sum_{j=1}^\infty j q_j}$, because

$$\lim_{n\to\infty} n\left[1-\sum_{i=1}^{n-1} q_i\right] = \lim_{n\to\infty} n\sum_{n=n}^{\infty} q_i \leq \lim_{n\to\infty}\sum_{i=n}^{\infty} i q_i = 0,$$

and the assertion follows:

$$\hat{\eta}(i) = \gamma\left[1-\sum_{j=1}^{i-1} q_j\right], \qquad i \geq 1.$$

□

Proof of Proposition 4.2. (a) For $i \geq 1$, the mean recurrence time from state i' to state 0 equals $M_{i'0}(\pi) = 2^i$ for any control strategy π. Similarly, $M_{i0}(\varphi^n) = 2^i$ if $i \geq n$. For $0 < i < n < \infty$ we have

$$M_{i0}(\varphi^n) = \frac{1}{2} + \frac{1}{2}(1 + M_{i+1,0}) = 1 + \frac{1}{2} + \frac{1}{4} + \cdots + \left(\frac{1}{2}\right)^{n-i-1} + \left(\frac{1}{2}\right)^{n-i} M_{n0}(\varphi^n) = 2 + 2^i - \left(\frac{1}{2}\right)^{n-i-1},$$

and $M_{i0}(\varphi^\infty) = 2$. Therefore,

$$\begin{aligned} M_{00}(\varphi^n) &= 1 + \sum_{i=1}^{\infty} \frac{3}{2}\left(\frac{1}{4}\right)^i \left[M_{i0}(\varphi^n) + 2^i\right] \\ &= 1 + \frac{3}{2}\sum_{i=1}^{n-1}\left(\frac{1}{4}\right)^i \left[2 + 2^i - \left(\frac{1}{2}\right)^{n-i-1}\right] + \frac{3}{2}\sum_{i=n}^{\infty}\left(\frac{1}{4}\right)^i 2^i + \frac{3}{2} \\ &= 5 - 3\left(\frac{1}{2}\right)^n + 2\left(\frac{1}{4}\right)^n < 5, \quad n = 1, 2, \ldots, \end{aligned}$$

$\lim_{n\to\infty} M_{00}(\varphi^n) = 5$, and the convergence is monotonic (increasing). Note that

$$\begin{aligned} M_{00}(\varphi^\infty) &= 1 + \sum_{i=1}^{\infty} \frac{3}{2}\left(\frac{1}{4}\right)^i [M_{i0}(\varphi^\infty) + M_{i'0}(\varphi^\infty)] \\ &= 1 + \frac{3}{2}\sum_{i=1}^{\infty}\left(\frac{1}{4}\right)^i [2 + 2^i] = \frac{7}{2}. \end{aligned}$$

Equations for $v_{\mathsf{P}_0}^{\varphi^n}$ $(n = 1, 2, \ldots, \infty)$ follow from Remark 4.3.

(b) Now let π be an arbitrary control strategy, $x \in \mathbf{X}$ be the initial state and estimate from above the mean recurrence time $M_{x0}(\pi)$ to state 0.

Clearly, $M_{i'0}(\pi) = 2^i$. Denoting $V^*(i) := \sup_\pi M_{i0}(\pi)$ for $i \geq 1$, we obtain the optimality equation

$$V^*(i) = 1 + \max\left\{\left(1 - \left(\frac{1}{2}\right)^i\right) 2^i; \;\; \frac{1}{2}V^*(i+1)\right\}, \quad i \geq 1,$$

corresponding to the following MDP with the expected total reward:

$$\tilde{\mathbf{X}} = \{0, 1, 2, \ldots\}, \; \tilde{\mathbf{A}} = \{1, 2\},$$

$$\tilde{p}(0|0, a) \equiv 1, \; \tilde{p}(0|i, 1) = 1, \; \tilde{p}(0|i, 2) = \tilde{p}(i+1|i, 2) = \frac{1}{2},$$

$$\tilde{c}(0, a) \equiv 0, \; \tilde{c}(i, 1) = 1 + \left(1 - \left(\frac{1}{2}\right)^i\right) 2^i = 2^i, \; \tilde{c}(i, 2) = 1.$$

The value of $\mathsf{E}_i^\pi\left[\sum_{t=1}^\infty \tilde{c}(X_{t-1},A_t)\right]$ for $i\ge 1$ in this model equals $M_{i0}(\pi)$. This model is positive and the objective is to be maximized. Using the corresponding version of Theorem 2.2 (i.e., change the sign of the function $c(\cdot,\cdot)$ and consider minimization of the expected total cost), we have $V^*(i)=\lim_{n\to\infty}V^n(i)$, where $V^n(\cdot)$ comes from the value iteration (2.6):

$$V^0(i)\equiv 0; \qquad V^n(0)\equiv 0;$$
$$V^{n+1}(i)=\max\{2^i,\ \ 1+\frac{1}{2}V^n(i+1)\},\quad i\ge 1.$$

It is straightforward to show by induction that $V^n(i)\le 2+2^i$ and $\lim_{n\to\infty}V^n(i)=V^*(i)=2+2^i$, $i\ge 1$.

Let us show that there are no uniformly optimal strategies in this auxiliary model. Suppose the initial state $x_0=i\ge 1$ is arbitrarily fixed. Without loss of generality, consider a Markov strategy π^{m}. (See Lemma 2.1.) If $\pi_t^{\mathrm{m}}(2|j)\equiv 1$, then

$$\mathsf{E}_i^{\pi^{\mathrm{m}}}\left[\sum_{t=1}^\infty \tilde{c}(X_{t-1},A_t)\right]=2<2+2^i=V^*(i).$$

When starting from $x_0=i\ge 1$, the values of $\pi_t^{\mathrm{m}}(2|j)$ are immaterial for $j\ne i+t-1$: the state at the decision epoch t can be either $j=i+t-1$ or 0, and in the latter case the action plays no role. Thus, suppose $\hat{t}\ge 1$ is the minimal time moment such that $\hat{t}=j-i+1$ and $\pi_{\hat{t}}^{\mathrm{m}}(2|j)=:q<1$. Now

$$\begin{aligned}\mathsf{E}_i^{\pi^{\mathrm{m}}}\left[\sum_{t=\hat{t}}^\infty \tilde{c}(X_{t-1},A_t)\right] &= \left(\frac{1}{2}\right)^{\hat{t}-1}\mathsf{E}_j^{\widetilde{\pi^{\mathrm{m}}}}\left[\sum_{t=1}^\infty \tilde{c}(X_{t-1},A_t)\right]\\ &\le \left(\frac{1}{2}\right)^{\hat{t}-1}\left\{q\left[\tilde{c}(j,2)+\frac{1}{2}V^*(j+1)\right]\right.\\ &\quad \left.+(1-q)\tilde{c}(j,1)\right\}\\ &= \left(\frac{1}{2}\right)^{\hat{t}-1}\{q[2+2^j]+(1-q)2^j\}\\ &< \left(\frac{1}{2}\right)^{\hat{t}-1}(2+2^j)=\left(\frac{1}{2}\right)^{\hat{t}-1}(2+2^{i+\hat{t}-1}),\end{aligned}$$

where $\widetilde{\pi^{\mathrm{m}}}_t(da|x):=\pi_{t+\hat{t}-1}^{\mathrm{m}}(da|x)$ is the shifted Markov strategy.

Therefore,

$$\begin{aligned}\mathsf{E}_i^{\pi^{\mathrm{m}}}\left[\sum_{t=1}^{\infty}\tilde{c}(X_{t-1},A_t)\right] &= \sum_{k=0}^{\hat{t}-2}\left(\frac{1}{2}\right)^k + \mathsf{E}_i^{\pi^{\mathrm{m}}}\left[\sum_{t=\hat{t}}^{\infty}\tilde{c}(X_{t-1},A_t)\right] \\ &< \frac{1-(1/2)^{\hat{t}-1}}{1/2} + \left(\frac{1}{2}\right)^{\hat{t}-1}(2+2^{i+\hat{t}-1}) \\ &= 2+2^i = V^*(i),\end{aligned}$$

and again the strategy π^{m} is not optimal.

To summarize, for $i \ge 1$, $M_{i'0}(\pi) \equiv 2^i$, $M_{i0}(\pi) < \sup_\pi M_{i0}(\pi) = 2+2^i$ and hence $M_{00}(\pi) < 1 + \frac{3}{2}\sum_{i=1}^{\infty}\left(\frac{1}{4}\right)^i[2^i+2+2^i] = 5$. As was shown above, $\sup_\pi M_{00}(\pi) = 5 = \lim_{n\to\infty} M_{00}(\varphi^n)$.

In particular, for any stationary strategy π^{s}, the whole state space $\mathbf{X}$ is the closed communicating class and $M_{00}(\pi^{\mathrm{s}}) < 5$. Therefore, the controlled process under π^{s} is positive recurrent and, for any initial distribution P_0, the stationary probability of state 0 $\hat{\eta}^{\pi^{\mathrm{s}}}(0) = \frac{1}{M_{00}(\pi^{\mathrm{s}})}$ [Kemeny*et al.* (1976), Prop.6.25] is strictly greater than $\frac{1}{5}$. Thus, in the case under consideration, $v_{\mathsf{P}_0}^{\pi^{\mathrm{s}}} = \hat{\eta}^{\pi^{\mathrm{s}}}(0) > \frac{1}{5}$. (See Remark 4.3.) Note also that $v_{\mathsf{P}_0}^{\varphi^n} = \frac{1}{M_{00}(\varphi^n)} \downarrow \frac{1}{5}$ when $n\to\infty$ and $v_{\mathsf{P}_0}^{\varphi^\infty} = \frac{1}{M_{00}(\varphi^\infty)} = \frac{2}{7}$.

(c) Consider now the discounted functional (3.1) with $\beta \in (0,1)$. Since only cost $c(0,a)=1$ is non-zero, it is obvious that $v_x^{*,\beta} < v_0^{*,\beta}$ for all $x \in \mathbf{X}\setminus\{0\}$. This strict inequality also follows from the optimality equation. Indeed,

$$v_{i'}^{*,\beta} = \frac{\beta\left(\frac{1}{2}\right)^i v_0^{*,\beta}}{1-\beta[1-\left(\frac{1}{2}\right)^i]}$$

and

$$v_i^{*,\beta} \le \beta\left(\frac{1}{2}\right)^i v_0^{*,\beta} + \beta\left[1-\left(\frac{1}{2}\right)^i\right]v_{i'}^{*,\beta} = \frac{\beta\left(\frac{1}{2}\right)^i v_0^{*,\beta}}{1-\beta[1-\left(\frac{1}{2}\right)^i]}.$$

But function $\frac{\beta\left(\frac{1}{2}\right)^i}{1-\beta[1-\left(\frac{1}{2}\right)^i]}$ decreases with i, so that, for all $x \in \mathbf{X}\setminus\{0\}$,

$$v_x^{*,\beta} \le \frac{\frac{\beta}{2}v_0^{*,\beta}}{1-\beta+\frac{\beta}{2}} < v_0^{*,\beta}.$$

Recall that $v_x^{*,\beta} \ge 0$.

Fix the initial state $x_0 = 0$ and an arbitrary control strategy π, which without loss of generality can be regarded as Markov (see Lemma 2.1) and denoted as π^{m}.

Let $T_0 = 0$, $T_1, \ldots$ be the sequence of time moments when $X_{T_n} = 0$. In part (a), it was proved that $\mathsf{E}_0^{\pi^{\rm m}}[T_1] = M_{00}(\pi^{\rm m}) < 5$, $\mathsf{E}_0^{\pi^{\rm m}}[T_2 - T_1] = M_{00}(\pi^{\rm m}) < 5$, etc. So $\mathsf{E}_0^{\pi^{\rm m}}[T_1] < 5$, $\mathsf{E}_0^{\pi^{\rm m}}[T_2] < 10$, ..., $\mathsf{E}_0^{\pi^{\rm m}}[T_n] < 5n$, ... Now

$$v_0^{\pi^{\rm m},\beta} = \mathsf{E}_0^{\pi^{\rm m}}[1 + \beta^{T_1} + \beta^{T_2} + \cdots]$$

and, from the Jensen inequality, we have

$$v_0^{\pi^{\rm m},\beta} \geq 1 + \sum_{n=1}^{\infty} \beta^{\mathsf{E}_0^{\pi^{\rm m}}[T_n]} > \sum_{n=0}^{\infty} \beta^{5n} = \frac{1}{1-\beta^5}.$$

According to the Tauberian/Abelian Theorem (see [Hernández-Lerma and Lasserre (1996a), Lemma 5.3.1]), since $c(\cdot,\cdot) \geq 0$,

$$\begin{aligned} v_0^{\pi^{\rm m}} &= \limsup_{T\to\infty} \frac{1}{T}\mathsf{E}_0^{\pi^{\rm m}}\left[\sum_{t=1}^{T} c(X_{t-1}, A_t)\right] \\ &\geq \limsup_{\beta\to 1-}(1-\beta)\mathsf{E}_0^{\pi^{\rm m}}\left[\sum_{t=1}^{\infty} \beta^{t-1} c(X_{t-1}, A_t)\right] \\ &= \limsup_{\beta\to 1-}(1-\beta) v_0^{\pi^{\rm m},\beta} \geq \lim_{\beta\to 1-}(1-\beta)\frac{1}{1-\beta^5} = \frac{1}{5}. \end{aligned}$$

If the initial state x is $i \geq 1$ or i' then the expectation of $\tilde{T}$, the first recurrence time to state 0 is smaller than $2 + 2^i$ (see part (a)) and, again using the Jensen inequality, we have

$$v_x^{\pi^{\rm m},\beta} = \mathsf{E}_x^{\pi^{\rm m}}\left[\beta^{\tilde{T}} v_0^{\widetilde{\pi^{\rm m}},\beta}\right] \geq \beta^{2+2^i} \frac{1}{1-\beta^5},$$

where $\widetilde{\pi^{\rm m}}$ is the shifted Markov strategy: $\widetilde{\pi^{\rm m}}_t(da|x) = \pi^{\rm m}_{\tilde{T}+t}(da|x)$. Thus, the similar argument as above leads to inequality $v_x^{\pi^{\rm m}} \geq \frac{1}{5}$ as well.

Finally, if P_0 is an arbitrary initial distribution, then

$$v_{\mathsf{P}_0}^{\pi^{\rm m},\beta} \geq \frac{1}{1-\beta^5}\left\{\sum_{i=1}^{\infty} [\mathsf{P}_0(i) + \mathsf{P}_0(i')]\, \beta^{2+2^i} + \mathsf{P}_0(0)\right\}$$

and the similar argument as above leads to inequality $v_{\mathsf{P}_0}^{\pi^{\rm m}} \geq \frac{1}{5}$ as well. Statement (c) is proved.

(d) We know from part (a) that, for the stationary selector φ^n, $M_{00}(\varphi^n) < 5$ and $M_{00}(\varphi^n) \uparrow 5$ as $n \to \infty$. In what follows, the initial distribution P_0 is arbitrarily fixed. As mentioned above, $v_{\mathsf{P}_0}^{\varphi^n} > \frac{1}{5}$ and $v_{\mathsf{P}_0}^{\varphi^n} \downarrow \frac{1}{5}$ as $n \to \infty$.

For $k = 1, 2, \dots$ let φ^{n_k} be a selector (4.17) such that $v_{\mathsf{P}_0}^{\varphi^{n_k}} < \frac{1}{5} + \frac{1}{2k}$, $1 \le n_1 < n_2 < \cdots$, and fix $N_k \ge 1$ such that

$$\frac{1}{T}\mathsf{E}_{\mathsf{P}_0}^{\varphi^{n_k}}\left[\sum_{t=1}^{T} c(X_{t-1}, A_t)\right] < v_{\mathsf{P}_0}^{\varphi^{n_k}} + \frac{1}{2k}$$

for all $T \ge N_k$. Let $\bar{N}_1 > N_1$ be such that

$$\frac{1}{\bar{N}_1}\mathsf{E}_{\mathsf{P}_0}^{\varphi^{n_1}}\left[\sum_{t=1}^{\bar{N}_1} c(X_{t-1}, A_t) + N_2\right] < v_{\mathsf{P}_0}^{\varphi^{n_1}} + \frac{1}{2}$$

and define $\bar{N}_k > N_k$, $n = 2, 3, \dots$, recursively by letting $\bar{N}_k$ be such that

$$\frac{1}{\bar{N}_k}\mathsf{E}_{\mathsf{P}_0}^{\varphi^{n_k}}\left[\sum_{t=1}^{\bar{N}_k} c(X_{t-1}, A_t) + \sum_{j=1}^{k-1} \bar{N}_j + N_{k+1}\right] < v_{\mathsf{P}_0}^{\varphi^{n_k}} + \frac{1}{2k}. \tag{A.4}$$

The selector φ^* of our interest is constructed in the following way.

Apply the stationary selector φ^{n_1} on the interval $T_0 := 0 < t \le T_1$, where the random time moment $T_1 := \min\{t \ge T_0 + \bar{N}_1 : X_t = 0\}$ is the first after $T_0 + \bar{N}_1$ when the controlled process (under $\varphi^* = \varphi^{n_1}$) equals 0.

Apply the stationary selector φ^{n_2} on the interval $T_1 < t \le T_2$, where the random time moment $T_2 := \min\{t \ge T_1 + \bar{N}_2 : X_t = 0\}$ is the first after $T_1 + \bar{N}_2$ when the controlled process (under $\varphi^* = \varphi^{n_2}$) equals 0.

And so on: for $k = 1, 2, \dots$

$$T_k := \min\{t \ge T_{k-1} + \bar{N}_k : X_t = 0\}; \ \varphi_t^*(x) = \varphi^{n_k}(x)\mathbb{I}\{T_{k-1} < t \le T_k\}.$$

(See Figure A.2). Note that, under φ^*,

$$c(X_{t-1}, A_t) = 0 \text{ for all } \ t = T_{k-1} + \bar{N}_k + 1,\ T_{k-1} + \bar{N}_k + 2, \dots, T_k, \quad k = 1, 2, \dots. \tag{A.5}$$

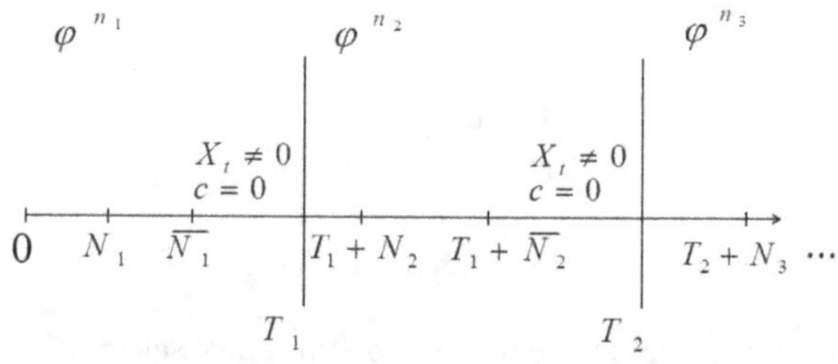

Fig. A.2 Construction of the selector φ^*.

Fix an arbitrary T such that $\sum_{i=1}^{k} \bar{N}_i < T \le \sum_{i=1}^{k+1} \bar{N}_i$, where $k > 1$, and prove that

$$\frac{1}{T}\mathsf{E}_{\mathsf{P}_0}^{\varphi^*}\left[\sum_{t=1}^{T} c(X_{t-1}, A_t)\right] < v_{\mathsf{P}_0}^{\varphi^{n_k}} + \frac{1}{2k} < \frac{1}{5} + \frac{1}{k}. \tag{A.6}$$

Obviously, $T \le T_k + \bar{N}_{k+1}$ because $\sum_{i=1}^{k} \bar{N}_i \le T_k$. We shall consider two cases:

(i) $T \le T_k + N_{k+1}$. Now $\sum_{t=1}^{T} c(X_{t-1}, A_t) \le \sum_{j=1}^{k-1} \bar{N}_j + \sum_{t=T_{k-1}+1}^{T_{k-1}+\bar{N}_k} c(X_{t-1}, A_t) + N_{k+1}$ (recall (A.5)). Therefore,

$$\begin{aligned}
&\frac{1}{T}\mathsf{E}_{\mathsf{P}_0}^{\varphi^*}\left[\sum_{t=1}^{T} c(X_{t-1}, A_t) \cdot I\{T \le T_k + N_{k+1}\}\right] \\
&\le \frac{1}{\bar{N}_k}\left\{\mathsf{E}_0^{\varphi^{n_k}}\left[\sum_{t=1}^{\bar{N}_k} c(X_{t-1}, A_t)\right] + \sum_{j=1}^{k-1} \bar{N}_j + N_{k+1}\right\} \\
&< v_{\mathsf{P}_0}^{\varphi^{n_k}} + \frac{1}{2k}
\end{aligned} \tag{A.7}$$

by (A.4).

(ii) $T_k + N_{k+1} < T \le T_k + \bar{N}_{k+1}$, below, we write this event as D for brevity. Now

$$\begin{aligned}
&\frac{1}{T}\mathsf{E}_{\mathsf{P}_0}^{\varphi^*}\left[\sum_{t=1}^{T} c(X_{t-1}, A_t) \cdot I\{D\}\right] \\
&= \frac{1}{T}\mathsf{E}_{\mathsf{P}_0}^{\varphi^*}\left[\left(\sum_{t=1}^{T_k} c(X_{t-1}, A_t) + \sum_{t=T_k+1}^{T} c(X_{t-1}, A_t)\right) I\{D\}\right] \\
&\le \frac{\bar{N}_k}{T}\mathsf{E}_0^{\varphi^{n_k}}\left[\frac{1}{\bar{N}_k}\sum_{i=1}^{k-1} \bar{N}_i + \sum_{t=1}^{\bar{N}_k} c(X_{t-1}, A_t)\right] \\
&\quad + \frac{T - \bar{N}_k}{T}\mathsf{E}_0^{\varphi^{n_{k+1}}}\left[\frac{1}{T - \bar{N}_k}\sum_{t=1}^{T-T_k} c(X_{t-1}, A_t)\mathbb{I}\{D\}\right]
\end{aligned}$$

Since $T_k \ge \bar{N}_k$ and, under assumption D, $T - \bar{N}_k \ge T - T_k > N_{k+1}$ ($P_{P_0}^{\varphi^*}$-a.s.), we conclude that

$$\begin{aligned}
&\mathsf{E}_0^{\varphi^{n_{k+1}}}\left[\frac{1}{T-\bar{N}_k}\sum_{t=1}^{T-T_k} c(X_{t-1},A_t)\right]\\
&\leq \mathsf{E}_0^{\varphi^{n_{k+1}}}\left[\frac{1}{T-\bar{N}_k}\sum_{t=1}^{T-\bar{N}_k} c(X_{t-1},A_t)\right]\\
&< v_{\mathsf{P}_0}^{\varphi^{n_{k+1}}}+\frac{1}{2(k+1)} \leq v_{\mathsf{P}_0}^{\varphi^{n_k}}+\frac{1}{2k}.
\end{aligned}$$

The second inequality is by the definition of N_{k+1}. The inequality

$$\mathsf{E}_0^{\varphi^{n_k}}\left[\frac{1}{\bar{N}_k}\sum_{i=1}^{k-1}\bar{N}_i+\sum_{t=0}^{\bar{N}_k} c(X_{t-1},A_t)\right] < v_{\mathsf{P}_0}^{\varphi^{n_k}}+\frac{1}{2k}$$

follows from the definition of $\bar{N}_k$. Therefore,

$$\frac{1}{T}\mathsf{E}_{\mathsf{P}_0}^{\varphi^*}\left[\sum_{t=1}^{T} c(X_{t-1},A_t)\cdot I\{D\}\right] < v_{\mathsf{P}_0}^{\varphi^{n_k}}+\frac{1}{2k}.$$

This inequality and inequality (A.7) complete the proof of (A.6) because $v_{\mathsf{P}_0}^{\varphi^{n_k}} < \frac{1}{5}+\frac{1}{2k}$.

Now statement (d) of Proposition 4.2 is obvious. □

Proof of Lemma 4.1. On step 2 of the algorithm, we have to determine $\rho_n(3)=\rho_n(4)$ and $h_n(3), h_n(4)$.

The value of $\rho_n(3)$ comes from equation $\rho_n(3) = 5(\frac{1}{4}+(\varphi^n(3))^2)+4(\frac{1}{4}+\varphi^n(3))+\rho_n(3)(\frac{1}{2}-\varphi^n(3)-(\varphi^n(3))^2)$:

$$\rho_n(3)=\frac{9+20(\varphi^n(3))^2+16\varphi^n(3)}{2+4(\varphi^n(3))^2+4\varphi^n(3)} \in (4,5). \tag{A.8}$$

The proof of inequality $\rho_n(3)<5$ is by contradiction: if $\rho_n(3)\geq 5$, then

$$\rho_n(3)=5(\frac{1}{4}+(\varphi^n(3))^2)+4(\frac{1}{4}+\varphi^n(3))+\rho_n(3)(\frac{1}{2}-\varphi^n(3)-(\varphi^n(3))^2)<\rho_n(3)$$

because $(\frac{1}{4}+(\varphi^n(3))^2,\ \frac{1}{4}+\varphi^n(3),\ \frac{1}{2}-\varphi^n(3)-(\varphi^n(3))^2)$ is a vector of probabilities. Similarly, $\rho_n(3)$ cannot be smaller than 4. Finally, the values of $h_n(3)$ and $h_n(4)$ come from equations

$$\begin{aligned}
&\rho_n(3)+h_n(3)=6+h_n(3)\left(\frac{1}{2}-\varphi^n(3)-(\varphi^n(3))^2\right);\\
&\rho_n(4)+h_n(4)=6+h_n(3).
\end{aligned}$$

On step 3 for $x = 3$, we need to minimize the expression

$$F(\alpha) = 5\left(\frac{1}{4} + \alpha^2\right) + 4\left(\frac{1}{4} + \alpha\right) + \rho_n(3)\left(\frac{1}{2} - \alpha - \alpha^2\right)$$
$$= \frac{9}{4} + \frac{1}{2}\rho_n(3) + (5 - \rho_n(3))\alpha^2 + (4 - \rho_n(3))\alpha.$$

(The value $a = \hat{0}$ will be considered later.) The minimum $\min_{\alpha \in [0,1/3]} F(\alpha)$ is provided by

$$\alpha^* = \min\left\{\frac{\rho_n(3) - 4}{2(5 - \rho_n(3))}, \quad 1/3\right\}$$

because $\frac{\rho_n(3)-4}{2(5-\rho_n(3))} \in (0, \infty)$ is the argmin$\{F(\alpha)\}$ and $F(\cdot)$ is the convex function.

If $\varphi^n(3) = 1/3$, then, according to (A.8), $\rho_n(3) = \frac{149}{34}$ and $\frac{\rho_n(3)-4}{2(5-\rho_n(3))} = \frac{13}{42} < 1/3$. Thus, $\alpha^* = \frac{13}{42} \neq 1/3 = \varphi^n(3)$.

If $\varphi^n(3) < 1/3$, then the case $\alpha^* = \varphi^n(3)$ is excluded because, after we substitute expression (A.8) into equality $\frac{\rho_n(3)-4}{2(5-\rho_n(3))} = \varphi^n(3)$, we obtain the single positive root $\varphi^n(3) = \frac{\sqrt{5}-1}{4}$ contradicting the assumption that $\varphi^n(3)$ is rational.

Therefore, in any case $\alpha^* \neq \varphi^n(3)$ and

$$\min_{\alpha \in [0,1/3]} F(\alpha) = F(\alpha^*)$$
$$< 5\left(\frac{1}{4} + (\varphi^n(3))^2\right) + 4\left(\frac{1}{4} + \varphi^n(3)\right) + \rho_n(3)\left(\frac{1}{2} - \varphi^n(3) - (\varphi^n(3))^2\right)$$
$$= \rho_n(3).$$

Moreover,

$$\sum_{y \in \mathbf{X}} \rho_n(y)p(y|3, \hat{0}) = \rho_n(4) = \rho_n(3) > F(\alpha^*),$$

so that $\varphi^{n+1}(3) = \alpha^* \neq \varphi^n(3)$.

The values $\varphi^{n+1}(1, 2, 4) = \tilde{\alpha}$ remain the same, the value of $\varphi^{n+1}(3) = \alpha^*$ is rational because the value of $\rho_n(3)$ is rational by (A.8).

The proof is completed. □

Proof of Proposition 4.3. (a) Let the initial state $x_0 = k \geq 1$ as well a strategy π be arbitrarily fixed. For $n = k, k+1, \ldots$ we introduce disjoint events $B_n = \{X_l = (n, 1) \text{ for some } l > 0\}$. Clearly, if $\mathsf{P}_k^\pi(B_n) > 0$ then $l = n - k + 1$. Event $B_0 := \Omega \setminus \{\cup_{n=k}^\infty B_n\}$ means that the controlled process X_l takes values $k, k+1, k+2, \ldots$, so that

$$\mathsf{E}_k^\pi\left[\sum_{\tau=1}^t c(X_{\tau-1}, A_\tau)|B_0\right] = 0.$$

For each $n \geq k$, if $\mathsf{P}_k^\pi(B_n) > 0$, then we have

$$\mathsf{E}_k^\pi\left[\sum_{\tau=1}^{t} c(X_{\tau-1}, A_\tau)|B_n\right] = n \geq k \geq 1$$

for all $t \geq l + 2n = n + 1$ because any sequence

$$X_l = (n, 1),\ X_{l+1} = (n, 2), \ldots, X_{l+2n-1} = (n, 2n),\ X_{l+2n} = 0$$

provides the accumulated cost equal to n. Thus,

$$\liminf_{T\to\infty} \frac{1}{T} \sum_{t=1}^{T} \mathsf{E}_k^\pi\left[\sum_{\tau=1}^{t} c(X_{\tau-1}, A_\tau)|B_n\right] = n > 0.$$

Finally,

$$\liminf_{T\to\infty} \frac{1}{T} \sum_{t=1}^{T} \mathsf{E}_k^\pi\left[\sum_{\tau=1}^{t} c(X_{\tau-1}, A_\tau)\right]$$

$$\geq \sum_{n=0,k,k+1,\ldots} \mathsf{P}_k^\pi(B_n) \liminf_{T\to\infty} \frac{1}{T} \sum_{t=1}^{T} \mathsf{E}_k^\pi\left[\sum_{\tau=1}^{t} c(X_{\tau-1}, A_\tau)|B_n\right] \geq 0.$$

(b) Consider the stationary selectors $\varphi^N(x) = \mathbb{I}\{x = N\}$, $N \geq 1$. It is not hard to calculate $v_1^{\varphi^N,\beta} = \frac{3\beta^{2N} - 2\beta^{3N} - \beta^N}{1-\beta}$ for $\beta \in (0, 1)$:

$$v_{(N,2N)}^{\varphi^N,\beta} = 2, \ldots, v_{(N,N+1)}^{\varphi^N,\beta} = \frac{2(1-\beta^N)}{1-\beta},\ v_{(N,N)}^{\varphi^N,\beta} = -1 + \frac{2\beta(1-\beta^N)}{1-\beta}, \ldots,$$
$$v_{(N,1)}^{\varphi^N,\beta} = \frac{-(1-\beta^N)}{1-\beta} + \frac{2\beta^N(1-\beta^N)}{1-\beta},$$
$$v_N^{\varphi^N,\beta} = \frac{-\beta(1-\beta^N)}{1-\beta} + \frac{2\beta^{N+1}(1-\beta^N)}{1-\beta}, \ldots$$
$$v_1^{\varphi^N,\beta} = \frac{-\beta^N(1-\beta^N)}{1-\beta} + \frac{2\beta^{2N}(1-\beta^N)}{1-\beta} = \frac{3\beta^{2N} - 2\beta^{3N} - \beta^N}{1-\beta}.$$

The function $g(y) := 3y^2 - 2y^3 - y$ decreases on the interval $\left[0, y^* := \frac{3-\sqrt{3}}{6}\right]$, and has the (local) minimal value $\min_{y\in[0,1]} g(y) = g(y^*) < 0$. Since the function $g(\cdot)$ is continuous, there exist $\varepsilon < 0$ and $\underline{\beta} \in (0, 1)$ such that, for all $\beta \in [\underline{\beta}, 1)$, the inequality $g(\beta y^*) \leq -\varepsilon$ holds. Now, for each β from the above interval, take $N(\beta) := \min\{n : \ \beta^n < y^*\}$, so that $\beta^{N(\beta)} \in [\beta y^*, y^*)$ and $g(\beta^{N(\beta)}) \leq g(\beta y^*) \leq -\varepsilon < 0$ since the function $g(\cdot)$ decreases on $[\beta y^*, y^*)$. Now $v_1^{*,\beta} \leq v_1^{\varphi^{N(\beta)},\beta} = \frac{g(\beta^{N(\beta)})}{1-\beta} < g(\beta y^*) \leq -\varepsilon$. □

Proof of Proposition 4.4. (a) From the equations for states 2 and 3

$$\rho(2) = 0.5\rho(1) + 0.5\rho(3); \quad \rho(3) = 0.5\rho(1) + 0.5\rho(2)$$

we conclude that $\rho(2) = \rho(3) = \rho(1) = \rho$.

If function $h(\cdot)$ satisfies equations (4.2), then function $h(\cdot) + r$ also satisfies those equations for each $r \in \mathbb{R}$. Hence, one can put $h(2) = 0$ without loss of generality.

From the equations for states 2 and 3

$$\rho = -9 + 0.5h(1) + 0.5h(3); \quad \rho + h(3) = 6 + 0.5h(1),$$

after we calculate the difference $h(3) = 15 - 0.5h(3)$, we see that $h(3) = 10$ and $\rho = -4 + 0.5h(1)$. Now for state 1 we have

$$\begin{aligned} \rho + h(1) = -4 + 1.5h(1) = \min \; &\{1.4 + 0.7h(1) + 0.1h(3) = 2.4 + 0.7h(1); \\ &0.2 + 0.1h(1) + 0.7h(3) = 7.2 + 0.1h(1)\}. \end{aligned}$$

Assume that the first expression is not larger than the second one:

$$2.4 + 0.7h(1) \le 7.2 + 0.1h(1) \Longrightarrow h(1) \le 8.$$

Then

$$-4 + 1.5h(1) = 2.4 + 0.7h(1) \Longrightarrow h(1) = 8 \text{ and } \rho = 0.$$

If we assume that the second expression is not larger than the first one, we similarly obtain $h(1) \ge 8$.

$$-4 + 1.5h(1) = 7.2 + 0.1h(1) \Longrightarrow h(1) = 8 \text{ and } \rho = 0.$$

All the minimums in (4.2) are provided by both $a^* = 1$ and $a^* = 2$.

(b) Under the stationary selectors φ^1 and φ^2, the controlled process is ergodic, and the presented values of the stationary probabilities come from equations

$$\left.\begin{aligned} &\hat{\eta}^{\varphi^1}(1) = 0.7\hat{\eta}^{\varphi^1}(1) + 0.5\hat{\eta}^{\varphi^1}(2) + 0.5\hat{\eta}^{\varphi^1}(3); \\ &\hat{\eta}^{\varphi^1}(2) = 0.2\hat{\eta}^{\varphi^1}(1) + 0.5\hat{\eta}^{\varphi^1}(3); \\ &\hat{\eta}^{\varphi^1}(1) + \hat{\eta}^{\varphi^1}(2) + \hat{\eta}^{\varphi^1}(3) = 1 \end{aligned}\right\}$$

and

$$\left.\begin{aligned} &\hat{\eta}^{\varphi^2}(1) = 0.1\hat{\eta}^{\varphi^2}(1) + 0.5\hat{\eta}^{\varphi^2}(2) + 0.5\hat{\eta}^{\varphi^2}(3); \\ &\hat{\eta}^{\varphi^2}(2) = 0.2\hat{\eta}^{\varphi^2}(1) + 0.5\hat{\eta}^{\varphi^2}(3); \\ &\hat{\eta}^{\varphi^2}(1) + \hat{\eta}^{\varphi^2}(2) + \hat{\eta}^{\varphi^2}(3) = 1. \end{aligned}\right\}$$

(c) The given formulae are valid if $T = 1$:

$$\mathsf{E}_x^{\varphi^1}[c(X_0, A_1)] = c(x, 1); \quad \mathsf{E}_x^{\varphi^2}[c(X_0, A_1)] = c(x, 2), \quad x = 1, 2, 3.$$

Suppose they are valid for some $T \geq 1$. Then straightforward calculations show that

$$\begin{aligned}
&\mathsf{E}_x^{\varphi^1}\left[\sum_{t=1}^{T+1} c(X_{t-1}, A_t)\right] \\
&= c(x,1) + p(1|x,1)\left\{\frac{4}{3} + \frac{1}{21}\left[-10\left(-\frac{1}{2}\right)^T - 18\left(\frac{1}{5}\right)^T\right]\right\} \\
&\quad + p(2|x,1)\left\{-\frac{20}{3} + \frac{1}{21}\left[110\left(-\frac{1}{2}\right)^T + 30\left(\frac{1}{5}\right)^T\right]\right\} \\
&\quad + p(3|x,1)\left\{\frac{10}{3} + \frac{1}{21}\left[-100\left(-\frac{1}{2}\right)^T + 30\left(\frac{1}{5}\right)^T\right]\right\} \\
&= \begin{cases} \frac{4}{3} + \frac{1}{21}\left[-10\left(-\frac{1}{2}\right)^{T+1} - 18\left(\frac{1}{5}\right)^{T+1}\right], & \text{if } x = 1; \\ -\frac{20}{3} + \frac{1}{21}\left[110\left(-\frac{1}{2}\right)^{T+1} + 30\left(\frac{1}{5}\right)^{T+1}\right], & \text{if } x = 2; \\ \frac{10}{3} + \frac{1}{21}\left[-100\left(-\frac{1}{2}\right)^{T+1} + 30\left(\frac{1}{5}\right)^{T+1}\right], & \text{if } x = 3 \end{cases}
\end{aligned}$$

and

$$\begin{aligned}
&\mathsf{E}_x^{\varphi^2}\left[\sum_{t=1}^{T+1} c(X_{t-1}, A_t)\right] \\
&= c(x,2) + p(1|x,2)\left\{\frac{4}{3} + \frac{1}{3}\left[50\left(-\frac{1}{2}\right)^T - 54\left(\frac{-2}{5}\right)^T\right]\right\} \\
&\quad + p(2|x,2)\left\{-\frac{20}{3} + \frac{1}{3}\left[-10\left(-\frac{1}{2}\right)^T + 30\left(\frac{-2}{5}\right)^T\right]\right\} \\
&\quad + p(3|x,2)\left\{\frac{10}{3} + \frac{1}{3}\left[-40\left(-\frac{1}{2}\right)^T + 30\left(\frac{-2}{5}\right)^T\right]\right\} \\
&= \begin{cases} \frac{4}{3} + \frac{1}{3}\left[50\left(-\frac{1}{2}\right)^{T+1} - 54\left(\frac{-2}{5}\right)^{T+1}\right], & \text{if } x = 1; \\ -\frac{20}{3} + \frac{1}{3}\left[-10\left(-\frac{1}{2}\right)^{T+1} + 30\left(\frac{-2}{5}\right)^{T+1}\right], & \text{if } x = 2; \\ \frac{10}{3} + \frac{1}{3}\left[-40\left(-\frac{1}{2}\right)^{T+1} + 30\left(\frac{-2}{5}\right)^{T+1}\right], & \text{if } x = 3. \end{cases}
\end{aligned}$$

□

Proof of Proposition 4.5. (a) Denote $F(x) := \begin{cases} \frac{4}{3}, & \text{if } x = 1; \\ -\frac{20}{3}, & \text{if } x = 2; \\ \frac{10}{3}, & \text{if } x = 3 \end{cases}$ and fix an arbitrary $\varepsilon < \frac{3}{4}$.

For $i = 1, 2, \dots$ fix $\delta_i, \tilde{\delta}_i \in (0,1)$ such that

$$15.2\,\delta_i < \frac{\varepsilon}{4}; \quad \tilde{\delta}_i < \frac{\varepsilon}{8}; \quad \frac{76}{3}\delta_i + \tilde{\delta}_{i+1} + 1.6\,\delta_i\tilde{\delta}_{i+1} < \frac{\varepsilon}{2^{i+1}}, \tag{A.9}$$

and take $\tau_i > 1$ such that, for $i = 1, 2, \dots$,

$$\mathsf{E}_x^{\varphi^1,\varphi^2}\left[\sum_{t=1}^{m} c(X_{t-1}, A_t)\right] \in (F(x) - \tilde{\delta}_i,\ F(x) + \tilde{\delta}_i) \tag{A.10}$$

for all $m \geq \tau_i - 1$ and, for $\hat{\eta}^{\varphi^1}(y)$ as in Proposition 4.4(b), independently of x

$$\mathsf{P}_x^{\varphi^1}(X_{\tau_i - 1} = y) \in (\hat{\eta}^{\varphi^1}(y) - \delta_i,\ \hat{\eta}^{\varphi^1}(y) + \delta_i). \tag{A.11}$$

This can be done according to Proposition 4.4(c) and because the distribution $\{\mathsf{P}_x^{\varphi^1}(X_t = y)\}_{y=1}^3$, as $t \to \infty$, converges to the stationary distribution $\{\hat{\eta}^{\varphi^1}(y)\}_{y=1}^3 = (\frac{15}{24}, \frac{5}{24}, \frac{4}{24})$ geometrically fast [Brémaud (2000), Ch.6, Thm. 3.4], [Meyn and Tweedie (2009), Thm. 16.2.1]. Recall that the controlled process X_t is ergodic under the stationary selector φ^1.

We shall prove the proposition for the sequence $T_i := \sum_{j=1}^{i} \tau_j$, $i = 1, 2, \dots$.

For each $i = 1, 2, \dots$, since $A_{T_i} = 2$ and X_t for $t = T_{i-1}, T_{i-1} + 1, \dots, T_i - 1$ is just the controlled process under the stationary selector φ^1, we have

$$\begin{aligned}
&\mathsf{E}_x^{\varphi}[c(X_{T_i-1}, A_{T_i})] \\
&= \sum_{y=1}^{3} \mathsf{P}_x^{\varphi}(X_{T_i-1} = y | X_{T_{i-1}} = z)\mathsf{P}_x^{\varphi}(X_{T_{i-1}} = z)c(y, 2) \\
&= \sum_{y=1}^{3} \mathsf{P}_z^{\varphi^1}(X_{\tau_i-1} = y)\mathsf{P}_x^{\varphi}(X_{T_{i-1}} = z)c(y, 2) \\
&< \left(\frac{15}{24} + \delta_i\right) 0.2 + \left(\frac{5}{24} - \delta_i\right)(-9) + \left(\frac{4}{24} + \delta_i\right) 6 \\
&= -\frac{3}{4} + 15.2\,\delta_i < -\frac{3}{4} + \frac{\varepsilon}{4}.
\end{aligned} \tag{A.12}$$

Here formula (A.11) was in use.

Further, by the similar argument,

$$
\begin{aligned}
&\mathsf{P}_x^{\varphi}(X_{T_i}=1)\\
&=\sum_{y=1}^{3}\mathsf{P}_x^{\varphi}(X_{T_i-1}=y|X_{T_{i-1}}=z)\mathsf{P}_x^{\varphi}(X_{T_{i-1}}=z)p(1|y,2)\\
&=\sum_{y=1}^{3}\mathsf{P}_z^{\varphi^1}(X_{\tau_i-1}=y)\mathsf{P}_x^{\varphi}(X_{T_{i-1}}=z)p(1|y,2)\\
&<\left(\frac{15}{24}+\delta_i\right)0.1+\left(\frac{5}{24}+\delta_i\right)0.5+\left(\frac{4}{24}+\delta_i\right)0.5\\
&=\frac{6}{24}+1.1\,\delta_i.
\end{aligned}
$$

Similarly,

$$
\begin{aligned}
\mathsf{P}_x^{\varphi}(X_{T_i}=2)&>\left(\frac{15}{24}-\delta_i\right)0.2+\left(\frac{4}{24}-\delta_i\right)0.5=\frac{5}{24}-0.7\,\delta_i;\\
\mathsf{P}_x^{\varphi}(X_{T_i}=3)&<\left(\frac{15}{24}+\delta_i\right)0.7+\left(\frac{5}{24}+\delta_i\right)0.5=\frac{13}{24}+1.2\,\delta_i.
\end{aligned}
$$

Next, by the similar argument, using the estimated probabilities and inequality (A.12), we see that

$$
\begin{aligned}
&\mathsf{E}_x^{\varphi}\left[\sum_{t=T_i}^{T_{i+1}-1}c(X_{t-1},A_t)\right]\\
&=\mathsf{E}_x^{\varphi}[c(X_{T_i-1},A_{T_i})]+\mathsf{E}_x^{\varphi}\left[\sum_{t=T_i+1}^{T_{i+1}-1}c(X_{t-1},A_t)\right]\\
&<-\frac{3}{4}+15.2\,\delta_i+\sum_{y=1}^{3}\mathsf{E}_y^{\varphi^1}\left[\sum_{t=1}^{\tau_{i+1}-1}c(X_{t-1},A_t)\right]\mathsf{P}_x^{\varphi}(X_{T_i}=y)\\
&<-\frac{3}{4}+15.2\,\delta_i+\left(\frac{4}{3}+\tilde{\delta}_{i+1}\right)\left(\frac{6}{24}+1.1\,\delta_i\right)\\
&\quad+\left(-\frac{20}{3}+\tilde{\delta}_{i+1}\right)\left(\frac{5}{24}-0.7\,\delta_i\right)+\left(\frac{10}{3}+\tilde{\delta}_{i+1}\right)\left(\frac{13}{24}+1.2\,\delta_i\right)\\
&=\frac{76}{3}\delta_i+\tilde{\delta}_{i+1}+1.6\,\delta_i\tilde{\delta}_{i+1}<\frac{\varepsilon}{2^{i+1}}.
\end{aligned}
$$

The second inequality is according to (A.10); the last inequality is by (A.9).

Since, by (A.10) at $i = 1$, $m = \tau_1 - 1$, and according to (A.9),

$$\begin{aligned}
&\mathsf{E}_x^\varphi\left[\sum_{t=1}^{T_{k+1}-1} c(X_{t-1}, A_t)\right] \\
&= \mathsf{E}_x^\varphi\left[\sum_{t=1}^{\tau_1-1} c(X_{t-1}, A_t)\right] + \sum_{i=1}^{k} \mathsf{E}_x^\varphi\left[\sum_{t=T_i}^{T_{i+1}-1} c(X_{t-1}, A_t)\right] \\
&< F(x) + \tilde{\delta}_1 + \frac{\varepsilon}{2} < F(x) + \frac{5\,\varepsilon}{8}
\end{aligned}$$

for all $k = 1, 2, \ldots$, we conclude that

$$\begin{aligned}
&\mathsf{E}_x^\varphi\left[\sum_{t=1}^{T_{k+1}} c(X_{t-1}, A_t)\right] \\
&< F(x) + \frac{5\,\varepsilon}{8} + \mathsf{E}_x^\varphi[c(X_{T_{k+1}-1}, A_{T_k+1})] < F(x) - \frac{3}{4} + \frac{7\,\varepsilon}{8}.
\end{aligned}$$

Here inequality (A.12) was in use.

But, according to (A.10) at $i = 1$, $m = T_{k+1}$ and using (A.9),

$$\mathsf{E}_x^{\varphi^1,\varphi^2}\left[\sum_{t=1}^{T_{k+1}} c(X_{t-1}, A_t)\right] > F(x) - \tilde{\delta}_1 > F(x) - \frac{\varepsilon}{8}.$$

Since $T_{k+1} \uparrow \infty$ as $k \to \infty$, we conclude that the selectors φ^1 and φ^2 do not overtake the selector φ:

$$\begin{aligned}
&\limsup_{T\to\infty}\left\{\mathsf{E}_x^{\varphi^1,\varphi^2}\left[\sum_{t=1}^{T} c(X_{t-1}, A_t)\right] - \mathsf{E}_x^\varphi\left[\sum_{t=1}^{T} c(X_{t-1}, A_t)\right]\right\} \\
&\geq \limsup_{k\to\infty}\left\{\mathsf{E}_x^{\varphi^1,\varphi^2}\left[\sum_{t=1}^{T_{k+1}} c(X_{t-1}, A_t)\right] - \mathsf{E}_x^\varphi\left[\sum_{t=1}^{T_{k+1}} c(X_{t-1}, A_t)\right]\right\} \\
&\geq \frac{3}{4} - \varepsilon > 0.
\end{aligned}$$

(b) The optimality equation (3.2) takes the form

$$\begin{aligned}
v(1) &= \min\{1.4 + \beta[0.7\,v(1) + 0.2\,v(2) + 0.1\,v(3)], \\
&\qquad 0.2 + \beta[0.1\,v(1) + 0.2\,v(2) + 0.7\,v(3)]\} \\
v(2) &= -9 + 0.5\beta[v(1) + v(3)], \\
v(3) &= 6 + 0.5\beta[v(1) + v(2)].
\end{aligned} \tag{A.13}$$

From the second and third equations we obtain:

$$v(3) = \frac{6 - 4.5\beta + (0.5\beta + 0.25\beta^2)v(1)}{1 - 0.25\beta^2}. \tag{A.14}$$

After we substitute this expression into the two other equations, we obtain the following:

$$\begin{aligned}[1 - 0.25\beta^2]v(1) = \min\{&1.4 - 1.2\beta - 0.2\beta^2 + v(1)\beta[0.7 + 0.15\beta - 0.1\beta^2];\\ &0.2 + 2.4\beta - 2.6\beta^2 + v(1)\beta[0.1 + 0.45\beta + 0.2\beta^2]\}.\end{aligned}$$

From the second line we obtain

$$v(1) = \frac{0.2 + 2.4\beta - 2.6\beta^2}{1 - 0.1\beta - 0.7\beta^2 - 0.2\beta^3} = \frac{4}{3} - \frac{26}{63}(1 - \beta) + o(1 - \beta),$$

where $\lim_{\varepsilon\to 0} \frac{o(\varepsilon)}{\varepsilon} = 0$. Now, the difference between the first and the second lines equals

$$1.2 - 3.6\beta + 2.4\beta^2 + v(1)\beta[0.6 - 0.3\beta - 0.3\beta^2] = \frac{3}{7}(1 - \beta)^2 + o((1 - \beta)^2) > 0,$$

meaning that, indeed, $v_1^{*,\beta} = v(1) = \frac{0.2+2.4\beta-2.6\beta^2}{1-0.1\beta-0.7\beta^2-0.2\beta^3} = v_1^{\varphi^2,\beta}$ for all β close enough to 1, and the stationary selector φ^2 is Blackwell optimal. The values of $v_2^{*,\beta} = v_2^{\varphi^2,\beta}$ and $v_3^{*,\beta} = v_3^{\varphi^2,\beta}$ can be easily calculated using the formulae (A.13) and (A.14), and it is straightforward to notice that

$$\lim_{\beta\to 1-} v_1^{\varphi^2,\beta} = \frac{4}{3}; \quad \lim_{\beta\to 1-} v_2^{\varphi^2,\beta} = \frac{-20}{3}; \quad \lim_{\beta\to 1-} v_3^{\varphi^2,\beta} = \frac{10}{3}.$$

For the stationary selector φ^1, using (A.13) and (A.14), we obtain

$$v_1^{\varphi^1,\beta} = \frac{1.4 - 1.2\beta - 0.2\beta^2}{1 - 0.7\beta - 0.4\beta^2 + 0.1\beta^3} = \frac{4}{3} - \frac{1}{18}(1 - \beta) + o(1 - \beta),$$

and again the values of $v_2^{\varphi^1,\beta}$ and $v_3^{\varphi^1,\beta}$ can be calculated using formulae (A.13) and (A.14). Since $\frac{26}{63} > \frac{1}{18}$, this calculation additionally confirms that selector φ^2 is Blackwell optimal, and φ^1 is not. Again, we have equalities

$$\lim_{\beta\to 1-} v_1^{\varphi^1,\beta} = \frac{4}{3}; \quad \lim_{\beta\to 1-} v_2^{\varphi^1,\beta} = \frac{-20}{3}; \quad \lim_{\beta\to 1-} v_3^{\varphi^1,\beta} = \frac{10}{3}.$$

Stationary selector φ^2 is bias optimal because it is Blackwell optimal [Puterman (1994), Thm. 10.1.5]. Stationary selector φ^1 is also bias optimal because, for each strategy π and all $x \in \mathbf{X}$,

$$\limsup_{\beta\to 1-}[v_x^{\varphi^1,\beta} - v_x^{\pi,\beta}] \le \limsup_{\beta\to 1-}[v_x^{\varphi^1,\beta} - v_x^{\varphi^2,\beta}] + \limsup_{\beta\to 1-}[v_x^{\varphi^2,\beta} - v_x^{\pi,\beta}] \le 0.$$

The first limit on the right equals zero, and the last inequality holds because selector φ^2 is Blackwell optimal.

The selector φ^1 is not 1-discount optimal because

$$\begin{aligned}\limsup_{\beta\to 1-}(1-\beta)^{-1}[v_x^{\varphi^1,\beta}-v_x^{\varphi^2,\beta}] &= \limsup_{\beta\to 1-}\frac{\left(\frac{26}{63}-\frac{1}{18}\right)(1-\beta)+o(1-\beta)}{1-\beta}\\ &= \frac{5}{14}>0.\end{aligned}$$

The proof is completed. □

Proof of Lemma 5.1. (a) According to Theorem 5.1, one can restrict to π-strategies. Since the jump intensity $q_x(a)$ is uniformly bounded, there is no explosion: $T_\infty=\infty$ $\mathsf{P}^\pi_{\mathsf{P}_0}$-a.s. (See also Theorem 5.3.) Next, for each $x_0\in\mathbf{X}^\Delta$, as explained in Subsection 5.1.1,

$$\begin{aligned}\mathsf{P}^\pi_{x_0}(X_1=\Delta) &\ge \varepsilon\int_{(0,\infty)} e^{-\int_{(0,s]}\int_{\mathbf{A}} q_{x_0}(a)\pi_1(da|x_0,u)du}ds\\ &\ge \varepsilon\int_{(0,\infty)} e^{-Ks}ds=\frac{\varepsilon}{K}.\end{aligned}$$

Therefore, $\mathsf{P}^\pi_{\mathsf{P}_0}(X_1\in\mathbf{X}^\Delta)\le 1-\frac{\varepsilon}{K}$ and, arguing by induction in the similar way, we obtain

$$\begin{aligned}\mathsf{P}^\pi_{\mathsf{P}_0}(X_n\in\mathbf{X}^\Delta) &= \mathsf{P}^\pi_{\mathsf{P}_0}(X_n\in\mathbf{X}^\Delta|X_{n-1}\in\mathbf{X}^\Delta)\mathsf{P}^\pi_{\mathsf{P}_0}(X_{n-1}\in\mathbf{X}^\Delta)\\ &\le\left(1-\frac{\varepsilon}{K}\right)\left(1-\frac{\varepsilon}{K}\right)^{n-1}=\left(1-\frac{\varepsilon}{K}\right)^n.\end{aligned}$$

Finally, since $\mathsf{E}^\pi_{\mathsf{P}_0}[\Theta_{n+1}|X_n\in\mathbf{X}^\Delta]\le\frac{1}{\varepsilon}$,

$$\begin{aligned}\int_{(0,\infty)}\bar{\mathsf{P}}^\pi_{\mathsf{P}_0}(t,\mathbf{X}^\Delta\times\mathbf{A})dt &= \sum_{n=0}^\infty \mathsf{E}^\pi_{\mathsf{P}_0}\left[\mathsf{E}^\pi_{\mathsf{P}_0}[\Theta_{n+1}\mathbb{I}\{X_n\in\mathbf{X}^\Delta\}|X_n\in\mathbf{X}^\Delta]\right]\\ &= \sum_{n=0}^\infty \mathsf{E}^\pi_{\mathsf{P}_0}\left[\mathbb{I}\{X_n\in\mathbf{X}^\Delta\}\mathsf{E}^\pi_{\mathsf{P}_0}[\Theta_{n+1}|X_n\in\mathbf{X}^\Delta]\right]\\ &\le \sum_{n=0}^\infty\frac{1}{\varepsilon}\mathsf{P}^\pi_{\mathsf{P}_0}(X_n\in\mathbf{X}^\Delta)\le\sum_{n=0}^\infty\frac{1}{\varepsilon}\left(1-\frac{\varepsilon}{K}\right)^n\\ &= \frac{K}{\varepsilon^2}<\infty\end{aligned}$$

and

$$|W^0(\pi,\mathsf{P}_0)|\le\int_{(0,\infty)}\sup_{(x,a)\in\mathbf{X}\times\mathbf{A}}|c(x,a)|\bar{\mathsf{P}}^\pi_{\mathsf{P}_0}(t,\mathbf{X}^\Delta\times\mathbf{A})dt\le L$$

with $L:=\frac{\sup_{(x,a)\in\mathbf{X}\times\mathbf{A}}|c(x,a)|K}{\varepsilon^2}$.

(b) Obviously, it is sufficient to consider $V(x) := \mathbb{I}\{x \in \mathbf{X}^\Delta\}$. Then $V(X(t)) = \mathbb{I}\{t < T_N\}$, where $N := \min\{n : X_n = \Delta\}$, is the step function.

Suppose

$$\exists \delta > 0 : \forall T \ \exists t > T : \ \mathsf{E}^S_{\mathsf{P}_0}[V(X(t))] = \mathsf{P}^S_{\mathsf{P}_0}(t < T_N) > \delta.$$

Then, for all $\tau \leq t$, $\mathsf{P}^S_{\mathsf{P}_0}(\tau < T_N) > \delta$ leading to

$$\begin{aligned} \mathsf{E}^S_{\mathsf{P}_0}\left[\int_0^\infty V(X(\tau))\,d\tau\right] &\geq \mathsf{E}^S_{\mathsf{P}_0}\left[\int_0^T \mathbb{I}\{\tau < T_N\}\,d\tau\right] \\ &= \int_0^T \mathsf{P}^S_{\mathsf{P}_0}(\tau < T_N)\,d\tau > T\delta \to \infty \end{aligned}$$

as $T \to \infty$. (Here the Fubini-Tonelli Theorem was in use.)

Clearly, $\mathsf{E}^S_{\mathsf{P}_0}\left[\int_0^\infty V(X(\tau))\,d\tau\right]$ is the value of the performance functional for the bounded cost rate $c(x,a) = V(x)$. The obtained equality $\mathsf{E}^S_{\mathsf{P}_0}\left[\int_0^\infty V(X(\tau))\,d\tau\right] = \infty$ contradicts Item (a), and the proof is completed. □

Proof of Lemma 5.2. The function $u(v,x) : [0,\infty) \times \mathbf{X} \to [0,1]$ is built by iterations

$$u^{n+1}(v,x) = \int_{(0,\infty)} \int_{\mathbf{X}} u^n(v+t,y)\tilde{q}(dy|x,v+t)e^{-\alpha t - \int_{(0,t]} q_x(s+v)ds}dt, \quad n = 0,1,2,\ldots \tag{A.15}$$

starting from $u^0(\cdot,\cdot)$. This sequence increases and, by induction, $0 \leq u^{n+1}(v,x) \leq 1$ because, when integrating by parts, one obtains

$$\begin{aligned} u^{n+1}(v,x) &\leq \int_{(0,\infty)} \int_{\mathbf{X}} \tilde{q}(dy|x,v+t)e^{-\alpha t - \int_{(0,t]} q_x(s+v)ds}dt \\ &= -e^{-\alpha t} \cdot e^{-\int_{(0,t]} q_x(s+v)ds}\Big|_0^\infty - \int_{(0,\infty)} \alpha e^{-\alpha t} e^{-\int_{(0,t]} q_x(s+v)ds}dt \\ &\leq 1. \end{aligned}$$

The limiting function $u(v,x) := \lim_{n\to\infty} v^n(v,x)$ satisfies equation (5.14): just pass to the limit in (A.15) using the Monotone Convergence Theorem.

The proof for equation (5.15) is the same after one rewrites inequality (5.17) as

$$u(x) \leq \frac{\int_{\mathbf{X}} u(y)\tilde{q}(dy|x)}{\alpha + q_x}.$$

□

Afterword

This book contains about 170 examples, mainly illustrating that the conditions imposed in the known theorems are important. Several meaningful examples, leading to unexpected and sometimes surprising answers, are also given, such as voting, optimal search, queues, and so on. Real-life applications of Markov Decision Processes are beyond the scope of this book. One can find such examples in [Kallenberg (2010); Piunovskiy and Zhang (2020); Puterman (1994)] and other monographs and articles.

Of course, there are many interesting examples which are not presented in the current book, simply because it is impossible to embrace the immensity of the subject. Let us only mention the following articles on queues:

- It is not always optimal to join the shortest queue [Whitt (1986)].
- If more jobs are admitted for service, then the network performance may be improved [Ye (2007)].
- As is shown in [Inoie *et al.* (2006)], grouping together separated queueing systems into one system may not provide improved performance.

Apart from the mentioned earlier areas, MDPs can be successfully used for control of moving objects and water resources, for quality control in a production line, for forest management, in finance, insurance, economy, ecology, gambling, sports, management, teaching/learning, reliability, medicine and so on. It is nearly impossible to name an area where MDPs cannot be applied.

Bibliography

Altman, E. and Shwartz, A. (1991a). Adaptive control of constrained Markov chains: criteria and policies, *Ann. Oper. Res.*, **28**, pp. 101–134.

Altman, E. and Shwartz, A. (1991b). Markov decision problems and state–action frequencies, *SIAM J. Control and Optim.*, **29**, pp. 786–809.

Altman, E. and Shwartz, A. (1993). Time-sharing policies for controlled Markov chains, *Operations Research*, **41**, pp. 1116–1124.

Altman, E. (1999). *Constrained Markov Decision Processes* (Chapman and Hall/CRC, Boca Raton, FL, USA).

Altman, E., Avrachenkov, K.E. and Filar, J.A. (2002). An asymptotic simplex method and Markov decision processes, in Petrosjan, L.A and Zenkevich, N.A. (eds.), *Proc. of the 10th Intern. Symp. on Dynamic Games, Vol.I*, (St. Petersburg State University, Institute of Chemistry, St. Petersburg, Russia), pp. 45–55.

Arapostathis, A., Borkar, V.S., Fernandez-Gaucherand, E., *et al.* (1993). Discrete-time controlled Markov processes with average cost criterion: a survey, *SIAM J. Control and Optim.*, **31**, pp. 282–344.

Avrachenkov, K.E., Filar, J. and Haviv, M. (2002). Singular perturbations of Markov chains and decision processes, in Feinberg, E. and Shwartz, A. (eds.), *Handbook of Markov Decision Processes*, (Kluwer, Boston, USA), pp. 113–150.

Ball, K. (2004). An elementary introduction to monotone transportation, *Geometric Aspects of Functional Analysis, Lecture Notes in Math.*, Vol. 1850, pp. 41–52.

Bäuerle, N. and Rieder, U. (2011). *Markov Decision Processes with Applications to Finance* (Springer-Verlag, Berlin, Germany).

Bäuerle, N. and Rieder, U. (2014). More risk-sensitive Markov decision processes, *Math. Oper. Res.*, **39**, pp. 105–120.

Bellman, R. (1957). *Dynamic Programming* (Princeton University Press, Princeton, NJ, USA).

Bertsekas, D. and Shreve, S. (1978). *Stochastic Optimal Control* (Academic Press, New York, USA).

Bertsekas, D. (1987). *Dynamic Programming: Deterministic and Stochastic Models* (Prentice-Hall, Englewood Cliffs, NJ, USA).

Bertsekas, D. (2001). *Dynamic Programming and Optimal Control, V.II* (Athena Scientific, Belmont, MA, USA).

Bertsekas, D. (2005). *Dynamic Programming and Optimal Control, V.I* (Athena Scientific, Belmont, MA, USA).

Bertsekas, D.P. (2018). Proper policies in infinite-state stochastic shortest path problems, *IEEE Trans. Aut. Control*, **63**, pp. 3787–3792.

Bertsekas, D.P. (2020). Proper policies in infinite-state stochastic shortest path problems. arXiv preprint arXiv:1711.10129.

Billingsley, P. (1983). The singular function of bold play, *American Scientist*, **71**, pp. 392–397.

Bishop, C.J., Feinberg, E.A. and Zhang, J. (2014). Examples concerning Abel and Cesàro limits, *J.Math. Anal. Appl.*, **420**, pp. 1654–1661.

Blackwell, D. (1962). Discrete dynamic programming, *Ann. Math. Stat.*, **33**, pp. 719–726.

Blackwell, D. (1965). Discounted dynamic programming, *Ann. Math. Stat.*, **36**, pp. 226–235.

Bogachev, V. (2007). *Measure Theory* (American Mathematical Society, Providence).

Boel, R. (1977). Martingales and dynamic programming, in *Markov Decision Theory, Proc. Adv. Sem., Netherlands, 1976*, (Math. Centre Tracts, No. 93, Math. Centr. Amsterdam, Netherlands), pp. 77–84.

Brau-Rojas, A., Cavazos-Cadena, R. and Fernández-Gaucherand, E. (1998). Controlled Markov chains with risk-sensitive criteria: some (counter) examples, *Proc of the 37-th IEEE Conf. on Decision and Control*, pp. 1853–1858.

Brémaud, P. (2000). *M*arkov Chains. Gibbs Fields, Monte Carlo Simulation, and Queues (Springer, NY-Berlin-Heidelberg).

Carmon, Y. and Shwartz, A. (2009). Markov decision processes with exponentially representable discounting, *Oper. Res. Letters*, **37**, pp. 51–55.

Cavazos-Cadena, R. (1991). A counterexample on the optimality equation in Markov decision chains with the average cost criterion, *Systems and Control Letters*, **16**, pp. 387–392.

Cavazos-Cadena, R., Feinberg, E. and Montes-de-Oca, R. (2000). A note on the existence of optimal policies in total reward dynamic programs with compact action sets, *Math. Oper. Res.*, **25**, pp. 657–666.

Cavazos-Cadena, R. and Montes-de-Oca, R. (2012). Sample-path optimality in average Markov decision chains under a double Lyapunov function condition, in Hernández-Hernández, D. and Minjárez-Sosa, J.A. (eds.), *Optimization, Control, and Applications of Stochastic Systems. in Honor of Onésimo Hernández-Lerma*, (Birkhäuser, Springer, NY), pp. 31–57.

Cavazos-Cadena, R., Montes-de-Oca, R. and Sladký, K. (2014). A counterexample on sample-path optimality in stable Markov decision chains with the average reward criterion, *J. Optim. Theory Appl.*, **163**, pp. 674–684.

Chen, R.W., Shepp, L.A. and Zame, A. (2004). A bold strategy is not always optimal in the presence of inflation, *J. Appl. Prob.*, **41**, pp. 587–592.

Chen, A.Y., Pollett, P., Zhang, H.J. and Cairns, B. (2005). Uniqueness criteria for continuous-time Markov chains with general transition structures, *Adv. Appl. Probab.*, **37**, pp. 1056–1074.

Chung, K.L. (1967). Markov Chains with Stationary Transition Probabilities (Springer, Berlin, Germany).

Crabill, T.B. (1972). Optimal control of a service facility with variable exponential service times and constant arrival rate, Manage. Scie., **18**, pp. 560–566.

Della Vecchia, E., Di Marco, S. and Jean-Marie, A. (2012). Illustrated review of convergence conditions of the value iteration algorithm and the rolling horizon procedure for average-cost MDPs, *Ann. Oper. Res.*, **199**, pp. 193–214.

Dekker, R. (1987). Counter examples for compact action Markov decision chains with average reward criteria, *Commun. Statist. Stochastic Models*, **3**, pp. 357–368.

Denardo, E.V. and Miller, B.L. (1968). An optimality condition for discrete dynamic programming with no discounting, *Ann. Math. Stat.*, **39**, pp. 1220–1227.

Denardo, E.V. and Rothblum, U.G. (1979). Overtaking optimality for Markov decision chains, *Math. Oper. Res.*, **4**, pp. 144–152.

Derman, C. (1964). On sequential control processes, *Ann. Math. Stat.*, **35**, pp. 341–349.

Derman, C. (1966). Denumerable state Markovian decisionprocesses-average cost criterion, *Ann. Math. Stat.*, **37**, pp. 1545–1553.

Derman, C. and Strauch, R. (1966). A note on memoryless rules for controlling sequential control processes. *Ann. Math. Stat.*, **37**, pp. 276–278.

Dubins, L.E. and Savage, L.J. (1965). *How to Gamble if You Must* (McGraw-Hill, New York, USA).

Dufour, F. and Piunovskiy, A.B. (2010). Multiobjective stopping problem for discrete-time Markov processes: convex analytic approach, *J. Appl. Probab.*, **47**, pp. 947–996.

Dufour, F. and Piunovskiy, A.B. (2013). The expected total cost criterion for Markov Decision Processes under constraints, *Adv. Appl. Probab.*, **45**, pp.837–859.

Dufour, F. and Prieto-Rumeau, T. (2024a). Absorbing Markov decision processes, *ESAIM:COCV*, **30**, N.5.

Dufour, F. and Prieto-Rumeau, T. (2024b). Absorbing Markov decision processes and their occupation measures, *SIAM J. Control and Optim.*, accepted.

Dynkin, E.B. and Yushkevich, A.A. (1979). *Controlled Markov Processes and their Applications* (Springer-Verlag, Berlin, Germany).

Fainberg, E.A. (1977). Finite controllable Markov chains, *Uspehi Mat. Nauk*, **32:3**, pp. 181–182, (in Russian).

Fainberg, E.A. (1980). An ε-optimal control of a finite Markov chain with an average reward criterion, *Theory Probab. Appl.*, **25**, pp. 70–81.

Feinberg, E.A. (1982). Controlled Markov processes with arbitrary numerical criteria, *Theory Probab, Appl.*, **27**, pp. 486–503.

Feinberg, E.A. (1987). Sufficient classes of strategies in discrete dynamic programming. I. Decomposition of randomized strategies and embedded models, *Theory Probab. Appl.*, **31**, pp. 658–668.

Feinberg, E.A. (1994). Constrained semi-Markov decision processes with average rewards, *Math. Meth. Oper. Res.*, **39**, pp. 257–288.

Feinberg, E.A. and Shwartz, A. (1994). Markov decision models with weighted discounted criteria, *Math. Oper. Res.*, **19**, pp. 152–168.

Feinberg, E.A. (1996). On measurability and representation of strategic measures in Markov decision processes, in Ferguson, T. (ed.), *Statistics, Probability and Game Theory: Papers in Honor of David Blackwell, IMS Lecture Notes Monographs Ser.*, **30**, pp. 29–43.

Feinberg, E.A. and Sonin, I.M. (1996). Notes on equivalent stationary policies in Markov decision processes with total rewards, *Math. Meth. Oper. Res.*, **44**, pp. 205–221.

Feinberg, E.A. (2002a). Optimal control of average reward constrained continuous-time finite Markov decision processes, *Proc of the 41-st IEEE Conf. on Decision and Control*, pp. 3805–3810.

Feinberg, E.A. (2002b). Total reward criteria, in Feinberg, E. and Shwartz, A. (eds.), *Handbook of Markov Decision Processes*, (Kluwer, Boston, USA), pp. 173–207.

Feinberg, E.A. and Piunovskiy, A.B. (2002). Nonatomic total rewards Markov decision processes with multiple criteria, *J.Math. Anal. Appl.*, **273**, pp. 93–111.

Feinberg, E.A. (2004). Continuous time discounted jump Markov decision processes: a discrete-event approach, *Math. Oper. Res.*, **29**, pp. 492–524.

Feinberg, E.A. and Piunovskiy, A.B. (2010). On strongly equivalent nonrandomized transition probabilities, *Theory Probab. Appl.*, **54**, pp. 300–307.

Feinberg, E.A. (2010). Reduction of discounted continuous-time MDPs with unbounded jump and reward rates to discrete-time total-reward MDPs, in Hernandez-Hernandez, D. and Minjarez-Sosa, A. (eds), *Optimization, Control, and Applications of Stochastic Systems*, (Birkhäuser, Springer), pp. 77–97.

Feinberg, E.A. and Rothblum, U. (2012). Splitting randomized stationary policies in total-reward Markov decision processes, *Math. Oper. Res.*, **37**, pp. 129–153.

Feinberg, E.A. and Piunovskiy, A.B. (2019). Sufficiency of deterministic policies for atomless discounted and uniformly absorbing MDPs with multiple criteria, *SIAM J. Control and Optim.*, **57**, pp. 163–191.

Feinberg, E.A., Mandava, M. and Shiryaev, A.N. (2021). Sufficiency of Markov policies for continuous-time jump Markov decision processes, *Math. Oper. Res.*, **47**, pp.1266–1286.

Fernández-Gaucherand, E., Ghosh, M.K. and Marcus, S.I. (1994). Controlled Markov processes on the infinite planning horizon: weighted and overtaking cost criteria, *ZOR – Methods and Models of Oper. Res.*, **39**, pp. 131–155.

Fisher, L. and Ross, S.M. (1968). An example in denumerable decision processes, *Ann. Math. Statistics*, **39**, pp. 674–675.

Flynn, J. (1974). Averaging vs. discounting in dynamic programming: a counterexample, *The Annals of Statistics*, **2**, pp. 411–413.

Flynn, J. (1976). Conditions for the equivalence of optimality criteria in dynamic programming, *The Annals of Statistics*, **4**, pp.936–953.

Flynn, J. (1980). On optimality criteria for dynamic programs with long finite horizons, *J.Math. Anal. Appl.*, **76**, pp. 202–208.

Frid, E.B. (1972). On optimal strategies in control problems with constraints, *Theory Probab. Appl.*, **17**, pp. 188–192.

Gairat, A. and Hordijk, A. (2000). Fluid approximation of a controlled multiclass tandem network, *Queueing Systems*, **35**, pp. 349-380.

Gelbaum, B.R. and Olmsted, J.M.H. (1964). *Counterexamples in Analysis* (Holden-Day, San Francisco, USA).

Goffman, C. and Pedrick, G. (1983). *First Course in Functional Analysis* (Chelsea, New York, USA).

Golubin, A.Y. (2003). A note on the convergence of policy iteration in Markov decision processes with compact action spaces, *Math. Oper. Res.*, **28**, pp. 194–200.

Guo, X. (2007). Constrained optimization for average cost continuous-time Markov decision processes, *IEEE Trans. Aut. Control*, **52**, pp. 1139–1143.

Guo, X. and Hernández-Lerma, O. (2009). *Continuous-Time Markov Decision Processes* (Springer-Verlag, Berlin-Heidelberg).

Guo, X. and Song, X. (2009). Mean-variance criteria for finite continuous-time Markov decision processes, *IEEE Trans. Aut. Control*, **54**, pp. 2151–2157.

Guo, X., Huang, Y. and Song, X. (2012). Linear programming and constrained average optimality for general continuous-time Markov decision processes in history-dependent policies, *SIAM J. Control and Optim.*, **50**, pp. 23–47.

Guo, X. and Zhang, Y. (2016). Optimality of mixed policies for average continuous-time Markov decision processes with constraints, *Math. Oper. Res.*, **41**, pp.1276–1296.

Guo, X. and Zhang, Y. (2017). Constrained total undiscounted continuous-time Markov decision processes, *Bernoulli*, **23**, pp.1694–1736.

Guo, X. and Zhang, Y. (2021). A useful technique for piecewise deterministic Markov decision processes, *Oper. Res. Letters*, **49**, pp. 55–61.

Guo, X., Huang, Y. and Zhang, Y. (2024). On average optimality for nonstationary Markov decision processes in Borel spaces, *Math. Oper. Res.*, accepted.

Haviv, M. (1996). On constrained Markov decision processes, *Oper. Res. Letters*, **19**, pp. 25–28.

Heath, D.C., Pruitt, W.E. and Sudderth, W.D. (1972). Subfair red-and-black with a limit, *Proc. of the AMS*, **35**, pp. 555–560.

Hernández-Lerma, O. and Lasserre, J.B. (1996a). *Discrete-Time Markov Control Processes. Basic Optimality Criteria* (Springer-Verlag, New York, USA).

Hernández-Lerma, O. and Lasserre, J.B. (1996b). Average optimality in Markov control processes via discounted-cost problems and linear programming, *SIAM J. Control and Optim.*, **34**, pp. 295–310.

Hernández-Lerma, O. and Vega-Amaya, O. (1998). Infinite-horizon Markov control processes with undiscounted cost criteria: from average to overtaking optimality, *Applicationes Mathematicae*, **25**, pp. 153–178.

Hernández-Lerma, O. and Lasserre, J.B. (1999). *Further Topics on Discrete-Time Markov Control Processes* (Springer-Verlag, New York, USA).

Hernández-Lerma, O., Laura-Guarachi, L.R., Mendoza-Palacios, S. and González-Sánchez, D. (2023). *An introduction to Optimal Control Theory* (Springer Nature, Switzerland).

Hinderer, K. (1977). Does value iteration for finite-stage discrete dynamic programs hold "always"? *Bonn, Math. Schr.*, **98**, pp.41–56.

Hordijk, A. and Tijms, H.C. (1972). A counterexample in discounted dynamic programming, *J. Math. Anal. Appl.*, **39**, pp. 455–457.

Hordijk, A. and Puterman, M.L. (1987). On the convergence of policy iteration in finite state undiscounted Markov decision processes: the unichain case, *Math. Oper. Res.*, **12**, pp. 163–176.

Hordijk, A. and Spieksma, F. (1989). Are limits of α-discounted optimal policies Blackwell optimal? A counterexample, *System and Control Letters*, **13**, pp.31–41.

Hordijk, A. and Yushkevich, A.A. (2002). Blackwell optimality, in Feinberg, E. and Shwartz, A. (eds.), *Handbook of Markov Decision Processes*, (Kluwer, Boston, USA), pp. 213–249.

Inoie, A., Kameda, H. and Touati, C. (2006). A paradox in optimal flow control of M/M/n queues, *Computers and Oper. Research*, **33**, pp.356–368.

Jaquette, S.C. (1976). A utility criterion for Markov decision processes, *Manage. Scie.*, **23**, pp.43–49.

Jaśkiewicz, A. and Nowak, A.S. (2019). Constrained Markov decision processes with expected total reward criteria, *SIAM J. Control and Optim.*, **57**, pp. 3118–3136.

Kallenberg, L.C.M. (1983). *Linear Programming and Finite Markovian Control Problems* Math. Centre Tracts. **148**, Amsterdam.

Kallenberg, L.C.M. (2010). *Markov Decision Processes, Lecture Notes* (University of Leiden, The Netherlands). NB: the numbers of statements in later editions may be different.

Kallianpur, G. (1980) *Stochastic Filtering Theory* (Springer-Verlag, NY-Heidelberg-Berlin).

Kaspi, H. and Mandelbaum, A. (1994). On Harris recurrence in continuous time, *Math. Oper. Res.*, **19**, pp. 211–222.

Kemeny, J.G., Snell, J.L. and Knapp, A.W. (1976). *Denumerable Markov Chains* (Springer-Verlag, New York, USA).

Kertz, R.P. and Nachman, D.C. (1979). Persistently optimal plans for nonstationary dynamic programming: the topology of weak convergence case, *The Annals of Probability*, **1**, pp. 811–826.

Kilgour, D.M. (1975). The sequential truel, *Intern. J. Game Theory*, **4**, pp. 151–174.

Krass, D., Filar, J.A. and Sinha, S.S. (1992). A weighted Markov decision process, *Operations Research*, **40**, pp. 1180–1187.

Kurushima, A., Piunovskiy, A., and Zhang, Y. (2018). Nowak's theorem on probability measures induced by strategies revisited, *Theory Probab Appl.*, **62**, pp. 328–334.

Langford, E., Schwertman, N., and Owens M. (2001). Is the property of being positively correlated transitive? *The American Statistician*, **55**, pp. 322–325.

Lasserre, J.B. (1994). Average optimal stationary policies and linear programming in countable space Markov decision processes, *J.Math. Anal. Appl.*, **183**, pp. 233–249.

Lewis, M.E. and Paul, A. (2019). Uniform turnpike theorems for finite Markov decision processes, *Math. Oper. Res.*, **44**, pp. 1145–1160.

Liggett, T.M. and Lippman, S.A. (1969). Stochastic games with perfect information and time average payoff, *SIAM Review*, **11**, pp. 604–607.

Lippman, S.A. (1969). Criterion equivalence in discrete dynamic programming, *Oper. Res.*, **17**, pp. 920–923.

Loeb, P. and Sun, Y. (2006). Purification of measure-valued maps, *Illinois J. of Mathematics*, **50**, pp. 747–762.

Luque-Vasquez, F. and Hernandez-Lerma, O. (1995). A counterexample on the semicontinuity minima, *Proc. of the American Mathem. Society*, **123**, pp. 3175–3176.

Magaril-Il'yaev, G.G. and Tikhomirov, V.M. (2003). *Convex Analysis: Theory and Applications* (AMS, Providence, RI, USA).

Maitra, A. (1965). Dynamic programming for countable state systems, *Sankhya, Ser.A*, **27**, pp. 241–248.

McCall, B.P. and McCall J.J. (2008). *The Economics of Search* (Routledge, New York, USA).

Meyn, S. and Tweedie, R.L. (2009). *M*arkov Chains and Stochastic Stability (Cambridge University Press, UK).

Mine, H. and Osaki, S. (1970). *Markovian Decision Processes* (American Elsevier, New York, USA).

Nowak, A.S. (1988). On the weak topology on the space of probability measures induced by policies, *Bull. Pol. Acad. Sci. Math.*, **36**, pp. 181–186.

Nowak, A.S. and Vega-Amaya, O. (1999). A counterexample on overtaking optimality, *Math. Meth. Oper. Res.*, **49**, pp. 435–439.

Oksendal, B. (2013). *Stochastic Differential Equations* (Springer-Verlag, Berlin-Heidelberg).

Ornstein, D. (1969). On the existence of stationary optimal strategies, *Proc. of the American Mathem. Society*, **20**, pp. 563–569.

Pang, G. and Day, M. (2007). Fluid limits of optimally controlled queueing networks, *J. Appl. Math. Stoch. Anal.*, vol.2007, 1–20.

Parrondo, J.M.R. and Dinis, L. (2004). Brownian motion and gambling: from ratchets to paradoxical games, *Contemporary Physics*, **45**, pp. 147–157.

Parthasarathy, K.R. (2005). *Probability Measures on Metric Spaces* (AMS Chelsea Publishing, Providence, RI, USA).

Piunovskiy, A.B. (1997). *Optimal Control of Random Sequences in Problems with Constraints* (Kluwer, Dordrecht, Netherlands).

Piunovskiy, A. and Mao, X. (2000). Constrained Markovian decision processes: the dynamic programming approach, *Oper. Res. Letters*, **27**, pp. 119–126.

Piunovskiy, A.B. (2006). Dynamic programming in constrained Markov decision processes, *Control and Cybernetics*, **35**, pp. 645–660.

Piunovskiy, A.B. (2009a). When Bellman's principle fails, *The Open Cybernetics and Systemics J.*, **3**, pp. 5–12.

Piunovskiy, A. (2009b). Random walk, birth-and-death process and their fluid approximations: absorbing case, *Math. Meth. Oper. Res.*, **70**, pp. 285–312.

Piunovskiy, A. and Zhang, Y. (2011). Accuracy of fluid approximation to controlled birth-and-death processes: absorbing case, *Math. Meth. Oper. Res.*, **73**, pp. 159–187.

Piunovskiy, A. and Zhang, Y. (2012). Fluid approximations to Markov decision processes with local transitions, in Hernandez-Hernandez, D. and Minjarez-Sosa, A. (eds), *Optimization, Control, and Applications of Stochastic Systems*, (Birkhäuser, Springer), pp. 225–238.

Piunovskiy, A.B. (2013). *Examples if Markov Decision Processes* (Imperial College Press, London).

Piunovskiy, A. (2018). Realizable strategies in continuous-time Markov decision processes, *SIAM J. Control and Optim.*, **56**, pp. 473–495.

Piunovskiy, A. and Zhang, Y. (2020). *Continuous-Time Markov Decision Processes* (Springer, Cham).

Piunovskiy, A. (2021). Controlled random walk: conjecture and counter-example, *Modern Trends In Controlled Stochastic Processes* (Springer, Cham), pp. 38–56.

Piunovskiy, A. (2023). Turnpikes in finite Markov decision processes and random walk, *Theory Probab. Appl.*, **68**, pp. 123–149.

Piunovskiy, A. and Zhang, Y. (2024a). Extreme occupation measures in Markov decision processes with an absorbing state, *SIAM J. Control and Optim.*, **62**, pp. 65–90.

Piunovskiy, A. and Zhang, Y. (2024b). On the continuity of the projection mapping from strategic measures to occupation measures in absorbing Markov decision processes, *Applied Mathematics and Optimization*, **89**, N.58.

Platzman, L.K. (1980). Optimal infinite-horizon undiscounted control of finite probabilistic systems, *SIAM J. Control and Optim.*, **18**, pp. 362–380.

Pollett, P.K. (2003). Integrals for continuous-time Markov chains, *Mathem. Biosciences*, **182**, pp. 213–225.

Priestley, H.A. (1990). *Introduction to Complex Analysis* (Oxford University Press, Oxford, UK).

Prieto-Rumeau, T. and Hernández-Lerma, O. (2012). *Selected Topics on Continuous-Time Controlled Markov Chains and Markov Games* (Imperial College Press, London, UK).

Puterman, M.L. (1994). *Markov Decision Processes* (Wiley, New York, USA).

Robinson, D.R. (1976). Markov decision chains with unbounded costs and applications to the control of queues, *Adv. Appl. Prob.*, **8**, pp. 159–176.

Rockafellar, R.T. (1970). *Convex Analysis* (Princeton, NJ, USA).

Rockafellar, R.T. (1987). *Conjugate Duality and Optimization* (SIAM, Philadelphia, PA, USA).

Ross, S.M. (1968). Non-discounted denumerable Markovian decision models, *Ann. Math. Stat.*, **39**, pp. 412–423.

Ross, S.M. (1970). *Applied Probability Models with Optimization Applications* (Dover Publications, New York, USA).

Ross, S.M. (1971). On the nonexistence of ε-optimal randomized stationary policies in average cost Markov decision models, *Ann. Math. Stat.*, **42**, pp. 1767–1768.

Ross, S.M. (1983). *Introduction to Stochastic Dynamic Programming* (Academic Press, San Diego, CA, USA).

Schäl, M. (1975a). On dynamic programming: compactness of the space of policies, *Stoch. Processes and their Appl.*, **3**, pp. 345–364.

Schäl, M. (1975b). Conditions for optimality in dynamic programming and for the limit of n-stage optimal policies to be optimal, *Z. Wahrscheinlichkeitstheorie verw. Gebiete*, **32**, pp. 179–196.

Schmidli, H. (2008). *Stochastic Control in Insurance* (Springer-Verlag, London, UK).

Schochetman, I.E. and Smith, R.L. (2005). Optimality criteria for deterministic discrete-time infinite horizon optimization. *Intern. J. of Mathematics and Mathematical Sciences*, **1**, pp. 57–80.

Schweitzer, P.J. (1987). A Brouwer fixed-point mapping approach to communicating Markov decision processes. *J. Math. Anal. Appl.*, **123**, pp. 117–130.

Sennott, L. (1989). Average cost optimal stationary policies in infinite state Markov decision processes with unbounded costs, *Oper. Res.*, **37**, pp. 626–633.

Sennott, L. (1991). Constrained discounted Markov decision chains, *Prob. in the Engin. and Inform. Sciences*, **5**, pp.463–475.

Sennott, L. (2002). Average reward optimization theory for denumerable state spaces, in Feinberg, E. and Shwartz, A. (eds.), *Handbook of Markov Decision Processes*, (Kluwer, Boston, USA), pp. 153–172.

Seth, K. (1977). Optimal service policies, just after idle periods in two-server heterogeneous queuing systems, *Oper. Res.*, **25**, pp. 356–360.

Spieksma, F. (2016). Kolmogorov forward equation and explosiveness in countable state Markov processes, *Ann. Oper. Res.* **241**, pp. 3–22.

Stoyanov, J.M. (1997). *Counterexamples in Probability* (Wiley, Chichester, UK).

Strauch, R.E. (1966). Negative dynamic programming, *Ann. Math. Stat.*, **37**, pp. 871–890.

Suhov, Y. and Kelbert, M. (2008). *Probability and Statistics by Example. V.II: Markov Chains* (Cambridge University Press, Cambridge, UK).

Szekely, G.J. (1986). *Paradoxes in Probability Theory and Mathematical Statistics* (Akademiai Kiado, Budapest, Hungary).

Wal, J. van der and Wessels, J. (1984). On the use of information in Markov decision processes, *Statistics and Decisions*, **2**, pp. 1–21.

Wilkins, J.E. (1972). The bold strategy in presence of house limit, *P*roc. Amer. Math. Soc., **32**, pp.567–570.

Whitt, W. (1986). Deciding which queue to join: some counterexamples, *Operations Research*, **34**, pp. 55–62.

Whittle, P. (1983). *Optimization over Time* (Wiley, Chichester, UK).
in batch production, *Prob. in the Engin. and Inform. Sci.*, **12**, pp. 351–371.

Ye, H-Q. (2007). A paradox for admission contro of multiclass queueing network with differentiated service, *J.* Appl. Prob., **44**, pp.321–331.

Zhang, Y. and Zheng, X. (2024). Further remarks on absorbing Markov decision processes, *Oper. Res. Letters*, **57**, 107191.

Index

Notation

$\mathbf{A}$	action space	1,3,353
A_t	action (as a random element)	2,3,5
$A(t)$	control process	356
a, a_t	action (as a variable, argument of a function, etc.)	2
$\mathcal{B}(\mathbf{X})$	Borel σ-algebra	1
$c_t(x,a), c(x,a)$	one-step cost, cost rate	2,3,62,357
$C(x)$	terminal cost	2,3
$\mathcal{D}_{\mathsf{P}_0}$, $\mathcal{D}_{x_0}$, $\mathcal{D}^f_{\mathsf{P}_0}$	spaces of occupation measures	68,69,185
$\mathsf{E}^{\pi}_{\mathsf{P}_0}$ ($\mathsf{E}^{\pi}_{x_0}, \mathsf{E}^{S}_{\mathsf{P}_0}, \mathbb{E}^{\sigma^{\mathrm{m}}}_{\mathsf{P}_0}$ etc)	mathematical expectation w.r.t. $\mathsf{P}^{\pi}_{\mathsf{P}_0}$ ($\mathsf{P}^{\pi}_{x_0}, \mathsf{P}^{S}_{\mathsf{P}_0}, \mathbb{E}^{\sigma^{\mathrm{m}}}_{\mathsf{P}_0}$ etc)	5,61,355
$\bar{f}^{T}_{\pi,\mathsf{P}_0}$	expected frequencies	305
$\mathbf{G}_t = \mathbf{H}_t \times \mathbf{A}$	space of trajectories	5
$G^A(S, \mathsf{P}_0)$	ratio-average cost	414
$\mathbf{H}, \mathbf{H}_t = \mathbf{X} \times (\mathbf{A} \times \mathbf{X})^t$	spaces of trajectories (histories)	3
$\mathbf{H}_n$	n-term histories	355
$h_t = (x_0, a_1, \ldots, x_t)$	history	5
$h(x)$	element of a canonical triplet	249
$h^{\beta}(x), h_{\alpha}(x)$	increments of the discounted Bellman function	255,372
$l(x,a)$	one-step cost	367
J	number of constraints	155,402
$m^{S,\alpha}_{\mathsf{P}_0,n}$	detailed occupation measure	359
$\mathbb{M}^{S,\alpha}_{\mathsf{P}_0}$	total occupation measure	359
M_{jj}	mean recurrence time	252
$\mathsf{P}_0(dx)$	initial distribution of X_0	2,4,353
$\mathsf{P}^{\pi}_{\mathsf{P}_0}, \mathsf{P}^{\pi}_{x_0}, \mathsf{P}^{\pi}_{h_\tau}$	strategic measures	4,5,7,61,
$\mathsf{P}^{S}_{\mathsf{P}_0}, \mathsf{P}^{\pi}_{x_0}, \mathsf{P}^{\varpi}_{\mathsf{P}_0}, \mathbb{P}^{\sigma^{\mathrm{m}}}_{\mathsf{P}_0}$	strategic measures	355,367
$\bar{\mathsf{P}}^{S}_{\mathsf{P}_0}(t, dy \times da), \bar{\mathsf{P}}^{S}_{\mathsf{P}_0}(t, dy)$	marginal distribution	357
$\mathcal{P}_{\mathsf{P}_0}, \mathcal{P}_{x_0}, \mathcal{P}^{\mathrm{N}}_{\mathsf{P}_0}, \mathcal{P}^{\mathrm{M}}_{\mathsf{P}_0}$	spaces of strategic measures	5,61,72
$p_t(dy\|x,a)$, $p(dy\|x,a)$	transition probability	2,3,62

$p_\alpha(dy\|x,a)$	transition probability	367
$q(dy\|x,a)$	transition rate	353
$q(dy\|x_{n-1},\pi_n,s)$	transition rate	355
$\tilde{q}(dy\|x,a)$	transition rate	353
$\tilde{q}(dy\|x_{n-1},\pi_n,s)$	transition rate	355
$q_x(a), q_{x_{n-1}}(\pi_n,s)$	jump intensity	353,355
$q(dy\|x,t), \tilde{q}(dy\|x,t)$	transition rates associated with a fixed natural Markov strategy	364
$q(dy\|x,\pi^{\mathrm{s}})$	Q-function	392
$q^V(dy\|x,\pi^{\mathrm{s}})$	V-transformed Q-function	392
$\bar{q}_x, \bar{q}$	supremum of $q_x(a)$	353,371
sp	span	22,293
Supp μ	support of measure μ	7
S	control strategy	355
S^*	(average) uniformly optimal strategy	358
t	time parameter	1
T	time horizon	1,3,61
T_0	(random) time to absorption	64,183
T_n	(random) jump epochs	354
T_∞	moment of explosion	354
$v^\pi_{\mathsf{P}_0}, v^\pi_{x_0}, v^\pi_{h_\tau}, v^{\pi,\beta}_{\mathsf{P}_0}$	performance functionals	3,5,7,61,62 183,249
$v_t(x)$, $v(x)$	solution to the optimality equation	3,6,63,183
$v^*_t(x)$	minimal possible cost starting from $X_t = x$ (Bellman function)	7
v^*_x, $v^*_0(x)$, $v^*_{h_\tau}$, $v^{*,\beta}_x$	Bellman function	7,62,183, 184,249
$v^n(x)$	approximation of the Bellman function via value iteration	65,184
$v^\infty(x)$	limiting function in value iteration	65,184
$V(x)$	solution to the optimality equation	362,369
$\mathcal{V}_{\mathsf{P}_0}$, $\mathcal{V}^{\mathrm{MN}}_{\mathsf{P}_0}$	performance spaces	21
$V^{\pi,T}_x, V^{(n)}(\varphi,i)$	performance functionals	8,425
V^T_x	Bellman function	8
W	realized cost (as a random element)	2,5,62

$W^{\alpha}(S, \mathsf{P}_0)$	performance functional (expected α-discounted cost)	357
$W^{A}(S, \mathsf{P}_0)$	average cost	357
$W^{*\alpha}(\mathsf{P}_0), W^{*A}(\mathsf{P}_0)$	optimal values of performance functionals	358
$W^{*\alpha}(x)$	Bellman function	358
$\mathbf{X}$	state space	1,3,353
$\mathbf{X}^{\Delta}$	states apart from the cemetery	359
X_t	state of the controlled process (as a random element)	1,3,5
$X(t)$	controlled process	355
x, x_t, y, y_t	state of the controlled process (as a variable, argument of a function, etc.)	2,4,353
x_{∞}	artificial state in CTMDP	355
Y_t^{π}	estimating process	17
α	discount factor	357
β	discount factor	183
Δ (or 0, or $\Box$)	absorbing state (cemetery)	64,183, 359,367
$\eta_{\mathsf{P}_0}^{\pi}$, $\eta_{x_0}^{\pi}$, $\eta_{\mathsf{P}_0}^{\sigma,\alpha}$	occupation measures	68,69, 185,367
$\hat{\eta}_{\mathsf{P}_0}^{\pi}$	marginal occupation measure	70
$\hat{\eta}^{\pi^{s}}$, $\hat{\eta}$	stationary (invariant) probability	252,256
$\eta(\cdot,\cdot)$	state-action frequency	253
$\eta(\cdot,\cdot), \tilde{\eta}(\cdot,\cdot)$	variables in linear programs	69,252,302, 370,399
θ_i	sojourn time (as a variable, argument of a function, etc.)	354
Θ_i	sojourn time (as a random element)	354
π	(relaxed) control strategy	4,355
π^*	(uniformly) optimal control strategy	2,5,7,183, 249
$\pi_t^{\mathrm{m}}(da\vert x)$	Markov (relaxed) strategy	4,356
$\pi^{\mathrm{s}}(da\vert x)$	stationary (relaxed) strategy	4,356
$\breve{\pi}$	natural Markov strategy	355
ϖ	randomized control strategy	355
ϖ^{m}	randomized Markov strategy	356
ϖ^{s}	randomized stationary strategy	356

$\Pi^{\rm All}$ $(\Pi^{\rm M}, \Pi^{\rm MN})$	set of all strategies (Markov strategies and Markov selectors)	4
$\Pi^{\rm S}$ $(\Pi^{\rm SN})$	set of all stationary strategies (stationary selectors)	4
Π^p_{x0}	set of proper strategies	128
$\Pi(\omega, t, \Gamma_A)$	$\mathbf{P}(\mathbf{A})$-valued process	382
$\langle \rho, h, \varphi^* \rangle$	canonical triplet	249
$\rho(x) = v^*_x$	Bellman function	254
$(\sigma^{\rm m}), \sigma$	(Markov) strategy	367,368
φ	selector (non-randomized, deterministic strategy)	4,356
$\varphi_t(x)$, $\varphi(x)$	Markov and stationary selectors, deterministic stationary strategies	4,356
Ω	sample space	61,354

List of the Main Statements

www.ingramcontent.com/pod-product-compliance
Lightning Source LLC
LaVergne TN
LVHW011656200926
848484LV00004B/958